Algorithms for Robotic
Motion and Manipulation

Algorithms for Robotic Motion and Manipulation

1996 Workshop on the Algorithmic
Foundations of Robotics

edited by

Jean-Paul Laumond
LAAS–CNRS
Toulouse, France

Mark Overmars
Utrecht University
Utrecht, The Netherlands

A K Peters
Wellesley, Massachusetts

96
The Workshop on the Algorithmic
Foundations of Robotics

Editorial, Sales, and Customer Service Office

A K Peters, Ltd.
289 Linden Street
Wellesley, MA 02181

Library of Congress Cataloging-in-Publication Data

Not available at press time

ISBN 1-56881-067-9

Cover illustration: Mechanical drawing of Leonardo da Vinci. From
Leonardo da Vinci, Cod. Atl., fol. 357 r-a.Reynal and Company, New York,
p. 498. Copyright in Italy by the Istituto Geograqfico De Agostini - Novara -
1956.

Printed in the United States of America
01 00 99 98 97   10 9 8 7 6 5 4 3 2 1

Contents

Geometric Algorithms

Visibility

Minimalism and Controllability

Foreword

Robotics research is often organized along functional approaches attacking basic issues in perception and modeling, task planning, motion planning, motion control, decisional architecture, and more. Each of these issues uses its own domain of knowledge, such as control theory, computational geometry, computer science, signal processing, with only few interactions between them. This diversity is one of the reasons that make the design of integrated robotics systems so hard.

The necessity of combining the use of tools from various domains requires a formal and abstract viewpoint of the robot-environment relationship. Exploring the algorithmic foundations of this relationship should be based on a common knowledge and understanding of the different theoretical issues; this study appears recently as the main objective of several research groups.

In February 1994 the first workshop on Algorithmic Foundations of Robotics was held in San Francisco. The response was very enthusiastic. Hence, it was decided to turn this into a biennial event. The second workshop was held in July 1996 in Toulouse, France. We deviated slightly from the previous format in the sense that this time the program consisted of both invited and contributing papers. The invited talks were given bu R. Alami (Toulouse), M. Erdmann (Pittsburgh), H. Hirukawa (Tsukuba), A. Jones (Seattle), L. Kavraki (Stanford), M. Lin (Chapel Hill), D. Manocha (Chapel Hill), M. Mason (Pittsburgh), M. Overmars (Utrecht), T. Poggio (Cambridge), S. Sastry (Berkeley), and C. Yap (New York). The contributing papers were selected from 27 high-level submissions, by a program committee consisting of, in addition to the editors of this volume, A. Bicchi (Pisa), S. Cameron (Oxford), B. Donald (Ithaca), K. Goldberg (Berkeley), D. Halperin (Stanford), J. C. Latombe (Stanford), M. Mason (Pittsburgh), T. Matsui (Tsukaba), P. Raghavan (Almaden), and R. Wilson (Albuquerque).

This book contains the full versions of both the invited and contributing papers. Aside from papers that treat core problems in robotics, like motion planning, sensor-based planning, manipulation, and assembly planning, we also invited some speakers to talk about the application of robotics algorithms in other domains, like molecular modeling, computer graphics, and image analysis.

The workshop has been supported by CNRS, the Région Midi–Pyrénées, and the Conseil Général de la Haute–Garonne. We thank the Public and International Relations Section of LAAS–CNRS and Jackie Som for her now legendary efficiency and kindness.

Jean-Paul Laumond (Toulouse)
Mark Overmars (Utrecht)

August 1996

Participants

Rachid Alami	LAAS-CNRS (rachid@laas.fr)
Robert-Paul Berretty	Utrecht University (rp@cs.ruu.nl)
Antonio Bicchi	Universita di Pisa (bicchi@piaggio.ccii.unipi.it)
Karl F. Bohringer	Cornell University (karl@cs.cornell.edu)
Jean-Daniel Boissonnat	INRIA Sophia Antipolis (boissonn@sophia.inria.fr)
Amy J. Briggs	Middlebury College (briggs@mail.middlebury.edu)
Joel Burdick	California Institute of Technology (jwb@robby.caltech.edu)
Stephen Cameron	Oxford University (cameron@comlab.ox.ac.uk)
John Canny	UC Berkeley (jfc@cs.berkeley.edu)
Howie Choset	Carnegie Mellon University (choset@robby.caltech.edu)
Bruce Donald	Cornell University (brd@cs.stanford.edu)
Ioannis Z. Emiris	INRIA Sophia Antipolis (Ioannis.Emiris@sophia.inria.fr)
Michael Erdmann	Carnegie Mellon University (me@h.gp.cs.cmu.edu)
Pierre Ferbach	EDF (Pierre.Ferbach@der.edfgdf.fr)
Thierry Fraichard	INRIA Rhone Alpes & GRAVIR (Thierry.Fraichard@imag.fr)
Andre Gasquet	MATRA MARCONI Space (gasquet@mms.matra-espace.fr)
Malik Ghallab	LAAS-CNRS (malik@laas.fr)
Georges Giralt	LAAS-CNRS (giralt@laas.fr)
Ken Goldberg	UC Berkeley (goldberg@ieor.berkeley.edu)
Leonidas J. Guibas	Stanford University (guibas@cs.stanford.edu)
Dan Halperin	Tel Aviv University (danha@math.tau.ac.il)
Hirohisa Hirukawa	Electrotechnical Laboratory (hirukawa@etl.go.jp)
Seth Hutchinson	University of Illinois (seth@uiuc.edu)
Alan Jones	The Boeing Co. (jones@redwood.rt.cs.boeing.com)
Leo Joskowicz	The Hebrew Univ. of Jerusalem (josko@cs.huji.ac.il)
Ammar Joukhadar	INRIA Rhone Alpes & GRAVIR (ammar.joukhadar@imag.fr)
Lydia Kavraki	Stanford University (kavraki@cs.stanford.edu)
Vijay Kumar	University of Pennsylvania (kumar@central.cis.upenn.edu)
Eelco de Lange	MATRA MARCONI Space (edelange@isis.matra-espace.fr)
Jean-Claude Latombe	Stanford University (latombe@cs.stanford.edu)
Christian Laugier	INRIA Rhone-Alpes & GRAVIR (Christian.Laugier@imag.fr)
Jean-Paul Laumond	LAAS-CNRS (jpl@laas.fr)
Steve LaValle	Stanford University (lavalle@cs.stanford.edu)
Ming C. Lin	University of N. Carolina (lin@cs.unc.edu)
Dinesh Manocha	University of N. Carolina (manocha@cs.unc.edu)
Alessia Marigo	Universita di Pisa (marigo@sesame.mathp6.jussieu.fr)
Matt Mason	Carnegie Mellon University (Matt_Mason@cs.cmu.edu)
Rajeev Motwani	Stanford University (rajeev@cs.stanford.edu)

Mark Overmars   Utrecht University (markov@cs.ruu.nl)
Eric Paulos   UC Berkeley (paulos@cs.berkeley.edu)
Tomaso Poggio   MIT (tp-temp@ai.mit.edu)
Jean Ponce   University of Illinois (ponce@cs.uiuc.edu)
John Reif   Duke University (reif@cs.duke.edu)
Daniela Rus   Dartmouth (rus@cs.dartmouth.edu)
Elisha Sacks   Purdue University (eps@cs.purdue.edu)
Shankar Sastry   Univ. of California. Berkeley (sastry@eecs.berkeley.edu)
Sepanta Sekhavat   LAAS-CNRS (sepanta@laas.fr)
Micha Sharir   Tel Aviv University (sharir@math.tau.ac.il)
Rajeev Sharma   Univ. of Illinois at Urbana-Champaign (rajeev@cs.uiuc.edu)
Thierry Siméon   LAAS-CNRS (nic@laas.fr)
Attawith Sudsang   University of Illinois (attawith@dizzy.ai.uiuc.edu)
Petr Švestka   Utrecht University (petr@cs.ruu.nl)
Marek Teichmann   New York University (teichman@cs.nyu.edu)
Carl Van Geem   RISC-Linz (Carl.Van.Geem@risc.uni-linz.ac.at)
Marilena Vendittelli   Univ. di Roma "La Sapienza" (venditt@labrob.ing.uniroma1.it)
Joris Vergeest   Delft University of Technology (j.s.m.vergeest@io.tudelft.nl)
Chantal Wentink   Utrecht University (chantal@cs.ruu.nl)
Chee K. Yap   New York Univ. and N. Univ. of Singapore (yap@cs.nyu.edu)
Milos Zefran   University of Pennsylvania (milos@grip.cis.upenn.edu)

Multi-robot Cooperation based on a Distributed and Incremental Plan Merging Paradigm

Rachid Alami, *LAAS/CNRS, 31077 Toulouse - France*

We present and discuss a generic cooperative scheme for multi-robot cooperation based on an incremental and distributed plan-merging process. Each robot, autonomously and incrementally builds and execute its own plans taking into account the multi-robot context. The robots are assumed to be able to collect the other robots current plans and goals and to produce plans which satisfy a set of constraints that will be discussed.

We discuss the properties of this cooperative scheme (coherence, detection of dead-lock situations) as well as the class of applications for which it is well suited. We show how this paradigm can be used in a hierarchical manner, and in contexts where planning is performed in parallel with plan execution. We also discuss the possibility to negotiate goals within this framework and how this paradigm "fills the gap" between centralized planning and distributed execution.

We finally illustrate this scheme through an implemented system which allows a fleet of autonomous mobile robots to perform load transfer tasks in a route network environment with a very limited centralized activity and important gains in system flexibility and robustness to execution contingencies.

1 Introduction

In the field of multi-agent cooperation, we distinguish two main issues:

- **C1:** the first issue involves goal/task decomposition and allocation to various agents

- **C2:** the second issue involves the simultaneous operation of several autonomous agents, each one seeking to achieve its own task or goal.

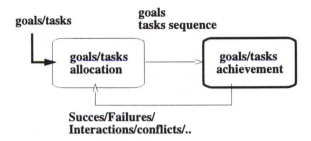

Figure 1: *The two main issues in multi-agent cooperation*

While several contributions have concentrated more particularly on one issue or the other, we claim that in numerous multi-agent applications, both issues appear and even "invoke" one another "recursively".

This is particularly true for autonomous multi-robot applications and, more generally, when the allocated tasks or goals cannot be directly "executed" but require further refinement.

The case of multi-robot applications is indeed directly concerned because the robots act in a same physical environment and because of the multiplicity of uncertainties.

Let us assume a set of autonomous robots, which have been given, through a centralized system or after a distributed planning process, a set of (partially ordered) tasks or goals. One can reasonably assume that this plan elaboration process is finished when the obtained tasks or goals have a sufficient range and are sufficiently independent to cause a substantial "selfish" robot activity. However, each robot, while seeking to achieve its task will have to compete for resources, to comply with other robots activities. Hence, several robots may find themselves in situations where they need to solve a new goal/task interaction leading to a

1

new goal/task allocation scheme.

Indeed, these two issues are somewhat different in nature, and should call for different resolution schemes. While the first (**C1**) is more oriented towards the collective search for a solution to a problem and calls for a purely deliberative activity, the second (**C2**) involves a more "compliant" behavior of the agents and integrates a closer interaction between deliberation and action.

While several generic approaches have been proposed in the literature concerning task or goal decomposition and allocation (Contract Nets [25], Partial Global Planning [9], distributed search [10], negotiation [14, 5, 12, 23, 6], motivational behaviors [21, 11]), cooperation for achieving independent goals have been mostly treated using task-specific or application-specific techniques [16, 20, 22, 27]

We argue that there is also a need for generic approaches to **C2**. One can of course make the agents respect a set of rules (e.g. traffic rules), or more generally "social behaviors" [24], which are specially devised to avoid as much as possible conflicts and to provide pre-defined solutions to various situations. However, this cannot be a general answer applicable to various domains.

We would like to devise a scheme which guarantees a coherent behavior of the agents in all situations (including the avalanche of situations which may occur after an execution failure) and a reliable detection of situations which call for a new task distribution process.

In the following, we propose a paradigm [3, 4] which, we believe, provides a generic framework for **C2** issues and clearly establishes a link with **C1** issues.

2   The *Plan-Merging* Paradigm

Let us assume that we have a set of autonomous robots equipped with a reliable inter-robot communication device which allows to broadcast a message to all robots or to send a message to a given robot.

Let us assume that each robot processes sequentially the goals it receives, taking as initial state the final state of its current plan. Doing so, it incrementally appends new sequences of actions to its current plan.

However, before executing any action, a robot has to ensure that it is valid in the current multi-robot context, i.e. that it is compatible with all the plans currently under execution by the other robots. This will be done by collecting all the other robot plans and by "merging" its own plan with them. This operation is "protected" by a mutual exclusion mechanism and is performed *without modifying* the other robots plans or inserting an action which may render one them invalid in order to allow the other robots to continue execution.

We call this operation, a *Plan-Merging Operation* (PMO) and its result a *Coordination Plan* (i.e. a plan valid in the current multi-robot context). Besides new actions for the robot such a plan will specify the necessary synchronization between these new actions and the other robots coordination plans.

Note that such an operation involves only communication and computation and concerns future (near term) robot actions. It can run in parallel with the execution of the current coordination plan.

2.1   The "global plan" and its properties

Everything works as if there was a *global plan* produced and maintained by the set of robots. In fact, no robot elaborates, stores or maintains a complete representation of such a *global plan*.

At any moment, a robot has its own *coordination plan* under execution. Such a *coordination plan* consists of a sequence of actions and events to be signaled to other robots as well as events which are planned to be signaled by other robots. Such events correspond to state changes in the multi-robot context and represent temporal constraints (precedence) between actions involved in different individual coordination plans.

At any moment, the "global plan" is the graph representing the union of all current robot coordination plans. Such a global plan is valid (i.e. it does not contain inconsistent temporal constraint) if it can be represented by a *directed acyclic graph (dag)*.

The key point here is how to devise a system composed of a set of robots which should, as much as possible, plan independently to achieve their tasks while maintaining such property of the global plan.

2.2   The *Plan-Merging* Protocol

Let us assume here that:

1. there exists a mean which allows a robot to get the right to perform a PMO while having the guarantee that it is the only robot doing so. This right should be thought of as a resource allocation[1]

2. there is a mean allowing a robot (which has obtained the right to perform a PMO) to ask for and obtain all the other robots coordination plans.

3. there exists a mean allowing robots to ask or inform one another about the occurrence of an event.

   Robots are assumed to plan from time to time (whenever it is necessary).

   When a robot is not planning and even when it is waiting to obtain the right to perform a PMO, it must be able to send its current coordination plan to another robot (which currently has the right to perform a PMO).

   When a robot has to plan, it uses the following protocol which we call the *Plan Merging Protocol* (see Figure 2):

1. It asks for the right to perform a PMO and waits until it obtains it together with the coordination plans of all the other robots.

2. It then builds the *dag* corresponding to the union of all coordination plans (including its own coordination plan)

3. It then tries to produce a new plan which can be inserted in the *dag*, *after* its current coordination plan. The new plan insertion may only add temporal constraints which impose that some of its

---

[1]A simple way to do it is, for instance, to maintain a "token" through communication

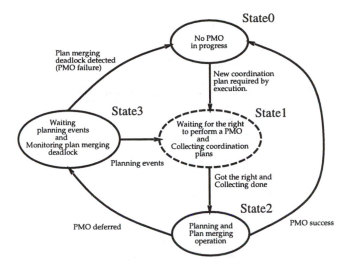

Figure 2: *The general protocol state graph*

actions must be executed after some time-points from other robots coordination plans.

Besides, the insertion must maintain the fact that the obtained global plan is still a *dag* (see Figure 3).

4. If it succeeds in producing the desired plan, the robot appends it to its current coordination plan.

5. And finally, it releases the right to perform a PMO.

When a robot executes its coordination plan, if it reaches a step with a temporal constraint linked to another robot time-point, it asks that robot if it has passed that time-point or not. Depending on the answer, the robot will wait until the other robot informs it or will immediately proceed.

2.3   Situations where PMO is deferred or where deadlock is detected

When a robot tries to perform a PMO, it may fail to produce a plan which satisfies the properties discussed earlier.

This may happen in two situations:

1. the goal can never be achieved. This can be detected if the robot cannot produce a plan even if it was alone in the environment.

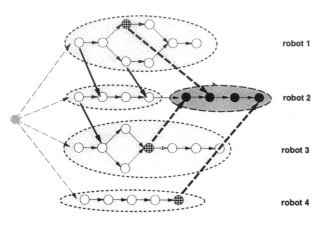

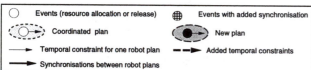

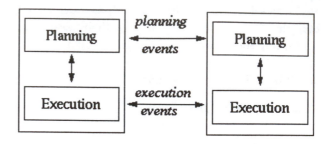

Figure 4: *Two event classes*

Figure 3: *Part of the "global plan"* dag *during the insertion of a new plan by robot 2.*

2. the robot can generate a plan but this plan cannot be inserted in the global plan. This means that the final state of another robot forbids it to insert its own plan.

   In such situation, the robot can simply abandon the PMO and decide to wait until the robots, that it has identified, have performed a new PMO which may possibly make them change the states preventing it to insert its plan.

Hence, we have introduced two types of events:

1. *execution events:* i.e. events which occur during plan execution and which allow robots to synchronize their execution.

2. *planning events:* i.e. events which occur whenever a robot performs a new PMO. These events can also be awaited for.

Note that, even when a robot fails in its PMO, it leaves the global plan in a correct state (it is still a *dag* and its execution can continue).

In order to detect deadlocks, a robot which finds itself in a situation where it has to wait for a *planning*

event from a particular robot, must inform it. Then, it becomes possible for a robot to monitor and detect *deadlock situations* by propagating and updating, a graph of robots waiting (directly or by transitivity) for *planning events* from itself. Indeed, a deadlock is detected when a robot finds itself in the list of robots waiting for itself.

When a deadlock occurs, it is necessary to take explicitly into account, in a *unique planning operation*, a conjunction of goals (which have been given separately to several robots).

This simply means that the global mission was too constrained to be solved using the Plan-Merging Paradigm. Here we must recall that we do not claim that the Plan-Merging paradigm can solve or help to solve multi-robot planning problems. The main point here is that the Plan-Merging paradigm is *safe* as it includes the detection of the deadlocks i.e. situations where a cooperation scheme of type **C1** should take place.

Note also that, in the case where only a small number of robots are involved in a deadlock, one can decide to allow the robot, which detected the deadlock, to plan for all the concerned robots . The Plan-Merging paradigm remains then applicable: the inserted plan will then concern several robots at a time.

A detailed discussion on the properties of the Plan-merging paradigm as well as on its ability to cope with execution failures can be found in [3].

The paradigm and the protocol presented so far is

generic. We believe that it can be used in numerous applications. Several instances of the general paradigm can be derived, based on different planners: action planners in the stream of STRIPS, as well as more specific task planners or motion planners.

One class of applications which seems particularly well suited is the control of a large number of autonomous mobile robots.

We present in the sequel an application in the case of a fleet of autonomous mobile robots.

3 A fleet of autonomous mobile robots

We have applied the Plan-Merging Paradigm in the framework of the MARTHA project[2] which deals with the control of a large fleet of autonomous mobile robots for the transportation of containers in harbors, airports and railway environments.

In such context, the dynamics of the environment, the impossibility to correctly estimate the duration of actions (the robots may be slowed down due to obstacle avoidance, and delays in load and un-load operations, etc..) prevent a central system from elaborating efficient and reliable detailed robot plans.

The use of the Plan-Merging paradigm allowed us to deal with several types of conflicts in a general and systematic way, and to limit the role of the central system to the assignment of tasks and routes to the robots (without specifying any trajectory or any synchronization between robots) taking only into account global traffic constraints.

3.1 Overall Architecture

A Martha system is composed of a Central Station (CS) and a set of autonomous mobile robots able to communicate with each other and with the Central Station.

The CS as well as the robots make use of the same description of the environment for several purposes dealing with mission specification, robot navigation or multi-robot conflict resolution.

---

[2]MARTHA: European ESPRIT Project No 6668. "Multiple Autonomous Robots for Transport and Handling Applications"

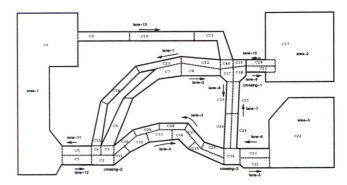

Figure 5: *An Environment: Geometrical Representation.*

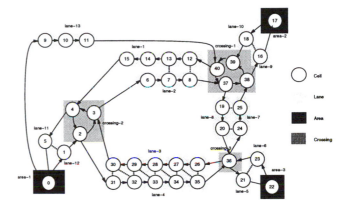

Figure 6: *Topological Representation.*

Environment Model An environment is a topological graph of *areas*, *routes* and *crossings*. The areas contains docking/undocking *stations*. The routes are composed of *lanes*; crossing and lanes are then composed of *cells* which have a nominal (but not exclusive) direction. Cells, areas and stations have a geometrical description (polygonal regions). An example of such an environment (geometrical and topological) is given in Figures 5 and 6.

Besides, one can have a geometrical description of known obstacles as well as complementary data for localization or docking purposes.

The robots heavily rely on this model to plan their routes and trajectories. Nevertheless, the real environment may also contain unknown obstacles/objects which have to be dealt with on-line by the robots (detection and avoidance if possible).

```
(mission (.
  (action 1 (goto (station 1))
            (using (lane 10)))
  (action 2 (dock))
  (action 3 (putdown))
  (action 4 (undock))
  (action 5 (goto (station 3))
            (using (lane 12) (lane 8)))
  (action 6 (dock))
  (action 7 (pick-up (container 5)))
  (action 8 (undock))
  (action 5 (goto (end-lane 0))
            (using (lane 9) (lane 0))) .))
```

Figure 7: *Example of a Mission sent by the Central Station*

Martha's Robots Missions Although one of the goal of the Martha project is to alleviate the burden on a Central Station (CS), one remains present. However, its role is mainly to plan the transshipment operations (which robot loads/unloads which container)[3] and the routes the robot *should* use (See Figure 7 for a mission example). The CS uses the topological model to plan these routes. The CS does not intervene in the robot plans coordination (such as in crossing or area), nor does it plan the precise trajectory which are executed by the robots. As a consequence, the communication bandwidth required between the robots and the CS is very low. Moreover, the computational power devoted by the CS to control the robot is far less important than the one used in a completely centralized application.

The robots receive their missions from the Central Station. From then on, each robot is on its own to perform the mission. It has to refine the mission, to plan its routes and then its trajectories, to coordinate the resulting plans and trajectories with other robots and to execute all these actions, monitoring critical situations (such as unknown obstacles) and reporting unrecoverable action failure to the CS (mostly those requiring an operator assistance).

---

[3] The transshipment operations planning problem, which remains under the responsibility of the CS is more or less a temporal allocation problem and is not presented in this paper.

3.2 A Plan-Merging Protocol for Multi-Robot Navigation

For the case of a number of mobile robots in a route network environment, we have devised a specific *Plan-Merging Protocol* (PMO) based on spatial resource allocation (see [4]). It is an instance of the general protocol described above, but in this context, *Plan-Merging Operation* is done for a limited list of required spatial resources: a set of cells which will be traversed during the plan to merge. The robot broadcasts the set or required cells, receives back the set of coordination plans from other robots which have already planned to use some of the mentioned cells, and then tries to perform a plan insertion which ensures that the union of the considered plans is a directed acyclic graph.

One of the most interesting attributes of this protocol is that it allows several PMOs to be performed simultaneously if they involve disjunctive resource sets. This is particularly useful when there are several local conflicts at the same time.

3.2.1 Plan-Merging for cell occupation:

In most situations, robot navigation and the associated Plan-merging procedure are performed by trying to maintain each cell of the environment occupied by at most one robot. This allows the robots to plan their trajectories independently, to compute the set of cells they will cross and to perform Plan-Merging at cell allocation level.

In order not to constrain unnecessarily the other robots, the allocation strategy makes a robot allocate one cell ahead when it moves along lanes, while it allocates all the cells necessary to traverse and leave crossings.

3.2.2 When reasoning about cells is not sufficient

While, most of the time, the robots may restrict their cooperation to cells allocation, there are situations where this is not enough. This happens when they have to cross large (non-structured) areas or when an unexpected obstacle, encountered in a lane or in a crossing, forces a set of robots to maneuver simultaneously in a set of cells. In such situations, a more detailed cooperation (using the same protocol but a different planner: the motion planner) takes place allowing robots to coordinate their actions at trajectory level. Figure 8 illustrates the results of a *PMO* at trajectory level leading to trajectory synchronizations: $R_i\text{-}W_j$ stands for a position where robot r_i should stop and wait that robot r_j has passed position $R_j\text{-}S_i$.

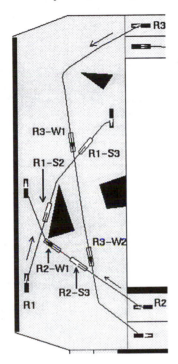

Figure 8: *The result of a PMO at trajectory level*

Thus, we have a hierarchy of PMOs:

- first, at the cell level, based on resource (cells) allocation

- then, depending on the context, at trajectory level: motion planning in a set of common cells determined by the first level

This hierarchy authorizes a "light" cooperation, when possible, and a more detailed one, when necessary.

3.3 Examples

We shall now illustrate the plan-merging paradigm and its capabilities with some sequences from our exper-

imentation with simulated robots in a route network with open areas. The first example presents a PMO at a crossing, the second, a PMO in an open area.

Example 1: Coordination at cell level (Figure 9).

- **Step 1:** This snapshot shows the involved cells of the environment.
  The robot destinations are the followings:

  - Robots 0 and 1 on the right go to cell $C8$ above the crossing using cell $C4$.
  - Robots 2 and 3 at the bottom right traverse the crossing to reach the left cell $C0$ using cells $C5, C4$ and $C2$.
  - Robot 6 goes from left to the right cell $C7$ using cells $C3$ and $C5$.
  - Robot 4 goes from up to the lower cell $C10$ using cells $C2$ and $C3$.

  The PMOs have occurred in the following order: robot-0 then robot-2 and then robot-6 in parallel with robot-1 (because robot-6 and robot-1 have disjunctive lists of resources) and finally robot-4.

- **Step 2:** The following synchronizations have been planned: robot-2 on robot-0 (which frees $C4$), robot-6 on robot-2 (which frees $C5$) and robot-1 on robot-2 (which frees $C4$), robot-4 on robot-2 (which frees c2) and robot-6 (which frees $C3$)

- **Step 3:** One should note that at this stage, robot-3 PMO fails and is deferred because robot-2 has not yet planned an action to free the cell $C0$.

Example 2: Trajectory level (Figure 10) The second example illustrates PMO at trajectories level in a large open area with two obstacles in the middle, and 10 docking/undocking stations. In such an environment, there are no cell allocations (the robots are all in the same cell), all synchronizations are made at trajectory level.

Figure 10 shows a situation where all the robots have planned and coordinated a complete trajectory. The trajectories displayed on the figure are the one which

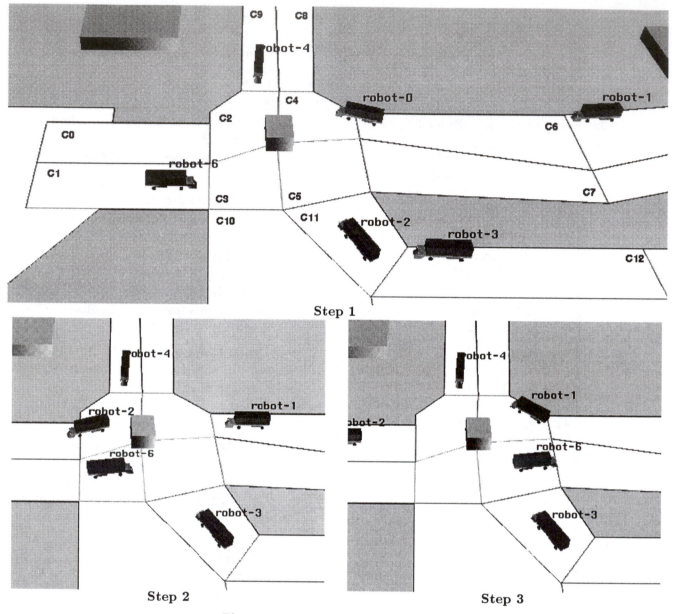

Figure 9: *Plan-merging at the cell level.*

have been sent by the robots for execution display.

• The robot destinations are: $r0$ goes to station 9, $r4$ to station 5, $r1$ to station 1, $r28$ to station 7, $r27$ to station 0, $r20$ to station 4.

• PMOs were done in the following order: $r1$, $r4$, $r27$, $r20$, $r28$ and $r0$.

• One can see the synchronizations established by the robots: $r4$ on $\{r1\}$, $r27$ on $\{r1, r4\}$, $r20$ on $\{r1, r27, r4\}$, $r28$ on $\{r27, r20, r1\}$, $r0$ on $\{r1, r4, r28, r27\}$.

The two examples, presented above, exhibit the following properties of the PMO:

• Planning and execution is done in parallel.

• Several robots may use the crossing simultane-

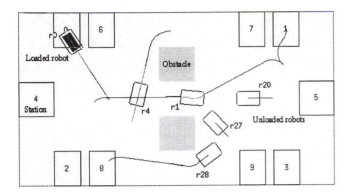

Figure 10: *Plan-merging at the trajectory level.*

ously.

- The example exhibits the two types of synchronization: *execution events* and *planning events* ($r2$ and $r3$ in the crossing example).

- Each robot produces and merges its plans iteratively, and the global plan for the use of the crossing is incrementally built through several PMOs performed by various robots.

3.4 Implementation and Results

We have developed a complete robot control system which includes all the features described. Its architecture is based on a generic control architecture for autonomous mobile robots developed at LAAS [7, 2]. It is instantiated in this case by adding an intermediate layer for performing Plan-Merging operations.

3.4.1 Emulation Testbed

For testing and demonstration purposes, the robot control system has been linked to a robot simulator.

Experiments have been run successfully on a dozen of workstations (each workstation running a complete robot simulator) communicating through Ethernet. A 3-d graphic server has been built in order to visualize the motions and the load operations performed by all the robots in a route network (Figure 9) or in-door (Figure 10) environments. The simulated robots where able to achieve navigation missions, performing hundreds of PMOs and solving local conflicts. Motion

planning and PMOs were sufficiently efficient to allow most often the robots to elaborate and merge their plans without stopping unless necessary.

Running a fleet of ten robots during thirty minutes on a large outdoor environment slightly more complex than the one presented on Figure 5 generates about 4024 messages between the robots which can be classified into 416 PMO requests, resulting into 3744 responses. 128 cooperation plans are exchanged resulting into 48 synchronizations between executable plans. 32 wait for planning are generated. 15 K-bytes of compressed data are exchanged over the robot network.

Another example also involving ten robots for the same duration, in a more constrained indoor environment resulted in 10168 messages exchanged (note the increase of the number of messages), including 928 PMO requests which led to 8352 responses. Due to the small number of cells, and large areas, 220 PMO request conflicts arose. 936 cooperation plans were exchanged, 176 cell synchronizations and 104 trajectories synchronizations were done. 219 plan dependencies were managed and led to 439 plan update messages. 100 K-bytes of compressed data were exchanged over the robot network.

3.4.2 Real Robots Testbed

Extensive experiments have also been performed using three laboratory robots (Figure 11).

The Hilare robots are equipped with two driving wheels, four free wheels, a VME rack supporting CPU boards of the Motorola 680x0 family, running under the VxWorks real-time system. The sensors used on each robot in this experiment are: an odometer and a gyroscope maintaining its position, a laser range finder used for absolute localization and obstacle modeling. In addition, one of the robots is equipped with a belt of sonars for obstacle detection. The problem of cooperation between sonars on several robots was not approached in this experiment. All robots use radio modems. A set of external cameras attached to the ceiling of the room completes the set of sensors. They are used to provide an absolute localization of the robots.

Our experiment room (which is about 10×7 meters large) has been structured into two areas including six

Figure 11: *The Three Hilare Robots in Mission*

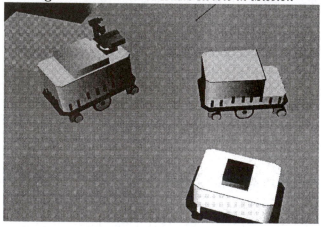

Figure 12: *The Same Mission Viewed with the Display Server*

docking stations and two lanes, according to the environment model presented in Section 3.1.

In this environment, we have conducted runs where the robots keep going for more than two hours. In a typical run, during one hour, one robot:

- covers a cumulated distance of 300 meters,

- exchanges 900 messages with the other robots,

- executes 250 coordination operations which yield to 70 synchronizations at the trajectory level and 20 at the resource level.

- the decisional level produces 1500 requests to the functional level.

The high number of coordinations observed here is a consequence of the small size of the environment compared to the size of our robots. But it fully demonstrates the capabilities of our decisional level.

4   Related work

There are numerous contributions dealing with multi-robot cooperation. However, the term "cooperation" has been used in several contexts with different meanings.

We will not consider here contributions to cooperation schemes at servo level (e.g. [17]) nor contributions which aim at building an "intelligent group" of simple robots (e.g. [18]). We will limit our analysis to contributions which involve an effective cooperation (at plan or program level) between several robots.

Several approaches have been proposed, such as generation of trajectories without collision (e.g. [8, 26]), traffic rules [13, 15], negotiation for dynamic task allocation [16, 5], and synchronization by programming [19, 22].

Inter-robot communication allows to exchange various information, positions, current status, future actions, etc [5, 22, 27] and to devise effective cooperation schemes.

Traffic rules have been proposed as a way to allow several robots to avoid collision and to synchronize their motion (with limited or even without communication). However, many aspects should be taken into account in order to build the set of traffic rules: the tasks, the environment, the robot features, and so on. This entails that the generated rules are valid only under the considered assumptions. If some of them are changed, the rules have to be modified or sometimes be regenerated completely. Besides, these systems are generally built heuristically and do not provide any guarantee such as deadlock detection.

Negotiation have been used for dynamic task [5] or resource [16] allocation to several robots on a situation-dependent basis.

Most contributions which make use of synchronization through communication are based on a pre-defined set of situations or on task dependent properties.

Indeed, most of the methods listed here, deal essentially with collision avoidance or motion coordination and cannot be directly applied to other contexts or tasks.

We claim that our Plan-Merging paradigm is a generic framework which can be applied in different contexts, using different planners (action planners as well as motion planners). It has some clean properties (and clear limitations) which should allow, depending on the application context, to provide a coherent behavior of the global system without having to encode explicitly all situations that may encoded.

5 Conclusion

We have argued that, in the field of multi-robot (and more generally multi-agent) cooperation, it is useful to distinguish between two main issues: **C1** goal/task decomposition and allocation, and **C2** cooperation while seeking to achieve loosely coupled goals. We have even claimed that in numerous multi-agent applications, both issues appear and even "invoke" one another "recursively".

We have then proposed a "generic" approach called *Plan-Merging Paradigm* which deals with **C2** issues and clearly establishes a link with **C1** issues.

The Plan-Merging paradigm has the following properties;

1. It makes possible for each robot to produce a co-ordination plan which is compatible with all plans executed by other robots.

2. No system is required to maintain the global state and the global plan permanently. Instead, each robot updates it from time to time by executing a PMO.

3. The PMO is safe, because it is robust to plan execution failures and allows to detect deadlocks.

We believe that it can be applied to a large variety of contexts and with different planners (from action planners to task or motion planners), and at different granularities.

Such a multi-robot cooperation scheme "fills the gap" between very high level planning (be it centralized or distributed) and distributed execution by a set of autonomous robots in a dynamic environment.

Indeed, it appears to be particularly well suited to the control of a large number of robots navigating in a route network. The application that we have implemented clearly exhibits its main features. It allowed us to make a large number of autonomous robots behave coherently and efficiently without creating a huge activity at the central system.

Besides the investigation of other classes of applications and the work on a more formal description of the proposed approach, our future work will concentrate on developing new cooperation schemes by embedding a multi-robot planning activity.

References

[1] L. Aguilar, R. Alami, S. Fleury, M. Herrb, F. Ingrand, F. Robert *Ten Autonomous Mobile Robots (and even more) in a Route Network Like Environment* In, *IEEE International Conference on Intelligent Robots and Systems*, IROS'95, Pittsburgh (USA), July 1995.

[2] R. Alami, R. Chatila, and B. Espiau. Designing an Intelligent Control Architecture for Autonomous Robots. in *International Conference on Advanced Robotics*, Tokyo (Japan), October 1993.

[3] R. Alami, F. Robert, F. Ingrand, and S. Suzuki. A paradigm for plan-merging and its use for multi-robot cooperation. In *IEEE International Conference on Systems, Man, and Cybernetics*, San Antonio, Texas (USA), October 1994.

[4] R. Alami, F. Robert, F. Ingrand, S. Suzuki. Multi-robot Cooperation through Incremental Plan-Merging In *IEEE International Conference on on*

Robotics and Automation, Nagoya (Japan), May 1995.

[5] H. Asama, K. Ozaki, et al. Negotiation between multiple mobile robots and an environment manager. In *International Conference on Advanced Robotic*, Pisa (Italy), June 1991.

[6] R. I. Brafman, Y. Shoham. Knowledge considerations in robotics and distribution of robotics tasks. *IJCAI 95*, Montréal (Canada), August 1995.

[7] R. Chatila, R. Alami, B. Degallaix, and H. Laruelle. Integrated planning and execution control of autonomous robot actions. In *IEEE International Conference on Robotics and Automation*, Nice, (France), May 1992.

[8] H. Chu and H.A. EiMaraghy. Real-time multi-robot path planner based on a heuristic approach. In *IEEE International Conference on Robotics and Automation*, Nice, (France), May 1992.

[9] E.H. Durfee, V. Lesser Partial Global Planning: A Coordination Framework for Distributed Hypothesis Formation In *IEEE Transactions on Systems, Man and Cybernetics*, Vol 21 (5), October 1991.

[10] E.H. Durfee, T. A. Montgomery Coordination as Distributed Search in a Hierarchical Behavior Space In *IEEE Transactions on Systems, Man and Cybernetics*, Vol 21 (6), December 1991.

[11] E. Ephrati, M. Perry, J.S. Rosenschein. Plan execution motivation in multi-agent systems. *AIPS'94*, Chicago, June 1994.

[12] G. Ferguson, J.F. Allen. Arguing about plans: plan representation and reasoning for mixed-initiative planning. *AIPS'94*, Chicago, June 1994.

[13] D.D. Grossman. Traffic control of multiple robot vehicles. *IEEE Journal of Robotics and Automation*, 4(5):491–497, October 1988.

[14] N.R. Jennings. Controlling cooperative problem solving in industrial multi-agent systems using joint intentions *Artificial Intelligence*, Vol 73 (1995) 195-240.

[15] S. Kato, S. Nishiyama, and J. Takeno. Coordinating mobile robots by applying traffic rules. In *IEEE International Conference on Intelligent Robots and Systems*, IROS'92, Raleigh, (USA), July 1992.

[16] C. Le Pape. A combination of centralized and distributed methods for multi-agent planning and scheduling. *IEEE International Conference on Robotics and Automation*, Cincinnati (USA), May 1990.

[17] Z.W. Luo, K. Ito, and M Ito. Multiple robot manipulators cooperative compliant manipulation on dynamical environments. In *IEEE International Conference on Intelligent Robots and Systems*, IROS'93, Yokohama, (Japan), July 1993.

[18] M. Mataric. Minimizing complexity in controlling a mobile robot population. In *IEEE International Conference on Robotics and Automation*, Nice, (France), May 1992.

[19] F.R. Noreils. Integrating multi-robot coordination in a mobile-robot control system. In *IEEE International Conference on Intelligent Robots and Systems*, IROS'90, Tsuchiura (Japan), July 1990.

[20] K. Ozaki, H. Asama, et al. Synchronized motion by multiple mobile robots using communication. *IEEE International Conference on Intelligent Robots and Systems*, IROS'93, Yokohama (Japan), July 1993.

[21] L.E. Parker. Heterogeneous multi-robot cooperation. *MIT Technical Report AITR-1465*, February 1994.

[22] S. Yuta, S. Premvuti. Coordination autonomous and centralized decision making to achieve cooperative behaviors between multiple mobile robots. *IEEE International Conference on Intelligent Robots and Systems*, IROS'92, Raleigh (USA), July 1992.

[23] J.S. Rosenschein, G. Zlotkin. Designing conventions for automated negotiation. *AI MAGAZINE, Volume 15, (1994)*

[24] Y. Shoham, M. Tennenholtz. On social laws for artificial societies: Off-Line design. *Artificial Intelligence, 73 (1995) 231-252*

[25] R.G. Smith. The Contract Net Protocol: High-level communication and control in a distributed problem solver. *IEEE Transactions on Computers*, Vol C-29 (12), December 1980.

[26] T. Tsubouchi and S. Arimoto. Behavior of a mobile robot navigated by an "iterated forecast and planning" scheme in the presence of multiple moving obstacles. In *IEEE International Conference on Robotics and Automation*, San Diego, (USA), May 1994.

[27] J. Wang. On sign-board based inter-robot communication in distributed robotic systems. In *IEEE International Conference on Robotics and Automation*, San Diego, (USA), May 1994.

Acknowledgments: This work was partially supported by the MARTHA (ESPRIT III) Project, the CNRS, the Région Midi-Pyrénées and the ECLA RO-COMI Project.

It is the fruit of a very intensive collaboration between numerous researchers: F. Robert, F. Ingrand, S. Fleury, M. Herrb, S. Qutub, L. Aguilar. I would like also to acknowledge the help and the effective involvement of M. Khatib, H. Bullata, S. Suzuki, J. Perret, T. Siméon, B. Dacre-Wright, M. Devy.

Robot Motion Planning: A Game-Theoretic Foundation

Steven M. LaValle, *Stanford University, Stanford, CA USA*

This paper proposes a dynamic game-theoretic framework that is used as an analytical tool and unifying perspective for a wide class of problems in motion planning. This approach is inspired by the foundation laid by configuration-space concepts for basic path planning. In the same manner that configuration-space concepts led to substantial progress in path planning, game-theoretic concepts provide a more general foundation which can incorporate any of the essential features of path planning, sensing uncertainty, decision theory, bounded-uncertainty analysis, stochastic optimal control, and traditional multiplayer games. By following this perspective, new modeling, analysis, algorithms, and computational results have been obtained for a variety of motion planning problems including those involving uncertainty in sensing and control, environment uncertainties, and the coordination of multiple robots.

1   Introduction

It is widely accepted that the configuration-space (C-space) representation has provided a powerful, unified foundation for the development and analysis of motion planning algorithms. In spite of this success, there has been little attempt to obtain further benefits by broadening this foundation into a common mathematical structure that encompasses many important, well-studied extensions of the basic planning problem. This paper proposes such a foundation by combining decision-theoretic concepts from areas such as dynamic game theory and stochastic optimal control with C-space concepts. The intent is not to provide an alternative formulation of motion planning, but instead to present an expanded foundation that is built on previous geometric concepts, while characterizing and unifying a broader class of problems.

The *basic* motion planning problem has been to determine a continuous, collision-free path that connects an initial configuration to a goal configuration. This problem was created by modularizing robotic tasks to isolate path planning from lower-level trajectory tracking. If the modularization is completely removed, many robotic tasks can be considered as a nonlinear control problem for which there are complicated constraints on the state space (which include the constraints due to static obstacles). Although basic motion planning has found many direct applications, there are fundamental limitations that have motivated many specific approaches to handle difficult extensions of the basic problem. Let *general* motion planning include complications such as sensing uncertainties, prediction uncertainties, nonholonomy, dynamics, performance criteria, and multiple robots with independent goals.

Several benefits arose from the use of the C-space representation for basic motion planning. The comparison of seemingly disparate approaches, such as cell decomposition methods, roadmap methods, and artificial potential field methods, was greatly facilitated through the use of C-space representations [22],[29]. Concepts such as *completeness* and *resolution completeness* were formulated in terms of configuration space, and hence applied to a wide class of problems and algorithms. Planning algorithms could also be generalized by utilizing the common C-space foundation. For instance, a randomized potential field planner was applied with only minor adaptations to a variety of problems ranging from multiple rigid robots to high degree-of-freedom manipulators [4]. Wide applicability is obtained because all problems are reduced to C-space terms.

The same basic philosophy can be preserved for general motion planning by using a broader mathematical foundation that provides the same types of benefits that configuration-space concepts provided for the basic planning problem. It is important to note, however, that this paper emphasizes a *mathematical foundation* as opposed to a particular model or computational approach. Algorithms and computed examples are presented in this paper for the purpose of demonstrating the power of this foundation, to encourage its use in future motion planning research. More details on the specific algorithms, analysis, and computed examples appear in [24] and [25]-[28], [37].

2   Mathematical Formulation

Before considering a formulation of general motion planning, first consider making small extensions to the basic motion planning problem. The basic problem is to find a continuous path $x : [0, t_f] \rightarrow C_{free}$ such that $x(0) = q_{init}$ and $x(t_f) = q_{goal}$. Recall that C_{free} implicitly incorporates all of the constraints due to the robot geometry and static obstacles in the workspace.

Suppose that there are nonholonomic constraints. To facilitate upcoming concepts, let C_{free} be renamed as a generic state space, $X = C_{free}$. It is well known that the nonholonomic constraints can be expressed as $\dot{x} = f(x(t), u(t))$, which constrains the allowable vector fields on X. Instead of directly choosing $x(t)$, one is forced to interact with the system using the input (or action) $u(t)$. This occurs, for example, when manipulating an object through pushing [31]. If $f(x(t), u(t)) = u(t)$, the original nonholonomic, basic motion planning problem is obtained since any desired, collision-free path in the state space can be obtained by selecting an appropriate input.

Suppose that optimality with respect to some criterion, such as path length or execution time, is important. A *loss functional* can be defined that evaluates any state trajectory and input:

$$L(x(\cdot), u(\cdot)) = \int_0^{t_f} l(x(t), u(t))dt + Q(x(t_f)). \quad (1)$$

The integrand $l(x(t), u(t))$ allows the specification of a cost that will accumulate during execution and will depend in general on the state trajectory and the input. The final term $Q(x(t_f))$ can indicate the importance of achieving the goal. The basic problem can be considered as a special form of optimal control [13]. Suppose $l(x(t), u(t)) \equiv 0$, and $Q(x(t_f)) = 0$ if $x(t_f) = q_{goal}$ and $Q(x(t_f)) = 1$ otherwise. This corresponds to the original case in which optimality is not important. The

space of possible inputs to the system in this case is partitioned into two classes: those that lead to the goal region, and those that fail.

Next, consider moving to a mathematical structure for the general motion planning problem, which is based on concepts from dynamic noncooperative game theory [1] and stochastic optimal control [20]. This structure will be formulated in discrete time to ease the specification of uncertainty aspects; however, continuous time can alternatively be used with some minor modifications. Thirteen components are first listed, and a discussion of each in relation to motion planning follows.

1. An index set, $\mathbf{N} = \{1, 2, \ldots, N\}$, of N decision makers

2. An index set, $\mathbf{K} = \{1, 2, \ldots, K\}$, that denotes the *stages* of the game

3. A set, X, called the *state space*. The state of the game, x_k, at stage k, belongs to X.

4. A set, U_k^i, defined for each $k \in \mathbf{K}$ and $i \in \mathbf{N}$, which is called the *action set* of the ith decision maker at stage k. The *action*, u_k^i, at stage k, belongs to U_k^i.

5. A set, Θ_k^a, defined for each $k \in \mathbf{K}$, which is called the *control action set for nature* at stage k. The *control action for nature*, θ_k^a, at stage k, belongs to Θ_k^a.

6. A function, $f_k : X \times U_k^1 \times \ldots \times U_k^N \times \Theta_k^a \rightarrow X$, defined for each $k \in \mathbf{K}$ so that

$$x_{k+1} = f_k(x_k, u_k^1, \ldots, u_k^N, \theta_k^a), \quad (2)$$

is a *state transition equation*.

7. A set, Y_k^i, defined for each $k \in \mathbf{K}$ and $i \in \mathbf{N}$, and called the *sensor space* of the ith decision maker at stage k, to which the sensed observation y_k^i belongs at stage k.

8. A set, $\Theta_k^{s,i}$, defined for each $i \in \mathbf{N}$ and $k \in \mathbf{K}$, which is called the *sensing action set for nature* at stage k. The *sensing action for nature*, $\theta_k^{s,i}$, at stage k, belongs to $\Theta_k^{s,i}$.

9. A function, h_k^i, defined for each $k \in \mathbf{K}$ and $i \in \mathbf{N}$, so that

$$y_k^i = h_k^i(x_k, \theta_k^{s,i}), \qquad (3)$$

which is the *observation equation* of the ith decision maker concerning the value of x_k.

10. A finite set, η_k^i, defined for each $k \in \mathbf{K}$ and $i \in \mathbf{N}$ as a subset of all actions and observations made by decision makers at any previous stage, $\{u_1^1, \ldots, u_{k-1}^N, y_1^1, \ldots, y_k^N\}$.

11. A set of all possible values for η_k^i, denoted by N_k^i, which is called the *information space* for the ith decision maker at stage k.

12. A set, Γ_k^i, of mappings $\gamma_k^i : N_k^i \to U_k^i$, which are the *strategies* available to the ith decision maker at stage k. The combined mapping $\gamma^i = \{\gamma_1^i, \gamma_2^i, \ldots, \gamma_K^i\}$ is a *strategy* for the ith decision maker, and the set Γ^i of all such mappings γ^i is the *strategy space* of the ith decision maker. A *game strategy*, γ, represents a simultaneous specification of the strategy for each decision maker, and the space of game strategies is denoted by $\Gamma = \Gamma^1 \times \cdots \times \Gamma^N$.

13. An (extended) real-valued functional $L^i : (X \times U_1^1 \times \ldots \times U_1^N) \times (X \times U_2^1 \times \ldots \times U_2^N) \times \ldots \times (X \times U_K^1 \times \ldots \times U_K^N) \times \Theta \to \Re^+$, defined for each $i \in \mathbf{N}$, and called the *loss functional* of the ith decision maker. The Cartesian product of all of nature's action spaces is represented here as Θ.

State transitions and control Item 1 defines the decision makers, which each typically refers to an independent, controllable robot. In general, however, any agent that is capable of making decisions and interfering with the other decision makers can be considered as a decision maker.

Item 2 defines stages that correspond to times at which decisions are made. For standard discrete-time analysis, decisions are made at each Δt time increment. The limiting case of $K = \infty$ can be defined. In general, decision making at regular intervals is not required. Suppose for instance that the decisions correspond to very high-level operations, which may have unpredictable completion times. This case is discussed in more detail in [37], for modeling the completion of a fine-motion operation. A continuum of stages can alternatively be considered, which results in a continuous-time differential game (e.g., [18]).

The state space is defined in Item 3. At the very least, the state space can be used to represent the free configuration space, $\mathcal{C}_{free}$. In the case of multiple robots, it can represent the composite configuration space that is formed by taking the Cartesian product of the configuration spaces of the individual robots. In general, however, the state space could incorporate additional information. For instance, dynamics can be included by expanding the state space to include configuration time derivatives. This corresponds to the standard use of state space representations in optimal control theory. The state space can also include any parameters that can be completely or partially controlled through the operation of the robot(s). In one application [28] the state space includes *environment modes* that characterize varying conditions in the environment that potentially affect the robot.

Item 4 defines the set of actions that are available to each decision maker at each stage.

Item 5 is used to model sources of uncertainty. Two common representations of uncertainty have been applied to motion planning problems. With a *nondeterministic* (or bounded-set) representation parameter uncertainties are restricted to lie within a specified set. A motion plan is then generated that is based on *worst-case* analysis (e.g., [6], [12], [23], [30]). With a *probabilistic* representation the parameter uncertainties are characterized with a probability density function (pdf). This often leads to the construction of motion plans through *average-case* or *expected-case* analysis (e.g., [5], [15]).

One key aspect of the proposed mathematical foundation is a general capacity to model uncertainties. This is accomplished by introducing a decision maker referred to as *nature*. It will be assumed that no one has complete control over actions that are chosen by nature; however, models can be constructed to partially predict nature's actions. Nature can introduce nondeterministic or probabilistic uncertainties into the game by applying either *control actions* or *sensing actions*. Item 5 defines the set of control actions that are available to nature, and Item 8 will define the set of sensing actions that are available to nature.

Item 6 defines how changes in state are effected. The state x_{k+1}, at stage $k + 1$, is obtained as a function of the previous state x_k and the actions chosen by all decision makers, including nature. If nature is omitted from the state transition equation, then perfect prediction of future states is possible, given the actions of the decision makers. Under nondeterministic uncertainty, a set of possible future states can be derived from the state transition equation as:

$F_{k+1}(x_k, u_k^1, \ldots, u_k^N) =$

$$\{f(x_k, u_k^1, \ldots, u_k^N, \theta_k^a) \in X | \theta_k^a \in \Theta_k^a\}. \quad (4)$$

Under probabilistic uncertainty, it is assumed that $p(\theta_k^a)$, is known. By using the state transition equation, the next state is represented by a pdf, $p(x_{k+1} | x_k, u_k^1, \ldots, u_k^N)$.

A control example   As an example of a state transition equation with uncertainty, consider characterizing the uncertainty model that is used for motion control in preimage planning research (e.g., [12], [23], [30]). Suppose there is a single decision maker that is a polygonal robot translating in the plane amidst polygonal obstacles. The action set defines commanded velocity directions, which can be specified by an orientation, yielding $U = [0, 2\pi)$ (for the case of a single decision maker, the superscripts will be dropped). The robot will attempt to move a fixed distance $\|v\|\Delta t$ (expressed in terms of a constant velocity modulus, $\|v\|$) in the direction specified by u_k. The action space of nature is a set of angular displacements θ_k^a, such that $-\epsilon_\theta \leq \theta_k^a \leq \epsilon_\theta$, for some maximum angle ϵ_θ. Under nondeterministic uncertainty, any action $\theta_k^a \in [-\epsilon_\theta, \epsilon_\theta]$ can be chosen by nature. When using probabilistic uncertainty, $p(\theta_k^a)$ could be a continuous pdf, which is zero outside of $[-\epsilon_\theta, \epsilon_\theta]$. If the robot chooses action u_k from state x_k, and nature chooses θ_k^a, then x_{k+1} is given by

$$f(x_k, u_k, \theta_k^a) = x_k + \|v\|\Delta t \begin{bmatrix} cos(u_k + \theta_k^a) \\ sin(u_k + \theta_k^a) \end{bmatrix}. \quad (5)$$

Sensing uncertainty   Items 7 through 11 characterize the information that can be used for the basis of decision making when there is not direct access to the state. This can be considered as a general form of the sensing problem in robotics. Each decision maker at each stage has a *sensor space*, Y_k^i (or observation space), which encodes information regarding the state that is observed during stage k. This type of projection is used in optimal control theory to define system

outputs, and has also been considered in robot sensing problems (see, for instance, [8]). In addition to a projection from the state space to the sensor space, this information is potentially corrupted by a sensing action, $\theta_k^{s,i}$, of nature, which is chosen from $\Theta_k^{s,i}$.

Under nondeterministic uncertainty, the possible current states from a single sensor observation are

$$F_k^i(y_k^i) = \{x_k \in X | y_k^i = h_k^i(x_k, \theta_k^{s,i}), \theta_k^{s,i} \in \Theta^{s,i}\}. \quad (6)$$

Under probabilistic uncertainty the current state, assuming only a single observation, is represented by a pdf, $p(x_k | y_k)$.

The sensing model can be generalized to include state history, $y_k^i = h_k^i(x_1, \ldots, x_k, \theta_k^{s,i})$.

A sensing example   Consider representing the sensing model used in [5], [12], [23]. Suppose that a single robot is equipped with a position sensor and a force sensor. Assume that the position sensor is calibrated in the configuration space, yielding values in $\Re^2$. The force sensor provides values in $[0, 2\pi) \cup \{\emptyset\}$, indicating either the direction of force when the robot is in contact with an obstacle, or no force (represented by $\emptyset$) when the robot is in the free space.

Independent portions of the observation equation are considered: h^p for the position sensor and h^f for the force sensor (which together form a three-dimensional vector-valued function). The sensing action of nature, θ_k^s, are partitioned into subvectors $\theta_k^{s,p}$ and $\theta_k^{s,f}$, which act on the position sensor and force sensor, respectively. The observation for the position sensor is $y_k^p = h^p(x_k, \theta_k^{s,p}) = x_k + \theta_k^{s,p}$. Under nondeterministic uncertainty, $\theta_k^{s,p}$ could be any value in $\Theta_k^{s,p}$. If probabilistic uncertainty is used, a pdf is presented, such as

$$p(\theta_k^{s,p}) = \begin{cases} \frac{2}{\pi \epsilon_p^2} & \text{for } \|\theta_k^{s,p}\| < \epsilon_p \\ 0 & \text{otherwise} \end{cases}. \quad (7)$$

In (7) a radius ϵ_p is specified, and $\theta_k^{s,p}$ is two-dimensional.

One of two possibilities is obtained for the force sensor: (1) a value in $[0, 2\pi)$, governed by $y_k^f = h^f(x_k, \theta_k^{s,f}) = \alpha(x_k) + \theta_k^{s,f}$, in which $x_k \in \mathcal{C}_{contact}$ (i.e., the position lies in the boundary of $\mathcal{C}_{free}$), and the true normal is given by $\alpha(x_k)$, or (2) an empty value,

$\emptyset$, when the robot is in $\mathcal{C}_{free}$. When the robot configuration lies in $\mathcal{C}_{contact}$ and probabilistic uncertainty is in use, then the pdf can be represented as

$$p(\theta_k^{s,f}) = \begin{cases} \frac{1}{2\epsilon_f} & \text{for } |\theta_k^{s,f}| < \epsilon_f \\ 0 & \text{otherwise} \end{cases}, \qquad (8)$$

for some positive prespecified constant $\epsilon_f < \frac{1}{2}\pi$.

Information spaces Items 10 and 11 characterize the history that is available for decision making. The relationship between sensor and action history and decision making has long been considered important in planning under uncertainty (e.g., [12], [22], [30]). Generally one would like to optimize the performance of a robot, while directly taking into account the complications due to limited sensing. By using the concept of information state, as considered in stochastic control and dynamic game theory, a useful characterization of this relationship is provided. When there is perfect state information, decisions can be made on the basis of state. However, with imperfect state information, the decisions are conditioned on information states. The information state concept is similar to the definition of knowledge states, considered in [10], and has also recently been proposed in [2].

In Item 10, the dimension of the information space can increase linearly with the number of stages; however, alternative representations are possible, and preferable in many cases. In the case of nondeterministic uncertainty, the information space can be alternatively represented as an algebra of subsets of X that are obtained by performing set intersections that maintain consistency with the history. With probabilistic uncertainty, the information space can be alternatively represented as a pdf on X that is obtained through the repeated application of Bayes' rule. Functional approximation can also be considered to produce a low-dimensional representation of the information space [24].

The strategy concept Item 12 defines a strategy for each decision maker. The computational goal is to design a strategy that will lead to the accomplishment of some robotic task. At a given stage, each decision maker conditions its actions on its information state. This represents a deterministic (or pure) strategy; however, a randomized (or mixed) strategy can

alternatively be defined. In this case a pdf of the form $p(u_k^i|\eta_k^i)$ is specified as the strategy, and actions are chosen by sampling.

Encoding preferences Item 13 defines a loss functional for each of the decision makers, which guides the selection of strategies. The loss can generally be based on actions taken by any decision maker at any stage, and on the state trajectory. In this general form, the loss functionals can also depend on nature.

One form that is often used in discrete-time optimal control theory is the stage-additive loss functional (for a single decision maker):
$$L(x_1, \ldots, x_{K+1}, u_1, \ldots, u_K) =$$

$$\sum_{k=1}^{K} l_k(x_k, u_k) + l_{K+1}(x_{K+1}), \qquad (9)$$

in which $l_k(x_k, u_k)$ represents a cost that can accumulate (such as time, distance, or energy), and $l_{K+1}(x_{K+1})$ is a final cost that could, for instance, penalize a strategy that does not terminate in a goal region.

The general task is to determine strategies that optimize the losses in some appropriate sense. In the case of a single decision maker without nature, the task is to select a strategy that minimizes L. In the case of nondeterministic actions from nature, the task is to select a strategy that minimizes the worst-case loss. In the probabilistic case, one natural choice is to minimize the expected loss. For cases in which there are multiple, independent decision makers, a number of different concepts may be appropriate. For instance, in a cooperative game in which there is a certain amount of trust, *Pareto optimality* may be appropriate [34]. In a noncooperative setting, a Nash equilibrium condition might be appropriate [1]. This corresponds to a game strategy that minimizes the loss of each decision maker, given that the strategies of the other decision makers cannot be changed.

3 Synthesizing and Extending Motion Planning Concepts

By following this game-theoretic perspective, modeling, analysis, algorithms, and computed examples have so far been obtained for three classes of problems: (1)

motion planning under uncertainty in sensing and control [24], [25], [26]; (2) motion planning under environment uncertainties [24], [28]; and (3) multiple-robot motion planning [24], [27]. For the first problem class, a general method for determining feedback strategies is developed by blending ideas from dynamic game theory with traditional preimage planning concepts. This generalizes classical preimages to *performance preimages* and preimage plans to *motion strategies with information feedback*. For the second problem class, robot strategies are analyzed and determined for situations in which the environment is changing, but not completely predictable. For the third problem class, dynamic game-theoretic concepts are applied to motion planning for multiple robots that have independent goals. Several versions of the formulation have been considered: fixed-path coordination, coordination on independent configuration-space roadmaps, and centralized planning.

This section highlights some of the key concepts, to illustrate the the utility of the mathematical representation. Section 4 discusses some selected algorithm issues and presents some illustrative, computed examples.

Modeling sources of uncertainty Several types of uncertainty will be discussed for the single-robot case. It is straightforward to extend the discussion to multiple robots. All types are modeled with nature, which can be assumed to be either nondeterministic or probabilistic.

Suppose that $X = \mathcal{C}_{free}$, and let $\mathbf{q}_k$ denote the configuration (or state) at stage k. The state transition equation (2) can be specialized to $\mathbf{q}_{k+1} = f_k(\mathbf{q}_k, u_k, \theta_k)$. This represents a generalization of the control model that was given in (5), and represents uncertainty in *configuration predictability*. Suppose further that the observation equation is of the form $y_k = h_k(\mathbf{q}_k, \theta_k^s)$. This represents a generalization of the sensing model that was given in (8), and represents uncertainty in *configuration sensing*.

Other sources of uncertainty can be considered in addition to configuration uncertainties. Suppose, for example, that $\mathcal{C}_{free}$ is not exactly known, but is instead known to be one of several possibilities. In this case there is uncertainty in the robot's environment. A set E can be used to index the alternatives, and

a state space is defined as some subset $X \subset \mathcal{C} \times E$ [28]. Thus, for every $e \in E$, a different free configuration space can be obtained. Let $[\mathbf{q}_k \ e_k]$ represent the state at stage k. A state transition equation can be defined in two portions. Suppose that the future configurations are obtained deterministically from $\mathbf{q}_{k+1} = f_k'(\mathbf{q}_k, u_k)$, and future environments are obtained from $e_{k+1} = f_k''(e_k, \theta_k^a)$. In this case nature causes uncertainty in *environment predictability*. More generally, the future environments can be conditioned on the robot's configuration (which occurs, for instance in a manipulation problem) and the action, to yield $e_{k+1} = f_k''(x_k, \theta_k^a)$, in which $x_k = [\mathbf{q}_k \ e_k]$.

If the current environment is unknown, then there is uncertainty in *environment sensing*, which is a problem that has been considered from several different perspectives (e.g., [7],[9],[17],[38]). This can be modeled by defining $y_k = h_k(x_k, \theta_k^s)$, in which $x_k = [\mathbf{q}_k \ e_k]$.

In general, sensing and predictability uncertainties can be defined for any state space, including those that include dynamics. Also, a set of parameters could characterize variations in the model, and used to form models of uncertainty in predictability and sensing, in the same way that E was used.

It has been assumed thus far that each decision maker knows all game components, including the loss functionals, of other decision makers. Another sensing model could be introduced that reflects imperfect information that each decision maker has about the game itself. Problems of this type are quite realistic, yet are very difficult to model [14], [16]. The information of each decision maker could be represented, for example, as a pdf over a set of possible games. To make appropriate decisions, each decision maker must speculate about the knowledge that other decision makers have regarding the game. This type of second-guessing can progress for an infinite number of layers, which leads to a formidable modeling task.

Forward projections In preimage planning research, the notion of a forward projection has been useful for characterizing robot execution when there is uncertainty in configuration predictability and sensing. This concept can be substantially generalized, and in Section 4 a computed example of a probabilistic forward projection is shown.

The forward projection in this section will characterize future states under the implementation of a strategy. Without uncertainties, this corresponds to providing the state trajectory that can be inferred from (2). Suppose that the strategy, γ_k is fixed for all k, and that there is perfect current-state information available at all times. Under nondeterministic uncertainty, a subset of X in which the system state will lie can be inferred. Consider the state at stage x_{k+2}, if x_k is known. From (4), it is already known that $x_{k+1} \in F_{k+1}(x_k, u_k)$, and $u_k = \gamma_k(x_k)$. The nondeterministic action of nature at stage $k+1$ must next be taken into account to yield $F_{k+2}(x_k, \gamma) =$

$$\{f(x_{k+1}, u_{k+1}, \theta^a_{k+1}) \in X | x_{k+1} \in F_{k+1}(x_k, \gamma), \ \theta^a_{k+1} \in \Theta^a\}. \quad (10)$$

This defines the forward projection at stage $k+2$ in terms of the projection at stage $k+1$. By induction, forward projections can be constructed to any future stage.

Forward projections can be analogously constructed for probabilistic uncertainty. For instance, the pdf at stage $k+2$ is $p(x_{k+2}|x_k, \gamma) =$

$$\int p(x_{k+2}|x_{k+1}, \gamma_{k+1}(x_{k+1}))p(x_{k+1}|x_k, \gamma_k(x_k))dx_{k+1}. \quad (11)$$

These are the forward projections for the cases of nondeterministic uncertainty and probabilistic uncertainty, with a single robot that has uncertainty only in predictability (i.e., the current state is known). Other forward projections, which include sensing uncertainty and multiple robots, are presented along with additional computed examples in [24].

Termination conditions The decision to halt the robot has been given careful attention in manipulation planning research, particularly in cases that involve configuration-sensing uncertainty. A motion plan might bring the robot into a goal region (*reachability*), but the robot may not halt if it does not realize that it is in the goal region (*recognizability*) [12]. The notion of a termination condition has been quite useful for formulating robot plans that tell the robot when to halt, based on its current, partial information [12], [23], [30]. The same concept can be introduced in a game-theoretic formulation by defining a binary-valued mapping (as part of a strategy),

$$TC_k : N_k \rightarrow \{true, false\}, \quad (12)$$

and enforcing the constraint that if $TC_k = true$, then $TC_{k+1} = true$. The *true* condition indicates that the robot should halt, and can be considered as a special action that can be considered by a decision maker (and hence incorporated into a strategy that uses information feedback). This termination condition, in the determination of an optimal strategy, is equivalent to an *optimal stopping rule*, which appears in optimal control theory [20], [24].

Performance preimages Recall that a classical preimage yields the set of places in the configuration space from which a goal will be achieved under the application of a fixed motion command. This principle can be significantly generalized within the game-theoretic framework to yield a *performance preimage*.

Assume that a strategy encodes a termination condition in addition to motion control. Suppose that there is nondeterministic uncertainty, which is standard in preimage planning research. Consider some subset of the reals, $\mathcal{R} \subseteq \Re$. The *performance preimage on X* is the subset of X that is given by

$$\breve{\pi}_x(\gamma, \mathcal{R}) = \{x_1 \in X | \breve{L}(x_1, \gamma) \in \mathcal{R}\}, \quad (13)$$

in which $\breve{L}(x_1, \gamma)$ represents the worst-case loss that could be obtained under the implementation of γ with an initial state x_1. The set $\breve{\pi}_x(\gamma, \mathcal{R}) \subseteq X$ indicates places in the state space from which if robot begins, the loss will lie in $\mathcal{R}$.

The state space, X, can be partitioned into *isoperformance classes* by defining an equivalence class $\breve{\pi}_x(\gamma, \{r\})$ for each $r \in [0, \infty)$. For a 0-1 loss functional (zero if the goal region is achieved), $\breve{\pi}_x(\gamma, \{0\})$ yields the classical preimage. With a general loss functional, and $\mathcal{R} = [0, m)$ a performance preimage is obtained that indicates all $x_1 \in X$ from which the goal will be achieved with a loss that is guaranteed to be less than m. If a termination condition is neglected, then $\breve{\pi}(g, \{0\})$ yields a *backprojection* similar to that in [12].

Suppose probabilistic uncertainty is considered instead of nondeterministic uncertainty. The performance preimage becomes

$$\bar{\pi}_x(\gamma, \mathcal{R}) = \{x_1 \in X | \bar{L}(x_1, \gamma) \in \mathcal{R}\}, \quad (14)$$

in which $\bar{L}(x_1, \gamma)$ represents the *expected* loss that is obtained under the implementation of γ from x_1. Suppose that $\mathcal{R} = [0, r]$ for some $r \geq 0$. The performance preimage yields places in X from which the expected performance will be less than or equal to r. If $\mathcal{R} = \{r\}$ for some point $r \geq 0$, then places in X are obtained in which equal expected performance will be obtained. With a 0-1 loss functional and ignoring the termination condition, the performance preimages can give isoprobability curves which are equivalent to the probabilistic backprojections in [5].

Performance preimages can also be defined on the information space to account for sensing uncertainty, and for multiple robots [24].

Decoupling multiple robots Consider the problem of coordinating multiple robots that have independent goals. Approaches to multiple-robot motion planning are often categorized as *centralized* or *decoupled*. A centralized approach typically constructs a path in a composite configuration space, which is formed by combining the configuration spaces of the individual robots (e.g., [3],[36]). A decoupled approach typically generates paths for each robot independently, and then considers the interactions between the robots (e.g., [11],[19],[33]). The suitability of one approach over the other is usually determined by the tradeoff between computational complexity associated with a given problem and the amount of completeness that is lost.

A variety of multiple-robot coordination problems can be formulated by defining appropriate state spaces [24]. Suppose there are a collection of N robots that share a common workspace and have free spaces $\mathcal{C}_{free}^1, \ldots, \mathcal{C}_{free}^N$. The state space can be defined as the Cartesian product

$$X = \mathcal{C}_{free}^1 \times \mathcal{C}_{free}^2 \times \cdots \times \mathcal{C}_{free}^N. \quad (15)$$

The subset of X in which two or more robots collide is avoided in a successful motion plan. The dimensionality of this composite space has previously prompted many approaches that decouple the problem. Motion plans are more or less constructed for each robot independently, and then combined to coordinate the robots.

In [25], two additional state space definitions are used. For fixed-path coordination, it is assumed that a collision-free path $\tau^i : [0, 1] \to \mathcal{C}_{free}^i$ is given for each robot, and the state space is defined as the Cartesian product $[0, 1]^N$. Instead of a single collision-free path,

suppose that each robot is given a network of collision-free paths, referred to as a *roadmap*. Let $\mathcal{R}^i$ denote a space that is formed by combining the domains of the roadmap paths for the ith robot. A roadmap coordination space can be defined as

$$X = \mathcal{R}^1 \times \mathcal{R}^2 \times \cdots \times \mathcal{R}^N. \quad (16)$$

In general, many other combinations of constrained spaces are possible to define the state, leading to a variety of ways to define decoupled planning problems.

Multiple-robot optimality Little concern has been given in previous research to optimality for multiple-robot motion planning problems. For a single robot, a scalar loss is optimized. Previous multiple-robot motion planning approaches that consider optimality project the vector of individual losses onto a scalar loss. As a result, these methods can fail to find many potentially useful motion plans.

There are many well-studied optimality concepts from game-theory and multiobjective optimization literature. An optimality concept will be briefly described for the multiple-robot planning problem that results in a small set of alternative strategies that are guaranteed to be less than or equivalent to (in terms of losses) than any other possible strategy.

For each robot, assume there are no uncertainties and define a loss functional of the form $L^i(x_{init}, x_{goal}, u^1, \ldots, u^N) =$

$$\int_0^T l^i(t, x^i(t), u^i(t))dt + \sum_{j \neq i} c^{ij}(x(\cdot)) + q^i(x^i(T)), \quad (17)$$

which maps to the extended reals, and

$$c^{ij}(x(\cdot)) = \begin{cases} 0 & \text{if } x(t) \in X_{valid} \text{ for all } t \\ \infty & \text{otherwise} \end{cases} \quad (18)$$

and

$$q^i(x^i(T)) = \begin{cases} 0 & \text{if } x^i(T) = x_{goal}^i \\ \infty & \text{otherwise} \end{cases}. \quad (19)$$

The variables x_{init} and x_{goal} represent the initial and goal configurations for all of the robots.

The integrand l^i represents a continuous cost function, which is a standard form that is used in optimal control theory. It is additionally required, however, that

$$l^i(t, x^i(t), u^i(t)) = 0 \quad \text{if } x^i(t) = x^i_{goal}. \quad (20)$$

This implies that no additional cost is received while the ith robot "waits" at x^i_{goal} until time T. The term (18) penalizes collisions between the robots. The subset $X_{valid} \subset X$ represents the (closed) set of all states at which no robots or obstacles are in collision. This has the effect of preventing any robots from considering game strategies that lead to collision. The term (19) represents the goal in terms of performance. If a robot, $\mathcal{A}^i$, fails to achieve its goal x^i_{goal}, then it receives infinite loss.

Suppose that the initial state is given. For each game strategy, γ, a vector of losses will be obtained. A partial ordering, $\preceq$, can be defined on the space of game strategies, Γ. For a pair of elements $\gamma, \gamma' \in \Gamma$ let $\gamma \preceq \gamma'$ if $L^i(\gamma) \leq L^i(\gamma')$ for every i. The minimal game strategies with respect to $\preceq$ are better than or equal to all other game strategies in Γ, and it is shown in [24] that very few minimal game strategies typically exist (ignoring those that produce equivalent losses).

These solutions can be generated using algorithms that are based on the dynamic programming principle. For the criterion (17) it is shown that minimal solutions are consistent with other well-established forms of optimality from optimization literature [24]. The minimal game strategies are equivalent to the *nondominated* strategies used in multiobjective optimization and *Pareto optimal* game strategies used in cooperative game theory. Furthermore, it can be shown that the minimal game strategies satisfy the Nash equilibrium condition from noncooperative game theory, which implies that for a game strategy $\gamma^* = \{\gamma^{1*} \ldots \gamma^{N*}\}$, the following holds for each i and each $\gamma^i \in \Gamma^i$:

$$L^i(\gamma^{1*}, \ldots, \gamma^{i*}, \ldots \gamma^{N*}) \leq L^i(\gamma^{1*}, \ldots, \gamma^i, \ldots \gamma^{N*}). \quad (21)$$

Moving obstacles and other nonstationary systems It has been assumed so far in this section that the system is not time-varying. From a control perspective, this corresponds to a *stationary* problem. Optimal solutions to problems of this type depend only on state (or the information state with sensing uncertainty) and not on time.

By allowing time-varying models, many interesting motion planning problems can be defined. Suppose, for instance, that several moving obstacles exist in the workspace. For a single-robot problem, this leads to a time-varying free configuration space $\mathcal{C}_{free}(t)$ [22], which can be approximated in discrete time as

$$\mathcal{C}_{free}[k] = \bigcap_{t \in [(k-1)\Delta t, k\Delta t)} \mathcal{C}_{free}(t). \quad (22)$$

In general, many game items from Section 2 can encode time-dependent models.

4 Algorithms and Computed Examples

This section briefly discusses one of several algorithms that have been developed using this game-theoretic foundation. One purpose is to describe an approach that was inspired by numerical optimal control research, and was straightforward to develop, given the mathematical framework. This section also presents a variety of computed examples that were obtained using various algorithms, to indicate the broad applicability of the concepts. A more thorough presentation of algorithms and computed examples appears in [24].

An algorithm that handles uncertainty in prediction Suppose that there is one robot with probabilistic uncertainty in predictability, perfect configuration sensing (i.e., the information space reduces to X), and the models are not time-varying. The task is to compute a strategy that is optimal in the expected sense, which is a challenging extension of the basic motion planning problem. The expected loss obtained by starting from stage k and implementing the portion of the optimal strategy $\{\gamma^*_k, \ldots, \gamma^*_K\}$ can be represented as

$$\bar{L}^*_k(x_k) = E\left\{ \sum_{i=k}^{K} l_i(x_i, \gamma^*_i(x_i)) + l_{K+1}(x_{K+1}) \right\}, \quad (23)$$

in which $E\{\}$ denotes expectation taken over the actions of nature.

The principle of optimality [20] states that $\bar{L}^*_k(x_k)$ can be obtained from $\bar{L}^*_{k+1}(x_{k+1})$ by the following recurrence:
$$\bar{L}^*_k(x_k) =$$
$$\min_{\gamma_k \in \Gamma_k} \left\{ l_k(x_k, u_k) + \int \bar{L}^*_{k+1}(x_{k+1}) p(x_{k+1} | x_k, u_k) dx_{k+1} \right\}. \quad (24)$$

Note that the integral is taken over states that can be reached using the state transition equation.

An optimal strategy is determined by successively building approximate representations of $\bar{L}_k^*$. Each dynamic programming iteration can be considered as the construction of an approximate representation of $\bar{L}_k^*$. A discretized representation is used to construct a good approximation of the continuous function $\bar{L}_k^*$ over the entire state space. The value for $\bar{L}_k^*(x_k)$ is obtained by computing the right side of (24) for various values of u_k and using linear interpolation. Other schemes, such as quadratic interpolation, can be used to improve numerical accuracy [21].

Note that $\bar{L}_K^*$ represents the cost of the optimal one-stage strategy from each state x_K. More generally, $\bar{L}_{K-i}^*$ represents the cost of the optimal $(i+1)$-stage strategy from each state x_{K-i}. For a motion planning problem, one is typically concerned only with strategies that require a finite number of stages before terminating in the goal region. For a positive $\delta \approx 0$ the dynamic programming iterations are terminated when $|\bar{L}_k^*(x_k) - \bar{L}_{k+1}^*(x_{k+1})| < \delta$ for all values in the state space. The resulting strategy is formed from the optimal actions and termination conditions in the final iteration. Note that no choice of K is necessary. Also, at each iteration of the dynamic programming algorithm, only the representation of $\bar{L}_{k+1}^*$ is retained while constructing $\bar{L}_k^*$; earlier representations can be discarded.

To execute a strategy, the robot uses the final cost-to-go representation (which is called $\bar{L}_1^*$) in a way similar to the use of a navigation function [4], [35]. The optimal action can be obtained from any real-valued location $x \in X$ though the use of (24) (or the appropriate dynamic programming equation), interpolation, and the approximate representation of $\bar{L}_1^*$. A real-valued initial state is given. Thus, the robot is not confined to move along the quantization grid that is used for determining the cost-to-go functions. The application of the optimal action will yield a new real-valued configuration for the robot. This form of iteration continues until the termination condition is met.

Let Q denote the number of cells per dimension in the representation of $\mathcal{C}_{free}$. Let n denote the dimension of the state space. Let $|U|$ denote the number of actions that are considered. Let $|\Theta|$ denote the number of actions that are considered by nature. The space complexity of the algorithm is $O(Q^n)$. For each itera-

tion of the dynamic programming, the time complexity is $O(Q^n |U| \, |\Theta|)$, and the number of iterations is proportional to the number of stages required for sample paths to reach the goal. The complexity is exponential in dimension, but efficient for fixed dimension. Execution times vary dramatically depending on the resolutions, but computation times typically range from a minute or two for a basic 2D problem up to several hours for a challenging 3D problem, on a typical workstation with little regard to code optimization. It is important to note, however, that this algorithm is not competing with known algorithms that solve the basic problem, since the algorithm described in this paper overcomes uncertainty in prediction and yields an optimal strategy.

Several variations of this algorithm, and other algorithms that apply to problems that were discussed in Section 3, are presented in [24].

Computed examples To indicate the broad applicability of the game-theoretic concepts, a variety of computed examples are presented. The scope of the mathematical framework is not limited to problems shown in this section; however, the examples were computed using algorithms that were developed by utilizing the game-theoretic framework [24].

Figures 1 and 2 show some computed results for problems that involve uncertainty in control (cases that additionally involve uncertainty in sensing are presented in [24]). The state transition equation and sensing models are the same as the examples given in Section 2.

Figures 1(a)-(c) show computed preimages for a classic peg-in-hole task with a fixed, downward motion command (i.e., u_k represents a fixed angle in (5)). Figure 1(a) shows a performance preimage (using a variation of the previously discussed computation technique) under nondeterministic uncertainty and a loss functional that returns 0 when the goal is achieved, and 1 otherwise. The curve shown in Figure 1(a) corresponds closely to the classical preimage that has been determined for this problem in previous manipulation planning research (e.g., [12], [22]). Figure 1(b) assumes probabilistic uncertainty, and shows probabilistic back-projections that are quite similar to those that appear in [5]. Figure 1(c) shows performance preimages for a case in which a Gaussian error model is used to

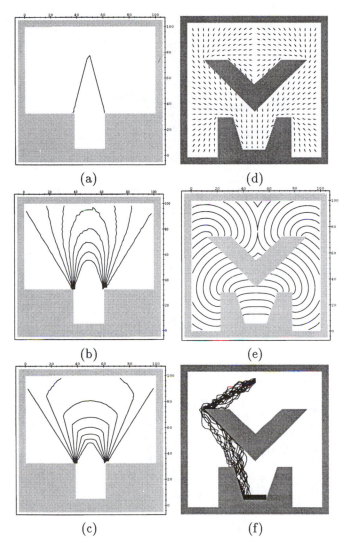

Figure 1: *Several computed performance preimages for the classic peg-in-hole problem: (a) a classical preimage; (b) a single-stage probabilistic preimage for a uniform state transition pdf; (c) a single-stage probabilistic preimage for a truncated Gaussian state transition pdf; and a computed optimal strategy for a different problem: (d) the state-feedback solution; (e) performance preimages; (f) simulated executions of the optimal strategy.*

represent the uncertainty in control, as opposed to a bounded uniform pdf as in Figure 1(b) and in [5].

Figure 1 shows a computed optimal strategy for a problem that involves probabilistic uncertainty in con-

trol and a loss functional that measures the time to achieve the goal. The goal is at the lower central part of the workspace. Figure 1(d) depicts the optimal strategy by showing the direction of the motion command $u_k = \gamma_k^*(x_k)$ at different locations in the state space. Figure 1(e) shows performance preimages under the implementation of the optimal strategy. Figure 1(f) shows 30 superimposed, simulated executions of the computed optimal strategy.

Figure 2 shows several stages of a computed forward projection under probabilistic uncertainty for a peg-in-hole problem and a fixed motion command. Initially, there is little uncertainty in configuration; however, as time progresses, the pdf on the state space becomes diffuse. Due to uncertainty reduction through compliance (which has been used in preimage planning research [32]), the pdf becomes flattened in the final stages. This type of simulation can provide useful information for experimenting with different uncertainty models and computed strategies.

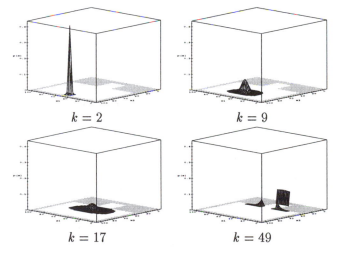

Figure 2: *The forward projection at several stages, with probabilistic uncertainty.*

Figures 3 and 4 show computed examples for problems that involve an environment that changes over time and is not completely predictable (more details appear in [28]). Figure 3(a) shows a problem for which there is a single rigid robot that can rotate in place or translate along its major axis. There are two doors that can become open or closed at various points in the future, and the behavior of the doors is modeled with a Markov process. The state space for this problem is the Cartesian product of the configuration space of the robot and a set of four possible combinations of open and closed doors. Figures 3(b) and (c) show two simulated executions under the implementation of a computed strategy that minimizes the expected time to reach the goal. Different trajectories are taken in different executions because the openings and closings of doors vary; however, both behaviors are obtained from the same strategy. Figure 3(d) shows a problem in which there is a nonholonomic car robot that is capable of only moving in a forward direction and has a limited turning radius. There are two regions in the workspace that are designated as service areas. In this case, the robot interacts with the environment by processing service requests that can occur at various points in the future (again modeled with a Markov process). Figures 3(e) and (f) show two simulated executions under the implementation of the strategy that minimizes the expected time to reach the goal region while there are no outstanding requests. Figure 4 shows a problem that involves a three degree-of-freedom manipulator that delivers one of two possible parts from one of two sources to one of two destinations. The execution of a strategy that minimizes the expected time that parts wait to be processed is depicted, assuming Markov models for resuests of parts, sources, and destinations. Other problems of this type are studied in [37].

Figures 5 and 6 show computed examples for problems that involve multiple robots. Each robot is constrained to move with bounded velocity along an independent roadmap, and a minimal strategy is depicted. These strategies were computed by applying the principle of optimality to a partially ordered space of strategies [24].

5 Conclusion

A dynamic game-theoretic framework has been proposed in this paper to serve as a mathematical foundation for a broad class of motion planning problems. Results obtained by following this perspective were summarized with the intent of indicating the general utility of this foundation. By no means is it intended to provide a general solution to a broad class of problems, but instead it provides a useful characterization upon which motion planning algorithms can be developed.

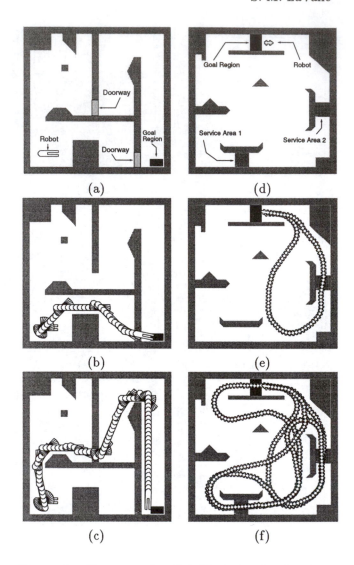

Figure 3: *Two examples that illustrate planning in a changing, partially-predictable environment.*

In this way, it can serve the same purpose that configuration space concepts served for basic path planning problems.

This foundation can provide several key advantages for future research: (1) A common, unified structure facilitates the comparison of techniques. Just as configuration space concepts provided a precise, ideal formulation of basic path planning, the dynamic game-theoretic concepts provide a formulation of the ideal (or optimal) strategies that can be achieved. For many difficult problems, tradeoffs are inevitably made to im-

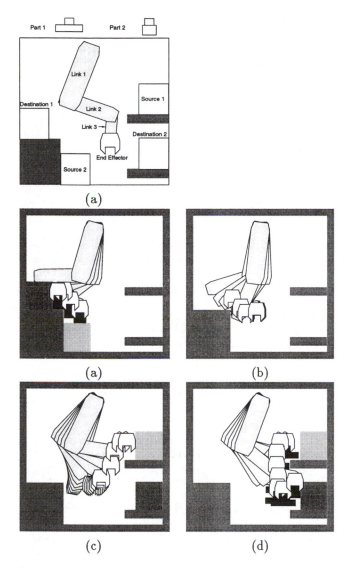

Figure 4: *A three degree-of-freedom manipulator with a constrained, rotating end-effector is in a workspace in which there are two parts, two sources, and two destinations that are modeled with a Markov process. The optimal strategy is shown.*

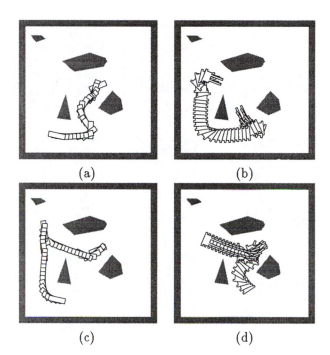

Figure 5: *Two strategies for a two-robot, roadmap-coordination problem.*

prove computational performance. As approximate or incomplete methods are proposed, it is useful for the purposes of analysis to have precise, ideal formulations. (2) Clear directions are provided along which the concepts and methods can be generalized. For example, the preimage and forward projection concepts have been shown to apply in very general settings by gen-

eralizing their definitions within the framework. This has provided a clear relationship between nondeterministic and probabilistic uncertainty models, and numerical navigation functions and preimages. (3) A variety of different models can be incrementally tested. One of the greatest difficulties in motion planning under uncertainties is determining appropriate models of uncertainty, while previous algorithms have often applied to very specific uncertainty models. The framework allows the substitutions of a variety of different models while many of the principles remain unchanged. This is particularly true of the numerical computation method briefly discussed in Section 4, which makes few restrictions on the models.

Although the framework has only been applied so far to three classes of motion planning problems, one important direction for future research will be to characterize and analyze additional problems. For example, problems that involve dynamics, sensing from a vision system, or complex manipulations, have yet to be considered. For many problems, specialized representations will undoubtedly be useful for developing algorithms. In many cases, useful concepts from the ex-

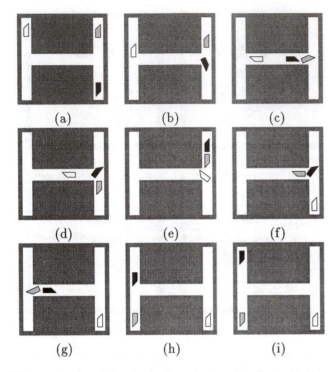

(a) (b) (c)

(d) (e) (f)

(g) (h) (i)

Figure 6: *A minimal solution strategy for three rotating robots on independent roadmaps.*

isting literature can be combined with the mathematical structure, such as in the case of using preimage planning research to develop the performance preimage. Such constructions are useful for developing algorithms, and are compatible with the dynamic game-theoretic concepts.

Acknowledgments

I thank Narendra Ahuja, Tamer Başar, Bruce Donald, Mike Erdmann, Ken Goldberg, Dan Koditschek, Jean-Claude Latombe, Jean Ponce, and Mark Spong, for their helpful comments and suggestions regarding this research. James Kuffner made helpful suggestions regarding this manuscript. I especially thank Seth Hutchinson and Rajeev Sharma who made significant contributions to the individual applications. This work was sponsored at the University of Illinois by a Beckman Institute research assistantship, and a Mavis Fellowship. Jean-Claude Latombe's research lab at Stanford University provided additional support.

References

[1] T. Başar and G. J. Olsder. *Dynamic Noncooperative Game Theory.* Academic Press, London, 1982.

[2] J. Barraquand and P. Ferbach. Motion planning with uncertainty: The information space approach. In *IEEE Int. Conf. Robot. & Autom.*, pages 1341–1348, 1995.

[3] J. Barraquand, B. Langlois, and J. C. Latombe. Numerical potential field techniques for robot path planning. *IEEE Trans. Syst., Man, Cybern.*, 22(2):224–241, 1992.

[4] J. Barraquand and J.-C. Latombe. A Monte-Carlo algorithm for path planning with many degrees of freedom. In *IEEE Int. Conf. Robot. & Autom.*, pages 1712–1717, 1990.

[5] R. C. Brost and A. D. Christiansen. Probabilistic analysis of manipulation tasks: A computational framework. *Int. J. Robot. Res.*, 15(1):1–23, February 1996.

[6] J. F. Canny. On computability of fine motion plans. In *IEEE Int. Conf. Robot. & Autom.*, pages 177–182, 1989.

[7] B. R. Donald. *Error Detection and Recovery for Robot Motion Planning with Uncertainty.* PhD thesis, Massachusetts Institute of Technology, Cambridge, MA, 1987.

[8] B. R. Donald and J. Jennings. Sensor interpretation and task-directed planning using perceptual equivalence classes. In *IEEE Int. Conf. Robot. & Autom.*, pages 190–197, Sacramento, CA, April 1991.

[9] A. Elfes. Using occupancy grids for mobile robot perception and navigation. *IEEE Computer*, 22(6):46–57, June 1989.

[10] M. Erdmann. Randomization for robot tasks: Using dynamic programming in the space of knowledge states. *Algorithmica*, 10:248–291, 1993.

[11] M. Erdmann and T. Lozano-Perez. On multiple moving objects. In *IEEE Int. Conf. Robot. & Autom.*, pages 1419–1424, 1986.

[12] M. A. Erdmann. On motion planning with uncertainty. Master's thesis, Massachusetts Institute of Technology, Cambridge, MA, August 1984.

[13] E. G. Gilbert and D. W. Johnson. Distance functions and their application to robot path planning in the presence of obstacles. *IEEE Trans. Robot. & Autom.*, 1(1):21–30, March 1985.

[14] P. J. Gmytrasiewicz, E. H. Durfee, and D. K. Wehe. A decision-theoretic approach to coordinating multi-agent interations. In *Proc. Int. Joint Conf. on Artif. Intell.*, pages 62–68, 1991.

[15] K. Y. Goldberg. *Stochastic Plans for Robotic Manipulation*. PhD thesis, Carnegie Mellon University, Pittsburgh, PA, August 1990.

[16] J. C. Harsanyi. Games with incomplete information played by Bayesian players. *Management Science*, 14(3):159–182, November 1967.

[17] H. Hu and M. Brady. A Bayesian approach to real-time obstacle avoidance for a mobile robot. *Autonomous Robots*, 1(1):69–92, 1994.

[18] R. Isaacs. *Differential Games*. Wiley, New York, NY, 1965.

[19] K. Kant and S. W. Zucker. Toward efficient trajectory planning: The path-velocity decomposition. *Int. J. Robot. Res.*, 5(3):72–89, 1986.

[20] P. R. Kumar and P. Varaiya. *Stochastic Systems*. Prentice-Hall, Englewood Cliffs, NJ, 1986.

[21] R. E. Larson and J. L. Casti. *Principles of Dynamic Programming, Part II*. Dekker, New York, NY, 1982.

[22] J.-C. Latombe. *Robot Motion Planning*. Kluwer Academic Publishers, Boston, MA, 1991.

[23] J.-C. Latombe, A. Lazanas, and S. Shekhar. Robot motion planning with uncertainty in control and sensing. *Artif. Intell.*, 52:1–47, 1991.

[24] S. M. LaValle. *A Game-Theoretic Framework for Robot Motion Planning*. PhD thesis, University of Illinois, Urbana, IL, July 1995.

[25] S. M. LaValle and S. A. Hutchinson. An objective-based stochastic framework for manipulation planning. In *Proc. IEEE/RSJ/GI Int'l Conf. on Intelligent Robots and Systems*, pages 1772–1779, September 1994.

[26] S. M. LaValle and S. A. Hutchinson. Evaluating motion strategies under nondeterministic or probabilistic uncertainties in sensing and control. In *Proc. IEEE Int'l Conf. Robot. & and Autom.*, pages 3034–3039, April 1996.

[27] S. M. LaValle and S. A. Hutchinson. Optimal motion planning for multiple robots having independent goals. In *Proc. IEEE Int'l Conf. Robot. & and Autom.*, pages 2847–2852, April 1996.

[28] S. M. LaValle and R. Sharma. Motion planning in stochastic environments: Applications and computational issues. In *IEEE Int'l Conf. on Robotics and Automation*, pages 3063–3068, 1995.

[29] T. Lozano-Pérez. Spatial planning: A configuration space approach. *IEEE Trans. on Comput.*, C-32(2):108–120, 1983.

[30] T. Lozano-Pérez, M. T. Mason, and R. H. Taylor. Automatic systhesis of fine-motion strategies for robots. *Int. J. Robot. Res.*, 3(1):3–24, 1984.

[31] K. M. Lynch and M. T. Mason. Pulling by pushing, slip with infinite friction, and perfectly rough surfaces. *Int. J. Robot. Res.*, 14(2):174–183, 1995.

[32] M. T. Mason. Compliance and force control for computer controlled manipulators. In B. Brady *et al.*, editor, *Robot Motion: Planning and Control*, pages 373–404. MIT Press, Cambridge, MA, 1982.

[33] P. A. O'Donnell and T. Lozano-Pérez. Deadlock-free and collision-free coordination of two robot manipulators. In *IEEE Int. Conf. Robot. & Autom.*, pages 484–489, 1989.

[34] G. Owen. *Game Theory*. Academic Press, New York, NY, 1982.

[35] E. Rimon and D. E. Koditschek. Exact robot navigation using artificial potential fields. *IEEE Trans. Robot. & Autom.*, 8(5):501–518, October 1992.

[36] J. T. Schwartz and M. Sharir. On the piano movers' problem: III. Coordinating the motion of several independent bodies. *Int. J. Robot. Res.*, 2(3):97–140, 1983.

[37] R. Sharma, S. M. LaValle, and S. A. Hutchinson. Optimizing robot motion strategies for assembly with stochastic models of the assembly process. *IEEE Trans. on Robotics and Automation*, 12(2):160–174, April 1996.

[38] A. Stentz. Optimal and efficient path planning for partially-known environments. In *IEEE Int. Conf. Robot. & Autom.*, pages 3310–3317, 1994.

Constrained Motion Planning: Applications in Mobile Robotics and in Maintenance Operations

Pierre Ferbach, *EDF - Direction des Etudes et Recherches, Chatou, France*
Jean-François Rit, *EDF - Direction des Etudes et Recherches, Chatou, France*

A general method for motion planning with geometric and kinematic constraints is presented. It is based on a Progressive Constraints approach. A path satisfying only some of the geometric constraints is first generated. The other constraints are then applied progressively. Concurrently, the initial path evolves so as to be admissible with the current level of constraints, and converges towards a solution of the original problem.

The Progressive Constraints approach is implemented in an iterative way. The original problem is replaced by a series of progressively constrained ones that are solved successively. Each path is obtained from a dynamic programming exploration procedure, able to deal with optimization criteria. This exploration procedure is inspired from the method presented in [2].

1 Introduction

Motion planning consists in finding a sequence of actions to move a system from an initial to a final position [9]. The motions are generally faced with two kinds of constraints. *Geometric constraints* restrict the set of admissible configurations q. *Kinematic constraints* restrict at each q the set of admissible velocities $\dot{q}$.

We present a general approach for motion planning with geometric and kinematic constraints, called the method of Progressive Constraints (PC). First, a path satisfying only some of the geometric constraints is generated. Then the kinematic constraints and the other geometric ones are introduced progressively. This is represented in the phase space by a continuous growth of the forbidden areas. The initial path evolves so as to adapt to their expansion, and to always link the initial and goal configurations with a path that satisfies the current constraints. In this way the current path can converge towards a solution. Such an approach was used in [1, 7] for planning motions of systems submitted to geometric constraints only.

We implemented a planner based upon this approach. The method is a variational technique. At each iteration, a path is generated that satisfies more severe constraints than the previous one. A new current path is obtained by exploring a neighborhood of the previous path. The exploration technique is inspired from the method presented in [2]. The implemented planner works for systems with high-dimensional (*i.e.*, > 4) configuration spaces.

It has been tested on classical models of mobile robots. Some of them are purely kinematic. In others, dynamic constraints are also taken into account. But by extending the system's configuration to the velocity parameters, the dynamic constraints act as geometric or kinematic constraints on the extended system.

The planner was also tested on examples inspired from maintenance operations. The systems consist of large objects moved by manipulators. They are faced with classical geometric constraints such as obstacle avoidance. They are also subject to kinematic constraints that result from the possible actions of manipulators. The objects considered are large and heavy. Therefore, the manipulators are not robotic arms, but powerful machines like cranes and traveling cranes. Technical aspects and severe safety regulations allow, at a time, only one specific action of a manipulator on an object, like lifting or rotating, but not both concurrently. This is a kinematic constraint.

In the situations described above, a maneuver corresponds to one specific action of a manipulator. A change of maneuver necessitates to operate or to change the manipulator. Therefore, a major concern is to find trajectories that obey the constraints and have as few maneuvers as possible. This is also true with mobile robots where maneuvering produces reversals.

Relation to other work: We mention here previous work on motion planning with kinematic constraints.

It belongs generally to the field of nonholonomic motion planning [11]. Nonholonomic constraints are a type of kinematic constraints.

Geometric methods determine the geometric characteristics of admissible paths, and then link two positions with a path having these characteristics [6, 14].

Control theoretic methods are based on a control system equation $\dot{q} = f(q, u)$ that models the system's motions according to the kinematic constraints. Motion planning problems can be expressed as optimal control problems: a functional of the control function is introduced, accounting for the admissibility and optimality of the corresponding path; the optimization problem can be solved using standard variational techniques [8]. Or, steering techniques can be used. They generally consist in introducing a control function with unspecified parameters, that are determined by integrating the control system equation with boundary conditions at the initial and final configurations [3, 13].

A family of methods decomposes the search in two phases [17, 10, 16]. A path satisfying only the geometric constraints is produced first, and then transformed into a path that obeys also the kinematic constraints.

Another class of motion planning methods are search based methods that build and explore a graph of states, in which edges in the graph correspond to feasible motions of the system. Barraquand and Latombe [2] used such a method for planning motions for mobile robots. A graph of states is concurrently built and explored by performing step motions from states already in the graph, according to a dynamic programming procedure that minimizes the number of reversals. The exploration expands from initial states and stops when it reaches a goal state. In [4] the method is extended to deal with the robot's dynamics and physical interactions with the terrain. In [12] it is used for planning pushing paths. Similar methods have been used for kinodynamic planning [5, 15]. These methods work only for low-dimensional systems because their time/space complexity is exponential in the dimension of the explored set. In [18] another search based method is presented, that works for high-dimensional systems: a preprocessing phase builds a graph of admissible configurations called a roadmap. Motions are then generated by connecting the initial and goal configurations to configurations of the graph, and then by searching the graph for an admissible path.

We reuse the method from [2] for systems with low-dimensional configuration spaces. We apply it to a new type of motion planning problems with kinematic constraints, in which objects are moved with devices that can only perform some specific actions, generally some translations and some rotations, and not concurrently. These problems are inspired from construction and maintenance operations in which heavy objects are moved.

For systems with high-dimensional configuration spaces, we propose an approach with Progressive Constraints (PC). A path that satisfies only some of the geometric constraints is first generated. Then, the rest of the constraints are introduced progressively, while the path evolves so as to obey the current level of constraints, and converges toward a solution of the original problem. The approach is implemented in an iterative planner: each iteration uses an exploration technique inspired from the method for low-dimensional cases [2], searching for a new current path that satisfies more severe constraints than the previous one. The planner for high-dimensional problems can be seen as a variational extension of the method for low-dimensional cases.

The Progressive Constraints approach is not complete since convergence is not guaranteed. Nevertheless, the experiments show that the proposed planner can solve complicated problems. And it can be applied to any system whose kinematic constraints can be modeled with an equation $\dot{q} = f(q, u)$ with a finite number of controls u.

A different iterative method in which constraints appear progressively is presented in [16]. In this work, the initial path satisfies all the geometric constraints. The kinematic constraints are imposed progressively, but they are added one by one. The number of iterations is therefore deterministic. Techniques that build on previous work [10, 18] are used for generating the intermediate paths. They require at each iteration the existence of a local planner that can deal with the current kinematic constraints. It is shown that the method is complete when the local planners satisfy a topological property.

Experiments: Simulation experiments are presented in this paper. They were done with C-programs running on a Silicon Graphics Indigo2 R4600 workstation, rated on the SPEC-MARKS benchmark with 72 SPECfp92, 109 SPECint92.

2 Nonlinear control system

2.1 System representation

We consider a system $\mathcal{R}$ whose kinematic constraints are modeled with a control system equation

$$\dot{q} = f(q, u), \quad u \in \mathcal{U}(q) \subset \mathcal{U}. \quad (1)$$

$q \in \mathcal{C}$ is a configuration, $\mathcal{C}$ the configuration space. u is the control parameter taken in a control space $\mathcal{U}(q)$ that depends on q. Kinematic constraints that are initially expressed with constraints of the form $G(q, \dot{q}) = 0, > 0$ or ≥ 0, can be transformed [2] into such an equation. $f(\cdot)$ is assumed to be smooth.

$\mathcal{U}$ is the set of possible controls. We consider only cases where it contains a finite number n_u of controls: $\mathcal{U} = \{u_1, \ldots, u_{n_u}\}$. If a system has initially an infinite number of controls, one has to keep only a finite number by taking care, if possible, that it does not preclude the existence of solution paths.

The system's state is represented by $\tilde{q} = (q, q')$, where q' (or $\dot{q}$) is the velocity of the system. The state space is the phase space Φ, assumed to be bounded.

A path is represented by a differentiable map:

$$\gamma : t \in [0, T] \mapsto \gamma(t) \in \mathcal{C},$$

whose derivative $\dot{\gamma}(\cdot)$ is piecewise continuous. A path obeying the kinematic constraints is obtained by integrating equation 1 from an initial configuration with a piecewise continuous control function:

$$u : t \in [0, T] \mapsto u(t) \in \mathcal{U}.$$

2.2 The planning problem

Constraints determine whether a state (q, q') is admissible or not. Geometric constraints apply only on q and define the free configuration space $\mathcal{C}_{free}$. Kinematic constraints also consider q'. Let $\Phi_{free} \subset \Phi$ denote the set of states that obey all the constraints. Figure 1 represents a phase space Φ in a case where $\mathcal{C}$ has dimension 1. The white areas represent Φ_{free}. In the problems we consider, Φ_{free} only consists of submanifolds of Φ since only a finite number of controls can be applied. Figure 1 corresponds to a more general situation.

The motion planning problem is expressed as follows: *Given an initial configuration* $q_{init} \in \mathcal{C}_{free}$, *and a set* $\mathcal{Goal} \subset \mathcal{C}_{free}$ *of goal configurations, find a control function* u *that defines an admissible path starting*

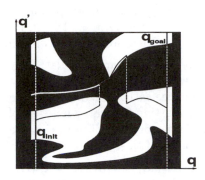

Figure 1: *A phase space Φ, with a path in Φ_{free} going from q_{init} to q_{goal}.*

from q_{init} *and ending in* $\mathcal{Goal}$. Also, since some solution paths are better than others, we may want to involve a criterion into the search, for finding good or optimal solutions.

The trajectory in Φ of an admissible motion lies in Φ_{free}. $q' = \dot{q}$ may be discontinuous when the control function is discontinuous. q is continuous. Not all trajectories in Φ_{free} correspond to a motion of the system: q' has to be equal to the derivative $\dot{q}$. Figure 1 depicts a path from q_{init} to q_{goal}. We consider that the area represented corresponds to positive values of q'. Then q can only increase along a path.

2.3 Bang motions

Given $u \in \mathcal{U}$ and $\delta t > 0$, applying u during time δt from some configuration is called a $(u, \delta t)$-*bang* [5]. We also use this term for the resulting step motion.

3 The low-dimensional case

The method presented in [2] is described now, in a slightly more general form. It can only be used in low-dimensional cases. It was applied in [2] to path planning for mobile robots. The method is in fact very general. We applied it to problems where large and heavy objects are moved. And the exploration technique we use in the Progressive Constraints method of Section 4 is inspired from it.

3.1 Method

The system's state space is explored with a procedure that concurrently builds and explores a graph $\mathcal{T}$ of states $(q, q') = (q, f(q, u))$, by performing bang motions from states already in $\mathcal{T}$. Dijkstra's algorithm is used, with a cost function used for optimizations.

The bang motions are calculated by integrating equation 1 with a fixed time step δt. The corresponding step motions satisfy the kinematic constraints. A $(u, \delta t)$-bang starting from a node (q, q') of $\mathcal{T}$ is admissible if its trajectory in $\mathcal{C}$ lies in $\mathcal{C}_{free}$, and if at any configuration q_b of this bang motion, $u \in \mathcal{U}(q_b)$.

The states $(q_{init}, f(q_{init}, u))$, $u \in \mathcal{U}(q_{init})$, are the roots of $\mathcal{T}$. The exploration stops when it reaches a state (q, q') with $q \in \mathcal{G}oal$. The solution path returned is then the sequence of configurations of the path in $\mathcal{T}$ going from a root to (q, q').

Because $\mathcal{T}$ is built by doing step motions, there is no guarantee that a state in $\mathcal{T}$ will reach $\mathcal{G}oal$. If necessary, we extend $\mathcal{G}oal$ to a neighborhood $\widetilde{\mathcal{G}oal}$ of it, that will be reached by the exploration.

A pruning technique is used for restricting the expansion of $\mathcal{T}$. Cells are put over Φ_{free}. Each node of $\mathcal{T}$ is associated to a cell containing it, and at most one state can be associated with each cell. Then, a state is discarded if it is too close to another state in $\mathcal{T}$.

In the experiments presented in 3.3, the optimization criterion was to minimize the number of maneuvers by minimizing the number of changes in the control applied. Therefore, in the Dijkstra explorations the incremental cost of a step motion was 1 if the control changed, and some very small positive number otherwise since we also wish to obtain short paths.

The cells we used in the implementation fit the optimization criterion. Instead of representing a node of $\mathcal{T}$ by (q, q') where $q' = f(q, u)$ for some control u, we represent it by (q, u). u is the control that led to this state. Then, small cells p covering $\mathcal{C}$ define cells $c = p \times \{u\}$ $(u \in \mathcal{U})$ covering Φ_{free}. We used a partition of $\mathcal{C}$ into small regular parallelepipeds p.

3.2   Asymptotic completeness

For $u \in \mathcal{U}$, let $\mathcal{C}_u$ be the set of configurations where the control u can be applied, $\mathcal{C}_u = \{q \in \mathcal{C}, u \in \mathcal{U}(q)\}$. Let Ψ denote a set of convex cells p covering $\mathcal{C}$, and d_Ψ the maximal Euclidean length of a line-segment included in such a cell. By slightly adapting the proof of similar results presented in [2], we have the following:

Let $\widetilde{\mathcal{G}oal}$ be a neighborhood of every configuration of

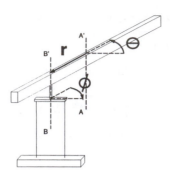

Figure 2: An object held under a rotating crane.

Goal. If $\mathcal{C}_{free}$ and all the $\mathcal{C}_u$, $u \in \mathcal{U}$, are open subsets of $\mathcal{C}$, and if any change of maneuver corresponds to a change in the control applied, then the above algorithm with goal set $\widetilde{\mathcal{G}oal}$ is guaranteed to generate a solution path with at most the minimum number of maneuvers of an admissible path from q_{init} to Goal, provided that δt has been set small enough, and that the algorithm is run with pruning cells $c = p \times \{u\} \in \Psi \times \mathcal{U}$ where d_Ψ is small enough.

This result does not provide adequate values for δt and d_Ψ. These are chosen in practice intuitively: a bang motion has to be small compared to the obstacle size, and d_Ψ has to be smaller than a bang motion length.

The algorithm needs the pruning technique since each node of $\mathcal{T}$ may have up to n_u successors. Also, it is only implementable if $\mathcal{C}$ has a low dimension since the number of cells of Ψ, given a fixed d_Ψ, increases exponentially with the dimension of $\mathcal{C}$.

3.3   Experimental results

The motion of an object with a rotating crane
We consider an object suspended under a rotating crane as depicted in Figure 2. Its configuration is $q = (r, \theta, \phi)$: θ is the angle of rotation of the upper beam around the vertical axis AA', ϕ is the angle of rotation of the lower part around BB', and r is the distance between AA' and BB'. The crane can act on r, θ, ϕ, but only separately. The object held is implicitly lifted at the beginning of a motion plan and laid down at the end. Its control system equation is

$$\dot{q} = f(q, u) = u, \quad u \in \{(\pm 1, 0, 0), (0, \pm 1, 0), (0, 0, \pm 1)\}.$$

We consider such a crane situated on top of a cylindrical building. The geometric constraints consist of avoiding collisions between the object and obstacles.

Figure 3 displays a computed path, as seen from the top of the building. The upper beam of the crane and the object are depicted. The computation time was $17.6s$, with 7700 collision tests done per second.

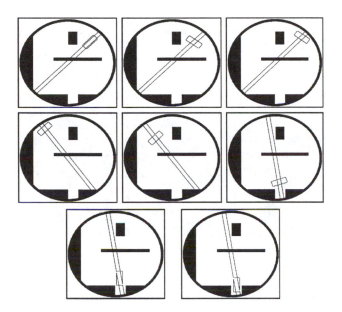

Figure 3: A rotating crane moving an object.

The motion of an object with a traveling crane
Consider the object in Figure 4, suspended under a traveling crane by means of vertical cables. This corresponds to a real situation where the heavy object is moved in a nuclear power plant. It has to be brought to the lower floor, through a hole in the upper floor. The system in Figure 4 is equivalent to the planar one of Figure 5. The object's configuration is $q = (x, z, \theta)$. (x, z) are the coordinates of the point A where cable L holds the object, and θ is the orientation of the segment AB. Let l be the distance between A and B.

The traveling crane can only perform the following separate actions at a given time: horizontal translation, vertical translation, R-tilting, L-tilting. An R-tilting is an action where the length of cable R varies. We consider a system in which when a cable's length varies, its abscissa adapts so that all cables remain vertical. An R-tilting is then a rotation around point A. In an L-tilting the length of L varies. The object is then

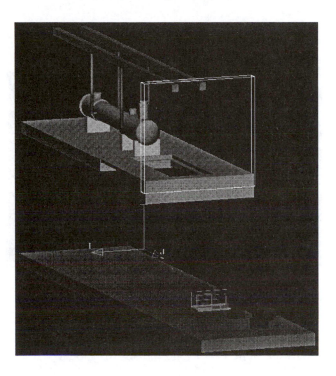

Figure 4: An object held under a traveling crane.

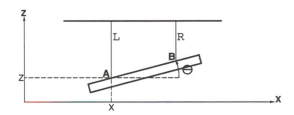

Figure 5: An equivalent planar system.

rotated around point B. For this system, equation 1 is

$$\dot{q} = f(q, u), \quad u \in \{u_1, u_2, \ldots, u_8\}.$$

u_1 and u_2 produce the horizontal translation, u_3 and u_4 the vertical translation, u_5 and u_6 the R-tilting, u_7 and u_8 the L-tilting. Thus $f(q, u_1) = (1, 0, 0)$, $f(q, u_2) = -f(q, u_1)$, $f(q, u_3) = (0, 1, 0)$, $f(q, u_4) = -f(q, u_3)$, $f(q, u_5) = (0, 0, 1)$, $f(q, u_6) = -f(q, u_5)$, $f(q, u_7) = (l.sin\theta, -l.cos\theta, 1)$, $f(q, u_8) = -f(q, u_7)$.

The geometric constraints are the collision avoidance and a bound on the slant of the object.

Figure 6 depicts a path obtained with the planner.

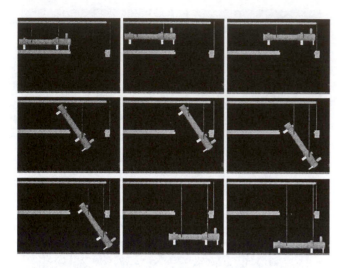

Figure 6: A traveling crane moving an object.

The computation time was $1580s$, and only 13 collision tests were done per second on this 3D model.

4   The high-dimensional case

We consider now problems where the previous method cannot be used because the dimension of $\mathcal{C}$ is higher than 4. The previous exploration technique will now be used in a variational framework based on the Progressive Constraints (PC) approach.

4.1   The method of Progressive Constraints

A planning problem $\mathcal{P}_+$ is said to be more constrained than a problem $\mathcal{P}_-$ if the free phase space of $\mathcal{P}_+$ is included in the one of $\mathcal{P}_-$. Let us introduce a distance d_Φ defined for two sets $\Omega_1, \Omega_2 \subset \Phi$ as

$$d_\Phi(\Omega_1, \Omega_2) = vol(\Omega_1 - \Omega_2) + vol(\Omega_2 - \Omega_1),$$

where $vol(A - B)$ denotes the volume of $A \cap B^C$. Φ is assumed to be a bounded set. We also introduce a distance d_Γ defined for two paths γ_1 and γ_2 as the maximum Euclidean distance between any configuration of one path and the other path: $d_\Gamma(\gamma_1, \gamma_2) =$

$$\sup\{ \sup_{q_1 \in \gamma_1} (\inf_{q_2 \in \gamma_2} \| q_1 - q_2 \|), \sup_{q_2 \in \gamma_2} (\inf_{q_1 \in \gamma_1} \| q_1 - q_2 \|)\}.$$

The PC method introduces constraints progressively, and makes an initial path converge toward a solution.

The forbidden areas of the phase space increase and tend toward those of the original problem. Concurrently, a path connecting the initial and goal configurations adapts so as to remain in the admissible regions of Φ, and converges toward a solution of the original problem. This is illustrated in Figure 7.

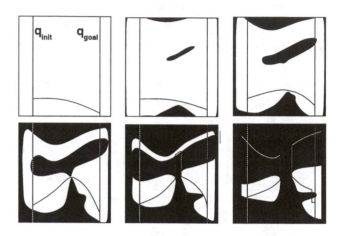

Figure 7: The Progressive Constraints approach.

Let $\epsilon \geq 0$ parameterize the level of constraints as follows. $\mathcal{P}^\epsilon$ denotes a planning problem with free phase space Φ_{free}^ϵ. $\mathcal{P}^0$ is the original problem, $\Phi_{free}^0 = \Phi_{free}$, and for ϵ greater than or equal to some $\epsilon_0 > 0$, $\mathcal{P}^\epsilon$ is only submitted to some of the geometric constraints of the original problem. Let us denote by $\mathcal{C}_{free}^\#$ the set of configurations that satisfy these initial geometric constraints. It can be any subset of $\mathcal{C}$ such that $\mathcal{C}_{free} \subset \mathcal{C}_{free}^\#$. The successive snapshots in Figure 7, with decreasing freespaces, correspond to decreasing values of ϵ. If $\epsilon < \epsilon'$, $\Phi_{free}^\epsilon \subset \Phi_{free}^{\epsilon'}$, i.e., $\mathcal{P}^\epsilon$ is more constrained than $\mathcal{P}^{\epsilon'}$. Φ_{free}^ϵ is assumed to evolve continuously with ϵ, with respect to the distance d_Φ. In particular, $d_\Phi(\Phi_{free}^\epsilon, \Phi_{free}) \to 0$ when $\epsilon \to 0$. Concurrently, a solution path γ^ϵ of $\mathcal{P}^\epsilon$ evolves continuously with respect to d_Γ, and converges toward a solution of the original problem when ϵ tends toward zero.

4.2   Progressiveness of the constraints

A state (q, q') is admissible if and only if

$$q \in \mathcal{C}_{free} \quad \text{and} \quad \exists u \in \mathcal{U}(q) \text{ such that } q' = f(q, u).$$

Let g and $h : [0, \epsilon_0) \to R^+$ be two continuous increasing functions such that $g(0) = h(0) = 0$ and $g(\epsilon)$

and $h(\epsilon) \to +\infty$ when $\epsilon \to \epsilon_0$ (e.g., $\frac{\epsilon}{\epsilon_0-\epsilon}$). For $\epsilon < \epsilon_0$ we define $\mathcal{P}^\epsilon$ by imposing the initial geometric constraints ($q \in \mathcal{C}_{free}^{\#}$), and by "relaxing by ϵ" the rest of the constraints: $(q, q') \in \Phi_{free}^\epsilon$ iff

$$q \in \mathcal{C}_{free}^{\#},$$
$$\exists q_{free} \in \mathcal{C}_{free} \text{ with } \|q - q_{free}\| \le g(\epsilon), \quad (2)$$
$$\exists u \in \mathcal{U}(q_{free}) \text{ with } q' \in \mathcal{B}(f(q_{free}, u), h(\epsilon)).$$

$\mathcal{B}(x, \alpha)$ is the spherical neighborhood of x with radius α. Slightly more general expressions would be obtained with functions $g(\cdot)$ and $h(\cdot)$ depending also on q, and converging uniformly towards zero with ϵ.

Then clearly $\Phi_{free}^\epsilon \subset \Phi_{free}^{\epsilon'}$ if $\epsilon < \epsilon'$, Φ_{free}^ϵ is d_Φ-continuous and $d_\Phi(\Phi_{free}^\epsilon, \Phi_{free}) \to 0$ when $\epsilon \to 0$. The PC approach starts from any initial path linking q_{init} to a goal configuration in $\mathcal{C}_{free}^{\#}$, and makes this path evolve while ϵ tends toward zero.

If no geometric constraint is imposed initially, $\mathcal{C}_{free}^{\#} = \mathcal{C}$. If all the geometric constraints are imposed from the beginning, we have $\mathcal{C}_{free}^{\#} = \mathcal{C}_{free}$. In that case, one can also impose $q_{free} = q$. Φ_{free}^ϵ is then the set of (q, q') such that

$$q \in \mathcal{C}_{free},$$
$$\text{and } \exists u \in \mathcal{U}(q) \text{ with } q' \in \mathcal{B}(f(q, u), h(\epsilon)). \quad (3)$$

In expression 2, the second line illustrates analytically how geometric constraints can be imposed progressively. This expression may however be difficult to verify. Geometric constraints are often expressed with equalities $F_{eq}(q) = 0$ and inequalities $F_{ineq}(q) \le 0$. These constraints can easily be relaxed with inequalities: conditions 2 can be replaced by

$$F_{eq}^{\#}(q) = 0, \quad F_{ineq}^{\#}(q) \le 0,$$
$$-g(\epsilon) \le F_{eq}(q) \le g(\epsilon), \quad F_{ineq}(q) \le g(\epsilon), \quad (4)$$
$$\exists u \in \mathcal{U}(q) \text{ with } q' \in \mathcal{B}(f(q, u), h(\epsilon)).$$

The first line represents the geometric constraints that are imposed from the beginning, and the second line the geometric ones that are relaxed.

4.3   Algorithm

The algorithm of the implemented planner deals with the progressiveness in a discretized way. First, a procedure *Geometric_Planner* generates a solution of the purely geometric initial problem $\mathcal{P}^{\epsilon_0}$. This procedure is a standard geometric planner. Then, several iterations are performed:

```
begin
    γ = Geometric_Planner;
    dε = dε_max;
    ε = ε₀;
    while  ε > ε_min
        ε = max(ε − dε, ε_min);
        γ' = Variation(γ, ε);
        if  γ' == (nil)
            ε = ε + dε;
            dε = dε/2;
        else
            γ = γ';
            dε = dε_max;
        endif;
    endwhile;
    γ' = (nil);
    if  ε_min > 0
        while  γ' == (nil)
            γ' = Variation(γ, 0);
        endwhile;
    endif;
end.
```

The algorithm tries to solve successive problems while making ϵ decrease by a predefined increment $d\epsilon_{max}$. At each iteration, a procedure $Variation(\gamma, \epsilon)$ searches for a path satisfying the new level ϵ of constraints in a neighborhood of the previous current path γ. *Variation* returns the obtained solution path if it is successful, and (nil) otherwise. It may fail because ϵ has been decreased too much since the previous iteration, so that there is no solution path in the explored neighborhood of the previous current path. This problem is circumvented by dividing the last decrease $d\epsilon$ of ϵ by 2 each time a search fails, i.e., by trying with a level of constraints that is closer to the previous one. And each time a search succeeds, $d\epsilon$ becomes equal to $d\epsilon_{max}$ again.

ϵ_{min} can be taken equal to zero. If it is strictly positive, it is an accuracy threshold at which we stop the iterations. In that case, it has to be such that once this level of precision is reached, the path γ is very close to a solution of the original problem. Such a solution can then be found by the *Variation* procedure receiving γ and $\epsilon = 0$ (the exact constraints) as input.

4.4   The *Variation* procedure

The algorithm of the *Variation* procedure is inspired from the one used in the low-dimensional case. A dynamic programming procedure builds and explores a graph $\mathcal{T}$ of states by performing step motions from states already in $\mathcal{T}$. Its cost function is used for optimizations. In our experiments we wanted to minimize the number of maneuvers.

The differences with the low-dimensional case are the following. *Variation* receives as input a path γ and a positive number $\epsilon < \epsilon_0$. The explorations are restricted to a neighborhood $\mathcal{V}$ of γ and are performed with "ϵ-approximations" of bang motions: instead of $dq(q, u, \delta t)$, the step motion corresponding to the control u applied at q during δt is taken at random in a neighborhood $\mathcal{B}(dq(\bar{q}, u, \delta t), h(\epsilon))$ of a $(u, \delta t)$-bang applied at a configuration $\bar{q}$. If the progressiveness of constraints occurs according to expression 2, $\bar{q}$ has to be some $q_{free} \in \mathcal{C}_{free}$ such that $\|q - q_{free}\| < g(\epsilon)$. In our experiments, when some geometric constraints were relaxed, it was done in the form of expression 4, thus we took $\bar{q} = q$. And when only kinematic constraints appear progressively (expression 3), $\bar{q}$ was also taken equal to q. In the planner we had $dq(\bar{q}, u, \delta t) = f(\bar{q}, u)\delta t$. Other integration techniques can be used. If it succeeds, *Variation*(γ, ϵ) returns a path that satisfies the level ϵ of constraints.

The states $(q_{init}, f(q_{init}, u))$, $u \in \mathcal{U}(q_{init})$, are the roots of $\mathcal{T}$. The exploration stops when it reaches a state (q, q') with $q \in Goal$. Like in the low-dimensional cases, $Goal$ may have to be extended to a neighborhood $\widetilde{Goal}$ of it, that will be reached by the exploration. And $\widetilde{Goal}$ can be set as close to $Goal$ as one wishes, by taking a smaller δt.

Again, a pruning technique restricts the expansion of $\mathcal{T}$. A set of cells defined this time in a random fashion is put over $\mathcal{V}$. Each node of $\mathcal{T}$ is associated to a cell containing it, and at most one state can be associated with each cell. A state is then discarded if it is too close to another state in $\mathcal{T}$.

The procedure *Variation*(γ, ϵ) may fail for two reasons. First, as we already mentioned, the last incremental decrease of ϵ may be too high. Or, the set of pruning cells may not be adequate: during the exploration, states are discarded that would have belonged to a solution path obtained without pruning. These failure causes are circumvented by repeating a search when one fails, after dividing $d\epsilon$ by 2, and with randomly defined cells and step motions.

We describe here the pruning cells we used. Let N, T be two positive integers and $s_{ref}, n_{ref}, t_{ref}$ positive numbers. The input path γ of the procedure *Variation* is represented by a sequence of S configurations $q_1, q_2, \ldots, q_S$ ($q_1 = q_{init}$, $q_S \in \widetilde{Goal}$). Let $(q_1^n, q_1^t), \ldots, (q_S^n, q_S^t)$ be S random couples of orthonormal vectors, newly generated at each iteration. Let $p[s, n, t] \subset \mathcal{C}$, $s \in \{1, \ldots, S\}, n \in \{-N, \ldots, N\}, t \in \{-T, \ldots, T\}$, be the set of q such that $\|q - q_s\| \leq s_{ref}$, $n \leq \frac{(q - q_s) \cdot q_s^n}{n_{ref}} < n + 1$, $t \leq \frac{(q - q_s) \cdot q_s^t}{t_{ref}} < t + 1$. It is the intersection of the sphere centered at q_s with radius s_{ref}, and of a generalized cylinder with a rectangle section whose edges are the vectors $n_{ref} q_s^n$ and $t_{ref} q_s^t$. Let $c_\epsilon[s, n, t, w]$, $w \in \{1, \ldots, n_u\}$ be the set of states $\tilde{q} = (q, q')$ such that $q \in p[s, n, t]$ and $\tilde{q}$ can be obtained from a step motion corresponding to an $(u_w, \delta t)$-bang, given a level ϵ of constraints. Let $\mathcal{V} \subset \Phi$ be the union of all the $c_\epsilon[s, n, t, w]$. Let $c'[s, n, t]$ be the set of states $\tilde{q} = (q, q')$ such that $q \in p[s, n, t]$. Let $\mathcal{V}' \subset \Phi$ be the union of all the $c'[s, n, t]$. $\mathcal{V}$ and $\mathcal{V}'$ are phase space neighborhoods of γ.

4.5   Experimental results

The planner was tested with models of mobile robots and on problems inspired from maintenance operations. The optimization criterion was the minimization of the number of maneuvers. For some of the systems, a change of maneuver corresponds to some changes in the control applied. Then we used the pruning cells $c_\epsilon[s, n, t, w]$ that fit the optimization criterion since w corresponds to the control applied. For other systems, the criterion deals only with the configurations occupied during a path. Then we used the cells $c'[s, n, t]$, and *Variation* was equivalent to exploring a configuration space neighborhood of the current path with the cells $p[s, n, t]$ for pruning.

Typically, we had $\epsilon_0 = 1$, $d\epsilon_{max} = 0.1$ and $\epsilon_{min} = 0.2$. And like in the low-dimensional case, in the Dijkstra explorations the incremental cost of a step motion was 1 at a change of maneuver, and some very small positive number otherwise.

4.5.1   Multibody mobile robots

We consider a tractor towing trailers (Figure 8).

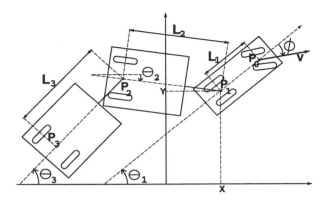

Figure 8: A tractor-trailer robot.

We call kinematic model of this vehicle the system with configuration $q_{kine} = (X, Y, \theta_1, \ldots, \theta_p)$. p is the number of bodies. X, Y are the coordinates of P_1, and θ_k is the orientation of the k-th body.

v is the velocity of P_0, and ϕ the steering angle. ϕ is bounded mechanically, and v for safety reasons:

$$|\phi| < \phi_{max}, \qquad |v| < v_{max}. \qquad (5)$$

Furthermore, in order to avoid dangerous skids, the car cannot be too fast when turning sharply. We assume that v and ϕ must be such that

$$slip(v, \phi) = |\phi.v| - s < 0. \qquad (6)$$

s is a positive parameter.

Nonholonomic kinematic constraints arise from the fact that the wheels can only roll or spin on the ground but not slide sideways. This leads [2] to the following control system equation $\dot{q}_{kine} = f_{kine}(q_{kine}, u_{kine}) =$

$$\begin{pmatrix} v.cos\phi.cos\theta_1 \\ v.cos\phi.sin\theta_1 \\ v.sin\phi/L_1 \\ \vdots \\ v.cos\phi.(\prod_{i=2}^{k-1} cos(\theta_i - \theta_{i-1})).sin(\theta_{k-1} - \theta_k)/L_k \\ \vdots \\ v.cos\phi.(\prod_{i=2}^{p-1} cos(\theta_i - \theta_{i-1})).sin(\theta_{p-1} - \theta_p)/L_p \end{pmatrix}$$

The control is $u_{kine} = (v, \phi)$.

In the kinematic models, v and ϕ can be discontinuous during a motion, which is unrealistic. Therefore we consider also models that account for dynamic constraints. We call them kinodynamic models.

They represent vehicles controlled by a driver who acts on the accelerator and brake, and on the steering wheel. The control is $u_{kino} = (a, \omega)$, where the control parameters are the acceleration $a = \dot{v}$ and the steering-wheel rotation speed $\omega = \dot{\phi}$. These are assumed to be bounded according to:

$$|\dot{v}| \le a_{max}, \qquad |\dot{\phi}| \le \omega_{max}. \qquad (7)$$

$q_{kino} = (X, Y, \theta_1, \ldots, \theta_p, v, \phi)$ is the configuration of a kinodynamic model. v and ϕ are now continuous. $q_{kino} = (q_{kine}, v, \phi)$. Constraints 5 and 6 are then geometric. The wheels are still only rolling or spinning on the ground. The control system equation for a kinodynamic model is

$$\dot{q}_{kino} = f_{kino}(q_{kino}, u_{kino}) = \begin{pmatrix} f_{kine}(q_{kino}) \\ a \\ w \end{pmatrix}.$$

A reversal corresponds to a change in the control parameter v for the kinematic models, and to a configuration where the configuration parameter v becomes zero for the kinodynamic models. Thus we used the pruning cells $c_\epsilon[s, n, t, w]$ in the first case and $c'[s, n, t]$ in the second case.

One kinematic tractor-trailer robot
We consider the kinematic model of a tractor towing trailers. The finite number of controls we consider are the $u = (v, \phi) \in \{-v_0, v_0\} \times \{-\phi_2, -\phi_1, 0, \phi_1, \phi_2\}$, where $v_0 > 0$, $\phi_1 = 36^o$ and $\phi_2 = 72^o$.

Four problems are considered: the tractor has 2 or 3 trailers, and has to move into a loading bay, either tractor first, with a forward motion, or tractor last, with a backward motion. Figure 9 depicts the two initial paths for the tractor with 3 trailers, and the initial and goal positions. The initial and goal positions are the same for the vehicle with only 2 trailers.

The collision avoidance was imposed from the beginning, in the generation of the initial paths. The kinematic constraints were then introduced progressively. Figure 10 depicts a path obtained.

The progressive part of the planner was run 50 times on each of the planning problems (2 or 3 trailers, forward (F) or backward (B) motion into the loading bay). Table 1 presents experimental results. For every series of 50 tests, t_{min} and t_{max} are the minimum and maximum duration needed. t_{90} is the time after which 90%

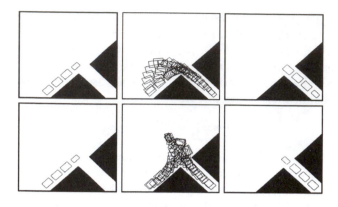

Figure 9: A tractor with 3 trailers. Initial paths.

Figure 10: A tractor towing 3 trailers. Solution path.

	2 trailers		3 trailers	
	F	B	F	B
computation time:				
t_{min}	58s	332s	104s	1h 1min
t_{90}	107s	480s	211s	3h 12min
t_{max}	144s	595s	520s	4h 16min
number of iterations:				
it_{min}	6	14	6	14
it_{max}	7	21	8	36
number of reversals:				
0	42%		80%	
1	58%	58%		56%
2-5		18%	14%	26%
6-10		14%	4%	18%
> 10		10%	2%	

Table 1: Statistics.

of the tests have produced a solution. About 11000 collision tests were done per second. it_{min} and it_{max} are the minimum and maximum number of times the procedure *Variation* was run. The numbers of reversals are indicated with percentages.

With the tractor with 3 trailers moving backwards into the loading bay, the computation times are very long because we used smaller and more pruning cells, so as to have finer explorations of the current paths' neighborhoods. Also, in this particular problem only, we put an upper bound on the maximum number of reversals a path could have in a graph $\mathcal{T}$. Otherwise, the implemented planner produced vibrating paths with many maneuvers. In fact, the smaller pruning cells are still not small enough for finding paths with few reversals without this bound. In the test series this upper bound was 10 reversals. Then, the procedure *Variation* failed when it would return a path with more than 10 reversals. In another test series with an upper bound of 1 reversal, 100% of the solution paths had only one reversal, but the computation times were about 30% longer since the procedure *Variation* failed more often.

Two kinodynamic tractor-trailer robots

We consider a system with two kinodynamic multibody mobile robots called a and b: a is a tractor with two trailers and b a tractor with one trailer. A configuration of the system is $q = (X^a, Y^a, \theta_1^a, \theta_2^a, \theta_3^a, v^a, \phi^a, X^b, Y^b, \theta_1^b, \theta_2^b, v^b, \phi^b)$. The explorations were performed by applying controls $u = (a^a, \omega^a, a^b, \omega^b)$ from $\{-a_{max}, a_{max}\} \times \{-\omega_{max}, \omega_{max}\} \times \{-a_{max}, a_{max}\} \times \{-\omega_{max}, \omega_{max}\}$, where $(a^a, \omega^a) = (\dot{v}^a, \dot{\phi}^a)$ and $(a^b, \omega^b) = (\dot{v}^b, \dot{\phi}^b)$.

All the geometric constraints were imposed from the beginning, and the kinematic ones then progressively.

Figure 11: Two kinodynamic tractor-trailer robots.

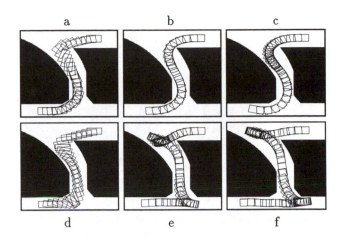

Figure 12: a. initial path; b. solution path; c. solution path with a lower ω_{max}; d. initial path; e. solution path; f. solution path with a lower a_{max}.

Figure 11 shows a solution path obtained after 15 iterations and $16min\ 47s$. 12000 collision tests were done per second. The system's model is kinodynamic, so the speeds and steering angles are continuous. During a reversal, a vehicle slows down, stops, accelerates.

One kinodynamic car-like robot
A kinodynamic model of one car-like robot is considered. The explorations were performed with controls $u = (a, \omega)$ from $\{-a_{max}, 0, a_{max}\} \times \{-\omega_{max}, 0, \omega_{max}\}$. All the geometric constraints were imposed in the initial paths.

A first planning problem is depicted in Figure 12. The initial position is in the right upper corner. In the initial and goal configurations, the car has a forward positive velocity. With the initial path of Figure 12a, solutions 12b,c were found, where the number of reversals is minimal. With the initial path of Figure 12d, solutions 12e,f were found.

Two successive positions correspond to a step motion with duration δt. In the path of Figure 12f, a_{max} was lower than in 12e: the car then needs more time and space to accelerate or decelerate. In the path of Figure 12c, ω_{max} was smaller than in 12b: the vehicle needs more time when turning the steering wheel.

The paths of Figures 12b,c,e,f were obtained in about $6s$ without progressiveness in the kinematic constraints: one iteration performed directly with $\epsilon = 0$ was sufficient because the neighborhood of the initial path explored contains a solution. 22000 collision tests were done per second.

Figure 13a displays an initial geometric path for another planning problem. The initial position is in the lower corner. The speed is null in the initial and goal configurations. Figure 13b depicts a solution path. In 13c the ground was very slippy: s was taken low in constraint 6. The car is then slower when turning. Again, in the paths of Figures 13b and c, one iteration at $\epsilon = 0$ was sufficient. The computation time was $5s$.

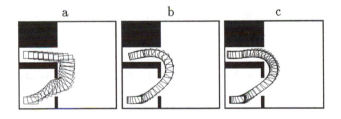

Figure 13: a. initial path; b. solution path; c. solution path on a slippy surface.

4.5.2 Motion of objects with a manipulator

The planner was used for generating motions for two objects moved with a traveling crane. The crane is shown in Figure 14 holding the first object. It can translate along the horizontal x-axis perpendicular to the beam. x represents its position on the x-axis. It can translate its lower part along the r-axis, the position is represented by r, and can rotate it around BB'

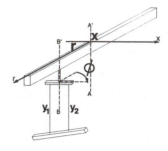

Figure 14: A traveling crane.

with the orientation angle ϕ. It can change the lengths y_1, y_2 of the vertical cables holding an object. The first object's position is represented by (x, r, ϕ, y_1, y_2). These parameters represent the configuration of the crane holding the object at this position. The second object is a vertical beam. It can be held by the two cables having the same length, and cannot be rotated. Its position is represented by (x_o, r_o, y_o), that are the $x, r, y_1 = y_2$ position parameters of the crane holding it at that position.

$q = (x, r, \phi, y_1, y_2, x_o, r_o, y_o)$ is a configuration of the system. The control system equation, corresponding to the permitted actions of the crane, is

$$\dot{q} = f(q, u) = u, \quad u \in \{u_1, u_2, \ldots, u_{16}\},$$

with the following controls and motions:

- $u_1 = (1, 0, 0, 0, 0, 0, 0, 0)$ and $u_2 = -u_1$ translate the first object in the x-axis direction.

- $u_3 = (0, 1, 0, 0, 0, 0, 0, 0)$ and $u_4 = -u_3$ translate the first object in the r-axis direction.

- $u_5 = (0, 0, 1, 0, 0, 0, 0, 0)$ and $u_6 = -u_5$ rotate the first object around BB'.

- $u_7 = (0, 0, 0, 1, 1, 0, 0, 0)$ and $u_8 = -u_7$ translate the first object vertically.

- $u_9 = (0, 0, 0, 1, -1, 0, 0, 0)$ and $u_{10} = -u_9$ produce the tilting of the first object. We assume that the cables' positions adapt so that they remain vertical and both at the same distance from BB'.

- $u_{11} = (0, 0, 0, 0, 0, 1, 0, 0)$ and $u_{12} = -u_{11}$ translate the second object in the x-axis direction.

- $u_{13} = (0, 0, 0, 0, 0, 0, 1, 0)$ and $u_{14} = -u_{13}$ translate the second object in the r-axis direction.

- $u_{15} = (0, 0, 0, 0, 0, 0, 0, 1)$ and $u_{16} = -u_{15}$ translate the second object vertically.

The geometric constraints consist of the collision avoidance. They also consist of manipulation constraints: only one object can be held by the crane, hence configurations in which none of the objects is lying in a stable position are forbidden.

Let $\mathcal{C}^{\#}_{free}$ be the set of configurations obeying the collision avoidance. $\mathcal{C}^{\#}_{free}$ is an 8-dimensional subset of $\mathcal{C}$. $\mathcal{C}_{free}$, however, consists of 6 and 7-dimensional submanifolds of $\mathcal{C}$, in which the first or second object is lying in a stable position, respectively.

Figure 16 displays a path obtained by introducing the geometric manipulation constraints and the kinematic constraints progressively, starting from an initial path depicted in Figure 15. The collision avoidance was imposed from the beginning. The computations took 16 iterations and $1h$ $16min$. 10000 collision tests were done per second. Figures 15 and 16 show two kinds of snapshots. One is seen from the top of the building and the x-axis is horizontal. The crane (not represented) is hanging above the obstacles depicted in black. The grey areas are holes with planar bottoms. The other snapshot is a vertical view taken from the left hand side of the building. The dotted lines represent the level of the holes bottom and of the normal floor. The objects are represented.

 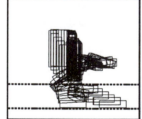

Figure 15: Initial path.

The same problem was solved with only the first object being moved. Then, the configuration is (x, r, ϕ, y_1, y_2), and only the controls $u_1, \ldots, u_{10}$ are considered. The solution was obtained by exploring a neighborhood of an initial path satisfying the collision avoidance, directly with the exact constraints ($\epsilon = 0$). The progressiveness was not necessary because

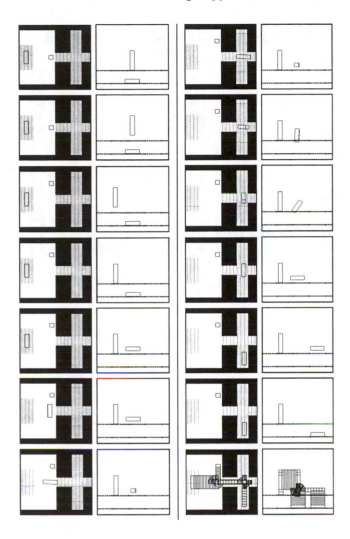

Figure 16: Solution path.

Let us illustrate this point, first from the geometric constraints point of view. Imagine a planning problem in which geometric and kinematic constraints appear progressively. In the phase space and in the configuration space, forbidden areas grow. In the PC approach, the initial path evolves continuously in $\mathcal{C}$ with respect to d_Γ, so as to adapt to the growth of forbidden areas. In some cases, this can lead to deadlock situations in $\mathcal{C}$, such as the one depicted in Figure 17: some forbidden regions of $\mathcal{C}$ join at configurations of the current path, making it impossible to find a new path close to the previous one and obeying more severe constraints. Such deadlock situations will be avoided with certainty if the freespaces in $\mathcal{C}$ keep the same homotopy classes while the forbidden regions in $\mathcal{C}$ grow. But this is not the case in general, and is difficult to verify a priori.

Figure 17: A deadlock situation.

the problem is simple enough. The computations took $3min\ 35s$. 16000 collision tests were done per second.

4.6   Discussion

Convergence

The experiments above show that the PC method can solve complicated problems for systems with high-dimensional configuration spaces. The underlying assumption of the PC approach, however, is that the current path can actually converge toward a solution, when the forbidden areas of the phase space expand. This assumption does not hold in general. The PC method is therefore not complete.

Similar deadlock situations can occur due to the kinematic constraints that appear progressively. Consider for example a kinodynamic model of a car-like robot, for which the initial path satisfying all the geometric constraints is the one of Figure 18a. The initial position is in the upper right corner. But imagine that the car has a high initial speed, and cannot decelerate enough for turning right without slipping. More precisely, for the geometric constraint $slip(v, \phi) = |\phi.v| - s < 0$ to be satisfied during a turn as in Figure 18a, the car has to violate the kinematic constraint $\dot{v} \leq a_{max}$. In this case, the initial path cannot evolve continuously and converge toward a solution. A solution path can be for example such as the one depicted in Figure 18b.

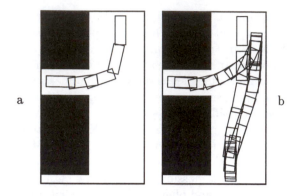

Figure 18: *Initial (a) and solution path (b) for a kinodynamic car-like robot.*

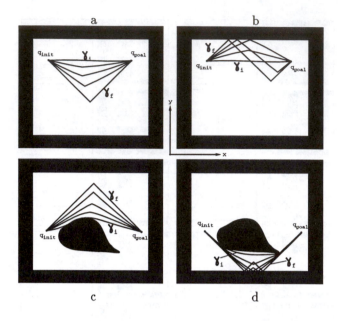

Figure 19: *Optimality is not guaranteed.*

In the cases depicted above, the current path is driven continuously into a deadlock. The evolution of the current path can also be stopped if forbidden areas appear suddenly in the phase space or configuration space, in the middle of previously free areas that the current path was crossing. Such deadlocks can occur or not, depending on the way the constraints are applied progressively.

Optimality

In the experiments we could obtain optimal or quasi optimal paths, that is, with few maneuvers. The optimization is possible through the cost function of the Dijkstra algorithm. The method is however not guaranteed to generate optimal paths.

A first reason is the pruning in the exploration technique. Nodes of the graph, that would have been part of an optimal path obtained without pruning, may be discarded during the exploration.

But independently of the technique used at each iteration for generating a new current path, the PC approach itself is not guaranteed to lead to an optimal path. Consider for example a point $q = (x, y)$ moving in a plane between obstacles, with the following kinematic constraints:

$$\dot{q} = f(q, u) = u, \qquad u \in \{(1, 1), (1, -1)\}.$$

Imagine that the obstacle avoidance constraints are imposed from the beginning in the generation of an initial path γ_i, and that the kinematic constraints appear progressively. And we wish to have paths with as few maneuvers as possible. Then, depending on the way

the current path evolves, even if during its evolution it remains locally optimal, the solution path γ_f is not guaranteed to be optimal. This is illustrated with Figures 19a,b. Also, the optimality of the final path γ_f depends on the initial path γ_i, as illustrated with Figures 19c,d, and with the experiments of Figure 12.

Pruning

Pruning is necessary for controlling the expansion of the graph $\mathcal{T}$ at every iteration. The size of the pruning cells influences the computation times and the optimization capacities of the planner.

Large cells enable short durations for the iterations since the number of nodes inserted into the graph $\mathcal{T}$ is at most the number of pruning cells. However, large cells may cause the failure of the procedure *Variation*, if the nodes of $\mathcal{T}$ do not spread enough through the explored neighborhood of the current path.

Small cells induce longer computation times through the higher number of nodes of $\mathcal{T}$, but enable to solve more difficult problems. Also, optimizations through the dynamic programming exploration procedure are more efficient with small cells.

5 Conclusion

We presented a new method for motion planning with geometric and kinematic constraints, that is able to solve complicated problems for systems with high-dimensional configuration spaces. First, a path satisfying only some of the geometric constraints is generated. This path is then transformed in an iterative way while constraints are introduced progressively. Each iteration consists of an exploration of a neighborhood of the current path, that can deal with optimization criteria.

The method works for any robot that is represented by a control system equation with a finite number of controls. It is not complete, but was able to solve difficult motion planning problems with multibody mobile robots, with or without dynamic constraints, and with objects moved by manipulators that can only perform specific actions at a time.

Acknowledgements

This research is part of P. Ferbach's PhD done with INRIA Sophia-Antipolis. Part of it was done at the Stanford University Computer Science Robotics laboratory.

References

[1] J. Barraquand, P. Ferbach, *Path Planning through Variational Dynamic Programming.* IEEE ICRA, 1994.

[2] J. Barraquand, J.-C. Latombe, *Nonholonomic Multibody Mobile Robots: Controllability and Motion Planning in the Presence of Obstacles.* Algorithmica, 10:121-155, Springer Verlag, 1993.

[3] L.G. Bushnell, D.M. Tilbury, S.S. Sastry, *Steering Three-Input Nonholonomic Systems: The Fire-Truck Example.* International Journal of Robotics Research, vol.14, no.4 , 1995.

[4] M. Cherif, Ch. Laugier, *Motion Planning of Autonomous Off-Road Vehicles under Physical Interaction Constraints.* IEEE ICRA, 1995.

[5] B.R. Donald, P.G. Xavier, J.F. Canny, J. Reif, *Kinodynamic Motion Planning.* Journal of the ACM, 40(5), 1993.

[6] L.E. Dubins, *On Curves of Minimal Length with a Constraint on Average Curvature and with Prescribed Initial and Terminal Positions and Tangents.* American Journal of Mathematics, vol.79, 1957.

[7] P. Ferbach, J. Barraquand, *A Penalty Function Method for Constrained Motion Planning.* IEEE ICRA, 1994.

[8] C. Fernandes, L. Gurvits, Z.X. Li, *Foundations of Nonholonomic Motion Planning.* Technical Report No.577, New York University Courant Institute of Mathematical Sciences, NY, 1991.

[9] J.-C. Latombe, *Robot Motion Planning.* Kluwer Academic, Boston, 1991.

[10] J.-P. Laumond, P.E. Jacobs, M. Taix, R.M. Murray, *A Motion Planner for Nonholonomic Mobile Robots.* IEEE Transactions on Robotics and Automation, vol.10, no.5, 1994.

[11] Z. Li, *Nonholonomic Motion Planning.* Kluwer Academic, Boston, 1992.

[12] K.M. Lynch, M.T. Mason, *Stable Pushing: Mechanics, Controllability, and Planning.* Workshop on the Algorithmic Foundations of Robotics, 1994.

[13] R.M. Murray, S.S. Sastry, *Steering Nonholonomic Systems Using Sinusoids.* IEEE CDC, 1990.

[14] J.A. Reeds, R.A. Shepp, *Optimal Paths for a Car that Goes both Forward and Backward.* Pacific Journal of Mathematics, vol.145, no.2, 1990.

[15] G. Sahar, J.M. Hollerbach, *Planning of Minimum-Time Trajectories for Robot Arms.* IEEE ICRA, 1985.

[16] S. Sekhavat, P. Švestka, J.-P. Laumond, M. Overmars, *Multi-Level Path Planning for Nonholonomic Robots Using Semi-Holonomic Subsystems.* Workshop on the Algorithmic Foundations of Robotics, 1996.

[17] H.J. Sussmann, W. Liu, *Limits of Highly Oscillatory Controls and the Approximation of General Paths by Admissible Trajectories.* Technical Report SYSCON-91-02, Rutgers Center for Systems and Control, 1991.

[18] P. Švestka, M.H. Overmars, *Coordinated Motion Planning for Multiple Car-Like Robots using Probabilistic Roadmaps.* IEEE ICRA, 1995.

Sensor Based Motion Planning:
The Hierarchical Generalized Voronoi Graph

Howie Choset, *Carnegie Mellon University, Pittsburgh, PA, USA*
Joel Burdick, *California Institute of Technology, Pasadena, CA, USA*

Abstract *The* hierarchical generalized Voronoi graph *(HGVG) is a* roadmap *that can serve as a basis for* sensor based robot motion planning. *A key feature of the HGVG is its incremental construction procedure that uses only* line of sight *distance information. This work describes basic properties of the HGVG and the procedure for its incremental construction using local range sensors. Simulations and experiments verify this approach.*

1 Introduction

Sensor based motion planning incorporates sensor information, reflecting the current state of the environment, into a robot's planning process, as opposed to *classical planning*, which assumes full knowledge of the world's geometry prior to planning. Sensor based planning is important for realistic deployment of robots because: (1) the robot often has no a priori knowledge of the world; (2) the robot may have only a coarse knowledge of the world because of limited computer memory; (3) the world model is bound to contain inaccuracies which can be overcome with sensor based planning strategies; and (4) the world is subject to unexpected occurrences or rapidly changing situations.

This work addresses two primary problems in sensor based motion planning for static environments. The first problem deals with the case of when a robot is given a target location and it always knows its current location (i.e., it has an on-board dead reckoning system). Assuming no a priori knowledge about its environment, the robot must find a collision-free path to the goal, based on sensory measurement. In the second problem, the robot is placed in a bounded environment with no a priori knowledge about that environment. Using only its on-board sensors and a dead reckoning system, the robot must build a complete *roadmap* of the bounded environment. A roadmap is a collection of one-dimensional curves that capture the important topological and geometric properties of the robot's environment. Roadmaps have the following properties: *accessibility*, *departability*, and *connectivity*. These properties imply that the planner can construct a path between any two points in a connected component of the robot's free space by first finding a collision free path onto the roadmap (accessibility), traversing the roadmap to the vicinity of the goal (connectivity), and then constructing a collision free path from a point on the roadmap to the goal (departability). The solution to this second problem automatically supplies a solution to the first problem, and so we will focus our attention on the second problem. The *hierarchical generalized Voronoi graph* (HGVG), defined in this work, is a roadmap which can be incrementally constructed using line of sight sensor data.

2 Relation to Previous Work

Many sensor based planners are heuristic and work well under a variety of conditions. Nevertheless, there are no proofs of correctness that guarantee a path can be found and furthermore, there do not exist well established thresholds for when heuristic algorithms fail. One class of heuristic algorithms is a behavioral based approach in which the robot is armed with a simple set of behaviors such as following a wall [2]. A hierarchy of cooperating behaviors forms more complicated behaviors such as exploration. An extension of this type of approach is called sequencing [11]. Since there are strong experimental results indicating the utility of these approaches, some of these algorithms may provide a future basis for provably correct sensor based planners.

There are many non-heuristic algorithms for which provably correct solutions exist in the plane (see [15] for an overview). For example, Lumelsky's "bug" algorithm [13] is one of the first provably correct sensor based schemes to work in the plane. However, this algorithm (like many described in [15]) requires knowledge of the goal's location during the planning process. Hence, the robot can't search for a goal "beacon." Furthermore, this algorithm simply returns a path from the start to the goal. The resulting path does not reflect the topology of the free space (the region of the environment not occupied by obstacles) and thus, it cannot be used to guide future robot excursions.

One approach to sensor based motion planning is to adapt the structure of a provably correct, or "complete," classical motion planning scheme to a sensor based implementation. Roadmaps are one of the complete classical methods. An example of a complete roadmap scheme is Canny and Lin's Opportunistic Path Planner (OPP) [4]. Rimon adapted this motion planning scheme for sensor based use [17], but, connectivity of the roadmap in [17] cannot be guaranteed without "active perception." Furthermore, from a practical point of view, there are two detractions to Rimon's method. First, to construct the roadmap, the robot must contain "interesting critical point" and "interesting saddle point" sensors, whose implementation is not well described. Second, a robust, detailed, and efficient procedure for constructing the roadmap edges from sensor data is not presented.

The HGVG is an extension of the *generalized Voronoi diagram* (GVD) into higher dimensions. The GVD is the locus of points equidistant to two or more obstacles which are convex sets in the plane. (The Voronoi diagram is the set of points equidistant to two or more points (sometimes termed sites) in the plane.) The GVD was first used for motion planning in [18]. Active research in applying the GVD to motion planning began with Ó'Dúnlaing and Yap [14], who considered motion planning for a disk in the plane. However, the method in [14] requires full knowledge of

the world's geometry prior to the planning event, and its retract methodology may not extend to non-planar problems. In [16], an incremental approach to create a Voronoi diagram-like structure, which is limited to the case of a plane, was introduced.

Prior work (e.g., [1]) describes the *Voronoi graph*, an extension of the Voronoi diagram into higher dimensions. The Voronoi graph is the locus of points in m dimensions equidistant to m point sites. The *generalized Voronoi graph (GVG)*, defined in Section 5, extends the Voronoi graph to the case of convex obstacles; that is, it is the set of points in m dimensions equidistant to m convex obstacles. Though the GVG introduced in this work appears to be new, a disconnected GVG-like structure for $SE(3)$ is described in [3].

The GVG can be thought of as the natural extension of the GVD into higher dimensions. However, unlike the GVD, the GVG is not necessarily connected in dimensions greater than two, and thus, in general, is not a roadmap. Therefore, we introduce additional structures, termed *higher order generalized Voronoi graphs* which are guaranteed to link the disconnected GVG components into a connected network. The resulting connected structure is the *hierarchical generalized Voronoi graph* (HGVG).

3 Contributions

Although the HGVG is applicable to classical motion planning, the intended use of the HGVG is for sensor based planning. When full knowledge of the world is available to the robot, we make no claim that the HGVG has any clear advantage over other one-dimensional roadmaps used in multi-dimensional configuration spaces, though our experience shows that the construction of this roadmap is efficient. However, since the HGVG is defined in terms of distance information, it lends itself to sensor based construction. We describe an incremental construction technique for the HGVG. Unlike other sensor based construction procedures, this procedure is proven to be complete (it is guaranteed to find the goal if it is reachable or to map the entire environment in finite time) and need not

Figure 1: *Mobile robot with sonar ring.*

require any artificial landmarks, complicated obstacle segmentation, nor abstract sensors. Additionally, this incremental construction technique can be applied to construct the edges of other roadmaps, such as Canny and Lin's Opportunistic Path Planner, Rimon's extension of the OPP, and wall following algorithms.

4 Distance Function

The HGVG and its properties are based on the following workspace distance function definitions. Assume that the robot is a point operating in a work space, $\mathcal{W}$, which is a subset of an m-dimensional Euclidean space. $\mathcal{W}$ is populated by convex obstacles $C_1, \ldots, C_n$. Non-convex obstacles are modeled as the union of convex shapes. It is assumed that the boundary of $\mathcal{W}$ is a collection of convex sets, which are members of the obstacle set $\{C_i\}$. The free space, $\mathcal{FS}$, is the subset of $\mathcal{W}$ not occupied by obstacles.

The distance between a point x and a convex set C_i is termed the single object distance function and is defined as

$$d_i^X(x) = \min_{c_0 \in C_i} \|x - c_0\|, \qquad (1)$$

where $\| \cdot \|$ is the two-norm in $\mathbb{R}^m$. The gradient of $d_i^X(x)$ is

$$\nabla d_i^X(x) = \frac{x - c_0}{\|x - c_0\|}. \qquad (2)$$

The vector $\nabla d_i^X(x)$ is a unit vector in the direction from c_0 to x, where c_0 is the nearest point to x in C_i.

To compute Equations (1) and (2) from sensor data, one need only know the distance and direction to the nearest point on C_i. For convex sets, the closest point is always unique.

Typically, the environment is populated with multiple obstacles, and thus we define a multi-object distance function as $D^X(x) = \min_i d_i^X(x)$.

The single object distance function measures the distance to obstacles that may not be within line of sight of the robot. We term this distance function the "X-ray" distance function because it assumes the robot has X-ray vision.

Alas, for the purposes of sensor based motion planning, the robot can only measure distances to obstacles it can see. To this end, we define the *single object visible distance function* which measures distance to nearby obstacles that are *within visible-line of sight*. A point c is *within line of sight* of x if there exists a straight line segment that connects x and c without penetrating any obstacle. That is, c is within line of sight of x if for all $t \in [0, 1]$, $(x(1 - t) + ct)$ is a point in $\mathcal{FS}$.

Let $\tilde{C}_i(x)$ be the set of points on an object C_i that are within line of sight of x, i.e.,

$$\tilde{C}_i(x) = \{c \in C_i : \forall t \in [0, 1], \, x(1 - t) + ct \in \mathcal{FS}\}.$$

Let $c_i = \operatorname{argmin} d_i^X(x)$, i.e., c_i is the closest point on the convex obstacle C_i to x. The *single object visible distance function* is the distance to obstacle C_i if c_i is within line of sight of x. In this case, C_i is within *visible*-line of sight of x. If C_i is not within visible-line of sight of x, then the distance to C_i is infinite. We denote the visible distance function as $d_i(x)$. If the visible distance function has a finite value at a point x, then its gradient is defined by Equation (2) and is denoted $\nabla d_i(x)$. The visible multi-object distance function is

$$D(x) = \min_i d_i(x). \qquad (3)$$

In this work, we use the visible distance functions. That is, our definition of the GVG and HGVG in terms of the visible distance function is unique.

An important characteristic of $d_i(x)$ and $\nabla d_i(x)$ is that they can be computed from sensor data. For example, consider a mobile robot with a ring of sonar sensors (Figure 1). The sensor measurement provides an approximate value of the distance function. The centerline of the sensor's measurement axis (which is orthogonal to the sensor face) approximates the distance gradient. Stereo vision, depth from focus, or laser range fingers can also be used to provide depth and gradients to points on surfaces in the surrounding environment. With this distance function, which makes use of line of sight information, we can define the HGVG, and related structures such as the generalized Voronoi graph.

5   The Generalized Voronoi Graph

Equidistant Faces. The basic building block of the GVD and GVG is the set of points equidistant to two sets C_i and C_j, which we term the *two-equidistant surface*,

$$\mathcal{S}_{ij} = \{x \in \mathcal{W} \backslash (C_i \bigcup C_j) : d_i(x) - d_j(x) = 0\}. \tag{4}$$

See Figure 2. Of particular interest is the subset of $\mathcal{S}_{ij}$ termed the *two-equidistant surjective surface*,

$$\mathcal{SS}_{ij} = \mathrm{cl}\{x \in \mathcal{S}_{ij} : \nabla d_i(x) \neq \nabla d_j(x)\}. \tag{5}$$

The two-equidistant surjective surface, $\mathcal{SS}_{ij}$, is the set of points which are equidistant to two objects such that $\nabla d_i \neq \nabla d_j$, i.e. the function $\nabla (d_i - d_j)(x)$ is surjective. This definition is required to deal with non-convex sets that are defined as the finite union of convex sets. See Figure 3. If $\mathcal{W}$ is solely populated with disjoint convex obstacles, then $\mathcal{SS}_{ij} = \mathcal{S}_{ij}$ for all i, j.

Using the Pre-image Theorem, it can be shown that $\mathcal{SS}_{ij}$ is an $(m-1)$-dimensional manifold in $\mathbb{R}^m$ (i.e., $\mathcal{SS}_{ij}$ has co-dimension one) [6].

Definition 1 (Two-Equidistant Face) The *two-equidistant face*,

$$\mathcal{F}_{ij} = \mathrm{cl}\{x \in \mathcal{SS}_{ij} :$$
$$d_i(x) = d_j(x) \leq d_h(x) \quad \forall h \neq i, j\}, \tag{6}$$

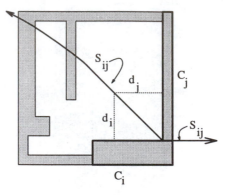

Figure 2: *The solid line represents $\mathcal{S}_{ij}$, the set of points equidistant to obstacles C_i and C_j. The dotted lines emphasize that at $x \in \mathcal{S}_{ij}$, $d_i(x) = d_j(x)$.*

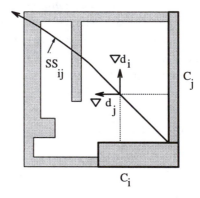

Figure 3: *The solid line represents $\mathcal{SS}_{ij}$, the set of points equidistant to obstacles C_i and C_j such that the two closest points are distinct. Note, it is also unbounded and only has one component.*

is the set of points equidistant to obstacles C_i and C_j, such that each point x in $\mathcal{SS}_{ij}$ is closer to C_i and C_j than to any other obstacle.

See Figure 4. In keeping with the conventions of the Voronoi diagram literature, a two-equidistant face is also termed a *generalized Voronoi face*.

The *two-Voronoi set*, $\mathcal{F}^2$, is the union of all two-equidistant faces, i.e.,

$$\mathcal{F}^2 = \bigcup_{i=1}^{n-1} \bigcup_{j=i+1}^{n} \mathcal{F}_{ij} \tag{7}$$

Since $\mathcal{F}^2$ is the set of points equidistant to the two or faces.

$$\begin{aligned}
\mathcal{F}_{i_1 i_2 \ldots i_k} &= \mathcal{F}_{i_1 i_2} \bigcap \cdots \bigcap \mathcal{F}_{i_1 i_k}, \\
&= \mathcal{F}_{i_1 i_2 \ldots i_{k-1}} \bigcap \mathcal{F}_{i_1 i_k}, \\
\mathcal{F}^k &= \bigcup_{i_1=1}^{n-k+1} \bigcup_{i_2=i_1+1}^{n-k+2} \cdots \bigcup_{i_k=i_{k-1}+1}^{n} \mathcal{F}_{i_1 i_2 \ldots i_k} \quad (8)
\end{aligned}$$

Generalized Voronoi Graph Definition. In m dimensions, a *generalized Voronoi edge* and a *generalized Voronoi vertex* are respectively an m-equidistant face, $\mathcal{F}_{i_1 \ldots i_m}$, and an $m+1$-equidistant face, $\mathcal{F}_{i_1 \ldots i_{m+1}}$. Generically (i.e, when equidistant faces transversally intersect), the Pre-image Theorem asserts that generalized Voronoi edges are one-dimensional and that the generalized Voronoi vertices are points. Since generalized Voronoi edges meet at generalized Voronoi vertices, generalized Voronoi vertices are termed *meet points*. Using these definitions, we can define:

Definition 2 (Generalized Voronoi Graph) *The generalized Voronoi graph* (GVG) *is the collection of generalized Voronoi edges and meet points. That is,*

$$GVG = (\mathcal{F}^m, \mathcal{F}^{m+1}). \quad (9)$$

The GVG's edges comprise the set of points equidistant to m objects, such that each point is closer to m objects than to any other object. *An important characteristic of the GVG is that it is defined in terms of a visible distance function, which can be computed readily from sensor data.* It is this feature that makes the GVG useful for sensor based motion planning. Our definition of the Voronoi graphs and diagrams in terms of a *visible* distance function is unique.

Example 1 *Figure 6 depicts a generalized Voronoi graph for a rectangular enclosure in $\mathbb{R}^3$. The GVG edges, delineated by solid lines, constitute the locus points equidistant to three obstacles, and the meet points are where the GVG edges intersect.* ◆

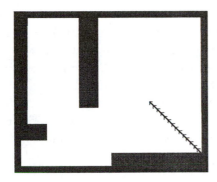

Figure 4: *The ticked solid line is the set of points equidistant and closest to obstacles C_i and C_j.*

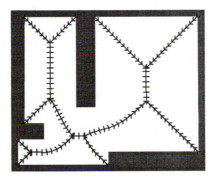

Figure 5: *The ticked solid lines is the set of points equidistant to two obstacles, such that each edge fragment is closest to the equidistant obstacles.*

more closest points on the boundary of $\mathcal{W}$, it is the generalized Voronoi diagram (GVD) of $\mathcal{W}$. See Figure 5. Since it can be shown that $\mathcal{F}_{ij}$ is an $(m-1)$-dimensional manifold with an $(m-2)$-dimensional boundary, $\mathcal{F}^2$ has co-dimension one, but is not necessarily a manifold.

To define the GVG, we continue to define lower dimensional subsets of $\mathcal{W}$. The k-*equidistant face*, $\mathcal{F}_{i_1 i_2 \ldots i_k}$, is the $(m-k+1)$-dimensional set of points that are equidistant to objects $C_{i_1}, C_{i_2}, \ldots, C_{i_k}$ such that each point is closer to objects $C_{i_1}, C_{i_2}, \ldots, C_{i_k}$ than to any other object and no two distance gradient vectors are collinear. The k-*Voronoi set*, $\mathcal{F}^k$, is simply the union of all $(m-k+1)$-dimensional k-equidistant

6   Incremental Construction of the GVG

A key feature of the GVG is that it can be incrementally constructed using line of sight range information.

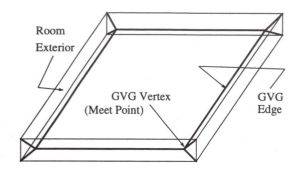

Figure 6: *The generalized Voronoi graph in a rectangular enclosure. The solid lines represent the GVG edges, which meet at GVG vertices, or "meet points."*

In the scenario in which the robot has no a priori information about the environment, the robot must construct a roadmap in an incremental manner because most environments do not contain one vantage point from which a robot can "see" the entire world, and thereby construct a roadmap from such a single vantage point.

The incremental construction techniques described in this section provide a rigorous approach to constructing the GVG using only line of sight sensory information. It is worth noting that these incremental construction procedures can be the basis of a numerical method to construct a roadmap when full geometry of the world is available. Furthermore, we have found that our incremental construction procedures are not only sensor implementable, but numerically efficient. Finally, the incremental construction techniques described in this section can be adapted to other sensor based planning methods such as the OPP (described in [4, 17]).

The HGVG's properties of accessibility, departability and connectivity translate to *incremental accessibility*, *incremental departability*, and *traceability*, respectively, in the incremental construction procedure. This section describes how to move onto (incremental accessibility) and trace along (traceability) the GVG *using only local information*. Incremental departability is described in Section 8. The algorithm is verified by simulations and experiments that are reviewed in Section 7.

6.1 Incremental Accessibility

A robot can access the GVG by following a path that is constructed using gradient ascent on the multi-object distance function $D(x)$, which is the distance to the nearest object from x. In [5], it was shown that D is not smooth, and thus does not have a conventional gradient. However, the multi-object distance function does exhibit a *generalized gradient* [10]. The generalized gradient of D was shown in [5] to be

$$\partial D(x) = \text{Co}\{\nabla d_i(x) : \ \forall i \in I(x)\}, \qquad (10)$$

where $I(x)$ is the set of indices where $d_i(x) = D(x)$, and where Co denotes convex hull.

Furthermore, it was shown in [5] that if $0 \in \text{int}(\partial D(x))$, where 0 is the origin of the tangent space at x, then x is a local maxima of D. It is worth noting that this local maxima is determined solely from first order information. Using this result and the following two lemmas, we can conclude that if x is a local maxima of D, then it is equidistant to $m + 1$ obstacles.

Lemma 1 *Given a set of n arbitrary vectors in $\mathbb{R}^m$, then $0 \in \text{int}(\text{Co}\{v_i \in \mathbb{R}^m : \ i = 1, \ldots, n\})$ if and only if $\{v_i \in \mathbb{R}^m : \ i = 1, \ldots, n\}$ positively span $\mathbb{R}^m$.*

Lemma 2 (Goldman and Tucker) *It requires a minimum of $(m + 1)$ vectors to positively span $\mathbb{R}^m$.*

6.2 Traceability

In an incremental context, the property of connectivity is interpreted as *traceability*. More specifically, traceability implies that using only local data, the robot can: (1) "trace" the GVG (or HGVG) edges; (2) determine all of the edges that emanate from a meet point; (3) change directions at a meet point, and thereby begin tracing new edges; and (4) determine when to terminate the tracing procedure.

Naively, one could trace an edge by repeated application of the accessibility method. That is, the robot would move a small distance along a given direction— either a fixed direction, or perhaps the tangent direction to the current edge. Gradient ascent would then be used to move back onto the local edge. The OPP

[4] method and its sensor based adaptation [17] use this strategy with a fixed stepping direction. However, gradient ascent can be a computationally expensive procedure because of its slow convergence. Also, the constant step direction leads to undesirable roadmap artifacts [5].

Our approach to edge construction borrows ideas from numerical continuation methods [12]. Continuation methods trace the roots of the expression $G(y, \lambda) = 0$ as the parameter λ is varied. The incremental construction of a GVG edge can be implemented as follows.

Let x be a point on the GVG. Choose local coordinates at x so that the first coordinate, z_1, lies in the direction of the tangent to the graph at x (see Figure 7). At x, let the hyperplane spanned by coordinates $z_2, \ldots, z_m$ be termed the "normal slice plane." We can thus decompose the local coordinates into $x = (y, \lambda)$, where $\lambda = z_1$ is termed the local "sweep" coordinate and $y = (z_2, \ldots, z_m)$ are the "slice" coordinates. Now define the function $G: \mathbb{R}^{m-1} \times \mathbb{R} \to \mathbb{R}^{m-1}$ as follows:

$$G(y, \lambda) = \begin{bmatrix} (d_1 - d_2)(y, \lambda) \\ (d_1 - d_3)(y, \lambda) \\ \vdots \\ (d_1 - d_m)(y, \lambda) \end{bmatrix} \qquad (11)$$

The function $G(y, \lambda)$ assumes a zero value only on the GVG. Hence, if the Jacobian of G, $\nabla_y G$, is surjective, then the implicit function theorem implies that the roots of $G(y, \lambda)$ locally define a generalized Voronoi edge as λ is varied.

By numerically tracing the roots of G, we can locally construct an edge. While there are a number of such techniques [12], we use an adaptation of a common predictor-corrector scheme. Assume that the robot is located at a point x on the GVG. The robot takes a "small" step, $\Delta \lambda$, in the z_1-direction (i.e., the tangent to the local GVG edge). This tangent is computed using line of sight information as follows:

Proposition 1 *The tangent to a GVG edge at x is defined by the vector orthogonal to the hyperplane that contains the m closest points: $c_1, \ldots, c_m$ of the m closest objects, $C_1, \ldots, C_m$.*

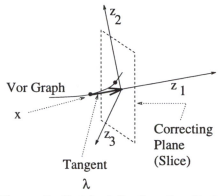

Figure 7: *Sketch of Continuation Method*

The proof of the above appears in [7].

In general, this "prediction" step will take the robot off the GVG. Next, a "correction" method is used to bring the robot back onto the GVG. If $\Delta \lambda$ is "small," then the graph will intersect a "correcting plane" (Figure 7), which is a plane parallel to the normal slice at distance $\Delta \lambda$. The correction step finds the location where the GVG intersects the correcting plane. (Figure 7)

The explicit correction procedure is an iterative Newton's Method. If y^k and λ^k are the kth estimates of y and λ, the $k + 1$st iteration is defined as

$$y^{k+1} = y^k - \left(\nabla_y G\right)^{-1} G(y^k, \lambda^k) \qquad (12)$$

where $\nabla_y G$ is evaluated at (y^k, λ^k). The following proposition, whose proof appears in [7], guarantees that Equation (12) is well defined:

Proposition 2 *(Equidistant Surface Full Rank Property) The matrix $\nabla_y G(y, \lambda)$ has full rank (i.e., has rank $(m-1)$) in a neighborhood of the GVG on the correcting plane.*

Practically speaking, this result states that the numerical procedure defined by Equation (12) will be robust for reasonable errors in robot position, sensor errors, and numerical round off.

There are several things worth noting about this method. First, to evaluate $G(y, \lambda)$ and $\nabla_y G(y, \lambda)$, one only needs to know the distance and direction to the m objects that are closest to the robot's current location—information that is easily obtained from local distance sensor data. Second, Newton methods are quadratic in their convergence, and thus they would be substantially faster than the naive gradient ascent techniques. Third, $\nabla_y G(y, \lambda)$ is an $(m-1) \times (m-1)$ matrix, and is thus typically quite small in size (e.g., a scalar for two-dimensional environments, or a 2×2 matrix for three-dimensional environments).

6.3   Terminating Conditions

So far, we have shown that the robot can access and trace a GVG edge. Due to the boundedness of the robot's environment, GVG edges must terminate. Most GVG edges have meet points for end points, but the some GVG edges may have other types of end points. A GVG edge may terminate on a *boundary point*, denoted $C_{i_1 \ldots i_m}$, where m obstacles intersect. A GVG edge may also terminate at a *floating boundary point*, denoted $FC_{i_1 \ldots i_m}$. At a floating boundary point, two distance function gradient vectors become collinear. The following proposition, whose proof appears in [9], guarantees that the only terminating conditions for a GVG edge are meet points, boundary points, and floating boundary points.

Proposition 3 *Given that equidistant faces transversally intersect in a bounded environment, if a generalized Voronoi edge is not a cycle (a GVG edge diffeomorphic to a circle), it must terminate: (1) at a generalized Voronoi vertex (a meet point), (2) on the boundary of the environment, or (3) at a point where two gradients of single object distance functions become collinear.*

This result is apparent by inspecting the definition of a GVG edge in $\mathbb{R}^3$:

$$\mathcal{F}_{ijk} = \mathrm{cl}\{x \in \mathcal{FS} : \nabla d_i(x) \neq \nabla d_j(x) \text{ and}$$
$$0 < d_i(x) = d_j(x) = d_k(x) < d_h(x)\} \quad (13)$$

The boundary of the GVG edge associated with the inequality $\nabla d_i(x) \neq \nabla d_j(x)$ is a floating boundary point (where two gradient vectors become collinear); the boundary of the GVG edge associated with the first inequality $0 < d_i(x)$ is a boundary point (where

distance to the environment is zero); and, the boundary associated with the last inequality is a meet point.

Incremental construction of the GVG is akin to a graph search where GVG edges are the "edges" and the meet points, boundary points, and floating boundary points are the "nodes." Once the robot has accessed a point on the GVG, it begins tracing an edge. If the robot encounters a meet point, it marks off the direction from where it came as explored, and then explores one of the other m edges that emanate from the meet point. It also marks off that direction as being explored. If the robot reaches another unvisited meet point, the above procedure is recursively repeated. When the robot hits a boundary point or a floating boundary point, it simply turns around and retraces its path to some previous meet point with unexplored directions. The robot terminates exploration of the GVG fragment when there are no more unexplored directions associated with any meet point. If the robot is looking for a particular destination whose coordinates are known, then the robot can invoke graph searching techniques, such as the A-star algorithm, to control the tracing procedure.

Meet Point Detection. Finding the meet points is essential to proper construction of the graph. While a meet point occurs when the robot is equidistant to $m+1$ objects, it is unreasonable to expect that a robot can exactly detect such points because of sensor error. Furthermore, since the robot is taking finite sized steps while tracing an edge, it is unlikely that the robot will pass exactly through an $(m+1)$-equidistant point. However, as shown in Figure 8, meet points can be robustly detected by watching for an abrupt change in the direction of the (negated) gradients to the m closest obstacles. Such a change will occur in the vicinity of a meet point.

Departing a Meet Point. Recall that the robot is equidistant to $m+1$ objects at a meet point. It must be able to identify and explore the $m+1$ generalized Voronoi edges that emanate from each meet point in order to completely construct the GVG. Each emanating edge corresponds to an m-wise combination of the $m+1$ closest objects. Assume that we wish to explore and trace the edge corresponding to objects

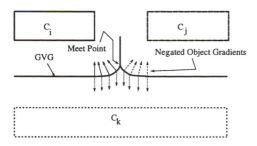

Figure 8: *Meet Point Detection*

$C_1, \ldots, C_m$. Proposition 1 yields the one-dimensional tangent space to the generalized Voronoi edge corresponding to these m objects. If v is a tangent vector computed from Proposition 1, the robot must determine if it should depart the meet point in the $+v$ or $-v$ direction. Let $d_1(x) = d_2(x) = \cdots = d_m(x) = d_{m+1}(x)$ be the distances to the $m+1$ closest objects at a meet point. If $\langle \nabla d_{m+1}, v \rangle > \langle \nabla d_i, v \rangle$ where $i \in \{1, \ldots m\}$, then the robot should move in direction $+v$, otherwise $-v$. Recall that a generalized Voronoi edge is closer to the m objects that define it, than any other object. This effects motion away from C_{m+1}.

So far, we have defined the GVG and demonstrated some of its properties. In particular, we showed that using line of sight information, a robot can access the GVG and incrementally trace out its edges. These properties are verified in the following section. A discussion of the properties of connectivity and departability are deferred to Section 8.

7 Simulations and Experiments

Planar Simulations. A planar simulator has validated this approach for a point or circularly symmetric robot operating in the plane. Figure 4 contains an example of a bounded environment in which our algorithm was tested. In this figure, the robot has accessed the GVG and traced one GVG edge. The ticked solid lines represent the planar GVG (also the GVD); these are the locus of points equidistant to the two nearest obstacles. The ticks point to the nearest obstacles. Figure 5 shows the final simulation result.

Three-Dimensional Simulator. A major advantage that the HGVG has over other methods is that it

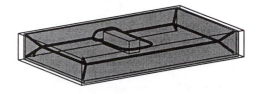

Figure 9: *GVG of a 3-dimensional box with a long box which is located off-center in the interior.*

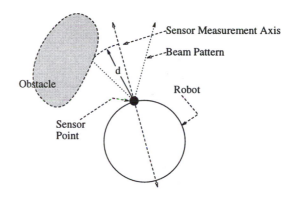

Figure 10: *Simplified distance measurement sensor model.*

is applicable in higher dimensional workspaces. To this end, we have implemented a three-dimensional simulator which traces GVG edges. The algorithm and data structure of the three-dimensional simulator is similar to that of the planar version. The distance function code used in this simulator was written by Brian Mirtich at Berkeley. Currently, the linking procedures (described in the next section) are under development. See Figure 9 for final results of GVG tracing.

Experimental Results. Another advantage of this approach is that the HGVG can be incrementally constructed from raw sensor data. Experiments were performed on a mobile robot with a ring of ultrasonic range sensors, depicted in Figure 1. We assume the sensors measure distance to nearby obstacles, along a fixed direction termed the *sensor measurement axis*. The sensor measurement axis is a function of the robot's position and orientation (See Figure 10). It is shown in [9] that the distances to individual obstacles correspond to local minima in the sensor array. The distance gradient is the unit vector pointing along the sensor measurement axis. An example is depicted in

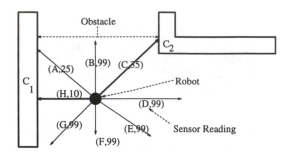

Figure 11: *Sensors H and Sensors C are associated with the distances to the two closest obstacles.*

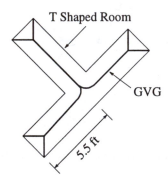

Figure 12: *T-shaped Room with Actual GVG.*

Figure 11 where a robot with eight sensors and their measurements is drawn.

The result of one experiment in a "T-shaped" room using the mobile robot is shown in Figure 12 (theoretical GVG) and in Figure 13 (experimental GVG). The small in Figure 13 squares denote the edge termination points, while the hatched squares represent meet points. For safety reasons, the robot does not trace the edge all the way to the wall's boundary. The octagon shown on the graph represents the point where the robot first accessed the GVG. The experimental GVG edges are jagged because the tangent is crudely approximated because of the angular inaccuracy of sonar distance sensors and the low resolution of sensor placement. However, the GVG is connected, and the edges are far away from the workspace boundary. Our experiments show that the actual GVG construction is quite robust even with crude distance sensors having large errors in distance localization.

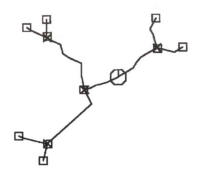

Figure 13: *Experimental GVG.*

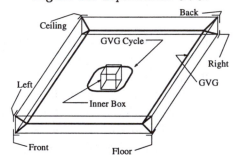

Figure 14: *Disconnected GVG.*

8 Hierarchical Generalized Voronoi Graph

Although the planar GVG and the particular GVG in Figure 9 are connected, the GVG is not guaranteed to be connected in work space dimensions greater than two. Figure 14 contains an example of a disconnected GVG with two connected components: (1) an outer GVG network similar to the one described in Example 1 and (2) an inner GVG network which forms a halo-like structure around the inner box. This section outlines our complete roadmap, which is the GVG augmented by additional structures that are termed *higher order generalized Voronoi graphs*. The higher order generalized Voronoi graphs are used to link disconnected GVG components.

Essentially, higher order generalized Voronoi graphs are like GVG's that are constrained to equidistant faces. In other words, we recursively invoke the GVG construction procedure on lower dimensional faces. For example, *a second order generalized Voronoi graph*, de-

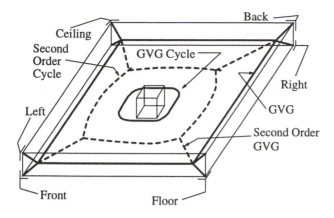

Figure 15: *The GVG^2 is drawn in dotted lines.*

noted GVG^2, is analogous to a GVG that is restricted to a two-equidistant face. An ith order generalized Voronoi graph, denoted GVG^i, is analogous to a GVG on an $(i-1)$st order two-equidistant face. The *hierarchical generalized Voronoi graph* (HGVG) is the GVG and all higher order generalized Voronoi graphs; each of these graphs can be defined in terms of line of sight information.

Connectivity of the GVD. The underlying philosophy of the HGVG is to exploit the connectivity property of the GVD. Recall from Section 5 that the GVD is the union of all of two-equidistant faces in $\mathcal{W}$. For the moment consider the case of $\mathcal{W} \subset \mathbb{R}^3$, where (1) the only higher order generalized Voronoi graph is the GVG^2, (2) a two-equidistant face is two-dimensional, and (3) a GVG edge (a three-equidistant face) is formed by the intersection of *three* two-equidistant faces. By definition, a GVG edge exists on the boundary of a two-equidistant face, and thus adjacent two-equidistant faces share a common GVG edge. If the GVG edges associated with each two-equidistant face are connected (i.e., the boundary of each two-equidistant face is connected), then the GVG is connected because the GVD is connected. A disconnected GVG may arise when a two-equidistant face contains disconnected boundary components, like the two-equidistant face defined by the floor and the ceiling depicted in Figure 14. The GVG^2 is used to connect the boundaries of two-equidistant faces with disconnected boundary components, and thereby connect all disconnected GVG components. In some cases, additional links are required.

Second Order Generalized Voronoi Region. A GVG^2 on a two-equidistant face, $\mathcal{F}_{ij}$, denoted $\text{GVG}^2\big|_{\mathcal{F}_{ij}}$, is a network of one-dimensional curves that divides a two-equidistant face, $\mathcal{F}_{ij}$, into sub-regions where there exists a common *second closest* object (C_i and C_j are the closest objects). See Figure 15. These regions are called *second order generalized Voronoi regions* and are defined as follows

$$\mathcal{F}_k\big|_{\mathcal{F}_{ij}} = \text{cl}\{x \in \mathcal{F}_{ij} : \forall h \neq i, j, k,$$
$$d_h(x) > d_k(x) > d_i(x) = d_j(x) > 0$$
$$\text{and } \nabla d_i(x) \neq \nabla d_j(x)\}, \quad (14)$$

where d_i is the visible distance function, defined in Section 4.

It can be shown that on each two-equidistant face, $\mathcal{F}_{ij}$, the $\text{GVG}^2\big|_{\mathcal{F}_{ij}}$ connects the boundaries of $\mathcal{F}_{ij}$ if and only if the boundaries of the individual second order generalized Voronoi regions are connected (or can be readily connected with a link). [9]

The rest of this section is now devoted to careful consideration of the structures on the boundary of a second order generalized Voronoi region. Furthermore, we show that these boundary structures are defined in terms of line of sight information and can be incrementally traced using range sensor data.

A rigorous proof enumerating all of the structures in the boundary of a second order generalized Voronoi region is contained in [9], but this result can be seen via inspection of Equation (14). The first inequality in Equation (14) is associated with the structures that exist on a common boundary of *two* second order generalized Voronoi regions on the same two-equidistant face. These structures are called the *second order two-equidistant face* (termed the GVG^2 equidistant edge in $\mathbb{R}^3$) and the *occluding two-face* (termed the occluding edge in $\mathbb{R}^3$).

GVG^2 Equidistant Edge. A GVG^2 equidistant edge, denoted $\mathcal{F}_{kl}\big|_{\mathcal{F}_{ij}}$, is the set of points on the boundary of adjacent second order generalized Voronoi regions where the distance to the second closest obstacle *continuously changes* as the robot crosses from one region to the other. In other words,

$$\mathcal{F}_{kl}\big|_{\mathcal{F}_{ij}} = \{x \in \mathcal{F}_{ij} \quad \text{such that}$$
$$\forall h,\ d_h(x) \geq d_k(x) = d_l(x) \geq d_i(x) = d_j(x) > 0\}. \tag{15}$$

The "equidistant" is included in this term to distinguish it from other GVG^2 edges such as occluding edges. In Figure 15, GVG^2 equidistant edges fully compose the GVG^2 on the two-equidistant face defined by the floor and ceiling.

Since the GVG^2 equidistant edge is defined in terms of the visible distance function, it can be incrementally constructed using only local range information. The incremental construction technique for GVG^2 equidistant edges is similar to that of GVG edges. That is, a GVG^2 equidistant edge is incrementally constructed by tracing the roots of

$$G_2(y, \lambda) = \begin{bmatrix} (d_1 - d_2)(y, \lambda) \\ (d_3 - d_4)(y, \lambda) \end{bmatrix}, \tag{16}$$

as the parameter λ is varied. In Equation 16, obstacles C_1 and C_2 are closest and obstacles C_3 and C_4 are second closest. Since the visible distance function makes up Equation (16), the incremental construction technique relies solely on line of sight information.

Analogous to the GVG, we continue our construction with lower dimensional subsets of $\mathcal{F}_{ij}$. The *second order three-equidistant face,*

$$\mathcal{F}_{klp}\big|_{\mathcal{F}_{ij}} = \mathcal{F}_{kl}\big|_{\mathcal{F}_{ij}} \bigcap \mathcal{F}_{lp}\big|_{\mathcal{F}_{ij}} \bigcap \mathcal{F}_{kp}\big|_{\mathcal{F}_{ij}},$$

is the set of points where C_k, C_l and C_p are *second* closest equidistant objects and C_i and C_j are the closest equidistant objects. In $\mathbb{R}^3$, the second order three-equidistant face are GVG^2 *equidistant vertices,* or *second order meet points.*

Occluding Edge. The occluding edge is the set of points on the boundary of adjacent second order generalized Voronoi regions where the distance to the second closest obstacle does *not continuously change*

as the robot crosses from one region to the other. See Example 2.

Boundary Edge, Floating Boundary Edge. The second inequality from Equation (14) is associated with a GVG edge (points where $d_i(x) = d_j(x) = d_k(x)$). Boundary edges (points where distance to the environment is zero) are associated with the next inequality and floating boundary edges (points where the gradients are collinear) are associated with the last inequality.

A key feature of all of these structures is that they can be incrementally constructed. A GVG edge can be incrementally traced using line of sight information (Section 6). Tracing a boundary edge is akin to wall following, and in a numerical context, it can be viewed as tracing the roots of

$$G_b(y, \lambda) = \begin{bmatrix} d_1(y, \lambda) \\ d_2(y, \lambda) \end{bmatrix}, \tag{17}$$

as the parameter λ is varied. Finally, it can be shown that a floating boundary edge is a straight line and need not require a complicated edge tracing procedure (but rather assumes the robot has a good dead reckoning sensor).

Connectivity. We have shown that the boundary of a second order generalized Voronoi region may contain: GVG edges, boundary edges, floating boundary edges, GVG^2 equidistant edges, and occluding edges. Now, we need to consider the connectivity of the boundary of a second order generalized Voronoi region.

In Figure 15 the GVG cycle is surrounded by a GVG^2 cycle which is connected to the outer GVG component. The existence of a cycle informs the robot that there is a need to make the link. The link is made by following the second closest obstacle distance gradient (or negated gradient) projected onto the tangent plane of two-equidistant face. With this link, the boundary of the second order generalized Voronoi region is connected.

In prior work [6], it is assumed that all GVG edges must have at least one meet point, and thus no GVG cycles can exist. This assumption is met in "cluttered" environments, which are the ones of most interest for sensor based planning. Nevertheless, even when

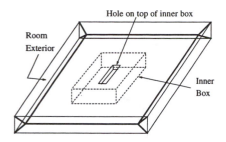

Figure 16: *Room with a box in the middle. The box, outlined with dotted lines, has an opening on top of it, delineated with solid lines.*

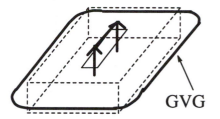

Figure 17: *Disconnected GVG components not within line of sight of each other.*

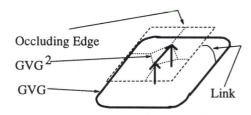

Figure 18: *Connected structure.*

this assumption is met, there are two other scenarios in which the boundary of a second order generalized Voronoi region may not be connected: when there is an occluding cycle (a connected second order generalized Voronoi region boundary component that has only occluding edges), and when there is a boundary cycle (a connected second order generalized Voronoi region boundary component that has only boundary edges).

A cycle of occluding edges (as is the case in Example 2) is detected by looking for discontinuities in distance sensor readings as the robot traverses a boundary component of a second order generalized Voronoi region. Forming a link to (from) an occluding cycle is done via gradient ascent (descent) of D. A final linking procedure is required for cycles formed by boundary edges. A cycle of boundary edges is detected by looking for local maxima of distance sensor readings as the robot traverses a boundary component of a second order generalized Voronoi region. Forming a link to (from) an boundary cycle is also achieved via gradient ascent (descent) of D.

Since the boundaries of the second order generalized Voronoi regions are connected (or can be readily connected with a link), the $GVG^2\big|_{\mathcal{F}_{ij}}$ connects disconnected GVG edges on a two-equidistant face. Since the GVD is connected, the GVG edges, combined with the GVG^2 edges and links, is connected. That is, the HGVG is connected.

Example 2 *Figure 16 is similar to Figure 14 except the box in the middle of the room contains an opening which can either be a through-hole, a dimple, or an* entrance to another internal environment. The GVG structure associated with the box and the hole (see Figure 17) contains two connected GVG components: one associated with the hole and ceiling, and one associated with the box, the floor, and the ceiling. Unfortunately, the two connected GVG components are not within line of sight of each other, and therefore an occluding edge is used to connect them.

The GVG structure associated with the hole is connected to the occluding edge using GVG^2 equidistant edges. Using a linking procedure, the GVG cycle is linked to the occluding edge. Now, the GVG is connected through a link, an occluding edge, and an equidistant GVG^2 edge. See Figure 18.

Finally, it can be shown that all points in the environment are within line of sight of the HGVG, and thus the HGVG has the property of *departability*. This line of sight departabilty property of the HGVG indicates that the HGVG has applications to sensor placement (e.g., the art gallery problem).

9 Conclusion and Future Work

We have developed a rigorous basis for sensor based motion planning for a robot, modeled as a point.

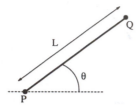

Figure 19: *The configuration of a rod is determined by the x and y coordinates of P and the orientation of the rod with respect to the horizontal.*

To this end, we defined the *hierarchical generalized Voronoi graph* (HGVG) to serve as a basis for robotic sensor based motion planning. The HGVG is a *roadmap*, which is a one-dimensional representation of an environment populated with obstacles, and has three key properties: (1) *accessibility*, (2) *connectivity*, and (3) *departability*. Simulations and experiments have validated this approach.

The ultimate goal of this work is to enable highly articulated robots equipped with sensors to explore unknown environments, via construction of a roadmap, and thus, many of the results of this work hold in spaces of dimension m. Nevertheless, the focus of this work is in dimension three where workspace distance measurements are available. Using these workspace distance measurements, the next step is to extend the definitions of the HGVG to the case of when the robot can be modeled as a line segment, sometimes called a rod (see Figure 19). The resulting roadmap is termed the *rod hierarchical generalized Voronoi graph* (rod-HGVG), the planar version of which has already been defined in [8].

The rod-HGVG is defined in terms of the *rod single object distance function* which is the shortest distance between a rod, R, at configuration q and a convex obstacle, C_i. (See Figure 20.) The rod distance function is denoted $D_i(q) = \min_{r \in q(R), c \in C_i} \|r - c\|$ where $q(R)$ is the set of points in the plane occupied by a rod, R. Since the rod-HGVG is defined in terms of a distance function, it can be incrementally generated using procedures similar to those described in Section 6.2.

The next step is to extend the results of the rod roadmap to that of a convex set, which in turn will be extended to the development of a roadmap for a chain

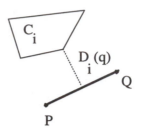

Figure 20: *The distance from the rod (thick solid line) to an obstacle is the distance (dotted line) between the nearest point on the rod to the obstacle and the nearest point on the obstacle to the rod.*

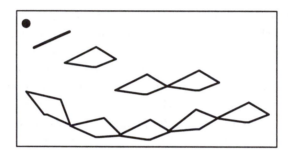

Figure 21: *Outline of future research.*

of convex sets which model a highly articulated robot.

We also believe that our focus on workspace distance as the underlying geometric foundation of these roadmaps will better enable future work on the analysis of sensor error and noise on planning performance. A parallel direction of current research focuses on sensor limitations such as sensor noise, effective sensor range, and sensor quantization. (Sensor quantization considers the discretization of range data.) Future research will focus on the use of robot vision to generate GVG edges in environments where range data is not readily available.

References

[1] D. Avis and B.K. Bhattacharya. Algorithms for Computing d-dimensional Voronoi Diagrams and Their Duals. *Advances in Computing Research*, 1:159–180, 1983.

[2] R.A. Brooks. A Robust Layered Control System for a Mobile Robot. *IEEE Journal on Robotics and Automation*, RA-2, March 1986.

[3] J.F. Canny and B. Donald. Simplified Voronoi Diagrams. *Discrete Comput. Geometry*, pages 219–236, 1988.

[4] J.F. Canny and M.C. Lin. An Opportunistic Global Path Planner. *Algorithmica*, 10:102–120, 1993.

[5] H. Choset and J.W. Burdick. Sensor Based Planning and Nonsmooth Analysis. In *Proc. IEEE Int. Conf. on Robotics and Automation*, pages 3034–3041, San Diego, CA, 1994.

[6] H. Choset and J.W. Burdick. Sensor Based Planning, Part I: The Generalized Voronoi Graph. In *Proc. IEEE Int. Conf. on Robotics and Automation*, Nagoya, Japan, 1995.

[7] H. Choset and J.W. Burdick. Sensor Based Planning, Part II: Incremental Construction of the Generalized Voronoi Graph. In *Proc. IEEE Int. Conf. on Robotics and Automation*, Nagoya, Japan, 1995.

[8] H. Choset and J.W. Burdick. Sensor Based Planning for a Planar Rod Robot. In *Proc. IEEE Int. Conf. on Robotics and Automation*, Minneapolis, MN, 1996.

[9] Howie Choset. *Sensor Based Motion Planning: The Hierarchical Generalized Voronoi Graph*. PhD thesis, California Institute of Technology, Pasadena, CA, 91125, 1996.

[10] F. H. Clarke. *Optimization and Nonsmooth Analysis*. Society of Industrial and Applied Mathematics, Philadelphia, PA, 1990.

[11] E. Gat and G. Dorais. Robot Navigation by Conditional Sequencing. In *Proc. IEEE Int. Conf. on Robotics and Automation*, pages 1293–1299, San Diego, CA, May 1994.

[12] H.B. Keller. *Lectures on Numerical Methods in Bifurcation Problems*. Tata Institute of Fundamental Research, Bombay, India, 1987.

[13] V. Lumelsky and A. Stepanov. Path Planning Strategies for Point Mobile Automaton Moving Amidst Unknown Obstacles of Arbitrary Shape. *Algorithmica*, 2:403–430, 1987.

[14] C. Ó'Dúnlaing and C.K. Yap. A "Retraction" Method for Planning the Motion of a Disc. *Algorithmica*, 6:104–111, 1985.

[15] N.S.V. Rao, S. Kareti, W. Shi, and S.S. Iyenagar. Robot Navigation in Unknown Terrains: Introductory Survey of Non-Heuristic Algorithms. *Oak Ridge National Laboratory Technical Report*, ORNL/TM-12410:1–58, July 1993.

[16] N.S.V. Rao, N. Stolzfus, and S.S. Iyengar. A Retraction Method for Learned Navigation in Unknown Terrains for a Circular Robot. *IEEE Transactions on Robotics and Automation*, 7:699–707, October 1991.

[17] E. Rimon and J.F. Canny. Construction of C-space Roadmaps Using Local Sensory Data — What Should the Sensors Look For? In *Proc. IEEE Int. Conf. on Robotics and Automation*, pages 117–124, San Diego, CA, 1994.

[18] P.F. Rowat. "Representing the Spatial Experience and Solving Spatial Problems in a Simulated Robot Environment". In *PhD. Thesis*, University of British Columbia, 1979.

Integrating Configuration Space and Sensor Space for Vision-based Robot Motion Planning

Rajeev Sharma, *The Pennsylvania State University, University Park, PA, USA*

Herry Sutanto, *University of Illinois, Urbana, IL, USA*

Visual feedback can play a crucial role in a dynamic robotic task such as the interception of a moving target. To utilize the feedback effectively, there is a need to develop robot motion planning techniques that also take into account properties of the sensed data. We propose a motion planning framework that achieves this with the help of a space called the Perceptual Control Manifold *or PCM defined on the product of the robot configuration space and an image-based feature space. We show how the task of intercepting a moving target can be mapped to the* PCM, *using image feature trajectories of the robot end-effector and the moving target. This leads to the generation of motion plans that satisfy various constraints and optimality criteria derived from the robot kinematics, the control system, and the sensing mechanism. Specific interception tasks are analyzed to illustrate this vision-based planning technique.*

1 Introduction

Sensor feedback is important in the flexible operation of a robot. The feedback could be critical in a dynamic manipulation task such as grasping a moving target, since the robot motion goal could be changing with time. Further, because of the temporal element in the definition of such tasks, effective motion planning is needed to drive the robot toward the moving goal. Such interception tasks would be involved, for example, in an assembly robot that picks a randomly placed object on a conveyor belt or in a space robot that acquires a moving target.

Computer vision holds great potential in providing the necessary feedback for the control of robot manipulator. One or more cameras that are either fixed or moving capture the image of the moving robot and a relevant portion of the surrounding scene to guide or "servo" the robot end-effector to the desired goal. This helps in overcoming uncertainties in modeling the robot and its environment, thereby increasing the scope of robot applications to include tasks that were not possible without sensor feedback, for example, welding. Most visual servo controllers are designed so that the error signal is defined directly in terms of some set of image features (image-based control), rather than in terms of the absolute position of the robot (position-based control) [3]. Visual servoing has been recently receiving substantial interest partly because of the increased performance and decreased cost of commercially available vision hardware [7, 9]. Several systems have been reported in which visual feedback is incorporated directly into the control feedback loop (e.g., [1, 15, 19, 23]).

However, sensors such as the video camera have limited range of operation and work well only when the objects in view are optimally configured with respect to the camera [16, 21]. Thus, to best utilize the sensor feedback, a robot motion plan should incorporate constraints from the sensor system as well as criteria for optimizing the quality of the sensor feedback. Unfortunately in most motion planning approaches, sensing is completely decoupled from motion planning. The role of vision in a motion planning framework, such as the one based on configuration space[13], would be to give an estimate of the current position of the robot in the configuration space with respect to the desired goal. For handling uncertainty in sensing of the robot position different motion planning approaches have been suggested. However, there is no general mechanism for including the variation of sensing parameters and sensor constraints into the motion plan. There is thus a growing need for extending the configuration space planning paradigm to bridge the gap between planning and sensing so that the motion plans can benefit by optimally utilizing the available sensing mechanism. In this paper we are concerned with developing such a motion planning approach.

63

We present a framework for motion planning that considers sensors as an integral part of the definition of the motion goal. This leads to vision-based plans that incorporate constraints and optimization criteria that involve both the robot and the sensor. The motion planning technique analyzed in this paper is based on the concept of *Perceptual Control Manifold* (*PCM*) that was introduced in [10, 17]. The *PCM* is a manifold defined on the product of the robot configuration space (defined in terms of the set of robot joint parameters) and image feature space (defined in terms of a set of image features from the image of the robot hand). These concepts are formalized in Section 2.

The task that we consider involves developing a gross-motion plan for "intercepting" a moving target, using the visual feedback from a fixed camera. The camera views both the end-effector of the robot and the moving target. The target trajectory is assumed to be estimated separately and is available to the planner. The interception task is used to illustrate how various constraints involving the robot joints and image features can be mapped to the *PCM*. A motion plan can then be developed using an appropriate optimization criterion. Alternatively, since all of the constraints are expressed in a common framework, this lends itself to reasoning about the goal, for example, determining the feasibility of interception, even before the actual motion plan is generated. Since a motion plan thus computed takes into account the sensor variations, an appropriate control law can be defined directly on the *PCM*. The constraints, such as image singularity avoidance help in achieving a better vision-based control.

2 Defining Interception Task on *PCM*

2.1 Background

The problem of motion planning of an articulated robot is usually defined in terms of the configuration space, $\mathcal{C}$ (or $\mathcal{C}$-space), which consists of a set of parameters corresponding to the joint variables of the robot manipulator. $\mathcal{C}$ is an n-dimensional manifold [13] for an n-DOF robot manipulator, *i.e.* $\mathcal{C} \equiv \mathcal{Q}_1 \times \mathcal{Q}_2 \times \ldots \mathcal{Q}_n \subseteq \mathbf{R}^n$, when $q_i \in \mathcal{Q}_i$ is a joint parameter (See Figure 1). The obstacles and other motion planning constraints are usually defined in terms of $\mathcal{C}$, followed by the application of an optimization criteria that yields a motion plan. However, since some of the task constraints

may be specified in terms of the cartesian task space, the actual motion planning may first involve mapping these constraints to $\mathcal{C}$. The task space of the robot is the set of positions and orientations that the robot end-effector or tool can attain. If the tool is a single rigid body moving in a three-dimensional workspace, the task space can be represented by $\mathcal{W} = \mathbf{R}^3 \times SO(3)$, where $\mathbf{R}^3$ is three dimension Cartesian space for robot position and $SO(3)$ is Special Orthogonal Group of the 3×3 matrices in $\mathbf{R}^9$ with orthonomal columns and rows and positive unit determinant for robot orientation. In many applications, the dimension of the task space can be reduced to reflect the number of degrees of freedom that are actually required to perform the task.

2.2 Image feature space, $\mathcal{S}$

In visual servo control, the robot configuration is related to a set of measurements which provide a feedback about the cartesian position of the end-effector using the images from one or more video cameras. We

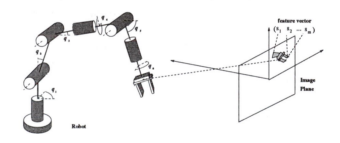

Figure 1: *Schematic diagram of a 6-DOF manipulator, and the mapping to the image feature space.*

assume that this feedback is defined in terms of measurable image parameters that we call *image features*, s_i (See Figure 1). Before planning the vision-based motion, a set of m image features must be chosen. Discussion of the issues related to feature selection for visual servo control applications can be found in [5, 20, 23]. The mapping from the set of positions and orientations of the robot tool to the corresponding image features can be computed using the projective geometry of the camera. Examples of commonly used projective geometry models include perspective, orthographic, or para-perspective projection models [2]. Since the cartesian position of the end-effector, in turn, can be considered to be a mapping from the configuration space of the robot, we can also define image features with a mapping from $\mathcal{C}$. Thus, an image feature can be defined as a function s_i which maps robot configurations to image feature values, $s_i : \mathcal{C} \rightarrow \mathcal{S}_i$. The set of all possible variations of the image features is termed *image feature space*, $\mathcal{S} \equiv \mathcal{S}_1 \times \mathcal{S}_2 \times \ldots \mathcal{S}_m$. A robot trajectory in configuration space will yield a trajectory in the image feature space.

Although we refer to vision, the discussion applies to any other sensor as well, and the term "image" is thus used for the generic sensor measurements. Moreover, we refer to *geometric* image features (e.g., position, size, distance, or surface area of objects in the image), as opposed to *photometric* features (pixel intensities, colors, etc.). Examples of image features used in visual servo control include image plane coordinates of a point [4, 15, 17, 19], length, orientation and other parameters of a line in the image [4, 5, 23], centroid, area, and other higher order moments of an image region [23] and composite features in [8].

The entire robot configuration space may not be observable by a particular image feature s_i either because its map may be out of s_i range or because it may be occluded by other objects. We define $\mathcal{C}_i^v$ as the maximum subset of $\mathcal{C}$ observable/visible by feature s_i. That is,

$$\mathcal{C}_i^v = \{\, \mathbf{q} \mid s_i(\mathbf{q}) \in \mathcal{S}_i \,\}$$

We limit our discussion to a subset of $\mathcal{C}$ which is observable by all image features under consideration, this set is denoted by $\mathcal{C}^v$.

We consider a hand/eye setup where the image features are derived from a stationary camera and mention its possible extension for other feedback modes in Section 5.

2.3 Perceptual Control Manifold, *PCM*

In order to include the image feature space, $\mathcal{S}$, in the planning space, we consider the $\mathcal{C} \times \mathcal{S}$ space, or $\mathcal{CS}$-space. We know that an n-dimensional configuration space $\mathcal{C}$ maps to an m-dimensional feature space $\mathcal{S}$. This mapping can be defined in terms of the vector-valued function $f : \mathcal{C} \rightarrow \mathcal{S}$, termed the *perceptual kinematic map* in [10, 17]. This mapping defines an n-dimensional manifold in $(n + m)$-dimensional space. We call this manifold the *perceptual control manifold*, or (*PCM*) and use it for developing a motion planning framework. The *PCM*, also called perceptual control

surface, was defined in [10, 17], but its use in motion planning was not explored.

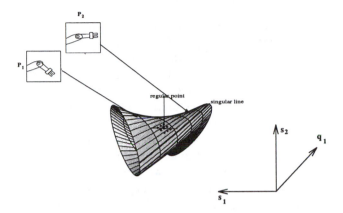

Figure 2: *Positions of the manipulator mapped into the* PCM.

For the robot in Figure 1, consider the variation of an image parameter, s_1, when a joint parameter, say q_1, is varied, while keeping the rest of the joints fixed. Without considering the joint limits for the time being, this would define an ellipse in the $\mathcal{Q}_1 \times \mathcal{S}_1$ space. Similarly, when two of the joints, say q_1 and q_2 are varied simultaneously, a hyper-ellipsoid will be defined in $\mathcal{Q}_1 \times \mathcal{Q}_2 \times \mathcal{S}_1 \times \mathcal{S}_2 \subseteq \mathbf{R}^4$. For ease of visualization, we project the corresponding *PCM* to $\mathcal{S}_1 \times \mathcal{S}_2 \times \mathcal{Q}_1 \subseteq \mathbf{R}^3$, as shown in Figure 2. Analogously, in higher dimensions, the *PCM* for a hand/eye setup is defined by varying all the joints and considering the parametric hypersurface defined in $\mathcal{C} \times \mathcal{S}$ space.

A given robot configuration maps to exactly one point on the *PCM*. The corresponding image features are not necessarily unique for a given position, but because the joint is also represented this leads to the uniqueness property that is needed for motion planning and control. Since the *PCM* represents both the control parameter and the sensor parameter, an appropriate control law can be defined on it [17]. Our concern in this paper is, however, on motion planning. An important characterization of the *PCM* for motion planning is the distinction between the *singular* and *regular* portions of the surface [10]. The singularities of the *PCM* need to be identified so that they may be avoided during manipulation. The singularity constraints can be incorporated into the motion planning along with other constraints that are defined on the

PCM as will be discussed later in the section.

2.4   Target Manifold, *TM*

The manipulation goal of the robot can be defined in terms of a *goal set* $GS \subset \mathcal{C} \times \mathcal{S}$, which represents the desired state of the robot at the end of the manipulation task. It can be specified either in terms of the cartesian position of the robot (as in position-based visual servoing) or in terms of the image features (as in image-based visual servoing). In either case the goal specification can be mapped as a set on the *PCM*. If the goal is time-varying, as in the dynamic manipulation task of intercepting a moving target, a useful structure called the *target manifold* or *TM* can be defined as follows. Imagine that we attach a disembodied end-effector to the moving target, holding it in a final grasping configuration. This would require that the end-effector is free to move wihout any kinematic constraints, as if it were detached from the robot arm. Consider the profile of the end-effector with time while it is grasping the target, defined in the $\mathcal{C} \times \mathcal{S}$ space. This is the target manifold, *TM*, and represents all possible positions where the target could be intercepted if there were no kinematic constraints or joint limits. The constraints for a particular robot arm can then be used to define

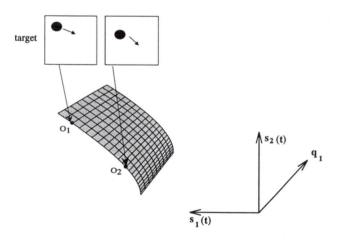

Figure 3: *Target Manifold (*TM*) associated with a moving target.*

the goal positions that can be actually achieved, as we will show in the next subsection. The *TM* is a function of time. Time can be implicitly represented and accounted for in motion planning on the *PCM* or it may be considered explicitly as an additional dimension, *T*,

for planning (see, for example, [12]). The latter case will involve the $\mathcal{C} \times \mathcal{S} \times T$ space, which we call the *temporal PCM patch* and use it for planning in Section 4.

Figure 3 gives an example of the target manifold for the task of catching a ball that moves along a parabolic trajectory. The surface is a parametric plot (as a function of time) of s_1, s_2, and q_1, which is the same dimension as the *PCM*. We note that to define the interception task, the value of q_1 is not important; hence, the target manifold for this particular case is just the parabolic surface stretched out along the axis for q_1. In general, the surface will be more complex and will represent a variety of positions that accomplish the same task. To simplify the discussion of the target manifold, we assume that a single configuration defines the interception goal. In practice, the task specification will give a set of acceptable grasping configurations [14] rather than a single configuration. Further, a particular visual tracking algorithm would be used to estimate the target trajectory. The resulting uncertainty in the estimation of the target trajectory can be expressed in terms of a region on the *PCM*.

2.5   Target Goal Set, *TGS*

For the interception to be feasible, the target should intersect the workspace of the robot manipulator. If the robot has a fixed base (we will discuss the case of a mobile manipulator in Section 5), then this condition

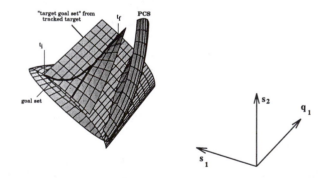

Figure 4: *Intersection of the* TM *and the* PCM *that gives the Target Goal Set (*TGS*).*

requires that the *TM* must have a non-empty intersection with the *PCM*. We call the intersection of the *TM* and the *PCM*, the *target goal set* or *TGS*. Thus for any interception goal to be feasible, there should be a path defined on the *PCM*, from the initial position on the

PCM to a point on the *TGS*. *TGS* is a function of time since *TM* is a function of time. When time is explicitly considered, $TGS \subset C \times S \times T$ and is defined on the temporal *PCM* patch. Figure 4 shows the goal set that is formed by intersecting the *TM* of Figure 3 with the *PCM* of Figure 2. The parameters t_i and t_f define the initial and final times for which interception can take place. Thus the entire manipulation task must be carried out in the interval $\tau \in [t_i, t_f]$. In general, because of joint limits and other constraints there may be several such intervals. The estimation of the time interval, τ, must also take into account the minimum time required for the manipulator to decelerate before reaching its joint limits. Notice that in Figure 4 the robot joint limits are expressed as gaps on the *PCM*; this and other motion planning constraints will be discussed next.

2.6 Expressing task constraints on the *PCM*

The interception task can be defined as a problem of trajectory planning on the *PCM* from the initial position of the manipulator to some goal position on the *TGS* . This motion planning requires the system to satisfy constraints presented by robot kinematics, the control system and the visual tracking mechanism. The aim is to get a feasible solution space for motion planning that satisfies these constraints, for example, in terms of the temporal *PCM* patch. The use of the *PCM* makes some of the sensor constraints easier to express compared to a potentially awkward *C*-space representation. An example, of such a constraint is image feature singularity [18]. This is considered in greater detail in Section 4 along with other constraints which involve robot kinematics and visual tracking. Constraints can be classified as being *hard* or *soft*. Hard constraints, e.g., joint velocity limits, must be satisfied in a motion plan. Soft constraints, e.g., image singularity avoidance, should preferably be satisfied, and can be included with the help of a cost function in motion planning.

2.7 Optimizing the interception trajectory

Once the necessary hard constraints have been applied to yield a feasible solution space defined on the *PCM*, any path on the *PCM*, from the point corresponding to the initial position of the robot to a point on the *TGS*, will give rise to a valid solution for the interception problem. However, the path chosen should be the one that optimizes a set of desired objectives. Some of these objectives could be related purely to the robot, e.g., minimizing joint movement, minimizing relative velocity of the robot end-effector and target at the time of interception. Some of these objectives could be related purely to the sensor, e.g., maximizing the variation in the image features. However, there are some objectives that involve both the robot and the sensor. These objectives directly effect the control and mainly depend on how the image features vary with an infinitesimal movement of the robot. The objective could be to steer away from image singularities or to find the robot trajectory such that it follows a singularity. These objectives will be discussed in detail in Section 4 with the help of specific examples. The *PCM* framework has the advantage that it allows a variety of optimization criteria to be expressed in a unified manner so that the optimal sensor-based plan can be generated for interception.

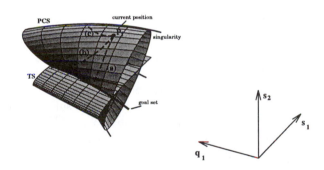

Figure 5: *Interception task defined as that of trajectory planning on the* PCM.

In Figure 5, P_1 represents the initial (current) position of the robot, and several different trajectories are defined on the *PCM*, that are based on the particular interception task. The goal set in the example is constructed by intersecting the *PCM* and the *TM*, as shown in Figures 2 and 3. The planning could consist of first deciding whether such a goal set exists, *i.e.* whether the task is feasible, and, if so, then determining the corresponding trajectory on the *PCM* for accomplishing the task.

2.8 Steps in the synthesis of the interception plan

The main steps involved in analyzing the interception task and synthesizing a motion plan that satisfies the given constraints and optimization criteria are summarized below.

1. *Computing the PCM.* For a given hand/eye setup and a set of image features compute the perceptual control manifold, *PCM*, using the robot kinematics and image projection equations. This computation is only done once and is then applicable to any manipulation task as long as the camera is stationary with respect to the robot base.

2. *Estimating the TM.* From the task description derive the final grasping condition; hence, derive the goal set in terms of the image features using the image projection of the robot end-effector. Using the predicted target trajectory from the trajectory estimator, compute the profile of the goal set, and hence the target manifold, *TM*.

3. *Estimating the TGS.* Obtain *TGS* as a function of time by intersecting the *PCM* and the *TM*.

4. *Applying the task constraints.* Define and apply the hard and soft constraints, which involve the robot kinematics, sensor data variation, the robot environment, and the task, to yield a subset of the *PCM* as the feasible solution space. This step is done explicitly only if there is a need to analyze the feasibility of the interception task; otherwise it is merged with the optimization step.

5. *Optimizing the motion plan.* Use the optimization criteria involving the robot, the sensor, and the task, to compute the motion plan in terms of the *PCM*, if such a plan is feasible. Since the plan involves both the configuration space and the image feature space, an appropriate image-based control law can be defined for accomplishing the interception task.

The above steps need not be executed sequentially. For example, some of the hard constraints (e.g., joint limits and obstacles) can be included in the first step while computing the *PCM*. The specific motion planning algorithms will be developed in Section 4 for the hand/eye setups that are described next.

3 Analyzing Specific Hand/Eye Setups

In this section we consider two setups that involve a robot manipulator and a camera (hand/eye setups), engaged in the task of intercepting a moving target. Assume that an estimated target trajectory has been given. The structures *PCM*, *TM*, and *TGS* will then be derived along with their relation to time. This will set the stage for applying the constraints and optimizing the robot trajectory for interception in Section 4.

3.1 Analysis of Setup 1

The first setup involves a two-link, planar robot arm with 1-DOF. The end-effector is confined to move only along the x-axis (imagine a folding door). The task involves intercepting a ball that moves in a linear trajectory along the x-axis toward the robot (see Figure 6). A constant negative acceleration is applied to the ball causing it to decelerate and eventually reverse its direction. The x position of the end-effector from the robot

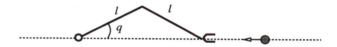

Figure 6: *Setup 1 involves a two link, 1-DOF robot for catching a moving ball.*

origin can be written as a function $x_{robot}(q) = 2\,l \cos q$, in which $q \in \mathcal{Q}$ is the angle between robot's first link and the positive x-axis, and l is the length of each link segment. The x position of the ball at time t is written as $x_{ball}(t) = \frac{1}{2}at^2 + v_0 t + x_0$, in which v_0 is the initial velocity, x_0 is initial position of the ball, and a is the acceleration (or deceleration). Unless the direction of the camera is parallel to the x-axis, changes in the x positions of end-effector and ball can be observed in the image. To simplify the presentation, we assume orthographic projection and that the camera is pointing at the robot base. The image feature, $s \in \mathcal{S}$, is chosen as the distance between the image of a point on the end-effector and the image of a point on the robot base.

The mapping of the joint space to this image feature is written as function $s_{robot}(q) = K \cos q$, in which s_{robot} is the distance of end-effector to the robot origin in the image, K is a constant that depends on camera parameters, and $q \in (-\frac{\pi}{2}, \frac{\pi}{2})$. The *PCM* of this robot is shown in Figure 7(a). This *PCM* remains unchanged over time since the camera is stationary; hence the *PCM* lifts as a surface in $\mathcal{C} \times \mathcal{S} \times T$ space, as shown in Figure 7(b).

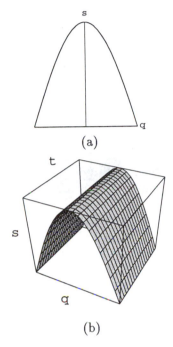

(a)

(b)

Figure 7: *(a) The* PCM *; (b) the* PCM *over time (*$\mathcal{C} \times \mathcal{S} \times T$ *space).*

Similarly, the distance of the ball to the robot origin in the image can be written as a parabolic equation, $s_{ball}(t) = At^2 + Bt + C$, in which A, B, and C are constants. This target manifold (*TM*) is shown in 8(a). The intersection of the *TM* with the *PCM* defines the target goal set (*TGS*) as shown in Figure 8(b) and Figure 9. The *TGS* curve can be written in a parametric form as:

$$\left[\begin{array}{c} s \\ q \\ t \end{array} \right] = \left[\begin{array}{c} At^2 + Bt + C \\ \cos^{-1}(\frac{At^2+Bt+C}{K}) \\ t \end{array} \right].$$

Further, the projection of the *TGS* into the $\mathcal{C} \times T$ plane can be defined in a parametric form as:

$$q(t) = \pm \cos^{-1}(\frac{At^2 + Bt + C}{K}).$$

and is shown in Figure 9(c).

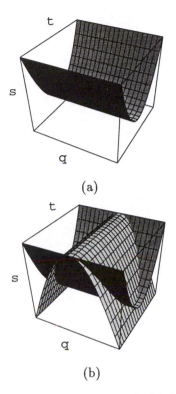

(a)

(b)

Figure 8: *(a) Target Manifold (*TM *); (b) the* TGS *defined as the intersection of the* PCM *and the* TM.

3.2 Analysis of Setup 2

The second setup involves a 2-DOF robot, two-link, planar robot arm (Figure 10) and a camera. The task is to intercept a ball that moves in the same plane as the robot. The ball follows a parabolic trajectory as it would for a idealized flight under the effect of gravity [19]. The position of the end-effector is given by

$$\mathbf{x}_{robot}(q_1, q_2) = \left[\begin{array}{c} l_1 \cos(q_1) + l_2 \cos(q_1 + q_2) \\ l_1 \sin(q_1) + l_2 \sin(q_1 + q_2) \end{array} \right],$$

which represent the x and y image coordinates of a point on the end-effector. Two image features are denoted as the vector $\mathbf{s} = (s_1, s_2) \in \mathcal{S}_1 \times \mathcal{S}_2$, and the joint parameters by the vector $\mathbf{q} = (q_1, q_2) \in \mathcal{Q}_1 \times \mathcal{Q}_2$. We again assume an orthographic projection for this example. The *PCM* of the robot can be expressed by the following parametric equations.

$$\mathbf{s}_{robot}(q_1, q_2) = \left[\begin{array}{c} A(l_1 \cos(q_1) + l_2 \cos(q_1 + q_2)) \\ B(l_1 \sin(q_1) + l_2 \sin(q_1 + q_2)) \end{array} \right],$$

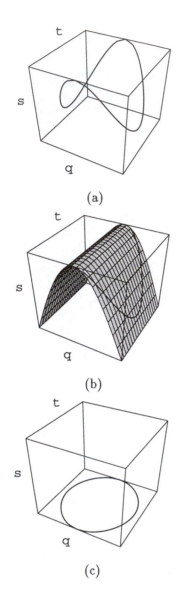

(a)

(b)

(c)

Figure 9: *(a) The* TGS *as curve in* $\mathcal{C} \times \mathcal{S} \times T$, *(b) the* TGS *overlaid on the* PCM, *and (c) projected into the* $\mathcal{C} \times T$ *plane.*

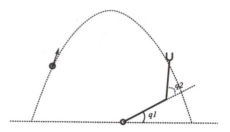

Figure 10: *Setup 2 involves a 2-DOF robot for intercepting a ball.*

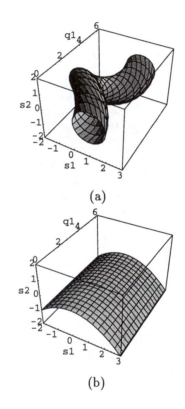

(a)

(b)

Figure 11: *(a) The* PCM *for the 2-DOF robot; (b) The* TM *of the ball. Both are projected into* $\mathcal{S}_1 \times \mathcal{S}_2 \times \mathcal{Q}_1$.

in which A and B are constants. The projection of this PCM into $\mathcal{S}_1 \times \mathcal{S}_2 \times \mathcal{Q}_1$ is shown in Figure 11(a). This projection is used to graphically illustrate the concept since we cannot view the 4-D PCM; the actual motion planning, however, will be done on without projecting the PCM.

The parabolic trajectory of the ball in the world is denoted as $\mathbf{x}_{ball}$, and its corresponding image feature,

s_{ball} is also a parabola. The latter forms the TM as defined in Section 2. The trajectories are

$$\mathbf{x}_{ball}(t) = \begin{bmatrix} x_0 + v_{x_0}t \\ y_0 + v_{y_0}t - \frac{1}{2}gt^2 \end{bmatrix}$$

and

$$s_{ball}(t) = \begin{bmatrix} C + Dt \\ E + Ft + Gt^2 \end{bmatrix},$$

in which (x_0, y_0) and (v_{x_0}, v_{y_0}) are initial position and velocity respectively, and C, D, E, F and G are constants. For illustration, we also project the *TM* to $S_1 \times S_2 \times Q_1$ as shown in Figure 11(b). Since the camera is stationary, the *PCM* stays the same for different t. Graphically, it is stretched along t dimension, creating a generalized cylinder. The intersection of the *PCM* and the *TM* defines the target goal set (*TGS*). The projection of this *TGS* into $S_1 \times S_2 \times Q_1$ space is shown as a curve in the projection of the *TM* (Figure 12). Further, the projection of the *TGS* into the $Q_1 \times Q_2 \times T$ is shown in Figure 13. The range of q_1 and q_2 is $[0, 2\pi]$, and t is in interval $[0, 6]$.

For the time, t, at which the *PCM* intersects the *TM*, there are two possible robot configurations that intercept the ball. These are shown as two curves along the t dimension. This gives rise to two separate *TGS* regions (shown in Figure 12), which is reflected as a discontinuity in the projected *TGS* into $Q_1 \times Q_2 \times T$ at time $t = 2$. Hence Figure 13(a) shows four different curves, two for $t \leq 2$ and two for $t > 2$. To show this more clearly, we remove t and project the *TGS* into $Q_1 \times Q_2$ plane in Figure 13(b).

4 Constraining and Optimizing Paths

The next step is to derive a path or trajectory from the initial robot position on the *PCM* to a point in *TGS*. This trajectory must satisfy the constraints defined for the robot, the sensor, and the task specification. The interception trajectory can then be computed using desired optimization criteria, applied to the feasible solution space in *PCM*.

4.1 Defining the task constraints on *PCM*

This section demonstrates how the *PCM* concept can accommodate several types of interception task constraints in a unified manner. The Setup 1 is used for illustrating the use of these constraints.

Robot kinematic constraint. Each of the mechanical robot joints has a maximum velocity bound; this can be described as a constraint between C and t. This constraint can be specified as:

$$|\frac{\partial q_i}{\partial t}| \leq V_{q_i}$$

in which V_{q_i} is the maximum velocity of joint i. Given an initial configuration, $\mathbf{q}_0$, this constraint defines a

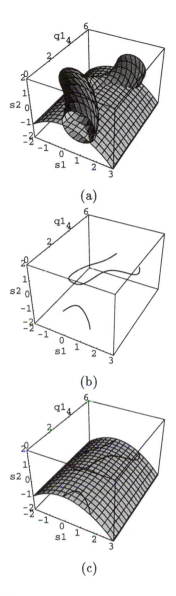

(a)

(b)

(c)

Figure 12: *(a) The intersection of the* PCM *and the* TM, *defining the* TGS. *(b) The projection of the* TGS *into* $S_1 \times S_2 \times Q_1$, *and (c) the same overlaid on the* TM.

hyper-pyramid in $C \times T$, which has the initial configuration, $\mathbf{q}_0$, as its apex, and $tan^{-1}(V_{q_i})$ as the half-angle in the $Q_i \times T$ plane. In $C \times S \times T$, the boundary of this constraint for Setup 1 is shown in Figure 14(a). Intersecting this constraint boundary and the *PCM* (see Figure 14(b)) yields the temporal *PCM* patch which is a feasible space satisfying the maximum joint velocity constraint (Figure 14(c) and (d)). The maximum joint

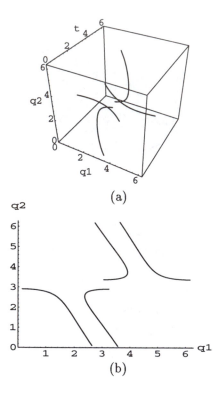

(a)

(b)

Figure 13: *(a) The* TGS *projected into* $Q_1 \times Q_2 \times T$*, and (b) into the* $Q_1 \times Q_2$*.*

velocity constraint is a hard constraint, meaning that it is impossible to intercept the target object outside the feasible temporal *PCM* patch defined above.

Image singularity constraint. For certain robot configurations relative to the camera, it may be the case that a robot movement cannot be detected by observing a given set of image features. This is known as an image singularity, and should be avoided since at these configurations, the ability to visually control the robot is lost. Given a set of image features, image singularity occurs at the robot configuration space where the determinant of its image Jacobian is zero ($J(\mathbf{q}, \mathbf{s}) = 0$, where $\mathbf{s}$ is the image feature vector). An image feature has a limited resolution due to the digital imaging equipment used to measure it. This limited resolution presents a minimum bound of the joint movement in order to be detectable as the image feature changes. Assuming a non-redundant system [18], this bound can be simplified as the minimum value, J_{min}, of the absolute value of the determinant of im-

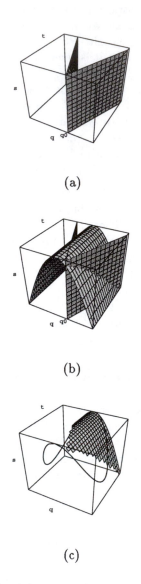

(a)

(b)

(c)

Figure 14: *Robot joint constraints expressed on the* PCM.

age Jacobian. Thus,

$$|J(\mathbf{q}, \mathbf{s})| > J_{min}.$$

For the robot in Setup 1, this constraint simply becomes $|\frac{ds}{dq}| > J_{min}$. Part of the *PCM* which satisfy this constraint is shown in Fig 15(a). Combining this constraint with maximum joint velocity constraint yields a feasible temporal *PCM* patch as shown in Figure 15(b). The image singularity constraint is considered soft; even though it should be avoided, it is still possible

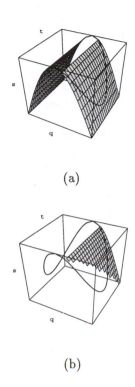

(a)

(b)

Figure 15: *(a) Image singularity constraint; (b) combining singularity and velocity constraints.*

to go beyond the temporal *PCM* patch defined by this constraint to intercept an object. This can be achieved by defining a corresponding cost for the "distance" to the image singularity during the optimization process [18].

Visual tracking constraints. An interception task requires the vision system to track image features of both the end-effector and the target. A feature that moves too fast with respect to the sampling rate of the tracking mechanism, will be difficult to track. To guarantee the stability of the visual tracking control mechanism, we define maximum velocity bounds for image feature values to reflect these constraints. This feature velocity constraint can be written as

$$|\frac{\partial s_i}{\partial t}| < V_{s_i},$$

in which V_{s_i} is the maximum velocity for feature s_i. This constraint does not apply to the features of the moving target because it is beyond the control of the

robot. Nevertheless, for an active vision system, this constraint can be incorporated into the camera planning stage so that the image features of the moving target can also be tracked reliably [16]. This constraint is also a soft constraint. The temporal *PCM* patch that satisfies this constraint is shown in Figure 16(a), that the parts of the parabolic cross section that have large $|\frac{\partial s}{\partial t}|$ are eliminated. The result of applying all three constraints is shown in Figure 16(b).

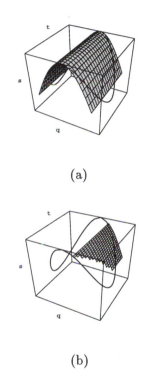

(a)

(b)

Figure 16: *(a) Visual tracking constraints; (b) Combining all three constraints.*

4.2 Optimizing interception plan using *PCM*

After deriving a temporal *PCM* patch that satisfies the interception constraints, all paths within this patch originated from initial configuration q_0 at t_0 to a point in the *TGS* are considered as feasible paths as long as they are monotonically increasing with time. The execution of an interception task can be improved by optimizing the path based on a desired set of objectives.

One of these objectives is to generate a plan whose

execution can be monitored easily. The execution of a task is monitored by sensors as sensor readings (image feature values). A path will be monitored easily if a deviation from the path can be detected quickly. This requires every point in the path to have high velocity of feature values with respect to an infinitesimal change in the robot configuration at that point [20]. In other words, this objective requires the path to be far from an image feature singularity, since at a singularity, a movement in configuration space does not reflect any change in the sensor reading. There could be another goal to minimize joint motion; this may lead to a path through singular configurations. The optimization process can accommodate these simultaneous goals by considering a objective or cost function that weighs the above objectives with respect to each other, leading to a plan that best magnifies the effect of movements in feature space, while minimizing joint movements.

There may be a certain task specification, which actually requires the plan to be generated along image singularities. Since robot motion along the image singularities does not change certain image features, this can be used for defining a robot manipulation plan. During plan execution, a deviation from the planned path will be reflected as a feature value change; hence, an appropriate control law can be defined for moving the robot along the image singularity. One instantiation of this path planning objective using visual positional feature is demonstrated in [6] as "visual compliance" motion.

Besides the above criteria, more general interception objectives can be defined with respect to the *PCM* framework. These are briefly described and illustrated for particular setups.

Optimizing robot movement and time of interception. A path on the temporal *PCM* patch that has the shortest projection into $\mathcal{C} \times T$ corresponds to an interception trajectory that optimizes both earliest time of interception and minimum robot movement. We can assign different weights for each dimension (joint or time dimension) to minimize movement of a particular joint or to minimize the time of interception. Figure 17(a) shows the resulting paths for the interception task with 3 different initial configurations ($q_0 = -1.5, -0.5$ and 1.0 radians) for Setup 1 (from Section 3.1) when this optimization criteria is applied.

Figure 18 shows the shortest paths for Setup 2 (from Section 3.2) when $q_0 = (5, 2)$ and $q_0 = (1, 1)$ (units are in radians). The resulting paths are shown as bold lines in the figures.

Minimum Robot Movement. Another common optimization criterion is minimizing the robot movement that accomplishes the interception task. This objective corresponds to finding a path on the temporal *PCM* patch with shortest projection into $\mathcal{C}$. This path reflects the situation in which the robot makes the smallest movement to a position where it can wait for the object to come. The Figure 17(b) shows minimum movement paths for 3 different initial configurations of the robot in Setup 1, and Figure 19 shows the minimum movement paths for the robot in Setup 2 with the same q_0 as previous cases.

Minimum Relative Velocity. Some tasks prohibit the object from making a sudden stop when intercepted. This objective translates to minimum relative velocity between the robot and the object at the time of interception. As indicated earlier, using the *PCM*, the planning is done directly on the image feature space, bypassing the cartesian space that defines the workspace of the robot. Therefore, actual velocities of the object and the end-effector may be unknown. However, some image features can be used to approximate them, such as, positional information in the image. Hence this objective can be achieved by generating a path for which the image feature velocities of the end-effector are the closest to the target velocity at the time of interception. For the end-effector of the robots in the two setups described in Section 3, the criterion involves getting the planning path to be tangential to the *TGS* curve at the interception point. Figure 17(c) and Figure 20 show paths that are derived with a minimum relative velocity objective for the two setups.

5 Discussion and Conclusions

For the analysis, we assumed that the target trajectory is estimated separately and is available to the planner. However, this is not a limitation of the approach. In fact, the uncertainty in the target position as well as the predicted trajectory can be factored into the *PCM* framework by defining an appropriate *TGS* region along with some uncertainty parameter. Another issue that we did not address in this paper was calibration. In the analysis we derived the *PCM* and the

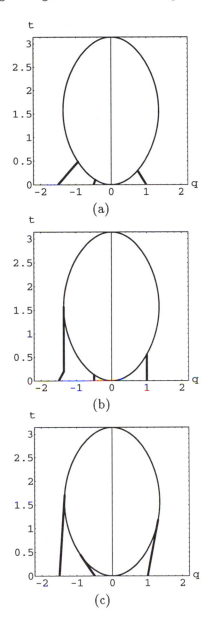

(a)

(b)

(c)

Figure 17: *Result of motion planning using different criteria: (a) optimizing both minimum robot movement and time of interception, (b) minimizing robot movement, and (c) minimizing relative velocity at interception. For each criterion, the resulting trajectories are shown from three different starting positions.*

TM in an algebraic form assuming the constants, e.g., A, B, C and K of the first setup, are known. There are two approaches that can be taken. The first would be

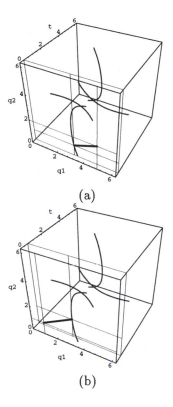

(a)

(b)

Figure 18: *Shortest paths from 2 different initial robot configurations.*

to calibrate the hand/eye system using a known calibration technique (e.g., [22]). The other alternative is to use a non-parametric approach, and use a learning technique such as the neural networks [11, 24] to build a representation of PCM. In fact, the algebraic approach that we followed to obtain the TGS, as intersection of manifolds, may be hard to generalize to higher dimensions. The non-parametric representation of the PCM may make the intersection easier to compute.

In this paper we consider the case of a stationary camera. A higher degree of mobility can be achieved if the camera is "active" and can be dynamically repositioned with respect to the robot [16]. The effect of camera movements can be assimilated into perturbations that alter the topology of the perceptual control manifold. This in turn can be used favorably for the dynamic reshaping of the PCM for obtaining a better motion plan.

Another possibility is to extend the motion planning technique to a mobile manipulator in which the cam-

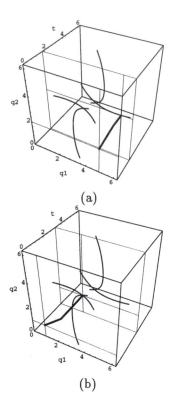

(a)

(b)

Figure 19: *Minimum movement paths from 2 different initial robot configurations.*

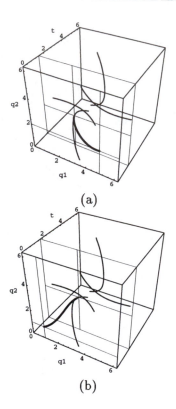

(a)

(b)

Figure 20: *Minimum relative velocity paths from 2 different initial robot configurations.*

era position is fixed with respect the robot base. The motion of the mobile platform does not change the shape of the *PCM*, and only effects the determination of *TGS*. The extra degrees of freedom of the motion of the robot base can, in fact, be factored into the motion planning for interception. Consider, for example, a baseball player catching a ball first positions himself relative to the ball trajectory before making a small corrective action just before catching the ball. Similar strategy can be defined for interception using the *PCM* framework.

In this paper we demonstrate how the *PCM* expands the notion of the configuration space to include a set of parameters (image features) from the sensor. This enables the sensor to become an integral part of motion planning; helping in expressing constraints and optimization criteria that best exploit the sensor feedback. The interception task helps in showing how the *PCM* can further incorporate the temporal element for motion planning. Specific hand/eye setups help in de-

veloping and illustrating the framework. The analysis shows the feasibility of using the *PCM* for motion planning. This can be viewed as a first step toward an integrated framework for sensor-based planning; further issues need to be addressed to make it a part of the flexible operation of a robot.

Acknowledgements

This research was supported in part by the U. S. Army Research Laboratory under Cooperative Agreement No. DAAL01-96-2-0003 and by the Beckman Institute through a Beckman Fellowship. We would like to thank Peter Cucka, Jean-Yves Hervé, Steve LaValle, Seth Hutchinson and Narayan Srinivasa for their help.

References

[1] P. Allen, B. Yoshimi, and A. Timcenko. Real-time visual servoing. In *Proc. DARPA Image Understanding Workshop*, pages 909–918, 1990.

[2] J. Aloimonos. Perspective approximations. *Image and Vision Computing*, 8(3):179–192, 1990.

[3] P. I. Corke. Visual control of robot manipulators—a review. In K. Hashimoto, editor, *Visual Servoing*, pages 1–32. World Scientific, 1993.

[4] B. Espiau, F. Chaumette, and P. Rives. A new approach to visual servoing in robotics. *IEEE Transactions on Robotics and Automation*, 8:313–326, 1992.

[5] J. T. Feddema, C. S. George Lee, and O. R. Mitchell. Weighted selection of image features for resolved rate visual feedback control. *IEEE Transactions on Robotics and Automation*, 7:31–47, 1991.

[6] A. Fox and S. Hutchinson. Exploiting visual constraints in the synthesis of uncertainty-tolerant motion plans. *IEEE Transactions on Robotics and Automation*, 11:56–71, 1995.

[7] G. Hager and S. Hutchinson, editors. *Proc. IEEE Workshop on Visual Servoing: Achievements, Applications and Open Problems*. Inst. of Electrical and Electronics Eng., Inc., 1994.

[8] G. D. Hager. Real-time feature tracking and projective invariance as a basis for hand-eye coordination. In *Proc. IEEE Conf. on Computer Vision and Pattern Recognition*, pages 533–539, 1994.

[9] K. Hashimoto. *Visual Servoing*. World Scientific, 1993.

[10] J-Y. Hervé, R. Sharma, and P. Cucka. The geometry of visual coordination. In *Proc. Ninth National Conference on Artificial Intelligence (AAAI-91)*, pages 732–737, July 1991.

[11] W. T. Miller III, R. S. Sutton, and P. J. Werbos. *Neural Networks for Control*. MIT Press, Cambridge, MA, 1990.

[12] K. Kant and S. W. Zucker. Toward efficient trajectory planning: The path-velocity decomposition. *International Journal of Robotics Research*, 5(3):72–89, 1986.

[13] J. C. Latombe. *Robot Motion Planning*. Kluwer Academic, Boston, MA, 1991.

[14] T. Lozano-Pérez, M. T. Mason, and R. H. Taylor. Automatic synthesis of fine-motion strategies for robots. *International Journal of Robotics Research*, 4:3–24, 1985.

[15] N. P. Papanikolopoulos, P. K. Khosla, and T. Kanade. Visual tracking of a moving target by a camera mounted on a robot: A combination of vision and control. *IEEE Transactions on Robotics and Automation*, 9(1):14–35, 1993.

[16] R. Sharma. Active vision for visual servoing: A review. In *IEEE Workshop on Visual Servoing: Achievements, Applications and Open Problems*, May 1994.

[17] R. Sharma, J-Y. Hervé, and P. Cucka. Dynamic robot manipulation using visual tracking. In *Proc. IEEE International Conference on Robotics and Automation*, pages 1844–1849, May 1992.

[18] R. Sharma and S. Hutchinson. Optimizing hand/eye configuration for visual-servo systems. In *Proc. IEEE International Conference on Robotics and Automation*, pages 172–177, May 1995.

[19] S. B. Skaar, W. H. Brockman, and R. Hanson. Camera-space manipulation. *International Journal of Robotics Research*, 6(4):20–32, 1987.

[20] H. Sutanto and R. Sharma. Global perfomance evaluation of image features for visual servo control. *Journal of Robotic Systems*, 13(4):243–258, April 1996.

[21] K. A. Tarabanis, P. K. Allen, and R. Y. Tsai. A survey of sensor planning in computer vision. *IEEE Transactions on Robotics and Automation*, 11:86–104, 1995.

[22] R. Y. Tsai and R. K. Lenz. A new technique for fully autonomous and efficient 3D robotics hand/eye calibration. *IEEE Transactions on Robotics and Automation*, 5:345–358, 1989.

[23] L. E. Weiss, A. C. Sanderson, and C. P. Neuman. Dynamic sensor-based control of robots with visual feedback. *IEEE Journal of Robotics and Automation*, 3:404–417, 1987.

[24] M. Zeller, R. Sharma, and K. Schulten. Vision-based motion planning of a pneumatic robot using a topology representing neural network. In *Proc. IEEE International Symposium on Intelligent Control*, 1996.

Multi-Level Path Planning for Nonholonomic Robots using Semi-Holonomic Subsystems

Sepanta Sekhavat, *LAAS/CNRS, Toulouse, France,*

Petr Švestka, *Department of Computer Science, Utrecht University, the Netherlands,*

Jean-Paul Laumond, *LAAS/CNRS, Toulouse, France,*

Mark Overmars, *Department of Computer Science, Utrecht University, the Netherlands*

We present a new and complete multi-level approach for solving path planning problems for nonholonomic robots. At the first level a path is found that disrespects (some of) the nonholonomic constraints. At each next level a new path is generated, by transformation of the path generated at the previous level. The transformation is such that more nonholonomic constraints are respected than at the previous level. At the final level all nonholonomic constraints are respected. We present two techniques for these transformations.

In the intermediate levels we plan paths for what we refer to as semi-holonomic subsystems. *Such a system is obtained by taking the real (physical) system, and removing some of its nonholonomic constraints.*

In this paper, we apply the scheme to car-like robots pulling trailers, that is, tractor-trailer robots. *In this case, the real system is the tractor-trailer robot, and the ignored constraints in the semi-holonomic subsystems are the kinematic ones on the trailers. These are the constraints of rolling without slipping, on the trailers wheels. Experimental results are given that illustrate the time-efficiency of the resulting planner. In particular, we show that using the multi-level scheme leads to significantly better performance (in computation time and path shape) than direct transformations to feasible paths.*

1 Introduction

Even in the absence of obstacles, planning motions for nonholonomic systems is not an easy task. So far there exists no general algorithm for planning the motions of any nonholonomic system, that guarantees to reach a given goal. The only existing results deal with approximation methods (i.e., methods that guarantee to reach a neighbourhood of the goal, e.g., [6, 2]) and exact methods for special classes of nonholonomic systems (e.g.,[6, 11, 13]); fortunately, these special classes contain several real robot models.

Obstacle avoidance adds a second level of difficulty : Not only must one take into account the constraints imposed by the kinematic nature of the system (i.e., linking the parameter derivatives), but also the constraints due to the obstacles (i.e., dealing with the configuration parameters of the system). It appears necessary to combine geometric techniques addressing the obstacle avoidance with control techniques addressing the nonholonomic motions. Such a combination is possible through subtle topological arguments ([7]).

Treating the holonomic constraints separately from the nonholonomic ones is nowadays almost a "classical" approach. It has resulted in planners for various nonholonomic robots ([8], [16], [19]). The idea is that the problem is solved in two separate steps. In the first, a collision-free path is computed without taking into account the nonholonomic constraints. Subsequently, in the second step, this geometric path is transformed into one that respects the nonholonomic constraints.

For relatively simple systems, that is, with few nonholonomic constraints, efficient path planners are obtained with the described scheme. This is for example the case for car-like robots ([8]). However, for systems with a higher degree of nonholonomy, it turns out that the second step is too time-consuming. This observation leads to the idea of further decomposition of the nonholonomic constraints, and introducing them separately.

Our paper, which develops this idea, is organised as follows : We first relate our planner to other works in nonholonomic motion planning (Section 2). Then, in order to introduce our *multi-level* planning scheme (Section 4), we discuss the concept of nonholonomy and we define what we refer to as *semi-holonomic subsystems* (Section 3). Our method consists of an initial search for a collision-free (but not necessarily feasible) path, and a number of subsequent transformation steps. Each such transformation step produces a path respecting more nonholonomic constraints than its input path. In Section 5 we present two general methods (the *PL* technique and *tube-PPP*) for the transformations. Section 6 is devoted to applying the multi-level planner to the particular example of a tractor towing a number of trailers. In order to use our planner for a given system, we need some specific local planners for the system and its semi-holonomic subsystems. The local planners that we use for the tractor-trailer problem (based on *RTR* and *sinusoidal inputs*) are presented in Section 6.2. A method for obtaining the first collision-free path (for a car towing fictive holonomic trailers) is given in Section 6.3. In Section 6.4 a general technique for smoothing the generated paths is presented. Finally, the experimental results of Section 7 aim to illustrate that decomposing the nonholonomic constraints, and introducing separately, clearly improves the computation time and the quality of the computed paths.

2   Previous work

In the past few years, there has been a great deal of interest in motion planning algorithms that generate collision-free paths for nonholonomic systems. The tractor-trailer system is one of the examples frequently used to illustrate different algorithms. For a given system, the first question we have to answer is : can a robot reach a given goal, while avoiding collisions with the obstacles of its environment? This is the *decision* problem. For *locally controllable systems* [1], the existence of a feasible path between two configurations in the interior of the *free configuration space* CS_{free} is equivalent to the existence of any path between them in the interior of CS_{free}. This has led to a family of algorithms, decomposing the search in two phases.

They first try to solve the geometric problem (i.e., the problem for the holonomic system that is geometrically equivalent to the nonholonomic one). Then they use the obtained path to build a feasible and collision-free one. So in the first phase the decision problem is solved, and only in the second phase the nonholonomic constraints are taken into account.

The first general result was presented by Sussmann and Liu [19], who proposed an algorithm constructing a sequence of feasible paths that *uniformly* converges to any given path. This guarantees that one can choose a feasible path *arbitrarily close* to a given collision-free path. The method uses high frequency sinusoidal inputs. Though this approach is general, it is quite hard to implement in practice. In [22], Tilbury et al. exploit this idea for a mobile robot with two trailers. Experimental results however show that the approach cannot be applied in practice, mainly because the convergence is very slow. Therefore this method has never been connected to a geometric planner in order to obtain a global planner that would take into account both environmental and kinematic constraints.

Another approach was developed in LAAS for car-like robots ([8]), using Reeds and Shepp works on optimal control to approximate the geometric path. In [12] Reeds and Shepp presented a finite family of paths composed of line segments and circle arcs containing a length-optimal path linking any two configurations (in absence of obstacles). The planner introduced in [8] replaces the collision-free geometric path by a sequence of Reeds and Shepp paths. This complete and fast planner was extended to the case of Hilare with one and two trailers, using near optimal paths numerically computed [9, 4] (so far the exact optimal paths for tractor-trailer system in absence of obstacles are unknown). The resulting planners are however neither complete nor time-efficient.

The same scheme was used for systems that can be put into the *chained form*. For these systems, Tilbury et al. [23] proposed different controls to steer the system from one configuration to another, in absence of obstacles. Sekhavat and Laumond prove in [15] that the *sinusoidal inputs* proposed by Tilbury et al. can be used in a complete algorithm transforming any collision-free path to a collision-free and feasible one. This algorithm was implemented for a car-like robot towing one or two trailers, which can be put into the chained form, and finds paths in reasonable times ([15]).

Another important class of motion planning algorithms consists of search based methods that build and explore a graph, with nodes being robot-configurations and edges corresponding to (simple) feasible paths. With increasing computation time, the graph tends to cover CS_{free}. Often heuristics are used to guide the search, in order to reduce the computation time required for obtaining a graph capturing sufficiently the connectivity of CS_{free}. These methods are of course penalised when the size and complexity of the search space grows.

The *Probabilistic Path Planner PPP* ([5, 20]) builds a graph of states in the free configuration space CS_{free}. The nodes are chosen randomly, and a local planner searches for paths linking pairs of admissible and mutually nearby configurations. Whenever such a (local) connection succeeds, a corresponding edge is added to the graph. Given path planning problems can be then solved by performing searches in this graph (or roadmap). The critical point of *PPP* when applied to nonholonomic robots is the speed of the nonholonomic local planner.

Local planners for tractor-trailer robots tend to be rather time-consuming. In [21] a local planner is presented and integrated into *PPP*, that uses exact closed form solutions for the kinematic parameters of a tractor-trailer robot. In [16] the local planner using sinusoidal inputs for chained form systems is used. For practical use, both local planners appear to be too expensive for capturing the connectivity of CS_{free}. For this reason, in [16] a two-level scheme is proposed. At the first level a geometric path is computed, not taking into account the nonholonomic constraints. At the second level a (real) solution is searched for within a neighbourhood of this geometric path. The multi-level algorithm proposed in this paper can in fact been seen as a generalisation of this two level scheme.

Barraquand and Latombe [1] propose a heuristic search approach to motion planning for nonholonomic robots. It consists of heuristically building and searching a graph whose nodes are small axis-parallel cells in configuration space. Two such cells are connected in the graph if there exists a basic path between two configurations in the respective cells. The completeness of this algorithm is guaranteed up to appropriate choice of certain parameters. The main drawback of this planner is that when the heuristics fail it requires an exhaustive search in the discretised configuration space. Nevertheless, in many cases the method produces nice paths (with minimum number of reversals) for car-like robots and tractors pulling one trailer.

Ferbach [3] builds on this method in his *progressive constraints* algorithm in order to solve the problem in higher dimensions. First a geometric path is computed. Then the nonholonomic constraints are introduced progressively in an iterative algorithm. Each iteration consists of exploring a neighbourhood of the path computed in the previous iteration, searching for a path that satisfies more accurate constraints. In fact, for an intermediate path the tangent vector at each point is not constrained to lie in a given tangent space (imposed by the nonholonomic constraints), but only in a larger space containing this tangent space. This "larger space" tends to the exact tangent space during iterations. The underlying assumption of this approach is that the current path can actually converge towards a nonholonomic solution, when the forbidden areas of the phase space[1] expand. This assumption is however not true in general, for which reason the algorithm is not complete. Nevertheless, smooth collision-free paths in non-trivial environments were obtained with this method for car-like robots towing two and three trailers. The fundamental difference between the *progressive constraints* planner and our *multi-level* planner is that, at each iteration of the algorithm, that the *progressive constraints* planner reduces the *size* of the phase space portion in which a solution is searched for, whereas the *multi-level* planner reduces the *dimension* of this space. Thanks to this, the number of interme-

---

[1]The *phase space* is the set of (c,c') where c is a configuration and c' a velocity (not necessarily adapted to c for any particular nonholonomic system).

diate paths constructed by the multi-level planner is deterministic and known at the very beginning of the search. Besides, there is no discretisation of the space and the elementary paths are exact (not computed by numerical integration).

Apart from the above general nonholonomic motion planners, some specific planners have been developed for special cases.

Dealing with tractor-trailer problem for example, we can cite [10] in which a rule-based incremental control was applied to the special problem of parallel parking.

Another example is [14] in which the goal is reached approximately (there is no control on the final position of trailers). This work concerns problems in which a path without reversals between the extremal configurations exists. It appears that in many such cases, paths can be planned just for the tractor (going only forwards) in a way that obstacle avoidance is guaranteed for the whole tractor-trailer system, when the trailers follow the tractor by motions induced by their nonholonomic constraints. This method is especially useful for industrial applications in which the exact position of trailers is not important, and the environment can be adapted to mobile robots that perform only forward motions. Although this method shows some resemblance with the first planning step of our multi-level planner, there is an important conceptual difference. Namely, whereas in [14] the existence of the trailers is ignored during the planning phase, in the first planning step of our multi-level planner only the nonholonomic constraints of the trailers are ignored, but not the trailers themselves. I.e., the first path will be computed for a fictive system composed of a real car pulling "holonomic" trailers, and the planning is done for this entire system. Through this, completeness of the multi-level planner is guaranteed. The concept of fictive (or semi-holonomic) systems is developed in the next section.

3 Nonholonomic systems and fictive simplifications

Holonomic constraints can be characterised by a set of equations involving only the system-state. This

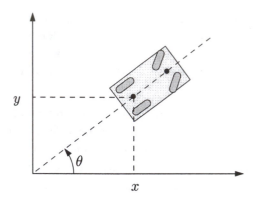

Figure 1: *A car-like robot.*

means that they only reduce the free configuration space CS_{free}, which makes it possible to apply any classical geometric approach (e.g., cell-decomposition, potential field) for tackling such problems. Nonholonomic constraints can only be characterised by a set of non-integrable equations involving, apart from the system-state, also the derivative (with respect to time) of this state. Because the equations cannot be integrated, the nonholonomic constraints do not reduce CS_{free}. However, they restrict the *direction* in which a CS motion is allowed. So, for a nonholonomic robot, not every collision-free path (that is, a path lying in the CS_{free}) is a feasible path.

For example, a car-like robot is nonholonomic. Physically, the nonholonomy is induced by the fact that the wheels cannot slide. This imposes a relation between the robot orientation and its admissible velocity. This relation can be expressed by the following non-integrable equation (see also Figure 1) :

$$\dot{x}\sin\theta - \dot{y}\cos\theta = 0$$

Now let us consider a car-like tractor pulling n trailers. A configuration c of this system can be represented by $n + 3$ parameters : $c = (x, y, \theta_0, \ldots, \theta_n)$. The position of the tractor is defined by x and y, the orientation of the tractor by θ_0, and the orientation of the kth trailer by θ_k. See also Figure 2. The orientation Φ of the tractors front wheels is not taken into account here.

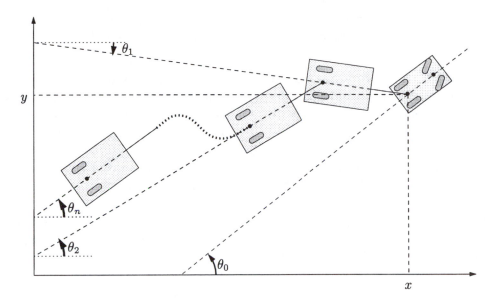

Figure 2: *A tractor-trailer robot (with n trailers) placed at configuration* $c = (x, y, \theta_0, \theta_1, \ldots, \theta_n)$.

For each trailer we have an equation analogous to the one above, linking the velocity of the (rear) axle midpoint and the orientation of the trailer. So the nonholonomic constraints of the tractor-trailer system are represented by $n+1$ equations of type $(0 \leq i \leq n)$:

$$E_i : \quad F_i(c, \dot{c}) = 0$$

These can be transformed to a non-integrable system of equations of the form :

$$\dot{c} = G(c, u)$$

where u is a control vector of dimension 2 (see [1]).

For such a system we can define $n+1$ fictive systems : $S_0, \ldots, S_n$. S_i is defined as the system respecting the nonholonomic constraints represented by the equations E_0 to E_i. We refer to S_i as the *semi-holonomic sub-system of degree i*.

Notice that the semi-holonomic subsystem, as explained above by the example of tractor-trailers, is a general concept. Indeed any nonholonomic system is defined by a set $\mathcal{E}$ of constraint equations of type E_i. So we can associate to it semi-holonomic subsystems by considering only subsets of $\mathcal{E}$.

4　The multi-level scheme using transformations between semi-holonomic subsystems

For simple systems, that is, with few nonholonomic constraints, efficient path planners are obtained by separating the holonomic constraints from the nonholonomic ones, and computing paths in two steps, according to the following (classical) scheme :

1. Compute a collision-free path without taking into account the nonholonomic constraints. If such a path does not exist, there is no solution to the problem.

2. Replace the (holonomic) path by a sequence of feasible ones.

The complexity of the transformation step (2) and the quality of the final path however depend on the initial path. There is a trade-off. The more the initial path respects the nonholonomic constraints, the faster will be the transformation, and the nicer will be the resulting path. However, more computation time is required to compute such initial paths. For robots with many nonholonomic constraints, it turns out that it pays off to spend more time on getting "good" initial paths.

The multi-level path planning scheme that we propose in this paper aims at this, by introducing a sequence of transformation steps, instead of just one. It uses the concept of semi-holonomic subsystems, as defined in the previous section. Given a real system S with $n + 1$ semi-holonomic subsystems $S_0, \ldots, S_n$, we first find a path P_0 for system S_0. Then, in n subsequent steps, we transform it to a path feasible for the real system S. At step i ($0 \leq i < n$), path P_i, feasible for S_i, is transformed to a path P_{i+1} feasible for S_{i+1}. Hence, the multi-level scheme that we propose is as follows :

1. Compute a collision-free path P_0, respecting the nonholonomic constraints of system S_0. If such a path does not exist, there is no solution to the problem.

2. **for** $i = 0$ **to** $n - 1$:
   Transform path P_i to a collision-free path P_{i+1}, respecting the nonholonomic constraints of system S_{i+1}

Of course, the key ingredient is the transformation of the paths in a way that they become feasible for the "next" system. In the following section we present two such techniques. Also, means for obtaining the initial path P_0 are required.

5 Obtaining initial paths and transformation techniques

The initial path P_0 is a path feasible for the (simple) system S_0, which has only one nonholonomic constraint (described by equation E_0). It is typically an easy task to compute such a path, with existing planners. For example, in the planner that we have implemented for tractor-trailer robots (as described in the Sections 6 and 7), an initial roadmap is computed with the *Probabilistic Path Planner* (See Section 6.3). This roadmap stores paths feasible for S_0, that is, paths

respecting the tractors nonholonomic constraints, but ignoring those of the trailer(s).

In the rest of this section, we focus on ways of performing the transformation steps in the multi-level algorithm. We assume that S_i is the (sub)system for which we want to transform a path P_{i-1} to a path P_i. Furthermore, we assume that L_i is a *local planner* for system S_i. That is, L_i is a function that takes two robot configurations as arguments, and returns a path connecting its arguments, that (in absence of obstacles) is feasible for system S_i. A local planner must posses a *topological property*, in order to guarantee *completeness* of the transformation step. We give this property in Section 5.3.

5.1 PL method

One way of performing the transformation of a path P_{i-1} to a path P_i is the *Pick and Link (PL)* method. Let us assume that the collision-free path P_{i-1} is parametrised by $s \in [0..1]$. $P_{i-1}(0)$ and $P_{i-1}(1)$ are the extremal configurations. We first try to join these two configurations using the local planner L_i . If the obtained path is collision-free then we have a feasible path avoiding obstacles, and the problem is solved. If not, we take an intermediate configuration (let us say $P_{i-1}(\frac{1}{2})$) on the collision-free path P_{i-1} and we apply recursively the same treatment to the portion of the collision-free path between $P_{i-1}(0)$ and $P_{i-1}(\frac{1}{2})$ and to the portion of the path between $P_{i-1}(\frac{1}{2})$ and $P_{i-1}(1)$. As the algorithm proceeds, the considered extremal configurations will lie closer and closer to each other. Thanks to the *topological property* of the local planner L_i, when the final configuration tends to the initial one, the length of the local path linking them tends to zero. This guarantees the convergence of the algorithm. For a serious demonstration see [15].

Strong points of this technique are its *completeness* and relative time-efficiency in cluttered regions of CS. The paths produced are however often very long and "ugly", and therefore require significant *smoothing* (See also Section 6.4). Furthermore, the completeness of the algorithm is only guaranteed if the input path P_{i-1} has a non-zero clearance from the obstacles.

5.2 Tube-PPP

The second transformation technique that we describe is based on the *Probabilistic Path Planner*, or *PPP* ([5, 20]).

PPP is conceptually quite simple. A roadmap R is constructed incrementally, by repeatedly generating a *random*[2] free configuration c, and trying to connect c to a number of previously generated configurations with the local planner. Whenever such a connection succeeds (that is, the computed local path is collision-free), the roadmap is extended with the corresponding local path. Once a roadmap has been constructed in the above manner, it can be used for retrieving feasible paths. We denote *PPP* with a specific local planner L by $PPP(L)$.

Given a path P we now define $T_\epsilon(P)$ to be the subset of CS that lies within distance ϵ of P, where ϵ is a (small) constant. We refer to $T_\epsilon(P)$ as the *CS-tube around* P. Formally :

$$T_\epsilon(P) = \{c \in CS | \exists \tilde{c} \in P : |\tilde{c} - c| \le \epsilon\}$$

The transformation is performed by executing $PPP(L_i)$ on $T_\epsilon(P_{i-1}) \cap CS_{free}$. That is, instead of picking the node configurations randomly from the whole CS_{free}, they are only picked from the free portion of $T_\epsilon(P_{i-1})$. The concept is illustrated in Figure 3. The start and goal configurations s and g are added as nodes at the very beginning, and the roadmap is extended in the standard way until s and g are graph-connected. Then, a graph search and concatenation of appropriate local paths gives a path P_i, feasible for the (sub)system S_i.

With respect to the *PL* transformation algorithm, *tube-PPP* has advantages as well as disadvantages. The paths produced are typically much shorter than those constructed by the *PL* technique. This means that less time has to be spent on smoothing them. Also, no minimal clearance between the (input) paths and the obstacles is required. However, the method is only *probabilistically* complete, and transformations of path segments where only concatenations of many short paths

---

[2]Heuristics have been developed for generating more nodes in certain interesting/difficult areas of CS_{free}.

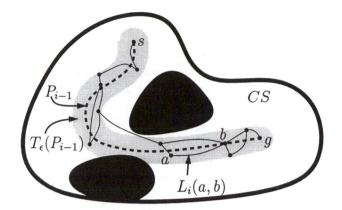

Figure 3: *A path P_{i-1} defines a tube in CS, in which a roadmap, feasible for system S_i, is constructed. Note that, although all nodes lie in the tube, the local paths are allowed to exit it.*

bring a solution tend to be more time-consuming than with the *PL* method.

Choosing the transformation techniques for the different transformation steps requires sensible choices to be made. In the implementation that we present in Section 7 we have based these choices on experiments.

5.3 Completeness of the transformation steps

Let us consider a configuration space CS equipped with a metric $d : CS \times CS \longrightarrow \Re+$. Let $B(c, r)$ be the ball of radius r around the configuration c. Let $\ell : CS \times CS \longrightarrow CS^{[0,1]}$ be a *local planner*; for two configurations $a, b \in CS$, $\ell(a, b)$ is a path $\ell_{a,b}(t), t \in [0, 1]$ such that $\ell_{a,b}(0) = a$ and $\ell_{a,b}(1) = b$.

Definition 1 *A local planner ℓ verifies the Topological Property (TP) if :*

$$\forall \epsilon > 0, \exists \eta > 0 \ such \ that \ \forall c_0 \in CS, \forall c \in B(c_0, \eta) :$$
$$\forall t \in [0, 1] : \ell_{c_0, c}(t) \in B(c_0, \epsilon)$$

Property : Let ℓ be a local planner that respects the topological property *TP*. The completeness of both the *PL* algorithm as *PPP* using ℓ is guaranteed. For proofs we refer to, respectively, [15] and [20].

6   Application to tractors with trailers

In this section we describe how the multi-level planning scheme can be applied to robots consisting of a car-like tractor, towing a number of trailers. First, in Section 6.1, we formally define tractor-trailer robots, and we give formulas that describe their nonholonomic constraints. We have seen, in Section 5, two transformation methods for the intermediate steps, requiring local planners that respect the topological property TP. In Section 6.2, we describe such local planners for tractor-trailer robots and their semi-holonomic subsystems. In addition to the transformation methods, we need an initial path. One way of obtaining such a path for S_0 is described in Section 6.3. Finally, in Section 6.4, we present a method for heuristically improving the quality of the paths produced by the transformation algorithms.

6.1   The tractor-trailer system

As described in Section 3, a configuration of a tractor with n trailers can be represented by $c = (x, y, \theta_0, \theta_1, \ldots, \theta_n)$. We have seen that the nonholonomic constraints of this system can be expressed by $n + 1$ equations that lead to a control system with two inputs. For example, we can choose as inputs the tangential and the angular velocities of the tractor. If we consider a real tractor, the mechanical stops on the front wheels constrain the motion of the vehicle to have a bounded curvature. For the motion planning problem, this imposes a relation between the two inputs. Rather than treating the problem in this way, we consider the angle Φ of the car front wheels (with respect to the x-axis) as a coordinate of the configuration $c = (x, y, \Phi, \theta_0, \theta_1, \ldots, \theta_n)$. Then the mechanical stops are simply geometric constraints on this coordinate. We use $\frac{1}{3}\pi$ as *maximal steering angle*. That is, when the front wheels are in a "straight" position, they cannot turn over an angle of more than $\frac{1}{3}\pi$ (clockwise or counterclockwise). This corresponds approximately to normal cars.

We now first introduce some notations and terminology. The tractor is denoted by $\mathcal{A}_0$, and the j-th trailer by $\mathcal{A}_j$. We refer to the midpoint between the rear wheels of $\mathcal{A}_0$ as the *rear point* of the tractor, and to the midpoint between the front wheels of $\mathcal{A}_0$ as the *front point* of the tractor. For the trailers, we refer to the midpoint between the wheels of $\mathcal{A}_j$ as *the rear point of trailer j*. The (fixed) distance between the rear point R_i of a trailer and the rear point R_{i-1} of the vehicle in front is now denoted by d_i. d_0 denotes the distance between the tractors rear and front point.

For a tractor pulling two trailers, the kinematic model of the system can be expressed by the following equations :

$$\begin{cases} \dot{x} = cos\theta_0 v_0 \\ \dot{y} = sin\theta_0 v_0 \\ \dot{\Phi} = \omega \\ \dot{\theta}_0 = \frac{1}{d_0} tan(\Phi) v_0 \\ \dot{\theta}_1 = \frac{1}{d_1} sin(\theta_0 - \theta_1) v_0 \\ \dot{\theta}_2 = \frac{1}{d_2} sin(\theta_1 - \theta_2) cos(\theta_0 - \theta_1) v_0 \end{cases}$$

The inputs here are v_0, the velocity of the car, and ω, the front wheels angular velocity (that is, the derivative (in time) of the tractors steering angle). For a realistic system, not only do we have to take into account some bounds for the steering angle Φ, but also for the angles between consecutive bodies composing the robot. We use $-\frac{1}{2}\pi$ and $\frac{1}{2}\pi$ as bounds.

The nonholonomic constraints of this system are imposed by the rolling without slipping of the wheels. So a natural way of building subsystems for a system composed by a tractor and n trailers is to ignore the existence of some of the wheels. We define the subsystem S_k $(0 \leq k \leq n)$ as in Section 3. That is, S_k is a car-like robot with the first k trailers nonholonomic (that is, with wheels), and the remaining $n - k$ trailers being holonomic. This means that the last $n - k$ trailers can rotate freely (within $[-\frac{1}{2}\pi, \frac{1}{2}\pi]$) around their linking point with the vehicle in front.

6.2   Local planners for the subsystems

For the both described transformation schemes we need a local planner respecting the topological property. Let us assume that we posses such a local planner ℓ for the system composed of the first $i + 1$ bodies of our robot (that is, for the tractor and the first i trailers). We can then define a local planner ℓ_i for the system S_i as follows : we first plan a path only for the tractor and the first i trailers with the local planner ℓ. Then we simply interpolate the linking angles of the remaining trailers between the extremal configurations. It is easy to prove that such a local planner ℓ_i always respects the topological property for S_i. So in order to have a suitable local planner for the subsystem S_i, we need a local planner for a tractor with i nonholonomic trailers, respecting the topological property. We will now introduce such local planners.

6.2.1   RTR-planner for a car-like robot

For a car-like robot we use a quite simple local planner ℓ. Given two (car-like) configurations, it constructs the shortest path connecting them that consists of a (constant curvature) curve, a straight path segment, and another (constant curvature) curve. We refer to this local planner as the *RTR local planner*, and we denote it by L_{RTR}. This local planner is quite a cheap local planner, in the sense that construction and collision checking of the local paths can be done very efficiently. Furthermore, it guarantees probabilistic completeness of *PPP* (See also [20]).

6.2.2   Sinusoidal planner for tractor-trailers

Briefly, the principle of the *sinusoidal local planner* is to transform the state coordinates into the chained form and to apply sinusoidal inputs to the transformed system.

In what is considered as the *classical* tractor-trailer system in the literature (and that we consider in this paper), each vehicle is hitched to the centre of the rear axle of the front vehicle. Such system can be put in a form called the "chained form", locally around almost any point of the configuration space. The singularities are the points where one of the angles between two vehicles is $\frac{1}{2}\pi$. If we consider a physical system for which $\frac{1}{2}\pi$ is an upper bound for the admissible values of these angles (this is the assumption we have made for our tractor-trailer), then locally around any point of the configuration space the system can be put in the chained form :

$$\left\{ \begin{array}{l} \dot{z}_1 = u_1 \\ \dot{z}_2 = u_2 \\ \dot{z}_3 = z_2.u_1 \\ . \\ . \\ . \\ \dot{z}_n = z_{n-1}.u_1 \end{array} \right.$$

For the classical system of tractor with n trailers, a general method of transformation into chained form is given by Sordalen [18]. See [17] for the conversion into chained form of the particular system of a tractor pulling two trailers.

For chained form systems, there exists several types of control that bring them from one point to another : piecewise constant inputs, polynomial inputs or sinusoidal inputs [23]. We need one of the corresponding local planners that respects the topological property.

The sinusoidal inputs for a chained form system of dimension n are defined as :

$$\left\{ \begin{array}{l} u_1 = a_0 + a_1 sin(\omega t) \\ u_2 = b_0 + b_1 cos(\omega t) + b_2 cos(2\omega t) + ... \\ \qquad + b_{n-2} cos((n-2)\omega t) \end{array} \right.$$

ω is fixed and $T = 2\frac{\pi}{\omega}$ is the integration time (the time required to steer the system from the start to the goal position). a_1 is chosen non-zero, and has influence on the shape of paths. The other parameters $(a_0, b_0, ..., b_{n-2})$ are functions of the extremal configurations.

We can prove that for a chained form system, a local planner using sinusoidal inputs respects the topological property TP ([15]).

6.3   Obtaining the initial paths for S_0

We have now all the ingredients required for transforming a S_i-path to a S_{i+1}-path, for any $i \geq 0$. This is however not enough. We must also have means for obtaining the initial S_0-paths, that is, paths for car-like robots towing a number of holonomic trailers. Path planning for such robots is however not a very challenging problem. One could first compute a fully holonomic path with, for example, a potential field method, and subsequently transform this path into a S_0-path with one of the transformation steps described in this paper.

In the planner that we implemented, and for which we will give experimental results in the next section, we however chose another solution. Given a particular scene, we construct a sufficient S_0-roadmap (that is, a roadmap storing paths feasible for system S_0) with *PPP*. As also described in Section 5.2, *PPP* builds a roadmap incrementally by probabilistically generating free configurations, and interconnecting these by a local planner, where possible. The local planner l_0 (based on *RTR*), as described in Section 6.2, is used. Given a constructed roadmap, solving a particular path planning problem (s, g) is done by connecting both the start configuration s and goal configuration g to the same connected component of the roadmap, with, for example, again the local planner. Of course, one may fail to compute such connections. In such a case, either there is no solution to the problem (that is, s and g lie in separate components of the free configuration space), or the roadmap is not yet sufficient, and has to be further extended.

6.4 Smoothing the intermediate and final paths

We have not yet said anything about the *quality* of the paths generated by the different steps in our algorithm, and neither about the influence that the quality of a S_i-path has on its transformation to a S_{i+1}-path.

For the initial and intermediate transformation steps we are interested in obtaining paths whose transformations are easy, that is, fast. Firstly, we recall that the *PL* method requires input-paths of a certain non-zero clearance for completeness. Furthermore, from experiments we noticed that the length of a path is of very large influence on the time-efficiency of its transformation. The shorter a path, the faster its transformation (on the average). So, during each step, we want generate paths that are short. Also, if the *PL* method is used, we must guarantee a non-zero clearance. The latter is quite easy (See also Section 7), so we now concentrate on the first criterion.

Both presented transformation algorithms are actu-ally very bad with respect to the resulting path length. The generated paths are typically very long, up a hundred times longer then necessary, and they can hardly be used directly. Fortunately, it appears to be quite easy to heuristically shorten the paths sufficiently in order to make their transformations possible (and fast). The probabilistic algorithm that we use for this cause is described below. Let P_i be a S_i-path, and L_i be a local planner for the system S_i. Repeatedly random segments of P_i are picked, and it is tried to replace these by shorter paths constructed by the local planner L_i. Of course, this is only possible in those cases where the local paths are collision-free. Formally :

> **loop** until . . .
>   Let Q be a *random* path segment of P_i,
>   with start-configuration s and end-configuration e.
>   Let Q_L be $L_i(s, e)$.
>   **if** Q_L is *collision-free* and $length(Q_L) < length(Q)$
>   **then** replace Q by Q_L in P_i.

This simple smoothing algorithm, that we refer to as *probabilistic path shortening*, works very well in practice. A problem however is the stop-criterion. In our experiments, we simply run the algorithm until it stops making (significant) progress. See again Section 7.1 for experimental results. We refer to [17] for other smoothing techniques for tractor-trailer robots.

7 An implementation of the algorithm and experimental results

We now describe an implementation of the multi-level algorithm for tractor-trailer robots with 2 trailers. The initial S_0-paths are obtained, as explained, by retrieval from a roadmap computed with *PPP*. For the $S_0 \rightarrow S_1$ transformation step we use *tube-PPP*, while for the $S_1 \rightarrow S_2$ transformation step we use the *PL* method. Because the *PL* method requires a non-zero clearance of its input-paths, we use a slightly grown robot (by a factor of 0.035) in the initial (*PPP*) step and the first (*tube-PPP*) transformation step. We based these choices on experiments. For the $S_0 \rightarrow S_1$ step both transformation algorithms seem to be of approximately the same speed, but the quality of the paths is better with *tube-PPP*. For the second step however, the *PL* method is much faster. We apply *probabilistic path shortening* to the intermediate and final paths.

We also give experimental results for the case where the $S_0 \rightarrow S_2$ transformation is performed in one single step (with the *PL* method). From these results it follows that a large gain is obtained by doing the transformations separately, according to the multi-level scheme that we propose in this paper.

We have performed experiments with our planner in 2 different environments, which we refer to as *Scene 1* and *Scene 2*. For both scenes, we have defined 3 different path planning problems (that is, 3 pairs of start and goal configurations), and we have measured the performance of our planner for solving these problems. We are mainly interested in the performance of the transformation steps (this is the difficult part). For this reason, we take, for each problem, a number of different S_0-paths solving it, and we measure the performance of the transformation steps for each of the (equivalent) S_0-paths. We will refer to these S_0-paths, as the *initial test paths*. Scenes 1 and 2 are shown in Figures 4 and 5, together with the initial test paths that we use. In Figures 6 and 7 we see, for each problem, a S_0-path and a (smoothed) S_2-path to which it has been transformed by the multi-level planner. Figure 8 shows a 3D visualisation of a path, feasible for system S_2, computed by our multi-level planner.

The set of equivalent (but different) S_0-paths is obtained by retrieval from independently constructed roadmaps by *PPP*. We think that the set of S_0-paths that we present for each problem is fairly representative for paths that *PPP*, in general, generates for S_0-robots.

7.1 Experimental results

All experiments have been performed on a Silicon Graphics Indigo[2] workstation rated with 96.5 SPECfp92 and 90.4 SPECint92 (136 MIPS). For both scenes we have first constructed a sufficient S_0-roadmap with *PPP*. For Scene 1 this took about 40 seconds, and the resulting roadmap had 550 nodes. For Scene 2 about 30 seconds where required, resulting in a roadmap with 650 nodes.

7.1.1 Results for the multi-level algorithm

For each initial test path $\mathcal{P}_0$, we measured the computation times required for :

1. Retrieving and smoothing $\mathcal{P}_0$ from the S_0-roadmap.

2. Transforming $\mathcal{P}_0$ to a S_1-path $\tilde{\mathcal{P}}_1$, with *tube-PPP*. Since *tube-PPP* is probabilistic, we give *averages* over 20 independent runs.

3. Transforming $\tilde{\mathcal{P}}_1$ to a "smooth" path $\overline{\mathcal{P}}_1$, with *probabilistic path shortening*. The algorithm was iterated until the point where no longer any significant length reduction was achieved.

4. Transforming $\mathcal{P}_1$ to a S_2-path $\tilde{\mathcal{P}}_2$, with the *PL* method.

5. Transforming $\tilde{\mathcal{P}}_2$ to a "smooth" path $\mathcal{P}_2$, with *probabilistic path shortening*. Again, the algorithm was iterated until the point where no longer any significant length reduction was achieved.

The results are presented in Tables 1 and 2. Each problem has its own row. For each problem, the average computation times over the different initial test paths (shown in the Figures 4 and 5) are given, for the above described mentioned computation phases of the algorithm. The most right column gives the total (average) computation times. The values between square brackets give the (average) *lengths*[3] of the corresponding paths.

We see that the total computation times are on the order of seconds for the two easy problems (Problem 1 in Scene 1 and Scene 2), and on the order of a few minutes for the other, more difficult, problems. When we look at the distribution of the computation times over the different steps of our algorithm, we see that most time is spent on the second transformation and,

---

[3]We measure the distance that the tractors rear point travels, considering the scene to be (tightly) bounded by a unit square.

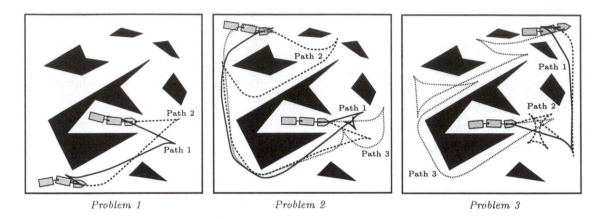

Problem 1   *Problem 2*   *Problem 3*

Figure 4: S_0-paths solving the problems 1, 2, and 3 in Scene 1.

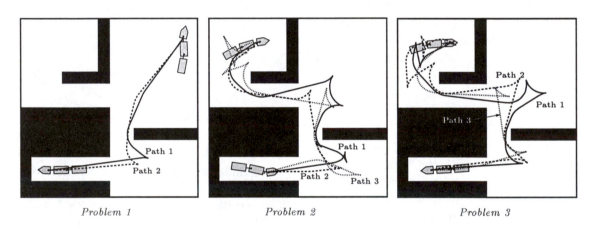

Problem 1   *Problem 2*   *Problem 3*

Figure 5: S_0-paths solving the problems 1, 2, and 3 in Scene 2.

especially, the final smoothing step. Hence, it appears to be much easier to obtain an arbitrary S_2-path than one which is (intuitively) nice. When we take into consideration the path lengths of the intermediate paths, we see very clearly that the length of the non-smoothed S_2-path $\tilde{\mathcal{P}}_2$ is of great influence on the final smoothing step. Also, not surprisingly, we see a strong correlation between the computation time of a (non-smoothed) S_2-path, and its length.

7.1.2 Other results

We have also done experiments with variations of the multi-level algorithm. Below, we give results obtained by skipping the S_1-level (Table 5). We have not

managed to smooth sufficiently all S_2-paths generated by the two-level planner (that is, the one without the S_1-level). We therefore do not give results for the final smoothing phase of this planner (the averages would be meaningless).

7.1.3 Comparison of the results

In order to have a comprehensive overview of the experimental data, we plot the average computation times for each of the 6 presented problems in separate charts (Figure 9). The (average) *cumulative* computation time is set against the computation phase of the algorithm. Again because of the smoothing problems for some of the S_2-paths obtained by the two-level plan-

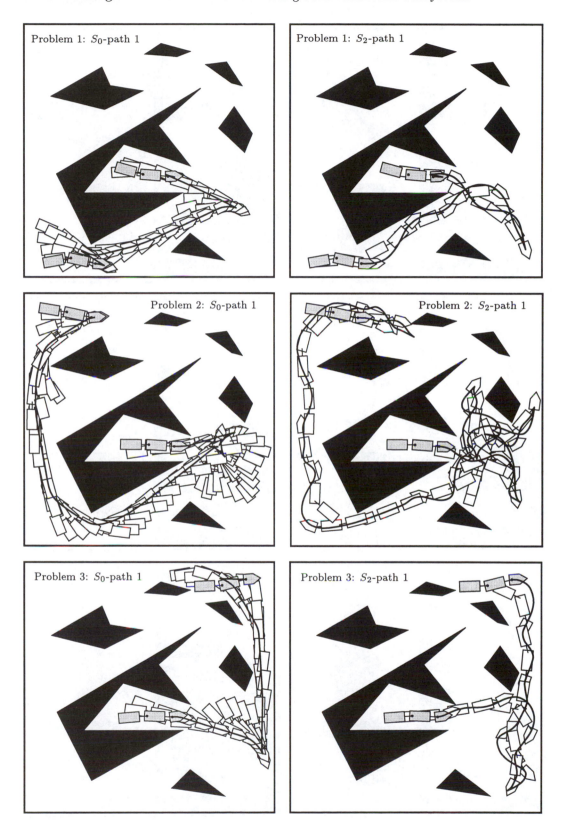

Figure 6: *For each problem in Scene 1, a S_0-path and a (corresponding) S_2-path solving the problem is shown.*

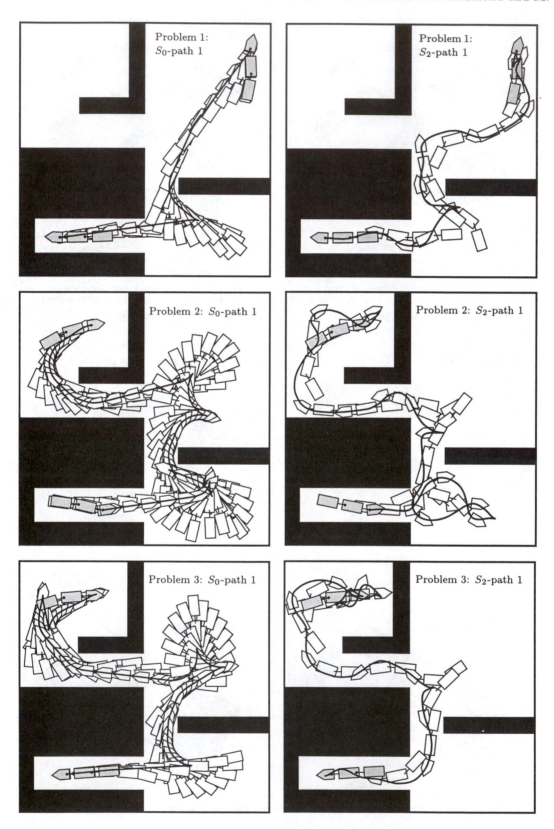

Figure 7: *For each problem in Scene 2, a S_0-path and a (corresponding) S_2-path solving the problem is shown.*

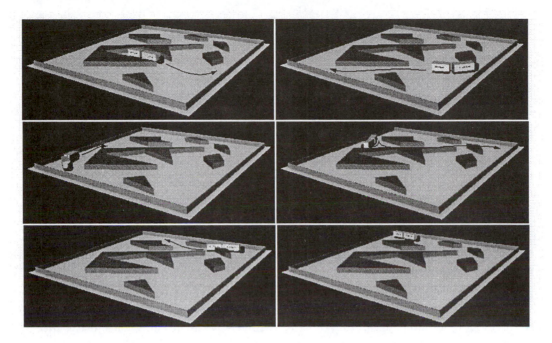

Figure 8: *A 3D visualisation of a S_2-path, computed by our planner. It is a transformation of Path 2 solving Problem 2 in Scene 1 (See Figure 4).*

Scene 1	$\rightarrow \mathcal{P}_0$	$\xrightarrow{P} \tilde{\mathcal{P}}_1$	$\xrightarrow{S_P} \mathcal{P}_1$	$\xrightarrow{L} \tilde{\mathcal{P}}_2$	$\xrightarrow{S_P} \mathcal{P}_2$	Total
Problem 1	0.3	1.3	0.6	2.6 [18.9]	5.4 [1.9]	10.1
Problem 2	2.0	5.9	2.9	20.6 [139.9]	76.0 [4.0]	107.4
Problem 3	1.4	14.9	3.6	25.9 [220.5]	481.6 [5.5]	527.4

Table 1: *Experimental results for Scene 1 with the multi-level algorithm.*

ner, the dotted plots stop at the $\tilde{\mathcal{P}}_2$ point.

For the simple problems, that is, the problems 1, we do not see much structure. For the other, more difficult, problems we do see clearly that the multi-level algorithm is better. Moreover, if we recall that most computation time is spent in the final smoothing phase and strongly depends on the length of the non-smoothed S_2-path $\tilde{\mathcal{P}}_2$, it is clear that the two-level algorithm performs very poorly indeed.

We have performed experiments with another two-level planner as well, namely one that transforms fully holonomic paths (that is, paths that also disrespect the tractors nonholonomic constraints) directly to S_2-paths, using the *PL* method. As could be expected, the results where considerably worse than even those for the $S_0 \rightarrow S_2$ planner. For sake of brevity, we do not include these results in our paper.

Summarising the experimental results, we can conclude that, over the whole, the *multi-level* algorithm performs better, with respect to computation time and path quality, than just *two-level* planning.

8   Conclusions and future work

We have presented a new *multi-level* approach to path planning for nonholonomic robots amidst static obstacles. At the first level an initial path is gener-

Scene 2	$\rightarrow \mathcal{P}_0$	$\xrightarrow{P} \tilde{\mathcal{P}}_1$	$\xrightarrow{S_P} \mathcal{P}_1$	$\xrightarrow{L} \tilde{\mathcal{P}}_2$	$\xrightarrow{S_P} \mathcal{P}_2$	Total
Problem 1	0.4	1.0	0.4	3.1 [24.6]	7.5 [1.9]	12.5
Problem 2	1.1	15.1	5.0	28.6 [226.4]	311.0 [6.1]	360.8
Problem 3	0.8	15.1	2.9	25.7 [250.9]	395.6 [4.7]	440.0

Table 2: *Experimental results for Scene 2 with the multi-level algorithm.*

Scene 1	$\mathcal{P}_0 \xrightarrow{L} \tilde{\mathcal{P}}_2$
Problem 1	13.5 [136.0]
Problem 2	72.0 [422.1]
Problem 3	92.6 [492.5]

Scene 2	$\mathcal{P}_0 \xrightarrow{L} \tilde{\mathcal{P}}_2$
Problem 1	3.4 [25.6]
Problem 2	134.0 [629.8]
Problem 3	79.1 [453.1]

Table 3: *Experimental results for Scenes 1 and 2 for direct $S_0 \rightarrow S_2$ transformations.*

ated that respects only some (easy) nonholonomic constraint(s). Then, at each subsequent level, this path is transformed to a more feasible path, that is, to a path that respects more nonholonomic constraints. The final path is fully feasible for the real robot. Two general transformation algorithms (*PL* and *tube-PPP*) have been described, as well means for obtaining initial paths (*PPP*). Also, a technique was presented for heuristically smoothing the intermediate and final paths.

The multi-level scheme is very general. It is applicable to any nonholonomic robot, provided that the robot is locally controllable, and that one has local planners for the corresponding subsystems. In order to obtain completeness, these local planners must respect a basic topological property.

We have applied the multi-level scheme to *tractor-trailer* robots (with two trailers). The resulting planner is *complete*, and, as experimental results show, quite *time-efficient*. Problems involving a tractor with two trailers in realistic environments are solved on the order of, at most, a few minutes. The experimental results also show that the multi-level algorithm performs significantly better than just two-level planning (that is, direct transformation of an initial path to a feasible one).

There remain various open questions and topics of future research. For example, it is not clear in general what should be the order in which the nonholonomic constraints are to be introduced. For tractor-trailer robots, we made this choice just intuitively. Also, from experimental results ([17]) we can observe that the running times of our algorithm vary considerably over different initial paths solving the same problem. Therefore, in our opinion, it is an important topic of future research to formulate criteria for initial paths, that indicate the difficulty of transforming these initial paths to feasible ones. Such criteria could be used for guiding the generation of the initial paths. Also, better stop-criteria for the smoothing algorithms would be desirable.

Finally, we want to mention the possibility of *multi-level roadmap generation*. Instead of taking an initial *path*, and, in a number of steps, transforming it to a feasible path, one could also take an initial *roadmap*, and perform the transformations on this whole roadmap. One then ends up with a fully feasible roadmap, that is, a roadmap from which paths, respecting all non-holonomic constraints of the robot, can directly be retrieved. Of course, the total computation times will, in non-trivial scenes, be very high (on the order of hours or days). However, if one is dealing with a truly static environment, an algorithm that within a few hours or days computes a roadmap from which fully feasible (and smoothed) paths are directly retrievable can be favourable to one that requires no preprocessing, but takes a few minutes to compute each single path.

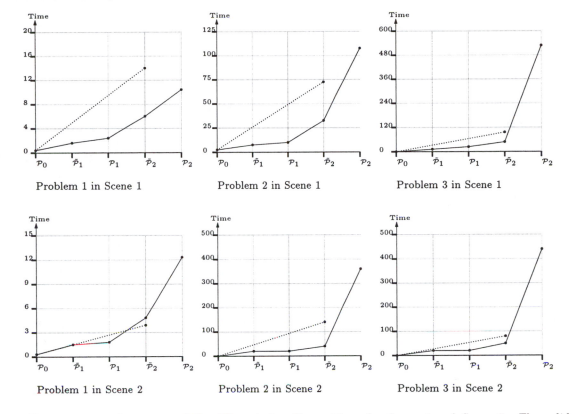

Problem 1 in Scene 1   Problem 2 in Scene 1   Problem 3 in Scene 1

Problem 1 in Scene 2   Problem 2 in Scene 2   Problem 3 in Scene 2

Figure 9: *The average performances of the different algorithm settings for Scene 1 and Scene 2. The solid polylines correspond to the* multi-level *planner, and the dotted polylines to the* two-level *planner.*

Acknowledgements

This research has been partially supported by the ES-PRIT III BRA Project 6546 (PROMotion) and by the Dutch Organisation for Scientific Research (N.W.O.).

We are very grateful to Geert-Jan Giezeman. All basic geometric operations have been performed by routines contained in the Plageo library, and also the software for our 3D path visualisation is due to Geert-Jan.

References

[1] J. Barraquand and J.-C. Latombe. Nonholonomic multibody mobile robots: Controllability and motion planning in the presence of obstacles. *Algorithmica*, 10:121–155, 1993.

[2] A. Bellaïche, J.-P. Laumond, and M. Chyba. Canonical nilpotent approximation of control systems : application to nonholonomic motion planning. In *Proc. 32nd IEEE Conf. on Decision and Control*, December 1993.

[3] P. Ferbach. A method of progressive constraints for nonholonomic motion planning. Technical report, Electricité de France. SDM Dept., Chatou, France, September 1995.

[4] C. Fernandes, L. Gurvits, and Z.X. Li. Optimal nonholonomic motion planning for a falling cat. In Z. Li and J.F. Canny, editors, *Nonholonomic Motion Planning*, Boston, USA, 1993. Kluwer Academic Publishers.

[5] L. Kavraki, P. Švestka, J.-C. Latombe, and M.H. Overmars. Probabilistic roadmaps for path planning in high dimensional configuration spaces. *To appear in IEEE Trans. Robot. Autom.*, 1995.

[6] G. Lafferriere and H. J. Sussmann. Motion planning for controllable systems without drift: A preliminary report. Technical Report SYSCON-90-04, Rutgers Center for Systems and Control, 1990.

[7] J.-P. Laumond. Singularities and topological aspects in nonholonomic motion planning. In Zexiang Li and J.F. Canny, editors, *Nonholonomic Motion Planning*, pages 149–199. Kluwer Academic Publishers, 1993.

[8] J.-P. Laumond, P.E. Jacobs, M. Taïx, and R.M. Murray. A motion planner for nonholonomic mobile robots. *IEEE Trans. Robot. Autom.*, 10(5), October 1994.

[9] J.-P. Laumond, S. Sekhavat, and M. Vaisset. Collision-free motion planning for a nonholonomic mobile robot with trailers. In *4th IFAC Symp. on Robot Control*, pages 171–177, Capri, Italy, September 1994.

[10] D. Luzeaux. Parking maneuvers and trajectory tracking. In *Proc. Int. Workshop on Advanced Motion Control*, Berkeley, CA, USA, 1994.

[11] R. Murray and S. Sastry. Steering nonholonomic systems using sinusoids. In *Proc. IEEE Conf. on Decision and Control*, pages 2097–2101, 1990.

[12] J.A. Reeds and R.A. Shepp. Optimal paths for a car that goes both forward and backward. *Pacific Journal of Mathematics*, 145(2):367–393, 1991.

[13] P. Rouchon, M. Fliess, J. Lévine, and P. Martin. Flatness and motion planning. In *Proc. European Control Conference*, pages 1518–1522, 1993.

[14] A. Sahai, M. Secor, and L. Bushnell. An obstacle avoidance algorithm for a car pulling many trailers with kingpin hitching. Technical Report UCB/ERL M94/10, University of California, Berkeley, CA, USA, March 1994.

[15] S. Sekhavat and J.-P. Laumond. Topological property of trajectories computed from sinusoidal inputs for nonholonomic chained form systems. In *Proc. IEEE Internat. Conf. on Robotics and Automation*, April 1996.

[16] S. Sekhavat, P. Švestka, J.-P. Laumond, and M.H. Overmars. Probabilistic path planning for tractor-trailer robots. Technical Report 96007, LAAS-CNRS, Toulouse, France, 1995.

[17] S. Sekhavat, P. Švestka, J.-P. Laumond, and M.H. Overmars. Multi-level path planning for nonholonomic robots using semi-holonomic subsystems. Technical Report UU-CS-1996-08, Dept. Comp. Sci., Utrecht Univ., Utrecht, the Netherlands, February 1996.

[18] O.J. Sordalen. Conversion of a car with n trailers into a chained form. In *Proc. IEEE Journal of Robotics and Automation*, pages 1382–13874, 1993.

[19] H.J. Sussmann and W. Liu. Limits of highly oscillatory controls and the approximation of general paths by admissible trajectories. Technical Report SYSCON-91-02, Rutgers Center for Systems and Control, February 1991.

[20] P. Švestka and M.H. Overmars. Motion planning for car-like robots using a probabilistic learning approach. *To appear in Intern. Journal of Rob. Research*, 1995.

[21] P. Švestka and J. Vleugels. Exact motion planning for tractor-trailer robots. In *Proc. IEEE Internat. Conf. on Robotics and Automation*, pages 2445–2450, Nagoya, Japan, 1995.

[22] D. Tilbury, J.-P. Laumond, R. Murray, S. Sastry, and G. Walsh. Steering car-like systems with trailers using sinusoids. In *Proc. 1992 IEEE Int. Conf. Robotics and Automation*, pages 1993–1998, Nice, France, May 1992.

[23] D. Tilbury, R. Murray, and S. Sastry. Trajectory generation for the n-trailer problem using goursat normal form. Technical Report UCB/ERL M93/12, Univ. California, Berkeley, CA, USA, February 1993.

Non-Uniform Discretization Approximations for Kinodynamic Motion Planning and its Applications

John Reif, *Duke University, Durham, NC, USA*

Hongyan Wang, *Duke University, Durham, NC, USA*

The first main result of this paper is a novel non-uniform approximation method for the kinodynamic motion-planning problem. *The kinodynamic motion-planning problem is to compute a collision-free, minimum-time trajectory for a robot whose accelerations and velocities are constrained. Previous approximation methods are all based on a uniform discretization in time space. On the contrary, our method employs a non-uniform discretization in configuration space (thus also a non-uniform one in time space). Compared to the previously best algorithm of Donald and Xavier, the running time of our algorithm reduces in terms of* $1/\varepsilon$*, roughly from* $O((1/\varepsilon)^{6d-1})$ *to* $O((1/\varepsilon)^{4d-2})$*, in computing a trajectory in a d-dimensional configuration space, such that the time length of the trajectory is within a factor of* $(1 + \varepsilon)$ *of the optimal. More importantly, our algorithm is able to take advantage of the obstacle arrangement, and is expected to perform much better than the analytical result in many cases when the obstacles are sparse or when the obstacles are unevenly distributed. This is because our non-uniform discretization has the property that it is coarser in regions that are farther away from all obstacles. So for the above mentioned situations, the total number of discretized grids will be significantly smaller.*

Our second main result is the first known polynomial-time approximation algorithm for the curvature-constrained shortest-path problem in 3 and higher dimensions. *We achieved this by showing that the approximation techniques for the kinodynamic motion-planning problem are applicable to this problem.*

1 Introduction

Nonholonomic motion planning involves planning a collision-free path (or trajectory) for a robot subject to *nonholonomic* constraints on its dynamics. A *holonomic* constraint is one that can be expressed as an equation of the robot's configuration parameters (a placement of a robot with k degrees of freedom can be uniquely specified by k such parameters), while a *nonholonomic* one can only be expressed as a *non-integrable* equation involving also the derivatives of the configuration parameters (see [20] for a more detailed discussion on nonholonomic constraints). Examples of nonholonomic constraints are bounds on velocities, accelerations or curvatures. Although there has been considerable recent work in the robotics literature (see [2, 3, 4, 5, 17, 21, 24, 32] and references therein) on nonholonomic motion-planning problems, relatively little theoretical work has been done on these problems. Nonholonomic motion planning is considerably harder than the holonomic one. For one thing, a robot with k degrees of freedom can not be described completely by k parameters. A complete description has to include the k parameters and their derivatives. The *configuration-space* approach, which is widely used for the holonomic motion-planning problem, does not apply to the problems with nonholonomic constraints, because such constraints are not expressed by the configuration-space representation. On the other hand, these problems bear major significance in robotic engineering. In reality, a robot arm has to move not only in a collision-free fashion, but also in conformation of the dynamic bounds due to limited force or torque from motors.

In this paper, we study two optimal nonholonomic motion-planning problems: the *kinodynamic motion-planning* problem and the *curvature-constrained shortest-path* problem. The kinodynamic motion-planning problem studies the problem of finding minimum-time trajectories for a robot whose motion is governed by Newtonian dynamics and whose acceleration and velocity are bounded. A *trajectory* of a robot with d degrees of freedom is a map Γ: $[0, T] \to \mathbb{R}^d \times \mathbb{R}^d$ given by $\Gamma(t) = (p(t), \dot{p}(t))$, where $p(t)$ and $\dot{p}(t)$ give the location and velocity at time t respectively, in d-dimensional configuration space. $\ddot{p}(t)$ is the acceleration function, which determines a trajectory uniquely, once an initial state is fixed. The constraints on dynamics are given by bounding the norms of the accelerations and velocities. The most studied norms are the L_∞ norm (called the *decoupled* case) and the L_2 norm (called the *coupled* case). In this paper, we study the coupled kinodynamic problem, which happens to be harder than the decoupled one. (The decoupled case is simpler because each dimension is independent of another, and a d-dimensional problem can be reduced to a 1-dimensional one.) We require that $\|\dot{p}(t)\|_2 \leq 1$ and $\|\ddot{p}(t)\|_2 \leq 1$, for $0 \leq t \leq T$.[1]

Given a continuous differentiable path $P : I \to \mathbb{R}^d$ parameterized by arc length $s \in I$, the *average curvature* of P in the interval $[s_1, s_2] \subseteq I$ is defined by $\|\dot{P}(s_1) - \dot{P}(s_2)\|_2/|s_1 - s_2|$. Our curvature constraint requires that the robot's path has an average curvature of at most 1 in every interval. The curvature-constrained shortest-path problem is to compute a shortest collision-free path for the robot such that the path satisfies the curvature constraint.

Our major contribution in this paper is a *non-uniform* discretization approximation method for the kinodynamic motion-planning problem. The discretization is non-uniform in the sense that it is coarser in regions which are farther from all obstacles. The intuition behind doing so is that in regions that are far away from all obstacles, even with a coarse discretization, we are still able to find an approximation trajectory which does not intersect obstacles. The non-uniform discretization on one hand reduces the search

space and the running time, and on the other hand still enables us to obtain a collision-free trajectory whose time length is within a given factor of the optimal time length. Non-uniform discretization is widely used in solving PDEs (see Miller *et al.* [22, 23] and references therein), but research in that area focuses on quite different issues. Applying the idea to kinodynamic motion planning was first suggested by Xavier [33] but without rigorous proof. As it happens, provably good bounds given by us are quite intriguing to obtain.

Our non-uniform discretization is based on a box decomposition of the configuration space. Other geometric algorithms using similar box decompositions include the work of Mitchell *et al.* [25] on ray-shooting problems and that of Hershberger and Suri [14] on 2D shortest-path problem without constraints. Using non-uniform discretization for geometric planning with nonholonomic constraints presents new challenge.

Instead of using a deterministic discretization, Kavraki *et al.* [18, 19] developed a *random sampling technique*, where in preprocessing, grid points (called *milestones*) are chosen *randomly* and are connected by feasible paths to form a network. The method is similar to ours in that both do a preprocessing to obtain a set of valid paths (or trajectories) which are used later in answering planning queries. However, they have to preprocess for each new environment, where in our method, the preprocessing is only done once for some fixed parameters, and can be used repeatedly for different environments.

1.1 Previous work on the kinodynamic motion-planning problem

With the exceptions of one and two dimensional cases (see Ó'Dúnlaing [26] and Canny *et al.* [6]), there are no exact solutions for the kinodynamic motion-planning problem. In fact, as an implication of the result of [7], the problem is at least NP-hard in 3 and higher dimensions. In light of this lower bound, most study has been focused on finding approximation solutions. The earlier approximation algorithms of Sahar and Hollerbach [29] did not guarantee goodness of their solutions. Moreover, the running time was exponential in resolution. Canny *et al.* [10] developed the

---

[1] In previous literature, $\|\dot{p}(t)\|_2 \leq v_{\max}$ and $\|\ddot{p}(t)\|_2 \leq a_{\max}$, where $v_{\max} > 0$ and $a_{\max} > 0$ are arbitrary. However, we can always scale $v_{\max}$ and $a_{\max}$ to 1, by scaling both time and the size of the configuration space; the proof is omitted in this version.

first provably-good, polynomial-time approximation algorithm for the decoupled kinodynamic case. Their work was followed up by a series of work, in which Donald and Xavier [8] improved the running time for the decoupled case, Heinzinger *et al.* [13] and Donald and Xavier [9] investigated the problem for open chain manipulators, and independent work of Donald and Xavier [9] and Reif and Tate [28] gave approximation algorithms for the coupled kinodynamic problem. The best known result for the coupled case, given by [9], is that given an $\varepsilon > 0$, one can compute in time $O(c(d)p(n, \varepsilon, d)L^d(1/\varepsilon)^{6d-1})$, an approximation trajectory whose time length is at most $(1+\varepsilon)$ times the time length of an optimal safe trajectory (roughly speaking, a safe trajectory is one that can be perturbed without intersecting obstacles), where d is the dimension, L is the size of the configuration space where the robot is confined to, n is the number of equations describing the configuration obstacles, $c(d)$ is a function depending soly on d, and $p(n, \varepsilon, d)$ is a lower-order polynomial in n, ε and d.

1.2 Previous work on the curvature-constrained shortest-path problem

Dubins [11] was perhaps the first to study the curvature-constrained shortest paths, who gave a characterization of the shortest paths in 2D in absence of obstacles. Reeds and Shepp [27] extended the obstacle-free characterization to robots that can make reversals. (Boissonnat *et al.* [3] gave an alternative proof for both cases, using ideas from control theory.) In presence of obstacles, Fortune and Wilfong [12] gave a $2^{poly(n,m)}$-time algorithm, where n is the number of vertices and m is the number of bits of precision with which all points are specified; their algorithm only decides whether a path exists, without necessarily finding one. Jacobs and Canny [16] gave an $O((\frac{n+L}{\varepsilon})^2 + (\frac{n+L}{\varepsilon})n^2 \log n)$-time algorithm that computes an approximation path whose length is at most $(1 + \varepsilon)$ times the length of an optimal path, where n is the number of obstacle vertices, and L is the total edge length of the obstacles. This running time was later improved significantly by Wang and Agarwal [31] to $O((n^2/\varepsilon^2) \log n)$. Agarwal *et al.* [1] studied a spe-

cial case when the boundaries of obstacles are also constrained to have curvature of at most 1. There has also been work on computing curvature-constrained paths when the robot is allowed to make reversals [2, 21, 24]. For the curvature-constrained shortest-path problem in 3D, Sussman [30] derived characterizations of the shortest paths in absence of obstacles, using control theory.

1.3 Models and results

Let B be a point robot in a d-dimensional configuration space W. Without loss of generality, we can assume that W is a d-dimensional cube, with each side of length L. A *state* x of B is a pair (LOC(x), VEC(x)), where LOC(x) is a point representing the location of B in the d-dimensional configuration space and VEC(x) is a vector representing the velocity of B. The accelerations and velocities of B are both bounded by 1 in L_2-norm. A trajectory is an $(\bar{a}, \bar{v})$-trajectory if at any time during the trajectory, the acceleration is bounded by $\bar{a}$ and the velocity is bounded by $\bar{v}$, both in L_2-norm. Let Ω be a set of configuration obstacles defined by a total of n constraints, where each constraint can be expressed by a polynomial of constant degree. A trajectory is *collision-free* if its path does not intersect the interior of Ω. A trajectory from a state x to another state y is *optimal*, if it is a collision-free $(1, 1)$-trajectory with a minimum time length, where the minimum is taken over all collision-free $(1, 1)$-trajectories from x to y. The kinodynamic motion-planning problem is to compute such optimal trajectories.

As we have seen that computing exact solutions is hard, we focus on developing fast approximation algorithms. To discuss about approximation solutions, the notion of a safe trajectory needs to be introduced. A point p has a *clearance* of μ if for any point $q \in \Omega$, $\|p - q\|_\infty \geq \mu$. A path is μ-*safe* if for any point p along the path, p has a clearance of μ. A trajectory is μ-*safe* if its associated path is μ-safe. A trajectory is an *optimal* μ-*safe* trajectory from x to y if its time length is the minimum over all μ-safe $(1, 1)$-trajectories from x to y. We will compute approximations to optimal safe trajectories.

The first result of this chapter is summarized in the following theorem.

Theorem 1.1 *Fix two real numbers $l, \varepsilon > 0$. After some pre-computation, we can achieve the following.*

Let W be a d-dimensional configuration space of size L. Let Ω be a set of configuration obstacles defined by a total of n polynomial constraints, each of a constant degree. Given any two states i and f, we can compute a collision-free $(1, 1)$-trajectory from i' to f', such that the time length of the trajectory is at most $(1+\varepsilon)$ times the time length of an optimal $3l$-safe $(1, 1)$-trajectory from i to f. Furthermore, $\|\text{LOC}(i') - \text{LOC}(i)\|_\infty \le \varepsilon l$, $\|\text{VEC}(i') - \text{VEC}(i)\|_\infty \le \varepsilon$, $\|\text{LOC}(f') - \text{LOC}(f)\|_\infty \le \varepsilon l$ and $\|\text{VEC}(f') - \text{VEC}(f)\|_\infty \le \varepsilon$. The running time of our algorithm is $O(nN + N \log N (1/\varepsilon)^{4d-2})$, where $N = O((L/l)^d)$.

Compared to the previously best algorithm of Donald and Xavier, the running time of our algorithm reduces in terms of $1/\varepsilon$, roughly from $O((1/\varepsilon)^{6d-1})$ to $O((1/\varepsilon)^{4d-2})$. More importantly, our algorithm is able to take advantage of the obstacle arrangement, and is expected to perform much better than the analytical result for many cases when the obstacles are sparse or when the obstacles are unevenly distributed. This is because our non-uniform discretization has the property that it is coarser in regions that are farther away from all obstacles. So for the above mentioned situations, the total number of discretized grids is significantly smaller.

By showing that the approximation techniques for the kinodynamic motion planning are applicable to the curvature-constrained shortest-path problem, we are able to obtain the second main result of this paper — the first known polynomial-time approximation algorithm for the curvature-constrained shortest-path problem in 3 and higher dimensions. The detailed model and result are presented in Section 3.

2 Kinodynamic Motion Planning

2.1 Preliminaries

For simplicity, we use $\|\cdot\|$ to denote $\|\cdot\|_2$.

Given a trajectory Γ, let $T(\Gamma)$ be the time length of Γ. Let $p_\Gamma(t), v_\Gamma(t), a_\Gamma(t)$, for $0 \le t \le T(\Gamma)$, be the location, velocity, and acceleration of Γ, respectively, at time t.

Let Π be a path and A a subset of the configuration space. We say that $d(\Pi, A) \le \rho$, for some real number $\rho \ge 0$, if for every point $p \in \Pi$, there is a point $q \in A$ such that $\|p - q\|_\infty \le \rho$, and $d(\Pi, A) = 0$ if $\Pi \in A$. Similarly, for two paths Π and Π', we say that $d(\Pi', \Pi) \le \rho$, if for every point $p \in \Pi'$, there is a point $q \in \Pi$ such that $\|p - q\|_\infty \le \rho$. Let Γ and Γ' be two trajectories. $d(\Gamma', \Gamma) \le \rho$ (resp. $d(\Gamma', A) \le \rho$), if $d(\Pi', \Pi) \le \rho$ (resp. $d(\Pi', A) \le \rho$), where Π' and Π are the paths of Γ' and Γ respectively, and A is a subset of the configuration space.

$x' \in ngb(x, \rho, \nu)$ if $\|\text{LOC}(x') - \text{LOC}(x)\|_\infty \le \rho$ and $\|\text{VEC}(x') - \text{VEC}(x)\|_\infty \le \nu$. Notice that if $x' \in ngb(x, \rho, \nu)$, then $x \in ngb(x', \rho, \nu)$.

The Time-rescaling Lemma of [15] states that the path Π of a $(\bar{a}, \bar{v})$-trajectory Γ, can be traversed by a $(\bar{a}/(1+\varepsilon)^2, \bar{v}/(1+\varepsilon))$-trajectory Γ' in time $(1+\varepsilon) T(\Gamma)$, for any $\varepsilon > 0$.

The following corollary states a result of the TC-graph method in [9]; see Appendix A for a brief description of the method.

Corollary 2.1 *Let W be a d-dimensional configuration space of size L. Let Γ be a $(1, 1)$-trajectory from state i to state f such that Γ lies inside W. Given any $\varepsilon > 0$ and $\rho = O(1)$, applying the TC-graph method with appropriate parameters, we can compute in time $O(L^d (1/\varepsilon)^{6d-1})$, a trajectory Γ' such that Γ' is from some $i' \in ngb(i, \varepsilon\rho/2, \varepsilon/2)$ to some $f' \in ngb(f, \varepsilon\rho/2, \varepsilon/2)$, that $T(\Gamma') \le (1 + \varepsilon) T(\Gamma)$, and that $d(\Gamma', W) \le \varepsilon\rho/2$.*

In the rest of the paper, we fix an $l > 0$ and an $\epsilon > 0$ unless otherwise stated, where l gives the size of the smallest box in the box decomposition and ϵ describes the fineness of discretization.

2.2 Overview of the algorithm

Following the framework of many motion-planning algorithms, our algorithm, termed as CS-graph

(Configuration-State space-graph) method, reduces the planning problem to that of computing and then searching on a graph. The nodes of the graph correspond to a set of non-uniform grid states (see Section 2.3). These grids are non-uniform in the sense that the discretization is coarser in regions that are farther away from all obstacles. Section 2.5 describes how to compute the edge set of the graph. Each edge corresponds to a collision-free, near-optimal trajectory between two nodes. In order to compute the edge set efficiently, we perform pre-computations. We derive a set of canonical trajectories (Section 2.4), which are sufficient for building the graph. These trajectories, once computed, can be used repeatedly for different problem instances with different obstacle arrangements. A technical lemma, the Correcting Lemma, which is essential to many of the proofs in preceding sections, is described in Section 2.6.

2.3   Non-uniform grids

Box decomposition. Let $\tilde{\Xi}(s)$ be a set $\{p = (p_1, \ldots, p_d) \,|\, -s/2 \le p_1, \ldots, p_d \le s/2\}$. $\tilde{\Xi}(s)$ is called a d-dimensional *canonical box* of size s. Each set of $\{p \,|\, p \in \tilde{\Xi}(s),$ and $p_j = -s/2\}$ and $\{p \,|\, p \in \tilde{\Xi}(s),$ and $p_j = s/2\}$, for $1 \le j \le d$, is called a *face* (of size s) of $\tilde{\Xi}(s)$. Ξ is a d-dimensional *box* of size s if it is a region which can be obtained from $\tilde{\Xi}(s)$ by translation, rotation and reflection. Similarly we can define the faces of Ξ. Notice that a face of a d-dimensional box Ξ of size s is really a $(d-1)$-dimensional box of size s. A d-dimensional box has $2d$ faces.

Given a d-dimensional box of size s, we can decompose it into 2^d boxes of size $s/2$. For each of them, we can further decompose it into 2^d boxes of size $s/4$ and so on. We stop further decomposing until certain conditions hold. We refer to this procedure, as well as the collection of *undecomposed* boxes obtained, as a *box decomposition*. Later, when we refer to a box of a decomposition, we mean an undecomposed box.

A box is *free* if it does not intersect the interior of Ω. (We consider the initial and final locations also as obstacle vertices.) A box is *occupied* if it is a subset of Ω. Otherwise a box is called *partially occupied*. A box

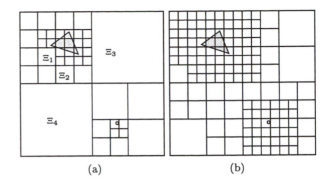

Figure 1: *(a) Box decomposition after the first stage; (b) After the second stage.*

B_1 is *adjacent* to a box B_2 (or B_1 and B_2 are *neighbors*) if B_1 and B_2 share common points. We perform box decomposition on W until the following conditions are satisfied: (C1) The size of a partially occupied box is l, (C2) If a free box is adjacent to a non-free box, or to a free box which has non-free boxes as neighbors, its size is l, and (C3) The decomposition is *balanced*, i.e., a free box can have adjacent free boxes whose sizes are either twice as large or half as small. Figure 1 shows an example of a box decomposition in 2D. The shaded triangle is an obstacle, and the shaded disk represents the initial location. (a) gives a box decomposition only satisfying C1. In (b), boxes like Ξ_1 and Ξ_2 are further decomposed to satisfy C2, and boxes like Ξ_3 and Ξ_4 are further decomposed so that the box decomposition becomes balanced.

Let N be the total number of boxes in the final box decomposition. In a rather straight-forward manner, the box decompsition can be computed in time $O(N \log(L/l))$. Notice that in the worst case, $N = O((L/l)^d)$, but N tends to be much smaller for cases where the obstacles are sparse.

Discretization. Let $\mathcal{B}$ be the box decomposition of W (with respect to Ω) satisfying C1–C3. For each face Δ of a box of $\mathcal{B}$ (recall that Δ is a $(d-1)$-dimensional box), we select $(2/\epsilon)^{d-1}$ uniformly spaced points on Δ, with spacing $\epsilon s/2$, where s is the size of Δ. The set of points obtained in this way is called the *CS-grid points* of $\mathcal{B}$. Notice that the number of grid points on each face is the same, while the spacing of grid points grows linearly with the size of the faces. Thus we obtain a non-uniform discretization of the configuration space.

Let

$$(1) \quad \mathcal{V} = \{v = (j_1\epsilon, \ldots, j_d\epsilon) \mid$$
$$j_1, \ldots, j_d \in \mathbb{Z}, \text{and } \|v\| \leq 1\}.$$

$\mathcal{V}$ is called the set of *grid velocities*. A state a is called a *CS-grid state* of $\mathcal{B}$ if $\text{LOC}(a)$ is a CS-grid point of $\mathcal{B}$ and $\text{VEC}(a) \in \mathcal{V}$. The grid states of $\mathcal{B}$ (including the initial and final state) are the nodes of the graph to be constructed.

For a state a that lies on a face Δ of a box, let $s(a)$ be the function returning the size of Δ. Define $ngb(a) = ngb(a, \epsilon s(a)/4, \epsilon/2)$. By the construction of the grid states, it is easy to show

Lemma 2.2 *If a is a state such that $\text{LOC}(a)$ lies on a face Δ of $\mathcal{B}$, there exists a CS-grid state a' such that $a \in ngb(a')$.*

2.4 Precomputing canonical trajectories

Extended boxes. Given a state a, define the *extended box* of a, $\xi(a)$, to be the smallest region comprised of boxes such that i) the boundary of the union of these boxes form a closed $(d-1)$-dimensional surface; and ii) for any point q that lies on the closed surface, $\|\text{LOC}(a) - q\|_\infty \geq s(a)/2$.

Let Δ be a face of $\mathcal{B}$, of size s, and let Ξ be the box that contains Δ (pick one arbitrarily if both of the two boxes containing Δ have size s). Decompose Δ into 2^{d-1} $(d-1)$-dimensional boxes, each of size $s/2$, and label them $\Delta_1, \ldots, \Delta_{2^{d-1}}$. Given any two states a and b, if both $\text{LOC}(a)$ and $\text{LOC}(b)$ lie on Δ_j, then $\xi(a)$ and $\xi(b)$ are the same. This implies that we can extend the definition of extended boxes for faces. Define $\xi(\Delta_j)$ to be the extended boxes of Δ_j, which is the same as the extended boxes of $\xi(a)$, for any state a such that $\text{LOC}(a)$ lies in Δ_j. $\xi(\Delta_j)$ is called an extended box of *size $s/2$*, since Δ_j is a face of size $s/2$.

An extended box (resp. a box) is said to have a *clearance* of μ, if for any point p that lies in the extended box (resp. the box), p has a clearance of μ. Notice

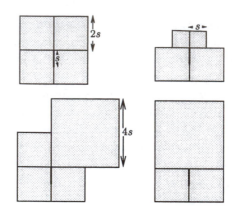

Figure 2: *Some canonical extended boxes of $\tilde{\Delta}(s)$, where $\tilde{\Delta}(s)$ is a line segment drawn in thick lines.*

that a free box has a clearance of s, if all its neighbors are free boxes and have a size of at least s. Using the box decomposition properties C1–C3, we can prove the following two lemmas.

Lemma 2.3 *An extended box of size $s \geq 2l$ has a clearance of at least $s/2$; an extended box of size l has a clearance of at least l.*

Lemma 2.4 *Let $\xi(\Delta_j)$ be an extended box of a face Δ_j of size $l/2$. If there exists a point p on Δ_j such that p has a clearance of $3l$, then the clearance of the extended box is at least l.*

Let $\tilde{\Delta}(s/2) = \{p = (p_1, \ldots, p_d) \mid p_1 = -s/2, \text{and } 0 \leq p_2, \ldots, p_d \leq s/2\}$. $\tilde{\Delta}(s/2)$ is called the *canonical face* of size $s/2$. We can transform Δ_j, Ξ and $\xi(\Delta_j)$, by translation, rotation and reflection, such that Ξ becomes $\tilde{\Xi}(s)$, the canonical box of size s, and Δ_j becomes $\tilde{\Delta}(s/2)$. The transformed extended box $\xi(\Delta_j)$ is called a *canonical extended box* of $\tilde{\Delta}(s/2)$, or a canonical extended box of size $s/2$. Since the box decomposition $\mathcal{B}$ is balanced, the number of canonical extended boxes of a fixed size is a function, depending only on the number of dimensions d, but not on the size, or the location. Let $\xi(s)$ be the set of canonical extended boxes of $\tilde{\Delta}(s)$. We can give an order to $\xi(s)$ and let $\xi(s, j)$ be the jth canonical extended box of $\tilde{\Delta}(s)$, for some valid j. Figure 2 gives some examples of canonical extended boxes in 2D.

Let $\mathcal{P}(s)$ be the set of $(1/\epsilon)^{d-1}$ uniformly spaced points on $\tilde{\Delta}(s)$, with spacing ϵs. Define $\mathcal{Q}(s,j)$ as follows. If Δ is a face of size s' belonging to the boundary of $\xi(s,j)$, include the set of $(2/\epsilon)^{d-1}$ uniformly spaced points with spacing $\epsilon s'/2$ on Δ in $\mathcal{Q}(s,j)$. Let $\mathcal{I}(s) = \mathcal{P}(s) \times \mathcal{V}$, and $\mathcal{F}(s,j) = \mathcal{Q}(s,j) \times \mathcal{V}$. A state in the set $\mathcal{I}(s)$ is called a *canonical starting state*, and a state in the set $\mathcal{F}(s,j)$ is called a *canonical ending state* of $\xi(s,j)$. It is easy to see that $|\mathcal{I}(s)| = O((1/\epsilon)^{2d-1})$. It also holds that $|\mathcal{F}(s,j)| = O(d(1/\epsilon)^{2d-1})$. This is because of the balanced property of the box decomposition. The boundary of $\xi(s,j)$ can contain at most $O(d)$ faces, each contributing $O((1/\epsilon)^{2d-1})$ states.

Computing connection tables. Let $\xi(s,j)$ be the jth canonical extended box of $\tilde{\Delta}(s)$, for some valid s and j. Let $CT(s,j)$ be the *connection table* of $\xi(s,j)$. The connection table contains pre-computed trajectories which connect canonical starting states and canonical ending states. In this section we describe how to compute the connection tables.

First we introduce our Correcting Lemma (whose proof can be found in Section 2.6), which states a result essential to the proofs of other lemmas.

Lemma 2.5 (Correcting Lemma) *Fix a constant* $c \geq 1$*, and let* $\rho_c = \bar{v}^2/(c\bar{a})$*, where* $\bar{a}, \bar{v} > 0$ *are arbitrary. Let* Γ *be an* $(\bar{a}, \bar{v})$*-trajectory from* i *to* f*. Given any* $\rho > 0$ *and* $\varepsilon > 0$*, and given any two states* g *and* h*, if* $\rho > \rho_c$*,* $\|\text{LOC}(f) - \text{LOC}(i)\|_\infty \geq \rho$*,* $g \in ngb(i, \varepsilon\rho/2, \varepsilon\bar{v}/2)$*, and* $h \in ngb(f, \varepsilon\rho/2, \varepsilon\bar{v}/2)$*, we can construct a trajectory* Γ' *from* g *to* h*, by correcting* Γ*, such that* Γ' *satisfies the following properties:* *(P1)* $T(\Gamma') = T(\Gamma)$*, (P2)* Γ' *is a* $((1+4c\sqrt{d}\varepsilon)^2\bar{a}, (1+4c\sqrt{d}\varepsilon)\bar{v})$*-trajectory, and (P3)* $d(\Gamma', \Gamma) \leq (17/16)\varepsilon\rho$*.*

Here onwards, we fix $c = \max(2/l, 1)$, where l is the size of the smallest box. In our later application of this lemma, it always holds that $\bar{v}^2/\bar{a} = 1$. Thus $\rho_c = 1/c$. This implies that $l/2 \geq \rho_c$. Also let $e = 4c\sqrt{d}$.

Consider a pair of states (i, f) that lie on the boundary of an extended box ζ. A $(1,1)$-trajectory from i to f is called *legal* if its path lies inside ζ. The pair (i, f) is *legal* if there exists at least a legal trajectory from i to f. We will see later that these legal trajectories

are the potential trajectories that we need to approximate. We call a pair of grid state (i, f) *good* if there exists at least a legal pair (g, h) such that $g \in ngb(i)$ and $h \in ngb(f)$. Roughly speaking, the next lemma (Existing Lemma) states that if (i, f) is a good pair, then there *exists* a trajectory Γ from i to f such that Γ approximates *any* legal trajectories for any legal pair (g, h), where $g \in ngb(i)$ and $h \in ngb(f)$. Thus for our purpose of approximation, it is enough to compute only the trajectories for good pairs. For a pair of legal states (a, b), let $\hat{T}(a, b)$ be the time length of the legal trajectory from a to b whose time length is the smallest among all legal trajectories from a to b.

Lemma 2.6 (Existing Lemma) *Let* $\xi(s,j)$ *be the* j*th canonical extended box of* $\tilde{\Delta}(s)$*, for some valid* s *and* j*. For any* $i \in \mathcal{I}(s)$ *and* $f \in \mathcal{F}(s,j)$*, if* (i, f) *is good, then there exists a trajectory* Γ *from* i *to* f*, such that the trajectory satisfies the following properties:* *(P1) For any* $i' \in ngb(i)$*, and* $f' \in ngb(f)$*, if* (i', f') *is legal,* $T(\Gamma) \leq \hat{T}(i', f')$*, (P2)* Γ *is a* $((1+e\epsilon)^2, (1+e\epsilon))$*-trajectory, and (P3)* $d(\Gamma, \xi(s,j)) \leq (17/8)\epsilon s$*.*

Proof: Let $g \in ngb(i)$ and $h \in ngb(f)$ be such that (g, h) is legal and that $\hat{T}(g, h)$ is the smallest among those of all such legal pairs; let Γ' be the $(1, 1)$-trajectory from g to h whose time length is $\hat{T}(g, h)$. We will show the existence of a trajectory satisfying P1–P3, by constructing one from Γ'. There are three cases depending on whether $s(f)$ (the size of the face where f lies) is s, $2s$ or $4s$. We will prove for the case when $s(f) = 4s$; this case gives the worst bounds. The other two cases can be handled in a similar way.

Let Γ be the trajectory obtained by correcting Γ'. Let $\bar{a} = 1$, $\bar{v} = 1$, and $\rho = 2s$ in the Correcting Lemma. Since $s \geq l/2$, $\rho = 2s \geq l > \rho_c$. We can show that $\|\text{LOC}(h) - \text{LOC}(g)\|_\infty \geq 2s = \rho$. Since $g \in ngb(i)$, $g \in ngb(i, \epsilon s(i)/4, \epsilon/2)$. Since $\rho = 2s = s(i)$, it is also true that $g \in ngb(i, \epsilon\rho/2, \epsilon/2)$. This implies that $i \in ngb(g, \epsilon\rho/2, \epsilon/2)$. Similarly, $f \in ngb(h, \epsilon\rho/2, \epsilon/2)$. Thus the conditions of the Correcting Lemma are satisfied. This implies that $T(\Gamma) \leq \hat{T}(g, h)$, satisfying P1. Γ is a $((1 + e\epsilon)^2, (1 + e\epsilon))$-trajectory, and $d(\Gamma, \Gamma') \leq (17/16)\epsilon\rho = (17/8)\epsilon s$. Since $d(\Gamma', \xi(s,j)) = 0$, $d(\Gamma, \xi(s,j)) \leq (17/8)\epsilon s$, proving P3. $\qquad \square$

For a good pair (i, f) (with respect to $\xi(s, j)$), our goal is to approximate (since we do not know how to compute exactly) such a trajectory Γ as stated in the Existing Lemma. The approximation is done in two steps. First, we apply the TC-graph method to compute a $((1+e\epsilon)^2, (1+e\epsilon))$-trajectory Γ' which is close to Γ time wise and spatial wise. However, the TC-graph method does not guarantee that Γ' is from i to f. Instead, we only know that Γ' is from some i' close to i to some f' close to f. Second, we correct Γ' to obtain a trajectory Γ'' which is really from i to f. By the Correcting Lemma, Γ'' approximates Γ', and thus approximates Γ. Also notice that Γ'' is a $((1+e\epsilon)^4, (1+e\epsilon)^2)$-trajectory. By using Corollary 2.1 with appropriate parameters, and the Correcting Lemma, we obtain

Lemma 2.7 (Loose Tracking Lemma) *Let $\xi(s, j)$ be the jth canonical extended box of $\tilde{\Delta}(s)$, for some valid s and j. Let $i \in \mathcal{I}(s)$ and $f \in \mathcal{F}(s, j)$. If (i, f) is good, then Γ'' computed as above satisfies the following properties: (P1) For any $g \in ngb(i)$ and $h \in ngb(f)$, if (g, h) is legal, $T(\Gamma'') \leq (1 + \epsilon) \hat{T}(g, h)$, (P2) Γ'' is a $((1 + e\epsilon)^4, (1 + e\epsilon)^2)$-trajectory, and (P3) $d(\Gamma'', \xi(s, j)) \leq 5\epsilon s$.*

Fix a canonical extended box $\xi(s, j)$, for some valid s and j. For each $i \in \mathcal{I}(s)$ and $f \in \mathcal{F}(s, j)$, we precompute a trajectory from i to f as described above. We store the trajectory (i.e., the initial state i, the final state f, the acceleration function, and the time length) in the connection table $CT(s, j)$. A connection table computed in this way satisfies the following property:

Lemma 2.8 (Connection Table Property) *Let $\xi(s, j)$ be the jth canonical extended box of $\tilde{\Delta}(s)$, for some valid s and j, and let $CT(s, j)$ be the connection table for $\xi(s, j)$. For any canonical starting state $i \in \mathcal{I}(s)$ and any canonical ending state $f \in \mathcal{F}(s, j)$, if (i, f) is good, then $CT(s, j)$ contains a trajectory Γ from i to f and Γ satisfies the following properties: (P1) For any $g \in ngb(i)$, and $h \in ngb(f)$, if (g, h) is good, $T(\Gamma) \leq (1 + \epsilon) \hat{T}(g, h)$, (P2) Γ is a $((1+e\epsilon)^4, (1+e\epsilon)^2)$-trajectory, and (P3) $d(\Gamma, \xi(s, j)) \leq 5\epsilon s$.*

Let $\xi(l/2, 0)$ be the canonical extended box that consists of 4 boxes of size l. Apart from $CT(l/2, 0)$, we

maintain a special connection table CT_0 for this canonical extended box. For each $i \in \mathcal{I}(l/2)$, we construct a TC-graph G rooted at i. We add a trajectory Γ from i to f in CT_0, if 1) f is a node of G, 2) f either lies inside $\xi(l/2, 0)$, or its L_∞ distance to the extended box is no more than $\epsilon l/2$, 3) there is a path in G from i to f (whose corresponding trajectory is Γ). By the property of the TC-graph method, we can show the following

Lemma 2.9 *Let $i \in \mathcal{I}(l/2)$. For any pair of states g and h, such that $g \in ngb(i, \epsilon l/2, \epsilon/2)$, that h lies inside $\xi(l/2, 0)$, and that there exists a $(1, 1)$-trajectory from g to h lying inside $\xi(l/2, 0)$, there exists a $(1, 1)$-trajectory in CT_0 from i to some f, such that $f \in ngb(h, \epsilon l/2, \epsilon/2)$, and that the time length of this trajectory is no more than $(1 + \epsilon)$ times the time length of the optimal $(1, 1)$-trajectory from g to h.*

The details of computing the connection tables are omitted in this version. By constructing a TC-graph and doing single-source-multiple-sink shortest-path search on the graph, it can be shown that computing a connection table $CT(s, j)$ can be done in $O(s^d (1/\epsilon)^{8d-2})$ time and space.

2.5 Approximation in presence of obstacles

The algorithm. In this section we present our approximation algorithm for computing collision-free near-optimal trajectories in presence of obstacles. Let W be a d-dimensional configuration space of size L, with a set Ω of obstacles. Let i and f be the initial and final states respectively. And let $\mathcal{B}$ be the box decomposition of W satisfying C1–C3. Our approximation algorithm constructs a weighted directed graph $G = (V, E)$, where V includes i, f and a subset of CS-grid states induced by $\mathcal{B}$. A CS-grid state a is included in V if i) $s(a) \geq 2l$; or ii) $s(a) = l$ and the clearance of $\xi(a)$ is at least l.

Let a and b be two states that lie within an extended box ζ. To see if there is a pre-computed trajectory from a to b, we transform ζ to its canonical form, and also a and b with it. If $\mathcal{T}$ is the transformation, we say that there is a pre-computed trajectory from a to b if there is a trajectory from $\mathcal{T} \circ a$ to $\mathcal{T} \circ b$ in the connection table of $\mathcal{T} \circ \zeta$. If Γ is the trajectory from $\mathcal{T} \circ a$ to

$\mathcal{T} \circ b$, then $\mathcal{T}^{-1} \circ \Gamma$ is the trajectory from a to b. It is easy to see that $\mathcal{T}^{-1} \circ \Gamma$ satisfies the connection table properties P1–P3, with respect to ζ.

We construct the edge set in the following way. For node i, add an edge from i to another node a, if i) i lies within $\xi(a)$; and ii) there is a pre-computed trajectory from a^- to i^-.[2] Similarly, for node f, add an edge from a node a to f, if i) f lies within $\xi(a)$; and ii) there is a pre-computed trajectory from a to f. For any other node a, add an edge from a to a another node b, if i) b lies on the boundary of $\xi(a)$; and ii) there is a pre-computed trajectory from a to b. The weight of each edge is equal to the time length of its corresponding trajectory.

Theorem 2.10 (Safe Tracking Theorem) *If there exists an optimal $3l$-safe $(1,1)$-trajectory Γ from state i to state f, then the graph G, constructed as above, contains a path whose corresponding trajectory Γ' satisfies the following properties: (P1) $T(\Gamma') \leq (1 + \epsilon) T(\Gamma)$, (P2) Γ' is a $((1 + e\epsilon)^4, (1 + e\epsilon)^2)$-trajectory, (P3) Γ' does not intersect the interior of Ω, and (P4) Γ' is from some $i' \in ngb(i, \epsilon l/2, \epsilon/2)$ to some $f' \in ngb(f, \epsilon l/2, \epsilon/2)$.*

Proof: Let Γ be divided into segments of trajectories, $\Gamma_{a_0 a_1} \| \Gamma_{a_1 a_2} \| \dots \| \Gamma_{a_{m-1} a_m}$, such that $a_0 = i$, $a_m = f$, and each a_j is the state where Γ first exits $\xi(a_{j-1})$, for $1 \leq j \leq m - 1$. Figure 3 gives an example in 2D. The optimal $(1,1)$-trajectory from i to f (drawn in thick curve) is divided into 10 segments (each dark circle represents an a_j, for $1 \leq j < 10$). Some of the extended boxes are shown as shaded regions. We consider the following parts, $\Gamma_{a_0 a_1}$, $\Gamma_{a_1 a_2} \| \dots \| \Gamma_{a_{m-2} a_{m-1}}$ and $\Gamma_{a_{m-1} a_m}$ separately.

Since each a_j, for $1 \leq j \leq m - 1$, is on some face of $\mathcal{B}$, by Lemma 2.2, we can find CS-grid state b_j, such that $a_j \in ngb(b_j)$. If $s(a_j) \geq 2l$, then $s(b_j) \geq 2l$ and b_j is in the node set V. Otherwise, since $\text{LOC}(a_j)$ has a clearance of $3l$, $\xi(a_j)$ has a clearance of at least l. Since $\xi(b_j) = \xi(a_j)$, $\xi(b_j)$ also has a clearance of at least l and b_j is in the node set V. We will show that the

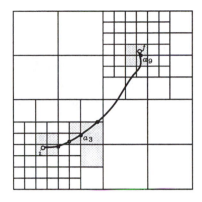

Figure 3: $\xi(a_0)$, $\xi(a_3)$, *and* $\xi(a_9)$.

graph path $(b_1, b_2) \| \dots \| (b_{m-2}, b_{m-1})$ corresponds to a trajectory satisfying P1–P3. To this end, we show that there is an edge from b_{j-1} to b_j whose corresponding trajectory $\Gamma_{b_{j-1} b_j}$ satisfies P2, P3, and $T(\Gamma_{b_{j-1} b_j}) \leq (1 + \epsilon) T(\Gamma_{a_{j-1} a_j})$, for $1 < j < m$.

By the way Γ is divided, $d(\Gamma_{a_{j-1} a_j}, \xi(a_{j-1})) = 0$. Thus each pair (a_{j-1}, a_j) is legal. This implies that each (b_{j-1}, b_j) is good. By the Connection Table Property, there is a pre-computed trajectory $\Gamma_{b_{j-1} b_j}$ from b_{j-1} to b_j. Thus an edge from b_{j-1} to b_j is added in the construction of G. Also by the Connection Table Property, $\Gamma_{b_{j-1} b_j}$ is a $((1 + e\epsilon)^4, (1 + e\epsilon)^2)$-trajectory, and its time length is no more than $(1 + \epsilon) T(\Gamma_{a_{j-1} a_j})$. Next we need to show that $\Gamma_{b_{j-1} b_j}$ is collision-free.

Let s be the size of $\xi(b_{j-1})$. We consider the three cases $s \geq 2l$, $s = l$ and $s = l/2$ separately. If $s \geq 2l$ (resp. $s = l$), the extended box $\xi(b_{j-1})$ has a clearance of $s/2$ (resp. l). On the other hand, $d(\Gamma_{b_{j-1} b_j}, \xi(b_{j-1})) \leq 5\epsilon s$. Thus if $\epsilon \leq 1/10$ is chosen small enough, $\Gamma_{b_{j-1} b_j}$ is collision-free. When $s = l/2$, $\xi(b_{j-1})$ has a clearance of at least l. This is because a_{j-1} is $3l$-safe. By Lemma 2.3, $\xi(a_{j-1})$ has a clearance of at least l. This implies that $\xi(b_{j-1})$ has a clearance of at least l, since $\xi(b_{j-1}) = \xi(a_{j-1})$. Since $d(\Gamma_{b_{j-1} b_j}, \xi(b_{j-1})) \leq (5/2)\epsilon l$, $\Gamma_{b_{j-1} b_j}$ is collision-free if ϵ is chosen small enough.

Now consider the part $\Gamma_{a_{m-1} a_m}$, which lies inside $\xi(a_{m-1})$. When constructing $\mathcal{B}$, $\text{LOC}(f)$ is considered as an obstacle vertex. Thus the box containing $\text{LOC}(f)$ and its neighbors all have size l. This

---

[2]For a state a, we use a^- to denote a state of $(\text{LOC}(a), -\text{VEC}(a))$.

means that $\xi(a_{m-1})$ is an extended box consisting of 4 boxes of size l. By Lemma 2.9, there exists in CT_0 a pre-computed trajectory $\Gamma_{b_{m-1}f}$ from b_{m-1} to some $f' \in ngb(f, \epsilon l/2, \epsilon/2)$. Thus an edge from b_{m-1} to f is added in the construction of G. Also $\Gamma_{b_{m-1}f}$ is a $(1,1)$-trajectory whose time length is no more than $(1 + \epsilon)\Gamma_{a_{m-1}a_m}$. Since $\text{LOC}(a_{m-1})$ has a clearance of $3l$, the clearance of $\xi(a_{m-1})$ is at least l. Thus $\Gamma_{b_{m-1}f}$ is also collision-free, since $d(\Gamma_{b_{m-1}f}, \xi(b_{m-1})) \leq \epsilon l/2$. We can show similar results for $\Gamma_{a_0 a_1}$. This completes the proof of this theorem. $\square$

Thus the approximation problem is transformed to one of searching for a shortest path on the graph G. Notice that in order for this algorithm to be correct, each edge introduced in G must correspond to a collision-free trajectory. This is guaranteed by the way we choose the node set V, and the connection table properties; the proof of this is similar to the one showing that each $\Gamma_{b_{j-1}b_j}$ is collision-free, used in the above theorem. Thus we do not have to explicitly check whether each trajectory segment is collision-free. The obtained shortest path corresponds to a $((1 + e\epsilon)^4, (1 + e\epsilon)^2)$-trajectory. Applying the Time-rescaling Lemma with a scaling factor of $(1 + e\epsilon)^2$, we can obtain the following

Corollary 2.11 *Let W be a d-dimensional configuration space of size L, and Ω a set of obstacles. Let Γ be an optimal $3l$-safe $(1,1)$-trajectory from a state i to a state f. We can compute a $(1,1)$-trajectory Γ' from some $i' \in ngb(i, \epsilon l/2, 3e\epsilon)$ to some $f' \in ngb(f, \epsilon l/2, 3e\epsilon)$ such that $T(\Gamma')$ is at most $(1 + 3e\epsilon)$ times $T(\Gamma)$.*

The time complexity. The running time of the algorithm consists of the following components.

(1) Time to generate the graph nodes (i.e., the grid states). If N is total number of boxes in the final decomposition, the time to perform the decomposition is $O(nN + N \log N)$. Since each box contributes at most $O((1/\epsilon)^{2d-1})$ grid states, the time to generate the grid states, after a decomposition, is $O(N(1/\epsilon)^{2d-1})$.

(2) Time to compute the graph edges. Since each node is connected to at most $O((1/\epsilon)^{2d-1})$ other nodes, the total number of edges is $O(N(1/\epsilon)^{4d-2})$. This

bounds the time to compute the edges, since it takes $O(1)$ time to compute an edge.

(3) Time to search for a shortest graph path. Using Dijkstra algorithm with the priority queue implemented with a binary heap, this time is $O((|V| + |E|) \log |V|)$, where $|V|$ are $|E|$ are the number of nodes and edges respectively. Plugging in our numbers, the time for searching is roughly $O(N \log N(1/\epsilon)^{4d-2})$.

(4) Time to rescale the obtained trajectory to a $(1,1)$-trajectory, which is $O(1)$.

Combining Corollary 2.11 and the time-complexity analysis, and choosing small enough $\epsilon = O(\varepsilon)$, we obtain Theorem 1.1.

2.6   Correcting a trajectory

Let Γ be an $(\bar{a}, \bar{v})$-trajectory from state i to state f. Given another pair of states g and h, we can construct a trajectory Γ' from g to h, by correcting Γ. Furthermore, we show that if g and h are close to i and f respectively, the correction is small.

For simplicity of illustration and analysis, we first look at the one dimensional case. Let $\delta a(t)$ be the corrective acceleration, i.e., $a_{\Gamma'}(t) = a_{\Gamma}(t) + \delta a(t)$, for $0 \leq t \leq T(\Gamma)$. Let $\delta v(t) = v_{\Gamma'}(t) - v_{\Gamma}(t)$, and $\delta p(t) = p_{\Gamma'}(t) - p_{\Gamma}(t)$. Also let $\Delta v_i = v_{\Gamma'}(0) - v_{\Gamma}(0) = \text{VEC}(g) - \text{VEC}(i)$, $\Delta v_f = v_{\Gamma'}(T) - v_{\Gamma}(T) = \text{VEC}(h) - \text{VEC}(f)$, $\Delta p_i = p_{\Gamma'}(0) - p_{\Gamma}(0) = \text{LOC}(g) - \text{LOC}(i)$, and $\Delta p_f = p_{\Gamma'}(T) - p_{\Gamma}(T) = \text{LOC}(h) - \text{LOC}(f)$. It can be shown that $\delta v(t) = \Delta v_i + \int_0^t \delta a(\mu) \, d\mu$ and

$$\delta p(t) = \Delta p_i + \Delta v_i t + \int_0^t \int_0^\mu \delta a(\nu) \, d\nu \, d\mu.$$ Our object

is to find $\delta a(t)$, such that $\Delta v_i + \int_0^T \delta a(\mu) \, d\mu = \Delta v_f$

and $\Delta p_i + \Delta v_i T + \int_0^T \int_0^\mu \delta a(\nu) \, d\nu \, d\mu = \Delta p_f$, where $T = T(\Gamma)$. We also want to keep the absolute values of $\delta v(t)$, $\delta a(t)$ and $\delta p(t)$ small.

Fix a constant $c \geq 1$ and let $\rho_c = \bar{v}^2/(c\bar{a})$. Assume that $|\text{LOC}(f) - \text{LOC}(i)| \geq \rho \geq \rho_c$, for some ρ, and let $\phi = \rho/\bar{v}$. Thus $T(\Gamma) \geq \phi$. Our correcting scheme is illustrated in Figure 4. Basically, we correct in three

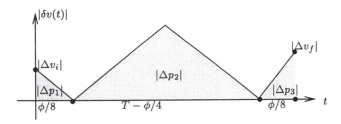

Figure 4: *The correcting scheme.*

phases. In the first phase, we use a constant corrective acceleration to make $\delta v(t)$ become 0, while in the last phase, we use a constant corrective acceleration to make $\delta v(t)$ become Δv_f. The middle phase is used to correct the distance. Notice that at the beginning and the end of the second phase, $\delta v(t) = 0$.

The time length of the first and the last phase is $\phi/8$. Let Δa_1 be the constant corrective acceleration used in the first phase. Thus $\Delta a_1 = -\frac{\Delta v_i}{\phi/8} = -8\Delta v_i/\phi$. If Δp_1 is the distance covered in this phase, then $\Delta p_1 = \Delta v_i \phi/16$. In this phase, $|\delta v(t)|$ is getting smaller and is upper bounded by $|\Delta v_i|$, while $|\delta p(t)|$ is upper bounded by $|\Delta p_i| + |\Delta p_1|$. Let Δa_3 be the constant corrective acceleration used in the last phase. Similarly, we have $\Delta a_3 = 8\Delta v_f/\phi$. $\Delta p_3 = \Delta v_f \phi/16$ is the distance covered in this phase. In this phase, $|\delta v(t)|$ is upper bounded by $|\Delta v_f|$. At the beginning of this phase, $\delta p(t) = \Delta p_f - \Delta p_3$. Thus $|\delta p(t)|$ is upper bounded $|\Delta p_f| + |\Delta p_3|$ during this phase.

Let $T' = T(\Gamma) - \phi/4$ be the time spent in the second phase. This phase is divided into two sub-phases of equal length. The corrective accelerations used in these two sub-phases have the same absolute value, but opposite directions. The effect is that at the end of this phase, $\delta v(t)$ becomes 0 again. If Δp_2 is the distance covered in this phase, $\Delta p_2 = \Delta p_f - \Delta p_i - \Delta p_1 - \Delta p_3$. Notice that $|\Delta p_2|$ and $\delta p(t)$ are both upper bounded by $|\Delta p_i| + |\Delta p_1| + |\Delta p_3| + |\Delta p_f|$. Let Δa_2 be the corrective acceleration used in the first sub-phase. Then $\Delta a_2 = 4\Delta p_2/(T')^2$. In this phase, $|\delta v(t)|$ is upper bounded by $2|\Delta p_2|/T'$.

Lemma 2.12 *For any $\varepsilon > 0$, and any two states g and h, if $g \in ngb(i, \varepsilon\rho/2, \varepsilon\bar{v}/2)$, and $h \in ngb(f, \varepsilon\rho/2, \varepsilon\bar{v}/2)$,*

where ρ is defined as above, then the trajectory Γ', constructed as above, satisfies the following properties: (P1) $T(\Gamma') = T(\Gamma)$, (P2) Γ' is a $((1 + 4c\varepsilon)^2 \bar{a}, (1 + 4c\varepsilon)\bar{v})$-trajectory, and (P3) $d(\Gamma', \Gamma) \le (17/16)\varepsilon\rho$.

Proof: P1 holds obviously. By simple calculations, we can show that, during the whole time period $T(\Gamma)$, the maximum difference in position is upper bounded by

$$|\Delta p_i| + |\Delta p_1| + |\Delta p_3| + |\Delta p_f|$$
$$\le \frac{1}{2}\varepsilon\rho + \frac{|\Delta v_i|\phi}{16} + \frac{|\Delta v_f|\phi}{16} + \frac{1}{2}\varepsilon\rho$$
$$\le \frac{17}{16}\varepsilon\rho,$$

proving P3. Also the maximum difference in velocity is upper bounded by

$$\max(|\Delta v_i|, |\Delta v_f|, \frac{2|\Delta p_2|}{T'}) \le \max(\frac{\varepsilon\bar{v}}{2}, \frac{2(17/16)\varepsilon\rho}{(1 - 1/4)\phi})$$
$$\le 4c\varepsilon\bar{v},$$

and the maximum difference in acceleration is upper bounded by

$$\max(|\Delta a_1|, |\Delta a_2|, |\Delta a_3|) \le \max(\frac{8|\Delta v_i|}{\phi}, \frac{4|\Delta p_2|}{(T')^2}, \frac{8|\Delta v_f|}{\phi})$$
$$= 8\varepsilon\bar{v}^2/\rho \le 8\varepsilon\bar{v}^2/\rho_c$$
$$\le 8\varepsilon c\bar{a}.$$

Thus the velocity of Γ' is upper bounded by $(1 + 4c\varepsilon)\bar{v}$ and the acceleration is bounded by $(1 + 4c\varepsilon)^2\bar{a}$, proving P2. $\square$

In higher dimensions, we can add corrective accelerations for each dimension separately. Since the time length of the trajectory is not changed, the corrections can be carried out simultaneously without affecting each other. Let $v_\Gamma^j(t)$ denote the velocity profile of trajectory Γ in the jth dimension, for $1 \le j \le d$. By the above lemma, we have $|v_{\Gamma'}^j(t)| \le 4c\varepsilon\bar{v} + |v_\Gamma^j(t)|$. By triangle inequality,

$$\|v_{\Gamma'}(t)\| \le \|v_\Gamma(t)\| + \|v_{\Gamma'}(t) - v_\Gamma(t)\|$$
$$\le \bar{v} + \sqrt{\sum_{j=1}^{d}(v_{\Gamma'}^j(t) - v_\Gamma^j(t))^2}$$
$$= (1 + 4c\sqrt{d}\varepsilon)\bar{v}.$$

Similarly for the acceleration, we can obtain that $\|a_{\Gamma'}(t)\| \leq (1 + 4c\sqrt{d}\varepsilon)^2 \bar{a}$. Combining everything together, we have proved the Correcting Lemma.

3 Curvature-Constrained Shortest Paths

3.1 Introduction

In the curvature-constrained shortest-path problem, a *state* X of the robot B is a pair $(\text{LOC}(X), \text{VEC}(X))$, where $\text{LOC}(X)$ is a point representing the location of B in the d-dimensional configuration space, and $\text{VEC}(X)$ is a *unit* vector representing its orientation (i.e., $\|\text{VEC}(X)\| = 1$). Let Ω be a set of configuration obstacles. A path is *collision-free* if it does not intersect the interior of Ω. Given an initial state I and a final state F, the curvature-constrained shortest-path problem is to compute a path from I to F whose path length is the minimum among all the paths from I to F which are collision-free and which have a maximum average curvature of 1.

It has been observed in [12] that the curvature-constrained shortest-path problem is a restricted case of the kinodynamic motion-planning problem, with the L_2 norms of the velocities fixed to be 1. However, if we require that the velocities of the approximation trajectories be fixed in L_2 norm, the techniques developed so far for the general kinodyanmic case can not be applied to this restricted case. As pointed out in [9], a necessary condition for the techniques to apply is that the set of *feasible instantaneous accelerations* (accelerations that can be applied without violating the dynamic constraints) spans d dimensions. On the other hand, if the velocities are fixed in L_2 norm, the set of instantaneous accelerations spans only $d - 1$ dimensions, because they have to be perpendicular to the instantaneous velocity.

Our contribution is that we look at the curvature-constrained shortest-path problem from a different view, which enables us to overcome the difficulty mentioned above. Basically, instead of requiring that the L_2 norms of the velocities be fixed to be 1, we only force them to fall within a small range close to 1. In this way, we are able to obtain a path whose maximum curvature is slightly larger than but can be arbitrarily close to 1.

In this section, we use lower-case letters to denote states as defined for the kinodynamic case, but upper-case letters (usually X, Y, U, V) to denote states whose velocities are 1 in L_2 norm (i.e., $\|\text{VEC}(X)\| = \|\text{VEC}(Y)\| = \|\text{VEC}(U)\| = \|\text{VEC}(V)\| = 1$). Given a path Π, let $L(\Pi)$ be its path length. A path is called a *c-constrained* path if its average curvature is at most c. A path is an *optimal* 1-constrained μ-safe path from state I to state F, if its path length is the minimum among all the 1-constrained μ-safe paths from I to F (see Section 1.3 for the definition of a safe path). We are interested in computing approximations to such optimal paths.

3.2 Approximating in absence of obstacles

The following lemma states a result similar to Corollary 2.1, but for the curvature-constrained case. This lemma enables us to apply the method developed in the previous sections to the curvature-constrained shortest-path problem.

Lemma 3.1 *Let W be a d-dimensional configuration space of size L. Let Π be a 1-constrained path from state X to state Y and lying inside W. Given any $\varepsilon > 0$ and $\rho = O(1)$, we can compute in time $O(L^d (1/\varepsilon)^{6d-1})$ a path Π', such that Π' satisfies the following properties: (P1) Π' is $(1 + \varepsilon)$-constrained, (P2) $L(\Pi') \leq (1 + \varepsilon)L(\Pi)$, (P3) $d(\Pi', W) \leq \varepsilon\rho/2$, and (P4) Π' is from some state $X' \in ngb(X, \varepsilon\rho/2, \varepsilon/2)$ to some state $Y' \in ngb(Y, \varepsilon\rho/2, \varepsilon/2)$.*

Proof: Let $\delta = \varepsilon/8$, $\eta_x = \varepsilon\rho/2$, and $\eta_v = \varepsilon/(8\sqrt{d})$.

We consider Π to be the path of a trajectory Γ, such that Γ is from $i = (\text{LOC}(X), \text{VEC}(X)/(1 + \delta))$ to $f = (\text{LOC}(Y), \text{VEC}(Y)/(1 + \delta))$, and that $\|v_\Gamma(t)\| = 1/(1 + \delta)$. Notice that $T(\Gamma) = (1 + \delta) L(\Pi)$. Since Π is a 1-constrained path, and at any time t, the curvature of Π is given by $\|a_\Gamma(t)\|/\|v_\Gamma(t)\|^2$, the acceleration of Γ is bounded by $\|a_\Gamma(t)\| \leq 1 \cdot \|v_\Gamma(t)\|^2 \leq 1/(1 + \delta)^2$. This implies that Γ is a $(1/(1 + \delta)^2, 1/(1 + \delta))$-trajectory. Let $\mathcal{A}_\delta$ be the set of accelerations whose L_2 norms are bounded by $1/(1 + \delta)^2$. This is the set of accelerations used by Γ.

Let $\mu = \kappa_l = \delta/(4\sqrt{d})$, and $\mathcal{A}_\mu$ the set of accelerations as defined in Appendix A (see (2)). Thus $\mathcal{A}_\mu$ has a uniform advantage of κ_l over $\mathcal{A}_\delta$. By the Tracking Lemma (see Appendix A), there exists a $\tau = O(\varepsilon)$ (satisfying (1)), and a τ-bang trajectory Γ' using $\mathcal{A}_\mu$, such that $T(\Gamma') = T(\Gamma)$, and that Γ' tracks Γ to a tolerance of (η_x, η_v). This implies that $\|v_{\Gamma'}(t) - v_\Gamma(t)\|_\infty \le \eta_v$. By simple calculations, we can show that

$$1 - \frac{\varepsilon}{4} \le \|v_{\Gamma'}(t)\| \le 1 + \frac{\varepsilon}{8}.$$

Let Π' be the path of trajectory Γ'. We will show that Π' satisfies P1–P4. Since $\|a_{\Gamma'}(t)\| \le 1$, this bounds the maximum curvature of Π' to be at most

$$\frac{1}{(1 - \varepsilon/4)^2} \le 1 + \varepsilon,$$

proving P1. Since $T(\Gamma') \le T(\Gamma) = (1 + \delta)L(\Pi)$, then

$$L(\Pi') \le (1 + \frac{\varepsilon}{8})T(\Gamma') \le (1 + \frac{\varepsilon}{8})(1 + \delta)L(\Pi)$$
$$\le (1 + \varepsilon)L(\Pi),$$

proving P2. P3 follows directly from that Γ' tracks Γ to a tolerance of (η_x, η_v), with $\eta_x = \varepsilon\rho/2$, and that Γ lies inside W.

Let i' and f' be the initial and final states of Γ', and X' and Y' be the initial and final states of Π'. Since X' is close to i' and i' is close to i, which is close to X, we can show that $\|\text{VEC}(X') - \text{VEC}(X)\|_\infty \le \varepsilon/2$. Similarly we can show that $\|\text{VEC}(Y') - \text{VEC}(Y)\| \le \varepsilon/2$. This proves P4 for Γ'.

Such a trajectory Γ' (and thus a path Π') can be obtained by the TC-graph method. In constructing the graph, an edge is added if i) its a (μ, τ)-bang; ii) the bang does not diverge from W by more than $\varepsilon\rho/2$; and iii) the L_2 norms of the velocities of the bang fall into $[1 - \varepsilon/4, 1 + \varepsilon/8]$. Since μ and τ are $O(\varepsilon)$, the number of edges in this graph is bounded by $O(L^d(1/\varepsilon)^{6d-1})$, and thus the running time. This completes the proof of the lemma. $\square$

Notice that we do not explicitly constrain the velocities of the tracking trajectory Γ'. This allows us to

apply the Tracking Lemma. However, it happens that $\|v_{\Gamma'}(t)\|$ falls into a small range of $[1 - \varepsilon/4, 1 + \varepsilon/8]$, by being able to track closely a trajectory Γ whose velocities are fixed to be 1 in L_2 norm. By upper bounding the accelerations and lower bounding the velocities of Γ', we are able to bound the curvature of Π'.

3.3 Approximating with obstacles

Let W be a d-dimensional configuration space of size L, with a set Ω of obstacles. Let $\mathcal{B}$ be the box decomposition of W, as described in Section 2.3. Let Π be an optimal $3l$-safe 1-constrained path from X to Y, where $\|\text{VEC}(X)\| = \|\text{VEC}(Y)\| = 1$.

As we did in the proof of the Safe Tracking Lemma, we can divide Π into segments of paths, $\Pi_{U_0 U_1}\|\ldots\|\Pi_{U_{m-1}U_m}$, such that $U_0 = X$, $U_m = Y$, and each U_i is the state where Π first exits $\xi(U_{i-1})$, for $1 \le i \le m$. Each segment $\Pi_{U_i U_{i+1}}$ is a 1-constrained path from U_i to U_{i+1} and lying inside $\xi(U_i)$. These are the paths we need to approximate.

We define the set of CS-grid states, the canonical extended boxes, the canonical starting states and the canonical ending states in the same way as we did in the previous sections, except that we let $\mathcal{V}$ be the set of *unit* vectors that are uniformly spaced with spacing δ. If $\delta = O(\epsilon)$ is chosen small enough, for any unit vector v, there is a unit vector $v' \in \mathcal{V}$ such that $\|v' - v\|_\infty \le \epsilon/2$. Thus $\mathcal{V}$ suffices for the curvature-constrained case, since $\|\text{VEC}(U_i)\| = 1$ for all U_i's along Π. Similar to what we did for the kinodynamic case, we can prove a version of the Correcting Lemma, the Loose Tracking Lemma and the Safe Tracking Theorem for the curvature-constrained case. Since $|\mathcal{V}| = O((1/\epsilon)^{d-1})$ (as opposed to $O((1/\epsilon)^d)$ for the kinodynamic case), the number of edges in the CS-graph is reduced to $O((1/\epsilon)^{4d-4})$.

Combining the above, and choosing small enough $\epsilon = O(\varepsilon)$, we can obtain

Theorem 3.2 *Fix two real numbers* $l, \varepsilon > 0$. *After some pre-computation, we can achieve the following.*

Let W *be a* d-*dimensional space of size* L *and let* Ω *be a set of obstacles with a total of* n *vertices. Given any*

two states X and Y (with $\|\text{VEC}(X)\| = \|\text{VEC}(Y)\| = 1$), we can compute a collision-free $(1 + \varepsilon)$-constrained path, from $X' \in ngb(X, \varepsilon l, \varepsilon)$ to $Y' \in ngb(Y, \varepsilon l, \varepsilon)$, such that the path length is at most $(1 + \varepsilon)$ times the length of an optimal 1-constrained 3l-safe path from X to Y. The running time of our algorithm is $O(nN + N \log N (1/\varepsilon)^{4d-4})$, where $N = O((L/l)^d)$.

Acknowledgements

Work is supported by NSF Grant NSF-IRI-91-00681, Rome Labs Contracts F30602-94-C-0037, ARPA/SISTO contracts N00014-91-J-1985, and N00014-92-C-0182 under subcontract KI-92-01-0182.

References

[1] P.K. Agarwal, P. Raghavan, and H. Tamaki. Motion planning for a steering-constrained robot through moderate obstacles. In *Proc. 27th Annual Symp. Theory of Computing*, pages 343–352, 1995.

[2] J. Barraquand and J.C. Latombe. Nonholonomic multibody mobile robots: Controllability and motion planning in the presence of obstacles. *Algorithmica*, 10:121–155, 1993.

[3] J.D. Boissonnat, A. Cerezo, and J. Leblond. Shortest paths of bounded curvature in the plane. In *Proc. IEEE Int. Conf. on Robotics and Automation*, pages 2315–2320, 1992.

[4] J.D. Boissonnat, A. Cerezo, and J. Leblond. A note on shortest paths in the plane subject to a constraint on the derivative of the curvature. Technical report, INRIA, Sophia Antipolis, France, 1994.

[5] X.N. Bui, J.D. Boissonnat, P. Soueres, and J.P. Laumond. Shortest path synthesis for Dubins nonholonomic robot. In *Proc. IEEE Int. Conf. on Robotics and Automation*, pages 2–7, 1994.

[6] J. Canny, A. Rege, and J. Reif. An exact algorithm for kinodynamic planning in the plane. *Discrete Comput. Geom.*, 6:461–484, 1991.

[7] J. Canny and J. Reif. New lower bound techniques for robot motion planning problems. In *Proc. 28th Symp. Foundations of Computer Science*, pages 49–60, 1987.

[8] B. Donald and P. Xavier. Provably good approximation algorithms for optimal kinodynamic planning:

Robots with decoupled dynamics bounds. *Algorithmica*, 14:443–479, 1995.

[9] B. Donald and P. Xavier. Provably good approximation algorithms for optimal kinodynamic planning for cartesian robots and open chain manipulators. *Algorithmica*, 14:480–530, 1995.

[10] B. Donald, P. Xavier, J. Canny and J. Reif. Kinodynamic motion planning. *J. of the ACM*, 40:1048–1066, 1993.

[11] L.E. Dubins. On curves of minimal length with a constraint on average curvature and with prescribed initial and terminal positions and tangents. *Amer. J. Math.*, 79:497–516, 1957.

[12] S. Fortune and G. Wilfong. Planning constrained motion. *Annals of Mathematics and Artificial Intelligence*, 3:21–82, 1991.

[13] G. Heinzinger, P. Jacobs, J. Canny, and B. Paden. Time-optimal trajectories for a robot manipulator: a provably good approximation algorithm. In *Proc. IEEE Int. Conf. on Robotics and Automation*, pages 150–155, 1990.

[14] J. Hershberger and S. Suri. Efficient computation of Euclidean shortest paths in the plane. In *Proc. 34th Symp. on Foundations of Computer Science*, pages 508–517, 1993.

[15] J.M. Hollerbach. Dynamic scaling of manipulator trajectories. In *Proc. of the Automatic Control Council*, pages 752–756, 1983.

[16] P. Jacobs and J. Canny. Planning smooth paths for mobile robots. In *Nonholonomic Motion Planning*. Kluwer Academic Publishers, 1992.

[17] P. Jacobs, J.P. Laumond, and M. Taix. Efficient motion planners for nonholonomic mobile robots. In *Proc. IEEE/RSJ Int. Workshop on Intelligent Robots and Systems*, pages 1229–1235, 1991.

[18] L. Kavraki and J.C. Latombe. Randomized preprocessing of configuration space for fast path planning. In *Proc. IEEE Int. Conf. on Robotics and Automation*, pages 2138–2145, 1994.

[19] L. Kavraki, J.C. Latombe, R. Motwani, and P. Raghavan. Randomized query processing in robot path planning. In *Proc. 27th Symp. on Theory of Computing*, 1995.

[20] J.C. Latombe. *Robot Motion Planning*. Kluwer Academic Publishers, Norwell, MA, 1991.

[21] J.P. Laumond, P.E. Jacobs, M. Taix, and R.M. Murray. A motion planner for nonholonomic mobile robots. *IEEE Trans. on Robotics and Automation*, 1993.

[22] G.L. Miller, D. Talmor, S. Teng, and N. Walkington. A Delaunay based numerical method for three dimensions: generation, formulation, and partition. In *Proc. 27th Symp. on Theory of Computing*, pages 683–692, 1995.

[23] G.L. Miller, S. Teng, W. Thurston, and S.A. Vavasis. Automatic mesh partitioning. In *Graphs Theory and Sparse Matrix Computation*, A. George, J. Gilbert, and J. Liu, editors, pages 57–84, Springer-Verlag, 1993.

[24] B. Mirtich and J. Canny. Using skeletons for nonholonomic path planning among obstacles. In *Proc. of the IEEE Int. Conf. on Robotics and Automation*, pages 2533–2540, 1992.

[25] J.S.B. Mitchell, D.M. Mount, and S. Suri. Query-sensitive ray shooting. In *Proc. 10th Symp. on Computational Geometry*, pages 359–368, 1994.

[26] C. Ó'Dúnlaing. Motion planning with inertial constraints. *Algorithmica*, 2(4):431–475, 1987.

[27] J.A. Reeds and L.A. Shepp. Optimal paths for a car that goes both forwards and backwards. *Pacific J. of Mathematics*, 145(2):367–393, 1990.

[28] J.H. Reif and S.R. Tate. Approximate kinodynamic planning using L_2-norm dynamic bounds. *Computers Math. Applic.*, 27(5):29–44, 1994.

[29] G. Sahar and J.M. Hollerbach. Planning of minimum-time trajectories for robot arms. Technical report, MIT, 1984.

[30] H.J. Sussmann. Shortest 3-dimensional paths with a prescribed curvature bound. In *Proc. 34th Conf. on Decision and Control*, pages 3306–3312, 1995.

[31] H. Wang and P.K. Agarwal. Approximation algorithms for curvature constrained shortest paths. In *Proc. 7th Annu. ACM-SIAM Sympos. Discrete Algorithms, to appear*, 1995.

[32] G. Wilfong. Shortest paths for autonomous vehicles. In *Proc. IEEE Int. Conf. on Robotics and Automation*, pages 15–20, 1989.

[33] P. Xavier. Private communication.

A The TC-Graph Method

For the sake of completeness, we describe the gists of the Tracking Lemma and the TC-graph method in this section.

A trajectory Γ' is said to *track* another trajectory Γ to a tolerance (η_x, η_v), if (1) $\|p_\Gamma(t) - p_{\Gamma'}(t)\|_\infty \le \eta_x$, and (2) $\|v_\Gamma(t) - v_{\Gamma'}(t)\|_\infty \le \eta_v$, for all $t \in [0, T]$. Notice that (1) implies that $d(\Gamma', \Gamma) \le \eta_x$.

A τ-*bang* is a trajectory segment of time duration τ, during which a *constant* acceleration is applied[3]. A τ-*bang trajectory* is a trajectory consisting of a sequence of τ-bangs. The set of bang trajectories is a restricted set of trajectories. But they suffice to track any non-restricted trajectories, if the acceleration set used by the bang trajectories has some *advantage* over the acceleration set used by the non-restricted trajectories. Next we define the notion of advantage more formally.

Let $\mathcal{P}$ and $\mathcal{Q}$ be two sets of accelerations. For any acceleration $a \in \mathcal{P}$, and any direction given by a d-length vector σ of 1's and -1's, if there exists an acceleration $b \in \mathcal{Q}$, such that $\sigma^j(b^j - a^j) \ge \kappa_l$, for $1 \le j \le d$, then $\mathcal{Q}$ is said to have a *uniform κ_l advantage* over $\mathcal{P}$ (α^j means the jth element of a vector α). The Tracking Lemma relates the parameter τ to the uniform advantage κ_l and the tracking tolerance (η_x, η_v).

Lemma A.1 (Tracking Lemma [9]) [4] *Let $\mathcal{P}$ and $\mathcal{Q}$ be two sets of accelerations such that, for each $a \in \mathcal{Q}$, $\|a\| \le 1$, and that $\mathcal{Q}$ has a uniform κ_l advantage over $\mathcal{P}$. Let Γ be a trajectory that uses $\mathcal{P}$. Let (η_x, η_v) be a tracking tolerance. There exists a time step τ, a τ-bang trajectory Γ' that uses $\mathcal{Q}$, such that Γ' tracks Γ to*

---

[3]Here we slightly abuse the notion of a *bang*. A bang usually means that its acceleration should be *extremal* in some sense.

[4]This is a simplified version of the Tracking Lemma presented in [9]. In [9], $\mathcal{Q}$ is a set of *instantaneous accelerations* that depends on the current state, and may be a subset of the set of allowed accelerations. This is because applying certain accelerations may violate the velocity constraint. In the context where we apply this Tracking Lemma, we do not explicitly bound the velocity of the tracking trajectory. This means that the set of instantaneous accelerations equals the set of all allowed accelerations.

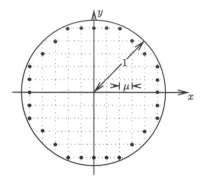

Figure 5: *The acceleration set $\mathcal{A}_\mu$ in 2D.*

tolerance (η_x, η_v). Moreover, it is sufficient that

(2) $$\tau = O(\min(\eta_v, \sqrt{\eta_x \kappa_l})).$$

Let $\mathcal{A}$ (resp. $\mathcal{A}_\varepsilon$) be the set of accelerations whose L_2 norms are bounded by 1 (resp. $1/(1+\varepsilon)^2$). Next we show how we can choose a finite set of accelerations such that this set has a uniform advantage over $\mathcal{A}_\varepsilon$. Given a parameter μ, a set of gridpoints of $\mathcal{A}$ with spacing μ is defined to be $\{a = (i_1\mu, \dots, i_d\mu) \mid i_1, \dots, i_d \in \mathbb{Z}$ and $a \in \mathcal{A}\}$. Let $\mathcal{A}_\mu$ be the set of *boundary* gridpoints (see Figure 5; the dark dots represent the boundary gridpoints in 2D). It is shown in [9], that if μ is chosen such that

(3) $$\mu \leq \kappa_l \leq \frac{\varepsilon}{4\sqrt{d}},$$

then $\mathcal{A}_\mu$ has a uniform advantage of κ_l over $\mathcal{A}_\varepsilon$. Notice that $|\mathcal{A}_\mu| = O(d(1/\mu)^{d-1})$.

A τ-bang (resp. τ-bang trajectory) is a (μ, τ)-*bang* (resp. (μ, τ)-*bang trajectory*) if the acceleration set used is $\mathcal{A}_\mu$. Given a tracking tolerance (η_x, η_v), and any $\varepsilon > 0$, there exist μ (satisfying (2)) and τ (satisfying (1)), such that for any $(1/(1+\varepsilon)^2, 1/(1+\varepsilon))$-trajectory Γ, there exists a (μ, τ)-bang trajectory Γ' (also a $(1,1)$-trajectory) that tracks Γ to a tolerance of (η_x, η_v).

To approximate any given $(1,1)$-trajectory Γ to within a factor of $(1+\varepsilon)$ in time length, we first time-rescale Γ to a $(1/(1+\varepsilon)^2, 1/(1+\varepsilon))$-trajectory Γ', with time length extended by a factor of $(1+\varepsilon)$. Since

there exists a (μ, τ)-bang trajectory Γ'' that tracks Γ' to a tolerance of (η_x, η_v), Γ'' approximates Γ in that $T(\Gamma'') \leq (1+\varepsilon)T(\Gamma)$. Γ'' does not track Γ to the tolerance of (η_x, η_v). However, $d(\Gamma'', \Gamma) \leq \eta_x$. This is because Γ' and Γ trace the same path and $d(\Gamma'', \Gamma') \leq \eta_x$. Also the two end states of Γ'' are close to the two end states of Γ respectively.

Having known the existence of a tracking bang trajectory, the next question is how to compute one. What makes it more difficult is that most of the time, the original $(1,1)$-trajectory Γ is not given except its initial and final states. The TC-graph method transforms this problem to that of finding a shortest path in a directed graph. (In the following, we only describe the TC-graph method in absence of obstacles, where we do not have to do collision-free checking.) The TC-graph G is roughly as follows: (a) The *root node* of G approximates the initial state of Γ; (b) A directed edge corresponds to a (μ, τ)-bang whose velocities are also bounded by 1 in L_2 norm. The weight of the edge is always τ; (c) The graph is generated and explored from the root node in a breadth-first manner, and the search terminates when either a node approximating the final state is found, or when no new nodes are generated.

Each node of G corresponds to a state. By carefully choosing the initial velocity, we can bound the number of possible velocities by $O((1/(\mu\tau))^d)$, and the number of possible locations by $O((L/(\mu\tau^2))^d)$, where L is the size of W. Thus the total number of nodes is bounded by $O((L/(\mu^2\tau^3))^d)$. Each node can have at most $O(d(1/\mu)^{d-1})$ out-going edges (because it can have at most this many choices of accelerations), thus the total number of graph edges is $O(L^d/(\mu^{3d-1}\tau^{3d}))$. This also bounds the running time.

If we set $\eta_x = \varepsilon\rho/2$ and $\eta_v = \varepsilon/2$, where $\rho = O(1)$, we obtain Corollary 2.1. Notice that since $\tau = O(\varepsilon)$ and $\mu = O(\varepsilon)$, the running time is $O(L^d(1/\varepsilon)^{6d-1})$.

Continuous Motion Plans for Robotic Systems with Changing Dynamic Behavior

Miloš Žefran, *University of Pennsylvania, Philadelphia, PA, USA*
Jaydev P. Desai, *University of Pennsylvania, Philadelphia, PA, USA*
Vijay Kumar, *University of Pennsylvania, Philadelphia, PA, USA*

The main objective of this paper is to address motion planning for systems in which the dynamic equations describing the evolution of the system change in different regions of the state space. We adopt the control theory point of view and focus on the planning of open loop trajectories that can be used as nominal inputs for control. Systems with changing dynamic behavior are characterized by: (a) equality and inequality constraints that partition the state space into regions (discrete states); and (b) trajectories that are governed by different dynamic equations as the system traverses different regions in the state space. The motion plan therefore consists of the sequence of regions (discrete states) as well as continuous trajectory (evolution of the continuous state) within each of the regions. Since the task may require that the system trajectories and the inputs are sufficiently smooth, we formulate the motion planning problem as an optimal control problem and achieve the smoothness by specifying an appropriate cost function.

We present a formal framework for describing systems with changing dynamic behavior borrowing from the literature on hybrid systems. We formulate the optimal control problem for such systems, develop a novel technique for simplifying this problem when the sequence of discrete states is known, and suggest a numerical method for dealing with inequality constraints. The approach is illustrated with two examples. We first consider the coordination between mobile manipulators carrying an object while avoiding obstacles. We show that the obstacle avoidance translates to inequality constraints on the state and the input. In this task no changes in the dynamic equations occur since no physical interaction between the manipulators and the obstacles takes place. In our second example, we study multi-fingered manipulation. In this case, the state space is partitioned into regions corresponding to different grasp configurations. We compute the motion plan for a task that requires a switch between two grasp configurations and obtain the optimal trajectory for the system as well as the optimal time when the switch should occur.

1 Introduction

Most of the early works on motion planning addressed the so called piano movers problem, formally defined by Schwartz and Sharir [36]. These methods rely purely on the geometry of the task and are not concerned with the physical interactions within the system. They are therefore usually referred to as geometric motion planning [21]. In recent years, minimalist approaches to manipulation are becoming increasingly prominent [13, 28]. This research led to important results in planning of planar manipulation tasks [12, 16, 27]. Most studies of minimalist manipulation are based on quasi-static analysis and are not concerned with the efficiency of the resulting motion. Of interest to the work presented here are also methods for nonholonomic motion planning [24, 31, 37], mostly inspired by developments in nonlinear control theory. The motion plans produced by these methods can be directly used for control.

Our work differs from other approaches in several important aspects. First, in contrast with the geometric motion planning, we take the control theory point of view and consider motion planning as a synthesis of open loop (or feed-forward) trajectories for control of the system. Further, we are ultimately interested in feedback control, hence we rely on the dynamic model of the system. Our study is therefore different from investigations of minimalist manipulation. Finally, we study systems that change dynamic equations as they evolve in the state space. This makes our study more general than, for example, the work on the nonholonomic motion planning.

For a system with changing dynamic behavior, the connected region in which the system is described by a single set of dynamic equations will be called a discrete state. The collection of discrete states defines a partition of the state space. Typically, such a partition can be described by a set of equalities and inequalities. The successful completion of a task necessitates transitions between discrete states and as the system passes from one discrete state to another its dynamic equations change. A typical example from robotics is a multi-fingered hand, where the set of fingers in contact with the object defines a discrete state. If no fingers are added or removed, the manipulation of an object relative to the wrist (continuous state) can be described by a system of ordinary differential equations [29] and the motion plan for such a continuous system can be generated by "traditional" techniques. Now consider the addition or removal of a finger in order to allow regrasping of the object. The occurrence of any such events causes a change in the system of ordinary differential equations describing the continuous state, and a change in the algebraic constraints [25]. Hence, the system switches from one region of the workspace (discrete state) to another[1].

The systems described above fall into the general class of hybrid systems. These are control systems that involve plants whose behavior can be modeled by continuous dynamics, discrete phenomena that may change the behavior of the plant, and control laws that may be continuous or discrete. There are different approaches to modeling of hybrid systems [18]. The most general model was proposed by Branicky *et al.* [2]. Of relevance to the work presented here are also the models of Brockett [4], and of Kohn and Nerode [30]. Motion planning for hybrid systems is a difficult task and good methods are yet to be developed. The hybrid systems framework of Branicky *et al.* [2] is used to prove the existence of optimal motion plans. Computational techniques for finding optimal hybrid control based on dynamic programming are proposed in [3], but they have prohibitive complexity and are thus not practical for solving complex motion planning problems.

This paper addresses a general method for generating continuous motion plans for a restricted class of hybrid systems that are encountered in robotics. In

robotic tasks, we require that the state variables, such as positions and velocities (and in some applications, accelerations and forces) must vary "smoothly"[2]. We are therefore interested in generating smooth motion plans that allow transitions between discrete states. Part of the motion plan should be a sequence of discrete states (for example, the walking or grasp gait). The main idea of the paper is to formulate the motion planning problem as an optimal control problem. By specifying the appropriate cost function, we can achieve smoothness. We can also prescribe constraints at the switching points. The resulting optimal control problem is very complex, so we simplify it by assuming that the sequence of discrete states is known. We develop a technique that uses this information to convert the problem with the unknown switching points to a problem in the standard form [5]. To solve the resulting optimal control problem, we use a numerical method derived from the calculus of variations. The method easily handles inequality constraints [39] and is well suited for problems discussed in this paper.

We demonstrate the technique on two examples. First, we investigate the problem of generating continuous motion plans for a team of cooperating manipulators mounted on nonholonomic carts and transporting an object in an environment with obstacles. We show that the coordination between mobile manipulators while avoiding obstacles can be appropriately modeled with inequality constraints on the state and the input, and show that we can efficiently solve the resulting optimal control problem. Although the trajectory for the system typically consists of constrained and unconstrained segments, in this task no changes in the dynamic equations occur since no physical interaction between the manipulators and the obstacles takes place. In other words, the whole state space forms a single discrete state. Our second example is representative of grasping and locomotion tasks: we generate smooth motion plans for the fine manipulation of a circular object with a two-fingered hand. In this case, the state space consists of discrete regions corresponding to different grasp configurations. As a result, the motion plan consists of a sequence of grasp configurations (discrete states) as well as a continuous trajectory (evolution of the continuous state) within each discrete

---

[1]In analogy to walking, the temporal sequence of the discrete states in grasping is called a grasp gait [7, 23].

[2]The degree of continuity that these variables must have depends on the application and the chosen model.

state. While the sequence of discrete states (grasp gait) is difficult to determine using our method, once this sequence is known it is possible to compute the optimal trajectory and the optimal times when the switches between the discrete states should occur. We argue that this often suffices for applications and that the sequence of discrete states can be inferred in advance for many robotics tasks.

2   Problem formulation

Let the continuous state space $\mathcal{X}$ of the dynamical system with n states be given by:

$$\mathcal{X} = \bigcup_{j=1}^{p} D_j, \tag{1}$$

where D_j are pairwise disjoint, connected subsets of $\mathbb{R}^n$. On each subset D_j, the system is described with system equations:

$$\dot{x} = F_j(x, u, t), \tag{2}$$

where $x \in \mathcal{X} \subset \mathbb{R}^n$ is the (continuous) state of the system, $u \in \mathbb{R}^m$ is the input and F_j is a (smooth) vector field. The sets D_j are called discrete states and they partition the state space into regions so that on each region the evolution of the system is governed by a (different) vector field F_j. We require that the continuous state changes continuously (but not smoothly) between the regions.

Since the sets D_j are disjoint, given a continuous state x, there is a unique set $D_{j(x)}$ such that $x \in D_{j(x)}$. We can therefore rewrite (2) as:

$$\dot{x} = f(x, u, t), \tag{3}$$

where $f(x, u, t) = F_{j(x)}(x, u, t)$.

Figure 1 schematically illustrates such a system. The continuous state space is partitioned into eight discrete states (regions). The shaded areas in the state space are regions which are not accessible (for example, where the robot would penetrate an obstacle). Three sample trajectories, denoted by α, β, and γ, are shown. Given a starting point A, and an end point B, it may be possible to go from A to B without changing the discrete state (system equations) following the trajectory α. Alternatively, the trajectory β exhibits a change from D_4 to D_5 and then a transition back to D_4. The optimal path between A and C may follow a straight line in

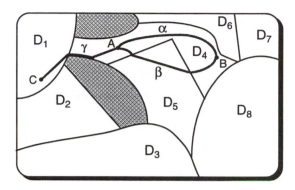

Figure 1: *A schematic of a system with changing dynamic behavior.*

D_4 until it hits a boundary, travel along the boundary until a state transition to D_1 occurs and then follow a straight line in D_1. Since along α the continuous state evolves within the same discrete state the trajectory is smooth. On the other hand, β and γ are only piecewise smooth.

We now define the motion planning problem for the system (3). Our premise is that a task often provides a way to evaluate different motion plans. We therefore regard the motion planning problem as an optimal control problem:

Problem: *Given a system described by (3), an initial state x_0 and a final state x_1, find a set of inputs, u, that minimizes the cost functional:*

$$J = \int_{T_0}^{T_1} L(x, u, t)\,dt. \tag{4}$$

To each trajectory $\gamma(x, u, t)$ of (3) that connects state x_0 with x_1, there corresponds a sequence of points $T_0 = t_0 < t_1 < \ldots < t_{N+1} = T_1$ and a sequence of indices $j_0, \ldots, j_N$ such that on the interval $[t_i, t_{i+1}]$, the trajectory belongs to the set D_{j_i} and at the time t_{i+1} it switches from the region D_{j_i} to $D_{j_{i+1}}$. (Times t_i thus correspond to switches in the discrete state of the system.) This implies that a solution of the optimal control problem consists of four components:

(a) The number of switches, N, between discrete states.

(b) The sequence, $\{D_{j_i}\}_{i=0}^{N}$, of discrete states (regions) that the system traverses as it moves along the optimal trajectory.

(c) The sequence, $\{t_i\}_{i=1}^N$, of times when the switches occur.

(d) The trajectories for the continuous state and the inputs on each interval $[t_i, t_{i+1}]$.

Finding all four components of the optimal solution is obviously quite a difficult task (see also [3, 20]). We can find an approximate solution by solving the problem hierarchically: first determining the sequence of the discrete states and then computing the trajectories for the continuous state.

3 Method for computing switching times

In tasks such as a multi-fingered hand grasping an object or a walking machine moving on the ground, the sequence of discrete states is usually known *a priori*. For the grasping task it can be obtained by investigating feasible grasp gaits [23]. In walking, the gait is usually computed in advance to avoid regions that are unsuitable for foot placement [1, 6]. Another instance where the sequence of discrete states is *a priori* known is planning of open-loop trajectories for transition between a free motion and position control to a constrained motion and force control [19, 32].

When the sequence of discrete states is given, the optimal control problem is significantly simplified and it reduces to finding the optimal switching times and the optimal trajectories for the continuous state. This problem is still difficult to solve, but we show that it can be reduced to a simpler problem. The main idea is to make the unknown switching times part of the state and introduce a new independent variable with respect to which the switching times are fixed. The resulting optimal control problem can then be solved using any existing method.

Assume that the number N and the sequence of discrete states $\{D_{j_i}\}_{i=0}^N$ through which the system evolves are known. Without loss of generality, we can assume that $T_0 = t_0 = 0$ and $T_1 = t_{N+1} = 1$. We can proceed in a similar way as in solving boundary value problems with an unknown terminal time [33]. The first step is to introduce new state variables $x_{n+1}, \ldots, x_{n+N}$ corresponding to the switching times t_i with

$$\begin{aligned} x_{n+i} &= t_i \\ \dot{x}_{n+i} &= 0. \end{aligned} \qquad (5)$$

We next introduce a new independent variable, s. The relation between s and t is linear, but the slope of the curve changes on each interval $[t_i, t_{i+1}]$. We establish piecewise linear correspondence between time t and the new independent variable s so that at every chosen fixed point s_i, $i = 0, \ldots, N+1$ (for convenience we choose values $s_i = i/(N+1)$, but any monotonically increasing sequence of N numbers on interval $[0, 1]$ would do), t equals t_i. Figure 2 illustrates this idea. As a result we obtain the following expressions:

$$t = \begin{cases} (N+1)x_{n+1}s, & 0 \leq s \leq \frac{1}{N+1} \\ \cdots \\ (N+1)(x_{n+i+1} - x_{n+i})s \\ +(i+1)x_{n+i} - ix_{n+i+1}, & \frac{i}{N+1} < s \leq \frac{i+1}{N+1} \\ \cdots \\ (N+1)(1 - x_{n+N})s \\ +(N+1)x_{n+N} - N, & \frac{N}{N+1} < s \leq 1. \end{cases}$$

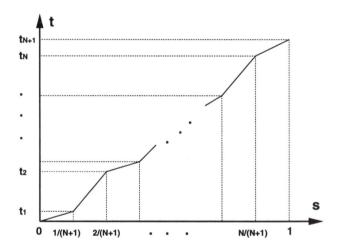

Figure 2: *A new independent variable has fixed values at the switching times t_i.*

With the new independent variable, the evolution equation on the interval $[t_i, t_{i+1}]$ that was given by:

$$\dot{x} = F_{j_i}(x, u, t), \qquad (6)$$

becomes:

$$x' = (N+1)(x_{n+i+1} - x_{n+i})\hat{F}_{j_i}(x, u, s), \qquad (7)$$

where $(.)'$ denotes the derivative of $(.)$ with respect to the new independent variable s and

$$\hat{F}(x, u, s) = F(x, u, t(s)).$$

If we denote by $\hat{x}$ the extended state vector:

$$\hat{x} = [x_1, \ldots, x_n, x_{n+1}, \ldots, x_{n+N}]^T,$$

we can define on each interval $\frac{i}{N+1} < s \le \frac{i+1}{N+1}$:

$$\hat{L}(\hat{x}, u, s) = (N+1)(x_{n+i+1} - x_{n+i})L(x, u, t(s)). \quad (8)$$

Finally, we can rewrite the functional (4) as:

$$J = \int_0^1 \hat{L}(\hat{x}, u, s)ds. \quad (9)$$

and the task is to minimize J in the extended state space. Points s_i at which the system described by the function $\hat{F}$ switches between the discrete states are known. In the optimal solution, $\hat{x}^*$, the last N components will be the optimal switching times $\{t_i\}_{i=1}^N$ for the original problem.

4 The optimal control problem

In this section, we present a numerical method that can be used to solve the optimal control problem for a system with inequality constraints. Consider a dynamic system described by the state equations

$$\dot{x} = f(x, u, t). \quad (10)$$

The vector x denotes the state of the system while u is the input vector. The state of the system is known at some initial time T_0 and is prescribed for some, possibly unknown, time T_1. We have a cost functional:

$$J = \int_{T_0}^{T_1} L(x, u, t)\, dt. \quad (11)$$

In addition, the following equality and inequality constraints must be satisfied by the state and the input:

$$g_i(x, u, t) = 0 \quad i = 1, \ldots, k \quad (12)$$

$$h_i(x, u, t) \le 0 \quad i = 1, \ldots, l. \quad (13)$$

The problem of optimal control is to find a piecewise continuous input vector u that brings the system from a known initial state x_0 to a desired final state x_f so that the constraints (12-13) are satisfied and the cost functional J is minimized.

4.1 The calculus of variations problem

In this subsection, we transform the optimal control problem into an equivalent unconstrained problem in the calculus of variations. First, each inequality constraint is converted into an equality constraint

$$h_i(x, u, t) \le 0 \Leftrightarrow \hat{h}_i = h_i(x, u, t) + \xi_i^2 = 0, \quad (14)$$

where ξ_i is an (unconstrained) slack variable. Next, define a Hamiltonian

$$H = L(x, u, t) + \lambda^T(\dot{x} - f(x, u, t)) + \phi^T g + \psi^T \hat{h}. \quad (15)$$

Note that the constraints g and h have been adjoined to the Hamiltonian with vectors of multipliers ϕ and ψ. Now define an extended state vector

$$X = [X_x^T, X_\mu^T, X_\lambda^T, X_\phi^T, X_\psi^T, X_\xi^T]^T, \quad (16)$$

where the individual components are defined by

$$X_x = x, \quad \dot{X}_\mu = u, \quad \dot{X}_\lambda = \lambda,$$
$$\dot{X}_\phi = \phi, \quad \dot{X}_\psi = \psi, \quad \dot{X}_\xi = \xi. \quad (17)$$

Consider the following variational problem:

$$\min J(X) = \int_{T_0}^{T_1} H(t, X, \dot{X})\, dt, \quad (18)$$

with the boundary conditions:

$$X_x(T_0) = x_0, \quad X_x(T_1) = x_f, \quad X_\mu(T_0) = 0,$$
$$X_\lambda(T_0) = 0, \quad X_\phi(T_0) = 0, \quad X_\psi(T_0) = 0, \quad (19)$$
$$X_\xi(T_0) = 0.$$

It can be shown that the critical solutions of the unconstrained variational problem (18) are exactly the extremals for the optimal control problem (10-13) obtained from the Minimum Principle [17, page 109].

4.2 Numerical method

To solve the variational problem (18) we use the numerical method by Gregory & Lin [17]. Consider the general variational calculus problem given by Equations (18-19). Define an admissible variation $Z(t)$ to be a piecewise smooth function such that if added to a trajectory $X(t)$ satisfying the boundary conditions at T_0 and T_1 the sum $X(t) + Z(t)$ satisfies the same boundary conditions. It can be shown [35] that if $X(t)$ is a solution of the variational problem (18-19), then the integral equation

$$\int_{T_0}^{T_1} [H_X^T Z + H_{\dot{X}}^T \dot{Z}]\, dt = 0, \quad (20)$$

holds for all admissible variations $Z(t)$. Here H_X and $H_{\dot{X}}$ denote partial derivatives with respect to X and $\dot{X}$ respectively.

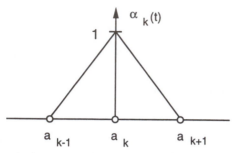

Figure 3: *Shape functions used in the numerical method.*

To numerically compute a critical solution, we first discretize the interval $[T_0, T_1]$ so that $T_0 = a_0 < a_1 < \cdots < a_M = T_1$. For simplicity, we assume that $a_i - a_{i-1} = h$ for $i = 1, \ldots, M$. Then we introduce a set of piecewise linear shape functions (Figure 3):

$$\alpha_k(t) = \begin{cases} \frac{t - a_{k-1}}{h} & \text{if } a_{k-1} < t \le a_k, \\ \frac{a_{k+1} - t}{h} & \text{if } a_k < t \le a_{k+1}, \\ 0 & \text{otherwise}, \quad (k = 0, \ldots, M). \end{cases}$$

Since (20) must be satisfied for any piecewise smooth admissible variation $Z(t)$, we can choose the functions $Z_i = \sum_{j=0}^{M} Z_{ij} \alpha_j(t)$. Using the central difference scheme to approximate the derivatives and the mean-value theorem to approximate the integral (20) on each subinterval $[a_{i-1}, a_i]$, the following set of vector equations in the unknown values X_k at points $k = 1, \ldots, M-1$ is obtained:

$$\begin{aligned} 0 = & \frac{h}{2} H_X \left(\frac{a_k + a_{k-1}}{2}, \frac{X_k + X_{k-1}}{2}, \frac{X_k - X_{k-1}}{h} \right) + \\ & H_{\dot{X}} \left(\frac{a_k + a_{k-1}}{2}, \frac{X_k + X_{k-1}}{2}, \frac{X_k - X_{k-1}}{h} \right) + \\ & \frac{h}{2} H_X \left(\frac{a_{k+1} + a_k}{2}, \frac{X_{k+1} + X_k}{2}, \frac{X_{k+1} - X_k}{h} \right) - \\ & H_{\dot{X}} \left(\frac{a_{k+1} + a_k}{2}, \frac{X_{k+1} + X_k}{2}, \frac{X_{k+1} - X_k}{h} \right). \quad (21) \end{aligned}$$

Note that the unknowns X_k are vectors. In all, we have $n(M + 1)$ unknowns. We also have $n(M - 1)$ equations (21). The remaining $2n$ equations are obtained from the boundary conditions, if they exist, or from the transversality conditions [35]. These are of the similar form as equations (21) and the reader should consult [17] for detailed derivations.

The resulting system of nonlinear equations is solved using the Newton-Raphson method. Each equation only depends on the three adjacent points. The matrix of the system of linear equations solved during each iteration is thus block-tridiagonal and the system can be solved very efficiently.

5 Coordination of Multiple Mobile Manipulators

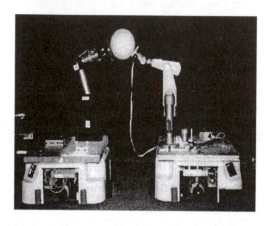

Figure 4: *A team of two cooperating mobile manipulators marching while grasping an object*

In this section we address the motion planning for multiple mobile manipulators. The basic task is to cooperatively grasp and transport an object from one point to another. In our experimental testbed shown in Figure 4, each mobile manipulator consists of a robot arm mounted on a TRC Labmate platform. The platform is a nonholonomic cart with two actuated degrees-of-freedom. The mobile platforms enable appropriate (and even optimal) positioning and configuring before grasping [38], and possible reconfiguration if necessary. We are particularly concerned with the problem of motion planning in the presence of obstacles. Because the mobile manipulators must maintain a stable grasp, it is necessary to plan a smooth trajectory that is free of collisions. For example, in Figure 5, the mobile manipulators must execute a "parallel-parking" like maneuver to avoid the obstacle while preventing excessive braking or large accelerations [10]. Similarly, as shown in Figure 6, it may be necessary to "change formations" to squeeze through narrow constrictions. From a dynamic viewpoint, it is desirable to generate a trajectory which requires a smooth variation of actuator forces while ensuring force-closure of the object. However, in this section, we focus on the problem of determining the optimal kinematic trajectory. The extension of results presented here to the case when the dynamics is incorporated is, although computationally intensive, conceptually straight forward.

Figure 5: *Executing a "parallel-parking" maneuver to circumvent an obstacle*

Figure 6: *Reconfiguring from a "march abreast" formation to a "march in a single file" formation*

5.1 Distance function

To formulate the problem of planning an obstacle-free path, we represent the obstacles as unilateral constraints in the state space. In turn, we need to be able to compute gradients of these constraints along different directions in the state space in order to employ the numerical method presented in Section 4.2. We use a distance function to accomplish this. The distance between two convex sets, S_A and S_B (if the links of the robot or the obstacles are concave, they can be represented as a union of convex sets), is given by

$$d(S_A, S_B) = min\{ \ ||x - y|| \ \ | \ x \in S_A, y \in S_B\}, \quad (22)$$

where $||x - y||$ stands for the Euclidean distance between the points x and y. The distance computation between a pair of convex sets in R^n is a well understood problem in computational geometry. Efficient algorithms for computing distances are described in [15, 26]. The method in [15] also affords analytical expressions for the gradient of the distance function. However, when the convex sets intersect, most algorithms ([15] included) do not compute the extent of the intersection. In order to compute the gradient, it is necessary to have a continuous distance function that returns a "negative distance" when the sets intersect,

a zero distance when the sets touch and a positive distance when the sets do not intersect.

We simplify the problem of computing the distance and its gradient by using a spherical representation [9, 14] for the links of the robot system and by representing the obstacles with convex polygons[3]. In the planar case, each link is represented by the circumscribing circle and the robot is a union of these circles. By decomposing each link into smaller polygons, we can describe the robot by smaller circumscribing circles and obtain a description with a finer resolution. The advantage of this approach is that the distance in Equation (22) can be easily computed.

In order to compute the distance between a robot link (a circle S_A) and an obstacle (a polygon S_B), we need to consider the two cases shown in Figures 7 and 8. In both cases, c is the center of the circle S_A, and o is the point on the polygon S_B that is closest to c. $\vec{V_c}$ and $\vec{V_o}$ are the corresponding position vectors.

1. Figure 7 shows the case when the center of S_A is outside S_B. The distance is given by:

$$d(S_A, S_B) = ||\vec{V_o} - \vec{V_c}|| - r_A, \quad (23)$$

where r_A is the radius of S_A.

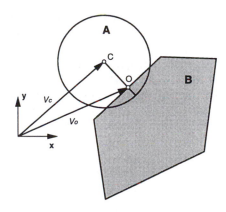

Figure 7: *Center outside the polygon.*

Since this expression is valid as long as the center of the circle S_A is outside S_B, it can be used to calculate the penetration of S_A into S_B as well as

---

[3]Equivalently we could have assumed that the robots are polygonal and they operate in a world of circular obstacles. We have chosen to approximate the robots since there are usually fewer robot links than obstacles.

the distance between S_A and S_B when they do not intersect.

2. Figure 8 shows the case when the center of S_A belongs to S_B. In this case, the negative distance is given by:

$$d(S_A, S_B) = (-||\vec{V_c} - \vec{V_o}|| - r_A). \qquad (24)$$

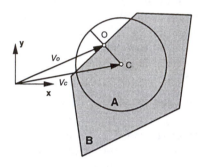

Figure 8: *Center within the polygon.*

Both computations return the distance $-r_A$ when c lies on the edge of a polygon. Additional details are available in [11].

Since r_A is a constant in the above equations, and V_c is a known function of the configuration space (or state space) variables, the simplified robot representation results in very simple expressions for the gradients[4].

5.2 Mathematical formulation

We consider the system of two cooperating agents shown in Figure 4. The manipulators are non redundant. The main task is to specify the trajectories for each platform and for the object, while ensuring that the two manipulators can cooperatively grasp the object. For this task it is sufficient to consider a kinematic model of the system [11].

Each platform is a nonholonomic cart with the state equations:

$$
\begin{aligned}
\dot{x}_1 &= u_1 \cos\theta_1 & \dot{x}_2 &= u_3 \cos\theta_2 \\
\dot{y}_1 &= u_1 \sin\theta_1 & \dot{y}_2 &= u_3 \sin\theta_2 \\
\dot{\theta}_1 &= u_2 & \dot{\theta}_2 &= u_4
\end{aligned}
\qquad (25)
$$

---

[4]There are special cases in which the gradients are not unique. In these cases, one must use generalized gradients [8] or choose the appropriate gradient depending on the history of the prior motion [11].

where (x_i, y_i) is the position of the reference point and θ_i is the orientation of the ith ($i = 1, 2$) cart. The nonholonomic constraint restricts the velocity of the cart so that the linear velocity of the reference point is parallel to the longitudinal axis of the vehicle. The inputs u_1 and u_3 are the linear velocities of the reference points while u_2 and u_4 are the angular velocities of the corresponding platforms.

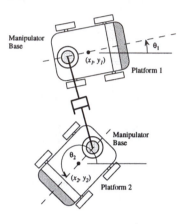

Figure 9: *Theoretical model of two cooperating mobile manipulators.*

We do not explicitly model the kinematics of the manipulators as we are not interested in the planning of motion for each manipulator. Instead, we imagine the bases of the manipulators being connected by a planar $R-P-R$ kinematic chain as shown in Figure 9. The revolute joints coincide with the base joints of the manipulators. The distance between the two revolute joints is denoted by l. The prismatic joint between the two revolute joints has limited range of motion. In this way, it is ensured that the mobile platforms do not drift too far apart or come too close. The limits on l model the limited reach of the manipulators and the need to keep the object in the dextrous workspace of each manipulator. The $R-P-R$ chain does not restrict the maneuverability of one platform relative to the other unless the prismatic joint is at a joint limit. The joint limit is modeled by unilateral constraints on the state:

$$l_{lower} \leq l \leq l_{upper}. \qquad (26)$$

The two drive wheels of the TRC Labmate platforms used in our experimental setup have a common axle so the platforms can turn without translating. But in a more general system this is not possible hence in the calculations we explicitly assume a finite value of the turning radius:

$$\rho_{min1}^2 u_2^2 - u_1^2 \leq 0, \qquad \rho_{min2}^2 u_4^2 - u_3^2 \leq 0. \qquad (27)$$

Here ρ_{min1} (ρ_{min2}) is the minimum radius of curvature for the path of the reference point on cart 1 (cart 2).

Finally, we require that the distance of the manipulators from the obstacles in the environment be positive at all times [5]. The distance constraint takes the form:

$$\epsilon - d_{ij} \leq 0 \quad i = 1, 2; \quad j = 1, \ldots, p \qquad (28)$$

where ϵ denotes the required clearance between the robot and the obstacle, p is the number of obstacles and d_{ij} denotes the distance of the manipulator i from the obstacle j.

We now treat the variational problem of determining the shortest path between the two given positions and orientations subject to the above constraints. The distance is measured in terms of the arc lengths traversed by the drive wheels of the two carts. The minimization of distance is mathematically equivalent to minimization of the integral of the square of the velocities. Thus we obtain the following cost function:

$$min \quad \frac{1}{2} \int_{T_0}^{T_1} \{(u_1^2 + u_3^2) + B^2(u_2^2 + u_4^2)\} dt$$

where B is a scaling factor that depends on the geometry of the vehicle.

In computations, we assume that there are two obstacles in the workspace. To reduce this problem to the standard form, we define the state vector:

$$x = [x_1, y_1, \theta_1, x_2, y_2, \theta_2]^T$$

and the extended state vector:

$$X = [x^T, X_\mu^T, X_\lambda^T, X_\psi^T, X_\xi^T]^T \qquad (29)$$

where

$$\dot{X}_\mu = [u_1 \ u_2 \ u_3 \ u_4]^T \qquad \dot{X}_\psi = [\psi_1 \ldots \psi_8]^T$$
$$\dot{X}_\lambda = [\lambda_1 \ldots \lambda_6]^T \qquad \dot{X}_\xi = [\xi_1 \ldots \xi_8]^T$$

The inequalities can be converted to equalities and written in the following form:

$$\hat{h}_1 = \rho_{min1}^2 u_2^2 - u_1^2 + \xi_1^2 \qquad \hat{h}_5 = \epsilon - d_{11} + \xi_5^2$$
$$\hat{h}_2 = \rho_{min2}^2 u_4^2 - u_3^2 + \xi_2^2 \qquad \hat{h}_6 = \epsilon - d_{21} + \xi_6^2$$
$$\hat{h}_3 = l_{lower} - l + \xi_3^2 \qquad \hat{h}_7 = \epsilon - d_{12} + \xi_7^2$$
$$\hat{h}_4 = l - l_{upper} + \xi_4^2 \qquad \hat{h}_8 = \epsilon - d_{22} + \xi_8^2$$

---

[5]The collisions between the two mobile manipulators are prevented by imposing the limits on the workspace of the two arms (modeled here by the limits on the prismatic joint).

At this point the Hamiltonian for the variational problem can be formulated.

With two obstacles, the final variational problem has 32 unknowns. Given a starting, x_0, and a final, x_1, position and orientation for the two platforms, the boundary conditions are:

$$\begin{aligned} x(T_0) = x_0, \quad x(T_1) = x_1, \quad X_\mu(T_0) = 0, \\ X_\lambda(T_0) = 0, \quad X_\psi(T_0) = 0, \quad X_\xi(T_0) = 0. \end{aligned} \qquad (30)$$

5.3 Examples

In the examples that follow (some experimental results are also presented in [10]), we used the following set of parameters to model the mobile platforms in Figure 4:

ρ_{min1}	ρ_{min2}	B	l_{lower}	l_{upper}	ϵ
0.05m	0.05m	0.1m	0.8m	1.2m	0.01m

We chose a total of 200 mesh points for our simulation and required a relative accuracy of 10^{-7} in all the variables for convergence.

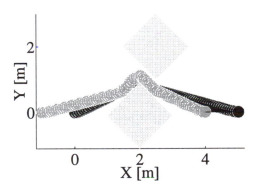

Figure 10: *The shortest distance paths around obstacles.*

In Figure 10, two mobile manipulators carrying an object follow a shortest distance trajectory while maneuvering around obstacles. The figure shows snapshots of the two platforms along the path at equal time intervals. The circumscribing circles "hug" the obstacle boundaries in the corridor. As expected from the cost function, there is a tendency for the platforms to travel, roughly speaking, along straight lines in the "obstacle free" region. When the platforms turn, the arcs have the same minimal turning radius as predicted by [34] for a single platform.

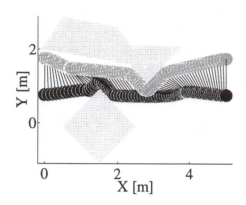

Figure 11: *The two mobile manipulators change formation in order to navigate the narrow corridor.*

Figure 11 shows the platforms moving amidst concave obstacles. The concave obstacle is represented as a union of two convex obstacles. The platforms reconfigure from a "marching abreast" formation to a "single file" formation (in which the object is between a leader and a follower) to avoid the obstacles. After passing through the constricted space the platforms change their formation back to the original formation. Once again, it is evident that during the maneuver the circumscribed circles pass very close to the obstacle boundaries.

The motion plans in these examples are optimal in the sense that they minimize the distance traveled by the wheels. However, the solution is not a global minimum and it depends on the initial guess. In fact, we can define homotopy classes of trajectories such that no trajectory in one class can be "deformed" into a trajectory in another class without passing through an obstacle. An initial guess that belongs to one homotopy class will never converge to a solution in another homotopy class.

The approach adopted here is in many ways superior to others discussed in the literature. The work that is closest in flavor is reported in [21, 22]. There, the planning problem is divided into three stages: (a) Find the holonomic path amid obstacles; (b) Find the nearest path that also satisfies the nonholonomic constraints; and (c) Optimize the obtained path. In our method, we concurrently solve the two stages (b) and (c). Further, we do not rely on analytical expressions for the shortest distance paths which might not be available

except in very simple examples.

Finally, we note that the generation of optimal trajectories for the examples shown here with a mesh of 200 points takes approximately 60 seconds on a Sparc 10 station.

6 Planar manipulation with a two-fingered hand

In the previous example, we were able to formulate the system equations without partitioning the continuous state space into discrete states. We now consider motion planning in the case when switches between multiple discrete states occur. We study an example of two fingers with limited workspace rotating a circular object in a horizontal plane. In [23], a similar example is used to study grasp gaits. Our approach is also in line with the conceptual framework proposed in [25] for motion planning for dextrous manipulation.

For our task we require that exactly one finger is on the object during any finite time interval (an example of such task would be a driver turning a steering wheel). This effectively partitions the continuous state space into two discrete states, each representing the case when one finger is in contact with the object and the other freely moves in the plane. There is also an intermediate third state when both fingers are in contact with the object, through which the system passes instantaneously at each switch. We assume that the object can rotate around a fixed axis passing through the center of the object and that the fingers can exert arbitrary forces on the object. The motion of each finger is restricted to a cone in the plane of the object. We can easily generalize the example to the case of a multi-fingered hand manipulating a planar object.

6.1 Mathematical formulation

The position of the object is given by its turning angle φ (Figure 12). The center of the object (and thus the pivot point) is at the origin of the global coordinate system and the radius of the object is equal to R. Positions of the two fingers in the plane are expressed in polar coordinates and are (r_1, θ_1) and (r_2, θ_2). The dynamics of the object is given by:

$$\begin{aligned} I\ddot{\varphi} = {} & -F_{1x}R\sin\theta_1 + F_{1y}R\cos\theta_1 \\ & -F_{2x}R\sin\theta_2 + F_{2y}R\cos\theta_2, \end{aligned} \quad (31)$$

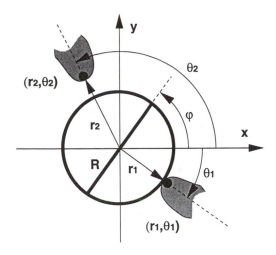

Figure 12: *Two fingers turning a circular object in a plane.*

where I is the moment of inertia of the object around the axis of rotation while F_1 and F_2 are the forces, expressed in the global coordinate frame, that the fingers 1 and 2 exert on the object.

The dynamic equations of the two fingers can be written in Cartesian space. For simplicity we assume that the fingers behave like point masses located at the finger tip. (This assumption is easily relaxed.) The dynamics of the finger i is thus given by:

$$
\begin{aligned}
-F_{ix} + u_{ix} &= m_i(\ddot{r}_i \cos\theta_i - 2\dot{r}_i\dot{\theta}_i \sin\theta_i \quad (32) \\
&\quad - r_i\dot{\theta}_i{}^2 \cos\theta_i - r_i\ddot{\theta}_i \sin\theta_i) \\
-F_{iy} + u_{iy} &= m_i(\ddot{r}_i \sin\theta_i + 2\dot{r}_i\dot{\theta}_i \cos\theta_i \\
&\quad - r_i\dot{\theta}_i{}^2 \sin\theta_i + r_i\ddot{\theta}_i \cos\theta_i),
\end{aligned}
$$

where m_i is the effective mass of the finger i, and u_i is the 2×1 vector of driving forces for finger i. By defining a state vector:

$$
x = [\varphi, \dot{\varphi}, r_1, \theta_1, \dot{r}_1, \dot{\theta}_1, r_2, \theta_2, \dot{r}_2, \dot{\theta}_2]^T \quad (33)
$$

the dynamic equations of the object and the two fingers can be transformed into the state-space form. The vectors u_1 and u_2 are the inputs to the system.

We assume that the workspace of the two fingers is limited. The workspace of each finger is a cone of angle 2α centered at the origin. The axis of the cone for the first finger corresponds to the half-line $\theta = 0$ while the axis of the cone for the second finger is the half-line $\theta = \pi$. The cones $\mathcal{W}_1$ and $\mathcal{W}_2$ representing

the workspace of the fingers 1 and 2, respectively, are thus given by:

$$
\begin{aligned}
\mathcal{W}_1 &= \{\theta \mid -\alpha \leq \theta \leq \alpha\} \\
\mathcal{W}_2 &= \{\theta \mid -\alpha + \pi \leq \theta \leq \alpha + \pi\}. \quad (34)
\end{aligned}
$$

For our task of rotating the object we require that exactly one finger is on the object during any finite time interval. It is easy to see that this requires partitioning of the state space into three regions:

$$
\begin{aligned}
D_1 &= \{x \mid \theta_1 \in \mathcal{W}_1, \theta_2 \in \mathcal{W}_2, r_1 > R, r_2 = R\} \\
D_2 &= \{x \mid \theta_1 \in \mathcal{W}_1, \theta_2 \in \mathcal{W}_2, r_1 = R, r_2 > R\} \\
D_3 &= \{x \mid \theta_1 \in \mathcal{W}_1, \theta_2 \in \mathcal{W}_2, r_1 = R, r_2 = R\}.
\end{aligned}
$$

Region D_1 corresponds to the case when the second finger is in contact with the object while the first finger does not touch the object. Region D_2 describes the opposite situation. In region D_3 both fingers are on the object. Because of the requirement that exactly one finger is on the object during any finite time interval, the system can not stay in D_3. This basically reduces the state space to $D_1 \cup D_2$, with the switch between the two corresponding to D_3.

The dynamic equations are the same in all three regions. Therefore, $f = f_1 = f_2$, where f is the system function obtained when the dynamic equations are rewritten in the state space. However, in the region D_1, the constraint $r_1 = 1$ forces $\dot{r}_1 = 0$. Similarly, the requirement $r_2 > 0$ implies $F_2 = 0$. Analogous equations hold for D_2.

Finally, we have to choose the cost functional for the optimal control problem. The choice depends on the desired properties of the optimal trajectories. For example, continuous force profiles can be obtained by minimizing the rate of change of forces. In this work we chose to minimize a measure of the energy necessary to move the two fingers:

$$
J = \int_0^1 L\,dt = \frac{1}{2}\int_0^1 (u_{1x}^2 + u_{1y}^2 + u_{2x}^2 + u_{2y}^2)\,dt. \quad (35)
$$

At $t = 0$, we have the initial conditions:

$$
\varphi = 0, \quad \theta_1 = 0, \quad \theta_2 = \pi, \quad r_1 = R, \quad r_2 = R,
$$

$$
\dot{\varphi} = 0, \quad \dot{\theta}_1 = 0, \quad \dot{\theta}_2 = 0, \quad \dot{r}_1 = 0, \quad \dot{r}_2 = 0.
$$

and we assume that the system immediately passes into the region D_1. We require the object to be rotated through $\pi/3$. Both fingers are prescribed to end their motion on the object, but we are not interested where. The final conditions are therefore:

$$\varphi = \tfrac{\pi}{3}, \quad r_1 = R, \quad r_2 = R,$$

$$\dot{\varphi} = 0, \quad \dot{r}_1 = 0, \quad \dot{r}_2 = 0, \quad \dot{\theta}_1 = 0, \quad \dot{\theta}_2 = 0.$$

In order to use the method from Section 3, we have to fix the number of transitions between the two regions in the state space. For simplicity, we assume that the system switches only once: we first turn the object with the second finger and then complete the rotation of the object with the first finger.

We now apply the technique outlined in Section 3. Let τ be the time when the system switches from D_1 to D_2. We extend the state vector x with τ and impose the state equation:

$$\dot{\tau} = 0.$$

Next, we define a new independent variable s with the equation:

$$t = \begin{cases} 2\tau s, & 0 < s \leq 0.5 \\ 2(1-\tau)s + 2\tau - 1, & 0.5 < s \leq 1.0. \end{cases} \tag{36}$$

The cost function L from Equation (35) becomes:

$$\hat{L}(\hat{x}, u, t) = \begin{cases} 2\tau L, & 0 < s \leq 0.5 \\ 2(1-\tau)L, & 0.5 < s \leq 1.0. \end{cases} \tag{37}$$

Similarly, if $f = f_1 = f_2$ is the system function for the independent variable t, the new system functions become:

$$\begin{aligned} \hat{f}_1 &= 2\tau f \\ \hat{f}_2 &= 2(1-\tau)f. \end{aligned} \tag{38}$$

In each region we also have to satisfy the following constraints:

$$\begin{array}{ll} D_1: & r_1 > R \qquad D_2: \quad r_1 = R \\ & r_2 = R \qquad\qquad\quad r_2 > R \\ & \dot{\varphi} = \dot{\theta}_2 \qquad\qquad\quad \dot{\varphi} = \dot{\theta}_1 \\ & F_1 = 0 \qquad\qquad\quad F_2 = 0 \end{array} \tag{39}$$

To solve the resulting optimal control problem we can now use the method from Section 4.2. First, the inequality constraints are converted to equality constraints with the slack variables. These are then adjoined to the cost function together with the rest of the equality constraints and the system equations to form a Hamiltonian. When the discretized equations are obtained, care has to be taken to use the right constraints

and system functions: for $s < 0.5$, the equations associated with D_1 are used while for $s > 0.5$, the equations corresponding to D_2 must be substituted. With all constraints appropriately adjoined, we obtain a Hamiltonian which contains 46 unknown functions.

6.2 Examples

In the examples, we chose $R = 1$ and $m_1 = m_2 = 1$. We show the results for two different choices of the workspace for the two fingers.

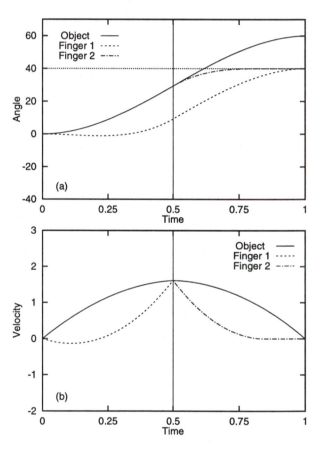

Figure 13: *(a) Angles φ, θ_1 and $\theta_2 - \pi$; and (b) velocities $\dot{\varphi}$, $\dot{\theta}_1$ and $\dot{\theta}_2$ for the workspace $\alpha = 40°$.*

Figure 13 shows the trajectories when the workspace is given by $\alpha = 40°$. We subtracted π from θ_2 to compare it with the other two angles. The new independent variable s is shown on the abscissa. Therefore, the switch between the two regions occurs at $s = 0.5$. For $\alpha = 40°$, the optimal value for the switching time is $\tau = 0.493$. As the figure shows, in the first half of the task the finger 2 rotates the object while the finger 1 is not in contact with the object. This can be clearly seen from the velocity plots, where $\dot{\theta}_2 = \dot{\varphi}$. The finger 1 moves along the surface of the object, but with lower speed. At $s = 0.5$ ($\tau = 0.493$), the object is rotated for 29.2° and the two fingers switch the roles. For $s > 0.5$, the finger 2 moves freely in space and continues its motion until it reaches the limit of the workspace. In the meantime, the finger 1 rotates the object to 60° and reaches the limit of its workspace. Both fingers thus terminate the motion on the edge of the workspace. Their final position is therefore different from the starting position which explains the asymmetry of the task and the difference in the angle through which each finger rotates the object.

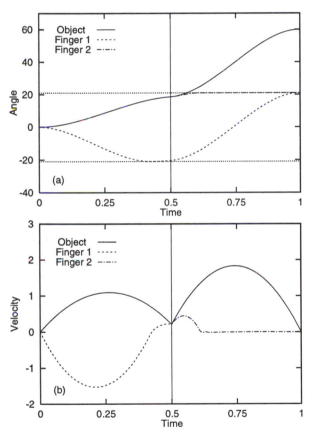

Figure 14: *(a) Angles φ, θ_1 and $\theta_2 - \pi$; and (b) velocities $\dot{\varphi}$, $\dot{\theta}_1$ and $\dot{\theta}_2$ for the workspace $\alpha = 21°$.*

Figure 14 shows the optimal trajectories for $\alpha = 21°$. In this case, the 60° rotation can be barely achieved with a single switch (the maximal achievable rotation

is 3α). For $s < 0.5$, the finger 1 not in contact with the object and it moves to the lower edge of the workspace. Meanwhile, the finger 2 rotates the object almost for the entire allowable range α. At $s = 0.5$ the finger 1 comes into contact with the object which is rotated for 18.6°. The finger is close to the lower edge of the workspace and can subsequently use most of the allowable range 2α to complete the rotation of the object. In this example, the switch between the discrete states D_1 and D_2 occurs at $\tau = 0.425$. Both fingers end their motion at the upper edge of the workspace.

We also make the following observations:

- The velocities of the contact points on the object and on the finger at the time of establishing or breaking the contact are equal so no impact occurs. Hence, the continuous state is continuous across the switches between discrete states.

- The positions of the fingers on the object at the switch are computed as part of the motion plan. In other words, the sequence of discrete states is assumed to be known, but the value of the continuous state at the switches is computed from the optimization.

- The workspace constraints of the two fingers are respected. We could further generalize the task by including frictional constraints.

7 Conclusion

Many robotic tasks lead to systems that change their dynamic equations as they evolve in the state space. Typical examples are grasping and walking. The state of such systems is composed of a discrete state, which determines the dynamic equations of the system, and a continuous state, that describes how the system evolves within the discrete state. The existing approaches to motion planning for systems with changing dynamic behavior concentrate either on planning of the sequence of discrete states to achieve the task or on planning and control of the system within each of the discrete states. This work is an attempt to bring these two paradigms together and view the motion planning as a single process.

We are interested in generating motion plans than can be directly used in control. The motion planning problem is thus formulated as an optimal control problem. In the paper, we addressed two aspects of computing the optimal motion plan:

1. Partition of the state space into discrete states can be described with a set of equality and inequality constraints. We hence presented a numerical method for solving optimal control problems subject to inequality constraints.

2. Changes in the dynamic equations of the system are difficult to address within the framework of optimal control. We showed how the optimal control problem with the unknown switching times can be simplified if the sequence of the discrete states for the task is given.

We demonstrated the approach on the examples of a team of two mobile manipulators manipulating an object in a cluttered environment and two fingers rotating a circular object. While the first example can be formulated as an optimal control problem with inequality constraints involving the input and the state, the second example requires partitioning of the state space into a finite number of disjoint regions. By choosing the cost function appropriately we can guarantee the desired level of smoothness across the discrete states in the velocities, accelerations or forces.

The main limitation of the proposed approach is that we must *a priori* know the sequence of discrete states. Future work focuses on methods for planning this sequence. However, the value of the continuous state at the switches between the discrete states (for example, placement of the legs on the ground during walking) can be determined from the optimization. The approach is therefore quite versatile and lends itself to a number of motion planning problems.

Acknowledgment

Financial support for this work has been provided by the NSF grants BCS 92-16691, MSS-91-57156, CISE/CDA 88-22719, ARPA Grant N00014-92-J-1647 and Army Grant DAAH04-96-1-0007. The first author is a fellow of the Institute for Research in Cognitive Sciences at the University of Pennsylvania.

References

[1] J.-D. Boissonnat, O. Devillers, and S. Lazard. Motion planning of legged robots. In K. Goldberg, D. Halperin, J.-C. Latombe, and R. Willson, editors, *Algorithmic Foundations of Robotics*, pages 49–67. A K Peters, Ltd., Wellesley, MA, 1995.

[2] M. S. Branicky, V. S. Borkar, and S. K. Mitter. A unified framework for hybrid control. In *Proceedings of the 33rd IEEE Conference on Decision and Control*, pages 4228–4234, Lake Buena Vista, FL, 1994.

[3] M. S. Branicky and S. K. Mitter. Algorithms for optimal hybrid control. In *Proceedings of the 34th IEEE Conference on Decision and Control*, pages 2661–2666, New Orleans, LA, 1995.

[4] R. W. Brockett. Hybrid models for motion control systems. In H. L. Trentelman and J. C. Willems, editors, *Essays in Control: Perspectives in the Theory and its Applications*, pages 29–53. Birkhäuser, Boston, 1993.

[5] A. E. Bryson and Y.-C. Ho. *Applied Optimal Control*. Hemisphere Publishing Co., New York, 1975.

[6] C. H. Chen and V. Kumar. Motion planning of walking robots in environments with uncertainty. In *Proceedings of 1996 International Conference on Robotics and Automation*, Minneapolis, MN, 1996.

[7] I.-M. Chen and J. W. Burdick. A qualitative test for N-finger force-closure grasps on planar objects with applications to manipulation and finger gaits. In *Proceedings of 1993 International Conference on Robotics and Automation*, pages 814–820, Atlanta, GA, 1993.

[8] F. H. Clarke. *Optimization and non-smooth analysis*. Wiley, New York, 1983.

[9] A. P. del Pobil and M. A. Serna. *Spatial representation and motion planning*. Lecture notes in Computer Science. Springer, 1995.

[10] J. Desai, C. C. Wang, M. Žefran, and V. Kumar. Motion planning for multiple mobile manipulators. In *Proc. of 1996 IEEE Int. Conf. on Robotics and Automation*, Minneapolis, MN, 1996.

[11] J. P. Desai and V. Kumar. Optimal motion plans for nonholonomic mobile manipulators in an environment with obstacles. Technical report, University of Pennsylvania, GRASP Laboratory, Philadelphia, 1996.

[12] B. R. Donald, J. Jennings, and D. Rus. Information invariants for distributed manipulation. In K. Goldberg, D. Halperin, J.-C. Latombe, and R. Willson, editors, *Algorithmic Foundations of Robotics*, pages 431–457. A K Peters, Ltd., Wellesley, MA, 1995.

[13] M. Erdmann and M. T. Mason. An exploration of sensorless manipulation. In *Proceedings of 1986 International Conference on Robotics and Automation*, pages 1569–1574, San Francisco, CA, April 1986.

[14] R. Featherstone. A hierarchical representation of the space occupancy of a robot mechanism. In J. P. Merlet and B. Ravani, editors, *Computational Kinematics '95*. Kluwer Academic Publishers, Boston, 1995.

[15] E. G. Gilbert, D. W. Johnson, and S. S. Keerthi. A fast procedure for computing the distance between complex objects in three-dimensional space. *IEEE Journal of Robotics and Automation*, 4(2):193–203, 1988.

[16] K. Goldberg. Orienting polygonal parts without sensors. *Algorithmica*, 10(2):201–225, 1993. Special Issue on Computational Robotics.

[17] J. Gregory and C. Lin. *Constrained optimization in the calculus of variations and optimal control theory*. Van Nostrand Reinhold, New York, 1992.

[18] R. L. Grossman, A. Nerode, A. P. Ravn, and H. Rischel (Eds.). *Hybrid systems*, volume 736 of *Lecture notes in computer science*. Springer-Verlag, New York, 1993.

[19] H.-P. Huang and N. H. McClamroch. Time-optimal control for a robot contour following problem. *IEEE Transactions on Robotics and Automation*, 4(2):140–149, 1988.

[20] W. Kohn, A. Nerode, J. B. Remmel, and Xiolin Ge. Multiple agent hybrid control: carrier manifolds and chattering approximations to optimal control. In *Proceedings of the 33rd IEEE Conference on Decision and Control*, pages 4221–4227, Lake Buena Vista, FL, 1994.

[21] J.-C. Latombe. *Robot motion planning*. Kluwer Academic Publishers, Boston, 1991.

[22] J.-P. Laumond, P. E. Jacobs, M. Taix, and R. M. Murray. A motion planner for nonholonomic mobile robots. *IEEE Transactions on Robotics and Automation*, 10(5):577–593, 1994.

[23] S. Leveroni and K. Salisbury. Reorienting objects with a robot hand using grasp gaits. In *International Symposium on Robotics Research*, pages 49–63, Munich, Germany, 1995.

[24] Z. Li and J. F. Canny. *Nonholonomic motion planning*. Kluwer Academic Publishers, Boston, 1993.

[25] Z. Li, J. F. Canny, and S. S. Sastry. On motion planning for dextrous manipulation, Part 1: The problem formulation. In *Proc. of 1989 International Conference on Robotics and Automation*, pages 775–780, 1989.

[26] M. C. Lin and J. F. Canny. A fast algorithm for incremental distance calculation. In *Proc. of 1991 International Conference on Robotics and Automation*, pages 1008–1014, Sacramento, CA, 1991.

[27] K. M. Lynch and M. T. Mason. Stable pushing: Mechanics, controllability and planning. In K. Goldberg, D. Halperin, J.-C. Latombe, and R. Willson, editors, *Algorithmic Foundations of Robotics*, pages 239–262. A K Peters, Ltd., Wellesley, MA, 1995.

[28] M. T. Mason. Mechanics and planning of manipulator pushing operations. *International Journal of Robotics Research*, 5(3), 1986.

[29] R. M. Murray, Z. Li, and S. S. Sastry. *A Mathematical Introduction to Robotic Manipulation*. CRC Press, 1994.

[30] A. Nerode and W. Kohn. Models for hybrid systems: Automata, topologies, stability. In R. L. Grossman, A. Nerode, A. P. Ravn, and H. Rischel, editors, *Hybrid systems*, pages 317–356. Springer-Verlag, New York, 1993.

[31] J. P. Ostrowski. *The Mechanics and Control of Undulatory Locomotion*. PhD thesis, California Institute of Technology, Pasadena, CA, 1995.

[32] P. R. Pagilla and M. Tomizuka. Control of mechanical systems subject to unilateral constraints. In *Proceedings of the 34th IEEE Conference on Decision and Control*, New Orleans, LA, 1995.

[33] W. H. Press, S. A. Teukolsky, W. T. Vetterling, and B. P. Flannery. *Numerical Recipes in C*. Cambridge University Press, Cambridge, 1988.

[34] J. A. Reeds and L. A. Shepp. Optimal paths for a car that goes both forwards and backwards. *Pacific. J. of Math.*, 145(2):367–393, 1990.

[35] H. Sagan. *Introduction to the calculus of variations*. McGraw-Hill, New York, 1969.

[36] J. T. Schwartz and M. Sharir. On the 'piano movers' problem: 1. The case of two-dimensional rigid polygonal body moving amidst polygonal barriers. *Communications on pure and applied mathematics*, 36:345–398, 1983.

[37] D. M. Tilbury. *Exterior differential systems and nonholonomic motion planning*. PhD thesis, University of California at Berkeley, Berkeley, CA, 1994.

[38] Y. Yamamoto. *Control and Coordination of Locomotion and Manipulation of Wheeled Mobile Manipulators*. PhD thesis, University of Pennsylvania, 1994.

[39] M. Žefran and V. Kumar. Optimal control of systems with unilateral constraints. In *Proceedings of 1995 International Conference on Robotics and Automation*, pages 2695–2700, Nagoya, Japan, 1995.

Collision Detection: Algorithms and Applications

Ming C. Lin, *U.S. Army Research Office and University of North Carolina, Chapel Hill, NC, USA*
Dinesh Manocha, *University of North Carolina, Chapel Hill, NC, USA*
Jon Cohen, *University of North Carolina, Chapel Hill, NC, USA*
Stefan Gottschalk, *University of North Carolina, Chapel Hill, NC, USA*

Fast and accurate collision detection between general geometric models is a fundamental problem in modeling, robotics, manufacturing and computer-simulated environments. Most of the earlier algorithm are either restricted to a class of geometric models, say convex polytopes, or are not fast enough for practical applications. We present an efficient and accurate algorithm for collision detection between general polygonal models in dynamic environments. The algorithm makes use of hierarchical representations along with frame to frame coherence to rapidly detect collisions. It is robust and has been implemented as part of public domain packages. In practice, it can accurately detect all the contacts between large complex geometries composed of hundreds of thousands of polygons at interactive rates.

1 Introduction

Collision detection is a fundamental problem in robotics, computer animation, physically-based modeling, molecular modeling and computer simulated environments. In these applications, an object's motion is constrained by collisions with other objects and by other dynamic constraints. The problem has been well studied in the literature.

A realistic simulation system, which couples geometric modeling and physical prototyping, can provide a useful toolset for applications in robotics, CAD/CAM design, molecular modeling, manufacturing design simulations, etc. Such systems create electronic representations of mechanical parts, tools, and machines, which need to be tested for interconnectivity, functionality, and reliability. A fundamental component of such a system is to model object interactions *precisely*. The interactions may involve objects in the simulation environment pushing, striking, or smashing other objects. *Detecting collisions and determining contact points* is a crucial step in portraying these interactions accurately.

The most challenging problem in a simulation, namely the collision phase, can be separated into three parts: collision detection, contact area determination, and collision response. In this paper, we address the first two elements by presenting general a purpose collision detection and contact area determination algorithm for simulations. The collision response is application dependent. The algorithm reports the contact area and thus enables the application to compute an appropriate response.

Our algorithm not only addresses interaction between a pair of general polygonal objects, but also large environments consisting of hundreds of moving parts, such as those encountered in the manufacturing plants. Furthermore, we do not assume the motions of the objects to be expressed as a closed form function of time. Our collision detection scheme is efficient and accurate (to the resolution of the models).

Given the geometric models, the algorithm precomputes the convex hull and a hierarchical representation of each model in terms of oriented bounding boxes. At runtime, it uses tight fitting axis-aligned bounding boxes to pair down the number of object pair interactions to only those pairs within *close proximity* [12]. For each pair of objects whose bounding boxes overlap, the algorithm checks whether their convex hulls are intersecting based on the closest feature pairs [22]. Finally for each object pair whose convex hulls overlap, it makes use of oriented bounding box hierarchy (OBBTree) to check for actual contact [18].

Organization: The rest of the paper is organized as follows: Section 2 reviews some of the previous work in collision detection. Section 3 outlines the algorithm for pruning the number of object pairs. We briefly describe the closest feature and contact determination algorithms in Section 4. Finally, we describe the implementation and performance on different applications in Section 5

2   Previous Work

Collision detection has been extensively studied in CAD, computer graphics, robotics, and computational geometry. Since collision detection is needed in a wide variety of situations, many different methods have been proposed. Most of them make specific assumptions about the objects of interest and design a solution based on object geometry or application domain.

Robotics literature deals with collision detection in the context of path planning. Using sophisticated mathematical tools, several algorithms have been developed that plan collision-free paths for a robot in restricted environments [8, 9]. However, in path planners based on potential field methods, collision detection and distance computation are still considered as major bottlenecks [21, 10].

Most computational geometry literature deals with collision detection of objects in a static environment. Objects are at a fixed location and orientation, and the algorithms determine whether they are intersecting [11, 13]. In most modeling and graphics applications, where many objects are in motion, such an approach would be inefficient. Moreover, the objects move only slightly from frame to frame and the collision detection scheme should take advantage of the information from the previous frame to initialize the computation for the current frame [3, 23]. Several solutions based on this idea of coherence have been proposed in [22]. Approaches that combine collision response with detection can be found in [3, 33, 25]. The methods in [32, 33] make use of the boundary representation to detect collisions.

Collision detection for multiple moving objects has recently become a popular research topic with the in-

creased interest in large-scaled virtual prototyping environments. For example, a vibratory parts feeder can contain up to hundreds of mechanical parts moving simultaneously under periodical force impulses in a vibratory bowl or tray. In a general simulation environment, there may be N moving objects and M stationary objects. Each of the N moving objects can collide with the other moving objects, as well as the stationary ones. Keeping track of $\binom{N}{2} + NM$ pairs of objects at every time step can become time consuming as N and M get large. To achieve interactive rates, the total number of pairwise intersection tests must be reduced before performing exact collision tests on the object pairs, which are in the close vicinity of each other. Several methods dealing with this situation are found in [7, 12, 16]. Most methods use some type of a hierarchical bounding box scheme. Objects are surrounded by bounding boxes. If the bounding boxes overlap, indicating the objects are near each other, a more precise collision test is applied.

As for curved models, algorithms based on interval arithmetic for collision detection are described in [15, 17]. These algorithms expect the motion of the objects to be expressed as a closed form function of time. Moreover, the performance of interval arithmetic based algorithms is too slow for interactive applications. Coherence based algorithms for curved models are presented in [24].

In many CAD applications, the input models are given as collections of polygons with no topology information. Such models are also known as 'polygon soups' and their boundaries may have cracks, T-joints, or may have non-manifold geometry. In general, no robust techniques are known for cleaning such models. Many of the algorithms described are not applicable to such models. Rather techniques based on hierarchical bounding volumes and and spatial decomposition are used on such models. Typical examples of bounding volumes include axis-aligned boxes (of which cubes are a special case) and spheres, and they are chosen for to the simplicity of finding collision between two such volumes. Hierarchical structures used for collision detection include cone trees, k-d trees and octrees [31], sphere trees [20, 30], R-trees and their variants

[5], trees based on S-bounds [7] etc. Other spatial representations are based on BSP's [27] and its extensions to multi-space partitions [32], spatial representations based on space-time bounds or four-dimensional testing [1, 6, 9, 20] and many more. All of these hierarchical methods do very well in performing "rejection tests", whenever two objects are far apart. However, when the two objects are in close proximity and can have multiple contacts, these algorithms either use subdivision techniques or check very large number of bounding volume pairs for potential contacts. In such cases, their performance slows down considerably and they become a major bottleneck in the simulation, as stated in [19].

3   Collision Detection between Multiple Moving Objects

We review our previous algorithm for multiple moving convex polytopes in complex environments. Coherence combined with incremental computation is a major theme of our algorithms. By exploiting coherence, we are able to incrementally trim down the number of pairwise object pairs and feature tests involved in each iteration.

Definition: *Temporal and geometric coherence* is the property that the state of the application does not change significantly between successive time steps or simulation frames. The objects move only slightly from frame to frame. This slight movement of the objects translates into geometric coherence, since their spatial relationship does not change much between frames.

For a configuration of N objects, the worst case running time for any collision detection algorithm is $O(N^2)$ where N is the number of objects. However, evidence suggests that these cases rarely occur in simulations [3, 12, 22, 28]. So our algorithm uses a *Sweep and Prune* technique to eliminate testing object pairs that are far apart, and later we show that the technique can be extended to eliminate testing features that are far apart between two colliding objects.

We use a bounding box based scheme to reduce the $O(N^2)$ bottleneck of testing all possible pairs of objects for collisions. In most realistic situations, an object has

to be tested against a small fraction of all objects in the environment for collision. For example, in a simulation of a vibratory parts feeder most objects are in close proximity to only a few other objects. It would be pointless and expensive to keep track of all possible interactions between objects at each time step.

Sorting the bounding boxes surrounding the objects is the key to our *Sweep and Prune* approach [12]. It is not intuitively obvious how to sort bounding boxes in 3-space to determine overlaps. We use a *dimension reduction* approach. If two bounding boxes collide in 3-D, then their orthogonal projections on the x, y, and z axes must overlap. The sweep and prune algorithm begins by projecting each 3-D bounding box surrounding an object onto the x, y, and z axes. Since the bounding boxes are axially-aligned, projecting them onto the coordinate axes results in intervals. We are interested in overlaps among these intervals, because a pair of bounding boxes can overlap *if and only if* their intervals overlap in all three dimensions.

We construct three lists, one for each dimension. Each list contains the values of the endpoints of the intervals in each corresponding dimension. By sorting these lists, we can determine which intervals overlap. In the general case, such a sort would take $O(N \log N)$ time, where N is the number of objects. We can reduce this time bound by keeping the sorted lists from the previous frame, updating only the interval endpoints. In environments where the objects make relatively small movements between frames, the lists will be nearly sorted, so we can re-sort using *insertion sort* in expected $O(N)$ time [26, 3]. Graphs 1 – 6 are timings taken from a multi-object simulation where we compare the performance of using fixed versus dynamic sized boxes [12]. Parameters such as the number of objects, the polygonal complexity of the objects, the velocity of the objects, etc. were varied as the graphs show.

4   Exact Collision Detection

Given two objects in close proximity, the algorithm initially checks whether their convex hulls are overlapping. It incrementally computes their closest feature

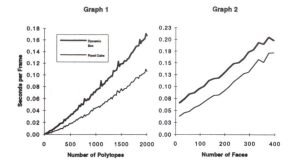

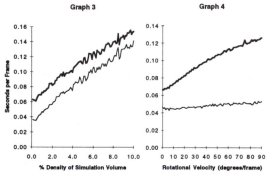

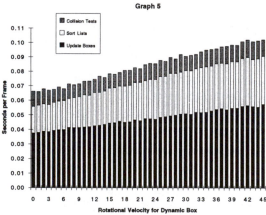

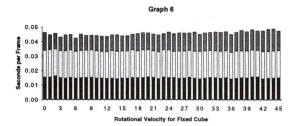

hierarchy of oriented bounding boxes to check for exact contact.

4.1 Collision Detection between Convex Polytopes

The algorithm computes the convex hull of all objects as part of pre-processing. Furthermore, it classifies the feature of convex hulls into *red* and *green* features. The red features correspond to the features of the original model and the green features are introduced by the convex hull computation.

We use the algorithm described in [22] to keep track of closest features for a pair of convex polytopes. The algorithm maintains a pair of closest features for each convex polytope pair and calculates the Euclidean distance between the features to detect collisions. This approach can be used in a static environment, but is especially well-suited for dynamic environments in which objects move in a sequence of small, discrete steps. The method takes advantage of coherence: the closest features change infrequently as the polytopes move along finely discretized paths. In most situations, the algorithm runs in *expected constant time* if the polytopes are not moving at large discrete steps (e.g. 180 degrees of rotation per step).

4.1.1 Voronoi Regions

Each convex polytope is pre-processed into a modified boundary representation. The polytope data structure has fields for its features (faces, edges, and vertices) and corresponding *Voronoi regions*. Each feature is described by its geometric parameters and its neighboring features, i.e. the topological information of incidences and adjacencies.

Definition: A *Voronoi region* associated with a feature is a set of points closer to that feature than any other [29].

The Voronoi regions form a partition of the space outside the polytope, and they form the generalized Voronoi diagram of the polytope. Note that the generalized Voronoi diagram of a convex polytope has linear number of features and consists of polyhedral regions. A *cell* is the data structure for a Voronoi region of a

pairs and check for overlap. If the convex hulls are intersecting, it checks whether the overlapping features belong to the original model or are introduced by the convex hull computation. Eventually it makes use of single feature. It has a set of constraint planes which bound the Voronoi region with pointers to the neighboring cells (which share a constraint plane with it) in its data structure. If a point lies on a constraint plane, then it is equi-distant from the two features which share this constraint plane in their Voronoi regions. For more details on this construction and its properties, please refer to [22].

4.1.2 Closest Feature Tests

Our method for finding closest feature pairs is based on Voronoi regions. We start with a candidate pair of features, one from each polytope, and check whether the closest points lie on these features. Since the polytopes and their faces are convex, this is a local test involving only the neighboring features of the current candidate features. If either feature fails the test, we step to a neighboring feature of one or both candidates, and try again. As the Euclidean distance between feature pairs must always decrease when a switch is made, cycling is impossible for non-penetrating objects. An example of the algorithm is given in Fig. 1.

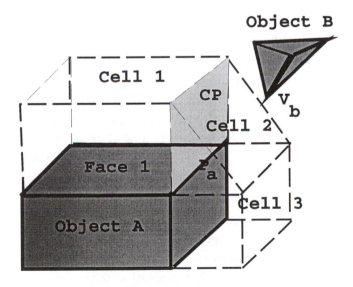

Figure 1: *A walk across Voronoi cells.*

Given a pair of features *Face 1* and *vertex V_b*, on objects A and B, as the closest features we test to see if vertex V_b lies within *Cell 1* of of *Face 1*. V_b violates the constraint plane imposed by *CP* of *Cell 1*. The

constraint plane *CP* has a pointer to its adjacent cell *Cell 2*, so the walk proceeds to test the containmentship of V_b within *Cell 2*. In similar fashion, vertex V_b has a cell of its own, and we see if the nearest point P_a on the edge to the vertex V_b lies within V_b's Voronoi cell. Basically, we are performing the containmentship tests of a point within a Voronoi region defined by the constraint planes of the region. The constraint plane causing the containmentship test to fail points the next direction for the algorithm to *advance* in the search of a new and *closer* feature. Eventually, we must reach the closest pair of features.

4.1.3 Penetration Detection for Convex Polytopes

The key to detecting penetrations lies in partitioning the *interior* as well as the exterior of the convex polytope. For internal partitioning, internal Voronoi regions can be used. The internal Voronoi regions can be constructed for any convex polytope by computing all the equi-distant hyperplanes between two or more facets on the polytope. However the general construction of the internal Voronoi regions is a non-trivial computation [29]. To detect a penetration – as opposed to knowing *all* the closest features – it is unnecessary to construct the exact internal Voronoi regions.

Rather we use an approximation, labeled as pseudo-internal Voronoi region. It is calculated by first computing the centroid of each convex polytope – the weighted average of all vertices. Then a hyperplane from each edge is extended towards the centroid. The extended hyperplane tapers to a point, forming a pyramid-type cone over each face. Each of the faces of the given polytope is now used as a constraint plane. If a candidate feature fails the constraint imposed by the face (indicating the closest feature pair lies possibly behind this face), the algorithm stepping "enters" inside of the polytope (as shown in Fig. 2). If at any time we find one point on a feature of one polytope is contained within the pseudo internal Voronoi region (i.e. this point satisfies the constraints posed by an pseudo internal Voronoi region of the other polytope), it corresponds to a penetration. A detailed discussion is presented in [28]

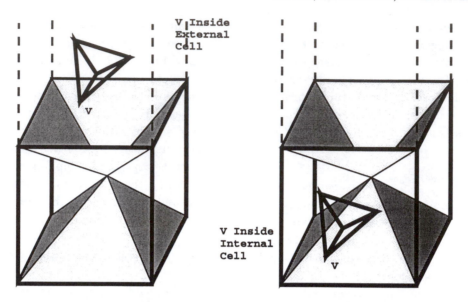

Figure 2: *Walk from external to internal Voronoi regions.*

4.1.4   Feature Classification

The algorithm highlighted above returns all pairs of overlapping features between the convex hulls. The resulting feature pairs can be classified into:

- Red-red feature overlap (as shown in Fig. 3(a)): This corresponds to an actual collision between the original models.

- Red-green feature overlap (as shown in Fig. 3(b)): This may or may not correspond to a collision. The algorithm presented in the next section is used to check for exact contact.

- Green-green feature overlap (as shown in Fig. 3(c)): Same as above. The algorithm presented in the next section is used to check for exact contact.

4.2   Exact Contact Determination

In this section, we present a robust, efficient and general purpose algorithm to compute all contacts between geometric models composed of polygons. The algorithms computes a hierarchical representation using *oriented bounding boxes (OBBs)*. An OBB is a rectangular bounding box at an arbitrary orientation in 3-space. The resulting hierarchical structure is referred to as an OBBTree. The idea of using OBBs is not new and many researchers have used them extensively to speed up ray tracing and interference detection computations [2]. In this paper, we briefly describe the algorithms for computing tight-fitting OBBs and checking them for overlap. More details are given in [18].

4.2.1   Building an OBBTree

In this section we describe algorithms for building an OBBTree. The tree construction has two components: first is the placement of a tight fitting OBB around a collection of polygons, and second is the grouping of nested OBB's into a tree hierarchy.

We want to approximate the collection of polygons with an OBB of similar dimensions and orientation. We triangulate all polygons composed of more than three edges. The OBB computation algorithm makes use of first and second order statistics summarizing the vertex coordinates. They are the mean, μ, and the covariance matrix, C, respectively [14]. If the vertices of the i'th triangle are the points p^i, q^i, and r^i, then the mean and covariance matrix can be expressed in vector notation as:

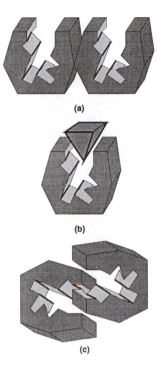

(a)

(b)

(c)

Figure 3: Feature classification based on the overlap between convex hulls

$$\mu = \frac{1}{3n} \sum_{i=0}^{n} (\mathbf{p}^i + \mathbf{q}^i + \mathbf{r}^i),$$

$$\mathbf{C}_{jk} = \frac{1}{3n} \sum_{i=0}^{n} (\overline{\mathbf{p}}_j^i \overline{\mathbf{p}}_k^i + \overline{\mathbf{q}}_j^i \overline{\mathbf{q}}_k^i + \overline{\mathbf{r}}_j^i \overline{\mathbf{r}}_k^i), \qquad 1 \leq j, k \leq 3$$

where n is the number of triangles, $\overline{\mathbf{p}}^i = \mathbf{p}^i - \mu$, $\overline{\mathbf{q}}^i = \mathbf{q}^i - \mu$, and $\overline{\mathbf{r}}^i = \mathbf{r}^i - \mu$. Each of them is a 3×1 vector, e.g. $\overline{\mathbf{p}}^i = (\overline{\mathbf{p}}_1^i, \overline{\mathbf{p}}_2^i, \overline{\mathbf{p}}_3^i)^T$ and $\mathbf{C}_{jk}$ are the elements of the 3 by 3 covariance matrix.

The eigenvectors of a symmetric matrix, such as $\mathbf{C}$, are mutually orthogonal. After normalizing them, they are used as a basis. We find the extremal vertices along each axis of this basis, and size the bounding box, oriented with the basis vectors, to bound those extremal vertices. Two of the three eigenvectors of the covariance matrix are the axes of maximum and of minimum variance, so they will tend to align the box with the geometry of a tube or a flat surface patch.

The basic failing of the above approach is that vertices on the interior of the model, which ought not in-

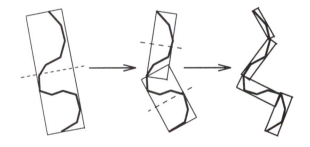

Figure 4: Building the OBBTree: recursively partition the bounded polygons and bound the resulting groups.

fluence the selection of a bounding box placement, can have an arbitrary impact on the eigenvectors. For example, a small but very dense planar patch of vertices in the interior of the model can cause the bounding box to align with it.

We improve the algorithm by using the convex hull of the vertices of the triangles. The convex hull is the smallest convex set containing all the points and efficient algorithms of $O(n \lg n)$ complexity and their robust implementations are available as public domain packages [4]. This is an improvement, but still suffers from a similar sampling problem: a small but very dense collection of nearly collinear vertices on the convex hull can cause the bounding box to align with that collection.

Given an algorithm to compute tight-fitting OBBs around a group of polygons, we need to represent them hierarchically. Most methods for building hierarchies fall into two categories: bottom-up and top-down. Bottom-up methods begin with a bounding volume for each polygon and merge volumes into larger volumes until the tree is complete. Top-down methods begin with a group of all polygons, and recursively subdivide until all leaf nodes are indivisible. In our current implementation, we have used a simple top-down approach.

Our subdivision rule is to split the longest axis of a box with a plane orthogonal to one of its axes, partitioning the polygons according to which side of the plane their center point lies on (a 2-D analog is shown in Figure 4). The subdivision coordinate along that axis was chosen to be that of the mean point, μ, of the vertices. If the longest axis cannot not be subdivided, the second longest axis is chosen. Otherwise, the shortest one is used. If the group of polygons cannot be partitioned along any axis by this criterion, then the group is considered indivisible.

4.2.2   Fast Overlap Test for OBBs

Given OBBTrees of two objects, the interference algorithm typically spends most of its time testing pairs of OBBs for overlap. A simple algorithm for testing the overlap status for two OBB's performs 144 edge-face tests. In practice, it is an expensive test. OBBs are convex polytopes and therefore, algorithms based on linear programming and closest features computation can be applied to check for overlap. However, they are relatively expensive.

One trivial test for disjointness is to project the boxes onto some axis (not necessarily a coordinate axis) in space. This is an 'axial projection.' Under this projection, each box forms an interval on the axis. If the intervals don't overlap, then the axis is called a 'separating axis' for the boxes, and the boxes must then

be disjoint. If the intervals do overlap, then the boxes may or may not be disjoint – further tests may be required. We make use of the *separating axis theorem* presented in [18] to check for overlaps. According to it, two convex polytopes in 3-D are disjoint iff there exists a separating axis orthogonal to a face of either polytope or orthogonal to an edge from each polytope. Each box has 3 unique face orientations, and 3 unique edge directions. This leads to 15 potential separating axes to test (3 faces from one box, 3 faces from the other box, and 9 pairwise combinations of edges). If the polytopes are disjoint, then a separating axis exists, and one of the 15 axes mentioned above will be a separating axis. If the polytopes are overlapping, then clearly no separating axis exists. So, testing the 15 given axes is a sufficient test for determining overlap status of two OBBs.

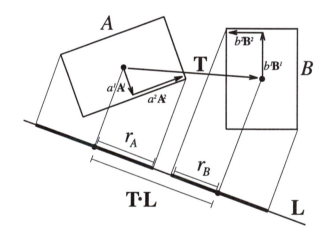

Figure 5: $\vec{L}$ is a separating axis for OBBs A and B because A and B become disjoint intervals under projection onto $\vec{L}$.

To perform the test, our strategy is to project the centers of the boxes onto the axis, and also to compute the radii of the intervals. If the distance between the box centers as projected onto the axis is greater than the sum of the radii, then the intervals (and the boxes as well) are disjoint. This is shown in 2D in Fig. 5. In practice, this corresponds to at most 200 arithmetic operations in the worst case [18]. Due to early exit (when the boxes are not overlapping), the algorithm takes about half the operations in practice. This algorithm for overlap detection between OBBs is about one order of magnitude faster than previous algorithms and implementations to check for overlap between OBBs.

5 Implementation and Performance

All these algorithms have been implemented and available as part of general purpose public domain packages. These are:

I_COLLIDE collision detection package available at **http://www.cs.unc.edu/~geom/I_COLLIDE.html**. It contains routines for the sweep and prune as well as the closest distance pair algorithm for convex polytopes. This package is applicable to environments, which can be described as union of convex polytopes. It has been widely used for dynamic simulation, architecture walkthrough and other applications.

RAPID interference detection package available at **http://www.cs.unc.edu/~geom/OBB/OBBT.html**. It contains routines for building the OBBTree data structure and fast overlap tests between two OBB-Trees. It has been used for virtual prototyping and simulation-based design applications. More details on their performance and robustness issues are given in [12, 18].

In practice, the algorithms based on OBBs asymptotically perform much better than hierarchies based on sphere trees or axis-aligned bounding boxes (like Octrees). The I_COLLIDE routines are able to compute all contacts between environments composed of hundred of convex polytopes at interactive rates (about 1/20 of a second). The OBBTree based interference detection algorithm (available as part of RAPID) has been applied to two complex synthetic environments to demonstrate its efficiency (as highlighted in Table 1). These figures are for an SGI Reality Engine (90 MHz R8000 CPU, 512 MB).

A simple dynamics engine exercised the collision detection system. At each time step, the contact polygons were found by the collision detection algorithm, an impulse was applied to the object at each contact before advancing the clock.

Scenario	Pipes	Torus
Environ Size	143690 pgns	98000 pgns
Object Size	143690 pgns	20000 pgns
Num of Steps	4008	1298
Num of Contacts	23905	2266
Num of Box-Box Tests	1704187	1055559
Num of Tri-Tri Tests	71589	7069
Time	16.9 secs	8.9 secs
Ave. Int. Detec. Time	**4.2 msecs**	**6.9 msecs**
Ave. Time per Box Test	7.9 usecs	7.3 usecs
Ave. Contacts per Step	6.0	1.7

Table 1: *Timings for simulations*

In the first scenario, the pipes model was used as both the environment and the dynamic object, as shown in Fig. 7. Both object and environment contain 140,000 polygons. The object is 15 times smaller in size than the environment. We simulated a gravitational field directed toward the center of the large cube of pipes, and permitted the smaller cube to fall inward, tumbling and bouncing. Its path contained 4008 discrete positions, and required 16.9 seconds to determine all 23905 contacts along the path. This is a challenging scenario because the smaller object is entirely embedded within the larger model. The models contain long thin triangles in the straight segments of the pipes, which cannot be efficiently approximated by sphere trees, octrees, and trees composed of axis-aligned bounding boxes , in general. It has no obvious groups or clusters, which are typically used by spatial partitioning algorithms like BSP's.

The other scenario has a complex wrinkled torus encircling a stalagmite in a dimpled, toothed landscape. Different steps from this simulation are shown in Fig. 8. The spikes in the landscape prevent large bounding boxes from touching the floor of the landscape, while the dimples provide numerous shallow concavities into which an object can enter. Likewise, the wrinkles and the twisting of the torus makes it impractical to decompose into convex polytopes, and difficult to efficiently apply bounding volumes. The wrinkled torus and the environment are also smooth enough to come into parallel close proximity, increasing the number of bounding volume overlap tests. Notice that the average number of box tests per step for the torus scenario is almost twice that of the pipes, even though the number of contacts is much lower.

We have also applied the RAPID library to detect collision between a moving torpedo on a pivot model (as shown in Fig. 6). These are parts of a torpedo storage and handling room of a submarine. The torpedo model is 4780 triangles. The pivot structure has 44921 triangles. There are multiple contacts along the length of the torpedo as it rests among the rollers. A typical collision query time for the scenario shown in Fig. 6 is 100 ms on a 200MHz R4400 CPU, 2GB SGI Reality Engine.

6 Acknowledgements

Thanks to Greg Angelini, Jim Boudreaux, and Ken Fast at Electric Boat for the model of torpedo storage and handling room. We would like to thank John Canny, Brian Mirtich and Krish Ponamgi for interesting discussions. Brian Mirtich and Krish Ponamgi also helped in the implementation of I_COLLIDE library. This research has been supported in part by a Sloan foundation fellowship, ARO Contract P-34982-MA, DARPA contract DABT63-93-C-0048, NSF grant CCR-9319957, NSF Grant CCR-9625217, ONR contract N00014-94-1-0738, NSF/ARPA Science and Technology Center for Computer Graphics & Scientific Visualization and NSF Prime contract No. 8920219.

References

[1] A.Garica-Alonso, N.Serrano, and J.Flaquer. Solving the collision detection problem. *IEEE Computer Graphics and Applications*, 13(3):36–43, 1994.

[2] J. Arvo and D. Kirk. A survey of ray tracing acceleration techniques. In *An Introduction to Ray Tracing*, pages 201–262, 1989.

[3] D. Baraff. *Dynamic simulation of non-penetrating rigid body simulation*. PhD thesis, Cornell University, 1992.

[4] B. Barber, D. Dobkin, and H. Huhdanpaa. The quickhull algorithm for convex hull. Technical Report GCG53, The Geometry Center, MN, 1993.

[5] N. Beckmann, H. Kriegel, R. Schneider, and B. Seeger. The r*-tree: An efficient and robust access method for points and rectangles. *Proc. SIGMOD Conf. on Management of Data*, pages 322–331, 1990.

[6] S. Cameron. Collision detection by four-dimensional intersection testing. *Proceedings of International Conference on Robotics and Automation*, pages 291–302, 1990.

[7] S. Cameron. Approximation hierarchies and s-bounds. In *Proceedings. Symposium on Solid Modeling Foundations and CAD/CAM Applications*, pages 129–137, Austin, TX, 1991.

[8] S. Cameron and R. K. Culley. Determining the minimum translational distance between two convex polyhedra. *Proceedings of International Conference on Robotics and Automation*, pages 591–596, 1986.

[9] J. F. Canny. Collision detection for moving polyhedra. *IEEE Trans. PAMI*, 8:200–209, 1986.

[10] H. Chang and T. Li. Assembly maintainability study with motion planning. In *Proceedings of International Conference on Robotics and Automation*, 1995.

[11] B. Chazelle and D. P. Dobkin. Intersection of convex objects in two and three dimensions. *J. ACM*, 34:1–27, 1987.

[12] J. Cohen, M. Lin, D. Manocha, and M. Ponamgi. I-collide: An interactive and exact collision detection system for large-scale environments. In *Proc. of ACM Interactive 3D Graphics Conference*, pages 189–196, 1995.

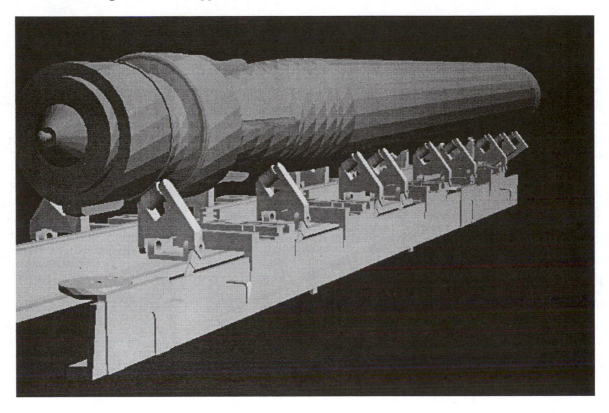

Figure 6: Interactive Interference Detection for a Torpedo (shown on the top) on a Pivot Structure – Torpedo has 4780 triangles; Pivot has 44921 triangles; Average time to perform collision query: 100 msec on SGI Reality Engine with 200MHz R4400 CPU

[13] D. P. Dobkin and D. G. Kirkpatrick. A linear algorithm for determining the separation of convex pol yhedra. *J. Algorithms*, 6:381–392, 1985.

[14] R.O. Duda and P.E. Hart. *Pattern Classification and Scene Analysis.* John Wiley and Sons, 1973.

[15] Tom Duff. Interval arithmetic and recursive subdivision for implicit functions and constructive solid geometry. *ACM Computer Graphics*, 26(2):131–139, 1992.

[16] P. Dworkin and D. Zeltzer. A new model for efficient dynamics simulation. *Proceedings Eurographics workshop on animation and simulation*, pages 175–184, 1993.

[17] J. Snyder et. al. Interval methods for multi-point collisions between time dependent curved surfaces. In *Proceedings of ACM Siggraph*, pages 321–334, 1993.

[18] S. Gottschalk, M. Lin, and D. Manocha. Obb-tree: A hierarchical structure for rapid interference detection. To Appear in Proc. of ACM Siggraph'96, 1996.

[19] J. K. Hahn. Realistic animation of rigid bodies. *Computer Graphics*, 22(4):pp. 299–308, 1988.

[20] P. M. Hubbard. Interactive collision detection. In *Proceedings of IEEE Symposium on Research Frontiers in Virtual Reality*, October 1993.

[21] J.C. Latombe. *Robot Motion Planning.* Kluwer Academic Publishers, 1991.

[22] M.C. Lin. *Efficient Collision Detection for Animation and Robotics.* PhD thesis, Department

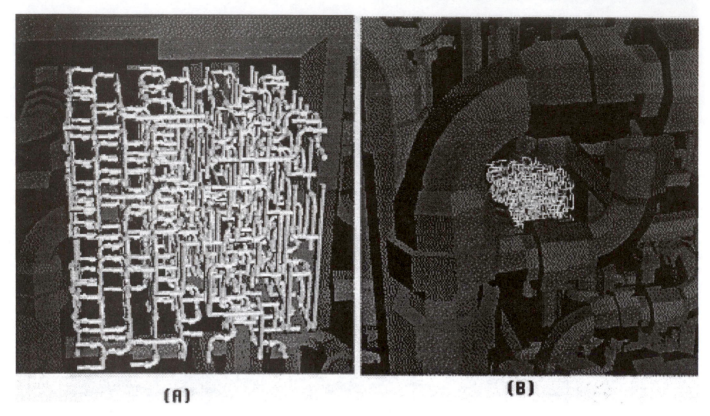

Figure 7: Interactive Interference Detection on Complex Interweaving Pipeline : 140, 000 polygons each; Average time to perform collision query: 4.2 msec on SGI Reality Engine with 90MHz R8000 CPU

of Electrical Engineering and Computer Science, University of California, Berkeley, December 1993.

[23] M.C. Lin and John F. Canny. Efficient algorithms for incremental distance computation. In *IEEE Conference on Robotics and Automation*, pages 1008–1014, 1991.

[24] M.C. Lin and Dinesh Manocha. Efficient contact determination between geometric models. *International Journal of Computational Geometry and Applications*, 1996. To appear.

[25] M. Moore and J. Wilhelms. Collision detection and response for computer animation. *Computer Graphics*, 22(4):289–298, 1988.

[26] M.Shamos and D.Hoey. Geometric intersection problems. *Proc. 17th An. IEEE Symp. Found. on Comput. Science*, pages 208–215, 1976.

[27] B. Naylor, J. Amanatides, and W. Thibault. Merging bsp trees yield polyhedral modeling results. In *Proc. of ACM Siggraph*, pages 115–124, 1990.

[28] M. Ponamgi, D. Manocha, and M. Lin. Incremental algorithms for collision detection between general solid models. In *Proc. of ACM/Siggraph Symposium on Solid Modeling*, pages 293–304, 1995.

[29] F.P. Preparata and M. I. Shamos. *Computational Geometry*. Springer-Verlag, New York, 1985.

[30] S. Quinlan. Efficient distance computation between non-convex objects. In *Proceedings of International Conference on Robotics and Automation*, pages 3324–3329, 1994.

[31] H. Samet. *Spatial Data Structures: Quadtree, Octrees and Other Hierarchical Methods.* Addison Wesley, 1989.

[32] W.Bouma and G.Vanecek. Collision detection and analysis in a physically based simulation. *Proceedings Eurographics workshop on animation and simulation*, pages 191–203, 1991.

[33] W.Bouma and G.Vanecek. Modeling contacts in a physically based simulation. *Second Symposium on Solid Modeling and Applications*, pages 409–419, 1993.

Figure 8: Interactive Interference Detection for a Complex Torus – Torus has 20000 polygons; Environment has 98000 polygons; Average time to perform collision detection: 6.9 msec on SGI Reality Engine

Largest Placements and Motion Planning of a Convex Polygon

Pankaj K. Agarwal, *Duke University, Durham, NC, USA*
Nina Amenta, *Xerox Palo Alto Research Center, Palo Alto, CA, USA*
Boris Aronov, *Polytechnic University, Brookly, NY, USA*
Micha Sharir, *Tel Aviv University, Tel Aviv, Israel and New York University, NY, USA*

We study two problems involving collision-free placements of a convex m-gon P in a planar polygonal environment: (i) We first show that the largest similar copy of P inside another convex polygon Q with n edges can be computed in $O(mn^2 \log n)$ time. We also show that the combinatorial complexity of the space of all similar copies of P inside Q is $O(mn^2)$, and that it can also be computed in $O(mn^2 \log n)$ time. (ii) We then consider the case where Q is an arbitrary polygonal environment with n edges. We give the first algorithm that constructs the entire free configuration space (the 3-dimensional space of all free placements of P in Q) in time that is near-quadratic in mn, which is nearly optimal in the worst case. The algorithm is also relatively simple. Previous solutions of the second problem were either incomplete, more expensive, or produced only part of the free configuration space. Combining our solution with parametric searching, we obtain an algorithm that finds the largest placement of P in Q in time that is also near-quadratic in mn. In addition, we describe an algorithm that preprocesses the computed free configuration space so that 'reachability' queries can be answered in polylogarithmic time.

1   Introduction

Problem statement. Let P be a (closed) convex m-gon. We consider two problems involving placements of P inside a (closed) planar polygonal *environment* Q bounded by a total of n edges. We allow P to translate, rotate, and sometimes also to scale. A *placement* of P is thus any congruent or similar copy of P (without reflections). A placement of P is *free* if it is fully contained in Q. A placement of a similar (resp. congruent) copy of P can be represented by four (resp. three) real parameters. The *free configuration space* $\mathcal{C}$ of P in Q is the space of all free placements of P in Q. In general, $\mathcal{C}$ is a four-dimensional space; if scaling is not

allowed, $\mathcal{C}$ is three-dimensional. There are two types of problems that we wish to consider in this setup:

Motion Planning: Assuming no scaling is allowed, construct the full configuration space $\mathcal{C}$. Also, preprocess $\mathcal{C}$ so that, given any two (congruent) placements of P, one can determine whether they lie in the same connected component of $\mathcal{C}$; that is, whether there exists a collision-free motion of P inside Q from one of these placements to the other.

Largest Placement: Allowing scaling, find the largest similar copy of P inside Q.

Previous results: General environment. Both problems are important basic problems in robotics and manufacturing, and have been studied intensively in computational geometry, starting about 15 years ago. Some of the initial results on this problem can be found in [6, 14, 21, 24]; these algorithms are either inefficient or consider only special cases (e.g., where P is assumed to be a line segment). The first significant development was made by Leven and Sharir [15], who showed that the combinatorial complexity of $\mathcal{C}$, when no scaling is allowed, is $O(mn\lambda_6(mn))$. Here $\lambda_s(q)$ is the maximum length of (q, s)-Davenport-Schinzel sequences [22], which is nearly linear in q for any fixed s. Thus the complexity of $\mathcal{C}$ is near-quadratic in mn. The goal then was to compute $\mathcal{C}$ in time that is also near-quadratic in mn. The first result in this direction is in [12], where an $O(mn\lambda_6(mn) \log mn)$-time algorithm was proposed. However, this algorithm turned out to have a technical difficulty: it computes a *superset* of all the vertices of $\mathcal{C}$ (where each such vertex is a free 'critical' placement of P in Q at which P makes three distinct contacts with the boundary of Q), and then aims to filter out the spurious (non-free) placements. However, this filtering was not handled correctly in some cases.

Two subsequent papers aimed to fix this problem. The first solution is given by Sharir and Toledo [23]. It processes Q into several range-searching data structures, and then it queries these structures with each placement of P produced by the algorithm of [12]. This allows us to detect and discard non-free placements, but the cost of each query is $O(m \log n)$, resulting in an overall complexity close to $O(m^3 n^2)$, which is significantly more expensive (when m is large, which is what we assume here). A second solution is given by Kedem et al. [13]. It remains within the time complexity $O(mn\lambda_6(mn) \log mn)$ of the algorithm of [12], but it may fail to produce the entire space $\mathcal{C}$. More precisely, given an initial free placement Z of P, the algorithm is guaranteed to construct the connected component of $\mathcal{C}$ that contains Z (which suffices for most motion planning applications), and may compute some other components, but it is not guaranteed to produce all components of $\mathcal{C}$. Consequently, their solution can determine in $O(mn\lambda_6(mn) \log mn)$ time whether there exists a collision-free path between two placements of P, but it cannot determine in time that is near-quadratic in mn whether a congruent copy of P can be placed inside Q.

The case in which scaling is allowed, and we seek the largest placement of P inside Q, has been studied in [8, 23]. Using generalized Delaunay triangulations induced by P in Q, Chew and Kedem [8] gave an $O(m^4 n^2 \alpha(n) \log n)$-time algorithm for computing a largest free similar placement of P in Q; here $\alpha(n)$ is the inverse Ackermann's function. Sharir and Toledo [23] proposed another algorithm that combines parametric searching [17] with a construction of the entire configuration space for the fixed-size case, as in the preceding paragraph; the running time of their algorithm is $O(m^2 n \lambda_4(mn) \log^3 mn \log\log mn)$, which is close to $O(m^3 n^2)$. If only translations and scalings are allowed, the largest homothetic placement of P inside Q can be computed in time $O(mn \log n)$, using the generalized Voronoi diagram of ∂Q induced by P [10, 16].

Previous results: Convex environment. One faces a much simpler situation when Q is also a convex polygon, although the resulting problems are still challenging and have an interesting geometric structure. They also have several applications (see, e.g, [5] for a computer vision application). The only previous attack on the problem, in which P is assumed to have a fixed size, is in [6]. It is shown there that one can determine in $O(mn^2)$ time whether P can be placed

inside Q. Computing a largest homothetic copy of P inside Q (i.e., allowing only translations and scalings) can be done in $O(m + n)$ time, using a linear-programming approach [23]. We are not aware of any previous work where P is also allowed to translate, rotate, and scale (except, of course, specializations of the algorithms mentioned above to the case where Q is convex; however, no improvements in the time bounds in such a specialization have been studied in these works).

New results and methods. We first study the problem where Q is convex, and we seek a largest placement of P inside Q. We show that such a placement can be computed in $O(mn^2 \log n)$ time. We also show that the combinatorial complexity of the space $\mathcal{C}$ of all similar placements of P inside Q is $O(mn^2)$, that this bound is tight in the worst case, and that $\mathcal{C}$ can also be computed in $O(mn^2 \log n)$ time. These results are obtained by a careful analysis of the structure of $\mathcal{C}$: we show that $\mathcal{C}$ can be represented as a convex polytope in 4-space with mn facets (this already implies that its complexity is $O(m^2 n^2)$). A more refined analysis yields the improved bounds noted above. To find a largest placement of P in Q, it suffices to consider a certain 2-dimensional projection of $\mathcal{C}$. We also analyze the structure of this projection, which has some additional properties, and give a simpler, $O(mn^2 \log n)$-time algorithm to compute this projection directly. Both algorithms are very simple to implement; they use some straightforward processing, followed by constructions of 3-dimensional convex hulls (for which optimal 'off-the-shelf' code is available).

We next study the case where Q is a general polygonal environment. We adapt a recent randomized algorithmic technique from [1, 3], to obtain a randomized algorithm that constructs (the boundary of) $\mathcal{C}$ in expected time $O(mn\lambda_6(mn) \log^2 mn)$. A somewhat more complex variant of the algorithm runs in expected $O(mn\lambda_6(mn) \log mn)$ time. This is the first correct solution whose running time is near quadratic in mn and which produces the entire configuration space $\mathcal{C}$. These algorithms (in particular the first one) are also relatively simple to implement (though not as simple as in the case of a convex Q, because here we need to process curved algebraic surfaces and arcs). Even for the task of computing only a portion of $\mathcal{C}$, our algorithms are simpler than the ones in [13, 23]. We can preprocess $\mathcal{C}$ in additional $O(mn\lambda_6(mn) \log^2 mn)$ time so that we can answer efficiently *reachability queries*: for any two placements of P, we can determine in $O(\log^2 mn)$ time whether there is a collision-free motion from one to the other (i.e., whether they lie in the same connected component of $\mathcal{C}$).

Using an approach based on parametric searching, similar to that of [23], we can find the largest similar placement of P in Q, in randomized expected time $O(mn\lambda_6(mn)\log^5 mn)$, thus improving significantly over the previous bounds in [8, 23]. We note that the parametric searching requires an 'oracle' procedure that has to determine, for a given size of P, whether the corresponding $\mathcal{C}$ is nonempty, which we can do using our algorithm for computing the entire $\mathcal{C}$. Notice that we cannot use the algorithm by Kedem et al. [13] here, because it may miss some of the components of $\mathcal{C}$.

2 Largest Placement of One Convex Polygon Inside Another

Let P be a convex polygon with m edges and Q a convex polygon with n edges. Our goal is to find a largest similar copy of P inside Q (allowing translation, rotation, and scaling of P); see Figure 1.

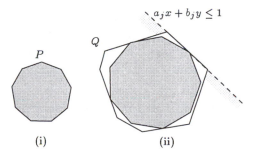

Figure 1: *(i) The polygon P; (ii) The polygon Q and a largest copy of P inside Q*

The geometric setup of the problem is as follows. We observe, following Baird [4], that similar placements of P can be parameterized nicely by referring to an arbitrarily chosen reference point $p \in P$. A placement is represented by a quadruple (s, t, u, v), where (u, v) is the position of p in the plane, and $s = \rho\cos\theta$, $t = \rho\sin\theta$, where P is rotated by θ and scaled by ρ, around p. The standard placement puts p at the origin, with $\rho = 1$, $\theta = u = v = 0$. Thus if (x, y) is a vertex of

P in the standard placement, its position at the placement (s, t, u, v) is $(sx - ty + u, tx + sy + v)$. Such a placement of P lies fully within Q if and only if every vertex (x_i, y_i) of P lies in every halfspace $a_j x + b_j y \leq 1$ containing Q and bounded by the line supporting an edge of Q (here we assume, without loss of generality, that Q contains the origin). That is, the placement (s, t, u, v) must satisfy the following system of mn linear inequalities:

$$a_j(sx_i - ty_i + u) + b_j(tx_i + sy_i + v) \leq 1$$

or

$$L_{i,j} : (a_j x_i + b_j y_i)s + (-a_j y_i + b_j x_i)t + a_j u + b_j v \leq 1.$$

In other words, the space $\mathcal{C}$ of all similar placements of P inside Q is a 4-dimensional convex polyhedron formed by the intersection of mn halfspaces (it is also easily seen to be bounded). This already implies that the combinatorial complexity of $\mathcal{C}$ is $O(m^2 n^2)$, and that it can be constructed in $O(m^2 n^2)$ time [19]. However, we will improve this bound in what follows, exploiting the fact that $\mathcal{C}$ is highly degenerate. The main results of this section are:

Theorem 1 *The vertices of the projection of $\mathcal{C}$ onto the st-plane can be computed in time $O(mn^2 \log n)$.*

Theorem 2 *The total number of vertices of $\mathcal{C}$ is $O(mn^2)$, and they can be computed in time $O(mn^2 \log n)$.*

Remark. Although Theorem 1 follows immediately from Theorem 2, we give a direct proof of Theorem 1, which is somewhat simpler and provides more geometric insight into the structure of the problem.

Proof of Theorem 1. We prove both theorems by applying the standard duality transform that maps a point $(\xi_1, \xi_2, \xi_3, \xi_4)$ to the hyperplane $\xi_1 s + \xi_2 t + \xi_3 u + \xi_4 v = 1$ and vice versa. We denote the coordinates in the dual space by s^*, t^*, u^*, v^*. The vertices of the polytope $\mathcal{D}$ dual to $\mathcal{C}$ are thus

$$w_{i,j} = (a_j x_i + b_j y_i, -a_j y_i + b_j x_i, a_j, b_j),$$

for $i = 1, \ldots, m$ and $j = 1, \ldots, n$. It is easy to verify that all the points $w_{i,j}$ are indeed extreme points of $\mathcal{D}$ (or, equivalently, that all the hyperplanes bounding the halfspaces $L_{i,j}$ contain facets of $\mathcal{C}$). Note that, for each fixed j (corresponding to an edge of Q), the convex hull

G_j of $\{w_{i,j}\}_{i=1}^m$ is a similar copy of P that lies in the 2-plane $\pi_j : u^* = a_j$, $v^* = b_j$. The dual polytope $\mathcal{D}$, then, is the convex hull of n similar copies of P, placed in parallel 2-planes in $\mathbb{R}^4$.

We exploit the well-known fact that projection in the primal is slicing in the dual. In more detail, let $\mathcal{C}_2$ denote the projection of $\mathcal{C}$ onto the st-plane $u = 0$, $v = 0$, as effected by the mapping $(s, t, u, v) \mapsto (s, t, 0, 0)$. Then a line $\alpha s + \beta t = 1$ in the st-plane is a supporting line of $\mathcal{C}_2$ if and only if the hyperplane $\alpha s + \beta t = 1$ is a supporting hyperplane of $\mathcal{C}$ in $\mathbb{R}^4$. This is equivalent, in the dual, to having the point $(\alpha, \beta, 0, 0)$ belong to the boundary of $\mathcal{D}$. Thus, computing $\mathcal{C}_2$ is equivalent to computing the cross section $\mathcal{D}_2$ of $\mathcal{D}$ with the 2-plane $u^* = 0$, $v^* = 0$.

Our strategy for computing $\mathcal{D}_2$ is first to compute $\mathcal{D}_3$, the cross section of $\mathcal{D}$ with the hyperplane $u^* = 0$, and then to slice $\mathcal{D}_3$ with the plane $v^* = 0$. Since it is trivial to intersect a three-dimensional polytope with a plane in time proportional to the complexity of the polytope, we only consider the construction of $\mathcal{D}_3$.

Without loss of generality, we can assume that none of the a_j's is 0. Then each of the polygons G_j lies outside the hyperplane $u^* = 0$. Hence, any vertex w of $\mathcal{D}_3$ must be an intersection of $u^* = 0$ with an edge of $\mathcal{D}$, connecting two vertices of a pair of distinct polygons, G_i and G_j, where G_i lies above $u^* = 0$ and G_j lies below. Moreover, w must also be a vertex of the intersection of the convex hull of $G_i \cup G_j$ with $u^* = 0$. So we can construct $\mathcal{D}_3$ by taking the convex hull, in $\mathbb{R}^4$, of every pair of polygons G_i, G_j, intersecting all of these sub-hulls with $u^* = 0$, and then taking the convex hull of the resulting intersections.

Let us consider the geometry of one such sub-hull. The two parallel 2-planes $u^* = a_i, v^* = b_i$ and $u^* = a_j, v^* = b_j$ lie in the common 3-plane $F_{i,j}$ defined by

$$(b_j - b_i)u^* + (a_i - a_j)v^* + (b_i a_j - b_j a_i) = 0$$

and so does the sub-hull determined by G_i, G_j. The three-dimensional geometry of $conv(G_i \cup G_j)$ in $F_{i,j}$ is as shown in Figure 2.

The intersection of $F_{i,j}$ with $u^* = 0$ is the 2-plane $u^* = 0$, $v^* = (b_i a_j - b_j a_i)/(a_i - a_j)$, which is also parallel to the two polygons G_i, G_j. By slicing the convex hull of the two parallel polygons with a parallel plane, we get a third parallel polygon $G_{i,j}$ which is the Minkowski sum of appropriately scaled copies of

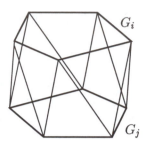

Figure 2: *Convex hull of parallel polygons*

G_i and G_j. This polygon has at most $2m$ vertices, and it is easy to compute directly from the vertices of G_i and G_j. Note that $G_{i,j}$ lies in both $F_{i,j}$ and in $u^* = 0$.

The 3-polytope $\mathcal{D}_3$ in $u^* = 0$ is the convex hull of all these polygons $G_{i,j}$. There are $O(n^2)$ such polygons, each with at most $2m$ vertices, so the total complexity of $\mathcal{D}_3$ is $O(mn^2)$ (which of course is also a consequence of the bound for the overall complexity of $\mathcal{D}$, as asserted in Theorem 2 and proven below).

The algorithm is simply to form the polygons $G_{i,j}$, take their three-dimensional convex hull, and intersect it with $v^* = 0$. Since the Minkowski sum of two convex polygons can be computed in linear time [11], we spend $O(mn^2)$ time in computing the polygons $G_{i,j}$. Their convex hull can be computed in $O(mn^2 \log n)$ time, using the divide-and-conquer algorithm of [18] (which has now only $O(\log n)$ recursive levels, because we start with the already available polygons $G_{i,j}$). Hence, the total running time is $O(mn^2 \log n)$. $\qquad \square$

Note that in practical terms, the implementation of this algorithm is a straightforward setup followed by a convex hull computation, which can be performed efficiently with publicly available software.

Proof of Theorem 2. We first consider the facets of $\mathcal{D}$ whose supporting hyperplanes are parallel to the 2-plane $u^* = 0, v^* = 0$. The equation of such a hyperplane h_F of a facet F has the form $\beta u^* + \gamma v^* + \delta = 0$. Hence, if h_F contains a vertex of some G_j, it must contain the entire polygon G_j. It then follows that F must be the convex hull of the union of two polygons G_i, G_j (as in the proof of Theorem 1 given above). The facet F is dual to the placement of P in which it is shrunk to a point and all its vertices are incident to the vertex of Q where edge i meets edge j (so that these two edges must be consecutive edges of Q). The number of such placements is n, and the complexity of each of the corresponding facets is $O(m)$, since it is the 3-dimensional convex hull of $2m$ points. (It is easily verified that each of these hulls is indeed a facet of $\mathcal{D}$.) It follows that the overall complexity of these facets of $\mathcal{D}$ is $O(mn)$. Constructing all these facets is easy to do in $O(mn)$ time.

Next, consider the facets of $\mathcal{D}$ whose supporting hyperplanes are not parallel to the 2-plane $u^* = 0$, $v^* = 0$. Let F be such a facet of $\mathcal{D}$, so that the equation of its containing hyperplane h can be written as $t^* = \alpha s^* + \beta u^* + \gamma v^* + \delta$. Then, for each $j = 1, \ldots, n$, the line ℓ_j of intersection between h and the 2-plane π_j containing G_j either touches G_j or is disjoint from G_j. The equation of ℓ_j is $t^* = \alpha s^* + \beta a_j + \gamma b_j + \delta$, $u^* = a_j$, $v^* = b_j$, Note that the coefficient α uniquely determines the vertex of G_j nearest to ℓ_j, for every j, unless α is a 'critical' value equal to the slope of an edge of some G_j. There are $\nu = mn$ such critical slopes α, corresponding to the orientations at which an edge of P is parallel to an edge of Q, and it is easy to compute them, in order, in time $O(mn \log n)$. Let $\alpha_1 < \alpha_2 < \cdots < \alpha_\nu$ be these critical slopes.

Let K be an open interval of α-coefficients between two successive critical slopes. Then, for each $j = 1, \ldots, n$, there exists a unique vertex $w_{i(K),j}$ of G_j, such that if h is any supporting hyperplane of $\mathcal{D}$, whose α-coefficient lies in K, then h can touch G_j, if at all, only at $w_{i(K),j}$. In other words, such an h is also a supporting hyperplane of $S_K = \{w_{i(K),j}\}_{j=1}^n$ (h must of course touch at least one of these vertices, and at least four if it contains a facet of $\mathcal{D}$). For two adjacent intervals K and K', the set $S_{K'}$ is obtained from S_K by replacing one vertex w by another vertex w' (both being adjacent vertices of some G_j). It easily follows that every facet F of $\mathcal{D}$ not parallel to $u^* = 0$, $v^* = 0$ is either a facet of $conv(S_K)$, for some interval K, or, if the α-coefficient of F is a critical value, a facet of $conv(S_K \cup S_{K'})$, for some pair of consecutive intervals K and K'. If the vertices of P and Q are in general position, these latter facets correspond to placements in which an edge of P is incident to an edge of Q. In fact, we can prove the following stronger claim. Assuming $\alpha_0 = -\infty$ and $\alpha_{\nu+1} = +\infty$, let K_i be the open interval (α_i, α_{i+1}) for $0 \le i \le \nu$, and let $w_i = S_{K_i} \setminus S_{K_{i-1}}$ for $1 \le i \le \nu$.

Lemma 1 *Every facet F of $\mathcal{D}$ that is not parallel to*

$u^* = 0$, $v^* = 0$ *is either a facet of the convex hull* $conv(S_{K_0})$ *or a facet of the convex hull* $conv(S_{K_{i-1}} \cup \{w_i\})$ *incident to w_i, for some $1 \le i \le \nu$.*

Proof: Let F be a facet of $\mathcal{D}$ that is not parallel to $u^* = 0$, $v^* = 0$ and that is not a facet of $conv(S_0)$. Let W be the set of vertices of F, and let $i \le \nu$ be the index such that the α-coefficient of the hyperplane supporting F lies in the (semi-open) interval $(\alpha_{i-1}, \alpha_i]$. Then, by the above argument, $W \subseteq S_{K_{i-1}} \cup \{w_i\}$. Suppose $j \le i$ is the largest index such that $w_j \in W$ (i.e., S_{K_j} is obtained from $S_{K_{j-1}}$ by inserting one of the points of W and deleting a point of $S_{K_{j-1}}$.) Then it is easily seen that $W \subseteq S_{K_{j-1}} \cup \{w_j\}$. Hence, F is a facet of $conv(S_{K_{j-1}} \cup \{w_j\})$ incident to w_j, as asserted. $\square$

This lemma suggests that we should compute $conv(S_{K_0})$ and, for each $1 \le i \le \nu$, we compute the facets of $conv(S_{K_{i-1}} \cup \{w_i\})$ incident to w_i. Since the hyperplanes containing the facets of $conv(S_{K_{i-1}} \cup \{w_i\})$ incident to w_i have only three degrees of freedom, this problem can be formulated as a three-dimensional convex hull problem, and can be solved in $O(n \log n)$ time; the number of these facets, as well as their overall complexity, is $O(n)$. Notice that the set S_{K_0} and the vertices w_i for $1 \le i \le \nu$ can be computed in $O(mn \log n)$. Repeating this algorithm for all $1 \le i \le \nu$ and computing $conv(S_{K_0})$, the algorithm produces a total of $O(mn^2)$ facets, of $O(mn^2)$ overall complexity, in time $O(mn^2 \log n)$.

These arguments already prove that the total number of facets of $\mathcal{D}$ is $O(mn^2)$, and that their overall complexity, and hence the overall complexity of $\mathcal{C}$, is $O(mn^2)$. Unfortunately, the algorithm might produce additional *spurious* facets, which are not facets of $\mathcal{D}$. Indeed, a facet F of $conv(S_{i-1} \cup \{\xi_i\})$ corresponds to a placement π of P such that there are at least 4 vertex-edge incidences between the vertices of P_π and the edges of Q, and F is spurious if $P_\pi \not\subseteq Q$. If the α-coefficient of F lies in the interval $K_{i-1} \cup K_i$, then it follows by definition that F cannot be spurious. However, if this α-coefficient lies in another interval K_j, for some $j \notin \{i-1, i\}$, then F may be spurious, because P_π may violate a constraint $L_{u,v}$ corresponding to some vertex $w_{u,v} \in S_j \setminus (S_{i-1} \cup S_i)$. An example of such a spurious facet is given in the full version of the paper.

Hence, to complete our algorithm, we need to detect and discard the facets of the hulls $conv(S_K)$ which are not facets of $\mathcal{D}$. This is accomplished as follows. We triangulate each computed facet F into $O(|F|)$ tetrahedra, using the bottom-vertex triangulation scheme described in [7]. Let Δ denote the set of resulting tetrahedra; $|\Delta| = O(mn^2)$. Let $\mathcal{D}^*$ be the bottom-vertex triangulation of the boundary of $\mathcal{D}$. We want to discard those tetrahedra of Δ that are not facets of $\mathcal{D}^*$. For a vertex w, let $\Delta_w \subseteq \Delta$ be the subset of tetrahedra incident to w, and let V_w be the set of vertices of the tetrahedra in Δ_w. It is easily verified that a tetrahedron $\Delta \in \Delta_w$ is a facet of $\mathcal{D}^*$ if and only if Δ is a tetrahedron in the bottom-vertex triangulation of the boundary of $conv(V_w)$, which is necessarily incident to w. We therefore compute the facets of $conv(V_w)$ that are incident to w, by the reduction, noted above, to a 3-dimensional convex hull construction, and then compute the bottom-vertex triangulation of each such facet. Note that these facets can be computed in $O(|V_w| \log n)$ time, since the vertices of V_w lie on only n 2-planes, so that the convex hull computation requires only $O(\log n)$ recursive levels; we omit the easy details. We can now discard those tetrahedra in Δ_w that do not lie on the boundary of $conv(V_w)$. Repeating this procedure for all vertices w of $\mathcal{D}$ gets rid of all spurious facets computed by the algorithm.

The running time of this step is $\sum_w O(|V_w| \log n)$, where the sum extends over all vertices w of $\mathcal{D}$. Since $\sum_w |V_w| = 4|\Delta| = O(mn^2)$, the total time spent is $O(mn^2 \log n)$. This completes the proof of Theorem 2. $\square$

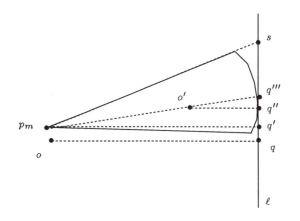

Figure 3: *Proof of claim*

An immediate corollary of Theorems 1 and 2 is the following.

Corollary 1 *The largest similar copy of P inside Q can be computed in $O(mn^2 \log n)$ time.*

We conclude this section by constructing a pair of polygons P and Q, with m and n vertices, respectively, such that there are $\Omega(mn^2)$ placements of P inside Q, each of which induces four incidences of the form (p, e), where p is a vertex of P and e is an edge of Q. This implies that the combinatorial bound of Theorem 2 is tight in the worst case.

The construction is depicted in Figure 4. Let n be of the form $2l + 2$, for some positive integer l, m an even integer, and o the origin. The first $n/2$ vertices $q_1, \ldots, q_{n/2}$ of Q are evenly distributed along the arc of the unit circle, centered at o, which goes from $-\pi/6$ to $\pi/6$ (in counterclockwise direction). The vertices $q_{n/2+1} \ldots q_n$ are evenly distributed along a tiny arc of a larger circle, say the circle with radius $10 + \varepsilon$ and center $(10, 0)$, and we let the tiny arc span the orientations between $\pi - \frac{\varepsilon}{2(10+\varepsilon)}$ and $\pi + \frac{\varepsilon}{2(10+\varepsilon)}$, so that its arc length is ε. The value of ε will be chosen sufficiently small, in a manner to be detailed in a moment.

We place one vertex p_m of P at the origin o and the remaining $m - 1$ vertices, equally spaced, on a circular arc of radius $1/4$, centered at $(3/4, 0)$, that spans the orientations between $-\frac{\pi}{40l}$ and $+\frac{\pi}{40l}$ from the center of that circle.

Claim: *If ε is chosen sufficiently small then the following holds. For every triple $n/2 + 1 \leq i \leq n$, $1 \leq j < n/2$, and $1 \leq k \leq m - 2$, there is a placement of P inside Q, using translation, rotation, and scaling, such that the vertex p_m of P coincides with the vertex q_i of Q, and such that the edge $p_k p_{k+1}$ of P coincides with the edge $q_j q_{j+1}$ of Q.*

Notice that every such placement of P induces four vertex-edge incidences between P and Q, and is thus a vertex of $\mathcal{C}$.

Proof: We consider the scaling, rotation, and translation of P that places $p_k p_{k+1}$ on the line ℓ supporting $q_j q_{j+1}$ and also places p_m at q_i.

As in Figure 3, let q be the center of edge $q_j q_{j+1}$; q is also the orthogonal projection of the origin o onto the line ℓ. Let q' be the projection of p_m, which is placed at q_i, onto ℓ. Let q'' be the projection onto ℓ of o', the center of the small circle whose boundary contains

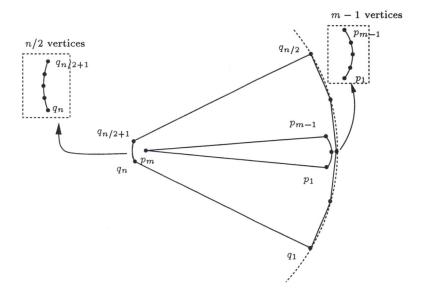

Figure 4: *Polygons P and Q for which there exist $\Omega(mn^2)$ similar placements of P in Q with four vertex-edge incidences per placement*

the points $p_1, \ldots, p_{m-1}$, which is appropriately shifted together with P. Let q''' be the intersection of the line from $p_m = q_i$ through o' with ℓ. Finally, let s be the intersection of the line supporting $p_m p_{m-1}$ (at this placement of P) with ℓ.

The distance from q to q' is at most ε. The angle $q''' p_m q'$ is the same as the angle $q''' o' q''$, which, by the construction of P, is at most $\frac{\pi}{40l}$. The angle $s p_m q'''$ is exactly $\frac{\pi}{40l}$. Since the distance from p_m to q' is at most $1 + \varepsilon$, the distance from q to s is

$$d(q, s) \le \varepsilon + (1 + \varepsilon) \tan \frac{\pi}{20l} .$$

Since the distance from q to q_{j+1} is $\sin \frac{\pi}{6l}$, ε can be chosen small enough so that

$$\varepsilon + (1 + \varepsilon) \tan \frac{\pi}{20l} < \sin \frac{\pi}{6l} ,$$

which then implies that this placement of P fully lies below the segment $p_m q_{j+1}$. An analogous argument shows that P lies above the segment $p_m q_j$, so P lies inside Q, as claimed.

We therefore obtain the following result.

Theorem 3 *There exist a convex m-gon P and another convex n-gon Q such that there are $\Omega(mn^2)$ placements of similar copies of P inside Q, each of which induces four vertex-edge incidences between P and Q.*

3 Placing a Convex Polygon in a General Polygonal Environment

We next consider the case where P is a convex m-gon translating and rotating in a general polygonal environment Q bounded by n edges (no scaling is allowed for the time being). As discussed in the introduction, the combinatorial complexity of the space $\mathcal{C}$ of free placements of P is $O(mn\lambda_6(mn))$, i.e., near-quadratic in mn, but there is no published algorithm that computes correctly the entire $\mathcal{C}$ in time that is close to this bound.

In this section, we propose a rather simple randomized technique for constructing the entire $\mathcal{C}$, with $O(mn\lambda_6(mn) \log^2 mn)$ expected running time. A somewhat more complicated randomized algorithm can compute $\mathcal{C}$ in expected time $O(mn\lambda_6(mn) \log mn)$. As it is usual for this type of algorithms, the expectation is over the random choices made by the algorithm, for any fixed input, and not over any assumed distribution of the input data.

In closing these introductory remarks, we note that the nice linear structure of the constraints defining $\mathcal{C}$ in the previous section does not exist here. Intuitively, this is because here we have a mixture of two types of constraints — those induced by contacts of vertices of P with edges of Q (as in the previous section) and those induced by contacts of edges of P with vertices of Q (which were absent in the previous analysis). One can choose coordinate frames in which either of these two types of constraints is linear, but then the other type is necessarily nonlinear.

Constructing $\mathcal{C}$. The space of all placements of P is three-dimensional, and can be parameterized by (x, y, θ) (or, preferably, by $(x, y, \tan\frac{\theta}{2})$), where (x, y) is the position of a reference point of P and θ is the angle by which P is rotated from some fixed reference orientation. For convenience, we make an assumption that simplifies our analysis. We triangulate the complement of Q, and from now on assume that it is the union of a set of n pairwise openly disjoint *triangular obstacles* (some of which may be unbounded); note that the new n is larger than the original n by a constant factor.

For each obstacle Δ, let $K(\Delta)$ denote the set of 'forbidden' placements of P at which it intersects the interior of Δ. These are open sets, and $\mathcal{C}$ is the complement of their union, so it suffices to compute (the boundary of) the union $K = \bigcup_\Delta K(\Delta)$. This leads to the following simple high-level description of our algorithm: Fix an obstacle Δ_0, and compute the the portion of ∂K that is contained in $\partial K(\Delta_0)$, which is the complement of $\bigcup_{\Delta \neq \Delta_0}(K(\Delta) \cap \partial K(\Delta_0))$. Hence, after applying this procedure to all obstacles Δ_0, we can 'glue' together these portions of ∂K to obtain (an appropriate discrete representation of) the entire boundary of K. We omit the details concerning the gluing process, as they are essentially the same as in the preceding algorithms [1, 3, 12].

Note that $\partial K(\Delta_0)$ consists of all (free or non-free) placements of P at which its boundary makes contact with $\partial\Delta_0$ and P and Δ_0 are openly disjoint. To simplify the algorithm, we partition $\partial K(\Delta_0)$ into $O(m)$ patches, each of which is the locus $\pi_{e,v}$ of all placements of P at which some fixed vertex v of P touches some fixed edge e of Δ_0, or some fixed edge e of P touches some fixed vertex v of Δ_0, and the interiors of P and Δ_0 are disjoint. If a patch is not xy-monotone, we further partition it into a constant number of xy-monotone patches. Such a partition is easy to obtain in $O(m)$ time. We refer to the patches $\pi_{e,v}$ as *contact surfaces*.

We thus obtain a collection of $O(mn)$ 2-dimensional contact surfaces $\pi_{e,v}$. For each such surface π, we com-pute the intersections $\Delta_\pi = \pi \cap K(\Delta)$, for all obstacles Δ, and construct, $K_\pi = \pi \setminus (\bigcup_\Delta \Delta_\pi)$, the complement of their union within π. K_π corresponds to placements at which v is in contact with e and P does not intersect the interior of any obstacle. Gluing these complements together will give us ∂K, as above. We refer to the sets Δ_π as *virtual π-obstacles*.

Let $\pi = \pi_{e,v}$ be a fixed contact surface. We can parametrize π by $(\rho, \tan\frac{\theta}{2})$, where ρ measures the displacement along e of its contact with v, and where θ is the orientation of P. For an obstacle Δ, constructing Δ_π is easy: Note that, for any fixed θ, the locus of placements contained in Δ_π with orientations θ is a line segment. (Indeed, the only motion available for P in this set is translation parallel to e; the set of such translations at which the two convex polygons P and Δ intersect is a line segment.) The critical θ's at which the combinatorial nature of an endpoint of this segment changes are such that the line parallel to e through some vertex of P passes through some vertex of Δ. There are $O(m)$ such orientations, and Δ_π can easily be obtained by sorting and processing these orientations in increasing order. Hence, $\partial\Delta_\pi$ consists of $O(m)$ arcs. As shown in [21], each such arc is a section of an algebraic curve of degree at most 4. Δ_π can be computed in $O(m \log m)$ time. The total time needed to produce the sets Δ_π, over all Δ, is thus $n \times O(m \log m) = O(mn \log m)$.

We compute $K_\pi = \pi \setminus (\bigcup_\Delta \Delta_\pi)$ using a randomized divide-and-conquer approach. We randomly divide the set of virtual π-obstacles into two equal subsets (so that every such partition occurs with equal probability), recursively compute the complement of their two unions in π, denoted by K_1, K_2, and compute $K_\pi = K_1 \cap K_2$ using a standard sweep-line technique. We assume, as is standard, an appropriate model of computation, in which various basic operations on the arcs forming the boundaries of the virtual obstacles (such as intersecting a pair of such arcs) can be done in $O(1)$ time. Since every intersection point of an edge of K_1 with an edge of K_2 is a vertex of K_π, the total time spent in dividing and in the merge step is $O((|K_\pi| + |K_1| + |K_2|) \log mn)$, where $|K_\pi|$, $|K_1|$ and $|K_2|$ are the number of vertices of these respective sets. If we let κ_π denote the total number of vertices in all the intermediate unions (of all recursive subproblems) produced by the algorithm, then the total running time of the algorithm (applied to a fixed π), including the time spent in computing the virtual π-obstacles, is $O((mn + \kappa_\pi) \log mn)$.

Applying this procedure to each of the $O(mn)$ contact surfaces independently and gluing the results together, we construct ∂K in time $O((m^2 n^2 + \sum_\pi \kappa_\pi) \log mn)$, where the summation is taken over all contact surfaces. We will prove below that the expected value of $\sum_\pi \kappa_\pi$ is $O(mn\lambda_6(mn) \log mn)$, which implies that the expected running time of the overall algorithm is $O(mn\lambda_6(mn) \log^2 mn)$. Hence, we can conclude:

Theorem 4 *Given a convex polygon P with m edges and a polygonal environment Q with a total of n edges, we can compute the entire free configuration space C by a randomized algorithm in expected time $O(mn\lambda_6(mn) \log^2 mn)$.*

Note that there is an alternative, randomized incremental approach to construct each K_π, in which we add the Δ_π's one at a time, in a random order, and maintain a 'trapezoidal' decomposition of the complement of their union; see [1, 3, 9]. The analysis of this technique is fairly standard, so we omit it here. The expected running time of this approach is only $O(mn\lambda_6(mn) \log mn)$, so this technique is slightly faster, but somewhat more complicated.

Bounding the expected value of $\sum_\pi \kappa_\pi$. In the remainder of this section, we establish the upper bound on the expected value of $\sum_\pi \kappa_\pi$, as stated above. For simplicity, assume that n, the total number of (triangular) obstacles, is of the form $2^h + 1$ for some integer h. Any vertex ζ that can appear on an intermediate union U produced by the algorithm, while computing K_π for some contact surface π, is an intersection of the boundaries of some pair of virtual π-obstacles (ignoring vertices of individual virtual π-obstacles, whose global number is easily shown to be $O(m^2 n^2)$). Therefore ζ represents a placement of P at which (a) P makes three simultaneous contacts with the obstacle boundaries, and (b) P is openly disjoint from the union of the obstacles Δ whose corresponding virtual π-obstacles participate in U, and from Δ_0, the obstacle for which π is a portion of $\partial K(\Delta_0)$.

A *triple-contact vertex* is a (not necessarily free) placement of P at which ∂P makes three simultaneous (vertex-edge or edge-vertex) contacts with ∂Q, so that P is locally free at a neighborhood of each contact. We say that any triple-contact vertex ζ has *level*

k (with respect to the full collection of obstacles) if removal of some k other obstacles (excluding the at most three that participate in the triple contact) causes ζ to become a free placement, relative to the remaining obstacles, and no set of fewer than k obstacles has this property. Note that level-0 vertices are exactly the vertices of C. Let F_k denote the number of level-k vertices for the given P and Q, and let $G(r)$ denote the expected number of level-0 vertices for P in an environment obtained by picking a random sample of r of the n obstacles, where any subset of r obstacles is chosen with equal probability. Fix a level-k vertex and let p_k denote the expected number of recursive subproblems of size r which contain it in the output. Then the expected value of $\sum_\pi \kappa_\pi$ is easily seen to be

$$\mathrm{E}\left[\sum_\pi \kappa_\pi\right] = \sum_{k=0}^{n-3} F_k \cdot p_k.$$

We first obtain a bound on p_k. Note that, throughout the execution of the algorithm, it encounters sets of virtual obstacles of cardinality 2^i, for $i = 0, \ldots, h$. Fix one such i. Consider any three obstacles, and fix a triple-contact vertex ζ of the free configuration space defined when only these three obstacles are present. What is the probability that ζ occurs during the execution of the algorithm, for any contact surface π, while processing subproblems involving $r = 2^i$ obstacles? The previous discussion implies that ζ lies at the intersection of three contact surfaces. Fix one of these contact surfaces π. Suppose ζ is a level-k vertex, with respect to the full set of obstacles. Then ζ appears in some fixed subproblem involving r obstacles in the construction carried out within π if and only if these r obstacles include the other two contact obstacles and do not include any of the k obstacles that "cover" ζ. Since every set of r obstacles (excluding the obstacle inducing π) has the same probability of being the set of input obstacles to our fixed subproblem, the probability of ζ appearing in the output of the subproblem is $\frac{\binom{n-3-k}{r-2}}{\binom{n-1}{r}}$. (Recall that we ignore vertices that are determined by fewer than three obstacles; these vertices show up as vertices of some virtual π-obstacle, so we already have a bound on their number, as above.) Hence,

$$p_k \leq \sum_{i=0}^{h} 3 \cdot 2^{h-i} \frac{\binom{n-3-k}{2^i-2}}{\binom{n-1}{2^i}}.$$

Here we used the fact that ζ may appear in the construction in the three different contact surfaces that define π, and that, in any fixed recursive construction within π, there are 2^{h-i} subproblems involving 2^i obstacles each. Hence,

$$
\begin{aligned}
\mathrm{E}\left[\sum_{\pi}\kappa_{\pi}\right] &\leq \sum_{k=0}^{n-3}\left(F_k\sum_{i=0}^{h}3\cdot 2^{h-i}\frac{\binom{n-3-k}{2^i-2}}{\binom{n-1}{2^i}}\right) \\
&= \sum_{i=0}^{h}3\cdot 2^{h-i}\sum_{k=0}^{n-3}\frac{\binom{n-3-k}{2^i-2}}{\binom{n-1}{2^i}}F_k . \quad (1)
\end{aligned}
$$

To bound this sum, we express $G(r)$ in terms of F_k. What is the probability that a level-k vertex ζ defined by three contacts, as above, is counted in $G(r)$? In other words, what is the probability that it corresponds to a vertex of the free configuration space, in the environment defined by r randomly selected obstacles? It is defined by three obstacles and "covered" by k other obstacles, so the probability is $\frac{\binom{n-3-k}{r-3}}{\binom{n}{r}}$. Thus, the expected number of free vertices (each defined by three obstacles) arising in the r-sample is

$$
G(r) = \sum_{k=0}^{n-3}\frac{\binom{n-3-k}{r-3}}{\binom{n}{r}}F_k .
$$

Putting $r = 2^i + 1$, we obtain

$$
\begin{aligned}
G(2^i + 1) &= \sum_{k=0}^{n-3}\frac{\binom{n-3-k}{2^i-2}}{\binom{n}{2^i+1}}F_k \\
&= \frac{2^i+1}{n}\sum_{k=0}^{n-3}\frac{\binom{n-3-k}{2^i-2}}{\binom{n-1}{2^i}}F_k . \quad (2)
\end{aligned}
$$

Substituting (2) into (1), we obtain

$$
\begin{aligned}
\mathrm{E}\left[\sum_{\pi}\kappa_{\pi}\right] &\leq \sum_{i=0}^{h}3\cdot 2^{h-i}\frac{n}{2^i+1}G(2^i+1) \\
&= O(n^2)\cdot\sum_{i=0}^{h}\frac{G(2^i+1)}{2^i(2^i+1)} .
\end{aligned}
$$

Notice that $G(2^i+1)$ is bounded by the combinatorial complexity of $\mathcal{C}$ for P moving amidst 2^i+1 obstacles, which, as noted above, is known to be $O(2^i m\lambda_6(2^i m))$, so the total expected output size of all subproblems is $O(mn\lambda_6(mn)\log n)$, as claimed.

Motion-planning queries for P. In order to answer reachability queries of the form "given two placements I and F of P, determine whether there is a collision-free path of P (inside Q) from I to F," we need to preprocess $\mathcal{C}$ into a data structure so that we can determine whether two query points lie in the same connected component of $\mathcal{C}$.

We first compute a refinement $\mathcal{C}^*$ of $\mathcal{C}$ and then preprocess $\mathcal{C}^*$ for point-location queries. We consider the collection of all boundary edges of the contact surfaces, and of the loci of points on these surfaces with vertical tangency (in the z-direction, where z is the parameter $\tan\frac{\theta}{2}$). For each of these curves γ, we draw a vertical segment from every point on γ in both the $(+z)$- and the $(-z)$-directions, until it intersects another contact surface. That is, we erect a vertical wall through γ within the connected component of $\mathcal{C}$ that contains γ. This gives the desired refinement $\mathcal{C}^*$. It can be shown that $\mathcal{C}^*$ is vertically convex (i.e., every line intersects any connected component of $\mathcal{C}^*$ in a connected interval), and that every two-dimensional face of the cross section of $\mathcal{C}^*$ with a plane parallel to the xz-plane is x-monotone. Following an argument similar to that in [1], one can show that the complexity of $\mathcal{C}^*$ is $O(mn\lambda_{10}(mn))$. This follows by showing that any pair of contact surfaces intersect at most 8 times within a fixed vertical wall. More details are given in the full version of the paper.

Two cells $f_1, f_2 \in \mathcal{C}^*$ lie in the same connected component of $\mathcal{C}$ if one can reach f_2 from f_1 by crossing only vertical walls of $\mathcal{C}^*$. We can now identify the faces of $\mathcal{C}^*$ that lie within the same connected component of $\mathcal{C}$ by a simple graph traversal of the edges of $\mathcal{C}^*$. The total time spent in this step is $O(mn\lambda_{10}(mn))$. Finally, we preprocess $\mathcal{C}^*$ for point-location queries, using the algorithm of Preparata and Tamassia, as described in [1]. Using this data structure, we can determine in $O(\log^2 mn)$ time whether two given placements lie in the same connnected component of $\mathcal{C}$. Hence, we can conclude:

Theorem 5 *Given a convex polygon P with m edges and a polygonal environment Q with a total of n edges, we can preprocess $\mathcal{C}$ in randomized expected time $O(mn\lambda_6(mn)\log^2 mn)$ into a data structure so that, for any two placements I and F of P, we can determine, in $O(\log^2 mn)$ time, whether there exists a collision-free motion of P from I to F.*

Finding the largest placement of P. As mentioned in the beginning of this section, we use the parametric-searching technique of Megiddo [17] to compute a largest collision-free similar placement of P inside Q. The parametric searching requires an 'oracle' procedure to determine, for a given scaling factor of P, whether the corresponding $\mathcal{C}$ is nonempty. Using Theorem 4, we can obtain an oracle that performs this task in expected time $O(mn\lambda_6(mn)\log^2 mn)$. An efficient implementation of the parametric searching, however, also requires a parallel algorithm for the oracle (in Valiant's comparisons model [25]). The only part of the above randomized algorithm that is difficult to parallelize is the sweep-line procedure used in the merge step, because a sweep-line algorithm is inherently sequential. We therefore perform the merge step in the parallel version using a different approach, based on segment trees, such as the one used in [2]. Omitting all further details from this version, we show that one can compute $\mathcal{C}$ in $O(\log^2 mn)$ parallel steps, using $O(mn\lambda_6(mn)\log mn)$ expected number of processors, in Valiant's comparison model. Megiddo showed that if the sequential algorithm for the oracle runs in time T_s and the parallel algorithm runs in time T_p using Π processors, then the parametric searching takes $O(T_p\Pi + T_sT_P\log\Pi)$ time. Hence, putting everything together, we can conclude:

Theorem 6 *Given a convex polygon P with m edges and a polygonal environment Q with a total of n edges, we can compute a largest free placement of P inside Q in randomized expected time $O(mn\lambda_6(mn)\log^5 mn)$.*

Open problems. We conclude this section by mentioning two open problems:

(i) What is the combinatorial complexity of the four-dimensional configuration space of all free placements of P in Q, when scaling is also allowed? Is it also near-quadratic in mn? (See the previous section for the case where Q is convex.)

(ii) How fast can one answer 'real' motion planning queries, where the output to a query should be a collision-free path connecting the two given placements (when such a path exists)? Can this be done in time that has a polylogarithmic overhead plus a cost that depends on the actual complexity of the path? The approach described above does not seem to yield such a performance.

Acknowledgements

Pankaj Agarwal has been supported by NSF Grant CCR-93–01259, an NYI award, and by matching funds from Xerox Corp. Nina Amenta has been supported by the Geometry Center, which is officially the Center for Computation and Visualization of Geometric Structures, supported by NSF/DMS-8920161. Boris Aronov has been supported by NSF Grant CCR-92-11541 and a Sloan Research Fellowship. Micha Sharir has been supported by NSF Grants CCR-94-24398 and CCR-93-11127, by a Max-Planck Research Award, and by grants from the U.S.-Israeli Binational Science Foundation, and the G.I.F., the German-Israeli Foundation for Scientific Research and Development.

The authors are grateful to Emo Welzl for helpful discussions, and to David Jacobs and Ronen Basri for bringing Baird's representation to our attention.

References

[1] P. Agarwal, B. Aronov, and M. Sharir, Computing envelopes in four dimensions with applications, *Proc. 10th ACM Symp. on Computationl Geometry* (1994), pp. 348–358.

[2] P. K. Agarwal, M. Sharir, and S. Toledo, Applications of parametric searching in geometric optimization, *J. Algorithms* 17 (1994), 292–318.

[3] B. Aronov and M. Sharir, The union of convex polyhedra in three dimensions, *Proc. 34th IEEE Symp. Found. Comput. Sci.* (1993), pp. 518–527.

[4] H. S. Baird, *Model-Based Image Matching Using Location*, Distinguished Dissertation Series, MIT Press, Cambridge, MA, 1984.

[5] R. Basri and D. Jacobs, Recognition using region correspondences, *Proc. 5th Int. Conf. Comput. Vision*, 1985, pp. 8-13.

[6] B. Chazelle, The polygon containment problem, in *Advances in Computing Research, Vol. 1: Computational Geometry* (F. P. Preparata, Ed.), JAI Press, London, England, 1983, pp. 1–33.

[7] B. Chazelle and J. Friedman, A deterministic view of random sampling and its use in geometry, *Combinatorica* 10 (1990), 229–249.

[8] L.P. Chew and K. Kedem, A convex polygon among polygonal obstacles: placement and high-clearance motion, *Comput. Geom. Theory Appls.* 3(2) (1993), 59–89.

[9] M. de Berg, K. Dobrindt, and O. Schwarzkopf, On lazy randomized incremental construction, *Discrete Comput. Geom.* 14 (1995), 261–286.

[10] S. Fortune, A fast algorit6hm for polygon containment by translation, *Proc. 12th Internat. Colloq. Automata, Languages and Programming*, 1985, pp. 189–198.

[11] L. Guibas, L. Ramshaw, and J. Stolfi, A kinetic framework for computational geometry, *Proc. 24th Annu. IEEE Sympos. Found. Comput. Sci.*, 1983, pp. 100–111.

[12] K. Kedem and M. Sharir, An efficient motion planning algorithm for a convex rigid polygonal object in 2-dimensional polygonal space, *Discrete Comput. Geom.* 5 (1990), 43–75.

[13] K. Kedem, M. Sharir and S. Toledo, On critical orientations in the Kedem-Sharir motion planning algorithm for a convex polygon in the plane, *Proc. 5th Canadian Conference on Computational Geometry* (1993), 204–209.

[14] D. Leven and M. Sharir, An efficient and simple motion planning algorithm for a ladder moving in two-dimensional space amidst polygonal barriers, *J. Algorithms* 8 (1987), 192–215.

[15] D. Leven and M. Sharir, On the number of critical free contacts of a convex polygonal object moving in two-dimensional polygonal space, *Discrete Comput. Geom.* 2 (1987), 255–270.

[16] D. Leven and M. Sharir, Planning a purely translational motion for a convex object in two–dimensional space using generalized Voronoi diagrams, *Discrete Comput. Geom.* 2 (1987), 9–31.

[17] N. Megiddo, Applying parallel computation algorithms in the design of serial algorithms, *J. ACM* 30 (1983), 852–865.

[18] F. Preparata and S. Hong, Convex hulls of finite sets of points in two and three dimensions, *Commun. ACM* 20 (1977), 87–93.

[19] F. P. Preparata and M. I. Shamos, *Computational Geometry: An Introduction*, Springer-Verlag, New York, 1985.

[20] F.P. Preparata and R. Tamassia, Efficient point location in a convex spatial cell-complex, *SIAM J. Comput.* 21 (1992), 267–280.

[21] J.T. Schwartz and M. Sharir, On the Piano Movers' problem: I. The case of a rigid polygonal body moving amidst polygonal barriers, *Comm. Pure and Appl. Math.* 36 (1983), 345–398.

[22] M. Sharir and P. K. Agarwal, *Davenport-Schinzel Sequences and Their Geometric Applications*, Cambridge University Press, New York, 1995.

[23] M. Sharir and S. Toledo, Extremal polygon containment problems, *Comput. Geom. Theory Appls.* 4 (1994), 99–118.

[24] S. Sifrony and M. Sharir, A new efficient motion planning algorithm for a rod in two-dimensional polygonal space, *Algorithmica* 2 (1987), 367–402.

[25] L. Valiant, Parallelism in comparison problems, *SIAM J. Comput.* 4(3) (1975), 348–355.

Dynamic Maintenance of Kinematic Structures

Dan Halperin, *Tel Aviv University, Tel Aviv, Israel*
Jean-Claude Latombe, *Stanford University, Stanford, CA, USA*
Rajeev Motwani, *Stanford University, Stanford, CA, USA*

We consider the following dynamic data structure problem. Given a collection of rigid bodies moving in 3-dimensional space and hinged together in a kinematic structure, our goal is to efficiently maintain a data structure that allows us to quickly answer range queries as the bodies move. This kinematic data structure problem arises in a variety of applications such as conformational search in molecular biology, simulation of hyper-redundant robots, collision detection, and computer animation. We study several models for dynamic maintenance of such structures and devise algorithms under these models. We obtain tight results on the worst-case, amortized, and randomized complexity of the data structure problem. For the offline version of the problem, we establish NP-hardness and provide efficient approximation algorithms.

1 Introduction

We study the following dynamic data structure problem: Given an articulated linkage in three-dimensional space, i.e., a collection of bodies (links) connected by joints, efficiently maintain a data structure allowing quick answers to range queries, as the bodies move (the joint parameters change). To be more concrete, let us assume that the data structure is a grid representation of the space occupied by the linkage, for some specified values of the joint parameters; for instance, it may store each grid cube intersected by the linkage in a hash-table. A query typically specifies a region in space (e.g., the region occupied by an obstacle) and asks whether the linkage intersects this region. Similarly, a query can ask whether the linkage lies within a certain distance from an obstacle, by enlarging the region occupied by the obstacle. The data structure is then used to quickly select small subsets of the bodies to which exact intersection/distance algorithms will be applied in order to answer the query. The initial construction of the data structure, for a given set of values of the joint parameters, is a well-studied problem that we will not address here; in fact, we do not assume a particular data structure, nor a specific technique to construct it, but we assume that the operations they support have certain costs. A more difficult, and novel, problem is to efficiently update this data structure as the joint parameters change. This problem is the subject of our paper.

To get a better understanding of our problem, consider a serial linkage with n links, such as a classical manipulator arm. We can construct n data substructures, e.g., n hash-tables each representing a grid occupancy by one link, in a coordinate system attached to the link. These rigid substructures need not be updated when joint parameters change, since each update can be viewed as a transformation of the coordinate system and stored as such. However, when a query is received, each substructure must be queried separately, after having computed the position of the query region relative to each of the links. This takes time proportional to the number n of substructures, a cost that we wish to avoid. Indeed, suppose that we receive a long sequence of updates that all modify the same joint parameter. In this limited case, a much better strategy is to decompose the linkage into two sublinkages at the joint being modified and build one data structure for each sublinkage. The cost of a query is then only twice the cost we would have incurred if we had computed a single data structure for the entire linkage. If we later receive another sequence of updates modifying another joint parameter, we can break and merge substructures accordingly. For a more involved sequence of updates and queries, this approach leads us to represent the entire linkage by a data structure that is a *dynamic* collection of substructures representing distinct sublinkages. We study strategies for maintaining the global data structure such that a sequence of updates and queries is answered in minimal time.

A possible application of our data structure is to improving the efficiency of path planners for robots with many links. Except in very limited cases, it is computationally infeasible to compute an explicit representation of the set of collision-free configurations – the free space – of such a robot. Instead, as discussed in [2], one can approximate the geometry and connectivity of this set by randomly sampling the configuration space. The configurations picked at random are checked for collision and the collision-free ones are retained as milestones. The distance (in the workspace) between the obstacles and the robot placed at a milestone is used to decide which pairs of milestones can safely be connected by straight paths (in the configuration space), yielding a network of milestones called a probabilistic roadmap. Several successful planners are based on this general scheme [3, 4, 19, 21, 22, 33]. However, they all spend most of their running time (typically, over 90%) computing distances or checking collision. Hence, faster distance computation will directly benefit these planners. Potential fields are also used by many planners, randomized [3] or otherwise [11], as a heuristic cost function guiding the search for paths of multi-link robots. These fields usually depend on the distance between the robot and the obstacles, e.g., they grow to infinity as this distance tends to 0 [23]. More efficient distance computation will speed up the potential field calculation. Note that different data structures have been proposed to represent the space occupied by robots for the purpose of collision checking and distance computation operations: single-level and hierarchical occupancy grids (e.g., [10, 20]), bounding boxes (e.g., [34]), and spherical approximations (e.g., [35]). The framework presented in this paper can be applied to all these representations.

Another application domain, molecular biology, motivated our kinematic data structure problem in the first place. The geometry of a molecule can be described by a collection of spheres, each representing an atom [8, 30]. As noted by Halperin and Overmars [18], the spheres fulfill certain properties that allow the construction of an efficient data structure (an occupancy grid represented by a hash-table) to answer range queries, where the range is a sphere whose radius is at most c times larger than the radius of the largest sphere in the original set, for some fixed constant c. For a molecule of n atoms, the structure uses $O(n)$ storage, requires $O(n)$ randomized preprocessing time, and allows a query to be answered in $O(1)$ time. This structure was proposed [18] for answering queries when the molecule is at a fixed configuration. However, due to possible rotation about bonds connecting atoms, certain molecules are highly flexible, e.g., drug ligands. Our results are relevant to the representation of such molecules [16].

The kinematic data structure problem is related to dynamic data structure problems previously studied in the computational geometry literature [6, 29, 31, 32]. However, in most of the earlier work, the dynamization is caused by addition or removal of objects. The novel aspect of our problem is that the set of objects is fixed and the dynamization is due to their motion. Another related problem is the representation of multibody systems (e.g., a collection of particles) for force calculation [1]. As in our problem, the bodies move but the set of bodies is fixed. However, due to the more involved mathematical form of the constraints on the relative motions of the bodies, prevailing solutions recompute the data structure from scratch at each step.

In the next section, Section 2, we present a formal model for our data structuring problem. In Section 3 we start by considering the case of path-like (serial) linkages and show that the complexity of this problem is $\Theta(\sqrt{n})$ for paths of length n when the measure of interest is the worst-case cost of an operation, i.e., we present an *optimal* strategy for maintaining the data structure so as to minimize over all input sequences of queries and updates the maximum time required by an operation. Our bounds apply unchanged to the amortized and randomized time measures. These results are extended to the case of trees in Section 4 based upon a novel tree decomposition strategy that may be of independent interest. We define a balance number κ for a tree T, and show that the worst-case and amortized complexity is $\Theta(\kappa)$. In some applications, such as sampling of configuration spaces in robot path planning or conformation search in molecular biology, the problem has an essentially offline nature, in that the entire sequence of queries and updates may be known in advance. In Sections 5 and 6 we discuss the *offline* version of the problem under two different cost measures. We show that already in the case of a path-like linkage, devising the best strategy is NP-hard, and we present efficient approximation algorithms for the problem. The approximation ratio achieved is bounded by 1.75 for one cost measure, and by $O(\log n)$ for the other. Some directions for further research are presented in Section 7. A particularly interesting direction of research is to apply the framework of competitive algorithms for online problems and we mention some preliminary results.

2 The Abstract Data Structure

In this section we give a formal description of our data structuring problem, henceforth referred to as the *kinematic data structure problem*. We are required to maintain a data structure D representing the space occupied by an articulated linkage L of n links with no closed loops. A real value, called the joint parameter, is associated with each joint and specifies the relative position of the two links connected by that joint. Since the linkage L is loop-free, it may be viewed as an abstract tree. It will be convenient in the later sections to switch to graph-theoretic terminology for the linkage, as follows. The linkage L is represented by a tree $T(V, E)$, where the links of L map to the vertices $V = \{v_1, v_2, \ldots, v_n\}$ and, for a joint connecting links i and j, there is an edge (v_i, v_j) in E connecting the corresponding vertices v_i and v_j. In the case of a serial linkage, the tree reduces to a single path.

The data structure D is a dynamic collection of substructures, each representing the space currently occupied by a connected subset of L, i.e., a subtree of T. Two distinct substructures represent disjoint subsets. Together, the substructures in D represent the entire linkage in that the corresponding subtrees are a decomposition of the tree T. The data structure D supports two operations, UPDATE and QUERY:

UPDATE(v_i, v_j, q) specifies a joint (edge) and requires changing the corresponding joint parameter to q.

QUERY requires a separate examination of all the substructures in D.

The details of the examination to be carried out by QUERY can vary from one task to another. A more general version of this operation would permit the specification of a subset of L and ask for the examination of only the substructures representing this subset. For simplification, we assume in this paper that QUERY always examines the entire structure. The goal of the algorithm maintaining D is to maintain the collection of substructures in D so as to be able to handle queries, given an input sequence with an arbitrary mix of UPDATE and QUERY operations.

As mentioned above, every data substructure in D corresponds to a subtree of T. The various subtrees in D may be viewed as a decomposition of T induced by the removal of a set of edges in E. Indeed, we will assume that at any time each edge in E is labeled as being BROKEN or MERGED, and that the decomposition of T is implicitly defined as being induced by the removal of the BROKEN edges. However, we do not actually remove any of the tree edges since in our model the topology of the tree or the linkage remains fixed over time. To update D, the algorithm uses two primitive operations, BREAK and MERGE, that may be applied to any edge (v_i, v_j) in T; if applied to a non-existent edge, the operations report failure.

BREAK(v_i, v_j): If v_i and v_j belong to the same substructure S of D, BREAK decomposes the sublinkage represented by S into two, by labeling the edge (v_i, v_j) as being BROKEN and partitioning S into two new substructures representing the resulting sublinkages. If v_i and v_j are not in the same substructure (i.e., (v_i, v_j) is already BROKEN), then BREAK leaves D unchanged.

MERGE(v_i, v_j): If v_i and v_j belong to two distinct substructures S_i and S_j, MERGE combines them into a single substructure representing the union of the two corresponding sublinkages connected by the joint corresponding to (v_i, v_j). If v_i and v_j are represented in the same substructure (i.e., (v_i, v_j) is already MERGED), then MERGE leaves D unchanged.

Essentially, a strategy for maintaining this data structure will control the partition of D into the various substructures so as to minimize the overall cost of processing the UPDATE and QUERY operations. For example, if the input sequence contains only an extremely small number of queries, the best strategy would be BREAK all edges and thereby decompose the tree into isolated vertices; conversely, when the input sequence contains only an extremely small number of updates, the best strategy would be to not BREAK any edge at all. Of course, in the applications we need to support a possibly arbitrary mix of these two operations and our paper is concerned with devising optimal strategies for adapting to that situation.

In order to be able to state precise results, we will assume throughout the rest of the paper that the cost of the primitive operations is as follows. Note that the cost models will change with the precise application, but our results should carry over after suitable modifications to the bounds obtained.

- BREAK(v_i, v_j) takes time $O(1)$ if the edge (v_i, v_j) is already BROKEN, and time $O(n_{ij})$ if it is MERGED, where n_{ij} is the size (number of vertices) of the subtree containing the two end-points v_i and v_j.

- MERGE(v_i, v_j) takes time $O(1)$ if the edge (v_i, v_j) is already MERGED, and time $O(n_{ij})$ if it is BROKEN, where n_{ij} is the size (number of vertices) of the resulting subtree with the two end-points v_i and v_j.

We will refer to the cost of BREAK and MERGE as described above as the TOTAL cost measure. An alternative cost measure, called MIN, is obtained by defining n_{ij} as the size of the smaller substructure (subtree) produced by BREAK or merged by MERGE. Depending on the application at hand, one of these measures may be more appropriate than the other. The MIN measure corresponds to the situation where a structure can be decomposed into two substructures by deleting the elements of the smaller substructure and then computing its representation, and two substructures can be MERGED into a common structure by inserting the elements of the smaller substructure into the larger substructure. Conversely, the TOTAL cost measure corresponds to the situation where the BREAK and MERGE operations involve completely destroying the old structures and recomputing the new structures from scratch.

The given costs correspond exactly to the data structure described by Halperin and Overmars [18]. This data structure is an occupancy grid stored in a hashtable representing a molecule modeled as a collection of spheres. The same data structure could be used to represent a linkage. The costs defined above assume that the links, as well as the query region, have approximately the same shape and size, and that there is not too much steric overlap between the links. The condition on the links is fulfilled by some modular robots [7, 12, 36]; when it is not satisfied, one may arbitrarily cut the larger links into smaller ones connected by fixed joints. The condition on the query region is often reasonable, since one is rarely interested

in computing the exact distance between the robot and an obstacle, when this distance is large. If the two conditions are not satisfied, the costs of the operations on the data structure may differ from those given above; then, the results presented in this paper would require to be reformulated appropriately.

The UPDATE operation can be implemented as follows: UPDATE(v_i, v_j, q) is implemented by first performing BREAK(v_i, v_j) and then storing the change q in the joint parameter at the edge (v_i, v_j). Thus, an UPDATE takes the same time as a BREAK. Of course, a series of updates not interspersed with any query can be accumulated and applied in a batch mode when a query finally arrives.

Finally, a QUERY takes time $O(k)$, where k is the number of BROKEN edges in T or the number of substructures in D. To justify this, we need to explain how a query is performed with our data structure. For clarity, let us assume that we are dealing with a serial linkage, i.e., the case where the corresponding tree is a path; the more general case can be dealt with in a similar fashion. Let S_1 denote the static structure (a hash table in the example above) containing (a representation of) the link v_1, let S_2 denote the next static structure along the path, and so on. We assume that the link v_1 is fixed in 3-space. Each static structure has a coordinate frame attached to it in which the links of the structure are described. The coordinate frame attached to S_1 is the universal frame in which the query regions will be given. For every pair of successive static structures S_i and S_{i+1}, we maintain a rigid transformation T_i which transforms points described in the frame of S_i so as to be described in the frame of S_{i+1}.

Given a query region Q, we query the structure S_1 with Q. Next, we transform Q into Q', using the transformation T_1, we query S_2 with Q', and so on. The final answer is easily deduced from the answers in all the structures S_i. For a path consisting of k static structures, the cost of the query is clearly $O(k)$.

Note that to update the joint value of an edge that lies between two static structures (i.e., not internal to any static structure), we simply update the transformation between the two structures. This takes $O(1)$ time, assuming that the update is applied to a BROKEN edge; as stated earlier, we will always BREAK an edge before applying an update.

3   Restriction to Paths

In this section we characterize the complexity of the update and query operations with respect to each of the following time measures: worst-case, amortized, and randomized. We begin by considering the case of paths and defer the extension to trees till the next section.

Theorem 1 *There is an algorithm for maintaining the kinematic data structure at a worst-case cost of $O(\sqrt{n})$ per operation, under both MIN and TOTAL cost measures.*

Proof: The idea is very simple: the algorithm chooses the initial state to be one where the BROKEN edges are spaced regularly along the path at intervals of $\sqrt{n}$. This state remains fixed throughout the processing of the input sequence. It is clear that any query can be answered in time $O(\sqrt{n})$ since that is the total number of BROKEN edges in the entire path. Further, any update operation has cost $O(\sqrt{n})$ under both MIN and TOTAL cost measures, since all subpaths are of length $\sqrt{n}$. Note that the algorithm will BREAK an edge for an update operation, but will then MERGE it right after that unless the edge is one of the initially BROKEN edges. The cost of the MERGE operation at most doubles the cost of the UPDATE operation.   □

As shown below, this bound is tight even for amortized and randomized time measures.

Theorem 2 *Any algorithm for maintaining the kinematic data structure must have a worst-case cost per operation that is $\Omega(\sqrt{n})$. The same holds for both amortized and randomized time measures, under both MIN and TOTAL cost measures.*

Proof: We first prove the worst-case lower bound using an adversarial approach. At any time, the adversary examines the state of the data structure being maintained by the algorithm. If the number of BROKEN edges exceeds $\sqrt{n}$, it requests a QUERY operation and this incurs a cost $\Omega(\sqrt{n})$. On the other hand, if the number of BROKEN edges is fewer than $\sqrt{n}$, then there exists a subpath with more than $\sqrt{n}$ vertices. In that case, the adversary inputs an operation which involves breaking the middle-most edge in this subpath. This costs $\Omega(\sqrt{n})$ regardless of whether we are using the MIN or the TOTAL cost measure.

Notice that this adversary is completely impervious to the strategy of the algorithm and can create an input sequence of arbitrary length where *each* operation costs $\Omega(\sqrt{n})$. Quite clearly then, the lower bound applies unchanged under the *amortized* time measure.

Finally, we extend our lower bound to the randomized case. Here we are allowing the algorithm to be randomized, and now the adversary can no longer look at the state of the data structure when choosing each operation in the input sequence. We modify the adversary strategy as follows: at each step, the adversary chooses to either supply an UPDATE operation or a QUERY operation, with equal probability; if it chooses an UPDATE operation, the edge to be updated is chosen uniformly at random. Suppose that the data structure has more than $\sqrt{n}$ BROKEN edges, then with probability $1/2$, the QUERY operation causes a cost of $\Omega(\sqrt{n})$. On the other hand, when the data structure has fewer than $\sqrt{n}$ BROKEN edges, the update operation is chosen with probability $1/2$ and the edge involved in this operation lies in a subpath of *expected* length $\Omega(\sqrt{n})$. It follows that the expected cost of each operation is $\Omega(\sqrt{n})$.   □

One way to get around the worst-case lower bound proved above is to consider the offline setting and the minimization of the total cost. Another approach would be to consider the online setting but employing competitive analysis, i.e., comparing the online algorithm's cost to the optimal offline cost. Both types of algorithms and analysis have their uses and applications. In Section 5 and 6 we will consider the offline setting in detail. The online setting is relatively open and we only have some preliminary results discussed in Section 7.

4   Generalization to Trees

To understand the case of trees, it is instructive to first examine the other extreme from paths, i.e., stars. Unlike in the case of paths, it is impossible to find a small number of edges whose removal decomposes the star into small subtrees and so it may seem that we will have to pay a significantly higher cost per operation. However, upon closer examination, it turns out that stars are much easier than paths provided we work with the MIN cost measure rather than the TOTAL cost measure (which will be considered later in Section 4.3).

One reason to focus on the MIN cost measure is that the cost of an UPDATE operation in a star is $O(1)$ for each edge, even if it is MERGED. Thus, for stars, the right solution is to not BREAK any edges at all, leading to $O(1)$ cost for a QUERY and $O(1)$ cost for an UPDATE.

Motivated by this insight, we make the following definition:

Definition 1 *In a tree T, the two subtrees resulting from the removal of an edge (v_i, v_j) are denoted T_i and T_j, according to which of these contains the two endpoints v_i and v_j. The* heaviness *of the edge is defined to be $\min\{|T_i|, |T_j|\}$, where $|T|$ denotes the size (number of vertices) of a tree T. An edge of heaviness k is said to be k'-heavy for any $k' \le k$.*

Basically, the "heaviness" of an edge is the cost of breaking or updating the edge. We extend the notion of heaviness to the entire tree.

Definition 2 *The* heaviness *of a tree T is the maximum heaviness of an edge in it. A tree T of heaviness k is said to be k'-heavy for all $k' \le k$.*

Based on this, we define the notion of balance number of a tree.

Definition 3 *The balance number κ of a tree T is the smallest integer k such that the removal of $k-1$ edges from T decomposes it into k subtrees $\tau_1, \tau_2, \ldots, \tau_k$ none of which is k-heavy. Such a decomposition is called a κ-balanced decomposition of T.*

Note that an edge that is not k-heavy in some τ_i could be k-heavy in the original tree T. In the above definition we are considering the heaviness of the edges in each τ_i with respect to that tree itself. Refer to Figure 1 for an example of a 4-balanced decomposition of a tree.

We now relate the balance number to the complexity of the kinematic data structure problem.

Theorem 3 *Let T be a tree with balance number κ. There is an algorithm that maintains the kinematic data structure at a worst-case cost of $O(\kappa)$ per operation, under the MIN cost measure.*

Proof: Let U be a set of at most $\kappa - 1$ edges in T which gives a κ-balanced decomposition into trees that are not κ-heavy. The idea is to keep the edges in U

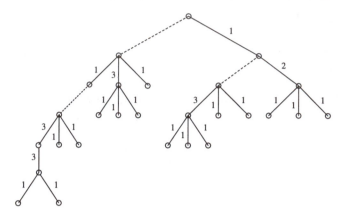

Figure 1: *A 4-balanced decomposition of a tree with balance number 4. The edges to be removed are shown as dashed lines and the resulting heaviness of each other edge is as shown.*

BROKEN. The cost of a QUERY is clearly at most κ. An UPDATE is also going to cost at most κ since in each of the induced subtrees, none of the edges are κ-heavy, □

This bound is tight, even for the amortized time measure.

Theorem 4 *Let T be a tree with balance number κ. Any algorithm for maintaining the kinematic data structure must have a worst-case cost per operation that is $\Omega(\kappa)$. The same holds for the amortized time measure.*

Proof: We first prove the worst-case lower bound using an adversarial approach. At any time, the adversary examines the state of the data structure being maintained by the algorithm. If the number of BROKEN edges is at least $\kappa - 1$, it inputs a QUERY operation and this incurs a cost $\Omega(\kappa)$. On the other hand, if the number of BROKEN edges is strictly less than $\kappa - 1$, then we claim that the resulting decomposition contains a subtree τ_j which is $(\kappa - 1)$-heavy. The claim follows from the observation that otherwise we would have at most $\kappa - 2$ deleted edges yielding a decomposition in which no resulting subtree is $(\kappa - 1)$-heavy, implying that the balance number of T is at most $\kappa - 1$ and thereby contradicting the assumption that T has balance number κ. By this claim, the adversary can identify a $(\kappa - 1)$-heavy subtree τ_j that must contain a $(\kappa - 1)$-heavy edge e — the adversary then inputs an UPDATE operation involving e and this costs $\Omega(\kappa)$.

Notice that this adversary is completely impervious to the strategy of the algorithm and can create an input sequence of an arbitrary length where *each* operation costs $\Omega(\kappa)$. Quite clearly then, the lower bound carries over to the *amortized* cost of operations without any changes. □

We have related the balance number κ to the data structure problem but we still have to devise an algorithm for computing the balance number κ and, in fact, for finding the $\kappa - 1$ edges that induce a balanced decomposition. This is not immediately obvious. The following lemma provides some insight.

Lemma 1 *For any tree T and any k, the set of k-heavy edges in T form a connected subtree of T.*

Proof: Let $e_1 = (v_1, v_2)$ and $e_2 = (v_3, v_4)$ be two k-heavy edges in T. Assume that v_2 is closer to e_2 than v_1, and that v_3 is closer to e_1 than v_4. We will show that the edges on the (unique) path from v_2 to v_3 are all k-heavy, and this will imply the desired result.

Suppose that the removal of the edge e_1 from T creates a subtree T_1 containing v_1 and a subtree T_2 containing v_2; similarly, the removal of the edge e_2 from T creates a subtree T_3 containing v_3 and a subtree T_4 containing v_4. Clearly, each of T_1, T_2, T_3, and T_4 has size at least k, since e_1 and e_2 are both assumed to be k-heavy.

Let $e = (v_5, v_6)$ be an edge on the path from v_2 to v_3, and suppose that the removal of the edge e from T creates a subtree T_5 containing v_5 and a subtree T_6 containing v_6. Clearly, T_1 is contained in T_5 and T_4 is contained in T_6, implying that both T_5 and T_6 have size at least k; therefore, e is k-heavy. □

4.1 Algorithm for Balanced Decomposition

We now describe a linear-time algorithm, Algorithm DFS-Decompose, for computing a κ-balanced decomposition of a tree with balance number κ. This algorithm assumes that the value of the balance number κ is provided along with the input tree; when κ is unknown, a binary search for the value of κ can be performed at the cost of increasing the running time to $O(n \log \kappa)$. Note that a slightly simpler version of this algorithm can be shown to compute a (2κ)-balanced decomposition of a tree with balance number κ; we omit the details.

Designate any arbitrary vertex r of T as its root. The algorithm is based on performing a depth-first search (dfs) of T starting at the root r. We assume that all edges are directed down from the root towards the leaves. The decision to delete an edge (u, v) directed from u to its child v is *usually* made when the dfs completes the traversal of all the vertices below u and is ready to leave u to move back up to the rest of the tree; in the sequel, we will refer to this as the *final return* to u. We assume throughout that the number of vertices remaining in the tree (as edges and subtrees are cut away) does not fall below 2κ; clearly, a tree with fewer than 2κ vertices cannot be κ-heavy and the algorithm can be terminated if the number of vertices ever falls below 2κ. Throughout the proof, the term "residual tree" will denote the subtree of T that remains at the point in time under consideration, with the removal of subtrees by deleting edges at earlier times in the dfs algorithm. The algorithm can track the total number of vertices remaining in the residual tree and can terminate whenever this falls below 2κ.

In the general case, suppose that the algorithm has just completed the traversal of the entire subtree rooted below a vertex u, and has just made a final return to u. Let the children of u be the vertices $v_1, \ldots, v_r$ and, for $1 \le i \le r$, let $e_i = (u, v_i)$. We will assume that the algorithm has recursively computed certain labels for each of the vertices $v_1, \ldots, v_r$. First, there is the label n_i for v_i which is the number of descendants of v_i (including itself) in the current residual tree. We assume that the children $v_1, \ldots, v_r$ are indexed in non-increasing order of n_i. In addition, some of these vertices may be "marked" and for each marked vertex, the algorithm associates a pointer to an edge p_i in the subtree rooted at that vertex and associates with the pointer an additional label m_i.

We will ensure in the algorithm that: *after the final return to a vertex u, the subtree rooted at u does not contain any κ-heavy edges.* This may have required deleting edges earlier while visiting the descendants of u. In fact, the algorithm's goal will be to try to ensure the following stronger condition: *after the final return to u, none of the subtrees rooted at a child v_i of u should have more than $\kappa - 1$ vertices.* It is easy to verify that if the root of a tree satisfies the second condition, then it satisfies the first condition, implying that the tree is not κ-heavy; however, the converse need not be true. As will soon become clear, the algorithm will sometimes be unable to achieve the second goal and then *exactly one* subtree (rooted at some v_i) with more than $\kappa - 1$ vertices will be allowed to remain, and that is precisely when a vertex v_i will be marked. At a marked vertex v_i, we store a pointer to an edge p_i directed from one of its descendants x_i to a vertex y_i such that y_i is the "lowest" descendant of v_i that has more than $\kappa - 1$ descendants (and hence is marked). The pointer's label m_i is the number of descendants of v_i that are non-descendants of y_i. The algorithm will ensure as an invariant that $m_i < \kappa - 1$.

If a child vertex v_i is marked, we were unable to meet our goal of ensuring that the final return to u should be with v_i's subtree having no more than $\kappa - 1$ vertices; however, we do not violate the requirement that upon final return to u the subtree rooted at u does not contain any κ-heavy edges. To understand how this is achieved, consider the decomposition of the edges of the subtree rooted at v_i into those contained in the tree rooted at y_i, and the remaining edges which span exactly m_i vertices. The former set of edges cannot contain any κ-heavy edges since by definition y_i is the lowest descendant of v_i with more than $\kappa - 1$ descendants and so cannot have any marked children. If the latter set of edges were to contain a κ-heavy edge, then we would delete the edge from u to v_i; observe that now the edges in the latter set could not be κ-heavy either since one of the two subtrees obtained by their deletion must have at most m_i vertices and $m_i < \kappa - 1$ by definition, implying that the subtree rooted at v_i is not κ-heavy. Note that, in general, it is entirely possible that an optimal solution would not delete the edge (u, v_i), preferring instead to delete an edge not contained in the subtree rooted at u to obtain a more fruitful deletion. As will soon become clear, our algorithm will only delete the edge (u, v_i) when absolutely essential for ensuring the invariants, and in that case we will be able to argue that some optimal solution must also delete this edge.

We are now ready to describe the recursive dfs-based computation. Suppose that at some point the algorithm has just made a final return to a vertex u. If $n_1 < \kappa - 1$, then the algorithm does not delete any of the edges $e_1, \ldots, e_r$; instead, it merely sets $n_u = 1 + \sum_{i=1}^{r} n_i$ and returns to u's parent. Otherwise, let s be the index such that $n_s \geq \kappa - 1$ and $n_{s+1} < \kappa - 1$; recall, the vertices v_i are in non-increasing order of number of descendants n_i. The algorithm has

two stages of deletion: the first stage, Stage A, is invoked only in the case $s \geq 2$; the second stage, Stage B, is applied in all cases, including the case where $s = 1$ and Stage A is applicable. During Stage A, the algorithm will cut away $s - 1$ of the subtrees of u, and thus this stage always produces a residual tree with $s = 1$ to which the algorithm then applies Stage B.

In Stage A, we have $s \geq 2$. Then out of the edges $e_1, \ldots, e_s$, the algorithm deletes all but one, say e_k. The index k is chosen as follows: if one of the vertices $v_1, \ldots, v_s$ is unmarked, then choose k such that v_k is an unmarked vertex; otherwise, all of them are marked, and the algorithm selects k such that m_k is the smallest amongst $m_1, \ldots, m_k$. Then, the algorithm deletes the edges $\{e_1, \ldots, e_s\} \setminus \{e_k\}$ and does not delete e_k or the edges $e_{s+1}, \ldots, e_r$. Notice that the resulting situation at u is the same as in the case where $s = 1$.

It remains to describe Stage B which the algorithm enters with $s = 1$. Since $s = 1$ implies that $n_i < \kappa - 1$ for $i > 1$, only v_1 can be a marked vertex. The algorithm computes $d = 1 + \sum_{i=2}^{r} n_i$ as the number of descendants of u (including itself) that are not descendants of v_1 (including itself). When v_1 is indeed a marked vertex, it defines $p = p_1$ and $m = m_1$; on the other hand, when v_1 is not marked, it defines $p = (u, v_1)$ and $m = 0$. If $m + d < \kappa - 1$, then the algorithm does not delete any edges and, before proceeding upwards from u, it labels u as follows: the label n_u is computed as $1 + \sum_{i=1}^{r} n_i$; u is marked; the pointer at u is to the edge p; and, the additional label $m_u = m + d$ is associated with this pointer. Conversely, if $m + d \geq \kappa - 1$, then the algorithm deletes the edge $e_1 = (u, v_1)$, updates n_u to be $1 + \sum_{i=2}^{r} n_i$, leaves u unmarked, and returns to u's parent.

It is easily verified that if u is marked, then the associated labels p_u and m_u satisfy the properties mentioned earlier. We remark that the algorithm does not really need to associate the pointer p with a marked vertex as the value of the pointer is never used; however, it will facilitate our analysis to be able to refer to the edge to which p points and that is the only reason why we ensure that the algorithm maintains such pointers.

This completes the description of the Algorithm DFS-Decompose. In Figure 2, we illustrate the execution of this algorithm on the tree shown in Figure 1.

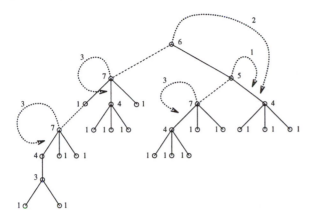

Figure 2: *Algorithm DFS-Decompose applied to a tree with balance number 4. The deleted edges are shown as dashed lines. For each vertex i we label it with the n_i value upon the return from that vertex to its parent. The pointers p_i at marked vertices are shown as dotted arrows labeled with the corresponding m_i values.*

4.2   Analyzing the Algorithm

Our goal in this section is to establish the following theorem.

Theorem 5 *Algorithm DFS-Decompose runs in time $O(n)$ on an input tree T with n nodes, and for any given integer κ, it returns a κ-balanced decomposition of T if it has balance number κ, and returns* FAILURE *otherwise.*

We analyze the performance and running time of Algorithm DFS-Decompose in the following sequence of lemmas. It should be clear that the lemmas combine to imply the proof of Theorem 5.

Lemma 2 *Algorithm DFS-Decompose runs in time $O(n)$.*

Proof: It is easy to verify that this algorithm runs in linear time. The depth-first search by itself runs in linear time. The additional work performed at a vertex u in maintaining vertex labels and choosing edges to delete can be charged to u and its incident edges such that only a constant amount of work is assigned to each vertex and edge.   □

Next, we show that that this algorithm produces a valid decomposition, i.e., none of the resulting subtrees are κ-heavy.

Lemma 3 *Algorithm DFS-Decompose produces a decomposition of the tree T such that none of the subtrees in the decomposition are κ-heavy.*

Proof: To this end, we establish the invariant that: *when the algorithm returns from a vertex u, the subtree rooted at u cannot be κ-heavy.* The invariant is easily seen to be true when the vertex u is unmarked, since then none of its subtrees are of size more than $\kappa - 2$. When u is marked, then consider the edge $p_u = (x, y)$ associated with u with corresponding weight label $m_u < \kappa - 1$. Since y is defined to be the lowest descendant of u that is marked, none of the subtrees of y are of size more than $\kappa - 2$, implying that none of the edges *below y* are κ-heavy since their deletion produces at least one subtree of size less than $\kappa - 1$. Further, the edge p_u itself and the edges below u that are not below y also cannot be κ-heavy. This is because the total number of vertices below u but not below y is $m_u < \kappa - 1$, and the deletion of these edges produces one subtree that does not contain vertices that are descendants of y.

Given the invariant, it is clear that when the algorithm returns from the root, the residual tree is not κ-heavy. However, we still need to show that the subtrees that were cut away from this residual tree are also not κ-heavy. Consider first the trees cut away upon a final return to a vertex u in Stage A, assuming that $s \geq 2$ and Stage A is invoked in the first place. Since each of these trees is rooted at a child v_i of u, and since the invariant also applies when the algorithm return from v_i to u, it is clear the cut subtrees are not κ-heavy. The only issue remaining is that of a subtree cut away in Stage B, assuming an edge is deleted at all in Stage B; the argument for this case is similar to that in the previous paragraph. The cut subtree is rooted at the marked vertex v_1 with pointer edge $p_1 = (x, y)$ and a label m_1. It is clear that none of the edges in the subtree rooted at y can be κ-heavy since no descendant of y has more than $\kappa - 2$ descendants of its own. The other edges cannot be κ-heavy since at least one of the subtrees resulting from their removal consists only of non-descendants of y; this subtree must have size at most $\kappa - 1$ since there are only $m_1 < \kappa - 1$ non-descendants of y in the subtree rooted at v_1.   □

Finally, the following lemma claims that the total number of edges deleted by this algorithm is no more than $\kappa - 1$. The proof of this lemma is fairly involved

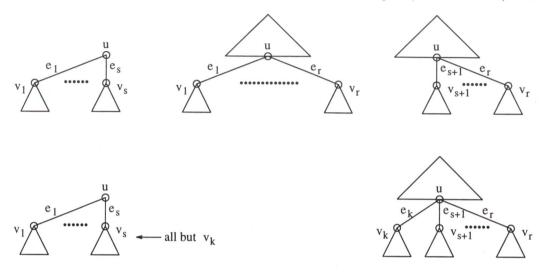

Figure 3: *An illustration for the subtrees referred to in the proof of Lemma 4. The top row shows the tree T (in middle), and the two subtrees T_1 (on left) and T_2 (on right). The bottom row shows the subtrees T_1' (on left) and T_2' (on right). Note that in T_1' the subtree rooted at v_k is not present.*

and is omitted due to lack of space; in fact, the rather complex nature of Algorithm DFS-Decompose is entirely due to the need to facilitate this proof of optimality. The complete proof can be found in the full version of this paper [17].

Lemma 4 *Algorithm DFS-Decompose produces a decomposition of T into at most κ subtrees.*

4.3   Extension to Trees with TOTAL Cost

We can show that essentially the same results as for paths carry over to the tree case with the TOTAL cost measure, provided we are willing to modify the notion of breaking an edge into that of breaking a vertex; clearly, this is not as satisfactory a resolution as in the case of the MIN measure with trees.

Note that now the star graph is a really bad case if we do not BREAK any edges, since each update has a linear cost. Also, breaking a sublinear number of edges is of no use since there will remain a subtree of linear size whose edges will require linear time to update, and of course breaking a linear number of edges will drive up the cost of a QUERY to a linear quantity.

The trick here is to modify our model slightly. Instead of breaking edges to decompose trees into small subtrees, we will use a different primitive operation that we will refer to as *breaking a vertex*. The idea

is to replace a vertex v by two copies of itself, say v_1 and v_2, with an edge between them and such that each edge incident on v is assigned to exactly one of the two copies v_1 and v_2. Then, breaking vertex v corresponds to breaking the (dummy) edge between v_1 and v_2. A possible physical interpretation of this is in terms of decomposing a link into a pair of sublinks. It can be argued that, at least in some applications, the resulting cost measures are an accurate reflection of the actual implementation discussed earlier.

It is now easy to see that there exists a choice of $O(\sqrt{n})$ vertices to break (and a choice of the assignment of incident edges to the two copies of a broken vertex) such that the tree decomposes into $O(\sqrt{n})$ subtrees, each of size $O(\sqrt{n})$. It is now possible to maintain the data structure at a cost of $O(\sqrt{n})$ per operation.

5   Offline Setting with TOTAL Cost

In this section, we focus on the offline version of the kinematic data structures when the linkage L has a path topology and using the TOTAL cost measure, henceforth referred to as the TOTAL problem. (We will consider the MIN version of this problem in the next section.) Interpreting the problem in geometric terms, we obtain that it is NP-complete. We also indicate briefly the known results for approximation to the problem.

5.1 The TOTAL Problem is NP-complete

Let $q_1, q_2, \ldots, q_m$ denote the ordered sequence of queries and recall that $v_1, v_2, \ldots, v_n$ denote the vertices in the path. We consider an $n \times m$ grid, lying inside a rectangle R, where each column stands for a query and each row stands for a vertex in the path. With a slight abuse of notation we will refer to the columns as $q_1, q_2, \ldots$ and to the rows as $v_1, v_2, \ldots$

The queries are interleaved with UPDATE operations. Suppose that between queries q_i and q_{i+1} we have an update operation UPDATE(v_j, v_{j+1}). In our geometric model this update operation translates into a point that lies at the intersection of the vertical grid line between columns q_i and q_{i+1} and the horizontal grid line between rows v_j and v_{j+1}. (Refer to Figure 4.) If there is more than one update operation between the queries q_i and q_{i+1}, they all translate into points on the same vertical grid line lying on the appropriate horizontal lines. Let $U = \{u_1, u_2, \ldots, u_N\}$ be the set of all the points corresponding to update operations.

R

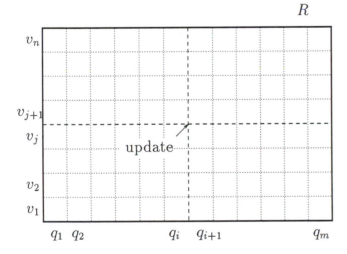

Figure 4: *A geometric interpretation of the TOTAL problem*

Consider an axis parallel rectangle whose height spans rows $v_i, v_{i+1}, \ldots, v_j$ and whose width spans columns $q_k, q_{k+1}, \ldots, q_l$. This rectangle represents the following situation: immediately before query q_k any BROKEN interior edge along the path $v_i, v_{i+1}, \ldots, v_j$ was MERGED, and each of the exterior edges of the path, namely the edges (v_{i-1}, v_i) and (v_j, v_{j+1}), was BROKEN (if it was previously MERGED), and all the

interior edges of this path remain MERGED until the completion of query q_l.

We claim that an optimal (i.e., minimum cost) solution for the problem corresponds to partitioning the entire grid rectangle R into rectangles $R_1, R_2, \ldots, R_t$ such that $\sum_{i=1}^{t}(h_i + w_i)$ is minimized, where h_i (w_i) is the length in unit grid size of the vertical (respectively, horizontal) edge of R_i, and such that no rectangle R_i contains a point of U in its interior.

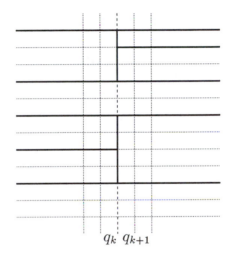

Figure 5: *The cost of restructuring between queries q_k and q_{k+1}*

To see this, consider first the intersection of a single column q_i with the rectangles R_i. The cost of the query q_i is the number of rectangles in the intersection, since we assumed that the cost of a query is equal to the number of substructures in the path. So we charge the cost of the query per rectangle R_i, to the portion of the lower horizontal edge of R_i that intersects the column q_i. Next consider the vertical grid line between the query columns q_k and q_{k+1}. (Refer to Figure 5.) The rectangle edges that appear on this vertical line correspond to the subpaths that have undergone change. The cost of these changes is exactly the length of these vertical rectangle edges. Recall that, if there are several update operations between queries q_k and q_{k+1} our planner executes them all at the same time, and the cost of restructuring a subpath of length s, by a collection of MERGE and BREAK operations is s.

The only constraint that our original problem imposes on the partitioning of R is that no rectangle in the partitioning contains a point of U in its interior. If a rectangle contains a point $u_j \in U$ in its interior, this implies that our data structure has not been updated by the update operation corresponding to the point u_j.

Lingas et al [27] have established the NP-completeness of partitioning an axis-parallel rectangle R with N point holes into axis-parallel rectangles with minimum total edge length, such that no rectangle in the partitioning contains a point hole in its interior. Hence, we obtain the following theorem

Theorem 6 *The TOTAL problem is NP-complete.*

5.2   Approximation Algorithms for TOTAL

A number of approximation algorithms have been proposed for the rectangular partition problem. Most of these algorithms rely on the connection between the rectangular partition as above and the so-called "guillotine" partition [9]. Finding the optimal guillotine partition is solvable in polynomial time, and it has been shown [9] that the optimal guillotine partition has edge length no greater than 1.75 times the length of the optimal rectangular partition.

Gonzalez and Zheng [9] give a 1.75 approximation bound for partitioning a rectangle with N point holes into axis-parallel rectangles, using dynamic programming. The running time of their algorithm is $O(N^5)$. Gonzalez, Razzazi, and Zheng [15] give a simple $O(N \log N)$ algorithm that obtains a 4-approximation for the same problem.

6   Offline Setting with MIN Cost

We now turn to the offline setting for path topologies under the MIN cost measure, which we call the MIN problem. As for the TOTAL measure, we interpret the MIN problem geometrically. We have a grid bounded inside a rectangle R. Each row corresponds to a vertex in the path; the order of the rows corresponds to the order of the elements in the kinematic chain. The columns correspond to queries and points on grid vertices correspond to updates.

Consider one column (i.e., one horizontal unit slab of the rectangle), say column q_i and suppose there are k horizontal line segments crossing this column besides R's edges. We interpret this as follows: at time q_i, the data structure consists of $k + 1$ static structures

describing rigid subpaths of the chain—therefore the cost of the query here is proportional to $k + 1$. The collection of all horizontal segments inside the rectangle and the length of one horizontal edge of the rectangle can be charged for all queries.

In between queries the structure may be reconfigured. The rebuilding of structures after a query depends on how they were organized just before the query. There are two ways to rearrange the structures:

- We can take any set of contiguous structures and rebuild them from scratch; the cost is obviously proportional to the number of elements in all these sets together and this will be expressed as a vertical line segment through all the rows corresponding to the elements.

- We can be more careful: in the rebuilding stage, recompute the new structures by removing elements from one structure and adding them to a neighboring structure. The cost is now proportional to the number of elements that are being moved around, and geometrically this is expressed as a vertical line segment through the rows corresponding to the moved elements. For an illustration, consider Figure 6: after query q_k the structure describing the vertices $v_i, \ldots, v_{i+4}$ is split into two. The cost of this restructuring is proportional to the shorter subpath.

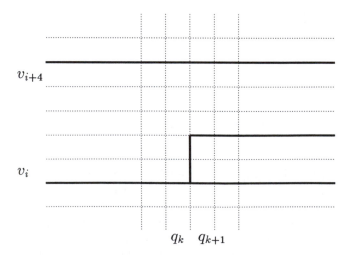

Figure 6: *A geometric interpretation of the MIN measure*

Definition 4 *A rectilinear polygon is a polygon with each side parallel to a coordinate axis.*

Definition 5 *A rectilinear polygon is vertically convex if every vertical line intersects it in at most one connected component.*

We will abbreviate *vertically convex rectilinear polygon* to VCRP.

Lemma 5 *In the geometric interpretation of the MIN problem, each structure is described by a VCRP.*

Proof: At each stage (column) a structure consists of a contiguous set of grid squares. The two restructuring rules above imply that there are no dangling edges inside the rectangle R. □

Now our partitioning has two constraints: it consists of VCRPs, and no VCRP contains a point hole in its interior.

Remark 1 *We have overloaded the vertical axis with two slightly different meanings: the cost per element of building from scratch and the cost of deletion or insertion. This leads to a constant factor error (2 in the model for molecules as described above). It may still be helpful to use this model for obtaining approximation algorithms.*

We now discuss the relation between the TOTAL and MIN measures. We show that the gain in using the MIN measure instead of the TOTAL measure is at most a factor of $O(\log N)$ in the total time to process a sequence of updates and queries, where N is the number of updates. We also show that there are instances of the problem where this gain is obtained. Our analysis will imply an $O(\log N)$ factor approximation algorithm for the MIN problem. We remark that we do not know whether the MIN problem is NP-complete.

Let R be the rectangular grid as above, containing N point holes $U = \{u_1, u_2, \ldots, u_N\}$ on interior vertices. For a given instance (R, U) of the problem, let $\mathcal{R}_{\mathrm{opt}}(R, U)$ denote the length of the minimum-length *rectangular* partitioning, and let $\mathcal{V}_{\mathrm{opt}}(R, U)$ denote the length of the minimum-length partitioning of R into VCRPs; in both cases we consider the edges of R to be a part of the partitioning. Also, in both cases the points of P need to lie on the boundaries of the partitioning objects.

Theorem 7 *Given a rectangular grid R and a set U of N points on grid vertices inside R,*

$$\mathcal{R}_{\mathrm{opt}}(R, U) = O(\mathcal{V}_{\mathrm{opt}}(R, U) \log N).$$

Moreover, there is a family of problems (depending on N) such that $\mathcal{R}_{\mathrm{opt}}(R, U) = \Omega(\mathcal{V}_{\mathrm{opt}}(R, U) \log N)$.

Proof: Let c be a grid vertex on the boundary of R that is closest to a point in U. We construct a minimum-length rectilinear Steiner tree S on the points in $U \cup \{c\}$. The total edge length of S is clearly not greater than $\mathcal{V}_{\mathrm{opt}}(R, U)$. This tree together with the boundary of R defines a degenerate rectilinear polygon Q without holes; it is degenerate in the sense that some edges bound the polygon on both sides. The total edge length of Q is no greater than twice $\mathcal{V}_{\mathrm{opt}}(R, U)$. We now turn Q into a simple polygon Q' by substituting each edge that bounds the polygon on two sides by two parallel edges very close to each other. The polygon Q' has at most $O(N)$ concave vertices; this is guaranteed for a minimum-length rectilinear Steiner tree, and that is why we did not use the optimal partition into VCRPs directly.

By a result of Levcopoulos and Lingas [26], for a simple rectilinear polygon with perimeter π and k concave vertices, there is a rectangular partitioning of length $O(\pi \log k)$. Hence Q' can be partitioned into rectangles with total length at most $O(\log N)$ times the length of Q. Thus, we have obtained a rectangular decomposition of (R, U) which is of length at most $O(\mathcal{V}_{\mathrm{opt}}(R, U) \log N)$. This implies the asserted upper bound.

To show that this bound is tight, we adapt a lower bound construction in Levcopoulos and Lingas [26]; see Figure 7 for an illustration. In a square grid of size $k \times k$, with the same unit resolution along both coordinates, we arrange the $2(k - 2) + 1$ points of U along a "staircase." A minimum length VCRP partition is obvious — it consists of the edges of the staircase (see Figure 7(b)), and has length $\Theta(k)$. Next, we consider a rectangular partition for the same setting, and we restrict our attention to the polygon W that is below the staircase (see Figure 7(c)). By the results of Levcopoulos and Lingas [26], any rectangular decomposition of W will be of length $\Omega(k \log k)$. □

Similarly, we obtain the following algorithmic result.

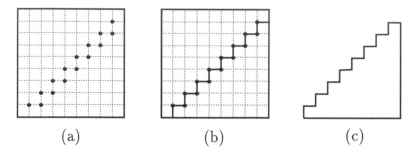

Figure 7: *An instance where the MIN measure is $\Omega(\log k)$ better than the TOTAL measure*

Theorem 8 *For a given maintenance problem (R, U), an $O(\log N)$-approximation for the MIN problem can be obtained in $O(N \log N)$ time, where N is the number of points in U.*

Proof: We apply the algorithm of Gonzalez et al [15] mentioned above. It is an $O(N \log N)$ algorithm that obtains a 4-approximation for the rectangular partition problem. By Theorem 7, the length of the resulting partition will be at most a factor $O(\log N)$ larger than the optimal partition into VCRP's. Thus it gives an $O(\log N)$-approximation for the MIN problem. □

7   Conclusion and Further Work

In this paper, we have initiated the study of a novel type of data structure for kinematic structures that efficiently supports intersection queries. We formulated an abstract model that captures a variety of settings ranging from collision detection for articulated robot arms to conformation search in molecular biology. Our results shed light on the complexity of efficient kinematic data structures for path and tree topologies, but a whole multitude of questions remain open at this point. We outline some of the issues worthy of further exploration.

- An important issue is that of an empirical testing of the ideas outlined in this paper. We are currently working on implementation of these kinematic data structures for conformation search in molecular biology and collision detection for articulated robot arms.

- In some application, particularly in randomized path planning and conformation search for molecular biology, there is considerable flexibility in the order in which the operations are performed on the kinematic data structure. This raises the issue of extending the offline results to the case where the algorithm can at least partially reorder the sequence of operations so as to minimize the total computational cost.

- In the offline setting, our results are concerned primarily with the path topology, where we show that the problem is NP-complete and there are some reasonably good approximation algorithms. Improving the approximation bounds obtained here seems rather difficult, while extending the results to tree topologies seems much more feasible.

- Another approach would be to consider the online setting employing competitive analysis, i.e., comparing the online algorithm's cost to the optimal offline cost. It is fairly easy to see that this problem is a special case of the *metrical task systems* formulation of Borodin, Linial, and Saks [5], but this leads to a fairly weak bound on the competitive ratio. We believe that the special structure of this problem should lead to significantly better competitive ratios. For example, the (artificial) situation where we fix the number of substructures at some value k corresponds exactly to the well-known k-server problem [28] for which a $(2k - 1)$-competitive algorithm is known [25]. In general, we can formulate the online version of our problem as a variant of the k-server problem that we call the *dynamic servers* problem. This is a server problem where the number of servers varies over time, and the online algorithm is required to pay a "rental" cost depending upon the number of servers in use at any given time. We will present our online model and results in a later paper.

- In Section 4.3 we outlined a strategy for maintaining the kinematic data structure for tree topologies under the TOTAL cost measure. This was based on applying the BREAK operation to vertices rather than edges. It would be interesting to further explore this notion and its validity in various applications.

- It is possible that in some applications even the topology of the linkage could undergo some change [12, 36]. It would be interesting to extend our results to such dynamic situations too.

Acknowledgements

This research was funded by the ARO MURI Grant DAAH04-96-1-007. The application to molecular biology is supported by a grant from Pfizer Central Research. The third author is also being supported by an Alfred P. Sloan Research Fellowship, an IBM Faculty Development Award, an OTL grant, and NSF Young Investigator Award CCR-9357849, with matching funds from IBM, Schlumberger Foundation, Shell Foundation, and Xerox Corporation.

References

[1] J. BARNES AND P. HUT, *A Hierarchical $O(N \log N)$ Force-Calculation Algorithm*, Nature, 324 (1986), pp. 446–449.

[2] J. BARRAQUAND, L. KAVRAKI, J.C. LATOMBE, T.Y. LI, R. MOTWANI, AND P. RAGHAVAN, *A Random Sampling Scheme for Path Planning*, to appear in *Robotics Research*, G. Giralt and G. Hirzinger (eds.), Springer Verlag (1996).

[3] J. BARRAQUAND AND J.C. LATOMBE, *Robot Motion Planning: A Distributed Representation Approach*, International Journal of Robotics Research, 10 (1991), pp. 628–649.

[4] P. BESSIÈRE, E. MAZER, AND J. AHUACTZIN, *Planning in a Continuous Space with Forbidden Regions: The Ariadne's Clew Algorithm*, in *Algorithmic Foundations of Robotics*, K. Goldberg et al (eds.), A.K. Peters, Wellesley, MA (1995), pp. 39–47.

[5] A. BORODIN, N. LINIAL, AND M.E. SAKS, *An Optimal On-Line Algorithm for Metrical Task Systems*, JACM, 39 (1992), pp. 745–763.

[6] Y.-J. CHIANG AND R. TAMASSIA, *Dynamic Algorithms in Computational Geometry*, in Proceedings of the IEEE, 80 (1992), pp. 1412–1434.

[7] G.S. CHIRIKJIAN AND J.W. BURDICK, *Kinematics of Hyper-Redundant Manipulators*, in Proceedings of the 2nd International Workshop on Advances in Robot Kinematics, 1990, pp. 392–399.

[8] M.L. CONNOLLY, *Solvent-accessible Surfaces of Proteins and Nucleic Acids*, Science, 221 (1983), pp. 709–713.

[9] D.-Z. DU, L.-Q. PAN, AND M.-T. SHING, *Minimum Edge Length Guillotine Rectangular Partition*, Technical Report MSRI 02418-86, MSRI, Berkeley, 1986.

[10] B. FAVERJON, *Obstacle Avoidance Using an Octree in the Configuration Space of a Manipulator*, in Proceedings of the IEEE International Conference on Robotics and Automation, 1984, pp. 504–512.

[11] B. FAVERJON AND P. TOURNASSOUD, *A Practical Approach to Motion Planning for Manipulators with Many Degrees of Freedom*, in *Robotics Research 5*, H. Miura and S. Arimoto (eds.), MIT Press, Cambridge, MA (1990), pp. 65–73.

[12] T. FUKUDA AND S. NAKAGAWA, *Dynamically Reconfigurable Robotic Systems*, in Proceedings of the IEEE International Conference on Robotics and Automation, 1988, pp. 1581–1586.

[13] N. GO AND H.A. SCHERAGA, *Ring Closure and Local Conformation Deformations of Chain Molecules*, Macromolecules, 2 (1980), pp. 178–187.

[14] T. GONZALEZ AND S.-I. ZHENG, *Improved Bounds for Rectangular and Guillotine Partitions*, Journal of Symbolic Computation, 7 (1989), pp. 591–610.

[15] T. GONZALEZ, M. RAZZAZI, AND S.-I. ZHENG, *An Efficient Divide-and-Conquer Approximation for Hyperrectangular Partitions*, in Proceedings of the 2nd Canadian Conference on Computational Geometry, 1990, pp. 214–217.

[16] D. HALPERIN, L. KAVRAKI, J.C. LATOMBE, R. MOTWANI, C. SHELTON, AND S. VENKATASUBRAMANIAN, *Geometric Manipulation of Flexible Molecules*, to appear in Proceedings of the

ACM Workshop on Applied Computational Geometry, 1996.

[17] D. HALPERIN, J.C. LATOMBE, AND R. MOTWANI, *Dynamic Maintenance of Kinematic Structures,* Full version (1996), *http://theory.stanford.edu/people/rajeev/postscripts/dynamic.ps.Z*

[18] D. HALPERIN AND M.H. OVERMARS, *Spheres, Molecules, and Hidden Surface Removal,* in Proceedings of the 10th ACM Symposium on Computational Geometry, 1994, pp. 113–122.

[19] T. HORSCH, F. SCHWARZ, AND H. TOLLE, *Motion Planning for Many Degrees of Freedom — Random Reflections at C-Space Obstacles,* in Proceedings of the IEEE International Conference on Robotics and Automation, 1994, pp. 3318–3323.

[20] S. KAMBHAMPATI AND L.S. DAVIS, *Multiresolution Path Planning for Mobile Robots,* IEEE Transactions on Robotics and Automation, 2 (1986), pp. 135–145.

[21] L. KAVRAKI AND J.C. LATOMBE, *Randomized Preprocessing of Configuration Space for Fast Path Planning,* in Proceedings of the IEEE International Conference on Robotics and Automation, 1994, pp. 2138–2145.

[22] L. KAVRAKI, P. ŠVESTKA, J.C. LATOMBE, AND M. OVERMARS, *Probabilistic Roadmaps for Fast Path Planning in High Dimensional Configuration Spaces,* to appear in IEEE Transactions on Robotics and Automation.

[23] O. KHATIB, *Real-Time Obstacle Avoidance for Manipulators and Mobile Robots,* International Journal of Robotics Research, 5 (1986), pp. 90–98.

[24] Y. KOGA, K. KONDO, J. KUFFNER, AND J.-C. LATOMBE, *Planning Motions with Intentions,* in Proceedings of SIGGRAPH, 1994, pp. 395–408.

[25] E. KOUTSOUPIAS AND C.H. PAPADIMITRIOU, *On the k-server conjecture,* in Proceedings of 26th Annual ACM Symposium on Theory of Computing, 1994, pp. 507–511.

[26] C. LEVCOPOULOS AND A. LINGAS, *Bounds on the Length of Convex Partitions of Polygons,* in Proceedings of the 4th Conference on the Foundations Software Technology and Theoretical Computer Science, Lecture Notes in Computer Science 181, Springer-Verlag, pp. 279–295.

[27] A. LINGAS, R.Y. PINTER, R.L. RIVEST, AND A. SHAMIR, *Minimum Edge Length Partitioning of Rectilinear Polygons,* in Proceedings of the 20th Annual Allerton Conference on Communication, Control, and Computing, 1985, pp. 53–63.

[28] M. MANASSE, L. MCGEOCH, AND D.D. SLEATOR, *Competitive Algorithms for Server Problems,* Journal of Algorithms, 11 (1990), pp. 208–230.

[29] K. MEHLHORN, *Multi-dimensional Searching and Computational Geometry,* Volume 3 of *Data Structures and Algorithms,* Springer-Verlag, New York (1985).

[30] P.G. MEZEY, *Molecular Surfaces,* in *Reviews in Computational Chemistry,* Volume I, K.B. Lipkowitz and D.B. Boyd (eds.), VCH Publishers (1990).

[31] K. MULMULEY, *Computational Geometry: An Introduction Through Randomized Algorithms,* Prentice-Hall, New York (1993).

[32] M.H. OVERMARS, *The Design of Dynamic Data Structures,* Lecture Notes in Computer Science 156, Springer-Verlag, Berlin (1983).

[33] M. OVERMARS AND P. ŠVESTKA, *A Probabilistic Learning Approach to Motion Planning,* in *Algorithmic Foundations of Robotics,* K. Goldberg et al (eds.), A.K. Peters, Wellesley, MA (1995), pp. 19–37

[34] M. PONAMGI, D. MANOCHA, AND M.C. LIN, *Incremental Algorithms for Collision Detection Between Solid Models,* in Proceedings of the 3rd ACM Symposium on Solid Modeling and Applications, 1995, pp. 293–304.

[35] S. QUINLAN, *Efficient Distance Computation Between Non-Convex Objects,* in Proceedings of the IEEE International Conference on Robotics and Automation, 1994, pp. 3324–3330

[36] M. YIM, *Locomotion with a Unit-Modular Reconfigurable Robot,* PhD thesis, Stanford Technical Report STAN-CS-94-1536, Department of Computer Science, Stanford University, December 1994.

Polyhedral Tracings and their Convolution

Julien Basch, *Stanford University, Stanford, CA, USA*

Leonidas J. Guibas, *Stanford University, Stanford, CA, USA*

G. D. Ramkumar, *Stanford University, Stanford, CA, USA*

Lyle Ramshaw, *Digital Equipment Corporation, Palo Alto, CA, USA*

1 Introduction

Over ten years ago, Guibas, Ramshaw, and Stolfi [8] introduced the *kinetic framework* for Computational Geometry in two dimensions. They augmented the standard boundary representation of planar shapes via polygons or closed curves with certain additional information, to make them into closed *tracings*. The advantage of tracings is that they admit of a multilinear calculus of operations that generalizes standard boolean operations on shapes such as union, intersection, set difference, etc. A particular accomplishment of the kinetic framework was to define the operation of *convolution* on planar tracings and to illustrate its use in a variety of algorithmic problems. Loosely speaking, the convolution $B * R$ of two tracings B (the barrier) and R (the robot) is another tracing that encodes the interaction of B with all possible translations of a copy of R mirrored through the origin. The convolution allows us to transform problems about two tracings B and R into problems about a point and a composite tracing, the convolution $B * R$. A related composition concept is familiar in robotics, under the names Minkowski sum or configuration space obstacle [15, 14]; but that concept lacks the advantageous multilinear structure of the convolution.

In two dimensions, a closed curve is normally defined via a continuous map from the unit circle to the plane. To augment the curve into a tracing, we make a value of this map be not only a point in the plane, but an associated direction as well, with the two related by a 'no-slipping' condition. Intuitively, we can think of a tracing as being drawn by a tiny car; the car can move forward, move backward, or turn while standing still, but may not skid sideways. A standard curve can be given an orientation at each point, corresponding to the sequence in which the points of the curve are traversed by the car. A tracing in addition specifies the direction the car is facing at each point, and that direction must be along the tangent to the curve whenever the latter is defined. The combination of {location, direction} is called a *state* in [8]. The direction of the car is the crucial new component that enables the definition of the convolution operation and gives it its desirable properties. The convolution is defined as a fiber product in [8], by matching states from the two factor tracings with the same direction, and then assembling those into a tracing.

Almost nothing is known about the convolution operation in dimensions higher than two, except in the special case of convex bodies, where it is equivalent to the Minkowski sum; this case for 3-D polyhedral objects has been treated in [9, 2, 12, 17]. Unfortunately the convex case is too special to suggest the proper extension of the concept of a tracing to three and higher dimensions. If, in two dimensions, one thinks of a tracing as a rule for assigning winding numbers to the elements (faces, but also edges and vertices) of the planar subdivision defined by the tracing, then we get a structure we call a *painting*. Paintings of arbitrary dimension were studied by Schapira [18] under the name of *constructible functions*. Using results from sheaf theory, Schapira was able to define a convolution operation on constructible functions and to show how it corresponds to the convolution of tracings discussed above. Nevertheless, Schapira's method does not give a definition of a tracing in higher dimensions, and as such it is not directly useful in algorithmic problems. A tracing can be thought of as a boundary representation of a painting; because of this, it can be cheaper to represent, since it need not encode features, such as self-intersections, that are artifacts of the embedding of the domain manifold into the range space, and not features of the mapping itself.

The main contribution of this paper is to define the notion of *polyhedral tracings*, which extends the classical notion of a polyhedron in exactly the same way a polygonal tracing extends the notion of a polygon. We also define the convolution $B * R$ of two polyhedral tracings B and R and show that it has the same desirable properties as in two dimensions. The technical challenge in getting this theory to work lies in how to extend normal polyhedra into tracings and, less obviously but equally importantly, how to orient the elements of the fiber product defining the convolution so that together they form another tracing manifold. We describe a data structure for representing polyhedral tracings based on the *quad-edge* data structure of Guibas and Stolfi [10] and give an efficient algorithm for computing the convolution of such tracings. Given tracings B and R of size m and n respectively, their convolution $Q = B * R$ can have size $\Theta(mn)$ in the worst case. If the actual size of Q is k, then we can compute Q in output-sensitive time $O(k\alpha(k)\log^3 k)$. This algorithm, which requires novel data structures, is based on a new method we have for detecting red-blue intersections between two families of segments, the red and the blue; within each family the segments need not be disjoint — in fact we require that they be connected. To keep the length of this paper within reason, the details of these data structures and the associated analysis will be reported in a separate paper [3]. We are currently working on extending to arbitrary dimensions and to more general types of tracings our current results on orienting fiber products, defining tracings and convolutions, and the relationship of tracings with paintings.

We view the importance of the work presented here to robotics as follows. For a polyhedral robot translating amidst polyhedral barriers in the Euclidean three dimensional space E^3 (or for polyhedral approximations of more general objects), the standard motion-planning solution involves the calculation of the free space, i.e. the locus of all placements of the robot where it does not collide with barriers. It is well-known that the free space is also a polyhedral domain and can be computed as the complement of the Minkowski sums of the barriers with the robot reflected through the origin. Once a description of the free space is available, one can develop algorithms for checking possible collisions during proposed robot motions, answering connectivity queries (the classical motion planning problem), rendering portions of the free space for visualization purposes, etc. In this paper we propose to accomplish the same goals, but using convolutions instead of the classical Minkowski sums.

To focus the issues, let us assume there is only one barrier B that the robot R must avoid. Why should we prefer the convolution $B * R$ over the Minkowski sum $B \oplus R$ (actually we should be using $-R$ in both cases)? There are two answers. First, the convolution gives us more information. Given a proposed placement of the robot, the convolution gives us information about the topology of the collision region between the barrier and the robot (its Euler-Poincaré characteristic) — not simply a bit on whether they collide or not; this is an advantage of the multilinear over the boolean formulation of the problem. More importantly, the convolution can be combinatorially much simpler than the Minkowski sum, while still allowing us to develop efficient algorithms for the kinds of motion queries described above. If, as above, B and R have sizes m and n, then the Minkowski sum of B and R can have size $\Theta(m^3 n^3)$, while the convolution is always of size at most $O(mn)$. Again, this is because the Minkowski sum has to explicitly represent features that correspond to self-intersections of the embedding of the convolution in $\mathcal{R}^3$. Thus, even when the Minkowski sum is what we really want, it is useful to think of the convolution as an *implicit representation* of it. The convolution is a structure that can be used to answer efficiently queries about the Minkowski sum, while being both smaller in size and easier to compute.

This paper is organized as follows. Section 2 introduces the general framework of tracings and paintings. Section 3 defines polyhedral tracings and introduces a data structure for representing them. Section 4 describes the convolution of two polyhedral tracings in terms of these data structures, while Section 5 sketches an algorithm to compute this convolution that runs in time nearly linear in input plus output size. Section 6 investigates in more detail the relationship between the Minkowski sum and the convolution, and Section 7 concludes.

2   Tracings and paintings

In this section, we generalize to higher dimensions the notion of *tracings* introduced in the kinetic framework, and relate that notion to Schapira's constructible functions, or *paintings*. We briefly describe his convolution operation on paintings and its relation to the Minkowski sum, and proceed to define an equivalent convolution operation on tracings. In this paper we restrict the discussion to the 3-D case: all manifolds below are 2-manifolds; S^2 denotes the 2-dimensional sphere, and E^3 denotes the oriented 3-dimensional Euclidean space. A basis of E^3 is called *direct* (resp. *indirect*) if it has positive (resp. negative) orientation. The *indicator* of a set $X \in E^3$ is the function whose value is 1 on X and 0 everywhere else.

For the definitions below to make sense, we need to place ourselves in a consistent framework. We may for instance follow Schapira, and assume that our manifolds are real analytic, that the sets are subanalytic, and that our functions are morphisms of real analytic manifolds (the focus of this paper, however, is on piecewise linear tracings).

2.1   Tracings

In the kinetic framework, a point on a curve is enriched into a state by the addition of a direction vector which can point either forward or backward along the curve. In three dimensions, we enrich a point on a surface into a state by adding a *whisker* vector at that point, which can point along either of the two directions normal to the surface.

More formally, a tracing R is given by an *oriented compact manifold without boundary* M_R, together with a pair of smooth functions: the location map $r : M_R \to E^3$, and the whisker map $\dot{r} : M_R \to S^2$. The two are related by a *no-slipping* condition:

$$\forall x \in M_R, \; \text{Im}(dr)_x \perp \dot{r}(x)$$

where $\text{Im}(dr)_x$ is the image of the differential of r at x. That is, the location map r is restricted locally to move in a direction orthogonal to $\dot{r}$.

The orientation of the manifold M_R can be thought of as a tiny rotating circle around each of its points. If the image under the location map of this circle around x is replaced by a rotating corkscrew, the direction in which this familiar object starts moving will be referred to as *locally outward* for M_R at x, while the opposite direction will be called *locally inward*.

There are two senses in which the whisker map $\dot{r}$ gives more information than the differential of the location map:

- When the tangent space $\text{Im}(dr)_x$ is a plane, the whisker provides exactly one bit of information, by selecting the normal to this plane which points locally inward or outward.

- The whisker map smoothly extends the notion of tangent plane to ridges and corners (where $\text{Im}(dr)_x$ is only a line or a point). Crossing a ridge causes the whisker vector to swing through an arc in S^2, while the whiskers associated with a corner fill some region of S^2.

2.2   Paintings

A painting ϕ on E^3 is a function that associates an integer to each point of E^3, in a "not too wild" fashion. The paintings we will deal with correspond to polyhedral subdivisions of E^3 with an associated constant integer value for each feature (vertex, edge, face, chamber) of the subdivision. For instance, there are two natural paintings that can be associated with a polyhedron P. The *closed painting* of P has value 1 everywhere inside the polyhedron and on its boundary, and zero elsewhere. The *open painting* of P is the same except for a value of 0 on the boundary. A painting ϕ is said to be *locally closed* at $p \in E^3$ if there exists an open neighborhood A of p on which $\max_{q \in A} \phi(q) = \phi(p)$. Likewise, we define *locally open* by replacing max with min. These definitions generalize the notions of open and closed sets to paintings. Since the overlay of spatial subdivisions is another spatial subdivision, algebraic operations (sum, product) on paintings can be defined in an obvious point-wise fashion.

Any tracing R has an associated painting ϕ, defined via the *winding number* function. We can compute the winding number at a point p of E^3 that does not lie on the range of the location map $r(M_R)$ of R as follows. Take a smooth path π from p to infinity, and count $+1$ or -1 each time the path crosses $r(M_R)$. We count the crossing as $+1$ if the 'velocity vector' of π at the crossing point is directed locally outwards (as defined by the local orientation), and -1 otherwise. The total count (the winding number) is independent of the path chosen, and thus defines a unique value except for those points that lie on $r(M_R)$. If p is such a point, one would like to perturb the location map by a tiny amount so as to move its image away from p, and proceed as before. One role of the whisker map $\dot{r}$ is to indicate the direction in which this perturbation should happen: the winding number is computed not with respect to r, but with respect to $r + \epsilon \dot{r}$, for some small enough $\epsilon > 0$. This perturbation transforms a ridge into a sector of a cylinder and a corner into a region of a small sphere centered at that corner. The winding number is thus established on every point of E^3 and defines a painting ϕ associated with R.

The distinction between open and closed sets is crucial in Schapira's theory of constructible functions. Tracings also make this distinction: it is one of their novel features as a boundary representation for solids. Consider a point $p \in E^3$ and assume, to simplify matters, that it has only one pre-image under r, say $x \in M_R$. If the whisker at p (i.e. $\dot{r}(x)$) points locally outward on the surface of r around p, the associated painting is locally closed at p, otherwise it is locally open.

2.3   Convolution

Any polyhedral painting ϕ can be triangulated, and thus it is not hard to see that it can be expressed as a finite weighted sum of indicators of closed, simply connected sets — such a decomposition, however, is not unique. Schapira defines the *integral* of ϕ, $\int \phi$, as the algebraic sum of these weights and shows that it is independent of the choice of decomposition. This integral has a familiar interpretation in certain cases. For example, if ϕ is the indicator of a closed set S, this integral is simply the Euler-Poincaré characteristic of the set, i.e., in three dimensions, the number of components, minus the number of tunnels, plus the number of internal holes of S.

The convolution of two paintings ϕ, ψ is then defined with respect to this integral $\int$, as

$$(\phi * \psi)(p) = \int_{q \in E^3} \phi(q)\psi(p - q)$$

and it relates to intersection problems in the following way [18]:

Theorem 2.1. *If ϕ, ψ are indicators of closed sets B, R, then the value of $\phi * \psi$ at a point $p \in E^3$ is the Euler-Poincaré characteristic of the intersection of B with R reflected through the origin and translated by p.*

We now wish to describe a convolution operation on tracings such that, given tracings B and R (blue and red) that define paintings ϕ and ψ, the convolution $Q = B * R$ defines the painting $\phi * \psi$. Disregarding the topology of tracings, and viewing them as bags of states as in [8], the convolution operation remains exactly the same as in two dimensions: points $x \in M_B$ and $y \in M_R$ define a state (x, y) in the convolution just when their whiskers match. In that case, the output feature keeps the same whisker and its location is $r(x) + b(y)$.

In our framework for tracings, the convolution domain is defined to be the following subset of $M_B \times M_R$:

$$M_Q = \{(x, y) \in M_B \times M_R \mid \dot{r}(x) = \dot{b}(y)\}$$

This is an instance of a general categorical construction known as a *fiber product* [13] applied to the two whisker maps $\dot{r}$ and $\dot{b}$. Under mild assumptions (*transversality*) on the pair $(\dot{r}, \dot{b})$, the fiber product M_Q is an oriented manifold of the same dimension as M_B and M_R, which makes the above define a valid tracing. The topology of the fiber product M_Q is not simply related to that of M_B and M_R; for instance, M_Q can have any number of components, even when both M_B and M_R are connected. Deriving a consistent local orientation rule for M_Q is a problem which is fully addressed in [4].

Note that when defining the winding number at a point in the image of the location map, the perturbed tracing can be viewed as the convolution of the original tracing with a ball of radius ϵ.

3 Polyhedral tracings

In this section, we define a polyhedral tracing as a specialized tracing defined in the framework of piecewise-linear geometry. We describe an efficient way to represent a polyhedral tracing using a *signed oriented quad-edge* structure, and show how an ordinary polyhedron can be made into a tracing using a sound set of conventions.

A polyhedral tracing is a tracing such that its domain manifold can be decomposed into (interior-disjoint) simply connected subsets of three types: vertex domains, edge domains, and face domains, with connectivity properties akin to a subdivision as defined in [10], and such that (Figure 1):

1. On a face domain, the whisker map is constant and the location map is a bijection whose range is a simple planar polygon,

2. On an edge domain, the location map range is a line segment (corresponding to a shared edge between the polygonal ranges of the two adjacent face domains), and the whisker map range is a great circle arc of length less than π on S^2, and

3. On a vertex domain, the location map is constant, while the whisker map is a bijection, and its range is a simple spherical polygon on S^2 entirely contained within a hemisphere [*the hemisphere assumption*].

Using a tracing to describe a polyhedron boundary has several advantages. First, a tracing can encode both open, closed, and mixed subsets of E^3, whereas no such distinction exists for a polyhedron. Second, a

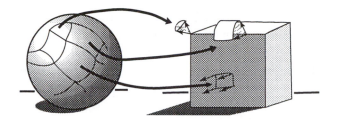

Figure 1: *A cube tracing indexed by a sphere. Note that the inverse image under the location map of points on faces, edges, and vertices of the cube are 0-dimensional, 1-dimensional, and 2-dimensional respectively.*

tracing can represent a larger variety of paintings with piecewise planar boundaries, and not just indicators of polyhedra. Third, the convolution of two (closed) polyhedra is in general not a polyhedron. Polyhedral tracings have in fact just the right expressiveness to allow the representation of convolutions of general non-convex polyhedra.

3.1 Enriched quad-edge structure

A quad-edge structure [10] is an algebra $(V, E, F, \text{Onext}, \text{Org}, \text{Lface}, \text{Sym})$, where V (resp. E, F) is the set of vertices (resp. edges, faces), with a number of relations between the operations which we do not review here. In such a structure, an edge has a *direction*, from its *origin* (Org) to its *destination* (Dest), and an *orientation*, from its *left face* (Lface) to its *right face* (Rface). The *symmetric* (Sym) of an edge is the same edge with both its orientation and direction reversed. The *next edge* (Onext) to an edge e whose origin is a vertex v is the edge with origin v whose Rface is the Lface of e. In the case of a connected orientable quad-edge, the quad-edge algebra is the union of two disjoint subalgebras: an edge with a given direction is present in both subalgebras, but with a distinct orientation. An oriented quad-edge structure is simply the choice of one of these subalgebras for each connected component. In this case, the direction of an edge determines its orientation and vice-versa, so that a face f has a ring of *directed* edges associated with it (those whose Lface is f).

We represent the domain manifold of a polyhedral

tracing by an oriented quad-edge structure (V, E, F). In this setting, a vertex and an edge in the structure also represent two-dimensional domains (vertex domains and edge domains) in the underlying manifold. Next, we encode the location and whisker maps by keeping only their values at vertices and faces respectively, via the two functions:

$$\lambda : \quad V \quad \mapsto \quad E^3$$
$$\mu : \quad F \quad \mapsto \quad S^2$$

where λ gives the location of each vertex, and μ gives the whisker at each face.

Obviously, this representation loses some information: the original domain manifold of the polyhedral tracing is known only up to a homeomorphism, but this level of precision is enough to define a painting. The location map can be smoothly reconstructed by linear interpolation from the data for vertices, to edges and faces, due to the assumption that the location map is a bijection on a face domain.

The reconstruction of the whisker map for edges and vertices from the whiskers of the faces is more delicate. Consider first the case of an edge. Its whiskers describe an arc joining the whiskers of its two adjacent faces. There are two such arcs, but only the smallest of the two satisfies the requirement that the whisker set of an edge be an arc of length less than π on S^2. For a vertex domain, the assumption that the whisker map is a bijection allows its specification using the values on its boundary, but, again, this boundary defines two polygons on the sphere. It is the hemisphere assumption that allows us to choose the smaller one without ambiguity. From this choice, the whisker map on a vertex domain is reconstructed up to an isomorphism.

3.2 Whisker and outward-pointing normal

Traditionally, when a quad-edge structure is used to represent a polyhedron boundary, a record for a vertex v stores its location $\lambda(v)$, and a record for a face f stores its outward-pointing normal $\nu(f)$. The outward-pointing normal is such that, when an observer is looking at a face f with $\nu(f)$ pointing towards her, the ring of oriented edges around f describes a counterclockwise cycle.

Due to the no-slipping condition, the outward-pointing normal to a face is either the same as, or opposite to, the whisker at that face. Thus, it is not necessary to store it in our structure, as it is possible to reconstruct it at the additional cost of only one extra bit per face. We define

$$\sigma : F \mapsto \{-1, +1\}$$

such that for each face f, $\nu(f) = \sigma(f)\mu(f)$, and store this sign bit with each face (hence the name "signed quad-edge").

This sign bit has an important meaning for the corresponding painting. If the sign at a face is $+1$, the whisker at that face points locally outwards, which defines a locally closed painting. If the sign is -1, the tracing defines a locally open painting. Storing this sign is redundant, as it can be reconstructed from the whisker and the orientation of the ring of edges around one face. However, this reconstruction is costly, as it requires tracing the entire ring of edges for each face.

3.3 From a polyhedron to a polyhedral tracing

In most applications, a polyhedron is defined by its boundary, represented by an oriented quad-edge structure, the location of each vertex, and the outward-pointing normal to each face. This structure can be easily converted into a polyhedral tracing, but there are two issues that require some discussion: what signs to choose for the faces, and how to enforce the hemisphere requirements for the vertex whisker maps.

The choice of signs for faces depends on whether the polyhedron is intended to represent a topologically closed or open set. While this distinction is rarely if ever considered in computational geometry, it may have some relevance in solid modeling. As we saw before, having the whisker map point locally outwards defines a locally closed painting. Thus, in order to represent a topologically closed polyhedron, one would set the sign bit of all faces to $+1$. To represent an open polyhedron, one would set the sign bit of all faces to -1. Once the signs are given, the 'tracification' boils down to replacing the outward-pointing normal data with the whisker data for each face (in effect multiplying the face normal by the sign). This clearly sets the values of the associated painting correctly on all faces, but it remains to check whether the edges and vertices have the correct values as well. We address this question while checking for the other requirements of a polyhedral tracing.

The choice of signs defines the whisker on each face. We take the convention that the whisker set of an edge is the shortest arc that joins the whiskers of its adjacent faces. Therefore, there is no special conversion necessary to represent an edge in the tracing. This convention mirrors a similar one used for turns in two dimensions [8]. It gets its full relevance from the open/closed interpretation of tracings. Indeed, with this choice, an edge adjacent to two closed faces is closed (i.e. the painting has the same value on the edge as in the polyhedron interior). An edge adjacent to two open faces is open.

The vertex case is more delicate. The path stroked by the whiskers of the edges whose origin is a vertex v might not fit in a hemisphere (Figure 2), or might even self-intersect (Figure 3), hence requiring a normalization process. There is no canonical way to normalize, but the simplest is probably to "triangulate" the vertex as follows: single out a face f adjacent to v, and introduce zero-length edges between f and each of the other faces adjacent to v. This cuts v into many vertices, each of which has degree three and defines a triangle on the sphere of directions. A triangle is not self-intersecting and is contained within a hemisphere, so this defines a valid tracing. Note that the edges introduced in this process are zero-length, but their whiskers span an arc on the sphere of directions (Figure 3).

It remains to see whether vertices have the desired value in the associated painting, i.e. whether they have value 1 if one settled for the closed polyhedron, and value 0 for the open polyhedron. This is indeed the case, provided that, for each vertex, there exists a hemisphere that contains all the whiskers to the faces adjacent to this vertex. Indeed, consider the closed case: for each vertex v the corresponding hemisphere gives us a direction (a whisker) with which the whiskers of all adjacent faces form an angle of less than $\pi/2$. Say that this direction is "up". Whichever normalization is used (if any is needed) at v, a convolution with a ball

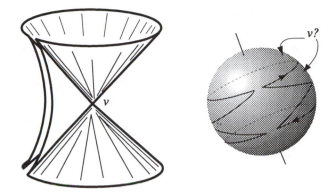

Figure 2: *A vertex v whose whisker map does not fit in any hemisphere. We can choose either of the two regions bounded by the spherical polygon on the right to be the range of the whisker map on the domain of v. Reversing this choice changes, by 1, the value of the associated painting at v.*

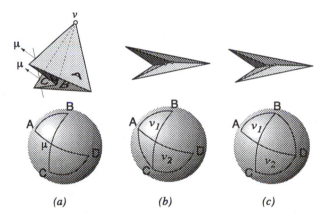

| (a) | (b) | (c) |

Figure 3: *(a) The simplest vertex that has a self intersecting path on the sphere of directions; (b,c) two ways to normalize it, cutting it into two vertices v_1, v_2 at the same location, by an adjunction of a zero-length edge. The top view shows the two vertices slightly displaced to emphasize the connectivity.*

of radius ϵ will locally pull up all the adjacent faces. Hence the winding number at v is the same as that of a point slightly inside the polyhedron, i.e. $+1$. If the hemisphere requirement is not satisfied at a vertex (Figure 2), it is unclear how to choose the whisker map so that the associated painting has value 1 at that vertex.

4   Convolution of polyhedral tracings

Since a polyhedral tracing is a special case of a tracing as defined in Section 2, the convolution operation is well defined. Moreover, the result is also a polyhedral tracing. Given the signed quad-edge representations of two input polyhedral tracings B and R, we show in this section how to obtain the convolution Q in the same representation. In the rest of the paper, we refer to features of B, R, and Q as *blue, red,* and *purple* respectively. The underlying manifold of the convolution is the fiber product of the whisker maps of B and R. It is described implicitly by the signed quad-edge structure of Q.

The features of the convolution are obtained by convolving pairs of matching features of B and R. A pair of features is said to *match* if the whisker sets of the two features have a non-empty intersection. In order to keep the notation simple, we assume that the intersection of the whisker sets of any two features is simply connected (and thus defines only one feature, whose whisker set is the intersection of the two original sets). In two dimensions, a forward move combined with a right turn becomes a backward move in the convolution. Likewise, in three dimensions, we expect that some faces will change sign in the convolution. A face f' in the convolution can arise when a face f of one tracing is convolved with a matching vertex v in the other. As f and f' have the same whisker, a change of sign requires a reversal of the edge ring (Figure 4). A face in the convolution is also obtained when a blue and a red whisker arc intersect on the sphere; their corresponding edges generate a parallelogram (the Minkowski sum of the two edges). The whisker of this new face is well defined, but its sign has to be determined.

4.1   Orienting the convolution

In the two-dimensional kinetic framework, the authors captured the reversals of sign in a very simple rule: forward moves and left turns have sign $+1$, backward moves and right turns have sign -1; furthermore, the sign of an output feature is the product of the signs of the corresponding input features. Guessing a similar

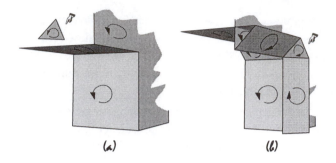

Figure 4: *In this convolution, the orientation of the little triangle is switched to create a consistent tracing. This reversal is captured by a negative sign assigned to the saddle vertex. Some pairs of edges whose whisker sets intersect create parallelogram faces in the convolution.*

sign rule in three dimensions, if it exists, is more difficult. We can deduce this sign rule from the general convolution definitions by calculating the orientation of the fiber product manifold on the domains corresponding to each feature of the convolution. The topology of this manifold also gives the connectivity of its quad-edge representation. We now proceed to define signs for edges and vertices for a tracing; we assume that these signs are also stored in the signed quad-edge structure.

1. The sign of an edge e, denoted $\sigma(e)$, is positive (resp. negative) if the triplet of vectors $(e\,\mathrm{Dest} - e\,\mathrm{Org}, e\,\mathrm{Lface}, e\,\mathrm{Rface})$ defines a direct (resp. indirect) basis.[1]

2. The sign of a vertex v, denoted $\sigma(v)$, is positive (resp. negative) if a traversal of the ring of edges around v via the Onext operator corresponds to a counterclockwise (resp. clockwise) traversal of the boundary of the whisker of v on the sphere of directions (from the point of view of an observer, facing from the outside, the hemisphere containing the whisker).

From the above definition, it follows that the sign of an edge e is the same as that of $e\,\mathrm{Sym}$. Also, when a tracing represents a closed polyhedron the above sign rule leads to giving convex edges a positive sign and

---

[1] A zero-length edge can be given an arbitrary sign

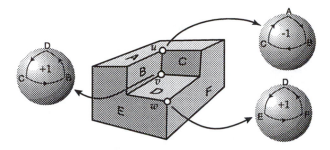

Figure 5: *The three simplest types of vertices with their ring of edges on the sphere of directions and their signs.*

concave edges a negative sign (much like left and right turns in polygonal tracings). Both convex and concave vertices have a positive sign, whereas a vertex like the one in Figure 4 has negative sign (Figure 5). If we view the boundary of the whisker set of a vertex as a 2-D tracing on the sphere of directions, then the sign of the vertex is exactly the winding number this tracing assigns to the region defining the whisker set. This is not a coincidence, and further work is needed to clarify and take advantage of this relationship.

4.2 Connectivity of the convolution

Let the vertex, edge, and face sets of B be V_b, E_b, and F_b, and those of R be V_r, E_r, and F_r. If features h_b and h_r of B and R match, the resulting convolution feature is denoted by the unordered pair (h_b, h_r). The assumption that every such pair produces at most one feature in the convolution makes the notation unambiguous. In an implementation, an output feature would be identified by a pointer, so that this assumption would be unnecessary. We denote $-e$ for e Sym, allowing expressions like $\sigma(v)e$, to refer to e or e Sym depending on the sign of v.

We now define the signed quad-edge structure of P by enumerating its features and their connectivity.

$$
\begin{aligned}
V(P) &= \{(v_b, v_r) \in V_b \times V_r \mid \mu(v_b) \cap \mu(v_r) \neq \emptyset\} \\
E(P) &= \{(e, v) \in E_b \times V_r \cup E_r \times V_b \mid \\
&\qquad \mu(e) \cap \mu(v) \neq \emptyset\} \\
F(P) &= \{(v, f) \in V_b \times F_r \cup V_r \times F_b \mid \mu(f) \in \mu(v)\} \\
&\cup \{(e_b, e_r) \in E_b \times E_r \mid \mu(e_b) \cap \mu(e_r) \neq \emptyset\}
\end{aligned}
$$

The pairs (e, e'), $(e\,\text{Sym}, e')$, $(e\,\text{Sym}, e')$, and $(e\,\text{Sym}, e'\,\text{Sym})$ refer to the same parallelogram face. Before we define the topological connectivity of the convolution and the geometry and signs of its features, we need to define a primitive Fedge (v, e). This primitive is defined only when $\mu(v)$ and $\mu(e)$ overlap.

Definition 4.1. If e is an edge whose whisker intersects the boundary of the whisker set of a vertex v, we denote by Fedge (v, e) the edge e' with $e'Org = v$, and such that e' is the first such edge crossed when going from $(\sigma(v)\,\sigma(e)\,e)$ Lface to $(\sigma(v)\,\sigma(e)\,e)$Rface on S^2.

The way to compute Fedge (v, e) is not addressed here. It will be a direct bi-product of the sweep-line algorithm described in the next section. The rules for constructing the polyhedral tracing representing the convolution are summarized in table 4.2.

Note that, just like in the two dimensional case, the topological adjacencies between features in the convolution are induced by those of the factor tracings and that the signs of its features are the products of the corresponding signs of the matching factor features. General results on tracings imply that the quad-edge structure defined above is consistent, and represents the tracing of the convolution. Except for the case of closed convex polyhedra, where it is the well-known Minkowski sum, the resulting convolution will in general self-intersect and have locally open faces (Figure 4 again).

4.3 Normalization and convolution size

When discussing output sensitive algorithms, there is an issue of what should be considered as the size of the output. This issue was discussed at length by Guibas and Seidel [9] for two-dimensional convolutions, and it is not surprising that it pops up again in three dimensions. In the present setting, two tracings that have different quad-edge representations are always considered distinct, although they might define the same painting (and the same "bag of states" in the vocabulary of the original kinetic framework). In this sense, the output described above is the only one that properly describes the exact topology of the fiber product manifold of the convolution.

Table 1: *Connectivity and geometry of the convolution of two polyhedral tracings.*

$$(v, e)\, \mathrm{Sym} \;=\; (v, e\, \mathrm{Sym})$$

$$(v, e)\, \mathrm{Org} \;=\; (v, \sigma\,(v)\, e\, \mathrm{Org})$$

$$(v, e)\, \mathrm{Onext} \;=\; \begin{cases} (v, (\sigma\,(v)\, e)\, \mathrm{Onext}^{\sigma(e)}) & \text{if } \mu\,(e\, \mathrm{Lface}) \in \mu\,(v) \\ ((\sigma\,(v)\, e)\, \mathrm{Org}, \mathrm{Fedge}\,(v, e)) & \text{otherwise} \end{cases}$$

$$(v, e)\, \mathrm{Lface} \;=\; \begin{cases} (v, (\sigma\,(e)\, e)\, \mathrm{Lface}) & \text{if } \mu\,(e\, \mathrm{Lface}) \in \mu\,(v) \\ (e, \mathrm{Fedge}\,(v, e)) & \text{otherwise} \end{cases}$$

$$\lambda\,(v_b, v_r) \;=\; \lambda\,(v_b) + \lambda\,(v_r)$$

$$\mu\,(e_b, e_r) \;=\; \mu\,(e_b) \cap \mu\,(e_r)$$

$$\mu\,(v, f) \;=\; \mu\,(f)$$

$$\sigma\,(v, f) \;=\; \sigma\,(v)\,\sigma\,(f)$$

$$\sigma\,(e_b, e_r) \;=\; \sigma\,(e_b)\,\sigma\,(e_r)$$

This view is however slightly hypocritical, as the objects we are ultimately interested in are paintings, and not one specific representation. As in the two dimensional case, it is possible to construct two tracings of sizes m, n, such that their convolution is of size $\Theta(mn)$, while $\Theta(m + n)$ features are sufficient to represent the same painting. Even worse, the normalization process used to obtain a polyhedral tracing from a polyhedron may also lead to variations in the output size: if one takes a polyhedron with n vertices like the one of Figure 3, and another polyhedron with m features crossing one of the normalization lines, the convolution may have a size that varies from linear to quadratic depending on which normalization line is used. It would be desirable to review the normalization process to avoid this inconsistent behavior.

5   An efficient algorithm for the convolution

In this section, we briefly describe an algorithm to compute the signed quad-edge structure of a convolution Q of two tracings B and R with a total of n vertices in time $O(k\alpha(n)\log^3 n)$, where k is the number of vertices of Q. We will assume that B and R all have faces

of sign $+1$, from which it follows that their whisker maps cover the sphere of directions S^2, and therefore $k \geq n$. The key algorithmic problem we must solve is to find the matching features of B and R. Geometrically, this reduces to finding the overlay, on the sphere of directions, of the whisker sets of the two tracings. This overlay will give us in fact the whisker sets of the convolution — its features will directly correspond to the elements of Q. The subdivision overlay problem requires us to locate all vertices of one subdivision into regions of the other, and to discover all pairs of intersecting red-blue edges. From the complexity point of view, the dominant cost of such an algorithm is the latter computation of all the bichromatic edge intersections. In the rest of the paper, we will assume that k actually represents the number of these bichromatic pairs. In the case where B and R are convex tracings (that is, when each whisker value appears only once per tracing) this problem has already been optimally solved [9] in $O(n + k)$ time. The difficulty with general tracings is that their whisker maps can multiply cover S^2 — they are effectively Riemann surfaces over S^2. During the overlay operation we want to discover all bichromatic edge intersections, but we cannot afford to pay for looking at crossings of edges of the same color (which would arise if we were to 'flatten out' each Riemann surface onto S^2).

Through an appropriate polar map, our problem is the same as the well-known problem of detecting red-blue intersections among a set of red and blue segments in the plane. This problem traditionally appears in two versions: general and disjoint (more precisely, the latter means that segments in each collection are required to have disjoint interiors). Neither is appropriate for our purposes: our collections need not be disjoint, as we pointed out above (monochromatic pseudo-intersections can arise from the overlay of different Riemann sheets), while at the same time a nearly linear red-blue merge algorithm seems highly unlikely for the general case. We show below how to exploit the fact that our segment collections are connected — that is, that the union of all points in the red segment collection forms a connected subset of the plane (and similarly for the blue). Agarwal and Sharir [1] investigated this version, and gave an $O(n\alpha(n)\log^2 n)$ algorithm to find *one* purple intersection when one exists, but their method could not be extended to find all purple intersections.

A classical Bentley-Ottmann sweep technique [5] can be applied for this problem, but it does not avoid the possibly quadratic cost of processing all blue-blue and red-red intersections (Figure 6). Our new algorithm, called HEAPSWEEP, revisits the sweep paradigm by relaxing the constraint that the segments be completely ordered along the sweep line. It uses a novel data structure that can be used to efficiently sweep the upper envelope of a family of segments in the plane by maintaining only a partial vertical ordering among the segments. We will report this algorithm and its detailed analysis elsewhere [3] — below we just indicate some of the ideas of the method.

The upper envelope of an arrangement of n line segments is a well-known concept in computational geometry [7] and it has size $O(n\alpha(n))$ (where $\alpha(n)$ is the familiar inverse Ackermann function). We propose a new (non-optimal) algorithm to compute this structure that can be used as a building block for a sweep-line procedure in the red-blue intersection problem. The idea is to perform a line sweep on the segments, keeping track of a partial (instead of exact) order of the

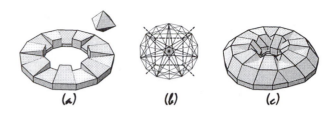

Figure 6: *(a) A diamond and a synchronizing gear with $n = 7$ teeth; (b) the arcs defined by the edges of the gear (in thin stroke) have $\Theta(n^2)$ intersections on the sphere, but only $\Theta(n)$ intersections with the features of the diamond (dashed stroke); (c) the convolution has only $\Theta(n)$ features.*

segment intersections on the sweep line in a heap-like data structure. Indeed, for the sweep line structure to be maintained in the totally ordered case, all segment intersections are events that have to be processed in the priority queue, leading to a possibly bad running time. In contrast, to maintain the heap-like structure, only intersections that involve a parent and child in the heap need to be scheduled and examined — we call these *tournament intersections*. No result is known about the maximum number of tournament intersections for standard heap implementations. In [3] we introduce several heap-like structures for which we can prove that the number of tournament intersections is roughly linear.

We now make use of the above as a basic component for an algorithm that computes the red-blue intersections in time $O(k\alpha(n)\log^3 n)$ in the connected version of the problem. At a given position, the sweep line is divided into *blue blocks* and *red blocks*. A block is a maximal interval on the line that intersect only one color of edges. All edges in a block are stored in a heap-like structure, so that the two extreme edges of the block are known at all times. Purple intersections can be discovered and scheduled by comparing the top and bottom segments of every pair of adjacent blocks, and can be processed in polylogarithmic time (see the schematic representation in figure 7-a).

Sometimes along the sweep, a blue block may encounter a red node that forces a split of the block into two blocks (Figure 7-b), a possibly expensive operation. Similarly, all arcs of a blue block may end at the

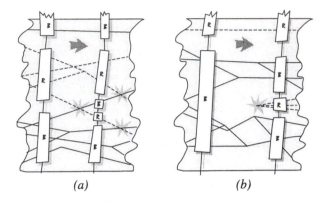

(a) (b)

Figure 7: *(a) A purple intersection: two new blocks are created, and new purple intersection events are scheduled. (b) A SPLIT of a blue block in two, triggered by an unexpected red node*

same node, without a new arc to replace them, requiring a costly merge of the two adjacent red blocks in order to maintain the prescribed sweep line structure (these two possibilities arise in the computation of the convolution as soon as the input polyhedra are allowed to be non-convex). The connectedness of our families is the crucial hypothesis that makes the amortized cost of these operations polylogarithmic.

Theorem 5.1. *Let B, R be two polyhedral tracings with n vertices in total, whose faces all have sign $+1$, and whose convolution Q has k features. The algorithm HEAPSWEEP computes the signed quad-edge structure of Q in time $O(k\alpha(n)\log^3 n)$*

6   Minkowski sum using convolution

Minkowski sums of polyhedra arise frequently in motion planning algorithms based on the configuration space approach [15, 14]. Though the case of computing the Minkowski sum has been well studied when both the polyhedra are convex [9, 6], non-trivial bounds are not known for the computation of Minkowski sum for general non-convex polyhedra [11]. We argue that the convolution of polyhedra provides a low-storage alternate representation which encodes much of the same (and often more) information as the Minkowski sum.

First let us address the size issue. If the two polyhedra are of sizes m and n their convolution is always

of size $O(mn)$, whereas the Minkowski sum can be $\Theta(m^3 n^3)$ in the worst case (an example where this happens can be found in [11]). The key savings in space and computation cost result from not computing or storing self-intersections of the convolution; these still give rise to features on the boundary of the Minkowski sum.

By Theorem 2.1, the winding number at x with respect to the convolution of closed tracings B and R is the Euler-Poincaré characteristic of the intersection of B with R reflected and translated by x. Thus the winding number captures information about how the robot and the barrier intersect. The intersection of two tracings representing simple closed polygons in the plane is always a set of simply connected components with Euler-Poincaré characteristic equal to the number of components. This means that the winding number with respect to the convolution is equal to the number of components in the corresponding intersection, and thus we can interpret this winding number as 'depth of collision.'

However, the correspondence between convolutions and Minkowski sums of polyhedral tracings is not as simple as in the case of polygonal tracings because the intersection of closed polyhedral tracings may not be simply connected. For example, imagine pushing two glasses close to each other until their mouths interpenetrate. The intersection is a torus of zero Euler-Poincaré characteristic (one component – minus one tunnel) even though it is a non-empty solid. Thus for polyhedral tracings, the region of the convolution of zero winding number is not the same as the complement of the Minkowski sum. However, the *outer cell* of the Minkowski sum is in fact the outer cell of the convolution in E^3.

6.1   Algorithmic issues

We expect that many algorithms using the Minkowski sum can be speeded up by using the combinatorially smaller convolution instead. For example, in order to render the outer face of the Minkowski sum from an external viewpoint, it is sufficient to render the faces of the convolution in any sequence. The depth buffer of current graphics hardware will automatically ensure that only the visible sections of the Minkowski sum are produced. Similarly, algorithms that interrogate the Minkowski sum in various ways, such as ray-shooting, winding number and distance computations, planar sectioning, extrema and bounding volume computations, etc., can all benefit from using the smaller convolution in lieu of the Minkowski sum. Note that the solution to several of these problems makes use of the sign (orientation) information stored with the convolution. For example, a winding number computation at a point p, based on connecting p to another point of known winding number and then algebraically summing the crossings with the convolution along the connecting path, makes use of the signs. We also anticipate that the convolution can be used in novel ways along the lines of [16]. We intend to develop some of these applications in greater detail in a future publication.

7 Conclusion and further work

In this paper we have presented the theory of polyhedral tracings and their convolutions. Polyhedral tracings generalize ordinary polyhedra by providing an explicit set of normals at all features of the polyhedron: at vertices and edges, as well as at faces. This additional structure is not costly; it can be adequately represented by keeping just one additional bit (a "sign") with each face of the polyhedron. It allows us to properly state which pieces of the boundary are part of the polyhedron, and which are not. Furthermore, it permits the definition of the convolution of two polyhedral tracings, which compactly encodes information about the topology of the intersection of the two polyhedra under all possible translations. The convolution can serve as an alternate representation of the Minkowski sum of the polyhedra, which has found numerous applications in motion planning problems. When we compare the two, we see that the convolution gives us more information about the relative placement of the two polyhedra, while at the same time being much smaller combinatorially in the worst case. And while computing the Minkowski sum can be a challenging problem

in three-dimensional computational geometry, our essentially optimal and output-sensitive convolution algorithm uses efficient planar data structures for its operation.

Clearly much remains to be done in further developing the theory of convolutions for polyhedral tracings and for more general instances. Other possible applications need to be investigated: object placement, distance computations, and more complex motion planning problems.

Acknowledgments

We wish to thank Rajeev Motwani and Dan Halperin for helpful discussions and Harish Devarajan, Scott Cohen, and Héloise Mallet for valuable comments on the manuscript. This work was supported by National Science Fundation grant NSF CCR-9215219, NSF/ARPA grant IRI-9306544, and US Army Grant 5-23542A.

References

[1] P. K. Agarwal and M. Sharir. Red-blue intersection detection algorithms, with applications to motion planning and collision detection. *SIAM J. Comput.*, 19(2):297–321, 1990.

[2] Chanderjit L. Bajaj and Myung-Soo Kim. Generation of configuration space obstacles: The case of moving algebraic curves. *Algorithmica*, 4:157–172, 1989.

[3] J. Basch, L.J. Guibas, and G.D. Ramkumar. Reporting red-blue intersections between connected sets of lines segments. *To appear in ESA'96*.

[4] J. Basch and L. Ramshaw. Orienting the fiber product of smooth manifolds. *In preparation*.

[5] J. L. Bentley and T. A. Ottmann. Algorithms for reporting and counting geometric intersections. *IEEE Trans. Comput.*, C-28:643–647, 1979.

[6] B. Chazelle, H. Edelsbrunner, L. Guibas, and M. Sharir. Algorithms for bichromatic line segment problems and polyhedral terrains. *Algorithmica*, 11:116–132, 1994.

[7] H. Edelsbrunner, L. J. Guibas, and M. Sharir. The complexity and construction of many faces in arrangements of lines and of segments. *Discrete Comput. Geom.*, 5:161–196, 1990.

[8] L. J. Guibas, L. Ramshaw, and J. Stolfi. A kinetic framework for computational geometry. In *Proc. 24th Annu. IEEE Sympos. Found. Comput. Sci.*, pages 100–111, 1983.

[9] L. J. Guibas and R. Seidel. Computing convolutions by reciprocal search. *Discrete Comput. Geom.*, 2:175–193, 1987.

[10] L. J. Guibas and J. Stolfi. Primitives for the manipulation of general subdivisions and the computation of Voronoi diagrams. *ACM Trans. Graph.*, 4:74–123, 1985.

[11] A. Kaul and M. A. O'Connor. Computing minkowski sums of regular polyhedra. Report RC 18891 (82557) 5/12/93, IBM T.J. Watson Research Center, Yorktown Heights, NY 10598, 1993.

[12] A. Kaul and J. Rossignac. Solid-interpolating deformations: construction and animation of PIPs. *Comput. & Graphics*, 16:107–116, 1992.

[13] Serge Lang. *Differential Manifolds*. Addison-Wesley, 1972.

[14] J.-C. Latombe. *Robot Motion Planning*. Kluwer Academic Publishers, Boston, 1991.

[15] T. Lozano-Pérez. Spatial planning: A configuration space approach. *IEEE Trans. Comput.*, C-32:108–120, 1983.

[16] G. D. Ramkumar. Algorithms to compute the convolution and minkowski sum outer-face of two simple polygons. In *Proc. 12th Annu. ACM Sympos. Comput. Geom.*, 1996.

[17] J. Rossignac and A. Kaul. Agrels and bips: Metamorphosis as a bézier curve in the space of polyhedra. In *Eurographics '94 Proceedings*, volume 13, pages 179–184, 1994.

[18] P. Schapira. Operations on constructible functions. In *Journal of pure and applied algebra* [8], pages 83–93.

Fast Construction of Near Optimal Probing Strategies

Eric Paulos, *University of California, Berkeley, CA, USA*
John Canny, *University of California, Berkeley, CA, USA*

We address the problem of defining and constructing optimal probing strategies for precisely localizing polygonal parts whose pose is approximately known. We demonstrate that this problem is dual to the well studied grasping problem of computing optimal finger placements as defined by Mishra et al. [17] and others [10, 16]. In addition we develop a quality measure for any given probing strategy based on a simple geometric construction in the displacement space of the polygon. Furthermore, we can determine a minimal set of probes that is guaranteed to be near optimal for constraining the position of the polygon. The size of the resulting set of probes is within $O(1)$ of the optimal number of probes and can be computed in $O(n \log^2 n)$ time. The result is a probing strategy useful in practice for refining part poses.

1 Introduction

In industrial manufacturing and automated assembly accuracy is money. Attaining and maintaining high precision can increase the cost of fixturing and feeding several fold [18]. The meaning of high verses low precision depends on the application, but for typical mechanical assembly, low precision tooling might provide accuracies in the tens of mils, while high precision would be around one mil or less (One mil = 10^{-3} inch = 25.4 microns). In this paper we study the *pose refinement problem*. In pose refinement, sensing is used as an inexpensive route to high precision part pose, assuming the pose is already known at low precision. Most research to date in computer vision and RISC sensing addresses the *pose acquisition problem*, where pose is determined with no knowledge of initial pose. The result of pose refinement is a high-precision estimate, but it differs from the problem of high-precision pose acquisition. Since initial pose is approximately known in pose refinement, it can be used to make judicious choices about sensor placement. The same accuracy can be achieved with fewer or less expensive sensors for pose refinement as compared to pose acquisition, which must deal with all possible poses.

Pose refinement is a problem suggested to us by our visits to several state-of-the-art manufacturing companies, especially PTI (Productivity Technologies Inc.) and HP Labs. In typical industrial workcells, it was pose refinement rather than pose estimation that was the dominant sensing task. There are two reasons for this:

1. Feeder economics. Vibratory feeders are an inexpensive way to provide many part types in known (albeit low precision) pose. Small parts can also be fed on tape, which is more expensive (a couple of cents per part) but still costs far less than a high-precision pallet. So the initial and ongoing costs of achieving low-precision pose without sensing are small.

2. Multi-step manufacturing. In typical manufacturing, there are not one but several sequential stages, including assembly stages, testing and packaging. A single step might mate two parts whose poses are known at high precision. But the assembly step itself introduces a small amount of uncertainty, and it is expensive to transport the partial assembly at high precision to the next assembly stage. A more economical solution is to use pose refinement at the next stage. So while there might

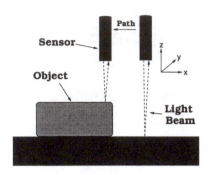

Figure 1: *A typical simple reflective sensor used for probing*

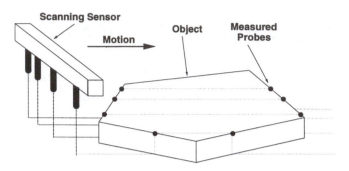

Figure 2: *Example of a fixed array scanning beam sensor*

be one pose acquisition step per part to get approximate initial pose, there will be several pose refinement steps for that part that start with the low precision output from the previous step and feeder, and increase the precision as needed for the next step.

So the arguments for pose refinement are (i) that it replaces the most expensive (high precision) fixturing and feeding steps (ii) it replaces the most frequent fixturing and feeding steps in multi-step assembly.

In this paper we employ simple light-beam sensors that act as line probes. A point light source and receiver define a line in space that is broken and unbroken by an object as it moves relative to the beam (see Figure 1). The positions of the object when the beam breaks give position readings, and three or more of these determine pose. Those readings are subject to error, and the pose estimate accuracy is limited by those errors and the sensor placements. This paper gives algorithms for choosing the probe placements to achieve near-optimal accuracy with a fixed number of probes, or to find a near-minimal number of probes to achieve a specified accuracy.

Problem Statement:

Given: *A polygonal part, its geometry, and an initial pose estimate of $O_i = (O_{ix}, O_{iy}, O_{i\theta})$ within $\sigma = |O - O_i|_m$ of the exact actual pose of the object, O.*

Solve: *Find the optimal set of point probes defined as the minimal number of probes and*

their placement necessary to reduce the uncertainty in the position of an object to better than some acceptable level. The probes are defined by a set of fixed points and vectors denoting the direction of travel for each probe. A probe returns boolean value with a positional error of at most ϵ, where $\epsilon \ll \sigma$, based upon the presence or absence of an object's edge at a particular point along the path of the probe.

With enough time, we could simply perform a large number of probes as shown in Figure 3. However, in real industrial robotic assembly workcell design, throughput is a heavily weighted criterion. Therefore, our aim is to produce the best possible probing strategies that conform to the imposed constraints. The probing strategies we obtain can be used by any line probe of the object's 2d projection. There are natural generalizations to higher dimensions, although they are not as efficient. A typical probe, shown in Figure 1, consists of a simple reflective light beam sensor that can easily detect the presence or absence of an object. Our algorithm also allows us to construct specialized optimal probing strategies such as those for a scanning array of probes as shown in Figure 2.

We achieve these near-optimal probes with a small number of actual probes by maximizing the utility of each sensor probe we place on the object. This in turn makes the problem tractable for a real robot in a high throughput automation system. These strategies are within a constant factor of the optimal probing strategy and can be solved in $O(n \log^2 n)$ time whereas the exact optimal solution is in NP-hard [7].

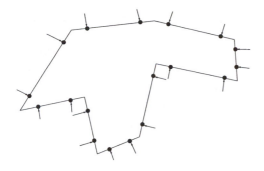

Figure 3: *A typical initial probe placement along the edges of an object*

2 Previous and Related Work

Since our probing strategy is dual to the grasping problem and uses recent results in set covering algorithms, we report advances and related works in all of these areas.

2.1 Work in Probing and Grasping

The importance of probing in terms of localizing and identifying objects with probes has been explored by several individuals. Cole and Yap [6] and Bernstein [1] developed algorithms for choosing probes to obtain the geometry of an unknown two-dimensional convex object. A generalization of this strategy for higher dimensions is presented by Dobkin *et al.* [8] while a non-convex version was developed by Boissonnat and Yvinec [2]. Also, Lindenbaum and Bruckstein [13] describe similar probing strategies for a geometric probe composed of two line probes rotating about a common axis point.

Development of efficient algorithms for scanning objects with probes for the purpose of identification and localization have been explored by Wallack *et al.* [22, 23]. Likewise, point probing strategies have been developed for insertion operations by Paulos and Canny [19].

Jia and Erdmann [12] demonstrate an elegant technique for choosing placements of simple binary sensors to discriminate objects in the plane. In fact they also employ recent work on hitting sets and set coverings in solving their problem. Our work differs mainly in

the type of problem that is solved. Jia and Erdmann choose fixed probes to discriminate individual object poses from a large set of possible poses. The problem we tackle in this paper is how to best choose moving probes to refine the pose of a known object.

We also explore the development of optimal grasping techniques since we will show our probing problem to be dual to this problem. The need for good grasp planning algorithms for arbitrary shapes has always been important for robotics and industrial automation. The problem of optimal finger placement has been addressed by Mishra *et al.* [17] who define easily computable quality metrics for grasps. Markenscoff and Papadimitriou [14] chose to optimize the grasp with respect to minimizing the forces needed to balance the object's weight through friction. Ponce and Faverjon [20] fix the number of fingers and solve a system of linear constrains in the positions of the fingers to optimally position them along the polygonal edges. A similar technique for three-dimensional polyhedral objects was developed by Ponce *et al.* [9]. Goldberg [11] also details a method for choosing grasps with a parallel jaw gripper when the initial pose of the object is unknown. Other optimizing grasps techniques based on simple geometric constructions have been developed by Brost [4] and later Mirtich, Canny, and Ferrari [10, 16].

2.2 Work in Set Coverings

We will show that finding the minimal set of probes is equivalent to solving the convex set covering problem. This problem is discussed by Clarkson [5] who describes a $O(cn \log^{O(1)} n)$ time randomized algorithm for finding covering sets of cardinality within $O(\log c)$ of the optimal set covering c.

More recent results by Brönniman and Goodrich [3] on the dual problem of finding minimal hitting sets improves on these bounds. They demonstrate an $O(n \log^2 n)$ algorithm that finds a hitting set of size $O(1)$ from the optimal set size. They employ work by Matoušek [15] using ϵ-nets.

3   Defining Optimality

When probing an object we would like to choose point probes that allow the minimum variation of the object pose. Our point probes inherently contain some known error so it is not enough to take k independent measurements to constrain k degrees of freedom. The placement of the probes effects the worst case object displacement. Therefore, we are interested in the relationship between the object displacements and the corresponding probe displacements. Our goal is to find a set of probe placements that minimizes the potential worst case object displacement.

3.1   Object Pose Definition

We define O as the actual pose of the object in two-dimensions as

$$O = (O_x, O_y, O_\theta)$$

From our problem statement, we note that we are interested not in locating an object, but in refining the position of a known object whose pose is known to some reasonable degree of accuracy. Our approach relies on this initial coarse accuracy pose information. We define our assumed initial pose as O_i and quantify a bound on the worse case displacement of the assumed pose from the actual pose as

$$\|O_i - O\|_m \leq \sigma$$

where $\|\cdot\|_m$ is defined to be the m-norm. At this point, we run into the usual problem of defining a metric on a space with distance and angular coordinates. There may be application-specific ways to weight the angular component, but a good default is to weight the angular component by the object's radius (i.e. the largest distance from any point in the object to its coordinate origin). With this choice, the metric bounds the maximum distance between any two corresponding points on the object at O and O_i. A typical value for σ would be tens of mils. Finally, while we are considering an m-norm for generality, the 2-norm would seem to be the most natural choice.

Recall that we will be using line probes to refine the position of the object. Therefore, for a given set of

probe measurements there will also be a set of valid poses for the object consistent with those sensor readings. We denote this object pose as $\bar{O}$ and define it to be an object pose chosen by an adversary consistent with some sensor readings given the object is at O. We define the difference between the actual object position and the adversary's choice as o.

$$o = O - \bar{O}$$

Recall that we are attempting to refine the position of the object so that σ will always be at least an order of magnitude larger than $\|o\|$. In terms of linear displacement the initial pose uncertainty, σ, is typically on the order of tens of mils while the uncertainty under probing is on the order of a mil or less.

$$\sigma \gg \|o\|$$

To quickly summarize we have O as the actual pose of the object, O_i as our initial pose estimate, and $\bar{O}$ as a pose that an adversary can choose that does not violate our sensor readings. That is, we cannot determine from the sensors if the object is at $\bar{O}$ or not. We will clarify later exactly how $\bar{O}$ is defined.

3.2   Probe Placement

We construct probes along the perimeter of the object and denote them as

$$p = (p_x, p_y, l_x, l_y)$$

where (p_x, p_y) is the point where the line probe touches the object when the object is at O_i, and (l_x, l_y) is the direction of the line probe. We must guarantee that no matter where the object actually is, this line probe always contacts the same edge. Assuming that the object radius was used to weight the angular component of the pose metric, this can be accomplished in a simple way: construct a strip about the line probe l whose boundaries are parallel to the line, and at distance σ from it; This strip represents the possible relative positions of the line probe for various actual object poses O. If the edge we are probing crosses the entire strip, it will always be probed correctly. If the edge crosses

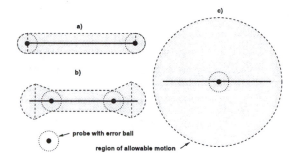

Figure 4: *Relation of placement of probes along an edge and the size of the allowable displacement regions*

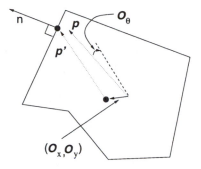

Figure 5: *Original probe p and probe p' resulting from an object displacement*

only part of the strip, then there is a possible O such that this line probe misses the edge completely. From now on, we assume that all line probes are chosen so that their σ strips touch only the edge of interest.

The initial probe placement consists of placing a pair of probes on each edge. Each probe is placed as near as possible to an endpoint of the edge, but subject to the strip constraint above. Our algorithm will choose a subset of these initial probes as the near optimal probe set. Our choice is based on the fact that we receive the most accurate pose information by probing near the vertices of an object. Observe that probes near the vertices give rise to large sensor displacements as a result of small rotational perturbations, while position information is the same anywhere on the edge. Figure 4 demonstrates how moving a set of probes with a given error out towards the vertices of an edge shrinks the size of of allowable displacements for that edge.

This set of probes is guaranteed to contain the optimal probe placement. Any edge-interior probes would give only redundant information in the worst case, and our probe choice is based on a worst-case analysis. A typical initial probe placement example is shown in Figure 3. The remaining problem is to determine a subset of these probes that still provide a substantial gain in object pose accuracy.

3.3 Probing Function

We place a coordinate system at the center of mass of the object. In addition we define the rotational displacement of the object to be about this c.o.m. axis. In

Figure 5 we depict the construction of the corresponding probe displacement for a given object displacement. This will define the probing function. In this figure n is a unit normal to the edge being probed, p is the initial probe location and p' is its location after the displacement O from the origin.

Recall that σ is very small allowing us to take small angle approximations and write

$$p'_k \approx p_k + (O_x, O_y) + p_k^\perp O_\theta$$

where $p^\perp = (x, y)^\perp = (-y, x)$ and k denotes the kth probe. It follows that the change in probe position is

$$
\begin{aligned}
\Delta p_k &= p'_k - p_k \\
&= (O_x, O_y) + p_k^\perp O_\theta
\end{aligned}
$$

The probe actually only gives us useful position information normal to the edge being probed. We could freely displace the object along the edge without changing that probe reading. Therefore, the change in probe position along the edge normal n_k can be written as $n_k \cdot \Delta p_k$. Observe that even if we approach an edge at an angle, when we detect the edge we can only claim that some point of the edge must intersect the detected point. This is equivalent to the information we receive if we approach normal to the edge. Therefore, the two probe approach techniques are equivalent and thus the choice of edge approach is independent and left as a final implementation detail. It *does* affect the σ-strip described earlier, and the amount of clearance from the edge endpoint needed to ensure the correct edge is detected.

We are now ready to define the probing function $P : \Re^3 \to \Re^k$ to be a real valued function which maps object positions into ideal probe outputs of the form $(P_1, P_2, \ldots, P_k)$. We define each element to be

$$P_k(O) = \Delta p_k \cdot n_k \tag{1}$$

Our probes will have a sensor error ϵ, typically a mil or so. We define the measured probes as $\bar{P} \in \Re^k$. Given a sensor error of ϵ, we observe that the measured probe values $\bar{P}$ must be consistent with the ideal probes given object pose O.

$$\|\bar{P} - P(O)\|_\infty \leq \epsilon \tag{2}$$

Similarly, any possible object position $\bar{O}$ that the adversary chooses must have all measured probes within ϵ of the given measurements.

$$\|P(\bar{O}) - \bar{P}\|_\infty \leq \epsilon \tag{3}$$

Using the triangle inequality on these last two expressions, we find that the O and $\bar{O}$ satisfy

$$\|P(O) - P(\bar{O})\|_\infty \leq 2\epsilon \tag{4}$$

and observe that for any $\bar{O}$ satisfying this inequality that there a is $\bar{P}$ satisfying Equations 2 and 3. Thus the combined bound is tight. We will call the set of object displacements $\bar{O}$ that satisfy this inequality $\mathcal{K}$.

Recall that the actual position of the object is defined as O and the possible interpreted object position for some sensor reading is $\bar{O}$ where $\bar{O}$ is any $\bar{O}$ satisfying Equation 4. We want to constrain the distance between the interpreted object position and the actual object position to be as small as possible. This in turn minimizes the worst case distance between the actual and measured poses, which is the ultimate goal of pose refinement. We represent the former quantity as

$$\|O - \bar{O}\|_m \tag{5}$$

We employ an adversarial argument and note that if an adversary is allowed to move the object to some valid $\bar{O}$ consistent with the sensor readings it will always choose the $\bar{O}$ such that Equation 5 is maximized. We express this as

$$\max_{\bar{O} \in \mathcal{K}} \|O - \bar{O}\|_m \tag{6}$$

However, we are allowed to choose the set of probes P. Furthermore, we desire a set of probes that will output drastically different values for different nearby object poses, thus allowing us to identify different poses easily. Essentially we would like to eliminate the possibilities of obtaining identical or near-identical sensor readings for an object in two different poses. We can write this as

$$\max_P \|P(O) - P(\bar{O})\|_\infty \tag{7}$$

or

$$\min_P \frac{1}{\|P(O) - P(\bar{O})\|_\infty} \tag{8}$$

Since both equations 6 and 8 scale linearly with $O - \bar{O}$, it is natural to combine them as a ratio which is then independent of the magnitude of $O - \bar{O}$:

$$\min_P \left(\max_{\bar{O} \in \mathcal{K}} \frac{\|O - \bar{O}\|_m}{\|P(O) - P(\bar{O})\|_\infty} \right) \tag{9}$$

Since $P(O)$ is linear in O we can simplify Equation 9 and arrive at our final optimality criterion and probe quality measurement Q.

$$Q(P) = \min_P \left(\max_{\bar{O} \in \mathcal{K}} \frac{\|O - \bar{O}\|_m}{\|P(O - \bar{O})\|_\infty} \right) \tag{10}$$

4 Displacement Space

Working in displacement space, we observe that there is a simple geometric construction of the optimality criterion as given in Equation 10. Displacement space, denoted $\mathbb{D}$ and $\mathbb{D} \in \Re^3$, is the space of all displacements in (x, y, θ) of the object O to be probed. Each probe sensor that we introduce imposes constraints on the allowable set of displacements of the object without violating the probe value.

Equation 4 from the previous section defines a pair of halfspaces in displacement space $\mathbb{D}$ for each probe p_k.

$$\|P_k(O) - P_k(\bar{O})\|_\infty \leq 2\epsilon$$
$$\|P_k(O - \bar{O})\|_\infty \leq 2\epsilon$$
$$\|P_k(o)\|_\infty \leq 2\epsilon$$
$$\|n_x o_x + n_y o_y + (p^\perp \cdot n) o_\theta\|_\infty \leq 2\epsilon$$

These two halfspace can be written as

$$n_x o_x + o_y o_y + (r^\perp \cdot n) o_\theta - 2\epsilon \;\leq\; 0 \qquad (11)$$

$$n_x o_x + o_y o_y + (r^\perp \cdot n) o_\theta + 2\epsilon \;\geq\; 0 \qquad (12)$$

The intersection of all $2k$ halfspaces constructed from k probes by definition represents a convex polytope in $\mathbb{D}$. We name this polytope S with the definition

$$S = \cap_{h \in \mathcal{H}(P)} h$$

where $\mathcal{H}(P)$ is the family of $2k$ halfspaces defined by the set of k probes P.

In displacement space this polytope S will have furthest outlying point which will occur in the non-degenerate case at a vertex of S. This furthest outlying point represents the largest object displacement from the assumed pose that still satisfies the given probe measurements. More formally we define this point as

$$\Gamma(S) = \sup_{q \in S} \|q\|_m$$

The distance to $\Gamma(S)$ is exactly the optimality criteria as defined in Equation 10. That follows because points in S fix the denominator of Equation 10, and $\Gamma(S)$ is chosen to maximize its numerator. We assume that P is fixed, so there is only optimization by the adversary over q. Recall that an adversary can choose the actual sensor readings $\bar{P}$ such that the object displacement $\Gamma(S)$ is a valid interpretation of $\bar{P}$. Hence, this is the largest displacement of the object undetectable by the given probing strategy.

For illustrative purposes we work through a simple example without rotation. In Figure 6 we show three probes on a triangle, an admittedly simple case, but enough to demonstrate our method. Notice that each probe in real space gives rise to a pair of parallel halfspaces in displacement space, $\mathbb{D}$. If we remove probe p_2, the area of polygon S in displacement space increases which represents the additional translational freedom that the object can undergo and still remain consistent with the remaining two sensor readings. Therefore, the added sensor p_2 is a useful addition since it decreases the area of S and reduces the distance to $\Gamma(S)$.

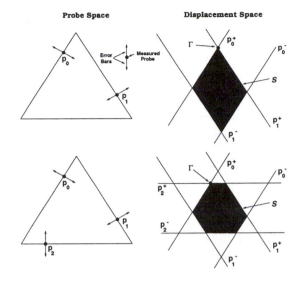

Figure 6: *Simple example of probe space and displacement space without considering rotation*

We are interested in probing strategies P' that have approximately the same quality metric P. Remember that every probe we remove from P removes a pair of halfspace in $\mathbb{D}$. This in turn changes the shape of S but not always the point $\Gamma(S)$ which defines the optimality criteria. Therefore, we would like to find other optimal probes with fewer probes. In particular, we would like to find

$$\min_{P' \subset P} |P'| : Q(P') \approx Q(P)$$

where $|\cdot|$ is simply the cardinality of the set P'.

We define S' to be the polytope defined by the intersection of the halfspaces defined by the probes P'. Observe that

$$S' \supseteq S$$

which implies that when we remove a probe, hence two halfspaces, we expect the furthest outlier to remain where it is or increase in distance from the origin giving

$$|\Gamma(S')| \geq |\Gamma(S)|$$

Rather than removing half-planes in an *ad hoc* manner such that $Q(S')$ remains essentially unchanged, we will dualize and solve for a minimal convex set covering for the corresponding points in the dual. These minimal set of points will be exactly dual to the minimal set of halfspaces in displacement space by definition of the minimal convex set cover problem. These resulting halfspaces in $\mathbb{D}$ correspond to minimal set of probes, as desired. We discuss this dualization in the next section.

Observe that the production of any such probing strategy is independent of the error, 2ϵ. This is true because we are interested in optimizing the ratio shown in Equation 10. One can also note that topology of the polytope S of the solution space is independent of ϵ which only serves as a scaling factor. That is, when we double ϵ we get the same polytope at twice the size (four times the volume). Therefore, without loss of generality we set ϵ to one for the duration of the paper.

5   Displacement Space Dual

A strong relationship to grasping is shown in this section. We show that finding the optimal k probe placements is equivalent to finding the optimal push-pull grasp for a set of k fingers. We recall that a push-pull grasp is defined as a grasp that employs fingers capable of exerting a pushing or pulling force at the contact.

We define $\mathbb{D}^D$ to be the dual of $\mathbb{D}$. We define the dual exactly in Table 1. In this mapping we show how points in $\mathbb{D}$ map to planes in $\mathbb{D}^D$ and similarly for planes in $\mathbb{D}$ to points in $\mathbb{D}^D$. We note that by definition the dual of $\mathbb{D}^D$ is $\mathbb{D}$, hence the duality operation is symmetric.

$\mathbb{D}$		$\mathbb{D}^D$
$p : (p_x, p_y, p_\theta)$	$\leftrightarrow$	$p^D : p_x x + p_y y + p_\theta \theta = 1$
$f : ax + by + c\theta = 1$	$\leftrightarrow$	$f^D : (a, b, c)$
$S = $ polytope	$\leftrightarrow$	$S^D = \{f^D : f \cap \text{Int}(S) = \emptyset\}$

Table 1: *Duality Mappings*

Observe that a polytope S defined as the intersection of a set of halfspaces h_k becomes the polytope S^D. We define $\text{Bound}(h_k)$ to be the plane on the boundary of the halfspace h_k. The polytope S^D can also be expressed as the convex hull of the union of dual points

$\text{Bound}(h_k)^D$.

$$S^D = \text{Conv}(\cup \text{Bound}(h_k)^D)$$

Let $r \in S$. The distance of r from the origin in $\mathbb{D}$ is simply

$$|r| = \sqrt{r_x^2 + r_y^2 + r_\theta^2}$$

The dual plane r^D in $\mathbb{D}^D$ by definition is represented as

$$r_x x + r_y y + r_\theta \theta = 1$$

This distance of the closest point on this plane to the origin in $\mathbb{D}^D$ is given by

$$|r^D| = \frac{1}{\sqrt{r_x^2 + r_y^2 + r_\theta^2}}$$

Setting $\alpha = \sqrt{r_x^2 + r_y^2 + r_\theta^2}$ we get that the distance of this point r from the origin in $\mathbb{D}$ is α and the minimal distance of the dual plane r^D from the origin in $\mathbb{D}^D$ is $\frac{1}{\alpha}$. Therefore,

$$|r| = \frac{1}{|r^D|}$$

Let f_c be the closest plane to the origin of $\mathbb{D}^D$ not intersecting $\text{Int}(S^D)$. The distance to f_c is the same as the distance to the closest point u_c in the boundary of S^D (which is contained in f_c). And its easy to see that $\Gamma(S)^D = f_c$ where $\Gamma(S)$ was defined earlier as the furthest outlying point in S.

The closest point to the origin in the boundary of a polytope lies on the largest inscribed sphere centered at the origin. Observe that $\Gamma(S)$ lies on the smallest circumscribing sphere of S in $\mathbb{D}$. Therefore, finding the smallest circumscribing sphere Σ for a polytope is equivalent to finding the largest inscribed sphere Σ^D of the dual polytope S^D. This follows from the relationship

$$|r_\Sigma| = \frac{1}{|r_{\Sigma^D}|}$$

Recall that the planes $\text{Bound}(h)$ through the halfspaces in $\mathbb{D}$ dualize to the points

$$\text{Bound}(h_k)^D = (n_x, n_y, p_k^\perp)$$

These points are equivalent to the wrenches due to unit pull finger forces acting at p. Recall that in the probing problem we obtain a pair of halfspaces for each probe. Hence, the optimal probing problem is equivalent to the optimal push-pull grasping problem. We use the optimal grasping criteria as defined by Mishra *et al.* [17] and others [10, 16] which is the set of finger placements such that the one-norm of the finger forces can resist the largest externally applied wrench on the object. We also define the optimal probe placement as the minimal number of probes and their placement necessary to reduce the uncertainty in the position of an object to better than some acceptable level. Using these metric definitions, we obtain the following result.

Theorem 1 *Finding the optimal placement of k probes is equivalent to finding the optimal push-pull frictionless grasp for a set of k fingers.*

6 Hitting Sets and Set Covers

Recall that the quality of that probing strategy is given directly by the radius of the maximally inscribed sphere in S^D. We would like to remove some vertices of S^D such that the radius of the maximally inscribed sphere does not decrease by much.

This problem can be posed as a convex set cover problem which states that given L and U a set of points in $\Re^d$, find $C \subseteq U$ with $L \subseteq Conv(U)$. Here we have L as the sphere of desired radius. This problem has been studied by several individuals. Recently, Clarkson [5] describes a randomized algorithm for computing the three-dimensional convex point set cover from an initial set of n points to within $O(\log c)$ of the optimal cover of c points. His algorithm has a running time of $O(cn \log^{O(1)} n)$.

Recent work by Brönniman and Goodrich [3] improve on both the running time and approximation to the optimal convex set covering. Their deterministic algorithm solves the equivalent problem of finding a minimal hitting set, where a hitting set is a subset $H \subseteq X$ such that H has a non-empty intersection with every set R in a collection of subsets of X. Their algorithm employs work by Matoušek [15] on ϵ-nets to obtain a

hitting set in $O(n \log^2 n)$ time that is within $O(1)$ of the optimal size hitting set. This set corresponds exactly to the optimal probe placement which we define as a set of c probes that reduce the uncertainty in the position of an object to at least some necessary level for the operation to be performed.

In our optimal construction we obtain pairs of halfspaces, hence pairs of points in the dual. However, in the Brönniman and Goodrich algorithm they are treated as two completely unrelated elements. This will result in near-optimal set sizes that are in the worst case twice as large as we could achieve by grouping the pairs. Alternatively, we can group them to obtain the near-optimal hitting set at a slight running time cost. This performance slowdown is a result of an increase in the VC-dimension [21] as a result of our pairing.

VC-dimension, named for Vapnik and Chervonenkis, is defined for a range space (X, R) with $P \subseteq X$ as the cardinality of the largest set P that is shattered by R. A set P is shattered by R if $\Pi_R(P)$ is the power-set of P.

To obtain an optimal probing strategy for an array of scanning sensors as shown in Figure 2, we identify the co-linear points in the displacement space and assign them labels such that the hitting set algorithm will include all or none of a set of co-linear points in the probe optimization selection. This also results in an increase in the VC-dimension which affects the running time but still finds a hitting set within $O(1)$ of the optimal one.

Our algorithm successfully handles other variations similar to the co-linear constraint for the scanning sensor without major modification. This makes it well adapted to situations where optimal probing strategies under special constrains are needed and not intuitive to observe.

The Lemma below summarizes much of the results of this paper.

Lemma 1 *A near optimal set of c point probes can be found for any polygonal object in $O(n \log^2 n)$ time. Furthermore, the size of the set c will be within $O(1)$ of the size of the optimal set of c point probes.*

7 Conclusion

We have demonstrated a fast method by which optimal probe placements can be obtained for any known polygonal object. More importantly, we can guarantee that the number of probes resulting from our technique is within a constant of the actual optimal number of probes necessary. These probing strategies refine the position of an object whose pose is approximately known. Furthermore, we claim that this is a real problem often encountered in industrial manufacturing. Our paper also shows that the problem of optimal probe placement is dual to the well studied push-pull grasping problem of positioning frictionless fingers on an object.

Acknowledgements

This research was supported in part by National Science Foundation Grant #IRI-9114446. The authors would like to thank Brian Mirtich and Ken Goldberg for several insights and suggestions towards the development of this paper. We would also like to thank the anonymous reviewers who provided numerous useful comments and suggestions.

References

[1] Herbert Bernstein. Determining the shape of a convex n-sided polygon by using 2nk+k tactile probes. In *Information Procssing Letters*, pages 225–260, 1986.

[2] J. Boissonnat and M. Yvinec. Probing a scene of nonconvex polyhedra. In *Algorithmica*, pages 321–342, 1992.

[3] H. Brönnimann and M.T. Goodrich. Almost optimal set covers in finite vc-dimension. In *Proc. 10th ACM Symp. on Computational Geometry (SCG)*, pages 293–302, 1994.

[4] Randy C. Brost. Automatic grasp planning in the presence of uncertainty. *International Journal of Robotics Research*, 7(1):3–17, 1988.

[5] Kenneth L. Clarkson. Algorithms for polytope covering and approximation. In *Proc. 3rd Workshop Algorithms Data Struct.*, volume 709, pages 246–252, 1993.

[6] Richard Cole and Chee Yap. Shape from probing. *Journal of Algorithms*, 8:19–38, 1987.

[7] G. Das and D. Joseph. The complexity of minimum nested polyhedra. In *Canadian Conference on Computational Geometry*, 1990.

[8] D. Dobkin, H. Edelsbrunner, and C. Yap. Probing convex polytopes. In *18th Anual ACM Symposium on Theory of Computing*, pages 424–432, 1986.

[9] B. Faverjon and J. Ponce. On computing two-finger force closure grasps of curved 2d objects. In *IEEE International Conference on Robotics and Automation*, pages 424–429, 1991.

[10] C. Ferrari and J.F. Canny. Planning optimal grasps. In *IEEE International Conference on Robotics and Automation*, pages 2290–2295, 1992.

[11] Ken Goldberg. Orienting polygonal parts without sensors. *Algorithmica. Special Issue on Computational Robotics*, Volume 10(3):201–225, 1993.

[12] Yan-Bin Jia and Michael Erdmann. The complexity of sensing by point sampling. *Algorithmic Foundations of Robotics*, pages 283–298, 1995.

[13] M. Lindenbaum and A. Bruckstein. Reconstructing a convex polygon from binary perspective projections. In *Pattern Recognition*, pages 1343–1350, 1990.

[14] Xanthippi Markenscoff and Christos H. Papadimitriou. Optimum grip of a polygon. *International Journal of Robotics Research*, 8(2):17–29, 1989.

[15] J. Matoušek. Construction of ε-nets. In *Discrete Computational Geometry*, pages 5:427–448, 1990.

[16] Brian Mirtich and John Canny. Easily computable optimum grasps in 2d and 3d. In *IEEE International Conference on Robotics and Automation*, 1994.

[17] B. Mishra, J. T. Schwartz, and M. Sharir. On the existence and synthesis of multifingered positive grips. *Algorithmica*, 2:541–558, 1987.

[18] James Nevins and Daniel Whitney. Computer controlled assembly. *Scientific America*, 238(2):62–74, February 1978.

[19] Eric Paulos and John Canny. Accurate insertion strategies using simple optical sensors. In *IEEE International Conference on Robotics and Automation*, pages 1656–1662, May 1994.

[20] J. Ponce and B. Faverjon. On computing three-finger force closure grasps of polyhedral objects. In *International Conference on Advanced Robotics*, pages 1018–1023, 1991.

[21] V. N. Vapnik and A. Ya. Červonenkis. On the uniform convergence of relative frequencies of events to their probabilities. In *Theory Probab. Appl.*, pages 16:264–280, 1971.

[22] Aaron Wallack and John Canny. Planning for modular and hybrid fixtures. In *IEEE International Conference on Robotics and Automation*, May 1994.

[23] Aaron Wallack, John Canny, and Dinesh Manocha. Object localization using crossbeam sensing. In *IEEE International Conference on Robotics and Automation*, pages 692–699, 1993.

Robust Geometric Algorithms for Sensor Planning

Amy J. Briggs, *Middlebury College, Middlebury, VT, USA*
Bruce R. Donald, *Cornell University, Ithaca, NY, USA*

We consider the problem of planning sensor strategies that enable a sensor to be automatically configured for robot tasks. In this paper we present robust and efficient algorithms for computing the regions from which a sensor has unobstructed or partially obstructed views of a target in a goal. We apply these algorithms to the Error Detection and Recovery problem of recognizing whether a goal or failure region has been achieved. Based on these methods and strategies for visually-cued camera control, we have built a robot surveillance system in which one mobile robot navigates to a viewing position from which it has an unobstructed view of a goal region, and then uses visual recognition to detect when a specific target has entered the room.

1 Introduction

This paper introduces a computational framework in which to study the problem of sensor configuration, and develops combinatorially precise algorithms for computing partial and complete visibility maps.

Many applications of automatic sensor configuration arise in the areas of cooperating robots and robot surveillance. In particular, our algorithms apply to the problems of intruder detection, execution monitoring, and robot reconnaissance. A natural extension of our work can be made to the problem of beacon placement for robot navigation. In Section 6 we discuss a demonstration surveillance system that we built using an implementation of one of the algorithms in this paper.

The Error Detection and Recovery Framework [11] provides a natural problem domain in which to apply our strategies. An *Error Detection and Recovery* (EDR) strategy is one that is guaranteed to achieve a specified goal when the goal is recognizably achievable, and signals failure otherwise. Our algorithms can be used in conjunction with an EDR planner to compute where a sensor should be placed in order to recognize success or failure of a motion plan. We explore this problem in Section 5.

In this paper we restrict our attention to visibility and recognizability problems in the plane. We show that even in the $2D$ case, the geometric computations are nontrivial and significant computational issues arise, making the $2D$ case a natural first consideration. Furthermore, planning motions for a mobile robot often reduces to a computation in $2D$: a mobile robot that maintains contact with the floor usually navigates among obstacles that can be modeled as swept polygons. When the $3D$ obstacles are projected to the floor, their $2D$ footprints yield a map in $2D$. For these reasons, this and much other work in the field of visual agents is in $2D$.

1.1 Error detection and recovery

Much of the early work in robotics focused on developing guaranteed plans for accomplishing tasks specified at a high level. Such task specifications might be of the form "mesh these two gears", or "place part A inside region B". It is not always possible, however, especially in the realm of assembly planning, to generate guaranteed plans. For example, errors in tolerancing of the parts might render an assembly infeasible. The Error Detection and Recovery (EDR) framework of Donald was developed to deal with these inadequacies of the guaranteed planning framework. EDR strategies will either achieve a goal if it is recognizably reachable, or signal failure. Given a geometrically-specified goal region G, an EDR strategy involves computing a failure region H and a motion plan that will terminate

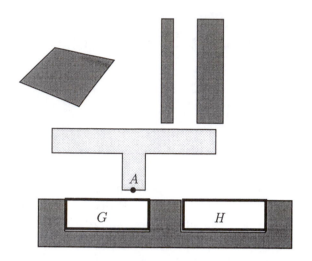

Figure 1: *An example setup for the problem of sensor configuration in EDR. A represents the target; its reference point is indicated by the black dot. G is the goal region and H is the failure region. The darkly shaded polygons are obstacles. The problem is to find a sensor placement from which $A \in G$ and $A \in H$ can be distinguished.*

recognizably either in G or H. The EDR framework guarantees that under generalized damper dynamics, the robot will eventually reach G or H. Furthermore, having entered G or H, it will never leave. Given this guarantee of reachability, we wish to strengthen it to a guarantee of recognizability: we want to know which of G and H has been attained. The visibility algorithms presented in Section 3 will be used in showing how a sensor can be configured to distinguish between a target in G and a target in H. Figure 1 gives an example of the problem we would like to solve.

1.2   Related Work

The sensor placement problem has previously been addressed by Nelson and Khosla [20] and Kutulakos, Dyer, and Lumelsky [16] for visual tracking and vision-guided exploration. Several researchers have explored the problem of optimal sensor placement. Cameron and Durrant-Whyte [6] and Hager and Mintz [13] present a Bayesian approach to optimal sensor placement.

Hutchinson [14] introduces the concept of a *visual*

constraint surface to control motion. The idea is to combine position, force, and visual sensing in order to produce error-tolerant motion strategies. His work builds on that of preimage planners by adding visual feedback to compensate for uncertainty. Details on the implementation of vision-based control are described by Hutchinson and Castaño [8].

Sharma and Hutchinson [24] define a measure of *robot motion observability* based on the relationship between differential changes in the position of the robot to the corresponding differential changes in the observed visual features. Lacroix, Grandjean, and Ghallab [17] describe a method for selecting view points and sensing tasks to confirm an identification hypothesis.

Cowan and Kovesi [9] study the problem of automatic camera placement for vision tasks. They consider the constraints on camera location imposed by resolution and focus requirements, visibility and view angle, and forbidden regions depending on the task. Given values bounding these constraints, they compute the set of camera locations affording complete visibility of a surface in $3D$. Laugier, Ijel and Troccàz [18] employ partial and complete visibility computations in selecting sensor locations to acquire information about the environment and to guide grasping movements. Zhang [29] considers the problem of optimally placing multiple sensors.

A different approach from the one taken in this paper to the incorporation of sensor planning in the EDR framework was first presented by Donald [11]. In that approach, an equivalence is established between sensing and motion in configuration space. *Active sensing* for a mobile robot is reduced to motion, by exploiting the similarity between visibility and generalized damper motions. In contrast, we present here a framework that is closer to actual sensors.

Research in the area of *art gallery theory* has introduced and addressed many problems pertaining to polygon visibility. The *art gallery problem* is to determine the minimum number of guards sufficient to guard the interior of a simple polygon (see [21] for more details). Sensor configuration planning addresses the related question of where sensors should be placed in order to monitor a region of interest. In this case we are interested in external visibility of a polygon rather than internal visibility. Furthermore, because we employ real sensors, considerations of uncertainty must be taken into account.

The questions of detecting polygon visibility and constructing visibility regions under a variety of assumptions is a rich area of past and ongoing research in computational geometry. We mention here a few of the papers most closely related to our problem. Suri and O'Rourke [25] give an $\Theta(n^4)$ algorithm for the problem of computing the locus of points weakly visible from a distinguished edge in an environment of line segments. Their lower bound of $\Omega(n^4)$ for explicitly constructing the boundary of the weak visibility region holds as well for our computation of recognizability regions under a weak visibility assumption. Bhattacharya, Kirkpatrick and Toussaint [2] introduce the concept of *sector visibility* of a polygon, and give $\Theta(n)$ and $\Omega(n \log n)$ bounds, depending on the size of the visibility wedge, for determining if a polygon is weakly externally visible. The problem of planar motion planning for a robot with bounded directional uncertainty is considered by de Berg *et al.* [10]. They give algorithms for constructing the regions from which goals may be reached, and show how the complexity of the regions depends on the magnitude of the uncertainty angle.

Teller [27] solves the weak polygon visibility problem for a special case in $3D$. Namely, he computes the antipenumbra (the volume from which some, but not all, of a light source can be seen) of a convex area light source shining through a sequence of convex areal holes in three dimensions. For an environment of total edge complexity n, he gives an $O(n^2)$ time algorithm for computing the piecewise-quadratic boundary of the antipenumbra, which will be non-convex and disconnected in general.

Tarabanis and Tsai [26] examine the question of complete visibility for general polyhedral environments in $3D$. For a feature polygon of size m and a polyhedral environment of size n, they present an $O(m^3n^3)$ algorithm for computing the locus of all viewpoints from which the fixed feature polygon can be entirely seen.

Guibas, Motwani and Raghavan consider an abstraction of the robot localization problem [12]. Given a simple polygon P (representing the map of a known environment) and a star-shaped polygon V (representing the portion of the map visible from the robot's position), the problem is to find a point or set of points in P from which the portion of P that is visible is congruent to V (*i.e.*, given V, the robot must determine its position in P). They give a method of preprocessing P so that subsequent queries V can be answered in optimal time in the size of the output.

1.3 Outline of paper

The remainder of the paper is organized as follows. In Section 2 we introduce our approach and define the notions of *recognizability* and *confusability*. Using point-to-point visibility as a model of detectability, we present in Section 3 our algorithms for computing recognizability regions for a target moving within a goal polygon in the plane. The computed regions can be restricted to account for the error characteristics of the sensor, as shown in Section 4. In Section 5 we apply these algorithms to the EDR framework, and show how to compute the set of sensor configurations so that readings that lead to confusion of G and H are avoided. Our experimental results using mobile robots in the Cornell Robotics and Vision Laboratory are presented in Section 6.

2 Preliminaries and definitions

We will start by introducing the notation used throughout this paper, and by formalizing the problems to be solved.

Our target configuration space is denoted C_r, and in this paper we consider two types of planar motion. We have $C_r = \mathbb{R}^2$ when the target has two translational degrees of freedom and a fixed orientation. When the target is allowed to both translate and rotate in the plane, the target configuration space is $\mathbb{R}^2 \times S^1$.

The sensor we employ is an idealized but physically realizable model of a point-and-shoot sensor, such as a laser range finder. A sensor configuration is specified by a *placement* and a viewing direction, or *aim*.

When in a particular configuration, the sensor returns a distance and normal reading to the nearest object, which is accurate to within some known bound. Such a ranging device has been developed in our robotics laboratory at Cornell, and has been used for both map-making and robot localization [4, 5].

We denote the space of sensor placements $C_{s_p} = \mathbb{R}^2$ and the space of sensor aims $C_{s_c} = S^1$. Our sensor configuration space is $C_s = C_{s_p} \times C_{s_c} = \mathbb{R}^2 \times S^1$. For a given sensor configuration $(\mathbf{p}, \psi)$, the sensor returns distance and normal readings for a subset of $\mathbb{R}^2$. We call this subset the *sensitive volume* of the sensor, and denote it by $SV(\mathbf{p}, \psi)$. In what follows, we restrict our attention to questions of visibility within the sensitive volume.

For a region X in the target object's configuration space, let $R(X)$ denote its *recognizability* region, that is, the set of all sensor placements from which the sensor can detect an object A in region X. Let $C(X, Y)$ denote the *confusability* region, that is, the set of all sensor placements from which the sensor cannot tell $A \in X$ and $A \in Y$ apart. To guarantee goal recognizability for an EDR strategy, we wish to find a sensor placement $\mathbf{p} \in C_{s_p}$ such that $\mathbf{p} \in R(G) \cap R(H) - C(G, H)$. Recognizability regions will be computed using the algorithms in the next section, and the computation of confusability regions will be addressed in Section 5.

3 Visibility algorithms

In Sections 3.1 and 3.2, we give algorithms for computing the recognizability region of a target in a goal under two different notions of visibility. Our algorithms explicitly construct the locus of points from which a target polygon is visible, and we analyze how these visibility regions change as the target moves.

3.1 Complete visibility

In this section we consider a simplified version of the target detection problem, in which the computed sensor placements are those that allow an unobstructed view of the target. We give algorithms for detecting

a stationary target, and for detecting a target at any position and orientation within a goal region.

Our result is the following:

Theorem 1 *The recognizability region of a target translating and rotating through a goal with k vertices can be computed in time $O(n\alpha(n) + nk)$ in an environment with n vertices in the complete visibility model.*

3.1.1 The complete visibility algorithm for a stationary target

We say that target A at configuration $\mathbf{q} = (x, y, \theta)$ is *completely visible* from sensor placement $\mathbf{p} \in C_{s_p}$ if for no point y on the boundary of the target does the segment $\overline{\mathbf{p}y}$ intersect an obstacle. Note that $\overline{\mathbf{p}y}$ may intersect $A_{\mathbf{q}}$. If $A_{\mathbf{q}}$ is completely visible from $\mathbf{p}$ in the presence of the obstacles then we say that the sensor at $\mathbf{p}$ has an *unobstructed view* of the target at configuration $\mathbf{q}$. Our algorithm assumes that no obstacle lies within the convex hull of $A_{\mathbf{q}}$.

The idea is that each obstacle casts *shadows* with respect to the target. Each shadow is a subset of the sensor placement space C_{s_p} from which the target is partially occluded. To compute the set of placements from which the target at configuration (x, y, θ) is completely visible, we use the following algorithm:

Complete visibility algorithm for a stationary target

1. Construct all local inner tangents between the obstacles and the target. Represent each tangent as a ray anchored on a vertex of the target.

2. Extend each tangent ray starting at the point of tangency with an obstacle until it hits an edge of the environment (an obstacle or the bounding polygon). We call these segments *visibility rays*.

3. Consider the arrangement of all the polygons in the environment, along with these visibility rays. Compute the single arrangement cell that contains the target polygon.

Complexity of the complete visibility algorithm

Let A be a polygon representing the target, and B_i be a polygonal obstacle. If A has m vertices and obstacle B_i has n_i vertices, we can compute the $O(n_i)$

local inner tangents between A and B_i in time $O(mn_i)$. For an environment of obstacles with n vertices overall, we can compute the $O(n)$ local inner tangents in time $O(mn)$. Computing a single cell in an arrangement is equivalent to computing the lower envelope of a set of line segments in the plane, which for a set of size n takes time $O(n\alpha(n))$, where $\alpha(n)$ is the inverse Ackerman function [22]. Thus, the overall time for computing the recognizability region of a stationary target in the complete visibility model is $O(n\alpha(n) + mn)$.

3.1.2 Complete visibility over a region

We now consider sensor placements with an unobstructed view of the target at any position or orientation within a goal region G. We model the target as a connected polygon, with a reference point that lies inside the polygon. Note that the target is said to "lie in the goal" if and only if its reference point lies in the goal. The idea is that a sensor placement is valid if and only if, as its reference point moves within the goal, the entire swept area covered by the target is visible.

Complete visibility algorithm for a translating and rotating target

Consider the case of a target rotating and translating through the goal. We want the set of placements from which no portion of the target's boundary is occluded by obstacles no matter what position or orientation the target has within the goal. We take the longest Euclidean distance from the target's reference point to a vertex of the target. We call this distance the *radius* of the target. Suppose the target has a radius of r and its reference point lies at $\mathbf{w}$. Then the disc of radius r centered at $\mathbf{w}$ is equivalent to the area covered by the target as it rotates around its reference point.

Hence, for target A with radius r, the Minkowski sum of the goal G with the disc of radius r represents the swept area covered by the target as it translates and rotates through the goal. Call this Minkowski sum the *swept goal region* $M(G, A)$. We now compute the set of sensor positions that have an unobstructed view of $M(G, A)$.

To compute the shadow boundaries, we introduce local inner tangents between each obstacle and the con-

vex hull of $M(G, A)$. This can be accomplished by simply computing all inner tangents between each obstacle and the disc of radius r at each vertex of G, then taking the outermost tangents at each obstacle. If inner tangent e is locally tangent at obstacle vertex v, then we again introduce a visibility ray that extends e away from vertex v. The arrangement of the visibility rays and the environment now partitions C_{s_p} into shadow regions and visibility regions, *i.e.*, regions from which $M(G, A)$ is partially occluded or entirely visible. But instead of computing the entire arrangement, we again note that it suffices to compute a single cell in the arrangement, namely the cell containing $M(G, A)$.

Complexity of the complete visibility algorithm over a region

Given the target radius r, we can compute $M(G, A)$ for a goal G of size k in time $O(k)$. If obstacle B_i has n_i vertices, we can compute the $O(n_i)$ local inner tangents between B_i and the convex hull of $M(G, A)$ in time $O(kn_i)$. For an environment of obstacles with n vertices overall, we can compute the $O(n)$ local inner tangents in time $O(kn)$. So the overall time for computing the recognizability region for a target rotating and translating through the goal in the complete visibility model is $O(n\alpha(n) + nk)$.

3.2 Partial visibility

We turn now to the question of computing recognizability regions in the *partial visibility* model. First we consider the problem of detecting a stationary target within a polygonal obstacle environment. We then apply these tools to the problem of detecting a translating target as it enters the goal.

We will show the following:

Theorem 2 *In the partial visibility model, the recognizability region of a target translating in a goal of size k can be computed in time $O(kmn^3(n+m))$ for an environment of complexity n and a target of complexity m.*

3.2.1 The partial visibility algorithm for a stationary target

We say that target A is *partially visible* from sensor placement $\mathbf{p} \in C_{s_p}$ if at least one point in the closure of A is visible from $\mathbf{p}$.

For target A at configuration $\mathbf{q} \in C_r$, we construct the partial visibility region using an approach similar to that given by Suri and O'Rourke for computing the region weakly visible from an edge [25]. Our algorithm is as follows:

Partial visibility algorithm for a stationary target

1. Construct the visibility graph for the entire environment, consisting of distinguished polygon A and obstacles B.

2. Extend each edge of the visibility graph maximally until both ends touch an edge of the environment. If neither of the endpoints of the extended visibility edge lie on the polygon A, discard the visibility edge. Otherwise, clip the edge at its intersection with A and call this piece a *visibility ray*.

3. For each vertex v in the environment, perform an angular sweep of the visibility rays incident to v. If A remains visible to v throughout the swept angle between two adjacent visibility rays anchored at v, then the triangular swept region is output as a *visibility triangle*.

The union of these visibility triangles forms the region from which A is partially visible. The complement of the union of triangles and the environment is a collection of holes in the visibility region, which we call *shadows*. Figure 3 shows the shadows for an example environment. This example demonstrates that in the partial visibility model, shadows are not necessarily bounded by tangents between an obstacle and the goal.

Complexity of the partial visibility algorithm

Suppose the obstacles and bounding polygon together have n vertices, and the target has m vertices. The visibility graph for this environment, the

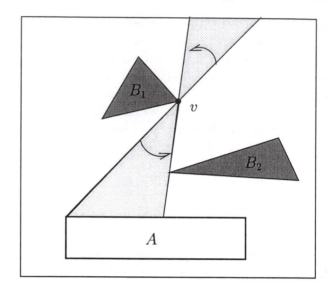

Figure 2: *An angular sweep between two visibility rays at vertex v. The lightly shaded regions are visibility triangles.*

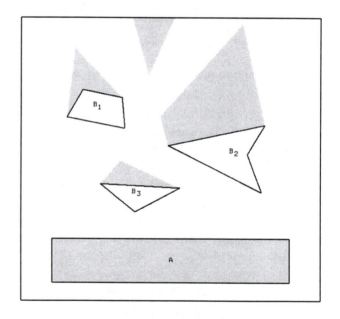

Figure 3: *The shadows cast by obstacles B_1, B_2, and B_3 are shown shaded. The complement of the shadows, the obstacles, and the target forms the partial visibility region of target A.*

basic data structure used in the algorithm, has size $O(n(n+m))$. Note that we are not interested in visibility edges between the vertices of the target itself. The extended visibility graph will, in practice, have fewer edges than the basic visibility graph, since we only keep the edges whose extensions intersect the target. Its worst-case complexity, however, remains $O(n(n+m))$. Each vertex of the environment has $O(n+m)$ visibility rays incident to it. Therefore each vertex contributes $O(n+m)$ visibility triangles, so we have $O(n(n+m))$ visibility triangles overall. In general, the union of these triangles has complexity $O(n^2(n+m)^2)$. As was mentioned in the paper by Suri and O'Rourke [25], the triangles can be output in constant time per triangle: Asano *et al.* have shown that the visibility edges at a vertex v can be obtained sorted by slope in linear time with Welzl's algorithm for computing the visibility graph [28, 1]. Thus, the overall time for explicitly computing the boundary of the partial visibility region for target A at any fixed configuration $\mathbf{q}$ is $O(n^2(n+m)^2)$. The region can be given as a union of triangles, without computing the boundary, in time $O(n(n+m))$.

3.2.2 Partial visibility over a region

The algorithm above solves the problem of detecting a stationary target in the partial visibility model. We now address the problem of maintaining line-of-sight contact with the target as it moves within the confines of a particular polygon, for example, as the target moves within the goal. How do the visibility triangles and shadows change as the target moves? To answer this question, we need to introduce some additional terminology. Let e be a visibility edge whose associated visibility ray intersects the target at point x. The endpoint of e lying closer to x (possibly x itself) is defined as the *anchor* vertex of e, while the further endpoint is called the *attachment* vertex of e. If a vertex of the shadow (considering the shadow as a polygon) lies in free space, *i.e.*, if it lies inside the bounding polygon and is not on the boundary of an obstacle, then we call it a *free* vertex of the shadow.

As the target translates, free shadow vertices trace out point conics if their generating edges are anchored on the target [3].

3.2.3 Swept shadows in the partial visibility model

We have shown how to compute shadows for any fixed target position, and have discussed how these shadows change as the target translates. In order to detect the target as it enters the goal, we must compute the shadows swept for all positions of the target in the goal. We define a *swept shadow* of the goal in the partial visibility model to be a maximal connected region of C_{s_p} such that for each point $\mathbf{p}$ in the region, there exists a configuration of the target in the goal from which the target is totally occluded.

We compute swept shadows for the target at a fixed orientation anywhere in the goal by translating the target polygon along the edges of the goal polygon. The boundary of a swept shadow is composed of obstacle segments and the curves (lines and conics) traced by free vertices. Discontinuities in the boundary of a swept shadow occur at *critical events*. We characterize the critical events as follows:

1. A moving visibility ray becomes aligned with a fixed edge of the visibility graph.

2. A free vertex of a shadow intersects an obstacle edge or the bounding polygon.

3. Two moving visibility rays bounding a shadow become parallel.

Below we present our algorithm for computing the partial visibility region of a target A as it translates through the goal at a known orientation θ. This gives us the set of all sensor placements from which at least one point on the boundary of the target A_θ can be seen, no matter where the target is in the goal.

Partial visibility algorithm for a translating target

1. Let e be any edge of goal G. Consider A_θ to be placed on one of the endpoints of e. Call this configuration $\mathbf{q}$. Construct the partial visibility region of target A at configuration $\mathbf{q}$.

2. Translate A_θ along e. As the shadows cast by the obstacles change, call the area swept out by a

shadow a *swept shadow*. Between critical events, the vertices of each shadow move along lines or conics. The equations of these curves can be computed algebraically given the positions of the obstacles in the environment and the visibility rays. Update the boundary of the swept shadows at critical events.

3. Translate A_θ along all other edges e_i, $1 \leq i \leq k$, of G, repeating step 2 for each edge.

4. Compute each swept shadow independently as described in the above steps. The complement of the union of all the swept shadows, the target, and the obstacles is the partial visibility region.

The output of the algorithm is the set of swept shadows. Note that the boundary of a swept shadow is piecewise linear and conic.

Complexity of the partial visibility algorithm over a region

The extended visibility edges bounding the shadows are all either external local tangents between an obstacle and the target, or internal local tangents between obstacles. Since the obstacles are fixed, the visibility edges between them remain fixed. As the target moves, the only visibility edges that move are those that are anchored on a vertex of the target.

With n vertices in the environment and m target vertices, there are $O(mn)$ moving visibility edges. As the target translates along an edge of the goal, a visibility edge anchored at target vertex a_i and attached at obstacle vertex b_j could become aligned with each of the $O(n)$ fixed visibility edges at obstacle vertex b_j. This gives $O(mn^2)$ critical events of the first type as the target translates along an edge of the goal. There are $O(m^2n^2)$ free vertices tracing out curves, which may intersect each of the $O(n)$ obstacle segments. This gives $O(m^2n^3)$ critical events of the second type. When the third type of critical event occurs, a free vertex disappears. There are $O(m^2n^2)$ of these events.

At a critical event of the first type, a visibility ray appears or disappears, causing a visibility triangle to appear or disappear. The total cost of handling all

updates of this type is $O(mn^3(n + m))$. Only local change is caused by events of the second type and third type.

Between critical events, we simply grow the shadows, either along lines or conics. Note that the shadows never shrink: A point $\mathbf{p} \in C_{s_p}$ is in a shadow with respect to a polygonal goal if there exists some target configuration such that the target is not at all visible from $\mathbf{p}$. The computation of swept shadows is done by translating the target polygon along the edges of the goal, updating the boundary at critical events. The total running time of the algorithm for a goal with k vertices is $O(kmn^3(n + m))$.

4 Uncertainty in sensor placement and aim

A real sensor cannot be configured exactly. Rather, it will be subject to both errors in placement and errors in aim. These errors depend on the sensor platform (*e.g.*, a mobile robot). Therefore we would like to compute sensor strategies that take uncertainty in sensor configuration into consideration. In this section, we sketch how the computation of visibility regions can be extended to handle this type of sensor error. Our approach does not address the problem of sensor measurement error.

Positional uncertainty characterizes the sensor placement error. Let ϵ_{pos} denote the worst-case positional uncertainty of the sensor. If the commanded sensor placement is $\mathbf{p}$, the actual sensor placement could be any position in the disc of radius ϵ_{pos} centered at $\mathbf{p}$. We handle positional uncertainty by growing the shadows by the uncertainty ball of radius ϵ_{pos}. The complement of the union of these grown shadows and the environment will be the visibility region that accounts for uncertainty in sensor position.

Directional uncertainty characterizes the sensor aim error. Let ϵ denote the maximum angular error of the sensor aim. That is, if the commanded sensing direction is ψ, the actual sensor heading could be any direction in the cone $(\psi - \epsilon, \psi + \epsilon)$. The effect of sensor directional uncertainty is that we must disallow angularly

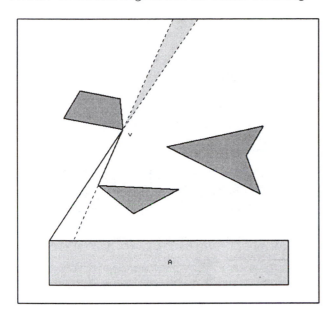

Figure 4: *A narrow visibility triangle anchored at vertex v is shown lightly shaded.*

narrow wedges of visibility. This type of uncertainty is most relevant in the case of partial visibility. See Figure 4 for an illustration of a narrow visibility triangle. This triangle does not become part of the visibility region when directional uncertainty is considered.

After we compute the visibility rays as described in Section 3.2.1, we visit each vertex in the environment, and combine adjacent visibility triangles that end on the same polygon. We make the following definitions:

1. The maximal union of adjacent visibility triangles anchored on a single vertex v and ending on the same polygon is called a *visibility polygon*. By construction, visibility polygons are simple.

2. The *core triangle* of a visibility polygon anchored at v is the maximal inscribed triangle whose apex is v.

If the angle at the apex of such a maximal visibility triangle is less than our angular uncertainty bound ϵ, we discard the polygon. Otherwise, we classify the maximal visibility triangle as an ϵ-*fat triangle*. After this processing, we now have $O(n(n+m))$ fat visibility triangles. We can now use a result of Matoušek

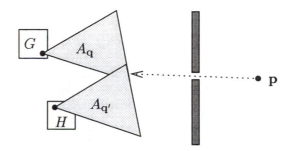

Figure 5: *A sensor reading that confuses $A_{\mathbf{q}} \in G$ and $A_{\mathbf{q'}} \in H$ is due to an edge of $A_{\mathbf{q}}$ being colinear with an edge of $A_{\mathbf{q'}}$. The darkly shaded rectangles are obstacles.*

et al. [19] on the union of fat triangles. Their result bounds the number of *holes* in a union of fat triangles. In our case, the "holes" are shadows in a union of visibility triangles. Their theorem states that for any fixed $\delta > 0$, and any family $\mathcal{F}$ of n δ-fat triangles, their union has $O(n/\delta^{O(1)})$ holes. When we restrict our visibility triangles to be at least ϵ-fat, we have at most $O((n(n+m))/\epsilon^{O(1)})$ shadows.

When ϵ is a fixed constant, we have at most $O(n(n+m))$ shadows. In effect, this means that considering directional uncertainty actually lowers the complexity of computing the recognizability region. Note that our construction yields a conservative approximation to the recognizability region under uncertainty.

The next section extends the sensor placement algorithms presented here to the domain of Error Detection and Recovery by avoiding placements that could give ambiguous readings.

5   Avoiding confusable placements

The set $C(G, H)$ is the set of all sensor placements that could lead to confusion of G and H. A placement $\mathbf{p}$ is in the confusable region if the only visible portion of the target polygon could be due to an edge of A in G or an edge of A in H. Note that a sensor reading that confuses a target $A_{\mathbf{q}}$ in G with a target $A_{\mathbf{q'}}$ in H is due to an edge of $A_{\mathbf{q}}$ being colinear with an edge of $A_{\mathbf{q'}}$. See Figure 5 for an example.

For each pair of edges (e_i, e_j) having the same orientation, we compute the *overlap* region $O(e_i, e_j) =$

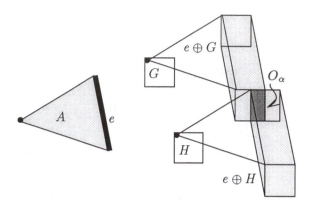

Figure 6: *Edge e at orientation α of target A is convolved with G and H. The darkly shaded region is the overlap O_α. Sensor readings in O_α can lead to confusion of G and H.*

$(e_i \oplus G) \cap (e_j \oplus H)$. We define O_α to be the union of all $O(e_i, e_j)$ for all pairs of edges (e_i, e_j) having orientation α. See Figure 6.

The confusable region is defined as

$$C(G, H) = \{\mathbf{p} \mid \exists \mathbf{q} \in G \cup H, \forall \psi : (SV(\mathbf{p}, \psi) \cap A_\mathbf{q})$$
$$\subseteq O(e_i, e_j) \text{ for some } (e_i, e_j).\}.$$

5.1 Discrete goal and failure regions

Before turning to the problem of handling polygonal goal and failure regions, we first consider the case in which the goal and failure regions are discrete points. Our technique for computing the set of good sensor placements is to first compute the set of overlap regions, and then compute the recognizability regions for the non-overlap portion of A in G and the non-overlap portion of A in H. The algorithm is as follows:

1. Compute all overlap regions $O(e_i, e_j)$ for all pairs of edges (e_i, e_j) having the same orientation. Note that in the case of point-sized goal and failure regions, the overlap regions consist of edge segments.

2. Perform the following steps for A in G and A in H:

   (a) Construct a new target A' by deleting the overlap segments from A. The new target

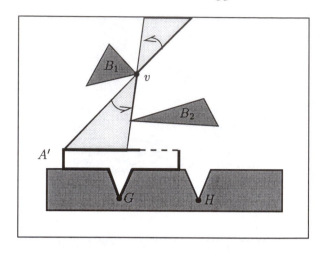

Figure 7: *An angular sweep between two visibility rays at vertex v. The lightly shaded regions are visibility triangles. The thick solid edges comprise A', and the dashed line is the overlap region.*

consists of a set of edge segments, where each edge segment has an associated outward-facing normal, so it is visible only from one side.

   (b) Compute the set of visibility triangles for target A' using the *partial visibility algorithm for a stationary target* as described in Section 3.2. Figure 7 gives an illustration of some visibility triangles.

   (c) Compute the union of the visibility triangles formed above. This is the partial visibility region for the non-overlap portion of A at this configuration.

3. Compute the intersection of the two visibility regions computed for A in G and A in H in steps 2(a)–2(c) above. This gives the set of all sensor placements from which both A in G and A in H can be recognized, but not confused.

5.2 Polygonal goal and failure regions

In the case of polygonal goal and failure regions, the computation of $R(G) \cap R(H) - C(G, H)$ is an incremental one. Recall that each overlap region is due to an edge of A in G being colinear with an edge of A in

H. In this case, the overlap region $(e_i \oplus G) \cap (e_j \oplus H)$ for parallel edges e_i and e_j is formed by a line sweeping through a region determined by G and H.

To determine the set of placements from which G and H can be distinguished but not confused, we do the following:

1. Compute the set of overlap regions O_α for all orientations α of the edges of A.

2. Place A at a vertex of G. Let $A'(\mathbf{q}) = A_{\mathbf{q}} - (A_{\mathbf{q}} \cap (\cup O_\alpha))$ be the set of edge segments of A at configuration $\mathbf{q}$ not intersecting any O_α.

3. Compute the partial visibility region of $A'(\mathbf{q})$ as it sweeps through G, as described in Section 3. Note that the endpoints of the edges of $A'(\mathbf{q})$ are not fixed, but vary during the sweep.

4. Repeat steps 2 and 3 for A sweeping through H.

5. Take the intersection of the regions computed for A sweeping through G and H, respectively.

The resulting region is the set of all placements from which A at any position in $G \cup H$ can be detected, but $A \in G$ and $A \in H$ can not be confused.

6   Experimental results

The algorithms for computing the complete and partial visibility regions of a polygon have both been implemented and used in conjunction with existing packages for graphics, geometric modeling, and plane sweep.

We used the implementation of the complete visibility algorithm to build a demonstration of robot surveillance using two of the mobile robots in the Cornell Robotics and Vision Laboratory. The autonomous mobile robots are called TOMMY and LILY. The task was for TOMMY to detect when LILY entered a particular doorway of the robotics lab. Initially TOMMY is at a position from which this doorway cannot be seen. Below we describe the various components of the system.

6.1   The visibility component

We constructed by hand a map of our lab, and used that map as the input environment to the complete visibility system. The map and the computed visibility region of the doorway are shown in Figure 8.

TOMMY's task was to monitor the doorway, which is marked in the Figure with "G". The dark gray regions are obstacles representing real objects in the room — chairs, desks, couches, bookshelves, *etc*. Given that most of the objects are regularly shaped and resting on the floor, the idea of using polygons as "footprints" of $3D$ objects turned out to give a good approximation of the $3D$ geometry. Given this map, our algorithms give us the exact placements from where the doorway can be monitored. The lightly shaded region in Figure 8 is the complete visibility region for this environment — the exact set of placements from where the doorway can be entirely seen with a sensing device such as a CCD camera.

6.2   Choosing a new placement

Based on the visibility region and the initial configuration of TOMMY, a new configuration is computed inside the visibility region. A motion plan to reach that new configuration is generated along with the distance from there to the goal.

In particular, we do the following to choose such a placement. We first shrink the visibility region to account for model and sensor uncertainty. The procedure to perform this shrinking returns a list of edges making up the *shrunk visibility region*. We now want to choose a new point inside this shrunk visibility region, one that is closest to the current position of the robot. We use the following heuristic to find such a point: we discretize the edges of the shrunk visibility region, obtaining a list of candidate points. We then sort this list of points by distance from the current position of the robot. Then test each of the points, searching for one that is reachable from the current position in a one-step motion. The first such point found is returned as the new configuration. If no such point is found, this is signaled. This could be due to two reasons: a point reachable in a one-step motion was missed due

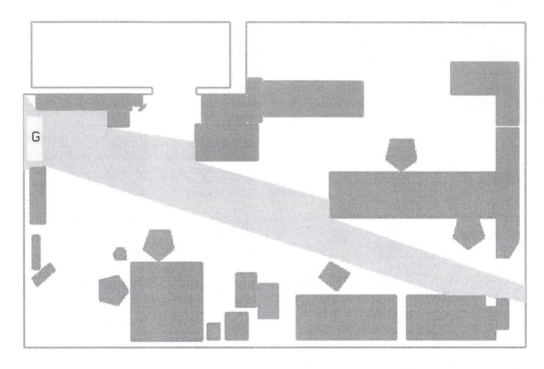

Figure 8: *The map and complete visibility region for the robotics lab.*

to the discretization being too coarse, or no one-step motion plan exists (*i.e.*, the robot would have to move around corners, or can not reach the visibility region at all). While the former case could easily be fixed by iteratively refining the discretization, the latter case requires the use of a full-fledged motion planner.

Figure 9 shows the shrunk visibility region and one of the starting points we used, as well as the new placement which was computed using the method described above.

6.3   Computing the viewing direction

The planner computes a viewing direction depending on the new placement and information obtained from the map. We fix a coordinate frame for the lab with the origin in one corner of the room. Then we compute the vector between the new computed placement and the centroid of the goal. The θ component of the new configuration is simply the angle of this vector. The final output from the planner is a vector containing the x-, y-, and θ-components of the new configuration,

along with the distance in world coordinates from this new configuration to the centroid of the goal.

6.4   Locating LILY

LILY's task is to move into the doorway and wait for TOMMY. LILY is run without a tether. She is programmed to translate a fixed distance and stop (in the center of the doorway). She then waits until her bump sensors are activated. When a bumper is pressed, LILY translates a fixed distance in reverse, rotates by 180 degrees, and then translates forward a fixed distance in order to leave the room.

Here is how the surveillance and recognition parts of the system work.

We first built a calibrated visual model of LILY. We used the Panasonic CCD camera mounted on TOMMY to take a picture of LILY from a known fixed distance (4 m). We then computed the intensity edges for that image using an implementation of Canny's edge detection algorithm [7]. The actual model of LILY that we

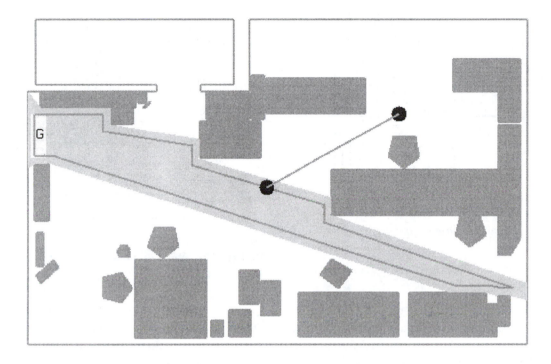

Figure 9: *The shrunk visibility region, the computed new configuration and the one-step motion.*

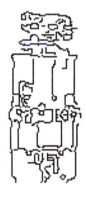

Figure 10: *Our model of* LILY.

created and used is shown in Figure 10. We did not alter the intensity edges that Canny's algorithm output, and experimentation demonstrated that our results are relatively insensitive to the particular image taken.

Based on the distance information from TOMMY's new configuration, the model edges are scaled to the expected size of LILY's image as seen from this configuration, using the fact that the image size is inversely proportional to the distance.

The video camera on TOMMY is used to repeatedly grab image frames, which along with the scaled model are input to a matcher that operates on edge images. The following loop is performed until a match is found:

1. Grab a frame.
2. Crop it, keeping only the portion of the image where LILY is expected to be.
3. Compute intensity edges for the cropped image.
4. Run the matcher to find an instance of the scaled model in the cropped image.

Figure 11 shows the intensity edges for a crop of one of the images that was grabbed with TOMMY's video-camera once TOMMY had moved into the computed configuration.

The matcher used in the experiment is based on the Hausdorff distance between sets of points and was written by William Rucklidge [23] and has been used extensively in the Cornell Robotics and Vision Laboratory

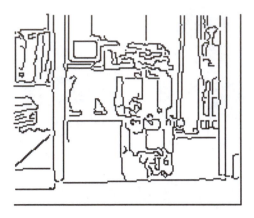

Figure 11: *Intensity edges for the cropped image.*

for image comparison, motion tracking, and visually-guided navigation [15].

The particular matcher used here is a translation-only matcher that uses a fractional measure of the Hausdorff distance. Matches are found by searching the $2D$ space of translations of the model, and computing the Hausdorff distance between the image and the translated model. A match occurs when the Hausdorff distance of a certain fraction of the points is below some specified threshold. All translations of the model that fit the image are returned.

The dark gray outline in Figure 12 shows all matches that were found between the scaled model of LILY and the image in Figure 11.

Based on where LILY is found in the image, TOMMY first performs a rotational correction so that LILY is centered in the image. An estimated value for the focal length of the camera was used to perform a rotation to correct for errors in dead reckoning. TOMMY then moves across the room to where LILY is using a simple guarded move.

6.5 Analysis

We videotaped several runs of the system. For these runs, we used two different starting positions for TOMMY, on different sides of the room, both from where the goal doorway could not be seen. We also demonstrated the robustness of the system by having people enter and leave through the doorway while

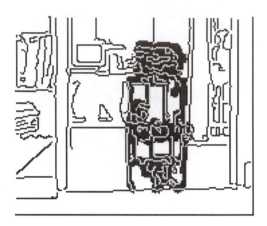

Figure 12: *Matches found between the image and the scaled model.*

TOMMY was monitoring it. The system performed consistently well. TOMMY never reported a false match — neither when the doorway was empty, nor when other people stood in the doorway. Once LILY was in position, the recognition component (on a SPARC 20 running Solaris) typically took 2–4 seconds to locate LILY. Disk access time accounted for some of this time (saving and loading image files) and could be eliminated by using a different file access strategy.

7   Conclusion

In this paper we explored the problem of automatic sensor placement and control. We presented methods for computing the placements from which a sensor has an unobstructed or partially obstructed view of a target region, enabling the sensor to observe the activity in that region. In particular, we have presented algorithms for computing the set of sensor placements affording complete or partial visibility of a stationary target, complete visibility of a target at any position or orientation within a goal, and partial visibility of a target translating through a goal at a known orientation. The algorithms account for uncertainty in sensor placement and aim.

The *Error Detection and Recovery (EDR)* system of Donald [11] provides a framework for constructing manipulation strategies when guaranteed plans cannot be found or do not exist. An EDR strategy attains the goal when the goal is recognizably reachable, and signals failure otherwise. Our results extend the guarantee of reachability to a guarantee of recognizability for the case of a polygon translating in the plane. In future work we plan to address the problem of planning sensing strategies when the target polygon may translate and rotate, resulting in unknown orientations of the target in G and H.

The implementation of the complete visibility algorithm was used as the planning component in a robot surveillance system employing both task-directed and visually-cued strategies. The system plans and executes sensing strategies that enable a mobile robot equipped with a CCD camera to monitor a particular region in a room, and then react when a specific visually-cued event occurs. Our experimental results demonstrate both the robustness and applicability of the visibility algorithms we have developed. They show that complete visibility of a goal region can be computed efficiently, and provides a good model of detectability in an uncluttered environment. We believe that this successful effort has validated our principled approach to planning robot sensing and control strategies.

Acknowledgements

Support for this work was provided in part by the National Science Foundation under grants No. IRI-8802390, IRI-9000532, IRI-9201699, and by a Presidential Young Investigator award to Bruce Donald, and in part by the Air Force Office of Sponsored Research, the Mathematical Sciences Institute, Intel Corporation, and AT&T Bell laboratories. The first author was additionally supported by an AT&T Bell Laboratories Graduate Fellowship sponsored by the AT&T Foundation.

Many of the ideas in this paper arose in discussions with Mike Erdmann, Tomás Lozano-Pérez, and Matt Mason. The robots and experimental devices were built in the Cornell Robotics and Vision Laboratory by Jim Jennings, Russell Brown, Jonathan Rees, Craig Becker, Mark Battisti, and Kevin Newman. William Rucklidge wrote the vision recognition system, and Mike Leventon and Daniel Scharstein wrote some of the vision code that was loaded onto the robots. Thanks especially to Daniel Scharstein for many discussions and insightful comments on this work.

References

[1] T. Asano, T. Asano, L. Guibas, J. Hershberger, and H. Imai. Visibility of disjoint polygons. *Algorithmica*, 1:49–63, 1986.

[2] B. Bhattacharya, D. G. Kirkpatrick, and G. T. Toussaint. Determining sector visibility of a polygon. In *Proc. ACM Symp. on Comp. Geom.*, pages 247–253, June 1989.

[3] A. J. Briggs. *Efficient Geometric Algorithms for Robot Sensing and Control*. PhD thesis, Department of Computer Science, Cornell University, Ithaca, NY, January 1995.

[4] R. G. Brown. *Algorithms for Mobile Robot Localization and Building Flexible, Robust, Easy to Use Mobile Robots*. PhD thesis, Department of Computer Science, Cornell University, Ithaca, NY, May 1995.

[5] R. G. Brown, L. P. Chew, and B. R. Donald. Localization and map-making algorithms for mobile robots. In *Proc. IASTED Int. Conf. on Robotics and Manufacturing*, pages 185–190, September 1993.

[6] A. J. Cameron and H. Durrant-Whyte. A Bayesian approach to optimal sensor placement. *IJRR*, 9(5):70–88, 1990.

[7] J. F. Canny. A computational approach to edge detection. *IEEE Transactions on Pattern Analysis and Machine Intelligence*, 8(6):34–43, 1986.

[8] A. Castaño and S. Hutchinson. Hybrid vision/position servo control of a robotic manipulator. In *Proc. IEEE ICRA*, pages 1264–1269, May 1992.

[9] C. K. Cowan and P. D. Kovesi. Automatic sensor placement from vision task requirements. *IEEE Transactions on Pattern Analysis and Machine Intelligence*, 10(3):407–416, May 1988.

[10] M. de Berg, L. Guibas, D. Halperin, M. Overmars, O. Schwarzkopf, M. Sharir, and M. Teillaud. Reaching a goal with directional uncertainty. In *Proc. Int. Symp. on Algorithms and Computation*, December 1993.

[11] B. R. Donald. *Error detection and recovery in Robotics*, volume 336 of *Lecture Notes in Computer Science*. Springer-Verlag, Berlin, 1989.

[12] L. J. Guibas, R. Motwani, and P. Raghavan. The robot localization problem. In *SODA*, 1992.

[13] G. Hager and M. Mintz. Computational methods for task-directed sensor data fusion and sensor planning. *IJRR*, 10(4):285–313, August 1991.

[14] S. Hutchinson. Exploiting visual constraints in robot motion planning. In *Proc. IEEE ICRA*, pages 1722–1727, April 1991.

[15] D. P. Huttenlocher, M. E. Leventon, and W. J. Rucklidge. Visually-guided navigation by comparing two-dimensional edge images. In *CVPR*, pages 842–847, 1994.

[16] K. N. Kutulakos, C. R. Dyer, and V. J. Lumelsky. Provable strategies for vision-guided exploration in three dimensions. In *Proc. IEEE ICRA*, pages 1365–1372, May 1994.

[17] S. Lacroix, P. Grandjean, and M. Ghallab. Perception planning for a multi-sensory interpretation machine. In *Proc. IEEE ICRA*, pages 1818–1824, May 1992.

[18] C. Laugier, A. Ijel, and J. Troccaz. Combining vision based information and partial geometric models in automatic grasping. In *Proc. IEEE ICRA*, pages 676–682, 1990.

[19] J. Matoušek, N. Miller, J. Pach, M. Sharir, S. Sifrony, and E. Welzl. Fat triangles determine linearly many holes. In *Proc. IEEE FOCS*, pages 49–58, 1991.

[20] B. Nelson and P. K. Khosla. Integrating sensor placement and visual tracking strategies. In *Proc. IEEE ICRA*, pages 1351–1356, May 1994.

[21] J. O'Rourke. *Art Gallery Theorems and Algorithms*. Oxford University Press, New York, 1987.

[22] R. Pollack, M. Sharir, and S. Sifrony. Separating two simple polygons by a sequence of translations. Technical Report 215, Department of Computer Science, New York University, Courant Institute of Mathematical Sciences, April 1986.

[23] W. J. Rucklidge. *Efficient Computation of the Minimum Hausdorff Distance for Visual Recognition*. PhD thesis, Department of Computer Science, Cornell University, Ithaca, NY, January 1995.

[24] R. Sharma and S. Hutchinson. On the observability of robot motion under active camera control. In *Proc. IEEE ICRA*, pages 162–167, May 1994.

[25] S. Suri and J. O'Rourke. Worst-case optimal algorithms for constructing visibility polygons with holes. In *Proc. ACM Symp. on Comp. Geom.*, pages 14–23, 1986.

[26] K. Tarabanis and R. Y. Tsai. Computing occlusion-free viewpoints. In *CVPR*, pages 802–807, 1992.

[27] S. J. Teller. Computing the antipenumbra of an area light source. *Computer Graphics (Proceedings SIGGRAPH '92)*, 26(2):139–148, July 1992.

[28] E. Welzl. Constructing the visibility graph for n line segments in $O(n^2)$ time. *Information Processing Letters*, 20:167–171, 1985.

[29] H. Zhang. Optimal sensor placement. In *Proc. IEEE ICRA*, pages 1825–1830, May 1992.

Visible Positions for a Car-like Robot amidst Obstacles

Marilena Vendittelli, *Università di Roma "La Sapienza", Rome, Italy*
Jean-Paul Laumond, *LAAS-CNRS, Toulouse, France*

This paper deals with the shortest paths for a car-like robot from an initial configuration (x_i, y_i, θ_i) to a goal position (x_g, y_g). We propose an efficient way to compute the shortest paths to polygonal obstacles. We then derive the "visibility" domain in presence of obstacles, i.e. the set of positions reachable from a starting configuration, by a collision-free shortest path unaffected by the presence of the obstacles.

1 Shortest paths for a car-like robot

1.1 History

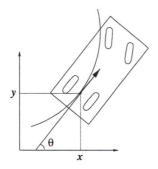

Figure 1: *Model of a car*

In the framework of the researches in nonholonomic motion planning for mobile robots, the case of the car-like robot has been the most investigated one (see e.g. [6] and the references therein). Numerous results are based on the knowledge of the shortest paths.

A car-like robot, the position and direction of which are defined by the coordinates (x, y) of the reference point and the angle θ between the abscissa axis and the main axis of the car (see Fig. 1), is completely specified as a point (x, y, θ) in the configuration space $\mathbf{R}^2 \times S^1$. Assuming the linear velocity constant, the motion is defined by the following control system

$$
\begin{cases}
\dot{x} = \cos\theta \cdot u_1 \\
\dot{y} = \sin\theta \cdot u_1 \\
\dot{\theta} = u_2
\end{cases}
$$

where $|u_1(t)| = 1$ and $|u_2(t)| \leq 1$ are, respectively, the linear and angular velocity of the car. The lower bound on the turning radius is supposed to be 1. This model corresponds to the motion in the plane of a particle subject to curvature constraints.

The study of the shortest paths for a car-like robot has already an history. The pioneering result has been achieved by Dubins who characterized the shape of the shortest paths for a particle subject to curvature constraints [4]. Starting from this result, Cockayne and Hall have computed the accessibility set for this model of particle [3] (i.e., the domain of the plane reachable by paths with bounded length). The particle model corresponds to a car-like robot moving forward with a constant velocity. More recently, Reeds and Shepp have provided a sufficient family of 48 shortest paths for a car-like system moving both forward and backward [10]: optimal paths are constituted by a finite sequence of at most five straight line segments or arcs of a circle of radius 1.

Then the problem has been revisited from a control theory point of view: Sussmann and Tang [13] and Boissonnat, Cerezo and Leblond [1] independently provided a new proof of Reeds and Shepp's result. In addition Sussmann and Tang reduced the sufficient family to 46 canonical paths.

At the same time, Souères and Laumond, using these results, computed a synthesis of the shortest paths, i.e.

a partition of the manifold $\mathbf{R}^2 \times S^1$ into cells reachable by only one type of shortest path (among the 46 ones) [11]. They also computed the exact shape of the shortest path metric, i.e., the shape of the domain in $\mathbf{R}^2 \times S^1$ reachable by paths with bounded length [7]. The projection of that domain on $\mathbf{R}^2$ corresponds to the accessibility domain in the plane; it has been computed by Souères, Fourquet and Laumond [12] who then extended the result of Cockayne and Hall [3] to the car moving also backward; the authors provide also a synthesis of the shortest path from an initial configuration to a point in the plane. On the other hand, with Bui and Boissonnat, they apply the same techniques to compute a synthesis for a car moving forward [2].

Finally, Moutarlier, Mirtich and Canny explored general tools to compute shortest paths to some submanifolds of $\mathbf{R}^2 \times S^1$ [9].

An intriguing question not yet investigated is: which is the set of positions reachable by a collision-free shortest path of the type considered in [12] and unaffected by the presence of the obstacles? The answer to this question is closely related to the visibility problems classically defined in the framework of Euclidean metric as will be illustrated in Section 4. We will refer to this problem as that of finding the set of positions visible from a start configuration in presence of obstacles.

1.2 Contribution

In this paper we consider a point robot moving amidst polygonal obstacles and we propose three geometric algorithms to:

- compute the shortest path from an initial configuration (x_i, y_i, θ_i) to a goal position (x_g, y_g)

- compute the shortest path from a configuration to a segment

- compute the domains of the plane visible from a start configuration in presence of obstacles.

Note that the results in [12] lead to an efficient algorithm to compute the shortest path to a position. We just propose an alternative geometric way for this computation (Section 2). Moreover the results presented

in [9] allow to design a procedure to compute the shortest path to a line segment. Again we propose here a geometric and more efficient algorithm to solve this problem (Section 3). Finally, the main contribution of the paper is to compute the set of positions visible from a start configuration in presence of obstacles (Section 4).

1.3 Shortest paths to a position

We briefly resume the results of Souères, Fourquet and Laumond [12] providing a synthesis of the optimal paths for the free final direction problem (i.e. the problem of finding the shortest path from an initial configuration (x_i, y_i, θ_i) to a point in the plane (O, x, y) with free direction). They prove that only three families of paths may be optimal to reach a position (x, y) from the origin $(0, 0, 0)$. Moreover they provide a partition of the set of positions $\mathbf{R}^2$ in domains reachable by a given family of paths (see Fig. 2). The paths

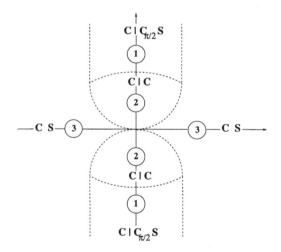

Figure 2: *Partition of the plane and corresponding families of paths*

are described using Sussmann's notation [13]: capital letters denote either a straight line segment (S) or an arc of a circle of radius 1 (C). The symbol | between two letters indicates the presence of a cusp. To specify the direction of rotation the letter C will be replaced by l for left turn and by r for right turn and the superscript $+$ $(-)$ will denote forward (backward) motion.

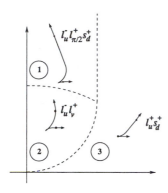

Figure 3: *Partition of the first quadrant and corresponding types of paths*

Subscripts represent positive real numbers giving the length of each piece. For any point not belonging to the y-axis, there is a *unique* shortest path constituted by:

- two arcs of a circle of minimum radius connected by a cusp and followed by a straight line segment (this family is denoted by $C|C_{\frac{\pi}{2}}S$) for points belonging to Region 1

- two arcs of a circle of minimum radius connected by a cusp (family $C|C$) for points belonging to Region 2

- an arc of circle of minimum radius followed by a straight line segment (family CS) for points belonging to Region 3.

If a path $\mathcal{P}$ is optimal for the free final direction problem, it is possible to build three other optimal paths $\mathcal{P}_x$, $\mathcal{P}_y$ and $-\mathcal{P}$ isometric to $\mathcal{P}$ and leading to points symmetric to the point reached by $\mathcal{P}$. They are obtained, respectively, by symmetry with respect to the x-axis, the y-axis and the origin of the coordinate axes. Assuming $\mathcal{P}$ known, the expression of the symmetric paths can be easily determined: the word describing $\mathcal{P}_x$ is obtained by permutating the "r" and the "l" in the word representing $\mathcal{P}$; the word describing $\mathcal{P}_y$ is obtained by permutating the "r" and the "l" and the superscript signs in the word representing $\mathcal{P}$; the word

representing $-\mathcal{P}$ is obtained by a combination of the two previous symmetric transformations.

Due to the symmetry properties, there exist exactly two shortest paths, belonging to the same family, reaching a given position on the y-axis. Moreover, using symmetry arguments (see Section 3.2), we can restrict ourselves to consider the first quadrant (Fig. 3). Note that any path is characterized by two free parameters. The coordinates of the points reached by each type of path in the first quadrant are [12]:

$$l_u^- l_{\frac{\pi}{2}}^+ s_d^+ \quad \begin{cases} x(u,d) = \cos u - (2+d)\sin u \\ y(u,d) = \sin u + (2+d)\cos u - 1 \end{cases} \quad (1)$$

with $0 \le u \le \arctan \frac{1}{2}$,

$$l_u^- l_v^+ \quad \begin{cases} x(u,v) = -2\sin u + \sin(u+v) \\ y(u,v) = 2\cos u - \cos(u+v) - 1 \end{cases} \quad (2)$$

with $0 \le u \le \frac{\pi}{6}$ and $0 \le v \le \frac{\pi}{2}$,

$$l_v^+ s_d^+ \quad \begin{cases} x(v,d) = \sin v + d\cos v \\ y(v,d) = -\cos v + d\sin v + 1 \end{cases} \quad (3)$$

with $0 \le v \le \frac{\pi}{2}$.

The boundary between the domain of points reachable by a path of the type $l_a^- l_{\frac{\pi}{2}}^+ s_d^+$ and the domain of the type $l_b^- l_v^+$ is obtained for $a = b$, $v = \frac{\pi}{2}$ and $d = 0$. It is an arc of the circle centered at $(0,-1)$ and radius $\sqrt{5}$.

The boundary between the domain of type $l_u^- l_{\frac{\pi}{2}}^+ s_d^+$ and the domain of type $l_v^+ s_d^+$ is obtained when $u = 0$ and $v = \frac{\pi}{2}$. It is the upper vertical half-line from the point $(1,1)$.

Between the domain of the type $l_u^- l_e^+$ and the domain of the type $l_b^+ s_d^+$ the boundary is obtained for $u = 0$, $b = e$ and $d = 0$. It is the arc of the unit circle centered at $(0,1)$, of length $\frac{\pi}{2}$ starting from the origin of the coordinate axes.

Figure 4 shows the sets of positions reachable by paths of length ℓ for seven different values of ℓ. They are obtained by expressing the length ℓ of the path in each region as a function of the two parameters characterizing the path and operating an appropriate change of variable (see Section 3.1).

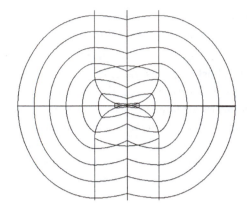

Figure 4: *Sets of reachable positions in the plane*

The isometric curves (also called "wave fronts") in Region 1 and 3 are involutes of a circle. In Region 2 the wave front is an arc of a circle of radius 2.

2   Geometric construction of the shortest paths to a position

From the results described in the previous section we can derive the following geometric constructions to compute the shortest path to a point P in the first quadrant:

- Point in Region 1. Type of path $l_u^- l_{\pi/2}^+ s_d^+$.
  First the tangent line $\mathcal{T}$ to the circle $\mathcal{C}_1$ (Fig. 5) of radius 1 and center $(0, -1)$, passing through the point $P(x, y)$ is traced, then the center of the circle $\mathcal{C}_2$ of radius 1, tangent to $\mathcal{C}_1$ and to $\mathcal{T}$, is computed. The value of the parameter u is given by the length of the arc $\widehat{OU}$. Parameter d is determined by the length of the segment $\overline{P'P}$.

- Point in Region 2. Type of path $l_u^- l_v^+$.
  Paths leading to points in Region 2 are computed by founding the center of the unit circle $\mathcal{C}_2$ tangent to the circle $\mathcal{C}_1$ and passing through the point P. As before the value of the parameter u is given by the length of the arc $\widehat{OU}$, while parameter v is equal to the length of the arc $\widehat{UP}$(see Fig. 6).

- Point in Region 3. Type of path $l_v^+ s_d^+$.
  Figure 7 shows that paths in Region 3 are obtained

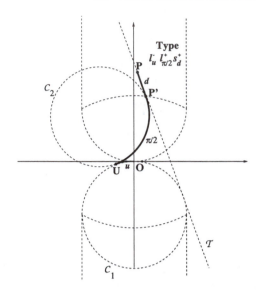

Figure 5: *The shortest path to a point in Region 1*

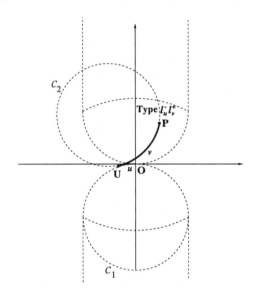

Figure 6: *The shortest path to a point in Region 2*

by drawing the line $\mathcal{T}$ tangent to the circle $\mathcal{C}_2$ and passing through P. Parameters v and d are given, respectively, by the length of the arc $\widehat{OV}$ and of the segment $\overline{VP}$.

Note that, in the worst case, only three comparisons between reals are requested to decide to which region the point P belongs.

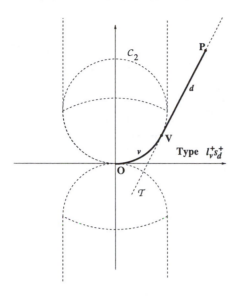

Figure 7: *The shortest path to a point in Region 3*

3 Shortest paths to a segment

Starting from the shape of the set of positions reachable by a path of length ℓ, as derived in [12] and shown in Fig. 4, we now address the problem of computing the shortest paths to a segment from the configuration $(0, 0, 0)$. The problem can be stated as that of finding the isodistance curve of minimum "radius" tangent to the line supporting the segment, that is finding the tangency point and, if it belongs to the segment, the shortest path leading to it. If the tangency point is not on the segment the problem is reduced to compare the length of the paths reaching the two extremities of the segment. Due to the triviality of this second case, we will describe the algorithm considering the case of the tangency point inside the segment. The problem is first attacked by assuming that the segment lies in the first quadrant. The general case will be solved by considering symmetry properties (Section 3.2).

3.1 First quadrant

As mentioned at the end of Section 1.3, the set of positions reachable by a path of length ℓ is described by three different curves in the three regions of the first quadrant. Noting that in Region 1 $\ell = u + \frac{\pi}{2} + d$, the change of variable $d = \ell - u - \frac{\pi}{2}$ in the system (1) leads

to [12]:

$$\left\{ \begin{array}{l} x(u, \ell) = \cos u - (2 + \ell - u - \frac{\pi}{2}) \sin u \\ y(u, \ell) = \sin u + (2 + \ell - u - \frac{\pi}{2}) \cos u - 1 \end{array} \right. \quad (4)$$

For ℓ constant this gives the equation of an involute of a circle.

The length of a path in Region 2 is $\ell = u + v$. Replacing the expression $u + v$ by ℓ in (2) we obtain:

$$\left\{ \begin{array}{l} x(u, \ell) = -2 \sin u + \sin \ell \\ y(u, \ell) = 2 \cos u - \cos \ell - 1 \end{array} \right. \quad (5)$$

Assuming ℓ constant, the parametric equation of a circle of radius 2 is obtained.

From the expression of the path length $\ell = v + d$ in the third region, the change of variable $d = \ell - v$ in (3) gives:

$$\left\{ \begin{array}{l} x(v, \ell) = \sin v + (\ell - v) \cos v \\ y(v, \ell) = -\cos v + (\ell - v) \sin v + 1 \end{array} \right. \quad (6)$$

The system (6) leads, for ℓ constant, to the equation of an involute of a circle.

Therefore the equation of the wave front in the first quadrant can be easily computed from the systems above. Its shape is shown in Fig. 8. According to the

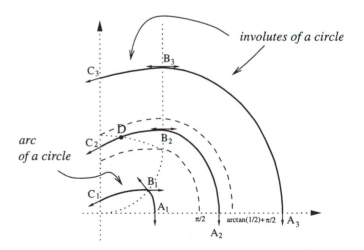

Figure 8: *Wave fronts in the first quadrant*

value of ℓ, the following description of the wave front structure can be given:

- $\ell \leq \frac{\pi}{2}$: concatenation of an involute of a circle and an arc of a circle

- $\frac{\pi}{2} < \ell < \arctan(\frac{1}{2}) + \frac{\pi}{2}$: concatenation of two involutes of a circle and an arc of a circle

- $\ell \geq \arctan(\frac{1}{2}) + \frac{\pi}{2}$: concatenation of two involutes of a circle.

In all cases the slope of the tangent line is monotonic in the first quadrant. The following table gives the angle of the half-tangent at the boundary points A_i, B_i, C_i ($i = 1, 2, 3$) in Fig. 8:

Case	$\ell \leq \frac{\pi}{2}$			
Points	A_1	B_1^-	B_1^+	C_1
Angle	$-\frac{\pi}{2}$	$v - \frac{\pi}{2}$	0	u
Case	$\frac{\pi}{2} \leq \ell \leq \arctan\frac{1}{2} + \frac{\pi}{2}$			
Points	A_2	B_2	C_2	
Angle	$-\frac{\pi}{2}$	0	u	
Case	$\ell > \arctan\frac{1}{2} + \frac{\pi}{2}$			
Points	A_3	B_3	C_3	
Angle	$-\frac{\pi}{2}$	0	u	

Note that at point D (connection of an involute of a circle and an arc of a circle), in Fig. 8, the first derivative is still continuous. Therefore looking at the slope of the line supporting the segment it is easy to decide in which region the tangency will occur.
We then solve the problem for each region:

- Regions 1 and 3
  The isodistance curves in regions 1 and 3 are, respectively, the involutes of the circles C_1 and C_2 (see Fig. 9) and have the nice property that the line $\mathcal{T}$ tangent to the circles and passing through the point T is perpendicular to the line tangent to the involute at the same point [14]. The segments of length d constituting the last part of the paths leading to regions 1 and 3 belongs to the line $\mathcal{T}$, as shown in the geometric construction of the shortest paths to points in these regions. The algorithm to find the shortest paths to the tangency points becomes then very easy: in Region 1 the parameter u is such that the robot final orientation is per-

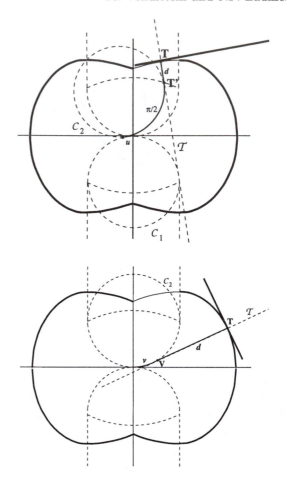

Figure 9: *Tangency in Region 1 (top) or in Region 3 (bottom)*

pendicular to the tangent line supporting the segment; therefore u is such that $\tan(u) = m$, where m is the slope of the tangent[1]. Parameter d is equal to the length of the segment $\overline{T'T}$ (Fig. 9, top). Analogously to the previous case, the parameter v in Region 3, is defined by $\tan(v) = -\frac{1}{m}$, where m has the same meaning as above[2]. Looking at Fig. 9 (bottom) it is easy to conclude that the length of the segment $\overline{VT}$ determines the value of the parameter d.

---

[1] This means that the car starts with a backward motion until it becomes parallel to the segment.

[2] In this case the robot final orientation is perpendicular to the segment.

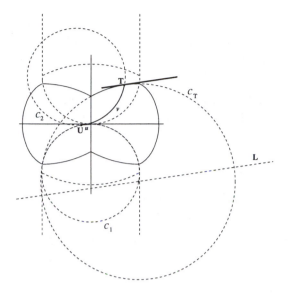

Figure 10: *Tangency point in Region 2*

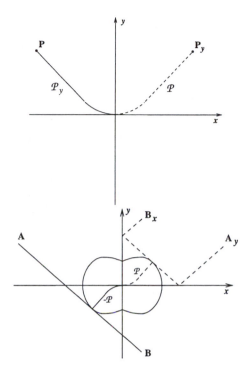

Figure 11: *Shortest path computation by symmetry*

- Region 2

  In Region 2 the wave front is an arc of the circle C_T of radius 2 the center of which is on the unit circle C_1 and on the line L parallel to the tangent line supporting the segment and located at a distance 2 from it, as shown in Fig. 10. Once found the center of this circle the tangency point is computed. Anyway, looking at the geometric construction shown in the figure it is possible to prove that the robot orientation at point U is parallel to the tangent line supporting the segment. The center of the circle C_2 can be then determined using this fact and the computation of the tangent point becomes trivial.

3.2 Symmetry property

We have solved the problem of finding the shortest path to a point (to a segment) in the first quadrant.

To solve the problem of the shortest path to a point P in the plane it suffices to consider the symmetric image of the point in the first quadrant, compute the shortest path attaining it and then apply to the path the same symmetric transformation, according to the rules given in Section 1.3. As an example (Fig. 11, top), if P belongs to the second quadrant, we compute

the shortest path $\mathcal{P}$ to the point P_y symmetric of P with respect to the y-axis. A symmetry with respect to the y-axis gives the path $\mathcal{P}_y$ to attain the point P in the second quadrant.

The problem of finding the shortest path to a segment in the plane is solved by computing the symmetric image in the first quadrant of the segment. Then the shortest path to this polygonal image is computed. The found path is then transformed by the appropriate symmetry.

Figure 11 (bottom) shows the case of a segment traversing the second, third and forth quadrant. The dashed polygonal line denotes its image in the first quadrant and the path $\mathcal{P}$ is the shortest path reaching the line. Since the path $\mathcal{P}$ reaches the line at a point belonging to the image of that part of the segment lying in the third quadrant, the shortest path to the segment is the image of $\mathcal{P}$ obtained by symmetry with respect to the origin.

The general rule can be informally summarized: the shortest path to an object (point, segment) in the plane, the image of which in the first quadrant has been computed by a symmetry, is obtained by applying the same symmetry to the shortest path to its image in the first quadrant. The same procedure will be used to find the shadow of a point (of a segment) in the plane.

Note that the computation of the shortest path to a straight line segment presented in this section is done in constant time. We can easily derive a $O(n)$ algorithm that gives the shortest distance from a given configuration to the obstacles in a polygonal environment constituted by n vertices. Any path of length smaller than such a minimal distance is guaranteed to be collision-free.

4 Visibility in presence of obstacles

As already mentioned in Section 1, the problem of finding the set of positions reachable by collision-free shortest paths of the kind considered in the paper can be related to that of the visibility defined in the framework of Euclidean metric. In such a framework the light rays are straight line segments. Here we consider a strange "light", the rays of which are supposed to follow the shortest paths in position we have dealt with. The definitions below are introduced according to this analogy; we are interested in building the "shadow" induced by the obstacles when lightened by a light of this kind.

The visibility notion is supported by the shape of the shortest paths. There is a main difference between the Euclidean case and the case addressed in the paper: while the rays of light are infinite half straight lines in the Euclidean case, they may be of *finite* length in our case. Moreover the "shadow" generated by a point may be 2-dimensional !...

Due to the symmetry properties with respect to the x-axis and the y-axis, we first restrict the study to the first quadrant: we compute a foliation supported by all the shortest paths reaching a point in the first quadrant; the method consists in *prolongating* all the finite length shortest paths. We then define a notion of shadow in a domain "extending" the first quadrant and we propose an efficient way to compute the shadow of a curve lying in this domain.

In a second step we address the general case (i.e., without any restriction on the location of the segment) through the symmetry properties and we show how to compute the visible sets.

As in the previous section we restrict ourselves to the case of a straight line segment. The extension to polygonal obstacles is trivial.

4.1 Visibility in the first quadrant

4.1.1 Foliation supported by the shortest paths reaching a position in the first quadrant

We have seen that a shortest path to a position lying in the first quadrant belongs to one of the three types $l_v^+ s_d^+$ (with $0 \le v \le \frac{\pi}{2}$), $l_u^- l_v^+$ (with $0 \le u \le \frac{\pi}{6}$ and $0 \le v \le \frac{\pi}{2}$) or $l_u^- l_{\frac{\pi}{2}}^+ s_d^+$ (with $0 \le u \le \arctan \frac{1}{2}$). Let $\mathcal{D}$ be the domain containing the first quadrant and bounded by the x-axis and the curve $l_{\frac{\pi}{6}}^- l_{\frac{\pi}{2}}^+ s_\infty$ (see Fig. 12). In this domain all the paths $l_u^- l_v^+$ are sub-paths of some $l_u^- l_{\frac{\pi}{2}}^+ s_d^+$ (with $0 \le u \le \frac{\pi}{6}$).

Lemma 1 *Let us consider all the paths $l_v^+ s_d^+$ ($0 < v < \frac{\pi}{2}$) and $l_u^- l_{\frac{\pi}{2}}^+ s_d^+$ ($0 \le u < \frac{\pi}{6}$) starting from the origin. When d tends to infinity the open sub-paths[3] s_d^+ in the family $l_v^+ s_d^+$ and $l_{\frac{\pi}{2}}^+ s_d^+$ in the family $l_u^- l_{\frac{\pi}{2}}^+ s_d^+$ realize a foliation[4] of $\mathcal{D}$.*

Proof: (Fig. 12) Let $\mathcal{L}_\alpha$, $\alpha \in I$ be the set of all the sub-paths defined in the Lemma. Each of them is characterized by the angle of its half-line s_∞^+ with the x-axis. By construction these angles vary continuously from 0 to $\frac{2\pi}{3}$. We may then index all the $\mathcal{L}_\alpha$ sub-paths by putting $I = \left]0, \frac{2\pi}{3}\right[$.

---

[3] i.e., without their endpoints.

[4] A *foliation* of a n-dimensional manifold $\mathcal{M}$ is a family $\mathcal{L}_\alpha, \alpha \in I$ of arcwise connected q-dimensional sub-manifolds ($q < n$), called *leaves*, of $\mathcal{M}$ such that:

- $\mathcal{L}_\alpha \cap \mathcal{L}_{\alpha'} = \emptyset$ if $\alpha \neq \alpha'$

- $\cup_{\alpha \in I} \mathcal{L}_\alpha = \mathcal{M}$

- every point in $\mathcal{M}$ has a local coordinate system such that $n - q$ coordinates are constant.

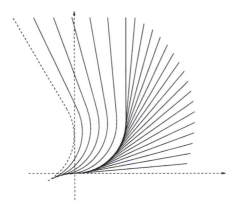

Figure 12: *Foliation of the domain* $\mathcal{D}$

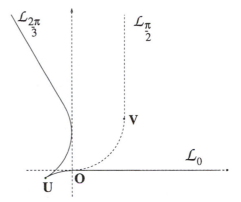

Figure 13: *Boundary of the domain* $\mathcal{D}$ *(solid line)*

The union of all the $\mathcal{L}_\alpha$, $\alpha \in I$, clearly spans the open domain $\mathcal{D}$.

The coordinates of a point that would belong to both distinct $\mathcal{L}_\alpha$ and $\mathcal{L}_{\alpha'}$ should be the common solution of two distinct parametric equation systems defining the optimal paths (see Section 1.3). We may check that it is impossible (this is the same argument that is used to prove the uniqueness of the shortest paths to points in the first quadrant; see [12]). Then $\mathcal{L}_\alpha \cap \mathcal{L}_{\alpha'} = \emptyset$ if $\alpha \neq \alpha'$.

Let us now introduce the following coordinate system in $\mathcal{D}$. Let P be a point in $\mathcal{D}$ with coordinates (x, y). The new abscissa of P is the index α of the curve $\mathcal{L}_\alpha$ that contains P; the new ordinate is the length ℓ of the path $l_v^+ s_d^+$, $l_u^- l_v^+$ or $l_u^- l_{\frac{\pi}{2}}^+ s_d^+$ reaching P. These new coordinates (α, ℓ) vary in $]0, \frac{2\pi}{3}[\times \mathbf{R}_+^\star$ and all the points on the same $\mathcal{L}_\alpha$ have the same abscissa. Therefore, this new coordinate system verifies the definition of a foliation of the 2-dimensional domain $\mathcal{D}$ by the 1-dimensional leaves $\mathcal{L}_\alpha$. $\square$ $\square$

4.1.2 Shadows in $\mathcal{D}$

Definition 1 *The shadow of a point P in $\mathcal{D}$ is the set of points reachable by a path $l_v^+ s_d^+$, $l_u^- l_v^+$ or $l_u^- l_{\frac{\pi}{2}}^+ s_d^+$ containing P. The shadow of a curve in $\mathcal{D}$ is the union of the shadows of all its points.*

The new coordinate system of $\mathcal{D}$ allows us to give an efficient way to characterize the shadow. Intuitively the light follows the leaves of the foliation. All the points "behind" a point-obstacle (i.e., their ordinate are greater than the ordinate of the point-obstacle) are not "enlightened". They are located on the vertical half line "above" the point obstacle in the coordinate system $]0, \frac{2\pi}{3}[\times \mathbf{R}_+^\star$.

Nevertheless the coordinate change is not continuous on the arc $\widehat{OV}$ (Fig. 13). This makes the formalization of the intuitive interpretation not immediate. To deal with the discontinuous cases we introduce two additional "leaves" bounding the closure $\overline{\mathcal{D}}$ of $\mathcal{D}$: $\mathcal{L}_{\frac{2\pi}{3}} = l_{\frac{\pi}{6}}^- l_{\frac{\pi}{2}}^+ s_\infty$ and $\mathcal{L}_0 = l_0^+ s_\infty$ (Fig. 13); moreover, we consider the following transformation f:

$$f : \overline{\mathcal{D}} \to [0, \tfrac{2\pi}{3}] \times \mathbf{R}^+$$

$$f(P(\alpha, \ell)) = \begin{cases} [\ell + \frac{\pi}{2}, \frac{2\pi}{3}] \times \{\ell\} & \text{if } \alpha = \frac{2\pi}{3} \text{ and } \ell < \frac{\pi}{6} \\ [\ell, \frac{\pi}{2}] \times \{\ell\} & \text{if } \alpha = \frac{\pi}{2} \text{ and } \ell < \frac{\pi}{2} \\ (\alpha, \ell) & \text{everywhere else} \end{cases}$$

This transformation is introduced to continuously prolongate the natural transformation induced by the coordinate system: the transformation by f of a finite length curve in $\mathbf{R}^2$ is a finite length connected curve in $[0, \frac{2\pi}{3}] \times \mathbf{R}^+$.

On the other hand, the shadow of a point P with coordinates $(\alpha, \ell) \in \overline{\mathcal{D}}$ is the set of points $P'(\alpha', \ell')$ such that :

$$\begin{array}{ll} \frac{\pi}{2} + \ell \leq \alpha' \leq \frac{2\pi}{3}, \ell' \geq \ell & \text{if } \alpha = \frac{2\pi}{3} \text{ and } \ell < \frac{\pi}{6} \\ \ell \leq \alpha' \leq \frac{\pi}{2}, \ell' \geq \ell & \text{if } \alpha = \frac{\pi}{2} \text{ and } \ell < \frac{\pi}{2} \\ \alpha' = \alpha, \ell' \geq \ell & \text{everywhere else} \end{array}$$

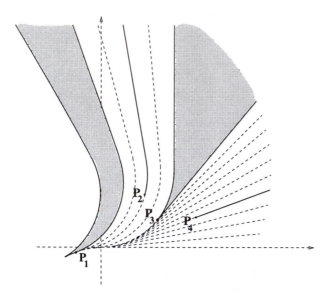

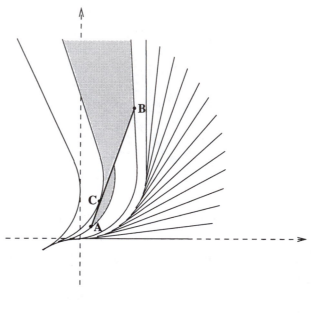

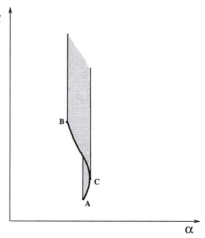

Figure 14: *The shadow of points in generic positions (P_2 and P_4), on the leaf $\mathcal{L}_{\frac{2\pi}{3}}$ (P_1) and on the leaf $\mathcal{L}_{\frac{\pi}{2}}$ (P_3)*

Note that the third case corresponds to a generic one (one-dimensional shadow, points P_2 and P_4 in Fig. 14), the two other cases occur, respectively, when P belongs to the arc $\widehat{OV}$ or to the arc $\widehat{UO}$ (two-dimensional shadow, points P_3 and P_1 in Fig. 14).

Combining the shape of the shadows of points and the definition of the transformation f, we get:

Property 1: The shadow of a finite length curve lying in $\overline{\mathcal{D}}$ is bounded in the coordinate system $[0, \frac{2\pi}{3}] \times \mathbf{R}^+$ of $\overline{\mathcal{D}}$ by two vertical half-lines and the lower envelop of the curve transformed by f.

Figure 15 illustrates this correspondence for a straight line segment.

4.1.3 Building the shadow of a straight line segment

A point in the domain $\overline{\mathcal{D}}$ may belong to the shadow of a segment only if the segment intersects the domain $\overline{\mathcal{D}}$.

The leaves of the foliation are either half-lines or arcs of a circle followed by an half-line. Therefore the number of intersection points between a given segment and

Figure 15: *The shadow of a segment (filled region) in the domain $\overline{\mathcal{D}}$ (top) and its image in the new coordinate system (bottom)*

a given leaf is at most two. The lower envelop of the image by f of any segment is then characterized by its endpoints and possibly a unique tangent point (see Fig. 15).

Based on these results an algorithm to build the shadow of a segment consists in:

- checking whether the segment is tangent to some leaf (to do that it suffices to check the existence of

a circle of radius 1 tangent to the line supporting
the segment and also tangent to the circle C_1 of
radius 1 centered at $(0, -1)$; this may happen only
if the segment, or a part of it, is in Region 2)

- computing the leaves containing the end points of
  the segment (this is done with the methods presented in Section 2)

- applying Property 1 (the shadow is the domain
  above the lower envelop of the segment transformed by f).

This computation is done in constant time.

The visibility notion has been restricted to the domain $\mathcal{D}$. To address a global point of view, we have
to deal with the symmetry properties and the non-
uniqueness of the shortest paths reaching positions located on the y-axis.

Each point in $\mathcal{D}$ is located on a leaf and may be
reached by a path of the type $l_v^+ s_d^+$, $l_u^- l_v^+$ or $l_u^- l_{\frac{\pi}{2}}^+ s_d^+$.
Nevertheless such a path is optimal if and only if the
point belongs to the first quadrant. This means for
instance that points belonging to the second quadrant
and to the shadow of an obstacle in $\mathcal{D}$ may perhaps
be reached by a collision-free shortest path lying in the
domain symmetric of $\mathcal{D}$ with respect to the y-axis. In
fact, the shadows in $\mathcal{D}$ should be "cut" along the x-axis
and the y-axis to keep only the part lying in the first
quadrant.

4.2   Building the visibility sets

Definition 2 *The visible set* Vis(P) *in presence of a
point-obstacle P is the set of endpoints of the shortest
paths that do not contain P. The visible set* Vis($\mathcal{C}$)
*in presence of a curve $\mathcal{C}$ is the intersection of all the
visible sets* Vis(P) *for $P \in \mathcal{C}$.*

Notice that this definition involves *shortest* paths
and deals with the whole plane while the Definition 1
of the shadow is restricted to $\mathcal{D}$ and deals with 3 types
of (not necessarily optimal) paths.

To make precise the construction of the visible sets,
we introduce a new domain $\tilde{\mathcal{D}}$ defined as the domain

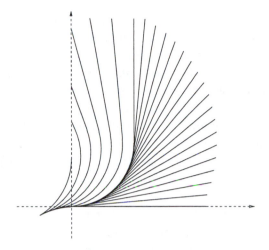

Figure 16: *Domain $\tilde{\mathcal{D}}$*

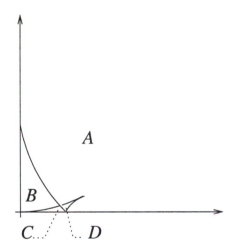

Figure 17: *Symmetric domains for shadow computation*

sweept by all the shortest paths reaching a point in
the first quadrant. $\tilde{\mathcal{D}}$ is shown in Fig. 16; it is clearly
included in $\overline{\mathcal{D}}$. A point in the first quadrant may belong
to the shadow of a segment only if the segment inter-
sects the domain $\tilde{\mathcal{D}}$. In order to compute the shadow of
a segment in the first quadrant, we consider the transformation of the sub-domain of $\tilde{\mathcal{D}}$ the abscissa of which
are negative by a first symmetry with respect to the y-
axis and then by a symmetry with respect to the x-axis.
The image of the sub-domain in the first quadrant is
shown in Fig. 17. It induces a partition of the first
quadrant into four domains. Generalizing the symme-

try properties described in Section 3.2 we have to apply the same symmetric transformation to the shadow generated by points lying in the image of the sub-domain. A point in the domain A induces a shadow only in the first quadrant; a point in B induces a shadow in the first and second quadrant; a point in C induces a shadow in the first, second and third quadrant; finally, a point in D induces a shadow in the first and third quadrant. The case of a segment (or a part of it) not lying in the first quadrant is treated using the symmetric transformations described in Section 3.2. Figure 18 shows the case of a segment in the first quadrant intersecting three of the four domains just described; filled areas represent the shadow induced by it.

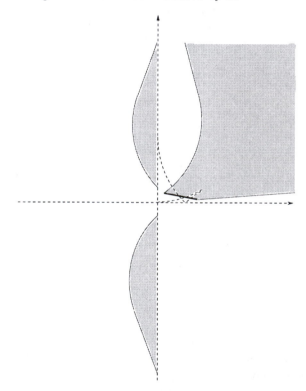

Figure 18: *Shadow of a segment intersecting domains A, B and D*

Therefore, from an algorithmic point of view, to compute the visible set in presence of an obstacle,

- we first compute the shadows of the various curve pieces belonging to the domain $\tilde{\mathcal{D}}$ (using the study in Section 4.1) and its three symmetric domains

- each (of the four) shadow is then cut along the y-axis and reduced to its part belonging to the corresponding quadrant

- the visible set is constituted by the complementary set of the union of these (four) restricted shadows.

It remains to check whether the points belonging to the frontier of a visible set are visible or not.

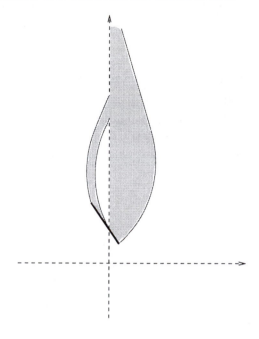

Figure 19: *All the points lying on the segments in dotted line of the y-axis are visible; points everywhere else on the frontier (solid lines) are not visible.*

Let us recall that there are exactly two shortest paths reaching a position on the y-axis. Let us consider a visible set such that a piece of the boundary of its closure is included on the y-axis, and let P be a point of this piece; one of the two shortest paths reaching P is collision-free (otherwise P would belong to two shadows lying in two adjacent quadrants, and then it would not belong to the frontier of the union of the shadows). Everywhere else, the points on the frontier of the visible set have a unique shortest path that belongs to the frontier of some shadow; such a path is in contact with the obstacle; then such frontier points do not belong to the visible set.

Therefore the visible set may be neither an open nor a closed set. Figure 19 gives an example of such a situation.

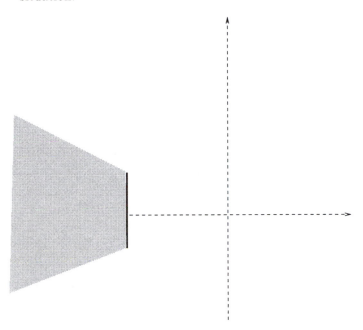

Figure 20: *A simple example : the obstacle is far from the origin.*

The following figures give various other examples of visible sets. Note that the domain not visible is quite "simple" and close to the "shadow" in the Euclidean case when the segment is located far from the origin (Figure 20); indeed the length of the shortest path in position converge to the Euclidean distance from the position to the origin when such a distance tends to infinity. The shape of the unvisible sets is "complicated" only when the segment is located in a neighborhood of the origin: such sets are not necessarily connected (Figures 21 and 22); moreover their connected components are not necessarily simply connected (Figure 23).

The complexity of the computation of the visible set in the presence of a segment is $O(1)$. Indeed, the computation consists in splitting the segment into (at most) three pieces, each belonging to a different quadrant (constant time). Then, for each of the piece, the intersection with the domains A, B, C and D in Fig. 17, is computed (constant time). Finally within each sub-

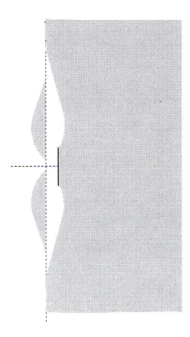

Figure 21: *The shadow may have three connected components.*

piece belonging to A, B, C or D one computes the associated shadow in constant time with the algorithm described in Section 4.1.3.

Conclusion

The algorithms induced by the study above have been implemented in C language and all the figures showing the shortest paths to a position or to a segment as well as the visible sets are the graphic output of the developed software. The efficiency of the algorithm for distance computation allows its integration into nonholonomic motion planning methods for a car-like robot. For instance, in the method developed in [8], the numerical computation of the distance function may be replaced by an analytical computation. Moreover such distance can be used in planning methods based on potential fields. The visible sets in position might be the building block to answer problems like the classical art gallery problems. Moreover it should be easy to devise a planning method for problems in which the final orientation is not relevant based on the construction of the visible sets.

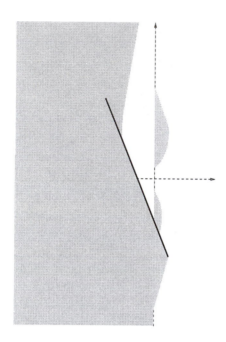

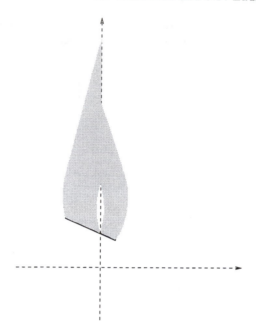

Figure 22: *Some part of the segment-obstacle may not belong to the frontier of the shadow.*

Figure 23: *The shadow may be bounded and not simply connected.*

Such developments as well as the extension of the visible sets in position to visible sets in $\mathbf{R}^2 \times S^1$ are under study.

Acknowledgment

This work has been done during the stay of the first author at LAAS in the context of the project ESPRIT 3 Basic Research #6546 (PROMotion).

We thank Philippe Moutarlier for helpful comments on the geometric properties of the shortest paths to a segment and Philippe Souères for useful discussions and comments.

Authors' e-mail: venditt@labrob.ing.uniroma1.it and jpl@laas.fr

References

[1] J.D. Boissonnat, A. Cerezo and J. Leblond, "Shortest paths of bounded curvature in the plane," in *IEEE Conf. on Robotics and Automation*, pp. 2315–2320, Nice, France, 1992.

[2] X-N. Bui, P. Souères, J.D. Boissonnat and J.P. Laumond, "The shortest path synthesis for a non-holonomic robot moving forward," Report INRIA 2153, Sophia-Antipolis, 1993.

[3] E.J. Cockayne and G.W.C. Hall, "Plane motion of a particle subject to curvature constraints," in *SIAM J. on Control*, 13 (1), 1975.

[4] L. E. Dubins, "On curves of minimal length with a constraint on average curvature and with prescribed initial and terminal positions and tangents," *American Journal of Mathematics*, vol. 79, pp. 497–516, 1957.

[5] J.P. Laumond, " Singularities and topological aspects in nonholonomic motion planning," in *Nonholonomic Motion Planning*, Zexiang Li and J.F. Canny Eds, The Kluwer International Series in Engineering and Computer Science 192, 1992.

[6] J.P. Laumond, P. Jacobs, M. Taïx, and R. Murray, "A motion planner for nonholonomic mobile robot," *IEEE Trans. on Robotics and Automation*, Vol. 10, 1994.

[7] J.P. Laumond and P. Souères, "Metric induced by the shortest paths for a car-like mobile robot," in *IEEE Int. Conf. on Intelligent Robots and Systems*, Yokohama, July 1993.

[8] B. Mirtich and J. Canny, "Using skeletons for nonholonomic path planning among obstacles," in *IEEE Int. Conf. on Robotics and Automation*, Nice, May 1992.

[9] P. Moutarlier, B. Mirtich and J. Canny, "Shortest paths for a car-like robot to manifolds in configuration space," in *International Journal of Robotics Research*, 15 (1), pp. 36–60, 1996.

[10] J. A. Reeds and R. A. Shepp, "Optimal paths for a car that goes both forward and backwards," *Pacific Journal of Mathematics*, 145 (2), pp. 367–393, 1990.

[11] P. Souères and J.P. Laumond, "Shortest path synthesis for a car-like robot," *European Control Conference*, Groningen, June 1993.

[12] P. Souères, J.Y. Fourquet and J.P. Laumond, "Set of reachable positions for a car", *IEEE Trans. on Automatic Control*, 1994.

[13] H.J. Sussmann and W. Tang, "Shortest paths for the Reeds and Shepp car : a worked out example of the use of geometric techniques in nonlinear optimal control," Report SYCON-91-10, Rutgers Univ., 1991.

[14] James and James, *Mathematics Dictionary*, Princeton, NJ: Van Nostrand, 1968

Sensorless Parts Feeding with a One Joint Robot

Srinivas Akella, *The Robotics Institute, Carnegie Mellon University, Pittsburgh, PA, USA*
Wesley H. Huang, *The Robotics Institute, Carnegie Mellon University, Pittsburgh, PA, USA*
Kevin M. Lynch, *Biorobotics Division, Mechanical Engineering Laboratory, Tsukuba, Japan*
Matthew T. Mason, *The Robotics Institute, Carnegie Mellon University, Pittsburgh, PA, USA*

A rigid object in the plane has three degrees of motion freedom, but it does not follow that a planar manipulator must have three independently actuated and controlled joints. As previous work has demonstrated, there are a variety of methods to perform manipulation tasks using fewer actuators than motion freedoms. The method explored in this paper is to use a single joint robot to push an object on a constant speed conveyor belt. This paper summarizes the approach, previously described in [3], and extends the approach to include the problem of orienting polygonal objects without a sensor.

1 Introduction

This paper describes an approach to planar manipulation called "1JOC" (One Joint Over Conveyor, pronounced "one jock") [3]. Initially the approach was conceived as a variation on the Adept Flex Feeder (see Figure 1), which is used to feed parts in automated factories. The Flex Feeder uses a system of conveyors to recirculate parts, presenting them in random orientation to a camera and robotic manipulator. Those parts that are in a graspable configuration may then be picked up by the robot and assembled into a product, placed in a pallet, or otherwise processed.

The question is whether, at least in parts feeding applications, we could replace the four-degree-of-freedom robotic manipulator with a simpler single-degree-of-freedom device. Figures 2 and 3 show a possible variation using a fence driven by a single revolute joint. By a sequence of pushing operations, punctuated by drift along the conveyor, the fence can move a part from an initially random pose to the entry point of a feeder track which carries the part to the next station.

A previous paper [3] focused on the simplest version:

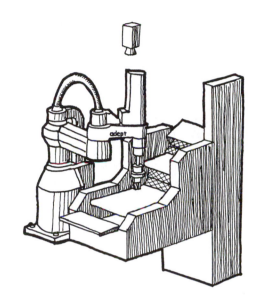

Figure 1: *The Adept Flex Feeder System. A SCARA robot picks parts off the middle of three conveyors. These three conveyors, along with an elevator bucket, circulate parts; an overhead camera looks down on the back-lit middle conveyor to determine the position and orientation of parts.*

a straight fence, collinear with the joint axis, working above a constant velocity conveyor. This paper introduces the *sensorless 1JOC* to address the issue of sensorless manipulation. Other variations include a 2JOC, multiple 1JOCs working in parallel, curved fences, and so on.

There are many different measures of manipulation. For example, a system with *small-time local controllability* could move the object along an arbitrary trajectory. A system with *global controllability* could move the object from an arbitrary start to an arbitrary goal. The 1JOC posesses neither of these properties. How-

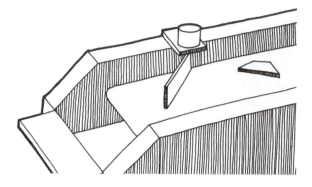

Figure 2: *The Flex Feeder with a rotatable fence.*

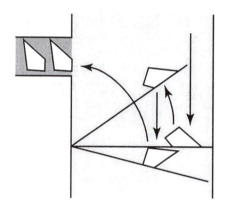

Figure 3: *We can feed a part by alternately pushing it with the fence and letting it drift along the conveyor.*

ever, the 1JOC can function as a parts feeding device. We formalize a measure of manipulation called the *feeding property*:

> A system has the *feeding property* over a set of parts $\mathcal{P}$ and set of initial configurations $\mathcal{I}$ if, given any part in $\mathcal{P}$, there is some output configuration $\mathbf{q}$ such that the system can move the part to $\mathbf{q}$ from any location in $\mathcal{I}$.

Our main results for the 1JOC are:

- 1JOC has the feeding property over the set of all polygons.

- For any polygon, a planner can determine a suitable goal, and can construct a sequence of pushes

and drifts from any initial configuration to the goal.

The feeding property is proven for an infinite length fence and infinitely wide conveyor. The planner takes bounds on fence length and conveyor width into account. We have also successfully demonstrated some plans in the laboratory.

Our primary result for the sensorless 1JOC is:

- The sensorless 1JOC can orient any polygon without sensing provided the polygon does not have a symmetric *push function* [13].

We have demonstrated an example plan on our system.

1.1 Previous Work

This paper touches on a number of topics, so that an adequate discussion of our intellectual precedents seems impossible in the space available. The interested reader should refer to [3, 11, 20].

Much of this work was inspired by industrial parts feeders such as bowl feeders and the APOS system (Hitakawa [16]). Related research includes dynamic parts orienting on a vibrating plate (Böhringer *et al.* [7], Swanson *et al.* [30]). Other forms of nonprehensile manipulation are parts orienting by tray-tilting (Erdmann and Mason [12]), tumbling (Sawasaki *et al.* [29]), pivoting (Aiyama *et al.* [2]), tapping (Higuchi [15], Huang *et al.* [17]), two pin manipulation (Abell and Erdmann [1]), and two palm manipulation (Erdmann [11], Zumel and Erdmann [32]).

Much of the work on underactuated manipulation has exploited dynamic coupling among freedoms. Research on underactuated manipulators includes that of Oriolo and Nakamura [26] and Arai and Tachi [5]. Arai and Khatib [4] demonstrated rolling of a cube on a paddle held by a PUMA. Their motion strategy was hand-crafted with the assumption of infinite friction at the rolling contact. In work closely related to the work reported here, Lynch and Mason [22, 20, 21] report automatic planning of rolling, throwing, and snatching tasks using a single degree of freedom robot. A good introduction to nonlinear control is given by Nijmeijer and van der Schaft [25], and nonholonomic robotic systems are discussed in the texts by Latombe [18] and Murray *et al.* [24].

Work on orienting parts using the task mechanics, with and without sensing, goes back to Grossman and Blasgen [14], whose system brought objects to a finite number of orientations in a tilted tray, where their orientation was determined by a tactile probe. Erdmann and Mason [12] developed an automatic sensorless tray tilting system based on the task mechanics. Peshkin and Sanderson [27] used results on the motion of a pushed object to find a sequence of fences to automatically orient a sliding part. Brokowski, Peshkin, and Goldberg [8] designed curved fence sections to eliminate uncertainty in the orientations of parts being oriented by the fences. Goldberg [13] developed a backchaining algorithm to orient polygonal parts up to symmetry using a frictionless parallel-jaw gripper. Wiegley *et al.* [31] developed a complete algorithm to find the shortest sequence of frictionless curved fences to orient a polygonal part.

Minimalism in robotics has also been studied by Donald *et al.* [10], Canny and Goldberg [9], and Böhringer *et al.* [6]. Raibert [28] and McGeer [23] constructed simple, elegant machines that use dynamics for stable locomotion.

2 Planar Manipulation with 1JOC

This section describes the 1JOC approach, sketches the proof that the 1JOC has the feedability property, and describes the planner. We adopt some conventions to simplify the presentation. We assume an origin coincident with the fence pivot, and we assume the belt's motion is in the $-y$ direction. The fence angle is measured with respect to the x axis. We will only use the right half of the belt, i.e. the half plane $x > 0$. The object's position along the fence is characterized by a *contact radius r*.

2.1 1JOC Primitives

We use four primitive actions, illustrated in Figure 4.

A *stable push* means that there is no motion of the object relative to the fence. A *turn* rotates the object, changing the edge in contact with the fence. It starts with a stable push to a fence angle of θ^+, followed by drift, until the object is caught at a fence angle of θ^-, followed by another stable push back to a fence angle of 0. A *jog* is a way of moving the object inward or outward along the fence, decreasing or increasing the contact radius r without changing the angle of the object. An inward jog is accomplished by slowly raising the fence a small angle θ, then quickly lowering it back to horizontal, and waiting while the object drifts back to the fence and settles back on the original edge. This moves the object inward by a distance $r(1 - \cos\theta)$. An outward jog is similar: the fence is quickly lowered by small angle θ, the object settles on the fence, the fence is slowly raised. The net effect is an outward motion of $r(\sec\theta - 1)$. The last primitive is a *convergent turn*, which is only required when an object must be turned through 180 degrees. A regular turn cannot get all the way to 180 degrees, but some objects will complete a full 180 degree turn, provided we allow the rotation to be completed after the catch.

The 1JOC manipulates objects as follows. Given a polygon and center of mass a goal is chosen, in most cases any stable edge aligned with the fence, at any radius. Now each time a polygon arrives, the camera identifies the initial orientation. The fence is held at zero degrees until the object comes to rest on the fence. The object is rotated to the desired edge by a sequence of turns (sometimes including a convergent turn) and jogs. A final jog brings the object to the desired contact radius. That this scenario always works, and just how to choose the parameters, are shown below.

2.2 Feedability

Here we sketch the proof of the feedability property.

- A stable edge is defined to be an edge that will remain in stable contact with the fence, with the fence at zero degrees plus or minus some small angle.

- Every polygon has at least one stable edge. This follows from the observation that the center of mass is in the interior of the polygon.

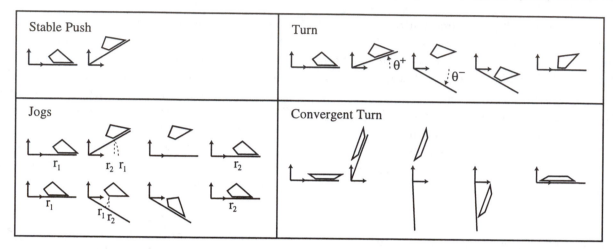

Figure 4: *Step by step illustrations of the 1JOC primitives.*

• A stable push is possible for any stable edge, provided that the contact radius is higher than some minimum. This minimum contact radius depends on the edge, on the object shape, the center of mass, the angular rate of the fence relative to the conveyor speed, and the coefficient of friction. Any stable push may be preceded by jogs to move the object out to the minimum contact radius.

• For most polygons, a counterclockwise rotation from one stable edge to another stable edge can be accomplished with turns. Every polygon has at most one stable edge from which it is impossible to rotate the polygon counterclockwise to another stable edge. There is one kind of polygon, with exactly two stable edges, parallel to each other, which requires a convergent turn.

• Given a polygon, we choose the goal configuration as follows. If there is a stable edge from which it is impossible to rotate counterclockwise to the next stable edge, then that edge is chosen as the goal edge. Otherwise any stable edge will serve. The goal contact radius is arbitrary.

The feeding property follows in a straightforward way. Given an arrival configuration, the fence can catch the polygon on a stable edge, use jogs as necessary to move the part far enough out on the fence for turns, use turns, and possibly a convergent turn, to rotate to

the goal edge, then use jogs to move the object to the goal radius.

2.3 Planning

Each 1JOC plan is a sequence of turns (possibly including a convergent turn) interleaved with jogs, rotating from the initial edge through a sequence of stable edges to the goal edge. We consider every sequence of stable edges from initial to goal edge that do not rotate more than 360 degrees. For every such sequence we must determine parameters for each primitive to minimize the total time required. Given a fence and conveyor of known dimensions, given the shape and center of mass of the polygon, and given the coefficient of friction, we formulate the constraints as follows:

• The contact radius must always exceed the minimum for any turn.

• The fence rotation must stay in valid ranges.

• The part must stay on the belt.

• The contact radius must not exceed the fence length.

• The fence does not contact the part during drift.

• The start and end configurations must match the given start and goal.

We use an approximation for the time required for the jogs, and find a plan minimizing the time, using a non-linear programming package GINO [19]. Details of the jogs are planned afterwards. A feasible solution to the above problem always exists, unless the belt is so narrow or the fence so short that the minimum contact radius cannot be satisfied for stable pushes. When this condition is satisfied, a plan with a maximum of three rotations exists, although it may not be the fastest plan.

3 Sensorless Parts Feeding

We have so far assumed that a camera gives us the initial position and orientation of the part after it first contacts the fence. The question we now ask is: Can we perform sensorless orienting and feeding with a one joint system over a conveyor? The answer is yes, with modifications to the 1JOC. We show that for our modified system, we can orient any object with an asymmetric *push function* [13]. We assume a rigid planar polygon with known shape and center of mass, quasistatic mechanics, uniform coefficient of support friction, and a frictionless fence.

The problem is to bring an object in an unknown initial orientation and position on the fence to a known orientation and position on the fence using a one joint system over a conveyor.

3.1 The Sensorless 1JOC

Our system to perform sensorless orienting and positioning consists of a frictionless fence, with a pivot in the center, with stops at both ends; we will refer to it as the *sensorless 1JOC* (see Figure 5). The pivot joint is not a simple revolute joint. Rather, the fence rolls without slipping on a circle, so that each point on the fence follows an evolute. If this pivot circle has diameter D, then when the fence has rotated 180 degrees CW (CCW), it will also have moved up the belt by D, and to the left (right) by $\pi D/2$. That makes 180 degree turns feasible—it leaves a space between the fence's dropping position and the fence's catching position, and an object being caught will not be too close to a stop.

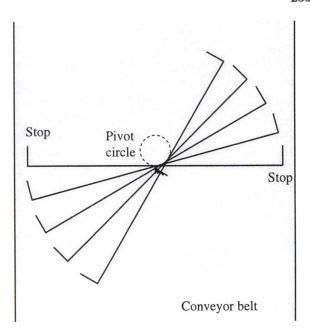

Figure 5: *The sensorless 1JOC.*

3.2 Sensorless Primitives

The primitives for sensorless orienting of polygons are (Figure 6):

1. *Catches:* A catch occurs when a part on the conveyor contacts the fence held stationary at 0 degrees and rotates onto a stable edge. An edge is stable if the perpendicular projection of the center of mass to the edge is in the interior of the edge segment. An edge is metastable if the center of mass projects to a vertex. If the polygon has metastable edges, slight rotations of the fence after the catch will destabilize metastable edges without affecting stable edges. Catches are used at the start of the orienting phase when the part drifts down the conveyor, and when the fence reacquires the part after each rotation by a stable push.

2. *Tilts:* A tilt is a CW or CCW rotation of the fence so that a part resting on a stable edge on the fence at 0 degrees slides without rotating to the right or left stop. A tilt eliminates positional uncertainty of the part by moving it into contact with a stop. To make sliding as fast as possible, we choose the

steepest tilt angle that does not cause the part to rotate. For a given stable edge, the object can be tilted through any angle for which the fence contact forces can balance the support friction forces. From the contacts that occur on the object once the slide is complete, we find the maximum tilt angle allowed for the edge. We find the smallest angle, up to 90 degrees, for the set of possible stable edges for the tilt and use that as the fence tilt angle. The angular velocity of the fence during downward motion should be small enough that the object does not lose contact with the fence.

3. *Stable pushes:* These actions are executed when the part is at one of the stops of the fence and we want to rotate the part CW or CCW. For a CW (CCW) rotation, the part has to be at the left (right) stop. This stable push differs from the stable pushes of earlier sections because we have more than just line contact, and there is zero friction. For a stable push, the contact forces due to the fence and stop should balance the frictional support force. We can show that by choosing a large enough fence angular velocity and a long enough fence, the contact force provided by the fence and the stop can balance the the support friction force for any pressure distribution.

These stable pushes allow the fence to rotate the part by up to 180 degrees in the CW or CCW direction. After the stable push, the fence comes to 0 degrees to catch the part, which then rotates to a stable edge.

3.3   Planning

A sensorless orienting plan begins with a catch to acquire the part, followed by a sequence of stages; each stage consists of tilt, a stable push, and a catch. The sequence of stages will take the part to a unique edge on the fence. The final step of the plan is a tilt to bring the part to a known stop of the fence. So at the end of the plan, the part is in a known orientation at a known position on the fence.

During a catch, the part rotation depends on the part shape, center of mass, and initial orientation. For

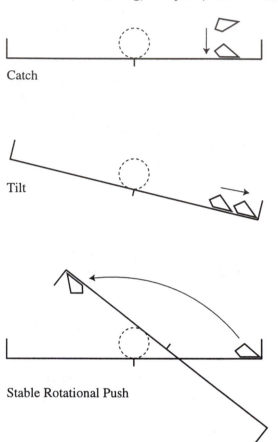

Figure 6: *Primitives used for sensorless orienting.*

a frictionless fence, the catch is effectively a normal push and the part rotation can be determined from the radius or push functions [13]. Since fence rotations in the range $[-180, 180]$ can rotate the part by an arbitrary amount, we can use Goldberg's backchaining algorithm [13] to generate a sequence of fence rotations (that is, the stable push angles). It follows that any part can be oriented up to symmetry in the push function and that a plan for an object with n stable edges has a length of $O(n)$ stages. The tilt directions are determined from the rotation direction of the succeeding stable push, and the tilt angles are determined from the set of edges that may be stable at each stage.

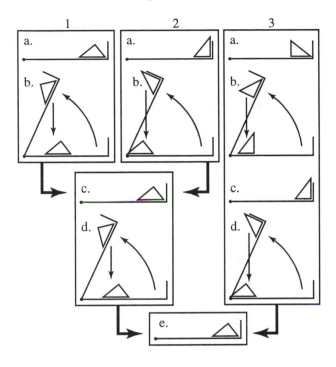

Figure 7: *Example sensorless feeding plan. Three different initial orientations converge to a single final orientation. For this plan only the right half of the sensorless 1JOC is used.*

3.4 Implementation

Most objects do not require rotations of 180 degrees, and do not require a frictionless fence. Our implementation of the sensorless 1JOC uses rotation about a fixed pivot point. We manually constructed a three stage plan to orient a right triangle, and tested it successfully. We used a conveyor velocity of 20 mm/sec and fence angular velocities of 20 degrees/sec and 12 degrees/sec for the stable pushes and tilts respectively.

The time for catches and tilts to proceed to completion, and the relative magnitudes of the conveyor velocity and fence angular velocity to guarantee stable pushes can be computed.

4 Conclusion

We have demonstrated two systems with one controlled degree of freedom to orient and feed parts. The original 1JOC system requires a sensor to give it the part's initial pose, while the variant system requires no sensing, except perhaps to know when a part has arrived. The sensorless variant has another advantage over the plain 1JOC. The sensorless variant appears to be more robust to errors in orientation and position of the fence and conveyor because of the uncertainty eliminating nature of the mechanical stops on the fence.

The sensorful 1JOC has advantages, too. Most obvious is that it can control the contact radius, and that it will often have shorter plans.

It may be possible to combine the two ideas, using the variant to bring the part to a known pose, followed by a plain 1JOC to reposition the part.

There are a number of other variations we would like to study. So far we have assumed singulated parts; we would like to explore the use of 1JOC to perform part singulation. Other variations of interest include three-dimensional parts, multiple part shapes, out-of-plane contact forces, faster motions, and pipelined 1JOCs.

Acknowledgments

Many thanks to Costa Nikou and Evan Shechter for implementations of the 1JOC on the Adept robot. We thank Mike Erdmann, Garth Zeglin, and Nina Zumel for early discussions about this paper. We thank Adept, NSK, and SONY for helping us learn about automation and for making equipment available. This work was supported by ARPA and NSF under grant IRI-9318496.

References

[1] T. Abell and M. A. Erdmann. Stably supported rotations of a planar polygon with two frictionless contacts. In *IEEE/RSJ International Conference on Intelligent Robots and Systems*, 1995.

[2] Y. Aiyama, M. Inaba, and H. Inoue. Pivoting: A new method of graspless manipulation of object by robot fingers. In *IEEE/RSJ International Conference on Intelligent Robots and Systems*, pages 136–143, Yokohama, Japan, 1993.

[3] S. Akella, W. Huang, K. M. Lynch, and M. T. Mason. Planar manipulation on a conveyor with a one joint robot. In *International Symposium on Robotics Research*, 1995.

[4] H. Arai and O. Khatib. Experiments with dynamic skills. In *1994 Japan–USA Symposium on Flexible Automation*, pages 81–84, 1994.

[5] H. Arai and S. Tachi. Position control system of a two degree of freedom manipulator with a passive joint. *IEEE Transactions on Industrial Electronics*, 38(1):15–20, Feb. 1991.

[6] K. Böhringer, R. Brown, B. Donald, J. Jennings, and D. Rus. Distributed robotic manipulation: Experiments in minimalism. In *International Symposium on Experimental Robotics*, 1995.

[7] K. F. Böhringer, V. Bhatt, and K. Y. Goldberg. Sensorless manipulation using transverse vibrations of a plate. In *IEEE International Conference on Robotics and Automation*, 1995.

[8] M. Brokowski, M. Peshkin, and K. Goldberg. Curved fences for part alignment. In *IEEE International Conference on Robotics and Automation*, pages 3:467–473, Atlanta, GA, 1993.

[9] J. F. Canny and K. Y. Goldberg. "RISC" industrial robotics: Recent results and open problems. In *IEEE International Conference on Robotics and Automation*, pages 1951–1958, 1994.

[10] B. R. Donald, J. Jennings, and D. Rus. Information invariants for cooperating autonomous mobile robots. In *International Symposium on Robotics Research*, Hidden Valley, PA, 1993. Cambridge, Mass: MIT Press.

[11] M. A. Erdmann. An exploration of nonprehensile two-palm manipulation: Planning and execution. In *International Symposium on Robotics Research*, 1995.

[12] M. A. Erdmann and M. T. Mason. An exploration of sensorless manipulation. *IEEE Transactions on Robotics and Automation*, 4(4):369–379, Aug. 1988.

[13] K. Y. Goldberg. Orienting polygonal parts without sensors. *Algorithmica*, 10:201–225, 1993.

[14] D. D. Grossman and M. W. Blasgen. Orienting mechanical parts by computer-controlled manipulator. *IEEE Transactions on Systems, Man, and Cybernetics*, 5(5), September 1975.

[15] T. Higuchi. Application of electromagnetic impulsive force to precise positioning tools in robot systems. In *International Symposium on Robotics Research*, pages 281–285. Cambridge, Mass: MIT Press, 1985.

[16] H. Hitakawa. Advanced parts orientation system has wide application. *Assembly Automation*, 8(3):147–150, 1988.

[17] W. Huang, E. P. Krotkov, and M. T. Mason. Impulsive manipulation. In *IEEE International Conference on Robotics and Automation*, 1995.

[18] J.-C. Latombe. *Robot Motion Planning*. Kluwer Academic Publishers, 1991.

[19] J. Liebman, L. Lasdon, L. Schrage, and A. Waren. *Modeling and Optimization with GINO*. The Scientific Press, 1986.

[20] K. M. Lynch. *Nonprehensile Robotic Manipulation: Controllability and Planning*. PhD thesis, Carnegie Mellon University, Mar. 1996. CMU-RI-TR-96-05.

[21] K. M. Lynch and M. T. Mason. Dynamic underactuated nonprehensile manipulation. In *IEEE/RSJ International Conference on Intelligent Robots and Systems*, 1996. To appear.

[22] M. T. Mason and K. M. Lynch. Dynamic manipulation. In *IEEE/RSJ International Conference on Intelligent Robots and Systems*, pages 152–159, Yokohama, Japan, 1993.

[23] T. McGeer. Passive dynamic walking. *International Journal of Robotics Research*, 9(2):62–82, 1990.

[24] R. M. Murray, Z. Li, and S. S. Sastry. *A Mathematical Introduction to Robotic Manipulation.* CRC Press, 1994.

[25] H. Nijmeijer and A. J. van der Schaft. *Nonlinear Dynamical Control Systems.* Springer-Verlag, 1990.

[26] G. Oriolo and Y. Nakamura. Control of mechanical systems with second-order nonholonomic constraints: Underactuated manipulators. In *Conference on Decision and Control*, pages 2398–2403, 1991.

[27] M. A. Peshkin and A. C. Sanderson. Planning robotic manipulation strategies for workpieces that slide. *IEEE Journal of Robotics and Automation*, 4(5):524–531, Oct. 1988.

[28] M. H. Raibert. *Legged Robots That Balance.* Cambridge: MIT Press, 1986.

[29] N. Sawasaki, M. Inaba, and H. Inoue. Tumbling objects using a multi-fingered robot. In *Proceedings of the 20th International Symposium on Industrial Robots and Robot Exhibition*, pages 609–616, Tokyo, Japan, 1989.

[30] P. J. Swanson, R. R. Burridge, and D. E. Koditschek. Global asymptotic stability of a passive juggler: A parts feeding strategy. In *IEEE International Conference on Robotics and Automation*, pages 1983–1988, 1995.

[31] J. Wiegley, K. Goldberg, M. Peshkin, and M. Brokowski. A complete algorithm for designing passive fences to orient parts. In *IEEE International Conference on Robotics and Automation*, pages 1133–1139, 1996.

[32] N. B. Zumel and M. A. Erdmann. Balancing of a planar bouncing object. In *IEEE International Conference on Robotics and Automation*, pages 2949–2954, San Diego, CA, 1994.

An Exploration of Nonprehensile Two-Palm Manipulation Using Two Zebras

Michael Erdmann, *Carnegie Mellon University, Pittsburgh, PA, USA*

1 Introduction

This paper describes our current research into *nonprehensile palm manipulation*. The term "palm" refers to the use of the entire device surface during manipulation, as opposed to use of the fingertips alone. The term "nonprehensile" means that the palms hold the object without wrapping themselves around it, as distinguished from a force/form closure grasp often employed by a fingered hand. Indeed, nonprehensile operations such as purposeful sliding and constrained dropping constitute important palm primitives.

We have implemented a system for orienting parts using two palms. The system consists of a planner and an executive. As input, the system expects a geometric description of a part, its center of mass, the coefficients of friction between the part and each of the palms, and a start and goal configuration of the part in stable contact with one of the palms. As output, the system computes and executes a sequence of palm motions designed to reorient the part from the specified start to the specified goal configuration.

1.1 System Specifics

For hardware we are using two Zebra Zero robots. We replaced the standard grippers on these arms with two flat surfaces; these are the "palms". Figure 1 shows the experimental setup. We used three different materials for the surfaces of these palms, namely aluminum, aluminum covered with duct tape, and aluminum covered with a hard impact-absorbing foam. These three surfaces provided us with a range of friction coefficients.

The system expects 2.5-dimensional polyhedral parts, specifically parts that are extruded convex polygonal shapes. The actual parts oriented consisted of several rectangular blocks as well as one irregularly-shaped

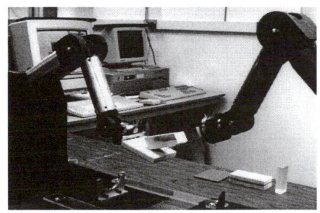

Figure 1: *The experimental setup consists of two Zebra robots, with flat palms attached to each wrist. The Zebras are shown manipulating an irregular block.*

block. The irregular block's two-dimensional cross-section consisted of a five-sided polygon with three stable edges, one unstable edge, and one marginally stable edge. Again, we used a variety of materials for the parts, namely hard foam, hard plastic, and wood.

The planner is planar, that is, it assumes that the parts have two translational and one rotational degrees of freedom. Hence the 3D parts oriented were extruded 2D shapes. The current version of the planner focuses on Type-B contacts, to use the classification of [Lozano-Pérez 1983, Section V]. In other words, the planner focuses on contacts involving the palm surfaces but not the boundaries of those surfaces. Viewed in 2D, this class of contacts includes both vertex and line contacts of the part with the palm lines. It excludes Type-A contacts, that is, again viewed in 2D, it excludes contacts between a line segment of the part and a vertex of the palms.

Finally, a word about sensing. The plans are sensorless. The planner takes careful account of object geometry, friction, and gravity, including knowledge of the object's center of mass. The resulting plans execute without any sensing of the part by the robots.

1.2   Contributions

This research makes two main contributions:

- **Automatic Two-Palm Manipulation.** We have demonstrated the feasibility of automatic nonprehensile palm manipulation, by implementing both a planner and an executive. The system reorients 2.5-dimensional polyhedral parts with two cooperating robot palms.

- **Contact Analysis Tools.** We have developed simple analysis tools for determining the different modes by which two palms can manipulate an object. These modes include holding the object, rotating the object, and slipping one palm or other against the surface of the object.

2   Examples

Figures 2 and 3 show two plans produced by our planner. We will refer to the left palm as "Palm 1" and the right palm as "Palm 2". The first plan reorients a rectangular block. The block was plastic, Palm 1 was aluminum, and Palm 2 was hard foam. The second plan reorients the irregular block mentioned earlier. The block was wood, Palm 1 was aluminum, and Palm 2 was aluminum covered with duct tape. The coefficients of friction may be found in the figure captions. The frames correspond to snapshots before and after actions planned by our planner.

Both plans have a similar overall character, but the precise actions differ. Both plans start with a TILT of Palm 1 to an angle that causes the block to slide to the right. The planner takes the tilt angle in each case to be 3 degrees greater than the angle of friction for that case. As part of the TILT, Palm 2 catches the block. With that, the reorienting process starts. Here are some highlights:

- **Slide.** A SLIDE operation consists of some action by the palms in which one palm slides relative to the part. Each plan happens to include one such SLIDE operation, beyond the initial tilt.

In Figure 2, Palm 2 drags the block to the right between Frames 5 and 6, moving tangentially to Palm 1. In Figure 3, Palm 2 drags the block down and to the right between Frames 2 and 3, again moving tangentially to Palm 1. The basic operation is the same in both cases, but the angles of the palms are different, in order to account for the different coefficients of friction.

A different plan for the wood block, shown in Figure 4, contains three different SLIDE actions. Between Frames 3 and 4, Palm 2 drags the block to the right, much as in the other two plans discussed. Between Frames 9 and 10, Palm 2 partially slips out from under the block, and between Frames 14 and 15, Palm 2 slides back down, reestablishing its configuration relative to the block. Despite how precarious these SLIDE operations may appear, especially the action between Frames 14 and 15, the Zebra robots executed all three SLIDE actions successfully.

- **Regrip.** Both plans contain a number of REGRIPS. A REGRIP is an action in which a palm rotates about its contact point with the part. The plans call for such REGRIP operations in order to strategically configure the palms for some subsequent action. For instance, between Frames 4 and 5 of Figure 2, Palm 2 reorients itself to a shallow angle in order to then drag the block to the right. Having dragged the block to the right, Palm 2 then reorients itself to a steeper angle in order to perform the rotation from Frame 7 to Frame 8. Similarly, the actions between Frames 3 and 15 of Figure 3 basically constitute a dance of REGRIP and ROTATE operations required to turn the block upright without permitting it to slip at the contacts or otherwise lose equilibrium.

- **Rotate.** No matter what else the palms do, ultimately they must ROTATE the blocks from their start to their goal configurations. In the plans shown, this is generally accomplished by moving Palm 2 so that its contact point traverses an arc of a circle about Palm 1's contact point. It is up to the planner to choose the amount of rotation so the part does not slip. Both plans contain a number of ROTATE operations. For instance, in Figure 2 there is a small ROTATE operation between Frames 2 and 3, and a large

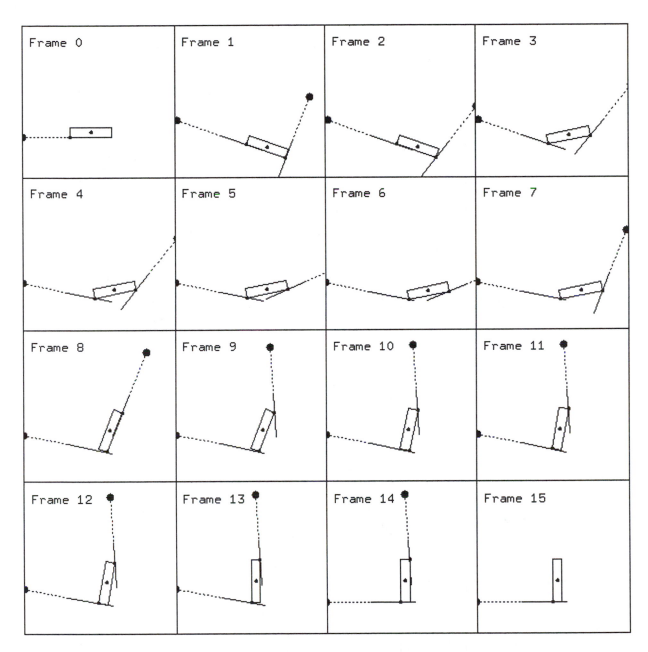

Figure 2: *Plan for reorienting a rectangular block using two robot palms. The plan was generated automatically, given a geometric description of the part and the coefficients of contact friction. Two Zebra Zero robots executed this plan, each acting as one palm. — Friction is 0.287 at the left palm and 0.810 at the right palm. The block dimensions are 83mm by 18mm. The palm lengths are drawn to scale. The big black dots are the locations of the robot wrist joints; the dashed lines depict unreachable parts of the robot hands, such as the force sensor housings. The small dots indicate the primary vertex of contact for each palm. — The friction coefficients were computed by measuring the friction angles to about one degree of accuracy, then taking tangents of those angles. The three digits shown do not reflect three digit accuracy.*

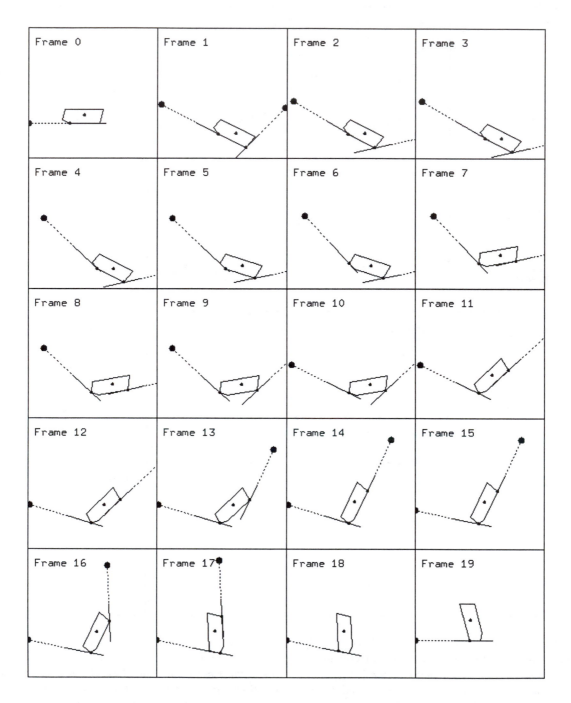

Figure 3: *Plan for reorienting a five-sided block using two palms. Friction at the left palm is 0.445 and friction at the right palm is 0.554.*

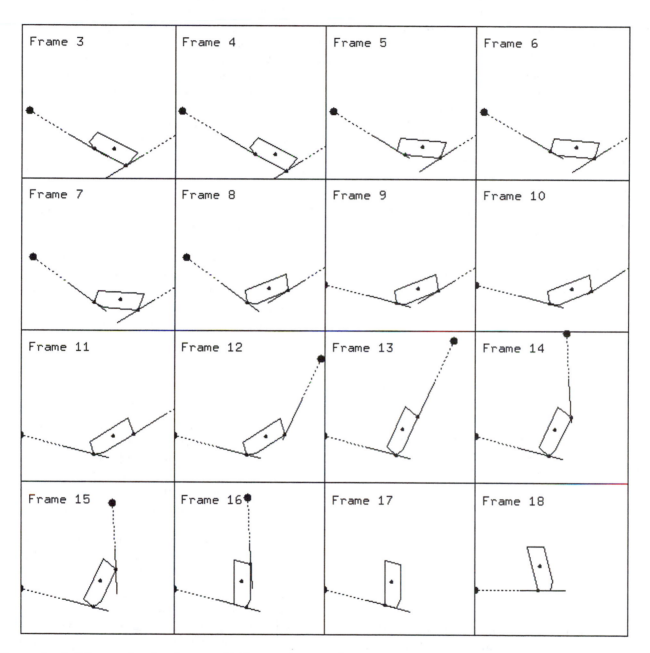

Figure 4: *A different plan for the same block as in Figure 3 (frames 0–2 are not shown; they are similar to the frames in the previous two plans). This plan includes three different* SLIDE *actions. These occur between Frames 3 and 4, between Frames 9 and 10, and between Frames 14 and 15.*

one between Frames 7 and 8. In Figure 3, there is a tiny ROTATE between Frames 4 and 5, just enough to bring the small unstable edge into flush contact with Palm 1. Two other rotations occur between Frames 6 and 7 and between Frames 10 and 11, each of which brings Palm 2 into flush edge contact with the block. (There are several more ROTATE actions subsequently.)

- **Release.** The two plans exhibit two different release modes. In Figure 2, Palm 2 releases the plastic block by doing a very careful dance with Palm 1. Before Palm 2 lets go of the block and moves away, the block and Palm 1 are in their final configurations (see Frame 14). In contrast, in Figure 3, Palm 2 actually pushes the block past its stable point, and allows the block to rotate to a new stable equilibrium. This happens between Frames 16 and 17. The action between these frames is a ROTATE. Palm 2 rotates the block about its contact with Palm 1. As the center of mass passes over the contact point, the block rotates on its own under the influence of gravity; Palm 2 eventually catches up. This operation is very delicate. The planner had to choose the angle of Palm 1 sufficiently steep to prevent the block from rotating too far and tipping over. After all, the goal edge is barely stable. On the other hand, choosing the tilt of Palm 1 too steeply would have caused the block to slide and thus slip between the two palms, again falling over. Finally, once the block rotates to its goal edge, Palm 2 can let go (Frame 18), and Palm 1 can then tilt to horizontal (Frame 19).

The heart of this paper is a discussion of how the planner analyzes the possible motions of a two-palm contact.

3   Motivation

We are motivated to study palm manipulation for three basic reasons: simplicity, naturalness, and foundation.

Simplicity.   Palm manipulation can reduce the mechanical complexity of devices needed to manipulate objects. We are interested in designing low degree of freedom devices for orienting and feeding parts to assembly workcells. In the past, feeder designs have suffered from the "frozen hardware" problem. Too much

of the mechanics of the task has been compiled into hardware, in the form of fixed feeder gates and orienting shapes. At the other extreme, general purpose robots have suffered from too much generality and too little reliable software. An important intermediate architecture consists of a sequence of simple manipulators under software control. Since the configurations and motions of the devices are under software control and since the devices are easy to program by virtue of having low degrees of freedom, last-minute part changes may be accommodated by the feeder with relative ease. For the devices, we envision a series of palms, each with one or two degrees of freedom, situated along a feeder belt.

Naturalness.   Nonprehensile palm manipulation is natural. Humans and animals often manipulate objects with a series of partial grasps, shuffling the object back and forth between various contacts. For instance, birds twirl seeds in their beaks, chipmunks roll acorns in their paws while shelling them with their teeth, and humans roll objects back and forth between their two palms in order to orient or inspect them. Turning over a large book is one everyday example. Lathering soap is another. These tasks include periods of prehensile grasps, but it is nonprehensile manipulation that provides the fluidity whereby one equilibrium grasp is dynamically transformed into another.

Even such a simple operation as *placing* an object must, by definition, end without prehension. Although one may transport the object using a force/form closure grasp, as one releases the object the interaction between the hand and the object becomes nonprehensile, often involving sliding contacts over an extended surface of the hand. The more quickly one wishes to release the object, the more significant this interaction. The placing operation effectively becomes a *slide-release* (see Section 5).

Moreover, the transport itself may involve a grasp that is merely an equilibrium grasp against gravity, without being prehensile. A waiter carrying a large tray on his or her palm is one example. A forklift raising a crate is another.

Other natural examples of palm manipulation include scooping, cradling, smacking, and flipping.

We wish to understand and harness these natural operations, with the aim of simplifying robot programming. By building planners and executors we highlight weaknesses and quantify competence, thereby increasing our understanding.

Foundation. Palm manipulation forms a key point in the spectrum of manipulation operations. At one end of this spectrum we have force/form closure grasps; at the other end we have free flight motions of thrown objects. Force/form closure grasps are characterized by complete control over the object and resultant insensitivity to environmental dynamics. Free flight motions are characterized by an impulse of initial control followed by strong sensitivity to environmental dynamics. In between these extremes lies a large class of nonprehensile manipulation operations. Research to date has explored only a small portion of this class. Some examples include quasi-static manipulation of an object on a planar support surface, such as pushing, tumbling, and pivoting. Other examples include controlled slip motion, reorienting by tipping, and object acquisition. The unstudied problems lie in two directions: (i) shaping and controlling constraint surfaces other than pure point-fingers or horizontal planes, and (ii) exploiting dynamics to take advantage of dynamic coupling, much as in the work on underactuated manipulators. Palm manipulation offers a simple domain in which to explore these two directions.

Hardware. Finally, we are motivated in our research by a fourth goal, to improve Manipulation hardware support. While other aspects of Robotics, such as Vision and Mobile Robotics have recently enjoyed the benefits of improved and relatively inexpensive technology, Manipulation has not. There are very few inexpensive arms or hands for researchers to use in developing ideas and algorithms. Many researchers make due with old arms, invest too much time building their own devices, or spend exorbitant amounts of money on fairly primitive robots. Vision and Mobile Robotics have had considerably more hardware support, in part because those fields have provided positive feedback to the hardware manufacturers. Better hardware has yielded better algorithms, with visible and tangible successes. We hope that demonstrations of automatic manipulation will help encourage and re-invigorate the robot arm industry. Admittedly, it is an elusive goal.

4 Related Work

Nonprehensile Manipulation

The importance of nonprehensile manipulation in robotics first makes its written appearance in Mason's Ph.D. thesis [Mason 1982] and the paper [Mason 1986]. Mason in turn points to work by Pingle, Paul, and Bolles from 1974, a conversation with Inoue, and general parts feeding approaches. The basic idea is for a robot to manipulate an object purposefully even though the robot does not have full control over the object. Mason looked at the problem of pushing objects in the plane. He analyzed the mechanics of pushing, including the classification of sliding and rolling contacts, and showed how pushing could be used to reduce an object's uncertainty. Mason's research spawned over a decade of further work on pushing. Peshkin analyzed the pushing problem [Peshkin 1986] and showed how to design a series of fences along a conveyor belt to orient parts on the conveyor [Peshkin and Sanderson 1988]. Peshkin's work was later extended by [Brokowski, Peshkin, and Goldberg 1993] to design curved fences that smoothly engaged and reoriented the parts on the conveyor. [Brost 1988] implemented a system that could orient parts through a series of pushing and squeezing operations. [Akella and Mason 1992] showed how to pose any polygon using only pushing strategies. [Lynch 1992] showed how to pose objects by pushing without relative slip, and [Lynch and Mason 1995] investigated various non-intuitive properties of pushing strategies. [Donald, Jennings, and Rus 1994] implemented a variety of pushing and reorienting strategies, using cooperating mobile robots to move large objects such as boxes and couches between rooms. Finally, in the context of nonprehensile finger-like manipulation, [Abell and Erdmann 1995] developed a planner for holding and reorienting planar parts stably, using only two fingers and gravity.

A number of researchers have investigated and implemented other ingenious forms of nonprehensile manipulation. [Sawasaki, Inaba, and Inoue 1993] considered the task of moving an object over a planar surface by tumbling it about various edges of contact. [Aiyama, Inaba, and Inoue 1993] considered the task of moving an object over a planar surface by pivoting it about various vertices, much as one might move a large refrigerator. These researchers observe that pushing, tumbling, and pivoting form a geometrically complete class of motions. Specifically, these operations preserve, respectively, plane, line, and vertex contact with a support surface. From this perspective, palm manipulation is a simple generalization of these operations, in which the support surface itself becomes moveable.

Palm-Like Manipulation

Probably the first to examine palm-like manipulation explicitly were Salisbury in one setting and Trinkle in another. [Salisbury 1987] proposed the idea of whole arm manipulation, in which an arm manipulates an object by making contact with the object anywhere on the arm's surface, as needed, rather than restricting contact to the fingertips of the hand. Around the same time, in a series of papers Trinkle and colleagues analyzed the mechanics of strategies for scooping up frictionless objects resting on a table, by enveloping them with two fingers [Trinkle 1992], [Trinkle, Abel, and Paul 1988], [Trinkle and Paul 1990].

Yun and colleagues looked at the problem of manipulating large objects by supporting them with two palms [Yun 1993], [Paljug and Yun 1993], [Paljug, Yun, and Kumar 1994]. These researchers developed and implemented control laws for stably moving and reorienting objects with non-sliding rolling contacts. Bicchi and colleagues analyzed the kinematics and manipulability of pairs of cooperating robots that manipulate objects with flat palm-like surfaces [Bicchi, Melchiorri, and Balluchi 1995], [Bicchi and Sorrentino 1995].

Slip

An important aspect of our system is the ability to purposefully slip one palm relative to the object being held, or, equivalently, to use one palm to drag or push the object relative to the other palm. There has been considerable prior work on slip. [Fearing 1984] and [Nguyen 1988] used slip to attain stable grasps on objects of unknown shape or pose. The works by Mason and Trinkle and their colleagues cited earlier made extensive use of slip to orient objects and reduce uncertainty. [Brock 1988] used slip to reorient grasped objects in the presence of gravity. [Carlisle, Goldberg, Rao, and Wiegley 1994] used slip to reorient grasped objects by pushing them against other objects. [Cole, Hsu, and Sastry 1992] developed a dynamic control law for regrasping objects using sliding. Perhaps the most extensive investigation of slip originated with

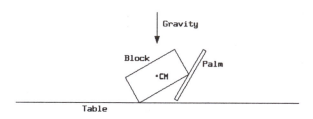

Figure 5: *Block in contact with a table and a palm.*

work by Kao and Cutkosky [Kao and Cutkosky 1989, 1992, 1993]. In one example, a robot rotates a card lying on a table, by slipping two fingers over the top surface of the card. Finally, [Yoshikawa, Yokokohji, and Nagayama 1993] analyze three-fingered grasping of three-dimensional objects using slip motions.

Friction

A key component of most grasping strategies, including our palm manipulation system, is the ability to account for friction. The heart of our planner is based on earlier work on friction found in [Erdmann 1994]. That paper also contains an extensive review of prior work on friction. Finally, the work reported in this paper has benefited greatly from the planar friction cone model, in particular the ideas underlying the work in the paper [Brost and Mason 1989].

5   Slide-Release

When two hands manipulate an object, there are three basic operations the hands perform: holding or rotating the object in an equilibrium grasp, allowing the object to fall in disequilibrium, and sliding part of one hand against the object. In this section we focus on the sliding operation.

5.1   Single-Handed Manipulation

Even single-hand manipulation gains scope and competence by judicious use of sliding. Imagine for instance placing a large dictionary on a table, using one hand.

Shortly before release, one edge of the book will be in contact with the table, while one's hand is holding an opposite edge. In all likelihood, one's fingers are supporting the bottom face of the book, while one's thumb is wrapped around the top side. In fact, the thumb is superfluous. Sooner or later one will release one's grip on the book, slide the fingers out from underneath, and allow the book to fall into place. This is a classic *slide-release*. It demonstrates one form of relative sliding, in which the hand slides relative to the object, while the object remains stationary in the world frame. In the example, the fingers slide out from under the book, while the book does not slide relative to the table. Of course, once the hand has slid all the way out from under the book, the book drops into place on the table, which is the whole point of a slide-release.

There is another form of sliding, dual to the one described, in which the hand moves and the object remains fixed to the hand, sliding in the world frame while tracking the hand. Such sliding can be used prior to a release to position the object in the world frame. In the example, we might move our hand so as to position the dictionary precisely on the table, and only thereafter release the book. Since we have a thumb, we would probably perform this fine positioning by grasping the raised end of the dictionary. However, again, the thumb is really unnecessary. One can both push and pull the dictionary simply by choosing carefully the angle with which one's palm or fingers support the raised end of the dictionary. Friction can substitute for the force closure provided by a thumb.

5.2   The Book-Placing Task

In the remainder of this section we will examine the book placing example in detail. We will develop a simple test for determining when sliding is possible and what type of sliding is possible. This test constitutes the central tool used by our two-palm planner for carving up its search space.

The basic problem is shown Figure 5. The figure depicts a block in contact with a horizontal table and a tilted palm. Gravity acts straight down. We assume that friction arises from dry Coulomb friction. We assume further that there is no difference between the static and dynamic coefficients of friction.

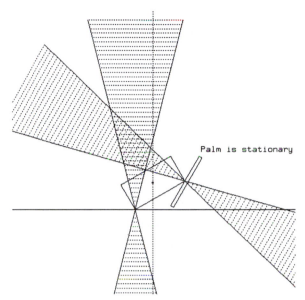

Figure 6: *Initial configuration of block, table, and palm, along with the two two-sided contact friction cones and the vertical center-of-mass line. Friction is 0.25 at each contact. Observe that the friction cones do not intersect anywhere along the center-of-mass vertical.*

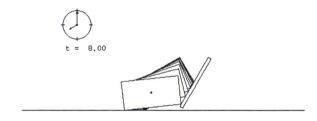

Figure 7: *Simulation of motion starting from the configuration in Figure 6. The block falls.*

5.3   Equilibrium Conditions

Let us look at the conditions under which equilibrium is possible. Figure 6 again shows the block in contact with the table and the palm, but now with the

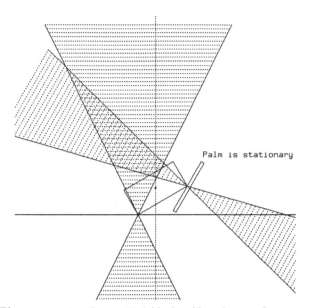

Figure 8: *Another initial block-table-palm configuration, along with the two-sided friction cones and the vertical center-of-mass line. In this case friction is still 0.25 at the palm contact, but it is now 0.5 at the table contact. Observe that in this example the friction cones do intersect along the center-of-mass vertical.*

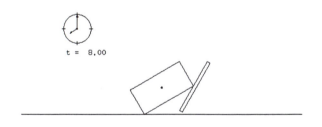

Figure 9: *Simulation of motion starting from the configuration in Figure 8. The block remains stationary.*

contact friction cones depicted as well. We have depicted the cones as two-sided friction cones in order to emphasize the lines of action of the reaction forces.

We have also drawn a dashed vertical line through the center of mass of the block. This line depicts the line of action of the force of gravity.

First, we ask whether the block will fall or remain stationary. It turns out the block will fall; Figure 7 shows the result after 8 centiseconds, as determined by our configuration space friction cone simulator.

In contrast, consider Figure 8. In that figure, the coefficient of friction between the block and the table is now $\mu_1 = 0.5$, larger than it was in Figure 6. This time the block remains stationary; it is in equilibrium. See Figure 9.

Mechanically, the difference between the examples of Figure 6 and Figure 8 is that the contact reaction forces can balance gravity in the second case but not the first. Geometrically, two conditions must be satisfied. First, the negative of the gravity vector must lie in the positive-span of the two contact reaction forces. Second, the lines of action of the two contact reaction forces and the gravity vector must all intersect at a common point. In other words, to use Brock's terminology [Brock 1988], the *force focus* of the three forces must exist. Together, the two conditions ensure that force-torque balance is physically possible.

We will focus mainly on the second condition in the remainder of this paper, that is, the *force focus test*. Nonetheless, we caution the reader not to forget the first condition, that is, the *positive-span test*.

Next, we ask ourselves whether we can tell instantly from Figures 6 and 8 if the force focus test is satisfied. The answer is yes. We need merely intersect the two two-sided friction cones and compare the resulting geometric object to the vertical line drawn through the object's center of mass. If the intersection of the two two-sided friction cones overlaps the vertical line, then it is possible to satisfy the force focus test. Conversely, if the intersection of the friction cones does not overlap the vertical line then it is not possible to satisfy the force focus test.

5.4 Sliding Contact Conditions

A robot expands its effective workspace when it can slide its hands relative to the objects it is manipulating. By strategically sliding the hands, the robot can reposition the objects in collision-free and mechanically advantageous locations on the hands.

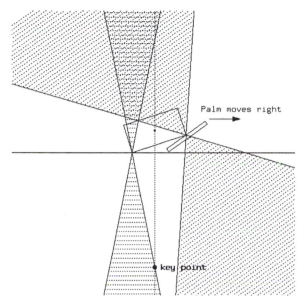

Figure 10: *Initial configuration of block, table, and palm. The palm is about to move straight to the right. Friction is 0.2 at the table contact and 0.8 at the palm contact. The "key point" is the intersection of the table's left friction cone edge with the center-of-mass vertical. Observe that the key point lies outside the palm's friction cone.*

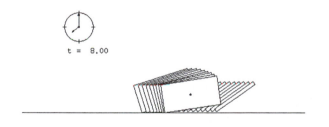

Figure 11: *Simulation of motion starting from the configuration in Figure 10. As the palm moves, the block slips and rotates down.*

Consider Figure 10. The figure again shows the vertical line through the object's center of mass and

the two two-sided friction cones. Suppose the palm moves to the right, tangential to the table. Will the block track the palm or fall? Figure 11 shows the simulation results after 8 centiseconds. We see that the block falls. In contrast, consider Figures 12 and 13. In that example the block tracks the hand.

The difference between these two examples may be explained and predicted using similar reasoning as we used with the equilibrium examples of Section 5.3, only we must be more careful in describing the source of the reaction forces. The reaction force at the sliding contact cannot arise arbitrarily from within its contact friction cone. Instead, the reaction force must be on an edge of the friction cone, namely the edge that opposes the sliding motion. The reaction force at the other contact, the non-sliding contact, can of course potentially arise from anywhere within its contact friction cone.

For the example of Figures 10 and 12, the reaction force at the block-table contact must lie on the left edge of its friction cone. In deciding whether quasi-static dragging is possible, we must therefore intersect the line through the left edge of the friction cone with the vertical line through the object's center of mass. If this intersection point lies within the two-sided friction cone anchored at the block-palm contact, then quasi-static sliding is indeed possible. Otherwise it is not.

The conditions for pushing the block to the left, or for sliding the palm relative to the block without disturbing the block, are similar.

5.5   Summary

To decide whether static equilibrium is possible one checks whether the two friction cones have a common overlap with the vertical line drawn through the center of mass. To decide whether quasi-static sliding is possible one checks whether an edge of one friction cone has a common overlap with the other friction cone on the vertical. The specific edge used depends on the particular sliding mode. The two tests are really the same, merely constrained differently to account for the different object motions.

5.6   Planning Tools

It is easy to generalize these equilibrium and sliding tests into planning tools. For a fixed block configuration and a fixed set of contacts, the truth value of an

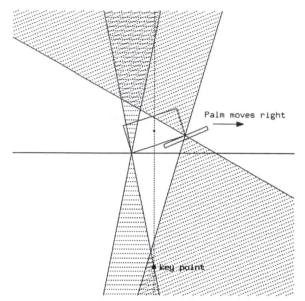

Figure 12: *Initial configuration of block, table, and palm. The palm is about to move straight to the right. Friction is 0.2 at the table contact and 0.8 at the palm contact. Observe that the palm is at a shallower angle than it was in Figure 10. Consequently, the "key point" now lies inside the palm's friction cone.*

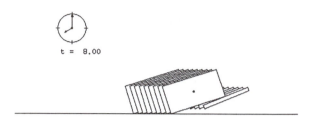

Figure 13: *Simulation of motion starting from the configuration in Figure 12. As the palm moves, the block tracks the palm without rotating.*

equilibrium or sliding test can only change at certain discrete orientations of the palm. These critical orientations are those angles at which a friction cone edge of one friction cone intersects a friction cone edge of the other friction cone at a point lying on the center-of-mass vertical (or lying at infinity). In short, the mechanics of the task splits the orientation space of the palm into a finite set of sectors yielding qualitatively similar object behaviors.

6   The Planner

6.1   Carving up the Configuration Space

The fundamental planning stage for the planner is a two-dimensional space that encodes the possible orientations ϕ_1 and ϕ_2 of the two palms for a fixed orientation θ of the object. The basic role of the two-palm planner is to carve up (ϕ_1, ϕ_2) space into regions within which the permissible palm motions are qualitatively identical. The planner does this by

first computing certain critical curves. The critical curves describe the palm orientations (ϕ_1, ϕ_2) at which two friction cone edges intersect at the same spot on the vertical center-of-mass line, or at which the positive-span test might change truth value. Finally, the planner creates a cylindrical decomposition based on these critical curves. The permissible behavior of the palms and the object are invariant across palm orientations within any region thus constructed.

Figure 14 shows two palms supporting a block in the presence of gravity and contact friction. The configuration is a critical configuration; the right edge of friction cone 1 intersects the vertical line at the same point as does the left edge of friction cone 2. If we rotate either palm slightly the resulting contact would become non-critical. Separating such generic contact states is a critical curve in (ϕ_1, ϕ_2) space. At each point on this curve, the right edge of Palm 1's friction cone meets the left edge of Palm 2's friction cone smack dab on the vertical through the block's center of mass. The planner constructs all such critical curves, for all possible contacts of the two palms with vertices of the block.

6.2   Planning Motions

The planner uses labelled subdivisions of the form shown in Figure 15 to plan motions of the palms to move the part from the start to the goal. The planner uses yet another critical event analysis, this time in θ,

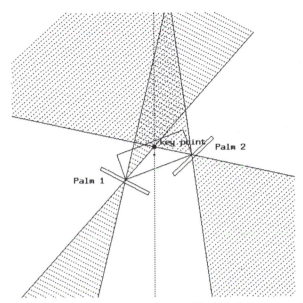

Figure 14: *Two palms holding a block. Friction at the left contact (Palm 1) is 0.25; friction at the right contact (Palm 2) is 0.70. The figure shows a singular configuration, in which Palm 1's right friction cone edge intersects Palm 2's left friction cone edge smack dab on the center-of-mass vertical.*

to construct (ϕ_1, ϕ_2) subdivisions at a small number of generic part orientations θ.

The planner searches the (θ, ϕ_1, ϕ_2) configuration space from start to goal by searching its collection of (ϕ_1, ϕ_2) subdivisions. Moving between different subdivisions is equivalent to rotating the part from one θ orientation to another. For a fixed part orientation, the palms can perform two types of motions: (i) they can rotate about their contact points, or (ii) they can slide left or right. The labelled subdivisions tell the planner which of these motions are legal. For instance, the region boundaries in the subdivision tell the planner how far it can reorient a palm before either changing region classification or making palm-edge contact with the part. And the region classification itself tells the planner whether it can slide a given palm left or right.

7 Future Directions

Although reasonably automatic, the system is far from truly autonomous. It is easy to design tasks for which the planner produces unexecutable plans. Low friction tasks are prime candidates. One way to increase the competence of our planner is to expand its mastery of mechanics. There are three directions future work should pursue:

1. Incorporate principled tests for contact stability.

2. Distinguish between static and dynamic coefficients of friction.

3. Build planners that can manipulate parts without requiring continuous stability or equilibrium.

Item (3) is probably the most interesting direction to explore. Our current planner has a modicum of nonequilibrium motion, namely the release operation discussed in Section 2. However, much more is possible. Zumel [Zumel and Erdmann 1996] is currently pursuing this direction, also in the domain of palm manipulation, focusing on tasks with very low friction. Additionally, [Mason and Lynch 1993] are pursuing this direction with their work on dynamic manipulation.

8 Acknowledgments

Many thanks to Craig Johnson for his excellent work with the Zebra Robots. This research benefited greatly from discussions with Tamara Abell, Bruce Donald, Kevin Lynch, Matt Mason, and Nina Zumel.

Financial support for this research was provided in part by Carnegie Mellon University through a Faculty Development Award, in part by an equipment grant from AT&T, and in part by the National Science Foundation through the following grants: NSF Presidential Young Investigator award IRI-9157643 and NSF Grants IRI-9213993 and IRI-9503648.

This paper is a revised version of two talks, one given at the *Seventh International Symposium on Robotics Research*, in Herrsching, Germany, October 21–24, 1995, the other given at the *Second Workshop on the Algorithmic Foundations of Robotics*, in Toulouse, France, July 3–5, 1996.

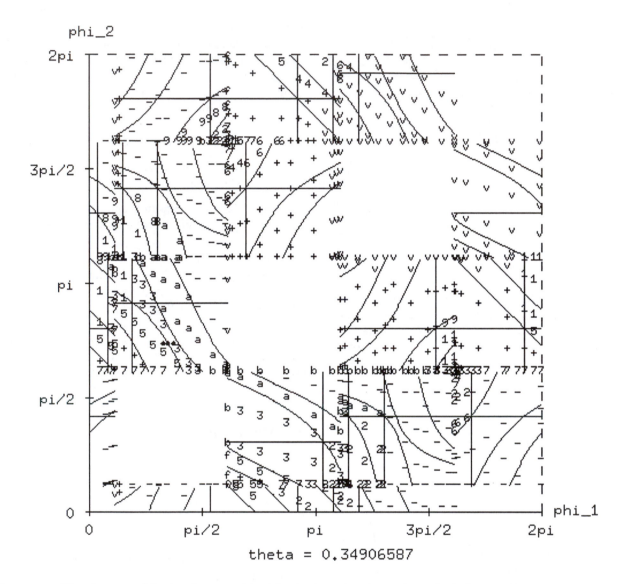

theta = 0.34906587

Figure 15: *The planner subdivides the space* (ϕ_1, ϕ_2) *of palm orientations for a fixed object orientation* θ *into regions of invariant contact mode, using a cylindrical decomposition. This figure shows the location of the representative state of each invariant region. The block orientation is fixed at 20 degrees, that is,* $\theta = 0.349$ *radians. The palms can make contact with any vertex of the block. Of course, not all contacts yield equilibrium states. — Each state is depicted by a letter. The letters are internal codes used by the planner to describe the types of motions possible at that state. For instance, the planner uses a one letter hexadecimal representation to encode the sliding modes possible when the block is in equilibrium. And a "v" means that the block is falling without rotating. — Note the three dots in the left central area. The middle dot corresponds to the critical configuration depicted in Figure 14. The underlying curve describes the orientations* (ϕ_1, ϕ_2) *at which the right edge of Palm 1's friction cone meets the left edge of Palm 2's friction cone smack dab on the center-of-mass vertical.*

References

[1] Abell, T., and Erdmann, M. 1995. Stably Supported Rotations of a Planar Polygon with Two Frictionless Contacts. *Proceedings of the 1995 IEEE/RSJ International Workshop on Intelligent Robots and Systems*, Pittsburgh, Pennsylvania, Vol 3, pp. 411–418.

[2] Aiyama, Y., Inaba, M., and Inoue, H. 1993. Pivoting: A New Method of Graspless Manipulation of Object by Robot Fingers. *Proceedings of the 1993 IEEE/RSJ International Workshop on Intelligent Robots and Systems*, Yokohama, Japan, pp. 136–143.

[3] Akella, S., and Mason, M. T. 1992. Posing Polygonal Objects in the Plane by Pushing. *Proceedings of the 1992 IEEE International Conference on Robotics and Automation*, Nice, France, pp. 2255–2262.

[4] Bicchi, A., Melchiorri, C., and Balluchi, D. 1995. On the Mobility and Manipulability of General Multiple Limb Robots. *IEEE Transactions on Robotics and Automation.* **11**(2):215–228.

[5] Bicchi, A., and Sorrentino, R. 1995. Dexterous Manipulation Through Rolling. *Proceedings of the 1995 IEEE International Conference on Robotics and Automation*, Nagoya, Japan, Vol 1, pp. 452–457.

[6] Brock, D. L. 1988. Enhancing the Dexterity of a Robot Hand Using Controlled Slip. *Proceedings of the 1988 IEEE International Conference on Robotics and Automation*, Philadelphia, Pennsylvania, pp. 249–251.

[7] Brokowski, M., Peshkin, M., and Goldberg, K. 1993. Curved Fences for Part Alignment. *Proceedings of the 1993 IEEE International Conference on Robotics and Automation*, Atlanta, Georgia, Vol 3, pp. 467–473.

[8] Brost, R. C. 1988. Automatic Grasp Planning in the Presence of Uncertainty. *International Journal of Robotics Research.* **7**(1):3–17.

[9] Brost, R. C., and Mason, M. T. 1989. Graphical Analysis of Planar Rigid-Body Dynamics. *Fifth International Symposium on Robotics Research*, Tokyo.

[10] Carlisle, B., Goldberg, K., Rao, A., and Wiegley, J. 1994. A Pivoting Gripper for Feeding Industrial Parts. *Proc. 1994 IEEE Intl. Conference on Robotics and Automation*, San Diego, California, pp. 1650–1655.

[11] Cole, A. A., Hsu, P., Sastry, S. S. 1992. Dynamic Control of Sliding by Robot Hands for Regrasping. *IEEE Transactions on Robotics and Automation.* **8**(1):42–52.

[12] Donald, B. R., Jennings, J., and Rus, D. 1994. Analyzing Teams of Cooperating Mobile Robots. *Proceedings of the 1994 IEEE International Conference on Robotics and Automation*, San Diego, California, pp. 1896–1903.

[13] Erdmann, M. A. 1994. On a Representation of Friction in Configuration Space. *International Journal of Robotics Research.* **13**(3):240–271.

[14] Fearing, R. S. 1984. Simplified Grasping and Manipulation with Dextrous Robot Hands. *Proceedings of the 1984 American Control Conference*, pp. 32–38.

[15] Kao, I., and Cutkosky, M. R. 1989. Dextrous Manipulation with Compliance and Sliding. *Fifth International Symposium on Robotics Research*, August 28–31, 1989, Tokyo.

[16] Kao, I., and Cutkosky, M. R. 1992. Quasistatic Manipulation with Compliance and Sliding. *International Journal of Robotics Research.* **11**(1):20–40.

[17] Kao, I., and Cutkosky, M. R. 1993. Comparison of Theoretical and Experimental Force/Motion Trajectories for Dextrous Manipulation With Sliding. *International Journal of Robotics Research.* **12**(6):529–534.

[18] Lozano-Pérez, T. 1983. Spatial Planning: A Configuration Space Approach. *IEEE Transactions on Computers.* **C-32**(2):108–120.

[19] Lynch, K. M. 1992. The Mechanics of Fine Manipulation by Pushing. *Proceedings of the 1992 IEEE International Conference on Robotics and Automation*, Nice, France, pp. 2269–2276.

[20] Lynch, K. M., and Mason, M. T. 1995. Pulling by Pushing, Slip with Infinite Friction, and Perfectly Rough Surfaces. *International Journal of Robotics Research.* **14**(2):174–183.

[21] Mason, M. T. 1982. Manipulator Grasping and Pushing Operations. AI-TR-690. Ph.D. thesis. Massachusetts Institute of Technology, Artificial Intelligence Laboratory.

[22] Mason, M. T. 1986. Mechanics and Planning of Manipulator Pushing Operations. *International Journal of Robotics Research.* **5**(3):53–71.

[23] Mason, M. T., and Lynch, K. M. 1993. Dynamic Manipulation. *Proceedings of the 1993 IEEE/RSJ International Workshop on Intelligent Robots and Systems*, Yokohama, Japan, pp. 152–159.

[24] Nguyen, V-D. 1988. Constructing Force-Closure Grasps. *International Journal of Robotics Research.* **7**(3):3–16.

[25] Paljug, E., and Yun, X. 1993. Experimental Results of Two Robot Arms Manipulating Large Objects. *Proceedings of the 1993 IEEE International Conference on Robotics and Automation*, Atlanta, Georgia, Vol 1, pp. 517–522.

[26] Paljug, E., Yun, X., and Kumar, V. 1994. Control of Rolling Contacts in Multi-Arm Manipulation. *IEEE Transactions on Robotics and Automation.* **10**(4):441–452.

[27] Peshkin, M. A. 1986. Planning Robotic Manipulation Strategies for Sliding Objects. Ph.D. thesis. Carnegie Mellon University, Physics Department.

[28] Peshkin, M. A., and Sanderson, A. C. 1988. Planning Robotic Manipulation Strategies for Workpieces that Slide. *IEEE Journal of Robotics and Automation.* **4**(5): 524–531.

[29] Pingle, K., Paul, R., and Bolles, R. 1974. *Programmable Assembly, Three Short Examples.* Film. Stanford Artificial Intelligence Laboratory.

[30] Salisbury, J. K. 1987. Whole Arm Manipulation. *Proceedings 4th International Symposium of Robotics Research*, pp. 183–189.

[31] Sawasaki, N., Inaba, M., and Inoue, H. 1989. Tumbling Objects Using Multi-Fingered Robot. *Proceedings of the 20th ISIR*, pp. 609-616.

[32] Trinkle, J. C. 1992. On the Stability and Instantaneous Velocity of Grasped Frictionless Objects. *IEEE Transactions on Robotics and Automation.* **8**(5):560–572.

[33] Trinkle, J. C., Abel, J. M., and Paul, R. P. 1988. An Investigation of Frictionless Enveloping Grasping in the Plane. *International Journal of Robotics Research.* **7**(3):33–51.

[34] Trinkle, J. C., and Paul, R. P. 1990. Planning for Dexterous Manipulation with Sliding Contacts. *International Journal of Robotics Research.* **9**(3):24–48.

[35] Yun, X. 1993. Object Handling Using Two Arms Without Grasping. *International Journal of Robotics Research.* **12**(1):99–106.

[36] Yoshikawa, T., Yokokohji, Y., and Nagayama, A. 1993. Object Handling by Three-Fingered Hands Using Slip Motion. *Proceedings of the 1993 IEEE/RSJ International Workshop on Intelligent Robots and Systems*, Yokohama, Japan, pp. 99–105.

[37] Zumel, N. B., and Erdmann, M. A. 1996. Nonprehensile Two Palm Manipulation with Non-Equilibrium Transitions between Stable States. *Proceedings of the 1996 IEEE International Conference on Robotics and Automation*, Minneapolis, Minnesota, pp. 3317–3323.

Upper and Lower Bounds for Programmable Vector Fields with Applications to MEMS and Vibratory Plate Parts Feeders

Karl-Friedrich Böhringer, Bruce Randall Donald, Noel C. MacDonald
Cornell University, Ithaca, New York, USA

Programmable vector fields can be used to control a variety of flexible planar parts feeders. These devices can exploit exotic actuation technologies such as arrayed, massively-parallel microfabricated motion pixels or transversely vibrating (macroscopic) plates. These new automation designs promise great flexibility, speed, and dexterity—we believe they may be employed to orient, singulate, sort, feed, and assemble parts. However, since they have only recently been invented, programming and controlling them for manipulation tasks is challenging. When a part is placed on our devices, the programmed vector field induces a force and moment upon it. Over time, the part may come to rest in a dynamic equilibrium state. By chaining together sequences of vector fields, the equilibrium states of a part in the field may be cascaded to obtain a desired final state. The resulting strategies require no sensing and enjoy efficient planning algorithms.

This paper begins by describing our experimental devices. In particular, we describe our progress in building the M-Chip (Manipulation Chip), a massively parallel array of programmable micro-motion pixels. As proof of concept, we demonstrate a prototype M-Chip containing over 11,000 silicon actuators in one square inch. Both the M-Chip, as well as macroscopic devices such as transversely vibrating plates, may be programmed with vector fields, and their behavior predicted and controlled using our equilibrium analysis. We demonstrate lower bounds (i.e., impossibility results) on what the devices cannot do, and results on a classification of control strategies yielding design criteria by which well-behaved manipulation strategies may be developed. We provide sufficient conditions for programmable fields to induce well-behaved equilibria on every part. We define composition operators to build complex strategies from simple ones, and show the resulting fields are also well-behaved. We discuss whether fields outside this class can be useful and free of pathology.

Using these tools, we describe new manipulation algorithms. In particular, we improve existing planning algorithms by a quadratic factor, and the plan-length by a linear factor. Using our new and improved strategies, we show how to simultaneously orient and pose any part, without sensing, from an arbitrary initial configuration. We relax earlier dynamic and mechanical assumptions to obtain more robust and flexible strategies.

Finally, we consider parts feeders that can only implement a very limited "vocabulary" of vector fields (as opposed to the pixel-wise programmability assumed above). We show how to plan and execute parts-posing and orienting strategies for these devices, but with a significant increase in planning complexity and some sacrifice in completeness guarantees. We discuss the tradeoff between mechanical complexity and planning complexity.

1 Introduction

Programmable vector fields can be used to control a variety of flexible planar parts feeders. These devices often exploit exotic actuation technologies such as arrayed, microfabricated motion pixels [9, 8, 7] or transversely vibrating plates [4]. These new automation designs promise great flexibility, speed, and dexterity—we believe they may be employed to orient, singulate, sort, feed, and assemble parts (see for example Figures 1 and 4). However, since they have only recently been invented, programming and controlling them for manipulation tasks is challenging. Our research goal is to develop a science base for manipulation using programmable vector fields.

When a part is placed on our devices, the programmed vector field induces a force and moment upon it. Over time, the part may come to rest in a dynamic equilibrium state. In principle, we have tremendous flexibility in choosing the vector field, since using

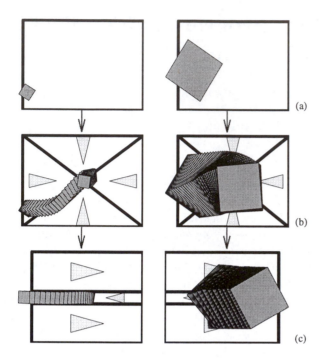

Figure 1: *Sensorless sorting using force vector fields: parts of different sizes are first centered and subsequently separated depending on their size.*

modern array technologies, the force field may be programmed pixel-wise. Hence, we have a lot of control over the resulting equilibrium states. By chaining together sequences of vector fields, the equilibria may be cascaded to obtain a desired final state—for example, this state may represent a unique orientation or pose of the part. A system with such a behavior exhibits the *feeding property* [2]:

> A system has the *feeding property* over a set of parts $\mathcal{P}$ and a set of initial configurations $\mathcal{I}$ if, given any part $P \in \mathcal{P}$, there is some output configuration $\mathbf{q}$ such that the system can move P to $\mathbf{q}$ from any location in $\mathcal{I}$.

Our work on programmable vector fields is related to nonprehensile manipulation [17, 44, 22, 20]: in either case, parts are manipulated without form or force closure.

This paper first describes our experimental devices, a technique for analyzing them called *equilibrium analysis*, lower bounds (i.e., impossibility results) on what the devices *cannot* do, and results on a classification

of control strategies yielding design criteria by which well-behaved manipulation strategies may be developed. Then we describe new manipulation algorithms using these tools. In particular, we improve existing planning algorithms by a quadratic factor, show how to simultaneously orient and pose a part, and we relax earlier dynamic and mechanical assumptions to obtain more robust and flexible strategies.

We pose the question *Which vector fields are suitable for manipulation strategies?* In particular, we ask whether the fields may be *classified.* That is: can we characterize all those vector fields in which every part has stable equilibria? While this question has been well-studied for a point mass in a field, the issue is more subtle when lifted to a body with finite area, due to the moment covector. To answer, we first demonstrate impossibility results, in the form of "lower bounds:" there exist perfectly plausible fields which induce *no* stable equilibrium in very simple parts.

Fortunately, there is also good news. We present conditions for fields to induce well-behaved equilibria when lifted, by exploiting the theory of potential fields. While potential fields have been widely used in robot control [30, 39, 38], micro-actuator arrays present us with the ability to *explicitly* program the applied force *at every point* in a vector field. Whereas previous work has developed control strategies with *artificial* potential fields, our fields are non-artificial (i.e., physical). This alone makes our application of potential field theory to micro-devices unique and novel. Moreover, such fields can be composed using addition, sequential composition, "parallel" composition by superposition of controls, or by a new kind of "morphing" of control signals which we will define.

Finally, the desire to implement complicated fields raises the question of control uncertainty. We close by describing how families of potential functions can be used to represent control uncertainty, and analyzed for their impact on equilibria, and we will give an outlook on still open problems and future work.

Because of limited space, we have abbreviated or omitted some of the proofs of our propositions. For a more detailed discussion please refer to the on-line version of our long paper at URL `http://www.cs.cornell.edu/home/karl/ProgVecFields`, or to [3].

2 Experimental Apparatus: Parts Feeders

It is often extremely costly to maintain part order throughout the manufacture cycle. For example, instead of keeping parts in pallets, they are often delivered in bags or boxes, whence they must be picked out and sorted. A parts feeder is a machine that orients such parts before they are fed to an assembly station. Currently, the design of parts feeders is a black art that is responsible for up to 30% of the cost and 50% of workcell failures [10, for example]. Thus although part feeding accounts for a large portion of assembly cost, there is not much scientific basis for automating the process.

The most common type of parts feeder is the *vibratory bowl feeder*, where parts in a bowl are vibrated using a rotary motion, so that they climb a helical track. As they climb, a sequence of baffles and cutouts in the track create a mechanical "filter" that causes parts in all but one orientation to fall back into the bowl for another attempt at running the gauntlet [10]. Sony's APOS parts feeder [27] uses an array of nests (silhouette traps) cut into a vibrating plate. The nests and the vibratory motion are designed so that the part will remain in the nest only in a particular orientation. By tilting the plate and letting parts flow across it, the nests eventually fill up with parts in the desired orientation. Although the vibratory motion is under software control, specialized mechanical nests must be designed for each part [36].

The reason for the success of vibratory bowl feeders and the Sony APOS system is the underlying principle of *sensorless manipulation* [21] that allows parts positioning and orienting without sensor feedback. This principle is even more important at small scales, because sensor data will be less accurate and more difficult to obtain. The APOS system or bowl feeders are unlikely to work in the micro domain: instead novel device designs for micro-manipulation tasks are required. The theory of sensorless manipulation is the science base for developing and controlling such devices.

Reducing the amount of required sensing is an example of *minimalism* [5, 14], which pursues the following agenda: For a given robot task, find the minimal configuration of resources required to solve the task. Minimalism is interesting because doing task A without

Figure 2: *A prototype* M-CHIP *fabricated in 1995. A large unidirectional actuator array (scanning electron microscopy). Each actuator is* $180 \times 240\,\mu m^2$ *in size. Detail from a* $1\,in^2$ *array with more than 11,000 actuators. For more pictures on device design and fabrication see URL* http://www.cs.cornell.edu/home /karl/MicroActuators.

resource B proves that B is somehow inessential to the information structure of the task. In robotics, minimalism has become increasingly influential. Raibert [37] showed that walking and running machines could be built without static stability. Erdmann and Mason [21] showed how to do dexterous manipulation without sensing. McGeer [34] built a biped, kneed walker without sensors, computers, or actuators. Canny and Goldberg [14] argue that minimalism has a long tradition in industrial manufacturing, and developed geometric algorithms for orienting parts using simple grippers and accurate, low cost light beams. Brooks [12] has developed online algorithms that rely less extensively on planning and world models. Donald et al. [17, 5] have built distributed teams of mobile robots that cooperate in manipulation without explicit communication. We intend to use these results for our experiments in micro-manipulation, and to examine how they relate to our theoretical proofs of minimalist systems.

2.1 Microfabricated Actuator Arrays

A wide variety of micromechanical structures (devices with features in the μm range) has been built recently by using processing techniques known from VLSI industry. However, the fabrication, control, and programming of micro-devices that can interact and actively change their environment remains challenging. Problems arise from

1. unknown material properties and the lack of adequate models for mechanisms at very small scales,

2. the limited range of motion and force that can be generated with microactuators,

3. the lack of sufficient sensor information with regard to manipulation tasks, and

4. design limitations and geometric tolerances due to the fabrication process.

MEMS manipulator arrays have been proposed by several MEMS researchers, among others Fujita et al. [24], Will et al. [31], or Suh et al. [42]. For an overview see [31] or [9, 8]. Our arrays (Figure 2) are fabricated using a SCREAM (Single-Crystal Reactive Etching and Metallization) process developed in the Cornell Nanofabrication Facility [43, 41]. The SCREAM process is low-temperature, and does not interfere with traditional VLSI [40]. Hence it opens the door to building monolithic microelectromechanical systems with integrated microactuators and control circuitry on the same wafer.

Our design is based on microfabricated torsional resonators [35]. Each unit device consists of a rectangular grid etched out of single-crystal silicon suspended by two rods that act as torsional springs (Figure 2). The grid is about $200\,\mu m$ long and extends $120\,\mu m$ on each side of the rod. The rods are $150\,\mu m$ long. The current asymmetric design has $5\,\mu m$ high protruding tips on one side of the grid that make contact with an object lying on top of the actuator. The other side of the actuator consists of a denser grid above an aluminum electrode. If a voltage is applied between silicon substrate and electrode, the dense grid above the electrode is pulled downward by the resulting electrostatic force. Simultaneously the other side of the device (with the tips) is deflected out of the plane by several μm. Hence an object can be lifted and pushed sideways by the actuator.

The fabrication process and mechanism analysis are described in more detail in [9, 8, 7].

2.2 Macroscopic Vibratory Parts Feeder

Böhringer et al. [4] have presented a device that uses the force field created by transverse vibrations of a

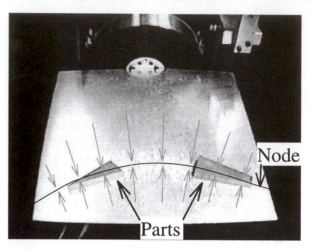

Figure 3: *Vibratory parts feeder: an aluminum plate (size 50 cm × 40 cm) exhibits a vibratory minimum. Parts are attracted to this* nodal line *and reach equilibrium there. See also URL* `http://www.cs.cornell.edu /home/karl/VibratoryAlign`*. Reproduced with permission from [4].*

plate to position and align parts. The device consists of an aluminum plate that is attached to a commercially available electrodynamic vibration generator,[1] with a linear travel of $0.02\,m$, and capable of producing a force of up to $500\,N$ (Figure 3). The input signal, specifying the waveform corresponding to the desired oscillations, is fed to a single coil armature, which moves in a constant field produced by a ceramic permanent magnet in a center gap configuration.

For low amplitudes and frequencies, the plate moves longitudinally with no perceptible transverse vibrations. However, as the frequency of oscillations is increased, transverse vibrations of the plate become more pronounced. The resulting motion is similar to the forced transverse vibration of a rectangular plate, clamped on one edge and free along the other three sides. This vibratory motion creates a force field in which particles are attracted to locations with minimal vibration, called the *nodal lines*. This field can be programmed by changing the frequency, or by employing clamps as programmable fixtures that create various vibratory nodes.

---

[1]Model VT-100G, Vibration Test Systems, Akron, OH, USA.

Figure 3 shows two parts, shaped like a triangle and a trapezoid, after they have reached their stable poses. To better illustrate the orienting effect, the curve showing the nodal line has been drawn by hand. *Nota bene*: This device can only use the finite manipulation grammar described in Section 6.2 since it can only generate a constrained set of vibratory patterns, and cannot implement radial strategies.

3 Equilibrium Analysis For Programmable Vector Fields

For the generation of manipulation strategies with programmable vector fields it is essential to be able to predict the motion of a part in the field. Particularly important is determining the stable equilibrium poses a part can reach in which all forces and moments are balanced. This *equilibrium analysis* was introduced in our short conference paper [8], where we presented a theory of manipulation for programmable vector fields, and an algorithm that generates manipulation strategies to orient polygonal parts without sensor feedback using a sequence of *squeeze fields*.

We now review the algorithm in [8] and give a detailed proof of its complexity bounds. The tools developed here are essential to understanding our new and improved results.

3.1 Squeeze Fields and Equilibria

In [8] we proposed a family of control strategies called *squeeze fields* and a planning algorithm for parts-orientation.

Definition 1 [8] *Assume l is a straight line through the origin. A squeeze field F is a two-dimensional force field defined as follows:*

1. *If $z \in \mathbb{R}^2$ lies on l then $f(z) = 0$.*
2. *If z does not lie on l then $f(z)$ is the unit vector normal to l and pointing towards l.*

We refer to the line l as the *squeeze line*, because l lies in the center of the squeeze field. See Figure 4 for examples of squeeze fields.

Assuming quasi-static motion, a small object will move perpendicularly towards the line l and come to rest there. We are interested in the motion of an arbitrarily shaped (not necessarily small) part P. Let us

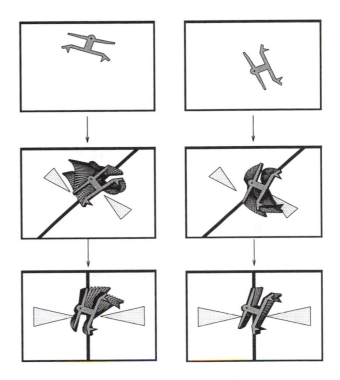

Figure 4: *Sensorless parts alignment using force vector fields: The part reaches unique orientation after two subsequent squeezes. There exist such alignment strategies for all polygonal parts. See URL* http://www.cs.cornell.edu /home/karl/MicroManipulation *for an animated simulation.*

call P_1, P_2 the regions of P that lie to the left and to the right of l, respectively, and C_1, C_2 their centers of mass. In a rest position both translational and rotational forces must be in equilibrium. We obtain the following two conditions:

I: The areas P_1 and P_2 must be equal.
II: The vector $C_2 - C_1$ must be normal to l.

P has a translational motion component normal to l if **I** does not hold. P has a rotational motion component if **II** does not hold. This assumes a uniform force distribution over the surface of P, which is a reasonable assumption for a flat part that is in contact with a large number of elastic actuators.

Definition 2 *A part P is in* translational equilibrium *if the forces acting on P are balanced. P is in* orientational equilibrium *if the moments acting on P are balanced. Total equilibrium is simultaneous translational* and orientational equilibrium.

Let (x_0, y_0, θ_0) be an equilibrium pose of P. (x_0, y_0) is the corresponding translation equilibrium, and θ_0 is the corresponding orientation equilibrium.

Note that conditions **I** and **II** do *not* imply that in equilibrium, the center of area of P has to coincide with the squeeze line l. For example, consider a large and a small square connected by a long rod of negligible width. If the rod is long enough, the center of area will lie outside of the large square. However, in equilibrium the squeeze line l will always intersect the large square.

Definition 3 A bisector *of a polygon P is a line that cuts P into two sections of equal size.*

Proposition 4 *Let P be a polygon whose interior is connected. There exist $O(k n^2)$ bisectors such that P is in equilibrium when placed in a squeeze field such that the bisector coincides with the squeeze line. n is the part complexity measured as the number of polygon vertices. k denotes the maximum number of polygon edges that a bisector can cross.*

If P is convex, then the number of bisectors is bounded by $O(n)$.

Proof: See URL http://www.cs.cornell.edu/home /karl/ProgVecFields, or [3].   □

For most part geometries, k is a small constant.[2] However in the worst-case, pathological parts can reach $k = O(n)$. A (e.g. rectilinear) spiral-shaped part would be an example for such a pathological case, because every bisector intersects $O(n)$ polygon edges.

3.2 Planning of Manipulation Strategies

In this section we present an algorithm for sensorless parts alignment with squeeze fields [8]. Recall from Section 3.1 that in squeeze fields, the equilibria for connected polygons are discrete (except for a neutrally stable translation parallel to the squeeze line which we will disregard for the remainder of Section 3).

To model our actuator arrays and vibratory devices, in [8] we made the following assumptions:

DENSITY: The generated forces can be described by a vector field, i.e. the individual microactuators are dense compared to the size of the moving part.

---
[2]In particular, in [8] we assumed that $k = O(1)$.

2PHASE: The motion of a part has two phases: (1) Pure translation towards l until the part is in translational equilibrium. (2) Motion in translational equilibrium until orientational equilibrium is reached.

Note that due to the elasticity and oscillation of the actuator surfaces, we can assume continuous area contact, and not just contact in three or a few points. If a part moves while in translational equilibrium, in general the motion is not a pure rotation, but also has a translational component. Therefore, relaxing assumption 2PHASE is one of the key results of this paper.

Definition 5 [8] *Let θ be the orientation of a connected polygon P in a squeeze field, and let us assume that condition **I** holds. The turn function $t : \theta \to \{-1, 0, 1\}$ describes the instantaneous rotational motion of P:*

$$t(\theta) = \begin{cases} 1 & \text{if } P \text{ will turn counterclockwise} \\ -1 & \text{if } P \text{ will turn clockwise} \\ 0 & \text{if } P \text{ is in total equilibrium (Fig. 5).} \end{cases}$$

This definition immediately implies the following lemma:

Lemma 6 [8] *Let P be a polygon with orientation θ in a squeeze field such that condition **I** holds. P is stable if $t(\theta) = 0$, $t(\theta+) \leq 0$, and $t(\theta-) \geq 0$. Otherwise P is unstable.*

Proof: Assume the part P is in a pose (x, y, θ) such that condition **I** is satisfied. This implies that the translational forces acting on P balance out. If in addition $t(\theta) = 0$, then the effective moment is zero, and P is in total equilibrium. Now consider a small perturbation $\delta\theta > 0$ of the orientation θ of P while condition **I** is still satisfied. For a stable equilibrium, the moment resulting from the perturbation $\delta\theta$ must not aggravate but rather counteract the perturbation. This is true if and only if $t(\theta + \delta\theta) \leq 0$ and $t(\theta - \delta\theta) \geq 0$.   □

Using this lemma we can identify all stable orientations, which allows us to construct the squeeze function [25] of P (see Figure 5c), i.e. the mapping from an initial orientation of P to the stable equilibrium orientation that it will reach in the squeeze field:

Lemma 7 [8] *Let P be a polygonal part on an actuator array A such that assumptions DENSITY and 2PHASE hold. Given the turn function t of P, its corresponding squeeze function $s : \mathbb{S}^1 \to \mathbb{S}^1$ is constructed as follows:*

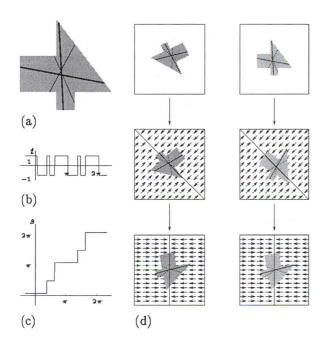

(a)

(b)

(c)   (d)

Figure 5: *(a) Polygonal part. Stable (thick line) and unstable (thin line) bisectors are also shown. (b) Turn function, which predicts the orientations of the stable and unstable bisectors. (c) Squeeze function, constructed from the turn function. (d) Alignment strategy for two arbitrary initial configurations. See URL* http://www.cs.cornell.edu /home/karl/Cinema *for an animated simulation.*

1. *All stable equilibrium orientations θ map identically to θ.*
2. *All unstable equilibrium orientations map (by convention) to the nearest counterclockwise stable orientation.*
3. *All orientations θ with t(θ) = 1 (−1) map to the nearest counterclockwise (clockwise) stable orientation.*

Then s describes the orientation transition of P induced by A.

Proof: Assume that part P initially is in pose (x, y, θ) in array A. Because of 2PHASE, we can assume that P translates towards the center line l until condition **I** is satisfied without changing its orientation θ. P will change its orientation until the moment is zero, i.e. $t = 0$: A positive moment $(t > 0)$ causes counterclockwise motion, and a negative moment $(t < 0)$ causes clockwise motion until the next root of t is reached. □

We conclude that any connected polygonal part, when put in a squeeze field, reaches one of a *finite* number of possible orientation equilibria [8]. The motion of the part and, in particular, the mapping between initial orientation and equilibrium orientation is described by the squeeze function, which is derived from the turn function as described in Lemma 7. Note that all squeeze functions derived from turn functions are monotone step-shaped functions.

Goldberg [25] has given an algorithm that automatically synthesizes a manipulation strategy to uniquely orient a part, given its squeeze function. While Goldberg's algorithm was designed for squeezes with a robotic parallel-jaw gripper, in fact, it is more general, and can be used for arbitrary monotone step-shaped squeeze functions. The output of Goldberg's algorithm is a sequence of angles that specify the required directions of the squeezes. Hence these angles specify the direction of the squeeze line in our force vector fields (for example the two-step strategy in Figures 4 and 5d).

It is important to note that the equilibria obtained by a MEMS squeeze field and by a parallel-jaw gripper will typically be different, even when the squeeze directions are identical. For example, to see this, consider squeezing a square-shaped part. Stable and unstable equilibria are switched. This shows that our mechanical analysis of equilibrium is different from that of the parallel-jaw gripper. Let us summarize these results:

Theorem 8 [8] *Let P be a polygon whose interior is connected. There exists an alignment strategy consisting of a sequence of squeeze fields that uniquely orients P up to symmetries.*

Corollary 9 *The alignment strategies of Theorem 8 have $O(k n^2)$ steps, and they may be computed in time $O(k^2 n^4)$, where k is the maximum number of edges that a bisector of P can cross. In the case where P is convex, the alignment strategy has $O(n)$ steps and can be computed in time $O(n^2)$.*

Proof: Proposition 4 states that a polygon with n vertices has $E = O(k n^2)$ stable orientation equilibria in a squeeze field ($O(n)$ if P is convex). This means that the image of its corresponding squeeze function is a set of E discrete values. Given such a squeeze function, Goldberg's algorithm [25] constructs alignment strategies with $O(E)$ steps. Planning complexity is $O(E^2)$. □

Goldberg's strategies [25] have the same complexity bounds for convex and non-convex parts, because when using squeeze grasps with a parallel-jaw gripper, only the convex hull of the part need be considered. This is not the case for programmable vector fields, where manipulation strategies for non-convex parts are more expensive. As described in Section 3.1, there could be parts that have $E = \Omega(k\,n^2)$ orientation equilibria in a squeeze field, which would imply alignment strategies of length $\Omega(k\,n^2)$ and planning complexity $\Omega(k^2\,n^4)$. In Section 6.1 we will present new and improved manipulation algorithms that reduce the number of equilibria to $E = O(k\,n)$.

This scheme may be generalized to the case where l is slightly curved, as in the "node" of the vibrating plate in Figure 3. See [4] for details. The remaining sections of this paper investigate using more exotic fields (not simple squeeze patterns) to

1. allow disconnected polygons,
2. relax assumption 2PHASE,
3. reduce the planning complexity,
4. reduce the number of equilibria,
5. reduce the execution complexity (strategy length), and
6. determine feasibility results and limitations for manipulation with general force fields.

3.3   Relaxing the 2PHASE Assumption

In Section 3.2, assumption 2PHASE allowed us to determine successive equilibrium positions in a sequence of squeezes, by a quasi-static analysis that decouples translational and rotational motion of the moving part. For any part, this provides a *unique* orientation equilibrium (after several steps). If 2PHASE is relaxed, we obtain a dynamic manipulation problem, in which we must determine the equilibria (x, θ) given by the part orientation θ and the offset x of its center of mass from the squeeze line. A stable equilibrium is a (x_i, θ_i) pair in $\mathbb{R} \times \mathbb{S}^1$ that acts as an *attractor* (the x offset in an equilibrium is, surprisingly, usually not 0). Again, we can compute these (x_i, θ_i) equilibrium pairs *exactly*, as outlined in Section 3.1.

Considering (x_i, θ_i) equilibrium pairs has another advantage. We can show that, even without 2PHASE, after two successive, orthogonal squeezes, the set of stable poses of any part can be reduced from $\mathcal{C} = \mathbb{R}^2 \times \mathbb{S}^1$ to a *finite* subset of $\mathcal{C}$ (the configuration space of part

P); see Claim 28 below. Subsequent squeezes will preserve the finiteness of the state space. This will significantly reduce the complexity of a task-level motion planner. Hence if assumption 2PHASE is relaxed, this idea still enables us to simplify the general motion planning problem (as formulated e.g. by Lozano-Pérez, Mason, and Taylor in [33]) to that of Erdmann and Mason [21]. Conversely, relaxing assumption 2PHASE raises the complexity from the "linear" planning scheme of Goldberg [25] to the forward-chaining searches of Erdmann and Mason [21], or Donald [16].

4   Lower Bounds: What Programmable Vector Fields <u>Cannot</u> Do

We now present "lower bounds" — constituting vector fields and parts with pathological behavior, making them unusable for manipulation. These counterexamples show that we must be careful in choosing programmable vector fields, and that, *a priori*, it is not obvious when a field is well-behaved.

In Section 3 we saw that in a vector field with a simple squeeze pattern (see again Figure 4), polygonal parts reach certain equilibrium poses. This raises the question of a *general classification of all those vector fields in which every part has stable equilibria*. There exist vector fields that do not have this property even though they are very similar to a simple squeeze.

Definition 10 *A skewed field f_S is a force vector field given by $f_S(x, y) = -\mathrm{sign}(x)(1, \epsilon)$, where $0 \neq \epsilon \in \mathbb{R}$.*

Proposition 11 *A skewed vector field induces no stable equilibrium on a disk-shaped part.*

Proof: Consider Figure 6, which shows a skewed field with $\epsilon = -\frac{2}{3}$: Only when the center of the disk coincides with the center of the squeeze pattern do the translational forces acting on the disk balance. But it will still experience a positive moment that will cause rotation.   □

Similarly we would like to identify the *class of all those parts that always reach stable equilibria* in particular vector fields. From Section 3 we know that connected polygons in simple squeeze fields satisfy this condition. This property relies on finite area contacts: it does not hold for point contacts. As a counterexample consider the part in Figure 7.

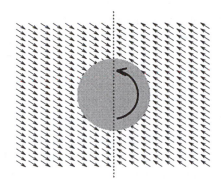

Figure 6: *Unstable part in the skewed squeeze field ($\epsilon = -\frac{2}{3}$). The disk with center on the squeeze line will keep rotating. Moreover, it has no stable equilibrium in this field.*

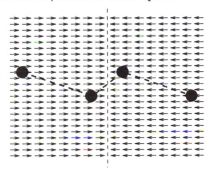

Figure 7: *S-shaped part with four rigidly connected point-contact "feet" in unstable total equilibrium (forces and moments balance). There exists no stable equilibrium position for this part in a vector field with a simple squeeze pattern. For an animation see URL* http://www.cs.cornell.edu /home/karl/MicroManipulation/Patho.

Proposition 12 *There exist parts that do not have stable equilibria in a simple squeeze field.*

Proof: The S-shaped part in Figure 7 has four rigidly connected "feet" with small contact surfaces. As the area of each of these four feet approaches zero, the part has *no* stable equilibrium in a simple squeeze field. There is only one orientation for the part in which both force and moment balances out, and this orientation is unstable. □

Finally, the *number of stable equilibria* of a given part influences both the planning complexity and the plan length of an alignment strategy. It also affects the resolution of the vector field that is necessary to perform a strategy accurately. Even though all parts

we have considered exhibit only one or two orientation equilibria, there exist no tight bounds on the maximum number of orientation equilibria in a unit squeeze field.

Proposition 13 *Let n be the number of vertices of a polygon P, and let k be the maximum number of edges that a bisector of P can cross:*

    A. *Regular polygons have n stable orientation equilibria in a squeeze field.*

    B. *Every connected polygon has $O(k n^2)$ stable orientation equilibria in a squeeze field.*

Proof:

    A. Because of their part symmetry, regular polygons have $2n$ equilibria. Half of them are stable, the other n are unstable.

    B. See Section 3.1.

□

As described in Section 3.1, there exist simple polygons with n vertices that can be bisected by a straight line in up to $O(k n^2)$ topologically different ways [6]. This suggests that there could be parts that have $\Omega(k n^2)$ orientation equilibria in a squeeze field, which would imply alignment strategies of length $\Omega(k n^2)$ and planning complexity $\Omega(k^2 n^4)$.

While the counterexample in Figure 7 may be plausibly avoided by prohibiting parts with "point contacts," the other examples (Figure 6 and Proposition 13B) are more problematic. In Section 5, we show how to choose programmable vector fields that exclude some of these pathological behaviors, by using the *theory of potential fields* to describe a class of force vector fields for which *all* polygonal parts have stable equilibria. In Section 6.1, we show how to combine these fields to obtain new fields in which all parts have only $O(k n)$ equilibria.

We believe parts with point contact (not having finite area contact) will behave badly in *all* vector fields. We can model a point contact with delta functions, such that e.g. for a point contact P_0 at (x_0, y_0):
$$\int_{P_0} f \, dA = \int f \, \delta(x_0, y_0) \, dA = f(x_0, y_0).$$ This model is frequently used in mechanics (see e.g. [19]). Point contact permits rapid, discontinuous changes in force and moment. Hence, bodies with point contact will tend to exhibit instabilities, as opposed to flat parts

that are in contact with a large number of (elastic) actuators. Finally, we believe that as the area contact—the size of the "feet" of a part—approaches zero, the part may become unstable. This represents a design constraint on parts which are to be manipulated using programmable planar parts feeders.

The lower bounds we demonstrate are indications of the pathologies that can arise when fields without potential or parts with point contact are permitted. Each of our counterexamples (Figures 6 and 7) is "generic" in that it can be generalized to a very large class of similar examples. However, these lower bounds are just a first step, and one wishes for examples that delineate the capabilities of programmable vector fields for planar parts manipulation even more precisely.

The separating field shown in Figure 1c is not a potential field, and that there exist parts that will spin forever, without equilibrium, in this field (this follows by generalizing the construction in Figure 6). However, for *specific parts*, such as those shown in Figure 1, this field is useful if we can pose the parts appropriately first (e.g., using the potential field shown in Figure 1b).

Finally, we may "surround" non-potential fields with potential fields to obtain reasonable behavior in some cases. Figure 1 shows how to "surround" a non-potential field in *time* by potential fields, to eliminate pathologies. Similarly, we can surround non-potential fields *spatially*. For example, if field 1c could be surrounded by a larger potential field, then after separation, parts can reach a stable equilibrium.

Non-potential fields can be used safely with the following methodology: Let $H \subset \mathcal{C} = \mathbb{R}^2 \times \mathbb{S}^1$ be the undesirable limit set. For example, H could be a limit cycle where the part spins forever. Let $\widehat{P}_V(H)$ be the weak pre-image [33, 15] of H under the field V. If we can ensure that the part starts in a configuration $z \notin \widehat{P}_V(H)$, it will not reach the unwanted limit cycle. For example, in Figure 1 the centering step (b) ensures that the part does not end up on the border between the two separating fields, where it would spin forever in step (c).

5 Completeness: Classification Using Potential Fields

We are interested in a *general classification of all those vector fields in which every part has stable equilibria.*

As motivation, recall that a skewed vector field, even though very similar to a regular squeeze field (see again Figure 4), induces *no* stable equilibrium in a disk-shaped part (Figure 6). In this section we give a family of vector fields that will be useful for manipulation tasks. These fields belong to a specific class of vector fields: the class of fields that have a potential.

We believe that fields without potential will often induce pathological behavior in many parts. Fields without potential admit paths along which a particle (point mass) will gain energy. Since mechanical parts are rigid aggregations of particles, this may induce unstable behavior in larger bodies. However, there are some cases where non-potential fields may be useful. For example, see Figure 1c, which is *not* a potential field. Such fields may be employed to *separate* but not to stabilize, pose, or orient parts. This strong statement devolves to our proof that fields like Figure 6 do not have well-behaved equilibria. Hence, they should only be employed when we want to induce an unstable system that will cast parts away from equilibrium, e.g. in order to sort or separate them.

Consider the class of vector fields on $\mathbb{R}^2$ that have a *potential*, i.e. fields f in which the work is independent of the path, or equivalently, the work on any closed path is zero, $\oint f \cdot ds = 0$. In a potential field each point (x, y) is assigned a real value $U(x, y)$ that can be interpreted as its potential energy. When U is smooth, then the vector field f associated with U is the gradient $-\nabla U$. In general, $U(x, y)$ is given, up to an additive constant, by the path integral $\int_\alpha f \cdot ds$ (when it exists and it is unique), where α is an arbitrary path from a fixed reference point (x_0, y_0) to (x, y).

An ideal point object is in stable equilibrium if and only if it is at a local minimum of U.

Definition 14 *Let f be a force vector field on $\mathbb{R}^2$, and let $\mathbf{p}$ be a point that is offset from a fixed reference point $\mathbf{q}$ by a vector $\mathbf{r}(\mathbf{p}) = \mathbf{p} - \mathbf{q}$. We define the generalized force F as the force and moment induced by f at point $\mathbf{p}$:*

$$F(\mathbf{p}) = (f(\mathbf{p}), \mathbf{r}(\mathbf{p}) \times f(\mathbf{p})) \qquad (1)$$

Let P be a part of arbitrary shape, and let $P_{\mathbf{z}}$ denote the part P in pose $\mathbf{z} = (x, y, \theta) \in \mathcal{C}$. We define the lifted force field f_P as the area integral of the force induced by f over $P_\mathbf{z}$:

$$f_P(\mathbf{z}) = \int_{P_\mathbf{z}} f \, dA \qquad (2)$$

The lifted generalized force field F_P is defined as the area integral of the force and moment induced by f over P in configuration $\mathbf{z}$:

$$F_P(\mathbf{z}) = \int_{P_\mathbf{z}} F \, dA$$
$$= \left(\int_{P_\mathbf{z}} f \, dA , \int_{P_\mathbf{z}} \mathbf{r} \times f \, dA \right) \qquad (3)$$

Hence, F_P is a vector field on $\mathcal{C}$. Finally, we define the lifted potential $U_P : \mathcal{C} \to \mathbb{R}$. U_P is the area integral of the potential U over P in configuration $\mathbf{z}$:

$$U_P(\mathbf{z}) = \int_{P_\mathbf{z}} U \, dA \qquad (4)$$

We now show that the category of potential fields is closed under the operation of lifting, and that U_P is the potential of F_P. Note that U need not be smooth.

Let $g : X \to Y$ and $h : Y \to Z$. Let $k : X \to Z$ be the function which is the composition of g and h, defined by $k(x) = h(g(x))$. In the following proposition, we use the notation $h(g)$ to denote k, the function composition of g and h.

Proposition 15 *Let f be a force field on $\mathbb{R}^2$ with potential U, and let P be a part of arbitrary shape. For the lifted generalized force field F_P and the lifted potential U_P the following equality holds: $U_P = \int_P U \, dA = \int_\alpha F_P \cdot d\mathbf{z} + c$, where α is an arbitrary path in $\mathcal{C}$ from a fixed reference point, and c is a constant.*

Proof: See URL http://www.cs.cornell.edu/home/karl/ProgVecFields, or [3]. □

So again, $U_P(x, y, \theta)$ can be interpreted as the potential energy of part P in configuration (x, y, θ). Therefore we obtain a lifted potential field U_P whose local minima are the stable equilibrium configurations in $\mathcal{C}$ for part P. Furthermore, potential fields are closed under addition and scaling. We can thus create and analyze more complex fields by looking at their components. In general, the theory of potential fields allows us to classify manipulation strategies with vector

fields, offering new insights into equilibrium analysis and providing the means to determine strategies with stable equilibria. For example, it allows us to show that orientation equilibrium in a simple squeeze field is equivalent to the stability of a homogeneous boat floating in water, provided its density is $\rho = \frac{1}{2}\rho_{\text{water}}$.

5.1 Examples: Classification of Force Fields

Example: Radial fields. A *radial field* is a vector field whose forces are directed towards a specific center point. It can be used to center a part in the plane. The field in Figure 1b can be understood as a radial field with a rather coarse discretization using only four different force directions. Note that this field has a potential.

As a specific example for radial fields, consider the *unit radial field R* which is defined by $R(\mathbf{z}) = -\mathbf{z}/\|\mathbf{z}\|$ for $\mathbf{z} \neq 0$, and $R(0) = 0$. Note that R has a discontinuity at the origin. A smooth radial field can be defined, for example, by $R'(\mathbf{z}) = -\mathbf{z}$. The corresponding potential fields are $U(\mathbf{z}) = \|\mathbf{z}\|$, and $U'(\mathbf{z}) = \frac{1}{2}\|\mathbf{z}\|^2$, respectively. Note that U is continuous (but not smooth), while U' is smooth.

Counterexample: Skewed squeeze fields. Consider again the *skewed squeeze field* in Figure 6. This is not a potential field, which explains why the disk-shaped part has no equilibrium: Note that for example the integral on a cyclic path along the boundary of the disk is non-zero.

Example: Morphing and combining vector fields. Our strategies from [8] (see Section 3) have *switch points* in time where the vector field changes discontinuously (Figure 4). This is because after one squeeze, for every part, the orientation equilibria form a finite set of possible configurations, but in general there exists no unique equilibrium (as shown in Section 3.2). Hence subsequent squeezes are needed to disambiguate the part orientation. Therefore these switches are necessary for strategies with squeeze patterns.

One may ask whether, using another class of potential field strategies, *unique* equilibria may be obtained without discrete switching. We believe that *continuously varying* vector fields of the form $(1-t)f + tg$, where $t \in [0,1]$ represents time, and f and g are squeezes, may lead to vector fields that have this property. Here "+" denotes point-wise addition of vector fields, and we will write "$f \rightsquigarrow g$" for the resulting continuously varying field. By restricting f and g to be fields with potentials U and V, we know that $U + V$ and $(1 - t)U + tV$ are potential fields, and hence we can guarantee that $f + g$ and $f \rightsquigarrow g$ are well-behaved strategies. These form the basis of our new algorithms in Section 6.

Let us formalize the previous paragraphs. If f is a vector field (in this case a squeeze pattern) that is applied to move part P, we define the *equilibrium set* $E_P(f)$ as the subset of the configuration space $\mathcal{C}$ for which P is in equilibrium. Let us write $f * g$ for a strategy that first applies vector field f, and then vector field g to move part P. $f + g$ can be understood as applying f and g simultaneously. We have shown that in general $E_P(f)$ is not finite, but for two *orthogonal* squeezes f and g, the discrete switching strategy $f * g$ yields a finite equilibrium set $E_P(f * g)$ (see Section 6.2, Claim 28). Furthermore, for some parts the equilibrium is unique up to symmetry.

We wish to explore the relationship between equilibria in simple vector fields $E_P(f)$ or $E_P(g)$, combined fields $E_P(f + g)$, discretely-switched fields $E_P(f * g)$, and continuously varying fields $E_P(f \rightsquigarrow g)$. For example, one may ask whether there exists a strategy with combined vector fields, or continuously varying fields, that, in just one step, reaches the same equilibrium as a discretely switched strategy requiring multiple steps. Finally, let $f_1 * f_2 * \cdots * f_s$ be a sequence of squeeze fields guaranteed to uniquely orient a part P under assumption 2PHASE. We wish to investigate how continuously varying strategies such as $f_1 \rightsquigarrow f_2 \rightsquigarrow \cdots \rightsquigarrow f_s$ can be employed to dynamically achieve the same equilibria even when 2PHASE is relaxed. The distributed actuation strategy $F * G$ is distributed in space, but not in time. The strategy $F + G$ is parallel with respect to space and time, since F and G are simultaneously "run." Research in this area could lead to a *theory of parallel distributed manipulation* that describes *spatially distributed* manipulation tasks that can be *parallelized over time and space* by superposition of controls.

5.2 Upward-Shaped Potential Fields

So far we have presented specific force fields that *always* (e.g. squeeze and radial fields) or *never* (e.g.

skewed squeeze fields) induce stable equilibria on certain classes of parts. We conclude this section with a criterion that provides a sufficient condition on force fields such that *all parts of a certain size reach a stable equilibrium.*

We have observed in Section 4 that *a priori* it is not obvious when a force field induces stable equilibria. Our Equilibrium Criterion will be based on two important properties:

1. The field has a potential. Potential fields do not allow closed paths (technically, limit cycles) along which the work is positive, which could induce infinite motion of a part.
2. The force field is "inward-directed," which implies that (assuming first-order dynamics) parts can never leave a certain region R. This useful property is a direct consequence of the definition of inward-directedness. An inward-directed force field corresponds to an "upward-shaped" potential, in which all paths that leave region R have an ascending slope.

We will require Property (1.) to hold for the entire force field, while Property (2.) devolves to a boundary condition.

5.2.1 Elementary Definitions

Definition 16 *Let* $\mathbf{z} \in \mathbb{R}^n$. *The ϵ-ball around* $\mathbf{z}$, *denoted* $B_\epsilon(\mathbf{z})$, *is the set* $\{\mathbf{r} \in \mathbb{R}^n \mid |\mathbf{r} - \mathbf{z}| < \epsilon\}$ *of all points within a distance ϵ of* $\mathbf{z}$.

Definition 17 (Lozano-Pérez [32]) *Let* A, B *be sets in* $\mathbb{R}^n$. *The Minkowski sum* $A \oplus B$ *of two sets A and B is defined as the set* $\{\mathbf{a} + \mathbf{b} \mid \mathbf{a} \in A, \mathbf{b} \in B\}$.

From these definitions it follows that for a region R with boundary ∂R, the set $\partial R \oplus B_d(\mathbf{0}) = \{\mathbf{r} + \mathbf{z} \mid \mathbf{r} \in \partial R$, and $|\mathbf{z}| \leq d\}$ comprises all points that are within a distance d from the boundary of R.

Definition 18 *Given a region* $R \subset \mathbb{R}^n$, *define the set* $CI(R, d) = R - (\partial R \oplus B_d(\mathbf{0}))$ *which is the region R shrunk by distance d. Note that $CI(R, d)$ is based upon the configuration space interior [32] of R for $B_d(\mathbf{0})$. Abusing terminology slightly, we call $CI(R, d)$ the configuration space interior of R in this paper.*

Definition 19 *The radius r_P of a part P is the maximum distance between an arbitrary point of P and the center of mass (COM) of P.*

5.2.2 Equilibrium Criterion

We are now able to state a general criterion for a force field f to induce stable equilibria on all parts in a region S. As mentioned at the beginning of Section 5.2, this criterion is based on two main conditions: (1) if f has a potential, limit cycles with positive energy gain are avoided inside S. (2) if f is "inward-directed" (see the definition below), parts cannot leave the region S.

In the following we give a general definition of inward-directed vector fields on a manifold Z. We then specialize the definition to the special instances of $Z = \mathcal{C} = \mathbb{R}^2 \times \mathbb{S}^1$ (the configuration space) and $Z = \mathbb{R}^2$, and give a sufficient, practical condition for inward-directed vector fields. We conclude with the presentation of the Equilibrium Criterion.

Definition 20 (Inward-Directed Force Fields) [3]
Let Z be an arbitrary smooth manifold, and let $Y \subset Z$ be a compact and smooth submanifold with boundary of Z. Assume that ∂Y has codimension 1 in Z, and that the boundary of Y is orientable. Let $q \in \partial Y$ be a point on the boundary of Y, and $V_q \in T_q Z$ be a tangent vector to Z at q.

We say V_q is inward-directed to ∂Y at q if there exists a sufficiently small $\epsilon > 0$ such that $q + \epsilon V_q \in Y$.

Let V be a vector field on Z. We say V is inward-directed to ∂Y if $V(q)$ is inward-directed to ∂Y at q for all $q \in \partial Y$.

Assume the set $S \subset \mathbb{R}^2$ is compact and smooth. Consider the part P when it is placed into the force field f such that its COM lies in S. The set of all such poses is a subset of the configuration space $\mathcal{C} = \mathbb{R}^2 \times \mathbb{S}^1$ which we call $\widetilde{S} = S \times \mathbb{S}^1$. The boundary of $\widetilde{S}$ is $\partial \widetilde{S} = \partial S \times \mathbb{S}^1$. Note that $\partial \widetilde{S}$ separates the interior $i\widetilde{S} = \widetilde{S} - \partial \widetilde{S}$ from the exterior $\mathcal{C} - \widetilde{S} = (\mathbb{R}^2 - S) \times \mathbb{S}^1$, and that $\partial \widetilde{S}$ is isomorphic to a torus $\mathbb{S}^1 \times \mathbb{S}^1$.

Now let $z = (x, y, \theta) \in \partial \widetilde{S}$, and let $F_z \in T_z \mathcal{C}$ represent the lifted generalized force acting on part P in pose z. F_z is *inward-directed* (w.r.t. $\partial \widetilde{S}$) if F_z points into the interior of $\widetilde{S}$. Note that this condition is equivalent to saying that the projection of F_z onto the tangent space at (x, y) to $\mathbb{R}^2$ points into S, because the rotational component of F_z is tangential to $\partial \widetilde{S}$. So for

---

[3]In this definition, for convenience we assume that Z is embedded in $\mathbb{R}^m$ for some m. This condition may be relaxed.

example, if $z = (x, y, \theta) \in \partial \widetilde{S}$, then $z' = (x, y, \theta') \in \partial \widetilde{S}$ for any θ'.

The following proposition gives a simple condition on a force field f that tells us if, for a given part P, its lifted generalized force field F_P is inward-directed:

Proposition 21 *Let P be a part with radius r whose COM is the reference point used to define its configuration space $\mathcal{C} = \mathbb{R}^2 \times \mathbb{S}^1$. Let f be a force vector field defined on a region $R \subset \mathbb{R}^2$, with F_P the corresponding lifted generalized force field. Let $S \subset \mathbb{R}^2$ be a convex, compact, and smooth subset of the configuration space interior of R, and $S \subset CI(R, r)$.*

Consider a point $q \in \partial S$ with outward normal n_q, and a ball $B_r(q)$ with radius r about q. If for every point $q \in \partial S$, and for every point s in the corresponding ball $B_r(q)$, the dot product $g(s) = f(s) \cdot n_q$ is less than 0, then the lifted generalized force field F_P is inward-directed to $\partial \widetilde{S}$ (note: $(\cdot)$ is the standard inner product).

Proof: Consider the part P in pose $z = (x, y, \theta) \in \partial \widetilde{S}$ such that $q = (x, y)$. P has radius r, hence it lies completely inside the ball $B_r(q)$, independent of its orientation θ. As we know that $g(p) = f(p) \cdot n_q < 0$ for all $p \in B_r(q)$, we can conclude that the integral of $g(p)$ over P is also less than 0: $\int_P g(p) \, dA = \int_P f(p) \cdot n_q \, dA = f_P \cdot n_q < 0$. This implies that for f_P, which is the translational component of F_P (see Definition 14), the vector $q + \epsilon f_P(z)$ lies inside S, if ϵ is positive and sufficiently small. As mentioned above in Section 5.2.2, this suffices to ensure that the vector $z + \epsilon F_P(z)$ lies inside $\widetilde{S}$. $\qquad \square$

Lemma 22 (Equilibrium Criterion) *Let P be a part with radius r, let f be a force field with potential U defined on a region $R \subset \mathbb{R}^2$, and let $S \subset R$ as specified in Proposition 21. Let us also assume that the motion of part P is governed by first-order dynamics.*

If the lifted force vector field F_P is inward-directed to $\partial \widetilde{S}$, then the part P will reach a stable equilibrium under f in $i\widetilde{S}$ whenever its COM is initially placed in S.

Proof: Assume that the COM of part P is placed at a point $(x, y) \in S$. This means that P is in some pose $z = (x, y, \theta) \in \widetilde{S}$. We now show that the COM of

P cannot leave S when initially placed inside S. We know that $\partial \widetilde{S}$ separates $i\widetilde{S}$ from $\mathcal{C} - \widetilde{S}$. Hence every path from z to some $z^* \in \mathcal{C} - \widetilde{S}$ must intersect $\partial \widetilde{S}$ at some point $z' \in \partial \widetilde{S}$. Now consider part P in pose z'. Under first-order dynamics, its velocity must be in direction of $F_P(z')$. Because F_P is inward-directed, the velocity of P must be towards $i\widetilde{S}$. In particular, this means that the COM will move into iS, hence P cannot leave S, and that there is no equilibrium on ∂S.

f, and hence (because of Proposition 15) F_P have potential U and U_P, respectively. Therefore limit cycles with energy gain are not possible. Furthermore, $U_P(\widetilde{S})$ is the continuous image of a compact set, $\widetilde{S}$. Therefore the image $U_P(\widetilde{S})$ is a compact subset of $\mathbb{R}$, hence has a minimum value attained by some point $s \in \widetilde{S}$. Since f is inward-directed, s must lie in $i\widetilde{S}$. This minimum is a stable equilibrium of P in f. □

Because of Lemma 22, the use of potential fields is invaluable for the analysis of effective and efficient manipulation strategies, as discussed in the following section. In particular, it is useful for proving the completeness of a manipulation planner.

6 New and Improved Manipulation Algorithms

The part alignment strategies in Section 3.2 have *switch points* in time where the vector field changes discontinuously (Figure 5). We can denote such a *switched strategy* by $f_1 * f_2 * \cdots * f_s$, where the f_i are vector fields. In Section 3.2 we recalled that a strategy to align a (non-convex) polygonal part with n vertices may need up to $O(kn^2)$ switches, and require $O(k^2 n^4)$ time in planning (k is the maximum number of polygon edges that a bisector can cross). To improve these bounds, we now consider a broader class of vector fields including simple squeeze patterns, radial, and combined fields as described in Section 5.

In Section 6.1 we show how, by using radial and combined vector fields, we can significantly reduce the complexity of the strategies from that of Section 3. In Section 6.2 we describe a general planning algorithm that works with a limited "grammar" of vector fields (and yields, correspondingly, less favorable complexity bounds).

6.1 Radial Strategies

Consider a part P in a force field f. Some force fields exhibit rotational symmetry properties that can be used to generate efficient manipulation strategies:

Property 23 *There exists a unique pivot point v of P such that P is in translational equilibrium if and only if v coincides with $\mathbf{0}$.*

Property 24 *There exists a unique pivot point v of P such that P is in (neutrally stable) orientational equilibrium if and only if v coincides with $\mathbf{0}$.*

We typically think of the pivot point v being a point of P; however, in generality, just like the center of mass of P, v does not need to lie within P, but instead is some fixed point relative to the reference frame of P. Now consider the part P in an ideal unit radial force vector field R as described in Section 5.

Proposition 25 *In a unit radial field R, Properties 23 and 24 hold.*

Proof: We fix the part P at an arbitrary orientation θ, and show that at this orientation P has a unique translational equilibrium $v(\theta)$. That is, placing $v(\theta)$ at the origin is necessary and sufficient for P to be in translational equilibrium at orientation θ. Second, we show that for any two distinct orientations θ and θ', $v(\theta) = v(\theta')$. We call this unique point v, dropping the orientation θ. Finally, we argue that whenever P is in translational equilibrium (i.e., v is at the origin), that P is neutrally stable w.r.t. orientation. This follows by the radial symmetry of R.

Consider the translational forces (but not the moments) acting on P in the radial field R. To do this, let us separate R into its x and y components, R_x and R_y, such that $R = (R_x, R_y)$. Assume for now that the orientation of P is fixed. If P is placed at a position $z_0 \in \mathbb{R}^2$, whose x-coordinate is sufficiently negative, the total force induced by R_x on P will point in the positive x direction. Symmetrically, placing P at a sufficiently large positive x coordinate will cause a force in the negative x direction. We claim that, by translating P rigidly with increasing x coordinate, this force decreases continuously and strictly monotonically, and hence has a unique root.

To verify this claim, consider a small area patch $\wp_0$ of P. A uniform translation t of $\wp_0$ in x direction can be described as $\wp(t) = \wp_0 \oplus (z_0 + t\hat{x})$ (with z_0 the initial position of the patch, $\hat{x}$ the unit vector in x direction, and $\oplus$ the Minkowski sum). The total force on $\wp(t)$ in x direction is $\int_{\wp(t)} R_x \, dA$. This force decreases continuously and strictly monotonically with t, because R_x is strictly monotone and continuous everywhere except on the x-axis, which has measure zero in $\mathbb{R}^2$. A similar argument applies for the y direction, and, because of the radial symmetry of R, for any direction.

If we choose the set S as a sufficiently large disk-shaped region around the origin and recall that R has a potential, we can apply Lemma 22 to conclude that there must exist at least one total equilibrium for P. Now assume that there exist two distinct equilibria $e_1 = (x_1, y_1, \theta_1)$ and $e_2 = (x_2, y_2, \theta_2)$ for P in R. We write "$P(e_i)$" to denote that P is in configuration e_i. Because of the radial symmetry of R, we can reorient $P(e_2)$ to $P(e_2')$ such that its orientation is equal to $P(e_1)$: $e_2' = (x_2', y_2', \theta_1)$, where $\begin{pmatrix} x_2' \\ y_2' \end{pmatrix} = M \begin{pmatrix} x_2 \\ y_2 \end{pmatrix}$, and M is a rotation matrix with angle $\theta_1 - \theta_2$. This reorientation does not affect the equilibrium. Note that P can be moved from e_1 to e_2' by a pure translation. From above we know that such a translation of P corresponds to a strictly monotone change in the translational forces acting on P. Hence we conclude that $P(e_1)$ and $P(e_2')$ cannot both be in translational equilibrium unless e_1 and e_2' are equal. This implies that e_1 and e_2 cannot both be equilibria of P in R unless they both have the same pivot point v. $\qquad \square$

Surprisingly, v need not be the center of area of P. For example, consider a large and a small square connected by a long rod of negligible width. The pivot point of this part will lie inside the larger square. But if the rod is long enough, the center of area will lie outside of the larger square. However, the following corollary holds:

Corollary 26 *For a part P in a continuous radial force field R' given by $R'(\mathbf{z}) = -\mathbf{z}$, the pivot point of P coincides with the center of area of P.*

Proof: The force acting on P in R' is given by $F = \int_P -\mathbf{z} \, dA$, which is also the formula for the (negated) center of area. $\qquad \square$

Now suppose that R is combined with a unit squeeze pattern S, which is scaled by a factor $\delta > 0$, resulting in $R + \delta S$. The squeeze component δS of this field will cause the part to align with the squeeze, similarly to the strategies in Section 3.2. But note that the radial component R keeps the part centered in the force field. Hence, by keeping R sufficiently large (or δ small), we can assume that the pivot point of P remains within an ϵ-ball of the center of R. This implies that assumption 2PHASE (see Section 3.2) is no longer necessary. Moreover, ϵ can be made arbitrarily small by an appropriate choice of δ.

Proposition 27 *Let P be a polygonal part with n vertices, and let k be the maximum number of edges that a bisector of P can cross. Let us assume that v, the pivot point of P, is in general position. There are at most $O(k\,n)$ stable equilibria in a field of the form $R + \delta S$ if δ is sufficiently small and positive.*

Proof: For a part in equilibrium in a pure radial field R (i.e., with $\delta = 0$), the pivot point v is essentially fixed at the origin. This is implied by Property 23. It is easy to see that Property 23 is not true in general for arbitrary fields of the form $R + \delta S$. Property 23 holds if $\delta = 0$, because then any orientation is an equilibrium when v is at the center of R. However, Property 24 does not hold if $\delta > 0$, because in general there does not exist a unique pivot point in squeeze fields (see Section 3.2).

We will conduct the combinatorial analysis of the orientation equilibria under the assumptions that (i) $\delta > 0$ and (ii) that v is fixed at the origin. Then we will relax the latter assumption (ii), and show that Property 23 holds, *approximately*, even in $R + \delta S$, for a sufficiently small $\delta > 0$. That is, we show that a sufficiently small δ can be chosen so that the combinatorial analysis is unaffected when assumption (ii) is relaxed.

First, we show that when δ is small but positive, and with v fixed at the center of R, there are only a linear number of orientation equilibria. (I.e., we constrain the pivot point v to remain fixed at the origin until further notice.) So let us assume that we are in a combined radial and small squeeze field $R + \delta S$.

Consider a ray $w(0)$ emanating from v. Assume w.l.o.g. that v is not a vertex of P, and that $w(0)$ intersects the edges $S(0) = \{e_1, \cdots, e_k\}$ of P in general position, $1 \le k \le n$. Parameterize the ray $w(\cdot)$ by its angle ϕ to obtain $w(\phi)$. As ϕ sweeps from 0 to 2π, each edge of P will enter and leave the *crossing* structure $S(\phi)$ exactly once. $S(\phi)$ is updated at *critical angles* where $w(\phi)$ intersects a vertex of P. Since there are n vertices, there are $O(n)$ critical angles, and hence $O(n)$ changes to $S(\phi)$ overall. Hence, since between critical angles $S(\phi)$ is constant, we see that $S(\phi)$ takes on $O(n)$ distinct values. Now place the squeeze line l to coincide with $w(\phi)$. For a given crossing structure $S(\phi) \cup S(\phi + \pi)$, satisfying conditions **I** and **II** as defined in Section 3.2 devolves to solving two algebraic equations of degree k, where k is the maximum number of edges intersected by the squeeze line as described in Section 3.1. This implies that between any two adjacent critical values there are only $O(k)$ orientations of l (given by $w(\phi)$) that satisfy conditions **I** and **II**. Hence, the overall number of orientations satisfying **I** and **II** is $O(kn)$.

If $\delta > 0$ the part P will be perturbed, so that Property 23 is only approximately satisfied. (That is, we now relax the assumption that v is constrained to be at the origin). However, we can ensure that v lies within an ϵ-ball around the origin (the center of the radial field). To see this, first consider P at some arbitrary configuration $\mathbf{z}$ in the squeeze field δS. The total squeeze force on $P_{\mathbf{z}}$ is given by the area integral $\delta S_P(\mathbf{z}) = \int_{P_{\mathbf{z}}} \delta S\, dA$. (Recall that S_P denotes the lifted force field of S; see Definition 14, Equation (2).) Now, δS_P is bounded above by $|\delta S_P| \le \delta A$, where A is the area of P (note that S is a unit squeeze field).

P is in equilibrium with respect to the radial field R if v is at the origin. Now consider the lifted force R_P when the pivot point of P is not at the origin. More specifically, Let $v_{\mathbf{z}}$ be the pivot point of $P_{\mathbf{z}}$, and let us define a function $\widehat{R}_P(d) = \min\{|R_P(\mathbf{z})|$ such that $|v_{\mathbf{z}}| = d\}$, i.e., $\widehat{R}_P(d)$ is the minimum magnitude of the lifted force acting on $P_{\mathbf{z}}$ when its pivot point $v_{\mathbf{z}}$ is at distance d from the origin.

By decomposing R_P into its x- and y-components, we can write $|R_P|$ as $\sqrt{R_{P,x}^2 + R_{P,y}^2}$. Because of the radial symmetry of R let us assume w.l.o.g. that $v_{\mathbf{z}} = (d, 0)$. From the proof of Proposition 25 we know that, for any given orientation of $P_{\mathbf{z}}$, the magnitude of $R_{P,x}$ increases continuously and strictly monotonically with increasing $d \ge 0$. Furthermore, $R_{P,y}$ is continuous in d, and $R_{P,y}(\mathbf{0}) = 0$, so $R_{P,y}^2$ is continuous and monotonically increasing for all d less than some sufficiently small $d_0 > 0$. Hence for any fixed orientation of $P_{\mathbf{z}}$,

R_P is a continuous and strictly monotonically increasing function for all $d \in [0, d_0]$. This implies that $\widehat{R}_P$ is also continuous and strictly monotone for sufficiently small $d \ge 0$.

Now consider $P_{\mathbf{z}}$ in equilibrium in the combined field $R + \delta S$, and again let d denote the distance between pivot point $v_{\mathbf{z}}$ and the origin. In equilibrium the lifted forces R_P and δS_P balance out, hence $\widehat{R}_P(d) \le |R_P| = |\delta S_P| \le \delta A$. Since $\widehat{R}_P$ is continuous and strictly monotone in d for sufficiently small d, we can ensure that d is less than a given ϵ, by chosing an appropriately small δ. This implies that $v_{\mathbf{z}}$ must lie within an ϵ-ball of the center of the radial field. In particular, we can make this ϵ-ball small enough so that the crossing structure $S(\phi)$ is not affected. Keeping d small also ensures that the torque τ_R about the pivot point induced by the radial field R is small, because τ_R is bounded by the product of d and R_P. This ensures that the equilibria of the squeeze field δS are not affected.

We conclude that the number of equilibria in a field $R + \delta S$ is bounded by $O(kn)$, for sufficiently small δ.
$\square$

In analogy to Section 3.2 we define the turn function $t : \mathbb{S}^1 \to \mathbb{S}^1$, which describes how the part will turn under a squeeze pattern, and hence yields the stable equilibrium configurations. Given the turn function t we can construct the corresponding squeeze function s as described in Section 3.2. With s as the input for Goldberg's alignment planner, we obtain strategies for unique part alignment (and positioning) of length $O(kn)$. They can be computed in time $O(k^2 n^2)$.

The result is a strategy for parts positioning of the form $(R + \delta S_1) * \cdots * (R + \delta S_{O(kn)})$. Compared to the old algorithm in Section 3.2 it improves the plan length by a factor of n, and the planning complexity is reduced by a factor of n^2. The planner is complete: For any polygonal part, there exists a strategy of the form $*_i(R + \delta S_i)$. Moreover, the algorithm is guaranteed to find a strategy for any input part. By appending a step which is merely the radial field R without a squeeze component, we are guaranteed that the part P will be uniquely posed (v is at the origin) as well as uniquely oriented. We can also show that the continuously varying "morphing" strategy $(R + \delta S_1) \rightsquigarrow \cdots \rightsquigarrow (R + \delta S_{O(kn)}) \rightsquigarrow R$ works in the same fashion to achieve the same unique equilibrium.

6.2 Manipulation Grammars

The development of devices that generate programmable vector fields is still in its infancy. The existing prototype devices exhibit only a limited range of programmability. For example, the prototype MEMS arrays described in Section 2.1 [9, 8, 7] currently have actuators in only four different directions, and the actuators are only row-wise controllable. Arrays with individually addressable actuators at various orientations are possible (see [9, 8, 31, 7, 42]) but require significant development effort. There are also limitations on the resolution of the devices given by fabrication constraints. For the vibrating plate device from Section 2.2 the fields are even more constrained by the vibrational modes of the plate.

We are interested in the capabilities of such constrained systems. In this section we give an algorithm that decides whether a part can be uniquely positioned using a given set of vector fields, and it synthesizes an optimal-length strategy if one exists. If we think of these vector fields as a vocabulary, we obtain a language of manipulation strategies. We are interested in those expressions in the language that correspond to a strategy for uniquely posing the part.

The elements of our "manipulation grammar" are (sequences of) vector fields that bring the part into a *finite* set of possible equilibrium positions (Figure 8). We call these (sequences of) vector fields *finite field operators*. Each field operator comes with the following guarantee: No matter where in $\mathbb{R}^2 \times \mathbb{S}^1$ the part starts off, it will always come to rest in one of E different total equilibria. That is: For any connected polygonal part P, either of these field operators is *always* guaranteed to reduce P to a *finite* set of equilibria in its configuration space $\mathcal{C} = \mathbb{R}^2 \times \mathbb{S}^1$.

From Section 6.1 we know that combined radial-squeeze patterns $R + \delta S$ have this property. However, there are other simple field operators that also have this finiteness property:

Claim 28 *Let f and $f_\perp$ be unit squeeze fields such that $f_\perp$ is orthogonal to f. Then the fields $f * f_\perp$ and $f + f_\perp$ induce a finite number of equilibria on every connected polygon P.*

Proof: First consider the field $f * f_\perp$, and w.l.o.g. assume that $f(x, y) = (-\text{sign}(x), 0)$. Also assume that

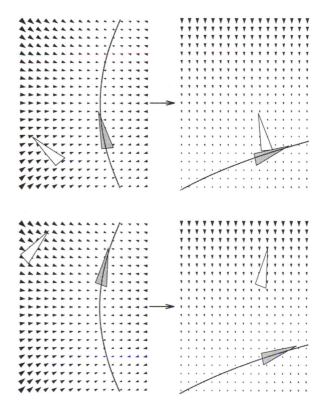

Figure 8: *Alignment vocabulary for a triangular part on a vibrating plate, consisting of two consecutive force fields with slightly curved nodal lines (attractors) which bring the part into (approximately) the same equilibria.*

the COM of P is the reference point used to define its configuration space $\mathcal{C} = \mathbb{R}^2 \times \mathbb{S}^1$. As discussed in Section 3.1, P will reach one of a finite number of orientation equilibria when placed in f or $f_\perp$. More specifically, when P is placed in f, there exists a finite set of equilibria $E_f = \{(x_i, \theta_i)\}$. Similarly for $f_\perp(x, y) = (0, -\text{sign}(y))$, there exists a finite set of equilibria $E_{f_\perp} = \{(y_j, \theta_j)\}$. Since the x-component of $f_\perp$ is zero, the x-coordinate of the reference point of P (the COM) remains constant while P is in $f_\perp$. Hence P will finally come to rest in a pose (x_k, y_k, θ_k), where $x_k \in \pi_1(E_f)$, $(y_k, \theta_k) \in E_{f_\perp}$, and π_1 is the canonical projection such that $\pi_1(x, y, \theta) = x$. Since E_f is finite, so is $\pi_1(E_f)$. $E(f_\perp)$ is also finite, therefore there exists only a finite number of such total equilibrium poses for $f * f_\perp$.

If P is placed into the field $f + f_\perp$, there exists a unique translational equilibrium (x, y) for every given, fixed orientation θ. In each of these translational equilibria, the squeeze lines of f and $f_\perp$ are both bisectors of P. Now consider the moment acting on P when P is in translational equilibrium as a function of θ. In analogy to Proposition 4 in Section 3.1 we can show that for any topological placement of the bisectors, this moment function has at most $O(k)$ roots, where k is the maximum number of edges a bisector of P can cross. This implies that there exist only $O(k n^2)$ distinct total equilibria for $f + f_\perp$. $\square$

We have seen in Sections 3 and 5 that for simple force fields such as e.g. squeeze or radial fields, we can predict the motion and the equilibria of a part with exact analytical methods. However, for arbitrary fields (e.g. the force fields described in Section 2.2 which are induced by vibrating plates) such algorithms may not exist. Instead we can employ approximate methods to predict the behavior of the part in the force field (for more details also see [18]). These methods are typically numerical computations that involve simulating the part from a specific initial pose, until it reaches equilibrium.[4] We call the cost for such a computation the *simulation complexity* $s(n)$. We write $s(n)$ because the simulation complexity will usually depend on the complexity of the part (i.e. its number of vertices n).

Proposition 29 *Consider a polygonal part P, and m finite field operators $\{F_i\}$, $1 \leq i \leq m$, each with at most E distinct equilibria in the configuration space C for P. There is an algorithm that generates an optimal-length strategy of the form $F_1 * F_2 * \cdots * F_m$ to uniquely pose P up to symmetries, if such a strategy exists. This algorithm runs in $O(m^2 E\, (s(n)+2^E))$ time, where $s(n)$ is the simulation complexity of P in F_i. If no such strategy exists, the algorithm will signal failure.*

Proof: Construct a transition table T of size $m^2 E$ that describes how the part P moves from an equilibrium of F_i to an equilibrium of F_j. This table can be constructed either by a dynamical analysis similar to Section 6.1, or by dynamic simulation. The time to construct this table is $O(m^2 E\, s(n))$, where $s(n)$ is the simulation complexity, which will typically depend on the complexity of the part.

---

[4]See for example URL http://www.cs.cornell.edu /home/karl/Cinema.

Using the table T, we can search for a strategy as follows: Define the *state* of the system as the set of possible equilibria a part is in, for a particular finite field operator F_i. There are m field operators and $O(E)$ equilibria for each of them, hence there are $O(m\, 2^E)$ distinct states. For each state there are m possible successor states as given by table T, and they can each be determined in $O(E)$ operations, which results in a graph with $O(m\, 2^E)$ nodes, $O(m^2\, 2^E)$ edges, and $O(m^2 E\, 2^E)$ operations for its construction. Finding a strategy, or deciding that it exists, then devolves to finding a path whose goal node is a state with a unique equilibrium. The total running time of this algorithm is $O(m^2 E\, (s(n) + 2^E))$. $\square$

Hence, as in [21], for any part we can decide whether a part can be uniquely posed using the vocabulary of field operators $\{F_i\}$ but (a) the planning time is exponential and (b) we do not know how to characterize the class of parts that can be oriented by $\{F_i\}$. However, the resulting strategies are optimal in size.

This result illustrates a tradeoff between mechanical complexity (the dexterity and controllability of field elements) and planning complexity (the computational difficulty of synthesizing a strategy). If one is willing to build a device capable of radial fields, then one reaps great benefits in planning and execution speed. On the other hand, we can still plan for simpler devices (see Figures 3 and 8), but the plan synthesis is more expensive, and we lose some completeness properties.

7 Conclusions and Open Problems

Universal Feeder-Orienter (UFO) Devices. It was shown in [8] that every connected polygonal part P with n vertices has a finite number of stable orientation equilibria when P is placed into a squeeze field S. Based on this property we were able to generate manipulation strategies for unique part alignment. We showed in Section 6.1 that by using a combined radial and squeeze field $R + \delta S$, the number of equilibria can be reduced to $O(k n)$. Using elliptic force fields $f(x, y) = (\alpha x, \beta y)$ such that $\alpha \neq \beta$ and $\alpha, \beta \neq 0$, this bound can be reduced to 2 [29]. Does there exist a *universal field* that, for every part P, has only one unique equilibrium (up to part symmetry)? Such a field could be used to build a *universal parts feeder* [1] that uniquely positions a part without the need of a clock, sensors, or programming.

We propose a combined radial and "gravitational" field $R + \delta G$ which might have this property. δ is a small positive constant, and G is defined as $G(x, y) = (0, -1)$. This device design is inspired by the "universal gripper in [1]. Such a field could be obtained from a MEMS array that implements a unit radial force field. Instead of rectangular actuators in a regular grid, triangular actuators could be laid out in a polar-coordinate grid. The array could then be tilted slightly to obtain the gravity component. Hence such a device would be relatively easy to build. Extensive simulations show that for every part we have tried, one unique total equilibrium is always obtained. We are working toward a rigorous proof of this experimental observation.

Magnitude Control. Consider an array in which the *magnitude* of the actuator forces cannot be controlled. Does there exist an array with constant magnitude in which all parts reach one unique equilibrium? Or can one prove that, without magnitude control, the number of distinct equilibria is always greater than one?

Geometric Filters. This paper focuses mainly on sensorless manipulation strategies for *unique positioning* of parts. Another important application of programmable vector fields are *geometric filters*. Figure 1 shows a simple filter that separates smaller and larger parts. We are interested in the question *Given n parts, does there exist a vector field that will separate them into specific equivalence classes?* For example, does there exist a field that moves small and large rectangles to the left, and triangles to the right? In particular, it would be interesting to know whether for any two different parts there exists a sequence of force fields that will separate them.

Performance Measures. Are there performance measures for how fast (in real time) an array will orient a part? In some sense the actuators are fighting each other (as we have observed experimentally) when the part approaches equilibrium. For squeeze grasps, one measure of "efficiency", albeit crude, might be the integral of the magnitude of the moment function, i.e., $\int_0^{2\pi} |M(\theta)| \, d\theta$. The issue is that if, for many poses,

$|M(\theta)|$ is very small, then the orientation process will be slow. Better measures are also desirable.

Uncertainty. In practice, neither the force vector field nor the part geometry will be exact, and both can only be characterized up to tolerances [15]. This is particularly important at micro scale. Within the framework of potential fields, we can express this uncertainty by considering not one single potential function U_P, but rather *families of potentials* that correspond to different values within the uncertainty range. Bounds on part and force tolerances will correspond to limits on the variation within these function families. An investigation of these limits will allow us to obtain upper error bounds for manipulation tasks under which a specific strategy will still achieve its goal.

A family of potential functions is a set $\{U_\alpha : \mathcal{C} \rightarrow \mathbb{R}\}_{\alpha \in J}$ where J is an index set. For example, we may start with a single potential function $U : \mathcal{C} \rightarrow \mathbb{R}$, and define a family of potential functions $\mathcal{F}(U, \epsilon, z)$ as $\{U_\alpha : \mathcal{C} \rightarrow \mathbb{R} \mid \|U_\alpha(p) - U(p)\|_z < \epsilon\}$ for some ϵ and norm z. This is analogous to defining a neighborhood in function space, using e.g. the compact-open topology.

When we differentiate a family of potential fields (using the gradient) we obtain a differential inclusion instead of a differential equation. So if $\mathcal{F}(u) = \mathcal{F}(u, \epsilon, z)$, then $\nabla \mathcal{F}(u) = \{\nabla U_\alpha\}_{\alpha \in J}$.

When considering families of potentials, the equilibrium may be known to lie only within a set E_i, although we may know that it is always a point in E_i. If the sets E_i are of a small diameter less than some $\epsilon > 0$, our algorithms could be extended to handle ϵ-approximations.

As a more general approach, we propose an algorithm based on *back-projections*: For a given part, let $B_{F_i}(G) \subset \mathcal{C} = \mathbb{R}^2 \times \mathbb{S}^1$ be the back-projection [33] of the set G under F_i, where $G \subset \mathcal{C}$, and F_i is a family of fields on $\mathbb{R}^2$. Then we wish to calculate a sequence of fields $F_1, F_2, \cdots, F_k$ such that $B_{F_1}(B_{F_2}(\cdots B_{F_k}(G) \cdots)) = \mathcal{C}$, where G is a single point in $\mathcal{C}$ (cf. [33, 21, 13, 15, 11]).

Output Sensitivity. We have seen in Sections 6.1 and 6.2 that the efficiency of planning and executing manipulation strategies critically depends on the number of equilibrium configurations. Expressing the planning and execution complexity as a function of the number of equilibria E, rather than the number of vertices n, is called *output sensitive analysis*. In practice, we have found that there are almost no parts with more than two distinct (orientation) equilibria, even in squeeze fields. This is far less than the $E = O(kn^2)$ upper bound derived in 3.1. If this observation can be supported by an exact or even statistical analysis of part shapes, it could lead to extremely good expected bounds on plan length and planning time, even for the less powerful strategies employing manipulation grammars (note that the complexity of the manipulation grammar algorithm in Proposition 29 is output-sensitive).

Discrete Force Fields. For the manipulation strategies described in this paper we assume that the force fields are continuous, i.e. that the generated forces are dense compared to the moving part (assumption DENSITY in Section 3.2). When manipulating very small parts on microactuator arrays, this condition may be only approximately satisfied. We are interested in the limitations of the continuous model, and we would like to know the conditions under which it is necessary to employ a different, discrete model of the array that takes into account individual actuators, as well as the gaps between actuators. In [9] we propose a model for the interaction between parts and arrays of individual actuators based on the theory of limit surfaces [26].

Resonance Properties. Is it possible to exploit the dynamic resonance properties of parts to tune the AC control of the array to perform efficient dynamic manipulation?

3D Force Fields. It may be possible to generate 3D force fields by using Lorentz electromagnetic forces. Tunable electric coils could be attached to various points of a 3D body, suspending the resulting object in a strong permanent magnetic field using magnetic levitation (the Lorentz effect) [28]. The tuning (control) of the electric coils could be effected as follows: Integrated control circuitry could be fabricated and co-located with the coils, and conceivably a power supply. The control could be globally effected using wireless communication, or, the control of each coil evolves in time until the part is reoriented as desired. The Lorentz forces could then be deactivated to bring the object to rest on the ground. Planning for such a 3D device might reduce to [23].

Acknowledgments

We would like to thank Tamara Lynn Abell, Vivek Bhatt, Bernard Chazelle, Paul Chew, Perry Cook, Ivelisse Rubio, Ken Steiglitz, Nick Trefethen, and Andy Yao for useful discussions and valuable comments. We are particularly grateful to Mike Erdmann for his suggestions and help for the proofs of Proposition 15 and Lemma 22, to Danny Halperin for sharing his insights into geometric and topological issues related to manipulation with force fields, to Lydia Kavraki for continuing discussions and creative new ideas concerning force vector fields, and Jean-Claude Latombe for his hospitality during our stay at the Stanford Robotics Laboratory.

Support is provided in part by the NSF under grants No. IRI-8802390, IRI-9000532, IRI-9201699, and by a Presidential Young Investigator award to Bruce Donald, in part by NSF/ARPA Special Grant for Experimental Research No. IRI-9403903, and in part by the AFOSR, the Mathematical Sciences Institute, Intel Corporation, and AT&T Bell laboratories. This work was supported by ARPA under contract DABT 63-69-C-0019. The device fabrication was performed at the Cornell Nanofabrication Facility (CNF), which is supported by NSF grant ECS-8619049, Cornell University, and Industrial Affiliates.

References

[1] T. L. Abell and M. Erdmann. A universal parts feeder, 1996. Personal communication / in preparation.

[2] S. Akella, W. H. Huang, K. M. Lynch, and M. T. Mason. Planar manipulation on a conveyor by a one joint robot with and without sensing. In *International Symposium of Robotics Research (ISRR)*, 1995.

[3] K.-F. Böhringer. *Manipulation with Programmable Vector Fields: Design, Fabrication, Control, and Programming of Massively-Parallel Microfabricated Actuator Arrays and Planar Vibratory Parts-Feeders*. PhD thesis, Cornell University, Department of Computer Science, Ithaca, NY 14853, Jan. 1997. Forthcoming.

[4] K.-F. Böhringer, V. Bhatt, and K. Y. Goldberg. Sensorless manipulation using transverse vibrations of a plate. In *Proc. IEEE Int. Conf. on Robotics and Automation (ICRA)*, pages 1989 – 1996, Nagoya, Japan, May 1995. http://www.cs .cornell.edu/home/karl/VibratoryAlign.

[5] K.-F. Böhringer, R. G. Brown, B. R. Donald, J. S. Jennings, and D. Rus. Distributed robotic manipulation: Experiments in minimalism. In *Fourth International Symposium on Experimental Robotics (ISER)*, Stanford, California, June 1995. http://www.cs.cornell.edu/home/brd.

[6] K.-F. Böhringer, B. R. Donald, and D. Halperin. The area bisectors of a polygon and force equilibria in programmable vector fields, 1996. In preparation.

[7] K.-F. Böhringer, B. R. Donald, and N. C. MacDonald. Single-crystal silicon actuator arrays for micro manipulation tasks. In *Proc. IEEE Workshop on Micro Electro Mechanical Systems (MEMS)*, San Diego, CA, Feb. 1996. http://www.cs.cornell.edu/home /karl/MicroActuators.

[8] K.-F. Böhringer, B. R. Donald, R. Mihailovich, and N. C. MacDonald. Sensorless manipulation using massively parallel microfabricated actuator arrays. In *Proc. IEEE Int. Conf. on Robotics and Automation (ICRA)*, pages 826–833, San Diego, CA, May 1994. http://www.cs.cornell.edu /home/karl/MicroManipulation.

[9] K.-F. Böhringer, B. R. Donald, R. Mihailovich, and N. C. MacDonald. A theory of manipulation and control for microfabricated actuator arrays. In *Proc. IEEE Workshop on Micro Electro Mechanical Systems (MEMS)*, pages 102–107, Oiso, Japan, Jan. 1994. http://www.cs.cornell.edu /home/karl/MicroActuators.

[10] G. Boothroyd, C. Poli, and L. E. Murch. *Automatic Assembly*. Marcel Dekker, Inc., 1982.

[11] A. J. Briggs. An efficient algorithm for one-step planar compliant motion planning with uncertainty. *Algorithmica*, 8(3), 1992.

[12] R. Brooks. A layered intelligent control system for a mobile robot. *IEEE Journal of Robotics and Automation*, RA(2), 1986.

[13] R. C. Brost. Automatic grasp planning in the presence of uncertainty. *Int. Journal of Robotics Research*, 7(1):3–17, 1988.

[14] J. Canny and K. Goldberg. "RISC" for industrial robotics: Recent results and open problems. In *Proc. IEEE Int. Conf. on Robotics and Automation (ICRA)*. IEEE, May 1994.

[15] B. R. Donald. *Error Detection and Recovery in Robotics*, volume 336 of *Lecture Notes in Computer Science*. Springer Verlag, Berlin, 1989.

[16] B. R. Donald. The complexity of planar compliant motion planning with uncertainty. *Algorithmica*, 5(3):353–382, 1990.

[17] B. R. Donald, J. Jennings, and D. Rus. Information invariants for distributed manipulation. In K. Goldberg, D. Halperin, J.-C. Latombe, and R. Wilson, editors, *International Workshop on Algorithmic Foundations of Robotics (WAFR)*, pages 431–459, Wellesley, MA, 1995. K. Peters.

[18] B. R. Donald and P. Xavier. Provably good approximation algorithms for optimal kinodynamic planning for cartesian robots and open chain manipulators. *Algorithmica*, 14(6):480–530, Nov. 1995.

[19] M. A. Erdmann. On a representation of friction in configuration space. *Int. Journal of Robotics Research*, 13(3):240–271, 1994.

[20] M. A. Erdmann. An exploration of nonprehensile two-palm manipulation: Planning and execution. Technical report, Carnegie Mellon University, Pittsburgh, PA, 1996.

[21] M. A. Erdmann and M. T. Mason. An exploration of sensorless manipulation. *IEEE Journal of Robotics and Automation*, 4(4), Aug. 1988.

[22] M. A. Erdmann and M. T. Mason. Nonprehensile manipulation. In *International Workshop on Algorithmic Foundations of Robotics (WAFR)*, Toulouse, France, July 1996.

[23] M. A. Erdmann, M. T. Mason, and G. Vaneček, Jr. Mechanical parts orienting: The case of a polyhedron on a table. *Algorithmica*, 10, 1993.

[24] H. Fujita. Group work of microactuators. In *International Advanced Robot Program Workshop on Micromachine Technologies and Systems*, pages 24–31, Tokyo, Japan, Oct. 1993.

[25] K. Y. Goldberg. Orienting polygonal parts without sensing. *Algorithmica*, 10(2/3/4):201–225, August/September/October 1993.

[26] S. Goyal and A. Ruina. Relation between load and motion for a rigid body sliding on a planar surface with dry friction: Limit surfaces, incipient and asymptotic motion. *Wear*, 1988.

[27] H. Hitakawa. Advanced parts orientation system has wide application. *Assembly Automation*, 8(3), 1988.

[28] R. Hollis and S. E. Salcudean. Lorentz levitation technology: A new approach to fine motion robotics, teleoperation, haptic interfaces, and vibration isolation. In *International Symposium of Robotics Research (ISRR)*, Hidden Valley, PA., Oct. 1993.

[29] L. E. Kavraki. On the number of equilibrium placements of mass distributions in elliptic potential fields. Technical Report STAN-CS-TR-95-1559, Department of Computer Science, Stanford University, Stanford, CA 94305, 1995.

[30] O. Khatib. Real time obstacle avoidance for manipulators and mobile robots. *Int. Journal of Robotics Research*, 5(1):90–99, Spring 1986.

[31] W. Liu and P. Will. Parts manipulation on an intelligent motion surface. In *IROS*, Pittsburgh, PA, 1995.

[32] T. Lozano-Pérez. Spacial planning: A configuration space approach. *IEEE Transactions on Computers*, C-32(2):108–120, Feb. 1983.

[33] T. Lozano-Pérez, M. Mason, and R. Taylor. Automatic synthesis of fine-motion strategies for robots. *Int. Journal of Robotics Research*, 3(1), 1984.

[34] T. McGeer. Passive dynamic walking. *Int. Journal of Robotics Research*, 1990.

[35] R. E. Mihailovich, Z. L. Zhang, K. A. Shaw, and N. C. MacDonald. Single-crystal silicon torsional resonators. In *Proc. IEEE Workshop on Micro Electro Mechanical Systems (MEMS)*, pages 155–160, Fort Lauderdale, FL, Feb. 1993.

[36] P. Moncevicz, M. Jakiela, and K. Ulrich. Orientation and insertion of randomly presented parts using vibratory agitation. In *ASME 3rd Conference on Flexible Assembly Systems*, September 1991.

[37] M. H. Raibert, J. K. Hodgins, R. R. Playter, and R. P. Ringrose. Animation of legged maneuvers: jumps, somersaults, and gait transitions. *Journal of the Robotics Society of Japan*, 11(3):333–341, 1993.

[38] J. Reif and H. Wang. Social potential fields: A distributed behavioral control for autonoomous robots. In K. Goldberg, D. Halperin, J.-C. Latombe, and R. Wilson, editors, *Algorithmic Foundations of Robotics*, pages 431–459. K. Peters, Wellesley, MA, 1995.

[39] E. Rimon and D. Koditschek. Exact robot navigation using artificial potential functions. *IEEE Transactions on Robotics and Automation*, 8(5), October 1992.

[40] K. A. Shaw and N. C. MacDonald. Integrating SCREAM micromechanical devices with integrated circuits. In *Proc. IEEE Workshop on Micro Electro Mechanical Systems (MEMS)*, San Diego, CA, Feb. 1996.

[41] K. A. Shaw, Z. L. Zhang, and N. C. MacDonald. SCREAM I: A single mask, single-crystal silicon process for microelectromechanical structures. In *Transducers — Digest Int. Conf. on Solid-State Sensors and Actuators*, Pacifico, Yokohama, Japan, June 1993.

[42] J. W. Suh, S. F. Glander, R. B. Darling, C. W. Storment, and G. T. A. Kovacs. Combined organic thermal and electrostatic omnidirectional ciliary microactuator array for object positioning and inspection. In *Proc. Solid State Sensor and Actuator Workshop*, Hilton Head, NC, June 1996.

[43] Z. L. Zhang and N. C. MacDonald. An RIE process for submicron, silicon electromechanical structures. *Journal of Micromechanics and Microengineering*, 2(1):31–38, Mar. 1992.

[44] N. B. Zumel and M. A. Erdmann. Nonprehensile two palm manipulation with non-equilibrium transitions between stable states. In *Proc. IEEE Int. Conf. on Robotics and Automation (ICRA)*, Minneapolis, MN, Apr. 1996.

Rolling Polyhedra on a Plane, Analysis of the Reachable Set

Yacine Chitour, *Centro E.Piaggio, University of Pisa, Pisa, Italy*

Alessia Marigo, *Centro E.Piaggio, University of Pisa, Pisa, Italy*

Domenico Prattichizzo, *Centro E.Piaggio, University of Pisa, Pisa, Italy*

Antonio Bicchi, *Centro E.Piaggio, University of Pisa, Pisa, Italy*

The problem of dexterous manipulation of objects, i.e. of arbitrary relocation and reorientation of rigid bodies by action of some mechanism, is considered. We build upon previous results on the possibility of implementing dexterous "robot hands" with few actuators, which can be afforded through the exploitation of nonholonomic rolling of regular surfaces. In this paper we focus on the manipulation of polyhedral objects, and prove a necessary and sufficient controllability–like result, which discloses some of the interesting aspects and perspectives of this problem.

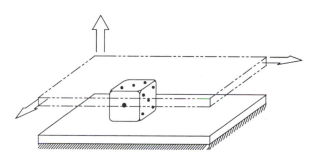

Figure 1: *A parallel–jaw gripper can manipulate polyhedral parts*

1   Introduction

The design of mechanisms for orienting and locating parts is of paramount importance in flexible automation and robotics. In some cases, the problem is that of reorienting a large number of parts coming in random positions and orientations, to a given posture within assembly tolerances. For such problems, industry most often uses *ad hoc* fixtures, such as vibrating part–feeders, fenced conveyor belts, etc.. The design and tuning of these devices is often time–consuming, and is only rewarding on large–size batch production. In other cases, where the typology of parts is more variate, more flexible manipulation means are preferable. In highly–flexible automation and robotics, the design of manipulation devices has been attacked by several different approaches, such as by developing dexterous multifingered hands ([7], [12]); using "pushing" or "tilting" actions ([10], [9]); "regrasping" ([15], [6]); and "finger gaiting" ([11], [4]).

Among these manipulation strategies, those using discontinuous contacts between the manipulator and the part are sometimes regarded as not reliable enough in real–world, unsteady environments. On the other hand, multifingered robot hands are often too costly, heavy, and complex, to be viable in many applications.

The idea of exploiting the nonholonomic nature of the constraint of rolling surfaces, so as to design a dexterous hand with few actuators, was presented in [3]. In that paper, building upon previous results of Li and Canny [8], it was conjectured that the position and orientation of an object with *regular* surface could be arbitrarily changed by rolling onto another regular surface by acting only on its relative angular velocities. The simple experiment of rolling a sphere on a plane surface, bringing it back to its initial position but with different orientation, is an intuitive explanation of this phenomenon. Based on the above controllability conjecture, a dexterous hand consisting of two parallel plates, with only three translational degrees of freedom, was devised and experimentally demonstrated (see fig. 1). Bicchi, Prattichizzo, and Sastry [2] confirmed later on the conjecture for all strictly convex objects with a regular surface of revolution, and discussed the differential–geometric aspects of planning and controlling the object motions.

The advantage of manipulation by rolling is that it accomplishes dexterity with very simple hardware, while it guarantees that the object is never "left alone" during manipulation. The intrinsic nonholonomic nature of rolling offers many difficulties to the planification and control of such devices, of which only few have been addressed so far.

Among the various open problems, the one we start considering in this paper is that of removing the limitation that manipulated objects should have regular (C^∞, analytic) surface. The main motivation of such an assumption is that for regular surfaces the powerful tools of differential geometry and nonlinear control theory are readily available. On the other hand, the assumption is rarely verified with industrial parts, which often have edges and vertices. Again, the simple experiment of rolling a die onto a plane without slipping, and bringing it back after any sufficiently rich path, shows that its orientation has changed in general, and hints to the fact that manipulation of parts with non–smooth (e.g., polyhedral) surface can be advantageously performed by rolling.

Some aspects of graspless manipulation of polyhedral objects by rolling have been considered already in the robotics literature (see e.g. [13], [1], [5]). However, a complete study on the analysis, planning, and control of rolling manipulation for polyhedral parts is far from being available, and indeed it comprehends many aspects, some of which appear to be non–trivial. In particular, the lack of a differentiable structure on the configuration space of a rolling polyhedron deprives us of most techniques used with regular surfaces. Moreover, peculiar phenomena may happen with polyhedra, which have no direct counterpart with regular objects. In this paper, we start such study by analysing the structure of the set of configurations reachable from a given one, and show that it may reveal extremely different structure depending on the polyhedron considered (see fig. 2 and 3).

Figure 2: *A polyhedron whose reachable set is everywhere dense*

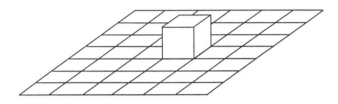

Figure 3: *A polyhedron whose reachable set is nowhere dense*

2 Problem formulation

Consider the simple device depicted in fig. 1, consisting of two plates, one of which is fixed, while the other can translate remaining parallel to the first. A part of known shape is put between the plates and successively moved by a combination of vertical and horizontal forces at the contacts that cause it to move. The goal is to bring the part from a given initial configuration (a point in $SE(3)$) to another desired one. A few considerations are in order:

- as the part is constrained to keep in touch with the two plates, to specify arbitrary desired configurations would require being able to move the lower plate vertically. With no loss of generality we only consider different configurations modulo a rigid translation of the whole mechanism;

- the surface of the part is considered to be piecewise flat, closed, and comprised of a finite number of faces, edges, and vertices;

- parts need not be convex. However, as the plates are assumed to be large w.r.t. the diameter of parts, we will only be concerned with the convex hull of the parts themselves. In what follows, we use the term "part" to refer to a piecewise flat convex surface, i.e., to a convex polyhedron;

- in general, three motions of a polyhedron on a plane are possible: by sliding on a face, tumbling about an edge, or pivoting about a vertex. Slippage is not considered desirable in this context, as it does not guarantee reliable enough results in manipulation. We assume therefore that high–friction, compliant materials are employed to cover the plates, and that the vertical degree–of–freedom of the upper plate is suitably used, so as to prevent slippage. By a similar concern, pivoting about a vertex is also banned. In fact, real–world parts never have perfectly sharp vertices, and the actual effects of pivoting about a vertex will strongly depend on the details of how "the corner is rounded" (recall that all the curvature of a polyhedron is concentrated at its vertices). On the other hand, tumbling about an edge is insensitive to the details of how the "edge" really looks like, since no curvature is concentrated on the edge of a polyhedron.

The only motions of the parts we will be concerned with are therefore comprised of a sequence of rotations about one of the edges of the face being in contact with the plate, by the amount that exactly brings another face in contact. This action on the parts will be referred to as an elementary tumble, or ET for short.

3 Definitions and properties

Let $\widetilde{P}$ be a convex polyhedron rolling on a plane P by ET's. We associate to $\widetilde{P}$ the following sets:

(a) $\widetilde{V} = \{v_1, \ldots, v_m\}$ is the set of vertices of $\widetilde{P}$ and $m = \mathrm{card}(\widetilde{V})$;

(b) $\widetilde{E} = \{e_1, \ldots, e_k\}$ is the set of edges of $\widetilde{P}$ and $k = \mathrm{card}(\widetilde{E})$;

(c) $\widetilde{F} = \{F_1, \ldots, F_l\}$ is the set of faces of $\widetilde{P}$ and $l = \mathrm{card}(\widetilde{F})$.

By the assumption of convexity, parts are topological spheres, hence for their Euler characteristic it holds $\chi = m - k + l = 2$.

The configuration space $\widetilde{M}$ of the system under investigation is the restriction of the space of rigid body configurations $SE(3)$ to those that have one face in contact with the plane P. One possible parameterization of this space is as follows.

Let Oxy be a fixed reference frame on the plane P. For each face F_i, $1 \leq i \leq l$, let c_i and u_i be two arbitrary distinct points fixed on F_i, for instance the center of gravity of F_i and one of its vertices. Let (x_i, y_i) be the coordinates of c_i, and θ_i be the oriented angle between Ox and $\vec{c_i u_i}$. A configuration of $\widetilde{P}$ on P is uniquely determined by the quadruple (x_i, y_i, θ_i, i), where $i \in \{1, \ldots, l\}$ is the index of the face in contact with P. The configuration set of our problem consists therefore of l copies of $SE(2)$, or explicitly

$$\widetilde{M} = \mathbb{R}^2 \times S^1 \times \widetilde{F}. \tag{1}$$

For $1 \leq i \leq l$, each copy of $SE(2)$ corresponds to the set of all the possible configurations for the face i, $\widetilde{M}_i$, i.e.

$$\widetilde{M}_i = \mathbb{R}^2 \times S^1 \times \{i\}. \tag{2}$$

The space $\widetilde{M}$ is endowed with the product metric associated to the metrics of the euclidean space $\mathbb{R}^2$, of the quotient space $S^1 = \mathbb{R}/2\pi\mathbb{Z}$ and of the discrete space $\widetilde{F}$, respectively. The latter is taken to be $\rho(F_i, F_j) = 1 - \delta_{ij}$, where δ_{ij} is the Kronecker symbol. Although very intuitive, this parameterization does not turn out to be the most convenient for our developments. We therefore introduce a slightly more technical description of $\widetilde{M}$ as the set of equivalence classes on a set $\widetilde{M}'$ by the relation $\sim$, where

- the set $\widetilde{M}'$ is defined as the subset of $\mathbb{R}^2 \times \widetilde{V} \times S^1 \times \widetilde{F}$ of points (x, y, v, θ, i) where i is the index of the face F_i in contact with P, v is any of the vertices of F_i (shortly $F_i \ni v$), (x, y) are the coordinates of v and θ is the oriented angle between $\vec{xx'}$ and $\vec{c_i v}$;

- two elements of $\widetilde{M}'$ are equivalent under the relation $\sim$ if $i = i'$ and $\theta' - \theta$ is equal to the oriented angle between $\vec{c_i v}$ and $\vec{c_i v'}$, for any fixed point c_i on face F_i

Note that corresponding to each configuration of the polyhedron, we have an equivalence class with n_{F_i} elements, where n_{F_i} is the number of vertices of the face F_i.

The actions we take on the configurations of the polyhedron are finite sequences of ET's, that will be referred to as "trips". The length of a trip is the number of ET's it is comprised of. The problem this paper is concerned with is to understand the structure induced on the configuration space by trips of arbitrary length. We therefore define reachability of a configuration as

Definition 1 *The configuration q_f is reachable from q_0 if there exists a trip steering $\widetilde{P}$ from q_0 to q_f. In this case, we write $q_0 \to q_f$.*

For every $q \in \widetilde{M}$, let $\widetilde{R}_q$ be the reachable set from q, i.e. the set of configurations that can be reached from q in a finite, but arbitrarily large number of ET's.

As mentioned in the introduction, the structure of the reachable set can be very diverse for different polyhedra. Note first that $\widetilde{R}_q$ is countable by its definition and therefore the inclusion

$$\widetilde{R}_q \subset \widetilde{M}$$

is strict. Introducing the canonical projections

$$\Pi_1 \;:\; \widetilde{M} \to \mathbb{R}^2,$$
$$\Pi_2 \;:\; \widetilde{M} \to S^1,$$

we have that $\Pi_1\left(\widetilde{R}_q\right)$ is trivially infinite and unbounded in $\mathbb{R}^2$. Various possibilities can occur: $\Pi_1\left(\widetilde{R}_q\right)$ (resp. $\Pi_2\left(\widetilde{R}_q\right)$) can be discrete in $\mathbb{R}^2$ (resp. finite in S^1) or can be dense in $\mathbb{R}^2$ (resp. idem in S^1). One can even distinguish differently dense structures for $\widetilde{R}_q$, among which are the following:

a) Density in $\widetilde{M}$:

$$\left(\mathbf{DM}\right) \begin{cases} \forall \varepsilon > 0, \forall q_f \in \widetilde{M}, \\ \exists q' \in \widetilde{R}_q \text{ such that} \\ q' \in B_\varepsilon(q_f), \end{cases}$$

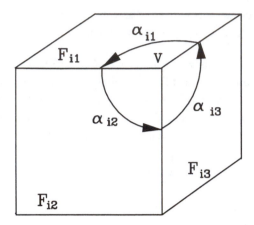

Figure 4: *The defect angle at vertex V is defined as $\beta_v = 2\pi - (\alpha_{i1} + \alpha_{i2} + \alpha_{i3})$*

b) Density in $\mathbb{R}^2 \times S^1$ for a given face i:

$$\left(\mathbf{DM}\right)_i \begin{cases} \forall \varepsilon > 0, \forall q_f \in \widetilde{M}_i, \\ \exists q' \in \widetilde{R}_q \text{ such that} \\ q' \in B_\varepsilon(q_f). \end{cases}$$

Here, $B_\varepsilon(\cdot)$ indicates a ball centered in its argument of radius ε in the suitable metric. Note that

$$\left(\mathbf{DM}\right) \Rightarrow \left(\mathbf{DM}\right)_i.$$

As usual, if the above properties hold for any initial configuration q, the properties will be said to hold globally.

In this paper, we explicitly consider two extreme cases of such behaviours represented in fig. 2 and 3, where we have respectively that

1. the reachable set is dense in $\widetilde{M}$ (see section 4);

2. $\Pi_1\left(\widetilde{R}_q\right)$ is a lattice of $\mathbb{R}^2$ and $\Pi_2\left(\widetilde{R}_q\right)$ is finite (see section 5).

The notion of "defect angle" β_v at a vertex will turn out to be crucial in the rest of this study.

Definition 2 *For each vertex $v \in \widetilde{V}$, let l_v be its valence, i.e. the number of faces of $\widetilde{P}$ which are adjacent to v, and name such faces as $F_{i_1}, \ldots, F_{i_{l_v}}$. Let α_{i_j},*

$1 \leq j \leq i_{l_v}$, be the angle at v corresponding to face F_{i_j}. The defect angle at v is then defined as

$$\beta_v = 2\pi - \sum_{j=1}^{l_v} \alpha_{i_j}. \tag{3}$$

The defect angle at a vertex (see fig. 4) is also known in the literature as the curvature concentrated at the vertex. Note that $0 < \beta_v < 2\pi$ since $\widetilde{P}$ is a convex polyhedron with null curvature on its faces. We also have the classical Euler relation given by

Proposition 1 (Euler relation) *Let $\widetilde{P}$ be a convex polyhedron and $\widetilde{V}$ the set of its vertices. Then,*

$$\sum_{v \in \widetilde{V}} \beta_v = 4\pi. \tag{4}$$

We state now two remarks that can also be seen as basic properties of the motion of a polyhedron on a plane. There is no proof because these remarks are elementary.

Remark 1 Let $v \in \widetilde{V}$ and suppose that $\widetilde{P}$ rests on P on a face F_i with $F_i \ni v$. By rolling clockwise $\widetilde{P}$ on all the faces containing v until coming back to F_i while keeping v immobile, $\widetilde{P}$ is rotated of an angle $2\pi - \beta_v$ around an axis Z_v orthogonal to P and passing through v, i.e., it moves from (x, y, v, θ, i) to $(x, y, v, \theta + \beta_v, i)$. We denote this trip by R_{β_v} and the analogous counterclockwise trip by $R_{-\beta_v}$. By repeating R_{β_v} clockwise or counterclockwise, we can go from (x, y, v, θ, i) to $(x, y, v, \theta + n\beta_v, i)$, $n \in \mathbb{Z}$.

If $\frac{\beta_v}{\pi}$ is irrational, then $\{n\beta_v\}_{n \in \mathbb{Z}}$ is dense in S^1, that is

$$\forall \varepsilon > 0, \; \forall \psi \in S^1, \; \exists n \in \mathbb{Z} : |n\beta_v - \psi| < \varepsilon. \tag{5}$$

We will refer to (5) as to the property of reorienting $\widetilde{P}$ "arbitrarily close" (AC for short) to any direction.

Remark 2 Suppose that a configuration $q_1 = (x_1, y_1, v, \theta_1, i)$ is brought in $q_1' = (x_1', y_1', v', \theta_1', i')$ by a certain trip T. Then, applying T to any configuration

$q = (x, y, v, \theta, i)$, we end up at $q' = (x', y', v', \theta', i')$, where

$$\begin{aligned}
(x', y') &= (x, y) + \\
&\quad \exp\left(i(\theta - \theta_1)\right)(x_1' - x_1, y_1' - y_1), \\
\theta' &= \theta_1' + (\theta - \theta_1).
\end{aligned}$$

For $i = 1, \ldots, l$, let $\widetilde{T}_i$ be the set of all the trips starting and finishing with F_i in contact. For any choice of $m - 1$ out of the m vertices of $\widetilde{P}$, labeled as $v_1, \cdots, v_{m-1}$, the following property holds:

Proposition 2 *For all trips $T \in \widetilde{T}_i$, there exist $m - 1$ integers $(n_i)_{1 \leq i \leq m-1}$ such that the total variation of orientation along T is given by:*

$$\Delta\theta_{|T} = \sum_{i=1}^{m-1} n_i \beta_{v_i}.$$

<u>Proof.</u> To each trip $T \in \widetilde{T}_i$, a closed continuous path γ_T can be associated as follows. Let $T = F_i \cdots F_j F_k \cdots F_i$. For all pairs of adjacent faces $F_j F_k$ with the edge e in common, take a continuous path γ_{jk} starting from the center of F_j and finishing at the center of F_k, that passes through the edge e only and through no vertex. The path γ_T is obtained by concatenating such γ_{ij} for all pairs of successive faces in T.

The polyhedron $\widetilde{P}$ is topologically equivalent to a two-dimensional sphere S^2 and is associated to $\widetilde{P}$, a curvature function K defined as follows ([14]):

$$K(x) = \begin{cases} 0 & \text{if } x \in \widetilde{P} \backslash \widetilde{V}, \\ \beta_{v_i} & \text{if } x = v_i, 1 \leq i \leq m. \end{cases}$$

Let $\gamma_1, \cdots, \gamma_{m-1}$ be a homology basis of $\widetilde{P} \backslash \widetilde{V}$ $\left(= S^2 \backslash \widetilde{V}\right)$ ([14]). Every path γ_T is therefore homologous to

$$\sum_{i=1}^{m-1} n_i \gamma_i, \quad n_i \in \mathbb{Z}, 1 \leq i \leq m - 1.$$

Each γ_i, $1 \leq i \leq m - 1$, can be taken as a simple continuous closed curve on $\widetilde{P} \backslash \widetilde{V}$ enclosing only v_i in one of the two connected components it defines. It is clear that any trip T associated with such a γ_i has the same effects on the polyhedron as the trip $R_{\beta_{v_i}}$. The variation of orientation along γ_T is equal to (Gauss–Bonnet theorem)

$$\Delta\theta_{|T} = \sum_{i=1}^{m-1} n_i \beta_{v_i}. \tag{6}$$

4   Density of the reachable set

The question whether a trip exists that can bring a given polyhedron AC to any configuration in $\widetilde{M}$, can be answered completely in terms of the curvature of the vertices of the polyhedron:

Theorem 1 *The set of reachable configurations of a polyhedron $\widetilde{P}$ is globally dense in $\widetilde{M}$ if and only if there exists a vertex $\bar{v}$ such that $\frac{\beta_{\bar{v}}}{\pi}$ is irrational.*

Proof.
($\Rightarrow$) The proof of the "if" part is subdivided as follows:

$$\frac{\beta_{\bar{v}}}{\pi} \text{ irrational} \overset{(ii)}{\Rightarrow} \left(\mathbf{DM}\right)_i \text{ holds}$$

$$\left(\mathbf{DM}\right)_i \text{ holds} \overset{(i)}{\Rightarrow} \left(\mathbf{DM}\right) \text{ holds} .$$

Proof of (i): By hypothesis, there exists a trip that brings the polyhedron AC to (x, y, θ, i) for some i and for any $(x, y, \theta) \in \mathbb{R}^2 \times S^1$. We want to show that a trip exists that approaches AC (x, y, θ, j), for all j's.

Let T be a trip that brings (x_1, y_1, θ_1, i) into (x_2, y_2, θ_2, j), for any fixed j. By hypothesis, we can go AC to (x', y', θ', i), where

$$\begin{aligned} (x', y') &= (x, y) - \\ &\quad \exp\left(i(\theta - \theta_2)\right)(x_2 - x_1, y_2 - y_1), \\ \theta' &= \theta - (\theta_2 - \theta_1). \end{aligned}$$

By remark 2, there exists a concatenation of trips that brings AC to (x, y, θ, j) from (x', y', θ', i).

Proof of (ii): Let $\bar{q} = (0, 0, \bar{v}, 0, i)$ be the initial configuration. We want to show that if $\frac{\beta_{\bar{v}}}{\pi}$ is irrational, then $\left(\mathbf{DM}\right)_i$ holds.

Let $\widetilde{R}_{\bar{v},i}$ the subset of $\widetilde{M}$ defined as

$$R_{\bar{v},i} = \{q \in \widetilde{R}_{\bar{q}} \mid q = (x, y, \bar{v}, \theta, i)\}.$$

Let us prove that the projection of $\widetilde{R}_{\bar{v},i}$ on $\mathbb{R}^2 \times S^1$ is everywhere dense. By remark 1, if some configuration $(x, y, \bar{v}, \theta, i)$ is reachable from $\bar{q}$, then for every $\psi \in S^1$, we can get AC to $(x, y, \bar{v}, \psi, i)$. Therefore, it is enough to prove that $\Pi_1(\widetilde{R}_{\bar{v},i})$ is everywhere dense. In turn, the previous property is a consequence of the following one:

$$\exists \delta > 0, \ \forall q \in \widetilde{R}_{\bar{v},i},$$
$$B_\delta\left(\Pi_1(q)\right) \subset clos\left(\Pi_1\left(\widetilde{R}_{\bar{v},i}\right)\right), \tag{7}$$

where $clos(.)$ stands for the set closure.

In order to prove (7), let us consider a vertex $v \in \widetilde{V}$, different from $\bar{v}$ and such that $\frac{\beta_v}{\pi}$ is irrational. The existence of the vertex v is insured by the Euler relation. From $\bar{q}$, we can surely reach a point $q_0 = (x_0, y_0, v, \theta_0, i')$ where $\vec{w} = (x_0, y_0) \neq 0$. The trip steering $\bar{q}$ to q_0 is denoted L and the reverse trip, L^{-1}.

Consider the trip $T_{\beta_{\bar{v}}, \beta_v}$ defined as

$$L^{-1} R_{\beta_v} L R_{\beta_{\bar{v}}}.$$

By remark 2, a simple computation shows that we reach from every point $q \in \widetilde{R}_{\bar{v},i}$ a point $q' \in \widetilde{R}_{\bar{v},i}$ such that $\Pi_1(q') = \Pi_1(q) + \vec{t}$, where

$$\vec{t} = \exp\left(i\beta_{\bar{v}}\right)\left(1 - \exp\left(i\beta_v\right)\right)\vec{w}. \tag{8}$$

In equation (8) and in $T_{\beta_{\bar{v}}, \beta_v}$, $\beta_{\bar{v}}$ and β_v can be replaced respectively by any of their multiples $m\beta_{\bar{v}}$ and $n\beta_v$, with $m, n \in \mathbb{Z}$.

Since $\frac{\beta_{\bar{v}}}{\pi}$ and $\frac{\beta_v}{\pi}$ are irrational, by remark 1 and by (8), we can therefore translate $\Pi_1(q)$ with a vector AC to any element of the set

$$\{\exp\left(i\theta\right)\left(1 - \exp\left(i\psi\right)\right)\vec{w} \mid (\theta, \psi) \in S^1 \times S^1\},$$

that is an open disc of radius $2\|\vec{w}\|$. Therefore, (7) is proved.

($\Leftarrow$) Assume now that there exist a vertex v and a face F_i with $F_i \ni v$ such that $\left(\mathbf{DM}\right)_i$ holds. We will show (a)$\Rightarrow$(c).

If $\frac{\beta_{v_i}}{\pi}$ is rational for $1 \leq i \leq m$, we have

$$\beta_{v_i} = \frac{p_i}{q_i}\pi.$$

Let q be the smallest common multiple of the q_i's. As a consequence of equation (6), the variation of orientation along γ_T is a entire multiple of $\frac{\pi}{q}$. This is a contradiction with $\left(\mathbf{DM}\right)_i$. Therefore there must exists $\bar{v} \in \widetilde{V}$ such that $\frac{\beta_{\bar{v}}}{\pi}$ is irrational. This ends the proof of Theorem 1.

5 Rolling a die

Upon examination, it is clear that the set of configurations of a unit cube $\widetilde{C}$, that are reachable from a given initial configuration $q_0 \in \widetilde{M}$, is a discrete set. Taking for instance $q_0 = (0,0,0,1)$, we have

$$\widetilde{R} \subset \mathbf{Z}^2 \times \widetilde{O} \times \widetilde{F},$$

where $\widetilde{O} = \{k\frac{\pi}{2}, \ k = 0,\ldots,3\}$. Let $(\vec{\imath},\vec{\jmath})$ denote an orthonormal frame of P that generates the square lattice determined by the motion of $\widetilde{C}$. If $\vec{k} = \vec{\imath} \wedge \vec{\jmath}$, then $(\vec{\imath},\vec{\jmath},\vec{k})$ is an orthonormal basis of $\mathbb{R}^3$.

Observe that, given any point $q \in \widetilde{M}$, there are 4 ET's, each of them corresponding to an edge of the face in contact with P. Furthermore, the restriction of these actions to $\widetilde{O} \times \widetilde{F}$ are well-defined as

r_1, the rotation with respect to $\vec{\jmath}$ of angle $\pi/2$,

r_2, the rotation with respect to $\vec{\imath}$ of angle $\pi/2$,

and their inverses r_1^{-1} and r_2^{-1}. Since these actions can be undertaken at every point of M, $\widetilde{O} \times \widetilde{F}$ can be seen as the group G generated by r_1 and r_2 with the composition as the multiplication law. One further representation of $\widetilde{M}$ is thus obtained as

$$\widetilde{M} = \mathbf{Z}^2 \times G. \tag{9}$$

The group G is the proper symmetry group of the cube $\widetilde{C}$ and has a simple description using the group S_3 of permutations of 3 elements. Let $\sigma \in S_3$ be represented by the triplet $\left(\sigma(\vec{\imath}),\sigma(\vec{\jmath}),\sigma(\vec{k})\right)$ and $\epsilon \in \mathbf{Z}_2^3$

by $(\epsilon_1,\epsilon_2,\epsilon_3)$, with $\epsilon_i^2 = 1$ for $i = 1,2,3$, where $\mathbf{Z}_2$ denotes the multiplicative group $\{-1,1\}$. Introduce the semi-direct product

$$\sigma \odot \epsilon = \left(\epsilon_1\sigma(\vec{\imath}),\epsilon_2\sigma(\vec{\jmath}),\epsilon_3\sigma(\vec{k})\right),$$

and let $G_1 = S_3 \odot \mathbf{Z}_2^3$. Note that the elements of G_1 transform the orthonormal basis $(\vec{\imath},\vec{\jmath},\vec{k})$ to another basis that is still orthonormal, but possibly with a different handedness. Then, G is the subgroup of G_1 of the elements transforming $(\vec{\imath},\vec{\jmath},\vec{k})$ to another basis with the same handedness, i.e. such that $\det\left(\sigma(\vec{\imath}),\sigma(\vec{\jmath}),\sigma(\vec{k})\right) = 1$. An element of $\widetilde{M}$ for the cube in the representation (9), is written as (m,n,g).

Define A_1 (resp. A_{-1}) as the subset of G corresponding to a product of an even (resp. odd) number of ET's. Note that A_1 is a subgroup of G and, by construction of the multiplication table of G, one gets $\mathrm{card}(A_1) = 12$. The subset A_{-1} has the same cardinality of A_1, however it is not a subgroup. Starting from a given face, there is an element in A_1 (resp. A_{-1}) that brings any face of the cube in contact with P. All elements of A_1 act on the cube so that it ends up with only two possible orientations, differing by π. The latter sentence applies to A_{-1} as well, but the set of possible final orientations under A_1 and A_{-1} are disjoint.

Consider next the infinite group $\widetilde{G} = \mathbf{Z}^2 \oplus G$ (isomorphic to $\widetilde{M}$) with the multiplication law "$\cdot$" defined by

$$(m,n,g) \cdot (m',n',g') = (m+m',n+n',g.g').$$

The group $\widetilde{G}$ is generated by the two ET $R_1 = (\vec{\imath},r_1)$ and $R_2 = (\vec{\jmath},r_2)$, and acts transitively on $\widetilde{M}$.

Define for $q = (m,n,g) \in \widetilde{M}$ and $g' \in G$ the homomorphisms h_1 and h_2 as

a) $h_1(q) = m + n \pmod 2$;

b) $h_2(g') = 1$ if $g' \in A_1$, -1 if $g' \in A_{-1}$.

We are now in a position to state our main results concerning the problem of rolling a die on a plane. The first proposition makes explicit the lattice structure and the restrictions on the reachable set of configurations:

Proposition 3 *Let* $q = (m, n, g) \in \widetilde{M}$. *Then*

$$\widetilde{R}_q = \Big\{ q' = (m + m', n + n', g') \in \widetilde{M}$$
$$\text{with } g' \in A_{h_1(q')h_2(g)} \Big\}.$$

The second result deals with bounds on the number of maneuvers necessary to reach an arbitrary configuration in $\widetilde{R}_q$:

Proposition 4 *Every* $q' = (m', n', g') \in \widetilde{R}_q$ *can be reached from* $q = (m, n, g) \in \widetilde{M}$ *by a trip of length* L *with*

$$L \geq |m' - m| + |n' - n| \qquad (10)$$
$$L \leq |m' - m| + |n' - n| + 6 \qquad (11)$$
$$L \leq \sup\{4, |m' - m|\} +$$
$$\qquad \sup\{4 + |n' - n|\} \qquad (12)$$

The proof of these propositions is based on examination of the multiplication table of the group G above defined, and is omitted for brevity.

Remark The lower and upper bounds (10) and (12) coincide in the region $|m' - m| \geq 4$, $|n' - n| \geq 4$. An optimal trip therefore exists for any reachable configuration in this region, and its length is exactly $|m' - m| + |n' - n|$.

6   Conclusions

In this paper we undertook the analysis of the set of configurations that a polyhedron can be brought to reach by rolling on a plane about its edges. The problem appears to be important to practical applications, such as that of automatic part manipulation, as well as theoretically stimulating. As a result of our analysis, we pointed out that the structure of the reachable set may show a much richer variety for different polyhedra than it results for different regular surfaces, which were analyzed previously. Results of this paper concern only two extreme cases, while several intermediate cases with different characteristics were not solved here (for instance, a right–angled box with different sides).

Among the many open problems that are left for future work, we point out that the criterion for density

of the reachable set of Theorem 1 is based on the irrationality of a quantity, whose actual value in physical problems can only be determined up to an error. It is clear therefore that it would be important to have more robust measures of how "manipulable" a given polyhedron is.

Acknowledgements

This research was supported in part by ESPRIT W.G. LEGRO, contr. no. 032/94/TS. Yacine Chitour is currently a Visiting Scientist in Pisa with the support of ERNET, the European Robotics Network, EU(HCM) contr. no. ERBCHRXCT930381. Alessia Marigo is currently Visiting Scholar at the Institute d'Analyse Algébrique, Université de Paris 6.

References

[1] Y. Aiyama, M. Inaba and H. Inoue. Pivoting: a new method of graspless manipulation of object by robot fingers. in *Proc. IEEE/RSJ Int. Conf. on Int. Robots and Systems, IROS'93*, 1993.

[2] A. Bicchi, D. Prattichizzo and S.S. Sastry. Planning motions of rolling surfaces. in *Proc. IEEE Conf. on Decision and Control*, 1995.

[3] A. Bicchi, and R. Sorrentino. Dexterous manipulation through rolling. in *Proc. IEEE Int. Conf. on Robotics and Automation*, 1995.

[4] I-M. Chen and J.W. Burdick. A qualitative test for n-finger force–closure grasps on planar objects with applications to manipulation and finger gaits. in *Proc. IEEE Int. Conf. on Robotics and Automation* 1993.

[5] M.A., Erdmann, M.T. Mason and G. Vaneček. Mechanical parts orienting: the case of a polyhedron on a table. *Algorithmica* volume 10, number 2, 1993.

[6] K. Goldberg. Feeding and sorting algorithms for the parallel –jaw gripper. in *Proc. 6th Int. Symposium on Robotics Research*, 1993.

[7] S.C. Jacobsen, S.G. Meek, and R.R. Fullmer. An adaptive myoelectric filter. in *6th IEEE Conf. Engineering in Med. & Biol. Soc.*, 1984.

[8] Z. Li and J. Canny. Motion of two rigid bodies with rolling constraint. *IEEE Trans. on Robotics and Automation,* volume 6, number 1, 1990.

[9] K. M. Lynch and M. T. Mason. Pulling by pushing, slip with infinite friction, and perfectly rough surfaces. *Int. J. of Robotics Research,* volume 14, number 2, 1995.

[10] M.A. Peshkin and A.C.Sanderson. Planning robotic manipulation strategies for workpieces that slide. *IEEE Journal of Robotics and Automation,* volume 4, number 5, 1988.

[11] D. Rus. Dexterous rotations of polyhedra. in *Proc. IEEE Int. Conf. on Robotics and Automation,* 1992.

[12] J.K. Salisbury, D.L. Brock and S.L. Chiu. Integrated language, sensing and control for a robot hand. in *Proc. 3rd ISRR,* MIT Press, Cambridge MA, 1985.

[13] N. Sawasaki M. Inaba and H. Inoue. Tumbling objects using a multifingered robot. in *Proc. 20th ISIR,* 1989.

[14] M. Spivak. *Differential geometry.* Publish or Perish, Houston, Texas, 1979.

[15] P. Tournassoud, T. Lozano–Perez and E. Mazer. Regrasping. in *Proc. IEEE Int. Conf. on Robotics and Automation,* 1987.

CAD Geometry Algorithms in a Large Industrial Enterprise

Alan K. Jones, *The Boeing Company, Seattle, WA, USA*

We survey CAD geometry algorithms currently in use at the Boeing company. Geometry creation, NC tool path generation, and visualization are well represented. Other areas, for example path planning for assemblability or maintainability, are not so widely used. We illustrate the principle that each design or manufacturing process, together with the algorithms that support it, is embedded in a network of upstream and downstream processes. Thus, each algorithm imposes constraints on, and is constrained by, many others. Finally, we discuss the difficulties of technology transfer. That is, how does one ensure that new mathematical techniques are actually used, and actually contribute to the profitability of the company?

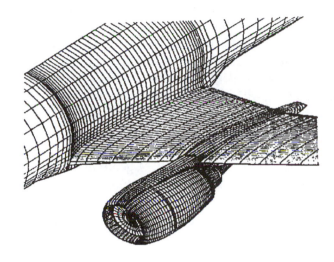

Figure 1: *A Discrete Grid on an Airplane*

1 Introduction

This paper is a brief survey of CAD geometry algorithms in the aerospace industry, and in Boeing in particular. Nothing in this work is to be construed as an official statement or position of The Boeing Company. Rather, it reflects the personal perspective and personal opinions of the author. The following background may help the reader appreciate what that perspective is.

Like any major corporation, Boeing is a big place, and not by any means a unified one. To begin with, there is a major split between the commercial and military sides of the house, the Commercial Airplane Group (BCAG) and the Defense and Space Group (BDSG). Within those groups, it has historically been organized around programs. In BCAG, for example, there are five major programs corresponding to the airplanes currently in production: 737, 747, 757, 767, 777, with internal cultures and practices developed over a period spanning three decades (747 in 1965, 777 in 1995). Even within a program, the complex process from design through manufacturing design to building, and then after-sale support is overseen by a number of historically separate organizations, each with their own point of view and way of doing things.

The author's organization, the Research & Technology arm of the Information and Support Services division, is a research, development, and consulting agency consisting of a few hundred specialists in computer science, mathematics, and related fields. We support the entire range of company activities, but the problems that we see at any given time tend to reflect the company's current priorities. For many years, our principal internal customers were the engineers and analysts supporting the design activity. Now, with the increased emphasis on high quality, cost effective production, the customers are just as likely to be from the manufacturing side.

Within BCAG at least, some de facto unity is imposed by the domination of computer-aided geometry over the last decade by a common CAD/CAM system and its vendors. Other products can be found there, but generally only if they provide capabilities not found in the common system. This standardization has been possible for BCAG because it has relatively few distinct programs, and is the prime contractor on each of them. The situation is different on the BDSG side. The culture there is of small, often "black" (secret) programs, isolated from each other for security reasons, each necessarily owning its own tools. Furthermore, it's very common for a BDSG group to subcontract for, or partner with other aerospace companies who impose their own CAD system on the project. A good rule of thumb is "BDSG has at least one of everything."

2 Taxonomy of Computer-Aided Geometry Software

First, in an attempt to impose order on the chaos, let's look at a few candidate organizing principles. One possible taxonomy for geometry software runs as follows

- Products of the primary vendor, Dassault Systèmes

- Other vendors (CADAM, Pro/E, etc.)

- Boeing additions to vendor software

- Boeing-developed systems

At the moment, there is a growing tendency not to seek solutions of the last type. It's related to the current fascination in the corporate world with COTS (Commercial Off The Shelf) everything, including software.

A second possible taxonomy runs as follows, roughly paralleling the product lifecycle process of design-build-maintain.

1. preliminary design / sizing / configuration

2. detail design (geometry creation and recording) – *the traditional CAD activity*

3. analysis (aerodynamic, structural, electromagnetic, and many others)

4. numerical control (NC) programming

5. design of assembly processes and tooling

6. other process planning issues

7. manufacturing process control

8. maintenance planning and documentation

Of course, the real industrial process is not a simple linear progression, but a collection of nested feedback loops. Items 2 and 3 in particular are in an especially tight loop.

Finally, there is a third, very loose taxonomy of mathematical and/or computing disciplines. I've listed these in approximate descending order of overall importance, or at least "market penetration", but these judgements are very subjective, and also very dynamic. KBE, for example, seems to be rapidly increasing in importance, and possibly in maturity and capability as well.

- graphics (See below)

- numerical analysis

  This is what my particular group at Boeing does. It's relevant here especially as applied to parametric spline curves and surfaces, and all the sorts of manipulations that need to be performed on them.

- robotics, particularly for NC applications

- Knowledge Based Engineering (KBE)

  Here, KBE is taken in the broadest sense, including some applications that might be more properly called parametric or generative geometry, though rule-based systems with inference engines also exist. We use a major commercial product as well as the internally developed system GENESIS [12].

- nonlinear optimization

  This is another strength of my local group at Boeing.

- classical computational geometry

  In my experience, algorithms like this seem to arise mostly in areas related to metrology. (But see Section 3.6 for a counterexample.) For example, Voronoi triangulations are a typical first step in trying to organize sets of measured data points taken off of surfaces in space, for metrological or surface reverse engineering tasks. (See Section 3.5.4.) Verifying mechanical tolerances can involve finding minimal containing circles to sets of discrete measured data points. (See Section 3.7.)

The rest of this paper is organized as follows. In the next section, we follow the life cycle taxonomy, and survey some of the geometry related tools and algorithms in use today, primarily though not exclusively on the commercial side. The search is mainly breadth-first, with just a a few topics explored in more depth. Then, we give examples of interesting and unexpected interactions among these algorithms. Finally, we abstract to the interactions between tools and people – issues of corporate culture.

3 Survey of Applications

First, before we begin, let's say a few words about two issues that pervade everything else – graphics and data storage.

Graphics

Visualization is the medium by which the modern user interacts with just about everything. It enables us to make best use of the unique pattern recognition capabilities of human brains. The best available tool in this area appears to be the FlyThru CAD visualizer described in [5] and [1], a Boeing-developed tool, retrofitted onto our primary CAD system and various other applications. More than just a graphics application, FlyThru also supports sophisticated searches and queries based on geometric location and various other categories of data. It was done in-house because we needed better performance than any then on the market, primarily because of the immense scale of operation associated with our products, in terms of sheer number of polygons, number of CAD models, etc. We'll have more to say on this issue of scale immediately below. Technically, FlyThru uses all the tricks one might expect – hierarchical levels of detail, frustum culling, optimization of display list traversal. One could say that it is the whole rather than the parts which is remarkable.

Data Representation and Storage

Questions of which algorithms to use for creating, manipulating and displaying geometric data are important, but equally important are questions of how and in what form those data are stored and shared between applications. Two key issues here are:

- **Scale.** It takes a huge amount of data to define the geometry of a single airplane, with $O(10^6)$ individual parts. A 777 airplane contains about 40,000 separate CAD models, and more than 15 gigabytes of disk space. A FlyThru rendering of the complete model would require $O(10^9)$ triangles. Thus, the development of FlyThru had to be a cooperative effort by graphics and database experts. Indeed, the data base aspect has been redesigned several times in order to handle the increasing amount of data and evolving patterns of access.

- **Time.** Our data must be accessible for 50-100 years.

  This is primarily for commercial reasons, to support our customers, but also to meet regulatory requirements, and for obvious legal reasons. The timespan chosen is not arbitrary or frivolous. DC-3's designed in 1936 are still flying, as are B-52's and 707's designed 40+ years ago. The model 747, designed c. 1965, is planned to be in production past 2010, and flying for decades beyond that. One obvious problem is that this time spans multiple generations of storage technology, but what concerns us at the moment is that it also spans multiple generations of mathematical technology. For example, electronic definitions in the 1960's were procedural definitions of individual surfaces, essentially just computerized versions of a conic lofting paradigm used in the industry since before World War II. In the 1990's, they are definitions of solids in terms of trimmed and joined rational B-spline (NURB) surfaces. No exact conversion is possible in either direction. One message is clear. The ability to migrate existing data into an enhanced, extended, or possibly totally different mathematical form is important, and to the extent possible we must plan for such eventual requirements from the beginning. One recent example comes to mind where this could, perhaps, have been done better. It was standard practice in some parts of the 777 program to isolate completed solid models (i.e., throw away the CSG history of the solid) in order to conserve disk space. It was a reasonable quick fix to an immediate problem. However, as soon as a new version of CAD software arrived with new, improved solid modeling capabilities, all the old polyhedral solids needed to be converted to exact solids. An operation that would have been more or less straightforward if more data had been saved became instead a difficult, labor-intensive exercise.

3.1   Preliminary Design

The preliminary design stage can be thought of as a microcosm of the full-scale design process characterized by simple, idealized geometry and relationships. On this level of abstraction, the number of variables being manipulated is sufficiently small that rule-based systems and numerical optimization algorithms can sometimes be used to guide the layout of major components of a design – wing and fuselage geometry, engine size and placement, landing gear size and placement, and type and location of major systems within the aircraft. A variety of sketching and synthesizing tools is used throughout the company, including at least one KBE-based configurator system.

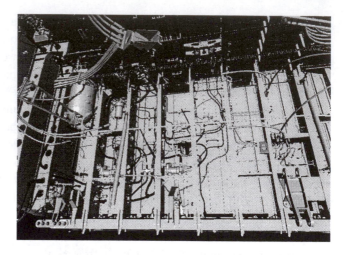

Figure 2: *Tubing on a Typical Airplane*

3.2   Detail Design

It is at the detail design stage where the exact, nominal geometry of the airplane is defined and recorded. CAD systems, at least those currently installed, are very good at the latter, and less well adapted to the former.

The 777 airplane, which was an all-new design, had a goal of 100% digital definition. The current design programs, derivatives of the 737 and 747 basic designs, are also being designed this way. This can pose a challenge if the designs on which derivatives are based are not already available in usable electronic form. A major effort in reverse engineering can be involved. See Section 3.5.4 below.

Not much of the process of creating new geometry is truly automated. Numerical optimization techniques are still not widely used, although their penetration is increasing. Rule-based generative design can be effective in some cases, for example in the routing of pipes and wire bundles, where complex design rules must be obeyed. For example, minimum separations are imposed between oxygen and fuel lines for safety reasons. The resulting simultaneous routing problem could be considered a complex variant of path generation with obstacles. Figure 2 shows the main landing gear wheel well of a typical airplane, with some of the tubing that must be routed through it.

Not much can be said about the geometry algorithms of our primary CAD system, because they're vendor proprietary, which means that even we don't know what they are. It is known that the internal representation for the most common form, namely parametric polynomial splines, is in terms of the power basis rather than B-splines, hence prone to some interesting numerical problems. The geometry construction algorithms themselves can give unexpected results, but generations of bandaids and workarounds have entered the common culture. Engineers are very good at that. On the positive side, it appears that the software architecture is now opening up, as the hardware environment already has. This should simplify the incorporation of third-party software point solutions in the future. Most of these observations are probably fairly typical of the current state of practice, and are more or less independent of the CAD vendor involved.

3.3   Analysis

Analysis comes in many flavors – aerodynamic, structural, electromagnetic (e.g., analyzing radar signatures for stealth aircraft designs), and mixtures of flavors – for example, aeroelastic analysis to predict the excitation of structural vibration modes by aerodynamic effects. All these analyses require some knowledge of airplane geometry, typically in a unique, idealized form. Thus, for aerodynamic analysis, one discretizes geometry into two- or three-dimensional grids. For structural analysis, one discretizes it into finite elements. See Figure 1, which is a surface grid on a generic transport airplane configuration.

On the aerodynamics side, for example, numerous packages are available from NASA and various commercial sources. In addition, there is the Aerodynamic Grid and Paneling System (AGPS) [6], [24], a Boeing-developed code which first evolved in the late 70's to bring spline geometry to aero preliminary design and analysis. It's interesting for two reasons. First, because the unusually rich command language of the command line driven version is in effect a geometry description language for parametric geometry of (idealized) airplanes and panelings. (That's of particular value for commercial transport programs, since the basic topol-

ogy of airliner design has been remarkably stable in recent decades.) Second, because of the rich grammar of mathematical function types that can be represented procedurally, including functional composition, Gordon blends, and recursive rectangular arrays. Portions of this mathematical structure are also available in the DT_NURBS spline library, developed by Boeing for the US Navy, and freely available to US companies [28].

Analysis engines themselves are out of scope for this survey. But once the results are computed, they need to be reduced to some useful form – visualized, or fed into some other analysis or process. Thus, it's not always just a question of graphics – a downstream computerized analysis will most likely need a functional form, not a picture, and if that analysis is used inside an optimization loop for design, it had better be a well-behaved one. Boeing has in-house expertise in fitting functions to scattered data in many dimensions, dating back, for example, to [10], but this is still not a pushbutton task. Finally, note that the same process applies when data are from wind tunnel or other tests, rather than simulations – not all analysis is digital! The additional complication in these cases is a random component in the error which calls for some statistical sophistication as well.

3.4   Numerical Control Programming

Programming of milling and drilling machines is done using standard CAD functions. Again, we don't always know the details of the algorithms involved, but they do have at least one interesting property which is explored in Section 4 below.

These traditional NC processes apply to tools that remove material. But there are machine tools that add material instead. For example, consider the case of a paint sprayer robot which must be programmed to produce a prespecified profile of coating thickness.

A more complex application of the material-adding type is tape layup for graphite-epoxy composite materials. Software for this task is provided by the vendors of the machine tools involved, though there was a Boeing effort a few years ago to explore the possibility of in-house codes. The fundamental geometry problem is to prescribe tape center lines along a doubly curved mold surface which are

1. approximate geodesics on the surface, so that the tape will lie flat without wrinkles

2. in approximately the right direction, to assure proper material properties of the finished part

3. spaced to avoid unacceptable gaps or overlaps between adjacent tape courses.

Complicating the analysis is the fact that, after the first layer of tape is laid down, the top surface of this layer becomes the new target surface on which the next layer must be laid. Its location in space, and its curvature properties, will be different from those of the original mold surface. In fact, both can be quite radically different if inserts and ply drop-offs are included in the design.

3.5 Design of Assembly Processes and Tooling

3.5.1 Assembly Planning

Assembly planning is a lively subject in academic circles, as shown for example in [3], [27], [30], [31], [32], and many others. For a commercial end user's perspective, see [16]. A typical goal is to generate automatically a feasible and efficient assembly sequence for a given assembly of parts. Feasibility is demonstrated by producing an approach path for each new part in the sequence. Numerous possible metrics for efficiency are proposed in the literature – minimizing number of hands required, or complexity of the assembly line tree, or whatever. In practice, planning like this is still done manually, using experience and intuition to balance multiple objectives. Manual verification of a plan's feasibility is very difficult without a visualization tool, but this is a natural application for a viewer like FlyThru; conversely, synthesizing the plan in the first place seems to be much easier for a human brain, with its massively parallel architecture, than for conventional software. (See also the discussion of disassembly planning in Section 3.8.)

3.5.2 Tolerance Stackup Analysis

There is one area of assembly analysis in which the objectives are precise and quantitative – tolerance stackup analysis. Not coincidentally, computerized geometry is used extensively in this area. The basic problem is this. We know that component parts are not of perfect, nominal geometry, but are free to vary within the prespecified tolerance zones which form part of their engineering definitions. If we define key characteristics in the final assembly (and specify how to measure each one), what can we say about the expected values of those characteristics in the assemblies actually produced? For a general introduction to the issues involved, see [4], [18].

In *worst case* tolerance analysis, one just calculates the resulting tolerance zones for the key characteristics. This is related to the work of [19], which seeks to produce assembly sequences which are worst-case feasible. Less pessimistically, *statistical* tolerance analysis assumes knowledge of the statistical distribution of parts within their respective tolerance zones, and computes the resulting distribution of key characteristics of the population of actual assemblies. This can be done either by straightforward Monte Carlo simulation, or by linearizing the geometrical relationships involved and exploiting the linearity of means and variances with a weighted RSS (root sum squares) computation.

Linearizing the local geometrical relationships involves a fairly deep understanding of the classification of mating feature relationships. Perhaps unsurprisingly, the mathematics involved turns out to be an infinitesimal, linearized version of what the mechanical engineering folklore has long known about the classification of lower kinematic pairs. See [8], [13].

Software for this type of analysis comes mainly from commercial vendors whose primary market is the US auto industry. Historically, its use within Boeing has been quite limited, although that appears to be changing rapidly. New developments in the field suggest that soon it will be possible to account for the effects of part flexibility on tolerance stackup [20], [23], and, conversely, for the effects of stackup on the kinematic functioning of mechanisms [17].

It's clear that there is a natural feedback loop connecting assembly tolerance analysis with assignment of individual part tolerances. The relationship between tolerances and assembly plans is probably less clear. This comes about because an assembly plan for real, as opposed to idealized, parts is more than just an assembly sequence. It also specifies at each step the *indexing scheme* by which the parts to be joined are actually located to each other – which features on part A are aligned with which feature on part B, and in what order. One option for alignment is to use a special jig or tool. So assembly planning can also include implicit specification of assembly tools.

3.5.3   Tool Design

Designing and fabricating tools can be just as complex and expensive as fabricating the end product. In this context, the word "tool" can mean anything from a hand-sized mold for forming small sheet metal parts to layup mandrels for large composite-material surfaces (see Section 3.4.) to massive assembly jigs.

Tool design is primarily a manual task using conventional CAD, but not always. For example, at least one project has had success with KBE in designing the substructures that support massive tools. The number and detailed design of the supports provided, not just their dimensions, are functions of the dimensions and weight of the tool. This illustrates the fact that parametric geometry isn't necessarily a continuous function of its parameters. Or, put another way, the topology of a piece of parametric geometry may itself be parametric.

Recently, work has been done on incorporating springback effects into mold designs. Traditionally, the term "springback" applies to metal forming – when the pressure is released, the workpiece "springs" back slightly from its fully bent position. With composite parts, a different but similar problem arises during the curing process. The problem in both cases is to modify the shape of the mold so that the actual part produced, including springback, is in fact the nominal part as designed. For a simple sheet metal part, this usu-

ally means just sharpening the angle of bend by a predictable amount. However, with more complex parts, or with composites rather than metal, a single handbook value isn't enough to characterize the problem. In these cases, springback can be computed by a finite element analysis, as a field of displacement vectors at discrete nodes over the entire surface. The geometry problem, then, is to modify the entire continuous surface definition to incorporate the corresponding springback corrections.

3.5.4   Surface Reverse Engineering

Older part and tool designs may have no electronic counterpart compatible with current systems and processes. Some may have no electronic design at all. In addition to the derivatives and redesign problems mentioned in Section 3.2 above, there is the issue of *tool scrapping*. Rarely used tools, which must still be maintained in case new parts are needed to support users in the field, incur tremendous expenses for storage, and eventually deteriorate anyway. Furthermore, the physical immobility of a very large tool in effect locks a company into a single manufacturing location, and a single supplier. Thus, it is sometimes necessary to reconstruct surface, or even solid geometry from clouds of data points digitized off existing artifacts. This is an area of tremendous activity in the academic and vendor community today, and of course many end-user companies have their own in-house codes. Many interesting issues enter into this problem, including:

1. manipulation and filtering of data points (thousands, millions, or more)

2. fitting scattered data with parametric surfaces

3. diagnosing local connectivity (topology) from discrete points

4. feature recognition – edges, creases, corners, and other discontinuities. An interesting complication is that not all detectable features should be reproduced. Scratches or hammer dings, for example, should probably be suppressed.

The vendor software we've seen does a good job of (1), but that's about all. In particular, capabilities like those described in [15] don't seem to be readily available.

3.6 Other Process Planning Issues

Keeping any modern factory running means careful scheduling and allocation of resources. Often those resources can be physical space, and we find the abstract methods of Operations Research useful in tackling problems with a geometrical flavor.

A problem of a more conventional geometric character provides a classic success story for "quick and dirty" industrial-grade mathematics. *Nesting* is the optimal placement of parts on two-dimensional sheet stock. The stock may be, for example, sheet metal, in which case the parts are free to rotate as well as translate. Conversely, it may be fabric for carpets or interior blankets, or perhaps some sort of composite material. In that case, grain restrictions will probably restrict the rotations to multiples of 90° or even 180°.

About four years ago it became clear that the COTS software then in use to do this job was too slow, and produced nests in which the users had no confidence. On the one hand, the automatically generated nests were sometimes clearly suboptimal, because it was obvious how to improve them. On the other hand, they contained unexpected features clearly not required by the solution.

A small internal team made an extensive search of the literature, and found a combination of off-the-shelf algorithms and clever programming which together produced much better performance:

1. By a sequence of local operations, approximate each part definition up to tolerance by an enclosing polygon with a small number of edges.

2. Cluster individual parts into enclosing rectangles with a small amount of waste, such that the rectangles are not too large for step 3. It is at this stage that rotations, if allowed, are tried. The location at which a candidate rotated part is added

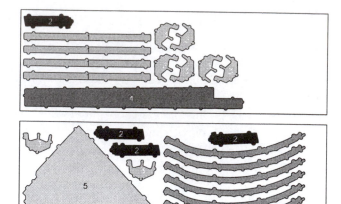

Figure 3: *Nesting of Parts on Sheet Metal*

to an existing cluster is found by optimization, using techniques of [26], over the appropriate configuration space polygon, which in turn is computed as in [2] and [21].

3. Apply the rectangle packing algorithm of [9].

See Figure 3 for a typical nest produced by this code.

3.7 Manufacturing Process Control

This section will actually treat two different but related subjects. The first is the geometrical side of statistical process control. The second is the geometrically elementary but still important problems that often come up in the real-time control of major assembly operations.

Manufacturing process control proper is mainly a statistical procedure. However, geometrical issues naturally arise when an objective of the process being controlled is to produce correct geometry of a part or assembly. Traditional practice for verifying geometrical correctness is based on the idea of hard gages, and go / no-go tests. More modern practice uses discrete points measured, either mechanically by special purpose robots called CMM (Coordinate Measuring Machines), or optically via laser range finders, CAT (Computer Aided Theodolites), photogrammetry, etc. Because these methods produce measurements rather than just yes/no results, they are more amenable to statistical analysis, but the underlying theory is not yet well developed, and the state of actual practice in the vendor community, as reflected in that of the national and international standards arena, is somewhat chaotic. The question of how many points to sample, and where, and what to do with them afterwards to verify a geometrical tolerance turns out to be an interesting mixture of statistics with computational geometry, as mentioned in Section 2. For an introduction to the questions that arise, and the kinds of algorithms involved in solving them, see [25].

Assembling very large structures – for example, joining body sections together, or joining wings to bodies – requires accurate location of very large parts with respect to each other. If this is done, for example, by CAT measurements, then how does their accuracy depend on the geometry of the relative locations of the three or more theodolites involved? And how should the setup be designed to maximize utility for a given task? Assuming, then, that several known tooling locations (or, failing that, many more unknown locations) on each part have been accurately measured, how do we use these data to construct the correct rigid motion that will match the coordinate frame of one component with that of the other? This is just the point-point alignment problem – find a rigid motion which best aligns two sets of corresponding labeled points. The least squares version of this problem has a simple, one-step solution using the matrix singular value decomposition (SVD) algorithm [11]. The problem becomes more difficult when we use a norm other than L_2, or add side conditions. For example, certain translational or rotational directions may be forbidden, or we may require the alignment to be *conservative*, that is, keep all the errors on the non-material side of the mating surfaces.

3.8 Maintenance Planning and Documentation

Part of the customer support activity is planning for maintenance. This involves a certain amount of *disassembly planning*. For example, if a part needs to be changed as part of some maintenance procedure,

there must be a plan for reaching it and then removing it. The same sort of tools are generally applicable to this problem as to the assembly planning described in Section 3.5.1. (Indeed, assembly planning is generally done in practice by producing a disassembly plan and then reversing it, since this is the most expedient means of specifying the desired fully assembled state.) Thus, the state of the art is essentially the same as that described in Section 3.5.1. Namely, software is more useful for verifying correctness of disassembly plans than for generating them in the first place. The only software that I know of in-house for this purpose is that described in [7], [29], which was developed by one of our engine vendors.

Another important, if unglamorous activity, is the creation of maintenance manuals and training materials. The PigsFly (Production Illustration Graphics System) application, built on top of FlyThru, has proved effective in cutting costs and flow times.

4   Technical Interactions

Computer aided geometry tools and processes are often designed and developed in isolation from each other, but as we may infer from the discussion of process flow above, they are seldom used that way. This means that there will inevitably be unanticipated interactions and incompatibilities among them. One consequence of this is this is that a discouragingly large fraction of all the computer programs ever written have been for the purpose of converting one file format into another. In the 1990's, the simple problems of moving geometry data around have been somewhat ameliorated by the prevalence of the IGES standard, and its successor, STEP. Thus, the interface problems that do crop up tend to be a little more complex. Here are some quick examples.

- Importing geometry into CAD for use in NC

  The algorithms used in our primary CAD system for NC programming of milling machines allocate the same number of GOTO points in each polynomial patch that a cutter path traverses. (This appears to be based on a quick procedure they have for providing an *a priori* upper bound of the number of points required per patch.) For this, and other reasons, surfaces as used in that system at Boeing must be strictly limited in the number of patches they contain. This means that surfaces produced by other CAD systems, which are generally not limited in this way, must be approximated up to tolerance by simpler surfaces before they can be imported. This is done by a Boeing-developed add-on application based on ideas related to, but significantly improved over [22].

- Designing surface geometry for use in NC

  It's possible to define cutter paths along plane cuts. However, this is very slow, so the preferred practice that has arisen is to use isoparameter curves instead wherever possible. This imposes a requirement on the surface creation stage, namely that individual surfaces should have smooth, evenly distributed isoparameter curves, and that such curves should match and join smoothly across surface boundaries. Thus, actual parametric continuity is preferred whenever possible.

- Design surfaces for efficient machinability

  For cost-effective milling, machinists want to use the largest cutting tool possible. This in turn imposes lower bounds on the minimum radius of curvature of the surfaces we design, which is an awkwardly nonlinear constraint. A Boeing-developed add-on application does this in simple cases.

- Exporting surface geometry from CAD

  When polynomial spline surfaces are exported from our primary CAD system, they come out in piecewise Bézier form, that is, as rectangular arrays of bipolynomial patches, with no explicit global parametrization or parametric continuity information. It's generally necessary to recreate these data, and repackage the surface in tensor product B-spline form before it can be used efficiently by other CAD systems. A Boeing-developed add-on application does this.

- Exporting CAD geometry to FlyThru

  The fast graphics of FlyThru depend on having polyhedral solids to work with. In early versions of our primary CAD system, this is the form in which solids where actually kept, so the connection was quick and easy. However, when a new version arrived which supported boundary representation (brep) solids based on general sculptured surfaces instead, this connection was suddenly broken. Workarounds were quickly found to reestablish it, of course, but this illustrates the maxim that an upgrade isn't always a good thing. Upgrading just one node in a graph of interrelated computer applications may not be a step forward. It may not even be possible.

The Solid Modeling Accuracy Problem

So far, we've talked about sharing curves and surfaces between processes. When brep solid models are being shared, the problem suddenly becomes much harder. All the algorithms that go into constructing solid models, especially the Boolean operations of union and intersection, and into interrogating the resulting new models later, must be logically/numerically consistent. That is, tests which are logically equivalent (in exact arithmetic) must give the same answers when actually made in the computer, using necessarily inexact arithmetic.

Various strategies have been identified for solving this problem in principle [14]. None has been rigorously applied, but, perhaps mostly through extensive trial and error, most individual CAD systems have matched their algorithms and choices of tolerances to get the failure rate down to the noise level. Problems do, however, tend to arise when a brep solid is built in one system and used in another. Valid solids on system A may not be recognized as valid by system B, even though all of their constituent parts appear to have transferred correctly. More insidiously, we may have the following failure of "commutativity". Let X and Y be two solids as represented in system A, and X_B and Y_B be the same solids as transferred to system B. Then we may have, for example,

$$(X \cap Y)_B \neq X_B \cap Y_B$$

That is, the following diagram may fail to commute.

$$
\begin{array}{ccc}
A & \xrightarrow{\cap} & A \\
\Downarrow & & \Downarrow \\
B & \xrightarrow{\cap} & B
\end{array}
$$

Organizations contributing to the STEP standard for exchange of product data are working this problem very aggressively. Boeing is in the forefront of applications testing, as a leading member of the PowerSTEP project. This is an effort to share geometry and related information between Boeing and its engine vendors – a high-leverage opportunity, since the airframe-engine interface is probably the most complex one in our entire business.

5 The Technology Transfer Problem

Analyzing industrial problems and creating new software, or even new mathematics, to deal with them is a challenging and rewarding task. However, it does not in and of itself provide any benefit to a company. To do that, we must ensure that the new tools are used, and used effectively, at the working levels of the organization. This is the problem of *technology transfer*.

Why is this a problem? After all, as the old proverb says, "If you build a better mousetrap, the world will beat a path to your door." Unfortunately, in real life, this does not always happen. It's necessary to get out and *sell* the mousetrap, if necessary door-to-door. And just convincing a customer to pay for research in the first place does not automatically mean that you have convinced him or her to use the results!

What sort of arguments do convince the potential user to try out a new mousetrap? First, note that two different sales must be made simultaneously – to the actual end users, and to their managers. Selling to these two different constituencies involves two different sets of challenges, and the form of the arguments used to convince them is superficially different, but the actual content is very similar. First, the potential user has to answer three questions.

- Can I see how to fit this into my work?

- Can I trust the results? (Recall the discussion of the former nesting tool in Section 3.6. Nothing is worse for a computer application than a perception that it is unreliable.)

- Can I see that the results will be worth the investment to climb the learning curve?

Second, and more importantly, the user needs to have a good "gut feeling" about the new tool. It's appropriate to quote here from the developers of FlyThru in [1].

> ...most (80% ?) of the challenge of getting users on board is emotional; that is, an easy to use software package was not enough. The success with the application problems – design quality improvements and reduced cycle-time had to be widely discussed and 'talked-up' to get new users trying the system.

5.1 The Management Level: Inertia of Scale

For the manager, the issue is usually phrased as an expanded version of the third question above, namely

- Can I make a business case for this?

That is, is it plausible that implementing this new technology will save more money than it costs? Note that this is *not* automatic, even when the new technology is completely successful on its own term.

Of course, whether or not a business case will be validated after the fact depends on how the goals are set. For example, a faster, more capable CAD system does not necessarily mean that the engineering design process itself will be made faster or cheaper. If each design iteration is made twice as fast, it may just mean that twice as many iterations will be made. But the overall result may be that the resulting designs are better, and that the costs of the manufacturing operation are lowered instead. All experienced managers recognize the biases and shortcomings of the business case approach; that's why a certain amount of emotional appeal, or "gut feel" is required, even on the highest levels.

Among the factors which can make a business case difficult to put together are the existing massive investments in

- Hardware environments tuned to their current uses and work levels.

- Educations of, and experience within the user community. (Changing a major CAD vendor can involve many person-months of formal training and/or lost productivity for each of thousands of users.)

- Standardized practices and workarounds. (In effect, this represents the learning curve for projects and organizations, whereas the preceding bullet represents the more widely recognized learning curve for individuals.)

Lastly, there is the problem that, as we saw in Section 4, new technology can be introduced in one process only as fast as all the other processes which interact with it can accommodate that change. As the number of processes and the number of potential interactions increase, so can resistance to change.

All of these factors create a certain amount of inertia, a resistance to change even when that change is known to be beneficial. Indeed, any large organization which attempts to maintain uniformity in its software and related processes naturally develops a huge inertia that tends to slow evolution of both the software and the processes. This is not entirely a bad thing. Changes in analysis and design tools used to make safety-critical decisions regarding an airplane, or any other product, should be made slowly and deliberately, and traceably. This is true even when government regulation does not mandate it.

5.2 The End User Level: Grass Roots Creativity

Engineers are bright, endlessly creative, and highly motivated people. It's axiomatic that the first thing they'll do with a new tool is to push it well beyond its intended domain of application. The second thing, of course, is that they'll break the tool. But that's how progress is made. To quote [1] again:

> Another challenge for developers of such systems in the future is to "Plan for success." If things work out, your system may need to be much more widely deployed than you had planned... Users will want to use the system in ways that they think they can not. Since [FlyThru] is a tool, it can be applied in ways that end-users might not anticipate. It also can be used in ways that the developers could not anticipate since they were not application engineers. Tracking and exploiting both of these situations has led to far more FlyThru process uses than had been originally imagined.

A corollary to the above is the following. If end users can be hard to sell on a solution they perceive as "technology push", they can be equally quick to "pull" technology from wherever it can be found to solve an immediate problem. If new startup companies or new products from established vendors appear, it's the end user community that's likely to know about it first. No matter what the official policy with respect to uniformity of tools and processes, a way will probably be found to bring a promising new tool in, at least for evaluation. Once a good tool is found, others quickly learn of it. Again from [1],

> After an initial use, most users would spread the word on their own. Our initial installation of 15 stations moved to over 100 in less than 4 months.

With the popularity in current business culture of concepts like "employee empowerment" (i.e., giving the power to make decisions to the people closest to the problems), this sort of thing can be expected to happen more and more often.

References

[1] R.M. Abarbanel, E.L. Brechner, and W.A. McNeely. FlyThru the Boeing 777, SigGraph, 1996.

[2] M. Adamowicz and A. Albano. Nesting two-dimensional shapes in rectangular modules. *Computer Aided Design*, 8:27–33, 1976.

[3] D. F. Baldwin, T. E. Abell, M.-C. M. Lui, T. De Fazio, and D. E. Whitney. An integrated computer aid for generating and evaluating assembly sequences for mechanical products. *IEEE Trans R&A*, 7:78–94, 1991.

[4] P. Bourdet and L. Mathieu, editors. *Proceedings of the 3rd CIRP Seminar on Computer Aided Tolerancing*. Ecole Normale Supérieure de Cachan, Cachan, France, April 1993.

[5] E. Brechner. Interactive walkthrough of large geometric databases. SigGraph short course, 1995.

[6] W. K. Capron and K. L. Smit. Advanced aerodynamic applications of an interactive geometry and visualization system. AIAA-91-0800, 29th Aerospace Sciences Meeting, Reno, Nevada, Jan. 7-10, 1991.

[7] H. Chang and T. Y. Li. Assembly maintainability using motion planning. In *Proceedings of IEEE International Conference on Robotics and Automation*, May 1995.

[8] André Clément, Alain Rivière, and Michel Temmerman. *Cotation Tridimensionnelle des Systèmes Mécaniques - Théorie & Pratique*. PYC Edition, Ivry-sur-Seine, France, 1994.

[9] R. D. Dietrich and S. J. Yakowitz. A rule-based approach to the trim-loss problem. *Int. J. Prod. Res.*, 29:401–415, 1991.

[10] D. Ferguson, R. Mastro, and R. Blakely. Modeling and analysis of aerodynamic data ii. practical experience. AIAA-89-2076, AIAA/AHS/ASEE Aircraft Design, System and Operations Conference, Seattle, WA, July 31-August 2, 1989.

[11] G. H. Golub and C. F. Van Loan. *Matrix Computations*. Johns Hopkins, Baltimore, 1989.

[12] J. A. Heisserman. Generative geometric design. *IEEE Computer Graphics and Applications*, 14, 1994.

[13] J. M. Hervé. Analyse structurelle des mécanismes par groupe des déplacements. *Mech. and Machine Theory*, 13:437–450, 1978.

[14] C. M. Hoffman. *Geometric & Solid Modeling: An Introduction*. Morgan Kaufmann, San Mateo, CA, 1989.

[15] H. Hoppe, T. DeRose, T. Duchamp, J. McDonald, and W. Stuetzle. Surface reconstruction from unorganized points. Tech. Report 91-12-03, Univ. of Washington, Dept. Of Computer Science & Engineering, Seattle, WA, 1991.

[16] A. K. Jones. Notes on assembly modeling. Tech. Report ISSTECH-95-012, Boeing Information & Support Services, Research & Technology, Seattle, WA, August 1995.

[17] L. Joskowicz, E. Sacks, and V. Srinivasan. Kinematic tolerance analysis. *Computer Aided Design*, 1996. (reprinted in these proceedings).

[18] F. Kimura, editor. *Proceedings of the 4th CIRP Seminar on Computer Aided Tolerancing*. University of Tokyo, Tokyo, Japan, April 1995.

[19] J.-C. Latombe, R. H. Wilson, and F. Cazals. Assembly sequencing with toleranced parts. Robotics Laboratory, Dept. of Computer Science, Stanford University, Stanford, CA, USA., 1996. (preprint).

[20] S. C. Liu, S. J. Hu, and T. C. Woo. Tolerance analysis for sheet metal assemblies. *Journal of Mechanical Design*, 118:1–6, 1996.

[21] T. Lozano-Perez. Spatial planning: A configuration space approach. *IEEE Trans. Comp.*, C-32:108–120, 1983.

[22] T. Lyche and K. Morken. Knot removal for parametric b-spline curves and surfaces. *Computer Aided Geometric Design*, 4:217–230, 1987.

[23] K. G. Merkley, K. W. Chase, and E. Perry. An introduction to tolerance analysis of flexible assemblies. NASTRAN User Group Meeting, July 1996.

[24] D. K. Snepp and R. C. Pomeroy. A geometry system for aerodynamic design. AIAA-87-2902, AIAA/AHS/ASEE Aircraft Design, Systems and Operations Meetings, St. Louis, MO., Sept. 14-16, 1987.

[25] V. Srinivasan and H. B. Voelcker, editors. *Proceedings of the 1993 International Forum on Dimensional Tolerancing and Metrology, CRTD-Vol. 27.* American Society of Mechanical Engineers, 1993.

[26] V. Torczon. On the convergence of the multidirectional search algorithm. *SIAM J. Optimization*, 1:123–145, 1991.

[27] R. H. Wilson and J.-C. Latombe. Geometric reasoning about mechanical assembly. *Artificial Intelligence*, 71:371–396, 1994.

[28] United States Navy, DT_NURBS home page, 1996. http://dtnet33-199.dt.navy.mil.

[29] GE R& D Ctr, Access Path Gen. home page, 1996. http://www.crd.ge.com/esl/cgsp/fact_sheet/path/index.html.

[30] MIT, Assembly Oriented Design home page, 1996. http://www.mit.edu:8001/afs/athena.mit.edu/user/k/r/krish1/www/assembly1.html.

[31] Sandia Labs, Archimedes Project home page, 1996. http://www.sandia.gov/2121/archimedes/archimedes.html.

[32] Stanford, Assembly-Planning Group home page, 1996. http://robotics.stanford.edu/users/assembly/assembly.html.

Freeform Shape Machining Using Minkowski Operations

Johan W.H. Tangelder, *Delft University of Technology, Delft, The Netherlands*

Joris S.M. Vergeest, *Delft University of Technology, Delft, The Netherlands*

Mark H. Overmars, *Utrecht University, Utrecht, The Netherlands*

In this paper we use Minkowski operations to describe freeform shape machining algorithms. Given a cutting tool, a toolholder and a stock-in-progress, that encloses the model to be machined, the computation of a tool path, for which the tool does not interfere the model and the toolholder does not interfere the stock-in-progress is described using the Minkowski addition. The computation of the stock-in-progress that is left, if the tool has followed the tool path, is described using the Minkowski subtraction. Grids of height values are described by real-valued functions on finite subsets of $\mathbb{Z}^2$, called numerical functions. We use Minkowski operations on these numerical functions to describe the well-know "remove as much material as possible" machining strategy and the well-know "slicing" machining strategy. As far as we know these strategies have not been described using Minkowski operations. Since the "slicing strategy" generates tool paths that are machined faster, an efficient implementation of the "slicing strategy" is described using Z-pyramids.

1 Introduction

The freeform shape machining problem includes both an interference avoidance problem and a volume processing problem. Given a cutting tool, a toolholder and a stock-in-progress, that encloses the model to be machined, only tool paths for which the tool does not interfere the model and the toolholder does not interfere the stock-in-progress, are allowed. The stock-in-progress that is left, if the tool has followed the tool path, is the result of a volume removal process by the tool from the stock-in-progress. We show that both this interference avoidance problem and this volume processing problem can be described using Minkowski operations.

In this paper we consider only removal of material by the cutting tool, whereas faster plans might be found by cutting off large chunks (rather than machining them away) in some cases. We also do not consider the stresses of possible breakage of the desired shape (such as a very thin wall, etc.) due to cutting forces.

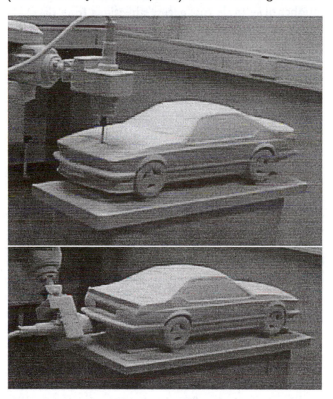

Figure 1: *The Sculpturing Robot system performs rapid prototyping of CAD-defined objects. The machining process consists of a number of stages. At each stage the milling tool approaches the upper face or one of the four side faces of the foam stock.*

For the freeform shape machining task we have installed a Sculpturing Robot system consisting of an industrial Manutec R15 robot with 6 rotational degrees of freedom and a turntable that can rotate around its z-axis, see figure 1. The robot holds a milling tool and a stock of foam is mounted on the turntable. The machining process shrinks the initial foam stock to a foam prototype of a CAD model. The machining process must be performed in such a way that a subsequent, more accurate machining process can be made. Hence, the obtained prototype should always enclose the freeform shape. Previously we have developed and implemented the SRPLAN1 robot motion planning and foam machining algorithm that uses only 5 predefined tool orientations [14]. Currently, we investigate methods to extend the shape domain and the size of the prototypes. These methods will be implemented in a new motion planner, that derives a limited number of suitable tool access directions from the CAD model geometry. For each tool access direction a robust and efficient freeform shape machining strategy will be implemented. In this paper we describe such strategies using Minkowski operations.

Researchers in the field of mathematical morphology have studied the Minkowski addition and subtraction extensively [5, 13]. In this field the Minkowski addition and subtraction are used to define the dilation and erosion image processing operations, respectively. Although mathematical morphology has initially been applied to digital image processing, the Minkowski addition and subtraction are defined generally for n-dimensional space. Therefore, with the Minkowski operations also volume processing can be described. E.g. blending solids using Minkowski operations has been described [11].

Also, the Minkowski addition and subtraction has been applied in the spatial planning problem [4, 10]. The spatial planning problem involves placing an object among other objects or move it without colliding with nearby objects. If the object can translate but not rotate the Minkowski addition and subtraction are used to compute the forbidden region and the feasible region, respectively. Avoiding interference between the tool and the model and between the toolholder and the stock-in-progress is a spatial planning problem.

In section 2 we formally define Minkowski operations on sets. In section 3 we describe the freeform shape machining problem using Minkowski operations on sets for a milling tool with a fixed orientation. Next, we focus on the description of free-form shape machining algorithms that represent the tool, the tool-

holder, the stock-in-progress and the model with grids of height values. In section 4 we formally define these grids of height values as numerical functions. We define Minkowski operations on these numerical functions. In section 5 we use Minkowski operations on numerical functions to describe the well-know "remove as much material as possible strategy" machining strategy (see e.g. [14], [18]) and the well-know "slicing" machining strategy (see e.g. [12]). As far as we know these strategies have not been described using Minkowski operations. [7] and [16] describe the relation between Minkowski operations and NC machining, but no toolholder is included in their analysis and no machining strategies are described. In section 6 we optimize the running time of the computation of the tool paths with the "slicing strategy" using Z-pyramids. We conclude in section 7.

2   Minkowski operations on sets

We have found in the literature several conventions to define Minkowski operations. We adopt the convention used by Serra [13].

The Minkowski addition of two sets A and B, each consisting of vectors in $I\!R^n$, the Euclidean space of dimension n, is constructed by translating A by each element of B and then taking the union of all the resulting translates, as illustrated in figure 2. Intuitively, the Minkowski addition can be considered as a dilation process where one set is expanded by the other. In more formal terms, the *Minkowski addition* is defined as

$$A \oplus B = \cup_{\mathbf{b} \in B} A + \mathbf{b},$$

where $A + \mathbf{b} = \{\mathbf{a} + \mathbf{b} \mid \mathbf{a} \in A\}$ denotes a translation of A by a vector $\mathbf{b}$. The Minkowski addition is commutative, i.e.,

$$A \oplus B = B \oplus A.$$

The Minkowski subtraction of two sets A and B is constructed by translating A by each element of B and then taking the intersection of all the resulting translates as illustrated in figure 3. Intuitively, the Minkowski subtraction can be considered as an erosion process where one set is eroded by the other. In more formal terms, the *Minkowski subtraction* is defined as

$$A \ominus B = \cap_{\mathbf{b} \in B} A + \mathbf{b}.$$

Let c denote set complementation. The Minkowski addition and subtraction are dual operations, i.e., dilating (eroding) a set A by a set B is equivalent with eroding (dilating) A^c by B. In more formal terms,

$$A \oplus B = (A^c \ominus B)^c$$

and

$$A \ominus B = (A^c \oplus B)^c.$$

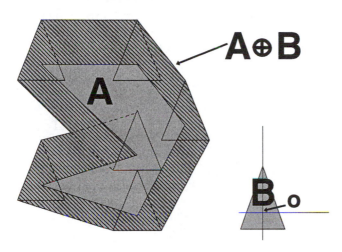

Figure 2: *The Minkowski addition $A \oplus B$ is constructed by translating A by each element of B and then taking the union of all the resulting translates. The triangles denote translates of the vertices of A.*

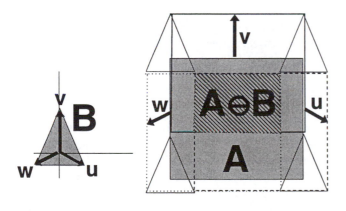

Figure 3: *The Minkowski subtraction $A \ominus B$ is constructed by translating A by each element of B and then taking the intersection of all the resulting translates.*

3 Freeform shape machining with a fixed tool orientation

We assume that a prototype $R \subset I\!R^3$, that encloses and approximates a freeform shape F, is to be obtained from a stock of material. The stock-in-progress S is a volume that shrinks from the initial stock to the prototype R due to the penetrating milling tool T. Collision between the toolholder H and S is not allowed. With these assumptions we state the freeform shape machining problem as follows.

Freeform shape machining problem. Let a freeform shape F and a stock-in-progress S be given in the same Cartesian coordinate system $\mathcal{S}$. Let the tool T and its toolholder H be given in the Cartesian coordinate system $\mathcal{T}$. Assume that T contains the origin $\mathbf{o}$ of $\mathcal{T}$. Assume that $T \cup H$ is a free-translating object in $\mathcal{S}$, i.e., T can translate but T can not rotate in $\mathcal{S}$. Further, assume that the orientations of the x, y and z-axes of $\mathcal{S}$ are identical to those of $\mathcal{T}$. Find a toolpath surface P for machining a prototype R of F, such that R is the minimal enclosing volume of F that can be obtained without interference between T and F and without interference between H and S, if the tool follows the toolpath surface.

In practice a toolpath will be extracted from the toolpath surface P with a finite accuracy. The toolpath is a curve on P such that the machined surface obtained by following the toolpath approximates the toolpath surface within that accuracy.

For a volume $V \subset I\!R^3$ let $-V = \{-v \mid v \in V\}$ be the symmetrical set of V. The following lemma describes the result of the freeform shape machining process if we would neglect interference between the stock-in-progress S and the toolholder H.

Lemma 1 *If interference between T and F is taken into account, but not yet the interference between S and H, then the final result of the milling process is $(F \oplus (-T)) \ominus T$.*

Proof: Sweeping volumes and forbidden volumes for positioning the origin $\mathbf{o}$ of $\mathcal{T}$ can be characterized with help of the Minkowski addition. Let a volume V be given in $\mathcal{S}$ and a volume W be given in $\mathcal{T}$. If $\mathbf{o}$ follows P the volume swept by W is

$$\cup_{\mathbf{p} \in P} W + \mathbf{p} = W \oplus P = P \oplus W.$$

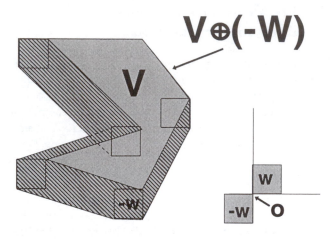

Figure 4: *V and W intersect, if* o $\in V \oplus (-W)$.

In robot motion planning (see, e.g. [8, 10]), it is a well-know fact that V interferes with W, if o is positioned such that

$$\mathbf{o} \in V \oplus (-W)$$

This is the forbidden region for o in S if collision between V and W should be avoided (see also figure 4). Hence, interference between the tool T and the shape F occurs if $\mathbf{o} \in F \oplus (-T)$, and interference between the stock-in-progress S and the toolholder H occurs if $\mathbf{o} \in S \oplus (-H)$.

If W contains o, then $V \ominus W = (V^c \oplus W)^c \subseteq V$ is the volume that is left of V after o has been positioned at every p outside V, while the volume swept by W, i.e., $V^c \oplus W$ is removed from V. If W does not contain o then $V \ominus W \subseteq V$ may be violated and $V \ominus W$ does not model the volume that is left of V. Since T contains the origin o of T the minimal volume that can be obtained with T and encloses F is $(F \oplus (-T)) \ominus T$, if we neglect interference between S and H.   □

In the field of mathematical morphology $(F \oplus (-T)) \ominus T$ is called the closing of F with respect to T. If T is a sphere centered at o with radius α then $(F \oplus (-T)) \ominus T$ is identical to the "filleted" blend of F [11]. It is also identical to the α-hull of F, as introduced by Edelsbrunner [2].

If we do not neglect interference between S and H the optimal result of the milling process can be specified as follows.

Result of the milling process. The optimal result R of the milling process is described iteratively

as follows. Let $S_0 \supseteq F$ denote the initial stock-in-progress. Let S_i denote the stock-in-progress after step i. S_{i+1} can be obtained from S_i by positioning o anywhere outside S_i without interference between F and T or between S_i and H. Hence, $S_{i+1} = S_i \cap (V_i \ominus T)$, where the toolpath surface P_i is the boundary of $V_i = (F \oplus (-T)) \cup (S_i \oplus (-H))$. Since $\forall i : F \subseteq S_{i+1} \subseteq S_i$, $\lim_{i\to\infty} S_i$ exists and $F \subseteq \lim_{i\to\infty} S_i$. Hence, we define R as follows. Let $R = S_i$ if there exists an i such that $S_i = S_{i+1}$. Otherwise let $R = \lim_{i\to\infty} S_i$. Figure 5 illustrates the first step of this description.

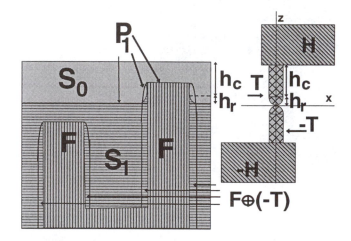

Figure 5: *First step of the milling process, where P_1 is the boundary of $V_1 = (F \oplus (-T)) \cup (S_0 \oplus (-H))$ and $S_1 = S_0 \cap (V_1 \ominus T)$.*

This "remove as much material as possible" strategy has been implemented without using Minkowski operations (see, e.g., [14, 18]).

From the S_i in this description, an interference free toolpath can be derived, but an interference free toolpath with which material is removed slice by slice can be extracted from a tool path surface P, derived directly from R as follows. We restrict ourselves to cutting tools that consist of a cylindric part with height h_c and a rotation symmetric remaining part with height h_r. In the case of a ball end tool the remaining part is a hemisphere with radius h_r and in the case of a flat end tool the remaining part is empty and $h_r = 0$. Without loss of generality, let the tool direction be in the $-z$ direction and the tooltip be positioned at the origin of T as indicated in figure 5. Further, we assume that T rotates around the z-axis of T. Let $R_{XY} = \{(x, y) \mid \exists z : (x, y, z) \in R\}$ and let for all $(x, y) \in R_{XY}$ $R(x, y) =$

$sup\{z \mid (x, y, z) \in R\}$. Let s_{max} be the height of S_0, $r_{min} = \inf\{R(x, y) \mid (x, y) \in R_{XY}$ the minimal height of R as measured along the z-axis. R can be obtained in $l = Round\_Up((s_{max} - (r_{min} + h_r))/h_c)$ stages. At each stage a slice is removed by following a plane parallel to the xy-plane or by following the boundary of $(F \oplus (-T)) \cup (R \oplus (-H))$. The slices are removed in a top down fashion such that in the first milling stage the tooltip is positioned at each point $(x, y) \in R_{XY}$ at height $\max(R(x, y), s_{max} - h_c - h_r)$. Hence, interference between H and S is avoided. Next, for $j = 2 \ldots l$ the tooltip is positioned in the jth milling stage at each point $(x, y) \in R_{XY}$ with $R(x, y) < s_{max} - (j-1)h_c - h_r$ in the xy-plane at height $\max(R(x, y), s_{max} - jh_c - h_r)$. If the jth slice is being milled, the stock-in-progress S is always identical to R above the height $s_{max} - (j-1)h_c$ and upon completion of milling this slice, S is identical to R above $s_{max} - jh_c$. Since H can hit S only above that slice, interference between S and H is avoided.

Also the "slicing" strategy has been implemented without using Minkowski operations (see, e.g. [12]).

In the next section we define Minkowski operations on numerical functions and in section 5 we describe the implementation of both algorithms using numerical functions on $\mathbb{Z}^2$.

4 Minkowski operations on numerical functions

In gray-scale morphology [5, 13] the definitions of Minkowski operations on sets are modified to real-valued functions defined on finite subsets of $\mathbb{R}^n$, the Euclidean space of dimension n, and on $\mathbb{Z}^n$, the Euclidean grid of dimension n. These functions are called numerical functions. In the next section we use numerical functions on $\mathbb{Z}^2$ and Minkowski operations to describe milling algorithms.

Let $D_f \subset \mathbb{R}^2$ denote the domain of the function f. Let the *translation* of f by a point (s, t) be denoted by $f_{s,t}$. $f_{s,t}$ is defined as

$$D_{f_{s,t}} = \{(x, y) \mid (x - s, y - t) \in D_f\}$$

and for all $(x, y) \in D_{f_{s,t}}$

$$f_{s,t}(x, y) = f(x - s, y - t).$$

Given a set S, the reflected set $\hat{S} = \{(x, y, -z) \mid (x, y, z) \in S\}$ is symmetrical to S with respect to the xy-plane.

The *maximum of a finite collection of numerical functions* F is defined on $\bigcup_{f \in F} D_f$ by $MAX(F)(x, y) = \max\{f(x, y) \mid f \in F \wedge (x, y) \in D_f\}$.

The *minimum of a finite collection of numerical functions* F is defined on $\bigcap_{f \in F} D_f$ by $MIN(F)(x, y) = \min\{f(x, y) \mid f \in F\}$.

The volume represented by a numerical function f on $\mathbb{R}^2$ is called its *umbra* $U(f)$. This is the set of all points that lie below f, i.e.,

$$U(f) = \{(x, y, z) \mid (x, y) \in D_f \wedge z \le f(x, y)\}.$$

In mathematical morphology the *Minkowski addition $f \oplus g$ of two numerical functions* f and g on $\mathbb{R}^2$ is defined as a dilation of f with g such that $U(f \oplus g) = U(f) \oplus U(g)$ as

$$D_{f \oplus g} = D_f \oplus D_g$$

and for all $(x, y) \in D_{f \oplus g}$

$$(f \oplus g)(x, y) = \sup_{(s,t) \in D_g} \{f_{s,t}(x, y) + g(s, t)\}.$$

In mathematical morphology the *Minkowski subtraction $f \ominus g$ of two numerical functions* is defined as an erosion of f with g such that $U(f \ominus g) = U(f) \ominus \hat{U}(g)$ as

$$D_{f \ominus g} = D_f \ominus D_g$$

and for all $(x, y) \in D_{f \ominus g}$

$$(f \ominus g)(x, y) = \inf_{(s,t) \in D_g} \{f_{s,t}(x, y) - g(s, t)\}.$$

Next, we consider numerical functions on $\mathbb{Z}^2$. Given a grid size d, we associate numerical functions on $\mathbb{Z}^2$ with volumes by a mapping M from numerical functions on $\mathbb{Z}^2$ to numerical functions on $\mathbb{R}^2$. The mapping M is given by

$$M(f)(x, y) = f(Round\_Down(x/d), Round\_Down(y/d)),$$

where

$$D_{M(f)} = \{(x, y) \in \mathbb{R}^2 \mid \exists (i, j) \in D_f : id \le x < (i + 1)d \wedge jd \le y < (j + 1)d\}.$$

Then the volume represented by f is $U(M(f))$.

Let $\oplus_z$ and $\ominus_z$ denote the definitions of $\oplus$ and $\ominus$ for numerical functions on $\mathbb{Z}^2$ instead of numerical functions on $\mathbb{R}^2$ as given above.

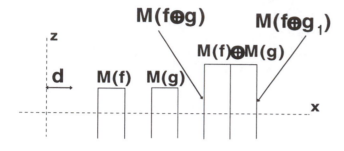

Figure 6: *Let f and g denote numerical functions on $\mathbb{Z}$. $M(f) \oplus M(g) \neq M(f \oplus g)$ and $M(f) \ominus M(g) \neq M(f \ominus g)$. The umbra of $M(f \oplus g)$ is constructed by translating the umbra of M(f) by each element of the umbra of M(g) and then taking the union of all the resulting translates. Hence, $M(f) \oplus M(g) = M(MAX(f \oplus g, f \oplus g_1))$, where g_x denotes the translate of g by x, i.e., $g_x(s) = g(s - x)$. Since the umbra of $M(f \ominus g)$ is constructed by translating the umbra of M(f) by each element of the umbra of M(g) and then taking the intersection of all the resulting translates, $M(f) \ominus M(g) = M(MIN(f \ominus g, f \ominus g_1))$.*

Note that for numerical functions on $\mathbb{Z}^2$ with finite domains, *sup* and *inf* are identical with *max* and *min*, respectively.

Figure 6 shows $M(f) \oplus M(g) \neq M(f \oplus_z g)$ and $M(f) \ominus M(g) \neq M(f \ominus_z g)$. In order to satisfy $M(f) \oplus M(g) = M(f \oplus_z g)$ and $M(f) \ominus M(g) = M(f \ominus_z g)$ we define Minkowski operations for numerical functions on $\mathbb{Z}^2$ as follows.

The *Minkowski addition* $f \oplus g$ *of two numerical functions f and g on $\mathbb{Z}^2$ is defined as*

$$f \oplus g =$$
$$MAX(f \oplus_z g, f \oplus_z g_{1,0}, f \oplus_z g_{0,1}, f \oplus_z g_{1,1})$$

and the *Minkowski subtraction* $f \oplus g$ *of two numerical functions f and g on $\mathbb{Z}^2$ is defined as*

$$f \ominus g =$$
$$MIN(f \ominus_z g, f \ominus_z g_{1,0}, f \ominus_z g_{0,1}, f \ominus_z g_{1,1}).$$

Given two numerical function f and g on $\mathbb{Z}^2$ with finite domains containing m and n elements, respec-

tively, computing $f \oplus g$ requires $O(mn)$ time. Also computing $f \ominus g$ requires $O(mn)$ time.

5   Toolpath computation

In this section we discuss the implementation of the algorithms described in section 3. Given a fixed tool orientation, we will use numerical functions $f : D_f \to \mathbb{R}$, where $D_f \subset \mathbb{Z}^2$ is a finite set, d is a grid size to approximate the stock-in-progress S, the freeform shape geometry F, the toolholder H and the milling tool T. We also use numerical functions on $\mathbb{Z}^2$ to describe toolpath surfaces. There exist several methods [1, 6] to extract toolpaths from these numerical functions.

Since the obtained prototype should always enclose F, we represent the actual stock-in-progress S by a minimal numerical function s^e such that the umbra of $M(s^e)$ encloses S. Since collision between F and T should be avoided we also represent F and $-T$ by such enclosing numerical functions f^e and t^e. We can choose the maximal value of t^e to be equal to zero. Also collision between S and H should be avoided. Hence, we use a minimal enclosing numerical function h^e to represent $-H$. We assume that the maximal value of h^e is equal to $-(h_r + h_c)$ as indicated in figure 7, i.e., $h^e(x, y) = -(h_r + h_c)$ for all $(x, y) \in D_{t^e}$. Because s^e must be updated such that it encloses the actual stock-in-progress we need also a maximal numerical function t^i such that the umbra of $M(t^i)$ is enclosed by $-T$. Since T is rotation symmetric around the z-axis of T, $-t^i$ approximates T and we can write $\hat{T} = -T$. Figure 7 illustrates these representations of T and H. Also a toolpath surface is represented by a numerical function p. The Minkowski addition and subtraction of these functions can be computed straightforwardly.

With these numerical functions the "remove as much material as possible" milling strategy can be described by the following algorithm that computes the final result r of the milling process and a set of toolpath surfaces p_i, where the umbra of p_i represents a volume V_i. Recall from section 3, that the toolpath surface P_i is the boundary of $V_i = (F \oplus (-T)) \cup (S_i \oplus (-H))$ and that $S_{i+1} = S_i \cap (V_i \ominus T)$.

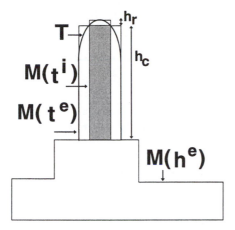

Figure 7: *The milling tool T is represented by the minimal numerical function t^e such that the umbra of $M(t^e)$ encloses $-T$ and the maximal numerical function t^i such that the umbra of $M(t^i)$ is enclosed by $-T$. $-H$ is represented by the minimal numerical function h^e such that the umbra of $M(h^e)$ encloses $-H$. $-\min\{t^i(j) \mid j \in D_{t_i}\}$ approximates h_r and $\max\{h^e(j) \mid j \in D_{h^e}\} - \min\{t^i(j) \mid j \in D_{t_i}\}$ approximates h_c.*

$i:=0$; $s^e:=$"the initial stock";
(* Use t^e to compute $F \oplus (-T)$ *)
$f^d := f^e \oplus t^e$;
REPEAT
   (* Use t^e to avoid collision *)
   $p_{i+1} := MAX(f^d, s^e \oplus h^e)$;
   (* Use t^i to update
   the stock-in-progress *)
   $s^e_{old} := s^e$; $s^e := MIN(s^e, p_{i+1} \ominus t^i)$;
   $i := i + 1$;
   (* Stop if no more material
   can be removed *)
UNTIL $s^e = s^e_{old}$;
(* s^e is the final result
of the milling process. *)
$r := s^e$;

Below, we describe an algorithm that computes "sliced" toolpath surfaces. With these toolpath surfaces the same final result r is obtained. If a toolpath surface p_i is contained in a slice, the toolholder can hit the stock-in-progress only above that slice. Therefore, the toolpath surfaces are computed efficiently slice by slice. Let s_{max} denote the maximal height of the initial stock s^e_0, let f_{min} denote the minimal height of f^e. Let

$D_{f^e} = D_{s^e}$ and for all $(x, y) \in D_{f^e}$ let $f^e(x, y) \leq s_{max}$. In this algorithm we use the following SLICE and SLI operations on numerical functions.

Let $SLICE^{to}_{from}(f)(x, y) = f(x, y)$, if $from \leq f(x, y) \leq to$. Otherwise let $SLICE^{to}_{from}(f)(x, y)$ be undefined.

Let $SLI^{to}_{from}(f)(x, y) = s_{max}$, if $f(x, y) > to$. Let $SLI^{to}_{from}(f)(x, y) = f(x, y)$, if $from \leq f(x, y) \leq to$. Let $SLI^{to}_{from}(f)(x, y) = from$, if $from > f(x, y)$.

The algorithm proceeds as follows.

$i:=0$; $s^e:=$"the initial stock";
$f^d := f^e \oplus t^e$;
(* l denotes the number of slices *)
$l := Round\_Up((s_{max} - (f_{min} + h_r))/h_c)$;
(* compute the boundaries of the first slice
and compute $F \oplus (-T)$ *)
$to := s_{max}$; $from := to - h_c$; $q := f^d$;
(* the top slice can be milled without
any toolholder collision check *)
(* extract the toolpath surface p_1
from $q = f^d$ *)
$p_1 := SLI^{to}_{from-h_r}(q)$;
(* update the stock-in-progress *)
$s^e := MIN(s^e, p_1 \ominus t^i)$; FOR $i := 1$ TO $l - 1$
LOOP
   FOR ALL $(i, j) \in D_{s^e \oplus t^i}$
   LOOP
      $q(i, j):=$
      $MAX(q, SLICE^{to}_{from}(s^e) \oplus h^e)(i, j)$;
   END LOOP;
   (* If we follow the toolpath surface q in
   the next slice, the toolholder
   does not collide with the
   stock-in-progress above that slice *)
   $to := from$; $from := to - h_c$;
   (* extract the toolpath surface p_{i+1}
   from q *)
   $p_{i+1} := SLI^{to-h_r}_{from-h_r}(q)$;
   (* update the stock-in-progress *)
   $s^e := MIN(s^e, p_{i+1} \ominus t^i)$;
END LOOP;
$r := s^e$;

We have checked the "remove as much material as possible" and the "slicing" algorithm for the 2D case by an implementation using numerical functions on $\mathbb{Z}$.

Figure 8 shows an example of a toolpath surface q which has been computed with the "slicing" algorithm.

If the initial stock is not a block, positioning the tool tip at some grid elements (k, l) may be of no use. Therefore, each path surface $p_i(k, l)$ should be marked to indicate whether the tool intersects the stock-in-progress. The actual path will be extracted from this marked toolpath surface.

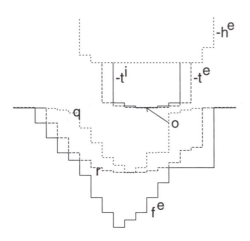

Figure 8: *Illustration of the algorithm in 2D. A toolpath "surface" q and the final result r computed by the slicing algorithm. If o is positioned on q then interference between the freeform shape F and the milling tool T, as well interference between the toolholder H and the stock-in-progress S is avoided.*

The computing time of this algorithm can be shown to be $O(mn)$ where m is the number of stock grid elements and n is the number of holder grid elements.

We expect that for our Sculpturing Robot system the machining time with the paths that are generated by the "slicing" strategy will be less than the machining time with the paths that are generated by the "remove as much material as possible" strategy. This can be explained as follows. The paths consist of a number of straight line segments. Our Sculpturing Robot system reduces the speed of the tool to zero at the end of each straight line segment. Since the paths that are generated by the "slicing" strategy contain far less line segments, we expect that its machining time will be shorter. We will implement both milling strategies as part of a new motion planner for the Sculpturing Robot system and compare the machining time and the com-

puting time of both strategies.

6 Efficient toolpath computation using Z-pyramids

The most CPU-intensive part of the "slicing" algorithm of the previous section is the computation of q by the inner loop

```
FOR ALL (i, j) ∈ D_{s^e ⊕ t^i}
LOOP
    q(i, j) := MAX(q, s ⊕ h^e)(i, j);
END LOOP;
```

where $s = SLICE^{to}_{from}(s^e)$. This computation can be optimized using a Z-pyramid representation [9] for q, s and h^e. The basic idea of the Z-pyramid is to use the original numerical function as finest level of the pyramid and then combine four z values at each level into one z value at the next coarser level by choosing the maximal z value. At the coarsest level of the pyramid there is one single z value which is the maximal value of the numerical function. Let a minimal Z-pyramid be defined by choosing at the next coarser level the minimal Z-value instead of the maximal Z-value.

In order to facilitate to read the description of a Z-pyramid below we define its numerical functions on the set of integer rectangles $\{[x1, x2] \times [y1, y2] \mid x1, x2, y1, y2 \in \mathbb{Z} \wedge x1 \leq x2 \wedge y1 \leq y2\}$.

Definition 1 *Let, for a numerical function f, its maximal Z-pyramid be given by k numerical functions f_i, $i = 1 \dots k$, where f_1 is a numerical function having one single value which is the maximal value of f and f_k corresponds with f. The Z-pyramid is derived from the original function as follows. Let $V(t, i)$ denote the integer interval starting at t that contains 2^{k-i} elements, i.e.,*

$$V(t, i) = [t, t + 2^{k-i} - 1].$$

Let f_k be defined as: $f_k([x, x] \times [y, y]) = f(x, y)$ for all $(x, y) \in D_f$. Let f_i, $i = 1 \dots k - 1$ be defined as
$f_i(V(2^{k-i}x, i) \times V(2^{k-i}y, i)) = \max(f_{i+1}(V(2^{k-i}x, i + 1) \times V(2^{k-i}y, i + 1)),$
$f_{i+1}(V(2^{k-i}x + 2^{k-(i+1)}, i + 1) \times V(2^{k-i}y, i + 1)),$
$f_{i+1}(V(2^{k-i}x, i + 1) \times V(2^{k-i}y + 2^{k-(i+1)}, i + 1)),$
$f_{i+1}(V(2^{k-i}x + 2^{k-(i+1)}, i + 1) \times V(2^{k-i}y + 2^{k-(i+1)}, i + 1)))$

Minimal Z-pyramids can be defined analogously. Let $<$ denote the usual partial ordering relation for numerical functions on $\mathbb{Z}^2$. Hence, for two numerical functions f_{i-1} and f_i in a maximal Z-pyramid holds $f_i \leq f_{i-1}$ and for two numerical functions f_{i-1} and f_i in a minimal Z-pyramid holds $f_i \geq f_{i-1}$.

Let a Z-pyramid representation for s and h^e be given. We assume that all these Z-pyramids have the same level (this can be achieved by adding levels with the same single Z value). From the Z-pyramid definition it follows that $s_i \leq s_{i-1}$ and $h_i^e \leq h_{i-1}^e$. From the definition of Minkowski addition follows $s_i \oplus h_i^e \leq s_{i-1} \oplus h_{i-1}^e$. Let a minimal Z-pyramid be given for q with the same level as the Z-pyramids for s and h_e. From the definition of minimal Z-pyramid follows $q_i \geq q_{i-1}$. Hence, q and its minimal Z-pyramid can be computed by starting the following recursive procedure at level 1 with appropriate values for $x1$, $x2$, $y1$ and $y2$.

```
PROCEDURE UPDATE_Q(x1,x2,y1,y2,l)
(* l denotes the Z-pyramid level *)
IS
BEGIN
w := s_l ⊕ h_l^e([x1, x2] × [y1, y2]);
IF w < q_l([x1, x2] × [y1, y2]) ∧ l < k
THEN
      (* updating q and its Z-pyramid
      is not necessary *)
      NULL;
ELSE
   (* update q_l and refine
   the computation of q *)
   q_l([x1, x2] × [y1, y2]) := w;
   (* Let "/" denote integer division,
   e.g. 5/2=2 *)
   UPDATE_Q
   (x1,(x1+x2)/2,y1,(y1+y2)/2,l+1);
   UPDATE_Q
   ((x1+x2)/2+1,x2,y1,(y1+y2)/2,l+1);
   UPDATE_Q
   (x1,(x1+x2)/2,(y1+y2)/2+1,y2,l+1);
   UPDATE_Q
   ((x1+x2)/2+1,x2,(y1+y2)/2+1,y2,l+1);
END IF;
END UPDATE_Q;
```

This algorithm has been tested for the 2D case with our implementation of numerical functions on $\mathbb{Z}$.

7 Conclusions and further research

We have shown that Minkowski operations form a powerful tool to specify the freeform shape machining problem and to describe and compare freeform shape machining algorithms. We have described the well-know "remove as much material as possible" machining strategy and the well-know "slicing" machining strategy using the Minkowski addition and subtraction. Since the latter strategy generates toolpaths, that can be followed by the tool in less time, we have described an efficient implementation of this strategy using Z-pyramids.

Further current research includes the selection of promising tool access directions [3, 15, 17], and the extraction of toolpaths from the computed toolpath surfaces [1, 6]. The "remove as much material as possible" strategy and the "slicing" strategy will be implemented using Z-pyramids as part of a new motion planner for the Sculpturing Robot system. The machining time and the computing time of both strategies will be compared.

Acknowledgements

We thank the staff of the Sculpturing Robot project for their support. Johan W.H. Tangelder was supported by the Delft University Research Fund. Mark H. Overmars was partially supported by the Netherlands Organization for Scientific Research (N.W.O).

References

[1] J.J. Cox, Y. Takezaki, H.R.P. Ferguson, K.E. Kohkonen, and E.L. Mulkay. Space-filling curves in tool-path applications. *Computer-Aided Design*, 26(3):215–224, 1994.

[2] H. Edelsbrunner and E.P. Mücke. Three-dimensional alpha shapes. *ACM Transactions on Graphics*, 13(1):43–72, 1994.

[3] J.G. Gan. Set-up orientations of workpieces for machining by three-axis numerical control machines. *Journal of Design and Manufacturing*, 2:59–69, 1992.

[4] P.K. Ghosh. A solution of polygon containment, spatial planning, and other related problems using Minkowski operations. *Computer Vision, Graphics and Image Processing*, 49:1–35, 1990.

[5] R.G. Giardina and R.D. Dougherty. *Morphological Methods in Image and Signal Processing*. Prentice Hall, 1988.

[6] J.G. Griffiths. Toolpath based on Hilbert's curve. *Computer-Aided Design*, 26(11):839–844, 1994.

[7] A. Kaul. Minkowski sums: A simulation tool for CAD/CAM. In G.A. Gabriele, editor, *Computers in Engineering - Volume 1*, pages 447–456, New York, 1992. ASME.

[8] J.C. Latombe. *Robot Motion Planning*. Kluwer Academic Publishers, Boston, 1991.

[9] D. Laur and P. Hanrahan. Hierarchical splatting: a progressive refinement algorithm for volume rendering. *Computer Graphics*, 25(4):285–288, 1991. (Proc. SIGGRAPH '91).

[10] T. Lozano-Pérez. Spatial planning: A configuration space approach. *IEEE Transaction on Computers*, C-32(2):108–120, 1983.

[11] J. Menon, R.J. Marisa, and J. Zagajac. More powerful solid modeling through ray representations. *IEEE Computer Graphics and Applications*, 14(3):22–35, 1994.

[12] T. Saito and T. Takahashi. NC machining with G-buffer method. *Computer Graphics*, 25(4):207–216, 1991. (Proc. SIGGRAPH '91).

[13] J. Serra. *Image Analysis and Mathematical Morphology*. Academic Press, 1982.

[14] J.W.H. Tangelder and J.S.M. Vergeest. Robust NC path generation for rapid shape prototyping. *Journal of Design and Manufacturing*, 4:281–292, 1994.

[15] J.W.H. Tangelder, J.S.M. Vergeest, and M.H. Overmars. Computation of voxel maps containing tool access directions for machining free-form shapes. In *Proceedings of the 1996 ASME Design Engineering Technical Conference and Computers in Engineering Conference*, New York, 1996. ASME.

[16] A.M Vepsäläinen. An application of morphological filters to NC-programming. In P.D. Gader, editor, *SPIE Vol. 1350 Image Algebra and Morphological Image Processing*, pages 177–183, Washington, 1990. SPIE - The international Society for Optical Engineering.

[17] T.C. Woo. Visibility maps and spherical algorithms. *Computer Aided Design*, 26(1):6–16, 1994.

[18] C.F. You and C.H. Chu. An automatic path generation method of NC rough cut machining from solid models. *Computers in Industry*, 26:161–173, 1995.

The Power of Friction:
Quantifying the "Goodness" of Frictional Grasps

Marek Teichmann, *Courant Institute, New York University, NY, USA*
Bud Mishra, *Courant Institute, New York University, NY, USA*

This paper attempts to quantify the trade-offs that exist among the number of fingers, coefficient of friction and the goodness of a grasp. In particular, we give a general framework for defining a grasp metric that takes friction into account. We contrast our approach to an older metric that was proposed by Ferrari and Canny [FC92] for the frictional case. In addition, we discuss several algorithmic solutions related to frictional grasps.

1 Introduction

A standard approach to grasp theory takes the point of view that it suffices to consider the situation where the fingers make only *frictionless point contacts* with the object to be grasped. Under such a model, referred to as a *positive grip* model, one then considers various grasp existence and synthesis problems. (See Mishra, Schwartz and Sharir [MSS87] for several key results under this model. For some related results, also see [MNP90, Ngu87, PSS+95].) This point of view is justified by the argument that presence of friction only improves the grasp and hence positive grips represent in some sense the most pathological situation. Another argument favoring the positive grip model is that even when there is uncertainty about how much friction is available, the grasps synthesized under this restricted model remain immune to such uncertainties. However, there have been relatively inadequate attempts to answer many questions that the above viewpoint raises.

For instance, given a frictionless closure grasp, precisely how much does the presence of friction improves that grasp? There are several qualitative results exploring the relation between frictional contacts and closure grasps. Markenscoff et. al. [MNP90] showed that in the presence of friction it is always possible to synthesize 3 and 4 finger closure grasps for planar and 3D objects, respectively. These numbers are also lower bounds. Recently Ponce, Sullivan, Sudsang and Boissonnat [PSS+95] have classified four finger frictional grasps of three dimensional objects using screw theory and propose methods for finding four finger equilibrium and closure grasps.

But in order to provide a more precise answer to the question posed earlier, one needs to define a quantitative measure for a goodness of a grasp and then provide a numerical measure of how a grasp improves in terms of the coefficient of Coulomb friction μ as it varies from 0 (frictionless situation) to ∞ (unbounded amount of friction). In the last few years, starting with the work of Kirkpatrick et. al. [KMY92] dealing with the positive closure grasps, there have been several "grasp metrics" proposed by many researchers[1] (see, for instance, [FC92, Mis94, Tei95a, Tei95b]). However, only the work of Ferrari and Canny directly addresses the problem of defining grasp metrics for the frictional contact models. While their approach is more appropriate for grippers and sliding fixtures, ours is more appropriate for multi-fingered hands and point fixtures.

Very recently, Bicchi [Bic95] has introduced a different approach to quantify the closure properties of grasping in the presence of friction. However it remains unclear how Bicchi's metric relates to the one proposed here.

We also consider a generalization of a problem first

---

[1]The paper by Mishra [Mis94] provides a unified framework for defining all of these metrics. The recent work of Teichmann [Tei95b, Tei95a] rectifies a problem with most of these metrics and proposes a class of metrics that do not depend on the reference frame.

addressed by Schwartz and Sharir [SS88].

> Given k "feasible gripping" points on an object and a positive real number $r > 0$, determine the least amount of friction $\mu^*(r)$ necessary in order that some subset of the gripping points will yield a good closure grasp of quality r or larger.

We provide a polynomial-time algorithm to solve this problem approximately.

2   The Formal Framework

We start with a highly idealized robot hand model. It consists of several independently movable force-sensing stiff fingers; this hand is used to grasp a rigid object B. The fingers are placed at points $\mathbf{p}$ of the boundary of B, which we shall denote by ∂B. Additionally, we make the following assumptions: (1) B is a full-bodied (i.e. no internal holes) compact subset of the Euclidean 3-space, and has a piece-wise smooth boundary ∂B. (2) For each finger-contact on the body, we may associate a nominal point of contact, $\mathbf{p} \in \partial B$. We let ∂B^* denote the set of points $\mathbf{p} \in \partial B$ such that the direction $\mathbf{n}(\mathbf{p})$ normal to ∂B at $\mathbf{p}$ is well-defined; by convention, we pick $\mathbf{n}(\mathbf{p})$ to be the unit normal pointing into the interior of B.

For each such point $\mathbf{p}$, we can define a wrench system $\{\boldsymbol{\Gamma}^{(1)}(\mathbf{p}), \boldsymbol{\Gamma}^{(2)}(\mathbf{p}), \ldots, \boldsymbol{\Gamma}^{(m)}(\mathbf{p})\}$, $(0 \leq m \leq 6)$, where the number and screw-axes of the wrench system depend on the contact type. Some of these wrenches can be *bisense* (i.e. can act in either sense) and the remaining wrenches, *unisense*.

In the case, where the contacts are frictionless, we call the corresponding grips 'positive grips.' The wrench system associated with each point is:

$$\boldsymbol{\Gamma}(\mathbf{p}) = \{[\mathbf{n}(\mathbf{p}), \mathbf{p} \times \mathbf{n}(\mathbf{p})]\}$$

In general, corresponding to a set of finger-contacts, we have a system of n wrenches,

$$\{\mathbf{w}_1, \ldots, \mathbf{w}_k, \mathbf{w}_{k+1}, \ldots, \mathbf{w}_n\},$$

the first k of which are bisense and the remaining last $n - k$ of the wrenches are unisense. Assume that the

magnitudes of these wrenches are given by the scalars f_i's

$$\{f_1, \ldots, f_k, f_{k+1}, \ldots, f_n\},$$

where $f_1, \ldots, f_k \in \mathbb{R}$ and $f_{k+1}, \ldots, f_n \in \mathbb{R}_{\geq}$, and not all the magnitudes are zero. We call such a system of wrenches and the wrench-magnitudes, a *grip*, G, and say that this grip G generates an external wrench $\mathbf{w} = [F_x, F_y, F_z, \tau_x, \tau_y, \tau_z] \in \mathbb{R}^6$, if

$$\mathbf{w} = \sum_{i=1}^{n} f_i \, \mathbf{w}_i.$$

This leads to the concept of a *closure grasp*:

Definition 2.1 *A set of gripping points on an object B to which corresponds a system of wrenches $\mathbf{w}_1, \ldots, \mathbf{w}_n$ (as before) is said to constitute a* force/torque closure grasp *(or simply, a* closure grasp*) if and only if any arbitrary external wrench can be generated by varying the magnitudes of the wrenches (subject to the constraints imposed by the senses of the wrenches).*

A necessary and sufficient condition for a closure grasp is that the (module) sum of the linear space spanned by the vectors $\mathbf{w}_1, \ldots, \mathbf{w}_k$ and the positive space spanned by the vectors $\mathbf{w}_{k+1}, \ldots, \mathbf{w}_n$ is the entire $\mathbb{R}^6$:

$$\text{lin}\,(\mathbf{w}_1, \ldots, \mathbf{w}_k) + \text{pos}\,(\mathbf{w}_{k+1}, \ldots, \mathbf{w}_n) = \mathbb{R}^6.$$

Denote, by L, the linear space $\text{lin}\,(\mathbf{w}_1, \ldots, \mathbf{w}_k)$, and, by $L^\perp$, the orthogonal complement of L in $\mathbb{R}^6$. Let π be the linear projection function of $\mathbb{R}^6$ onto $L^\perp$ whose kernel is L. A necessary and sufficient condition for a closure grasp is equivalently given as follows:

$$\mathbf{0} \in \text{int conv}\,(\pi\mathbf{w}_{k+1}, \ldots, \pi\mathbf{w}_n)$$

in $L^\perp$. In case of positive grips ($k = 0$), we have, $\mathbf{0} \in \text{int conv}\,(\mathbf{w}_1, \ldots, \mathbf{w}_n)$.

A simple linear time algorithm for finding *at least* one set of force targets that can generate a given external wrench has been presented in [MSS87]. Also, as the external wrench is varied in the course of a manipulation task, this algorithm updates the force targets in constant time. A more extensive discussion on the structure of the wrench map and how this structure can be exploited in grasp synthesis appears in Teichmann's thesis [Tei95a].

2.1 Introducing Friction

The number of fingers necessary for a closure grasp can be reduced in the presence of friction. Usually the Coulomb friction model for the surface contacts is assumed (see, for example, [KR86]). A friction cone is associated with each contact, and the line of action of the force transmitted through the contact must lie within this cone.

Let $\mu > 0$ be the friction coefficient. Closure with n fingers under friction occurs when any arbitrary external wrench $\mathbf{w}$ can be expressed as

$$\mathbf{w} = \sum_{i=1}^{n} f_i \left[\mathbf{\Gamma}(\mathbf{p}_i) + t_i (\mathbf{n}_i^\perp, \mathbf{p}_i \times \mathbf{n}_i^\perp) \right], \qquad (1)$$

where $f_i \geq 0$ and $\mathbf{n}_i^\perp$ is a unit normal to $\mathbf{n}(\mathbf{p}_i)$ with $0 \leq t_i \leq \mu$, $(1 \leq i \leq n)$. In other words, the actual force applied at point $\mathbf{p}_i$ is a non-negative multiple of $\mathbf{n}(\mathbf{p}_i) + t_i \mathbf{n}_i^\perp$ and belongs to the friction cone $\mathcal{C}_\mu(\mathbf{p}_i)$, a set of vectors forming an angle of at most $\arctan(\mu)$ with $\mathbf{n}(\mathbf{p}_i)$. Define the set of forces of unit magnitude in the friction cone

$$\mathcal{F}_\mu(\mathbf{p}) = \{ \mathbf{f} \in \mathcal{C}_\mu(\mathbf{p}) : \|\mathbf{f}\| = 1 \} .$$

We now show that the standard techniques for testing the closure properties still apply. We would like to say that since the finger can apply a force anywhere in the friction cone, we can include the image by pos $\mathbf{\Gamma}$ of the entire friction cone in the set of forces that can be generated at the contact point. Let $\mathbf{\Gamma}(\mathbf{p}, \mathbf{f}) = [\mathbf{f}, \mathbf{p} \times \mathbf{f}]$ and define

$$\mathbf{\Gamma}_\mu(\mathbf{p}) = \mathbf{\Gamma}(\mathbf{p}, \mathcal{F}_\mu(\mathbf{p})). \qquad (2)$$

Note that $\mathbf{\Gamma}_0(\mathbf{p}) = \{\mathbf{\Gamma}(\mathbf{p})\}$. We require the force applied from any direction to be of unit magnitude so that $f\mathbf{\Gamma}_\mu(\mathbf{p})$ represents the effect of applying a force of magnitude f at $\mathbf{p}$. In contrast, Ferrari and Canny [FC92]), require that the normal component of the force have unit magnitude. Their model yields larger grasp qualities and is only appropriate for grippers and sliding fixtures. In contrast, note that our model is better suited for point contact models, usually associated with multi-fingered robot hands and point fixtures.

To generate an external wrench $\mathbf{w}$, we would like to select points from $\bigcup_i \mathrm{pos}\, \mathbf{\Gamma}_\mu(\mathbf{p}_i)$ each point corresponding to a finger position, direction and magnitude of applied force. There are two potential difficulties. First, to a point in wrench space may correspond two or more fingers. The second difficulty is that we are allowed to select *at most one* point (actually ray) for each pos $\mathbf{\Gamma}_\mu(\mathbf{p}_i)$ since a finger can apply a force only in one direction at any one time. We show that this is always possible.

For $S \subseteq \mathbb{R}^d$, denote by $[0,1]S$ the set $\{\lambda s : s \in S, 0 \leq \lambda \leq 1\}$.

Lemma 2.1 *Both* pos $\mathbf{\Gamma}_\mu(\mathbf{p})$ *and* $[0,1]\mathbf{\Gamma}_\mu(\mathbf{p}, \mathbf{n})$ *are convex.*

Proof. To show the first statement, let $\mathbf{w}_i = [\mathbf{f}_i, \mathbf{p}_i \times \mathbf{f}_i] \in \mathrm{pos}\, \mathbf{\Gamma}_\mu(\mathbf{p})$, $i = 1, 2$. For $\lambda_1, \lambda_2 \geq 0$ with $\lambda_1 + \lambda_2 = 1$,

$$\lambda_1 \mathbf{w}_1 + \lambda_2 \mathbf{w}_2 = [\lambda_1 \mathbf{f}_1 + \lambda_2 \mathbf{f}_2, \mathbf{p} \times (\lambda_1 \mathbf{f}_1 + \lambda_2 \mathbf{f}_2)].$$

But $\lambda_1 \mathbf{f}_1 + \lambda_2 \mathbf{f}_2 \in \mathcal{F}_\mu(\mathbf{p})$ by convexity of the friction cone, which proves the first statement. Since $\lambda \mathbf{\Gamma}_\mu(\mathbf{p}, \mathbf{n}) = \mathbf{\Gamma}_\mu(\mathbf{p}, \lambda \mathbf{n})$, if in addition $\|\mathbf{f}_i\| \leq 1$, then we have $\|\lambda_1 \mathbf{f}_1 + \lambda_2 \mathbf{f}_2\| \leq \lambda_1 + \lambda_2 = 1$ by the triangle inequality and the second statement follows. $\square$

In two dimensions, the friction cone $\mathcal{C}_\mu(\mathbf{p})$ is bounded by two rays emanating from the origin and its image in wrench space $\mathbf{\Gamma}_\mu(\mathbf{p})$ is a planar cone ($\mathbf{\Gamma}$ being linear) and also bounded by two rays. For three dimensional objects however, the friction cone is a circular cone in three dimensions. Its image under $\mathbf{\Gamma}$ is a convex cone with apex at the origin, but is no longer linear. An early analysis of the image of a friction cone was done by Ji [Ji87].

We state two generalized versions of Carathéodory's theorem, which we call *thickened* versions[2].

Theorem 2.2 *Consider a family of convex sets A_1, A_2, ..., A_n in $\mathbb{R}^d$ and a point $x \in \mathrm{conv}\,(A_1, \ldots, A_n)$. Then there exist $m = \min\{n, d+1\}$ points*

---

[2] A different generalization called a 'multiplied version' is due to Bárány.

$y_1, y_2, \ldots, y_m$, *each belonging to a unique* A_i *such that* $x \in$ conv $\{y_1, \ldots, y_m\}$.

Proof. By Carathéodory's theorem, there exist $d + 1$ points $y_1, y_2 \ldots, y_{d+1} \in \bigcup_{i=1}^{n} A_i$ such that

$$x = \sum_{i=1}^{d+1} \lambda_i y_i \text{ with } 0 \leq \lambda_i \leq 1. \qquad (3)$$

Assume without loss of generality that $p_1, p_2 \in A_1$ and $\lambda_1, \lambda_2 \neq 0$. Then

$$
\begin{aligned}
x &= \lambda_1 y_1 + \lambda_2 y_2 + \sum_{i=3}^{d+1} \lambda_i y_i \\
&= (\lambda_1 + \lambda_2)\left(\frac{\lambda_1 y_1 + \lambda_2 y_2}{\lambda_1 + \lambda_2}\right) + \sum_{i=3}^{d+1} \lambda_i y_i \\
&= (\lambda_1 + \lambda_2) q + \sum_{i=3}^{d+1} \lambda_i y_i
\end{aligned}
$$

for $q = \left(\frac{\lambda_1}{\lambda_1 + \lambda_2}\right) y_1 + \left(\frac{\lambda_2}{\lambda_1 + \lambda_2}\right) y_2 \in A_1$ by convexity of A_1. Now x lies in the convex hull of d points which resides in an affine space of dimension $d - 1$. In effect, the hyperplane containing $y_3, \ldots, y_{d+1}$ is rotated around these points until it contained x. This process can be continued until the requirement of the theorem is met. $\square$

This theorem gives a constructive procedure to compute the y_i's, given an appropriate description of the convex sets A_i. The algorithms takes total time of $O(d^2)$, given the initial $d + 1$ points classified by which set they belong to. These points can be found using techniques from section 5. We also have the following

Theorem 2.3 *Consider a family of convex sets* A_1, A_2, $\ldots, A_n$ *in* $\mathbb{R}^d$ *and a point* $x \in$ conv $(0, A_1, \ldots, A_n)$. *Then there exist* $m = \min\{n, d\}$ *points* $y_1, y_2, \ldots, y_m$, *each belonging to a unique* A_i *such that* $x \in$ conv $\{0, y_1, \ldots, y_m\}$.

Proof. Similar to that of Theorem 2.2 using a version of Carathéodory's theorem which allows us to specify a point (0 in this case) such that $x \in$ conv $\{0, y_1, \ldots, y_d\}$. See for example the paper by Bárány [Bár82] for (a more general version of) this theorem. $\square$

Applying theorem 2.3 to the sets pos $\Gamma_\mu(\mathbf{p}_i)$, which are convex by lemma 2.1, we conclude that only one point from each is necessary to generate an external wrench (which is called x in the theorem.) Hence $\bigcup_i$ pos $\Gamma_\mu(\mathbf{p}_i)$ does indeed represent the set of possible wrenches the fingers can generate. Therefore, to test for closure in the frictional setting, we can check if the origin is in the interior of the convex hull of the friction cones associated with the finger contacts. This is however computationally non-trivial, as pos $\Gamma_\mu(\mathbf{p}_i)$ has non-linear boundaries, at least for three dimensional objects. In practice, an approximation is usually used. (See [Tei95a].) The test for closure can then be implemented as in the non-frictional case using such an approximation.

3 Efficiency of a Closure Grasp

In the description of closure grasps, there is an implicit unrealistic assumption that *the magnitudes of finger forces are in no way constrained*. In particular, it is quite likely that a closure grasp may resist any arbitrary external wrench; but it may only do so by applying an unrealistically large force at a finger in response to a fairly small external wrench in some direction.

In order to alleviate this problem, we may assume that certain additional constraint is imposed on the magnitudes of the finger forces—the *"finger force constraint"* being expressible as

$$
\begin{aligned}
\chi \; &: \; \mathbb{R}^n \to \{0, 1\} \\
&: \; (f_1, \ldots, f_n) \mapsto \begin{cases} 1, & \text{if the "constraint" holds;} \\ 0, & \text{otherwise.} \end{cases}
\end{aligned}
$$

Let $G = \{\mathbf{p}_1, \ldots, \mathbf{p}_n\}$ be a grasp. The set of external wrenches that can be generated by the grasp, subject to the finger force constraint, χ, is given by $\mathcal{W}_\chi$, called the *"feasible wrench set:"*

$$
\begin{aligned}
&\mathcal{W}_\chi(\Gamma_\mu(\mathbf{p}_1), \ldots, \Gamma_\mu(\mathbf{p}_n)) \\
&= \left\{ \mathbf{w} = \sum_{i=1}^{n} f_i \Gamma_\mu(\mathbf{p}_i) : \chi(f_1, \ldots, f_n) = 1 \right\} \subseteq \mathbb{R}^6.
\end{aligned}
$$

We also call $-\mathcal{W}_\chi$ the *"resistable wrench set,"* the set of external wrenches that can be resisted by the grasp.

Some natural finger force constraints that one may impose are of the following kinds:

Convex Constraint: This constraint allows us to bound on the total force applied by all the fingers.

$$\chi_{con} : f_1 \geq 0, \ldots, f_n \geq 0 \text{ and } \sum_{i=1}^{n} f_i \leq 1.$$

Thus, $\mathcal{W}_{\chi_{con}} = \text{conv}\{[0,1]\Gamma_\mu(\mathbf{p}_i) : 1 \leq i \leq n\}$.

Max Constraint: It corresponds to the case where each finger can apply a bounded force, independently of the others.

$$\chi_{max} : f_1 \geq 0, \ldots, f_n \geq 0 \text{ and } \max_{i \in \{1,\ldots,n\}} f_i \leq 1.$$

$\mathcal{W}_{\chi_{max}}$ is given by the Minkowski sum of the vectors $\Gamma_\mu(\mathbf{p}_1), \ldots, \Gamma_\mu(\mathbf{p}_n)$:

$$\mathcal{W}_{\chi_{max}} = \bigoplus_{i=1}^{n} \{[0,1]\Gamma_\mu(\mathbf{p}_i) : 1 \leq i \leq n\}.$$

Hybrid Constraint: This constraint corresponds for example to the case where we have several robot arms and wish to bound the total forces on each arm separately. Let $P_1, P_2, \ldots, P_l$ be a partition of the indices $\{1, \ldots, n\}$. Then

$$\chi_{hyb} : f_1 \geq 0, \ldots, f_n \geq 0 \text{ and } \sum_{i \in P_j} f_i \leq 1, 1 \leq j \leq l.$$

Thus

$$\mathcal{W}_{\chi_{hyb}} = \bigoplus_{j=1}^{l} \text{conv}\{[0,1]\Gamma_\mu(\mathbf{p}_i) : i \in P_j\}$$

$$= \text{conv} \bigoplus_{j=1}^{l} \{[0,1]\Gamma_\mu(\mathbf{p}_i) : i \in P_j\}$$

For $\mu > 0$, we need to show that with the new constraints on the finger forces, we can still express an external wrench $\mathbf{v} \in -\mathcal{W}_\chi(\Gamma_\mu(\mathbf{p}_1), \ldots, \Gamma_\mu(\mathbf{p}_n))$ as a sum $\mathbf{v} = \sum_{i=1}^{n} f_i \mathbf{w}_i$ with $\chi(f_1, \ldots, f_n) = 1$ and $\mathbf{w}_i \in \Gamma_\mu(\mathbf{p}_i)$.

For the case of χ_{con}, this follows from theorem 2.3 applied to $[0,1]\Gamma_\mu(\mathbf{p}_i)$, $i = 1, \ldots, n$, which are convex

sets by the second statement of lemma 2.1. For the case of χ_{max}, this follows from the definition of Minkowski sum. A combination of the reasonings used in these two cases, can be used to prove a similar result for χ_{hyb} also.

Note that analogous versions for non-frictional and frictional situations have been introduced earlier: Convex Constraint with no friction [KMY92], Convex and Max Constraints with friction [FC92] and a general unified framework including arbitrary contact models [Mis94]. However, the precise frictional models introduced here and the combined hybrid constraints are new.

3.1 Grasp Quality Measures

For any set $X \subseteq \mathbb{R}^d$, let the *residual ball* of X refer to the maximal ball $\mathcal{B}(X)$ centered at the origin 0 such that $\mathcal{B}(X)$ is fully contained inside the convex hull of X. The *residual radius* of X, denoted $r(X)$ is the radius of this residual ball $\mathcal{B}(X)$. Note that

$$\mathcal{W}_{\chi_{con}} \subseteq \mathcal{W}_{\chi_{hyb}} \subseteq \mathcal{W}_{\chi_{max}} \subseteq n\,\mathcal{W}_{\chi_{con}}$$

and $r_{\chi_{con}} \leq r_{\chi_{hyb}} \leq r_{\chi_{max}} \leq n\,r_{\chi_{con}}$. The corresponding residual radii are our grasp metrics under the appropriate constraints[3].

3.2 The Case of Friction

Allowing friction at the finger contacts causes the set of applicable forces to increase. Thus it is not surprising that grasp strength, as measured by the measures described here also increases. In fact some grasps which are not closure without friction, become closure grasps for a sufficiently large coefficient of friction. This is the case for example of two fingers placed at antipodal points on a planar object. Figure 1 shows a four finger closure grasp in the plane, call it G_4, and a graph showing grasp strength as measured by r_{con} as a function of

---

[3]We note that, for example in the frictionless case, the complexity of the Minkowski sum of the segments $\overline{0\Gamma(\mathbf{p}_i)}$ is much larger than the convex hull of those segments since every segment contributes more than one vertex on the Minkowski sum. This Minkowski sum is known as a *zonotope*. In the six dimensional case, the upper bound on its complexity is $O(n^5)$, while for the convex hull, it is $O(n^3)$.

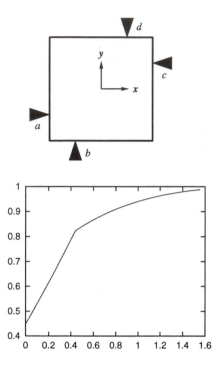

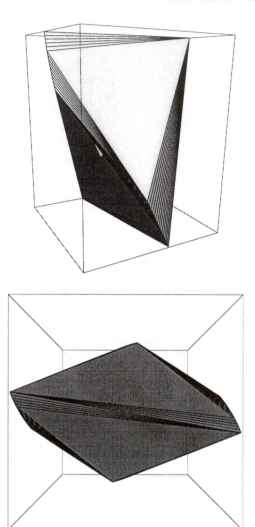

Figure 1: A four finger grasp and its strength as a function of the friction angle in radians.

the *friction angle*, the maximum angle a force can deviate from the object normal at the contact point, while maintaining contact. Corresponding to this grasp G_4, we also show in figure 2 a 'side' view and a 'top' view (from a point on the positive torque axis) of the set conv $\mathbf{\Gamma}(G_4)^4$.

4   Computational Issues

4.1   Approximating Friction cones

The main difficulty in computing these measures, not to mention performing grasp optimization based on them, is that conv $\mathbf{\Gamma}_\mu(\mathbf{p})$ has non-linear boundary for $\mu > 0$. It is customary in the robotics literature to approximate the boundary of the friction cone by a polyhedral cone. This is sufficient for detecting the presence or absence of closure. For the purpose of computing grasp strength,

---

[4]The figures For this section were generated using our residual radius implementation. Our software uses the QHULL library and the GEOMVIEW program of the Geometry Center, University of Minnesota.

Figure 2: The grasp of figure 1 for friction coefficient 0.2.

a course approximation causes the grasp strength to be underestimated. In this section we quantify this statement.

We have already defined $\mathbf{\Gamma}_\mu$ in equation (2) to be the image by $\mathbf{\Gamma}$ of all *unit* forces a finger can apply at a given point. This set lies on $\mathcal{S}^2 \oplus \mathbb{R}^3$ (or on $\mathcal{S}^1 \oplus \mathbb{R}$ for planar objects). We can approximate this set as follows. We start by approximating the friction cone $\mathcal{C}_\mu$ by N_1 regularly spaced rays. Similarly, we approximate the part of the truncated friction cone $\mathcal{F}_\mu$ that lies on $\mathcal{S}^2$. We obtain a N_2-vertex polyhedron S_{N_2} approximating a unit sphere. This polyhedron is then intersected with the approximation of the unbounded friction cone; the total complexity of the intersection is $O(N_1 + N_2)$. Let us denote by $\mathcal{F}_\mu(N_1, N_2)$ the resulting approximation. Also let $\mathbf{\Gamma}_\mu(N_1, N_2)$ denote the image of $\mathcal{F}_\mu(N_1, N_2)$ under the wrench map. After taking all such approximations for each finger, one can compute an approximate value $\tilde{r}$ of the grasp strength simply by solving at most M linear programming problems as in Kirkpatrick *et al.*[KMY92]. The values of N_1, N_2, and M will depend on the quality of the approximation desired. An alternative method is to compute the convex hull H of the set of points forming the vertices of $\mathbf{\Gamma}_\mu(N_1, N_2)$ for each finger, and finding the facet closest to the origin of H. Since there are $O(n(N_1 + N_2))$ points, computing the convex hull takes $O(n^3(N_1 + N_2)^3)$ time [Cha93]. Finding the facet closest to the origin is then trivial, given an appropriate representation of the hull.

We now give a lemma which gives a bound on the quality of the approximation of the friction cones. Let r_{μ, N_1, N_2} be an approximation of the residual radius of the grasp when the approximated friction cones $\mathcal{F}_\mu(N_1, N_2)$ are used, and $r_{\mu, N_1, N_2, M}$ be an approximation of the residual radius of the grasp when the approximated friction cones $\mathcal{F}_\mu(N_1, N_2)$ are used and the residual radius of these cones is further approximated using Kirkpatrick *et al.*'s scheme. Finally, $\tilde{r}$ will denote either of the two, according to the context.

Lemma 4.1 *Given* μ, N_1, N_2, *we have*

$$\frac{r_{\mu(1-\varepsilon)}}{1 + \delta_1} \le r_{\mu, N_1, N_2} \le r_\mu,$$

with $\delta_1 = 18/N_2$ *and* $\epsilon = \pi^2/2N_1^2$.

Proof. Consider a fixed finger contact point $\mathbf{p}$, and let $\mathcal{F}_\mu = \mathcal{F}_\mu(\mathbf{p})$ and $\mathcal{F}_\mu(N_1, N_2)$ be the corresponding friction cone and the approximated friction cone, as described above. First note that the infinite friction cone $\mathcal{C}_\mu$ can be approximated by a regular polyhedral cone C_μ such that $\mathcal{C}_{\mu\cos(\pi/N_1)} \subset \mathcal{C}_\mu$. But $\cos(\pi/N_1) \ge (1 - \frac{\pi^2}{2N_1^2}) = 1 - \epsilon$.

Next, we intersect the approximated infinite polyhedral cone with the approximated unit sphere S_{N_2}.

This sphere is approximated using N_2 points by techniques similar to those in Kirkpatrick *et al.* [KMY92], and the residual radius of this approximation is at least $1 - \frac{9}{N_2} = 1 - \frac{1}{2}\delta_1$. We conclude that

$$\frac{[0,1]\mathcal{F}_{\mu(1-\varepsilon)}}{1 + \delta_1} \subset (1 - \frac{1}{2}\delta_1)[0,1]\mathcal{F}_{\mu(1-\varepsilon)}$$
$$\subset [0,1]\mathcal{F}(N_1, N_2) \subset [0,1]\mathcal{F}_\mu$$

since $\frac{1}{1 - \frac{1}{2}\delta_1} < 1 + \delta_1$ for $\delta_1 < 1$, from which the lemma follows. $\square$

Corollary 4.2 *Given* μ, N_1, N_2, M, *we have*

$$\frac{r_{\mu(1-\varepsilon)}}{1 + \delta_3} \le r_{\mu, N_1, N_2, M} \le r_\mu,$$

with $\delta_3 = 2(\delta_1 + \delta_2)$, $\delta_2 = 36(\frac{36}{M})^{2/5}$ *and* $\epsilon = \pi^2/2N_1^2$.

Proof. Using techniques from [KMY92], we can approximate the residual radius of the convex hull of all approximated images of friction cones $\mathbf{\Gamma}_\mu(N_1, N_2)$ by performing M linear programs with an approximation factor of $1 - \frac{1}{2}\delta_2$. Thus

$$r_{\mu, N_1, N_2}(1 - \frac{1}{2}\delta_2) \le r_{\mu, N_1, N_2, M} \le r_{\mu, N_1, N_2}.$$

By lemma 4.1,

$$\frac{r_{\mu(1-\varepsilon)}}{1 + \delta_1} \le r_{\mu, N_1, N_2}$$
$$(1 - \frac{1}{2}\delta_2)\frac{r_{\mu(1-\varepsilon)}}{1 + \delta_1} \le r_{\mu, N_1, N_2, M}$$
$$\frac{r_{\mu(1-\varepsilon)}}{(1 + \delta_2)(1 + \delta_1)} \le r_{\mu, N_1, N_2, M}$$
$$\frac{r_{\mu(1-\varepsilon)}}{1 + 2(\delta_1 + \delta_2)} \le r_{\mu, N_1, N_2, M}$$

for $\delta_1, \delta_2 < 1/2$. $\square$

4.2 Finding the smallest friction coefficient guaranteeing a given grasp quality

Using the above lemmas, we can solve the following problem:

Given n "feasible gripping" points on an object and a positive real number $r > 0$, one can compute an approximate value for the coefficient of friction $\tilde{\mu}$ which guarantees that some subset of the gripping points will yield a good closure grasp of quality r or larger. Furthermore, if $\mu^*(r)$ represents the necessary optimal value, then

$$\tilde{\mu} - \mu^*((1+\delta)r) \leq \tilde{\mu}\varepsilon.$$

Of course, $\mu^*(r) \leq \tilde{\mu}$. The value $\tilde{\mu}$ satisfying the approximation requirements can be computed in time linear in n and polynomial in $1/\varepsilon$ and $1/\delta$.

We proceed as follows. We first choose N_1, N_2 and M to satisfy the approximation requirements. Then by binary search on $\tilde{\mu}$, we find $\tilde{\mu}$ such that

$$r \leq \tilde{r} \leq (1+\delta_4)r.$$

Then by lemma 4.1 or corollary 4.2 we have (letting δ_* represent either δ_1 or δ_3 resp. according to the algorithm used):

$$\frac{r_{\tilde{\mu}(1-\varepsilon)}}{1+\delta_*} \leq \tilde{r} \leq (1+\delta_4)r$$
$$r_{\tilde{\mu}(1-\varepsilon)} \leq (1+\delta_4)(1+\delta_*)r$$
$$\tilde{\mu}(1-\varepsilon) \leq \mu^*((1+\delta_4)(1+\delta_*)r)$$
$$1 - \frac{\mu^*((1+\delta_4)(1+\delta_*)r)}{\tilde{\mu}} \leq \varepsilon$$
$$\frac{\tilde{\mu} - \mu^*((1+\delta_4)(1+\delta_*)r)}{\tilde{\mu}} \leq \varepsilon$$
$$\frac{\tilde{\mu} - \mu^*([1+2(\delta_* + \delta_4)]r)}{\tilde{\mu}} \leq \varepsilon.$$

We simply let $\delta = 2(\delta_* + \delta_4)$ for the result.

There are at most $O(\log(1/\delta_4))$ binary search steps. In the convex hull version of the algorithm, at each step, the convex hull operations take $O(n^3(N_1 + N_2)^3)$ time, giving a total running time of $O(\log(1/\delta_4)n^3(N_1+N_2)^3)$. This translates to a running time of $O(\log \frac{1}{\delta_4} n^3(\frac{1}{\delta_1} + \frac{1}{\sqrt{\epsilon}})^3)$.

If instead we use the Kirpatrick *et al.* approximation scheme for computing the residual radius, each step takes $O(M\, \mathrm{LP}(n(N_1 + N_2)))$ time, for a total running time of $O(\log(1/\delta_4)M\, \mathrm{LP}(n(N_1 + N_2)))$. Here $\mathrm{LP}(n)$ is the time required to solve a linear program of size

n in fixed dimension. Since $M = 36^{5/2}36/\delta_2^{5/2}$, this translates to a running time of $O(n \log \frac{1}{\delta_4} \frac{1}{\delta_2^{5/2}}(\frac{1}{\delta_1} + \frac{1}{\sqrt{\epsilon}}))$. Here, we use the linear time linear programming algorithm of [CM93]. Unfortunately the constant here is quite large, due both to δ_2 and the linear programming algorithm.

4.3   Analysis of Ferrari and Canny model

In the model of Ferrari and Canny [FC92], the component normal to the object surface of the force applied by the finger has a unit magnitude. Let $\mathcal{F}'_\mu$ be the friction cone defined in this manner. It is the boundary of the set defined by the infinite cone $\mathcal{C}_\mu$ intersected with a plane located at unit distance from the origin, and perpendicular to the object normal. As mentioned earlier, this approach yields large grasp qualities as the friction angle increases. This can be seen in figure 3. The dotted curve corresponds to the model of Ferrari and Canny, while the continuous curve represents our model and is the same as that of figure 1. In this ex-

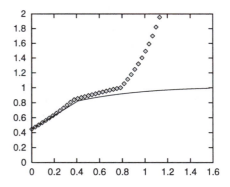

Figure 3: The grasp strength of G_4 as a function of the friction angle in radians under the two models.

ample, when the friction angle is larger than about 45 degrees, the convex hull structure changes drastically. The grasp quality grow unboundedly as the angle approaches 90 degrees.

To approximate $\mathcal{F}'_\mu$, we simply replace S_{N_2} by this plane in the approximations described previously. Let r'_μ be the residual radius of the grasp under this model, and r'_{μ,N_1} be the approximation of r'_μ when $\mathcal{F}'_\mu$ is approximated in a fashion analogous to $\mathcal{F}_{\mu,N_1,N_2}$. Also let $r'_{\mu,N_1,M}$ be the approximation of r'_μ as above where the residual radius r'_{μ,N_1} is further approximated using Kirkpatrick *et al.*'s scheme.

Then we can show that

$$r'_{\mu(1-\epsilon)} \leq r'_{\mu,N_1}$$

and

$$\frac{r'_{\mu(1-\epsilon)}}{1+\delta_2} \leq r'_{\mu,N_1,M} \leq r'_\mu$$

The algorithms for solving the problem of section 4.2 are also similar, and the running times are the same with the dependence on N_2 and δ_1 removed.

5 Generating Force Targets

Consider an external wrench $\mathbf{v}$ being applied to B which is being grasped by m fingers with frictional contacts. Mishra et. al. [MSS87] give a linear time procedure that given a set W of vectors $\mathbf{w}_i = \mathbf{\Gamma}(\mathbf{p}_i)$ in $\mathbb{R}^6$, $i = 1, \ldots, m$, finds a subset of 6 of them such that $\mathbf{v}$ can be expressed as a positive combination of those vectors. This can be used for finding force targets for frictionless fingers. The coefficients returned by their procedure however do not necessarily satisfy the finger force constraints.

We now describe a linear time procedure for finding force targets in the case of χ_{con} constraint. The problem discussed earlier can be handled as a result of the following

Theorem 5.1 *Given a set of points* $p_1, \ldots, p_m \in \mathbb{R}^d$, *with* $0 \in \text{conv}\{p_1, \ldots, p_m\}$, *and a point* $v \in \text{conv}\{p_1, \ldots, p_m\}$. *Then one can find* $\lambda_1, \ldots, \lambda_m$, *with* $0 \leq \lambda_i \leq 1$, *and* $\sum_{i=1}^m \lambda_i = 1$ *at most d of which are non-zero, such that* $v = \sum_{i=1}^m \lambda_i p_i$. $\square$

The theorem has a constructive counterpart, leading to an algorithm based on linear programming and some simple techniques to compute barycentric coordinates. The idea is to find a simplex containing v with one vertex at the origin 0. This is done by finding the facet of the convex hull of the p_i's which intersects the ray $\overrightarrow{0v}$, say at w. This can be done using linear programming [MSS87] in linear time. If the facet has more than d vertices, a similar procedure is repeated recursively with the ray $\overrightarrow{p_j w}$, where p_j is some vertex of the facet. We then obtain d vertices which, together with 0, form the required simplex. The coefficients λ_i, will simply be the barycentric coordinates of v in this simplex. The details appear in [Tei95a]. We use this algorithm for the case where $d = 6$ and the p_i's correspond to the wrenches $\mathbf{w}_i$'s and v to the external wrench $\mathbf{v}$. In the frictional case, one can use the above algorithm applied to an approximation of the friction cones as in section § 3, then "merge" any two points selected from the friction cones as per theorem 2.3. This might be sufficient in practice.

Acknowledgment

We wish to thank the anonymous referees for pointing out an error in our earlier interpretation of the Ferrari-Canny metric.

The research presented here was supported by a NYU Technology Transfer Grant (6-459-614) and an NSF IRIS grant: IRI-9414862.

References

[Bár82] I. Bárány. A Generalization of Carathéodory's Theorem. *Discrete Math.*, 40:141–152, 1982.

[Bic95] A. Bicchi. On the Closure Properties of Robotic Grasping. *International Journal of Robotics Research*, 14(4):319–334, 1995.

[Cha93] B. Chazelle. An Optimal Convex Hull Algorithm in Any Fixed Dimension. *Discrete Comput. Geom.*, 10:377–409, 1993.

[CM93] B. Chazelle and J. Matoušek. On Linear-time Deterministic Algorithms for Optimization Problems in Fixed Dimension. In *Proc. 4th ACM-SIAM Sympos. Discrete Algorithms*, pages 281–290, 1993.

[Ji87] Z. Ji. *Dexterous Hands: Optimizing Grasps by Design and Planning*. Ph.D. Thesis, Stanford Univ., Stanford, CA, 1987.

[FC92] C. Ferrari and J. Canny. Planning Optimal Grasps. In *Proceedings of the IEEE International Conference on Robotics and Automation*, pages 2290–2295, Nice, France, 1992.

[KR86] J. Kerr and B. Roth. Analysis of Multifingered Hands. *International Journal of Robotics Research*, 4(4):3–17, 1986.

[KMY92] D. Kirkpatrick, B. Mishra, and C.-K. Yap. Quantitative Steinitz's Theorems with Applications to Multifingered Grasping. *Discrete Comput. Geom.*, 7:295–318, 1992. Also in *Proc. 22nd Ann. ACM Symp. Theory Comp.*, pages 341–351, 1990.

[MNP90] X. Markenscoff, L. Ni, and C.H. Papadimitriou. The Geometry of Grasping. *The International Journal of Robotics Research*, 9(1), 1990.

[Mis94] B. Mishra. Grasp Metrics: Optimality and Complexity. In K. Goldberg, D. Halperin, J.C. Latombe, and R. Wilson, editors, *Algorithmic Foundations of Robotics*, pages 137–166. A.K. Peters, 1994.

[MSS87] B. Mishra, J.T. Schwartz, and M. Sharir. On the Existence and Synthesis of Multifinger Positive Grips. *Algorithmica*, 2:541–558, 1987.

[Ngu87] V.-D. Nguyen. Constructing Force-Closure Grasps in 3D. In *Proceedings of the IEEE International Conference on Robotics and Automation*, pages 240–245, 1987.

[PSS+95] J. Ponce, S. Sullivan, A. Sudsang, J.-D. Boissonnat, and J.-P. Merlet. Algorithms for Computing Force-closure Grasps of Polyhedral Objects. In K. Goldberg, D. Halperin, J.C. Latombe, and R. Wilson, editors, *Algorithmic Foundations of Robotics*, pages 167–184. A.K. Peters, 1995.

[SS88] J.T. Schwartz and M. Sharir. Finding effective 'force targets' for two dimensional multifinger frictional grasp. Unpublished Manuscript, 1988.

[Tei95a] M. Teichmann. *Grasping and Fixturing: a Geometric Study and an Implementation*. Ph.D. Thesis, New York Univ, 1995.

[Tei95b] M. Teichmann. A Grasp Metric Invariant under Rigid Motions. In *Proceedings of the IEEE International Conference on Robotics and Automation*, 1996.

Algorithms for Fixture Design

Chantal Wentink, *Utrecht University, Utrecht, the Netherlands*
A. Frank van der Stappen, *Utrecht University, Utrecht, the Netherlands*
Mark Overmars, *Utrecht University, Utrecht, the Netherlands*

Manufacturing and assembly processes often require objects to be held in such a way that they can resist all external wrenches. The problem of "fixture planning" is to compute, for a given object and a set of fixturing elements, the set of placements of the fixturing elements that constrain all finite and infinitesimal motions of the object (due to applied wrenches). As fixturing problems occur frequently in manufacturing and assembly, it becomes costly to build a dedicated fixturing solution for each different problem. Modular fixturing toolkits offer the advantage of reusability of the fixturing elements and have therefore gained considerable popularity. A modular fixturing toolkit consists of a fixturing table with a rectangular grid of holes, and a set of fixturing elements whose positions are restricted to the holes in the table. Several recent publications in the field of fixture planning aim at exploring the power of these modular fixturing toolkits. We give an overview of modular and non-modular fixture planning for various types of objects and sets of fixturing elements.

1   Introduction

Many manufacturing operations, such as machining, assembly, and inspection, require constraints on the motions of parts or subassemblies of parts [4, 8]. The concept of *form-closure* is over a century old [24] and refers to constraining, despite the application of an external wrench (force and moment), all motions of a rigid object (including infinitesimal motions) by a set of contacts on the object; any motion of an object in form-closure has to violate the rigidity of the contacts. Therefore, the problem is to compute contact locations on a given part shape that achieve form-closure.

In this paper, we are interested in immobilizing

Figure 1: *A polygon that is not fixtured with four point-contacts and the same polygon that is fixtured.*

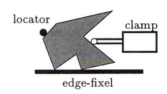

Figure 2: *A polygon in form closure with one edge-contact and two point-contacts. A point-fixel can reach into concavities, whereas an edge-fixel cannot.*

planar objects, in particular polygons. We give an overview of fixturability results and algorithms to compute form closure configurations under different fixture models. We refer to the set of contacts achieving form-closure as a *fixture*. We assume that the contacts are frictionless; note that this is a conservative assumption since any fixture computed assuming zero friction also holds in presence of friction. By the *fixture model*, we imply the set of allowable contact types. The conceptually simplest model is that of point contacts. See Figure 1 for an example of a fixture in a model with four point-contacts. It has been known since Reuleaux that fixturing a planar object requires at least four frictionless contacts. Mishra, Schwartz, and Sharir [20] and Markenscoff, Ni and Papadimitriou [18] independently proved that four point-contacts are also sufficient.

While point-contacts are conceptually simple, they

321

angle-fixel

Figure 3: *A polygon in form closure with two edge-fixels (one angle-fixel) and a point-contact.*

are not always easy to achieve in practice. The reason is that for form-closure, these point-contacts have to be capable of resisting arbitrary wrenches and therefore they have to be backed by bulky supports. This in turn implies difficulty of placing point-contacts at some points on the boundary of an object, in particular, at narrow concavaties. Hence, it is important to look for other possible practical fixture models in order to reduce the number of point-contacts. In everyday life, we frequently lean an object against a flat surface, such as a table or a wall, to constrain its motions. In the planar world, the analog of a wall is a supporting line. In this paper we also consider fixture models that include *edge-contacts* which offer straight-lines of support. Notice that an edge-contact can touch the object only along its convex hull. The object simply rests against it; there is no reaching into concavaties. See for an example Figure 2.

The first part of this overview is concerned with fixturability results. We identify classes of parts that are fixturable under different fixture models including point-contacts and one or two edge-contacts. Since it is not only important to know what parts can be fixtured under a model, but also what parts can not be fixtured, we also give negative results on existence of solutions.

The second part of the paper deals with *modular fixtures* which is a subject of considerable popular interest in the manufacturing industry for the past ten years or so [1, 2, 13, 14]. Basically, this involves a regular square grid of lattice holes together with fixture elements (or *fixels* [5]) that are constrained by the grid; the object

rests against these fixels which constrain its motions. Custom-built fixtures being expensive, the major benefit of modular fixtures stems from their reconfigurability; it is often necessary to fixture an object only for short periods of time after which the same set of fixels can be used to fixture different objects. Another advantage of modular fixtures is their easy assembly and disassembly.

Since research in computing form-closures generally involved point-contacts, it is not surprising that most fixels were designed to achieve point-contact. The simplest fixel is the *locator* which is a circular object centered at a lattice hole. Since achieving contact with four circles constrained to a grid is in general impossible ([30] shows this even for three circles), it is clear that we need a fixel that takes care of the slack. Such a fixel is called a *clamp* which has a fixed portion, the clamp *body*, attached to a movable rod, the clamp *plunger*, that can translate between certain limits along a grid line. The end of the plunger is the clamp *tip* which makes contact with the object. A clamp can be configured so that the motion of the plunger is parallel to either one of the axes; it is termed *horizontal* or *vertical* accordingly. See Figure 2 for an example of a clamp. An *edge-fixel* is simply a bar-like object of appropriate dimensions fixed to the lattice offering a straight-edge of support. We assume that an edge-fixel is at least as long as the longest edge in the convex hull of the object. Two edge-contacts are achieved by the use of a so-called *angle-fixel*, see for example Figure 3. The angle-fixel consist of two edge-fixels that are connected by a joint (an adjustable angle-fixel) or connected to each other with a fixed angle.

Wallack and Canny [28] consider an interesting model of modular fixtures which uses four locators and no clamps; instead, the slack is countered by mounting the part on a split horizontal lattice and allowing the one half to slide horizontally relative to the other. They call such a fixture device a vise. See Figure 4 for an example of this fixture model.

We will give fixturability results for models using the different fixture elements described above and also give an overview of algorithms to compute all possible form

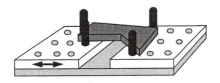

Figure 4: *A polygon in form closure on a vise.*

closure configurations on a grid. It is important to generate all solutions rather than just one. The reason is that additional conditions make some solutions preferable over others. By generating all solutions we can compare them and choose the one most suited for the particular problem. We consider both output-sensitive and non output-sensitive algorithms. An algorithm is called output-sensitive if its running time depends on the number of answers (form closure configurations) it computes. A non output-sensitive algorithm has a fixed running time. This is independent from the number of form closure configurations. Usually one prefers output-sensitive algorithms, but it is not always possible to find them.

This overview is structured as follows. We first give a short introduction on the notions of form closure and immobility in Section 2. Then, after introducing some preliminaries in Section 3, we discuss in Section 4 the different types of fixture models that are considered in the rest of the paper. In Section 5 we give an overview of (non-) fixturability results in different models, where we are not restricted to a grid. We also show how to compute one form closure configuration efficiently. Section 6 considers the case of modular fixtures. Finally, we conclude with possible extensions of the algorithms and open problems in Section 7.

2 Form Closure and Immobility

The terminology used in fixturing literature over the past years has not always been consistent. Here we discuss a few different notions on immobility of objects and show how our definition of *form closure*, which we will use in the rest of this paper, fits in. Several authors have discussed form closure and immobility [3, 11, 23, 25, 26]. Initially Reuleaux [24] used the

term force closure to describe immobilization (equilibrium) of an object that requires the application of an externally applied wrench. He defined form closure on a body as an equilibrium that is maintained despite the application of any possible externally applied wrench (force and moment). The method that we use (and that was first described by Reuleaux) is an instantaneous analysis and describes only the constraints on infinitesimal motions. This is equivalent to the notion of 1st order immobility introduced by Rimon and Burdick [25]. 2nd order immobility analysis includes the curvatures of the object and fixel surfaces into the determination of object immobility and therefore leads to other results [10, 25]. Since most of the literature focuses on 1st order immobility, we will define this as form closure. (So whenever we use the term form closure, we mean 1st order immobility.) This notion is a sufficient condition for immobility, but not a necessary one as pointed out by Rimon and Burdick [25]. When taking into account 2nd order immobility, it turns out that three contacts suffice to immobilize almost all two-dimensional objects. The three contact normals, however, must intersect in a single point which is usually hard to accomplish in practice. From a practical point of view it seems better to use 1st order immobility. Furthermore forces that will be applied on the point-contacts can be arbitrary large if we use only three point-contacts (if, for example, we drill a hole at the point where the three normals intersect each other)

3 Geometric Interpretation

Recall that a fixture is said to provide *form-closure* if it precludes all (planar) motion, translations and rotations. Let us start by examining what motions are ruled out by single point- or edge-contacts. Let from now on the object to be fixtured be a polygon P (not necessarily convex). A point-contact is a contact of a point-like fixture element (locator or clamp) with an edge or a vertex of P, whereas an edge-contact is a contact of an edge of a fixture element with an edge or one or more vertices of the object P. A wrench applied to P will make it translate or rotate. We will consider rotations only, with the understanding that translations in a direction are simply rotations about a point at infinity along the perpendicular direction. In this manner an (infinitesimal) motion of a polygon can be represented by a point in the plane, denoting the center of rotation of this motion; together with the direction of the rotation: clockwise (+) or counter clockwise (-). A point at infinity thus represents an (infinitesimal) translational motion.

Denote the boundary of the input polygon P by ∂P; n is the number of edges forming ∂P. Let $CH(P)$ denote the convex hull of P. Let the polygon edge containing a point a on its boundary be denoted by $E(a)$; the directed line perpendicular to $E(a)$ through a pointing to the interior of P is $l(a)$. We distinguish the following cases (See Figure 5).

Point contact at interior of an edge. This is the fundamental contact and the motions allowed by other types of contacts can be deduced by composing those allowed by elementary point-contacts.

Consider a point-contact at point a in the interior of edge $E(a)$ as shown in Figure 5. The allowed motions are defined by the line $l(a)$. If the object rotates in a clockwise (positive) direction with a point-contact at a, its center-of-rotation (COR) will have to lie in the region to the right of (and including) $l(a)$. Furthermore, any point in this closed right-half-plane is a possible COR for a positive rotation. Similarly, the COR's for counter-clockwise or negative rotations lie in the closed half-plane to the left of $l(a)$. These are all and the only constraints imposed by the point-contact at a. For future reference, we call these *half-plane constraints* imposed by a.

Point contact at concave vertex. This is shown in the upper-right of Figure 5; a is the concave vertex. Imagine two points a_1, a_2 infinitesimally close to a along the two edges defining it. The motions allowed by a can be determined by intersecting the motions allowed by the point-on-edge pairs $(a_1, E(a_1))$ and $(a_2, E(a_2))$ which may be individually analyzed as above.

Consider rotating a line l through a from $l(a_1)$ to $l(a_2)$ in the clockwise direction. The wedge constraints defined by a is the intersection of each of the half-plane constraints obtained along the sweep. Therefore, the

motions allowed by a are described by the intersection of two half plane constraints, this can be represented by a wedge. The result is shown in the figure. These will be called *wedge constraints* defined by a.

Edge contact at a vertex of $CH(P)$. An edge-fixel can be in contact with a vertex of the convex hull of $P, CH(P)$. Consider the line $l(a)$ perpendicular to the edge-fixel and directed towards the side of the fixel that contains the polygon. All points in the half plane to the left of $l(a)$ can be COR's for counterclockwise rotations. Similarly the points in the right half-plane are COR's for clockwise rotations of P. Thus, this type of contact can be described by one half-plane constraint.

Edge contact at an edge of $CH(P)$. An edge-fixel can be in contact with the polygon along an edge e the polygon or at two vertices a_3, a_4 (adjacent on the convex hull). The latter case is shown in the Figure 5; the former case can be similarly analyzed considering a_3, a_4 to be the end-vertices of e. Consider the lines $l(a_3), l(a_4)$ perpendicular to the edge-contact and confine the two half-plane constraints to get the result shown. The closed left half-plane at $l(a_3)$ allows for negative rotations while the right half plane of $l(a_4)$ allows for positive rotations. The open infinite "slab" in the middle, shown shaded in the figure, denoted $slab(a_3, a_4)$, disallows all motions and is a crucial entity in future analysis. The constraint imposed will be called a *slab constraint*, which is actually a combination of two half-plane constraints.

We do not give constraints for a point-contact at a concave vertex, because in practice, these contacts are not used. These contacts are not stable and will lead to deformation of the part if the forces become too large.

If we want to know if an object is in form closure, we have to construct the regions associated with the contacts that are imposed and confine them. If there is no region that allows for positive (or negative) rotations for every constraint, we have a form closure configuration, otherwise (infinitely small) motions are possible. The method of analysing possible motions in this geometrical (graphical) way was introduced by Reuleaux [24] in the case of point-contacts. It should be noted

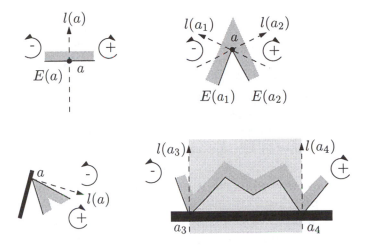

Figure 5: *Motions possible under simple contacts.*

Figure 6: *A rectangle that is not fixed with one point-contact and one edge-contact.*

that two (or more) parallel contact normals intersect in a point at infinity. See for an example Figure 6, in this case all positive and negative rotations seem to be excluded, but the intersection point at infinity still gives a possible center of rotation. Since this point is at infinity, a rotation means translation, thus the rectangle can still be translated to the left or to the right.

4 Fixture Models

In this paper we will consider four different ways of fixturing objects: with four point-contacts, with one edge-contact and two point-contacts and two edge-contacts and one point-contact (of which two perpendicular edge-contacts and one point-contact is a special case). All these models are minimal in the sense that for most polygons we need all the fixels in the model to achieve a form closure configuration. We can make a

comparison of the different fixture models using a number of criteria. The main criteria that we discuss here are classes of fixturable objects (which we will call the strength of the model, the model being stronger if more objects can be fixtured), loading the part, accessability, efficiency of computation of fixtures and stability of the fixtures. We also compare the number of contacts that are made in the model in terms of half-plane constraints. In the rest of the paper we will assume that whenever we use one or more edge-contacts, one of these edge-contacts will be an edge-contact at an edge of $CH(P)$.

Four point contacts. This is the basic model in which all polygons can be fixtured. In the following we will discuss why we also look at other fixture models that seem to be less strong, but do have advantages over the four-point model. A model with four point-contacts is preferred if we need to have a large accessability of the part, since edge-contacts will constrain accessability of P more than point-contacts. The maximum number of half plane constraints that we can get in this model is eight, if we place all four contacts at concave vertices of the polygon.

One edge and two point contacts. An edge-fixel placed against an edge of $CH(P)$ imposes two half-plane constraints on P. The two point contacts can result in a maximum of four half-plane constraints if we place them at concave vertices. The model is not as strong as a model with four point-contacts, because we cannot fixture some parts that have parallel edges, as will be shown below. The reason for this is that the two point-contacts as imposed by the edge are co-linear. The advantage of the system however is that the computation of all possible fixtures in a modular model can be done more efficiently, because P can only be in a fixed number of different orientations, which is not the case when we only have point-contacts. Intuitively, the stability of a fixture with an edge-fixel is better when forces are applied to it, because the force will be distributed along the edge and not only act on two points. We also believe that it is easier to load an object into the fixture even when the initial orientation of P is uncertain. If we first load the polygon against the edge-fixel, we already have fixed the orientation of P and can then slide it against the locator and finally apply the clamp.

Two perpendicular edge-contacts and one point contact. In this case, again we can not fixture some parts that have parallel edges. Although the model can impose 5 half-plane constraints (or 6 in the case that two perpendicular edges of P are placed against the edge-fixels), we can not fixture every polygon because there is a dependancy between the three half-plane constraints that are imposed by the edge-fixels (the directions of the contact normals and the contact point with the second edge, when we have fixed the edge against the first edge depend on each other). Once we have determined the edge of $CH(P)$ to place against the horizontal edge-fixel, we have fixed three of the half-plane constraints already. The computation of all possible fixtures is fairly easy. The number of orientations for P is the same as in the model with one edge-contact and two point-contacts. In addition, for each orientation there is only the position for the point-contact to consider. Loading the part seems to be quite easy in the model and intuitively the stability of the fixtures is good.

Two edge-contacts and one point contact. With this model we have more degrees of freedom than in the case of two perpendicular edge-fixels, since we can vary the angle between the two edge-fixels. As a result, it turns out that we can fixture all polygons. In the model we can have maximal 6 half-plane constraints, two for each edge-contact and two for the point-contact (if it is placed at a concave vertex). In addition however, these are not always the placements that will give us a form closure configuration (e.g. a rectangle). But, if this is not the case, we can find a form closure configuration with a different placement of the part (if we assume that there is no grid). A disadvantage of the model is that the edges should be positionable in any orientation, which is not always achievable in practise.

So, in conclusion we can say that the models with four point-contacts and with two edges and one point-contact are the strongest but they have the disadvantages that form closure configurations are harder to compute, they are more difficult to load and less stable.

5   Fixturability without a Grid

In this chapter we examine what kind of objects are fixturable with each of the toolkits mentioned in the previous section. We do not yet take modularity constraints into account. For every model we show what is known about fixturability of classes of objects and algorithms to compute one form closure configuration. We will frequently use a property of the contact points of the maximal inscribed circle of P with the boundary of P as is stated in the following lemma. The lemma follows from Markenscoff et al. [18].

Lemma 1 *[18] Let P be a polygon without pairs of parallel edges. Let $MIC(P)$ be any maximal inscribed circle of P and c the center of $MIC(P)$. Then the three vectors $\overrightarrow{ca_i}$, $a_i \in (MIC(P) \cap \partial P)$, positively span $\mathbf{R}^2$.*

The time to compute the maximal inscribed circle of a polygon P ($MIC(P)$) is $O(n)$ both for convex and non-convex polygons [7]. This can be done by

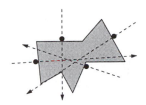

Figure 7: *A polygon that is fixtured with four point-contacts.*

computation of the medial axis of P, the center of the $MIC(P)$ is a vertex of the medial axis of P. We also use the following lemma that can easily be proved.

Lemma 2 *Given a set $V = v_1, v_2, v_3$ of three vectors that positively span $\mathbf{R}^2$. For every fourth vector v_4 in $\mathbf{R}^2$, we can define a new set of vectors consisting of v_4 and two vectors from V, such that this new set positively spans $\mathbf{R}^2$.*

5.1 Four Point-contacts

Since the graphical method of Reuleaux from Section 3 is not always the easiest test for automatically verifying for form closure, other methods for testing form closure with four point contacts are commonly used. A point-contact-vector on an object in two dimensions can be described by three variables; the position of the contact and its torque. Four point contacts provide form closure if and only if the four associated point-contact-vectors positively span $\mathbf{R}^3$. This can be verified using matrix computations [28] or force-sphere analysis [6]. Markenscoff et al. [18] proved the following Theorem.

Theorem 3 *[18] For any object O that is not a circle there is a set of four point contacts (possibly at a concave vertex) that provides form closure.*

The proof uses the properties of the maximum inscribed circle ($MIC(P)$) as stated in Lemma 1. The proof utilizes point contacts that are either on or close to the points of contact of a maximum inscribed circle of the object. It distinguishes two cases.

First there is the case where there are three contact points of $MIC(P)$ with P. In this case one of

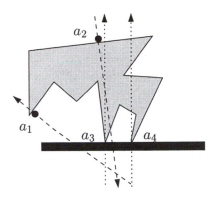

Figure 8: *A polygon fixtured with one edge-fixel and two point-fixels.*

these points is replaced by two points that are close to each other and on either side of the contact point of $MIC(P)$ and P. It is then shown that the point-contact-vectors at these four points positively span $\mathbf{R}^3$. The second case is the case where (some of the) $MIC(P)$'s touch only two points of P. This only happens when the $MIC(P)$ touches two parallel edges of P. In that case the MIC can be slided along these two edges, while its size remains the same. Thus we do not have a unique MIC. The intersection points of $MIC(P)$ and P are antipodal. If we assume that there is a third point on some $MIC(P)$ that touches P, we can always find a fourth point that will give us a form closure configuration. If there is no such a third contact point all the $MIC(P)$'s touch the boundary only in two points. In this case we can move the $MIC(P)$ to the left and the right in order to obtain two couples of forces that will give a so-called form closure by couples.

The proof is constructive in the sense that it does not only show the fixturability of objects in the four-point model, but it also indicates how to compute a form closure configuration for every two-dimensional object, by computing its MIC, leading to an algorithm with a running time of $O(n)$.

5.2 One Edge-fixel and Two Point-fixels

The problem of immobilizing a polygon with one edge-contact (where the edge-fixel is placed against a convex-hull-edge of P) and two point-contacts was analysed by Overmars et al. [22]. In Section 3 we have described what motions are constrained by an edge-edge contact and a point-edge contact. To constrain all motions of the object the set of allowable centers of rotation must be empty. The following lemma gives a more useful condition for form closure in this model. Let a_1 and a_2 be two points on an edge of P, and a_3 and a_4 two adjacent vertices on $CH(P)$.

Lemma 4 *[22] An object P is in form-closure with point-contacts a_1, a_2 and edge-contact (a_3, a_4) if and only if*

1. *the three vectors along $l(a_1), l(a_2), l(a_3)$ positively span $\mathbf{R}^2$, and*

2. *the intersection point of $l(a_1)$ and $l(a_2)$ lies in the interior of $slab(a_3, a_4)$.*

See Figure 8 for an example of an object that is in form closure with one edge-contact and two point-contacts.

Overmars et al. showed that if no edges of P are parallel then a form closure configuration always exists [22]. Their proof goes as follows. If we have the maximal inscribed circle of a convex polygon without parallel edges, there are always (at least) three intersection points (a_1, a_2, a_3) of P and $MIC(P)$, that lie in the interior of edges of P and whose contact normals positively span $\mathbf{R}^2$ (Lemma 1). If we place one of these edges against the edge-fixel, we obtain form closure by placing the two point-contacts at the other two intersection points of P and $MIC(P)$. This is true because the contact normals to these two points will intersect at the center (c) of $MIC(P)$ and the slab defined by the edge-contact has c strictly in its interior. From Lemma 4 we can now conclude that this is a form closure configuration. See for an example Figure 9.

A similar proof can be given if P is not convex. If at least one of the edges of $CH(P)$ is on the $MIC(P)$ we can use the same form closure construction as for a convex polygon, since the edge-fixel can then be placed against this edge of $CH(P)$. If none of the intersection points of P and $MIC(P)$ is on a convex hull edge of P we cannot place any of these edges against the edge-fixel. Instead, we grow the circle until we touch an edge

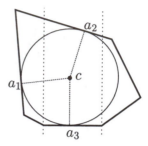

Figure 9: *The tangent points of the largest inscribed circle of P with P give us the edge- and point-contacts for form-closure.*

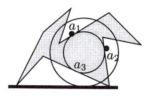

Figure 10: *Growing the maximal inscribed circle of a non-convex polygon, until it hits an edge of $CH(P)$ gives a form closure configuration.*

of $CH(P)$. We can then place this edge against the edge-fixel and choose two points of (a_1, a_2, a_3) whose contact normals will span $\mathbf{R}^2$ together with the direction of the edge-contact normal (corresponding to the edge-contact) using Lemma 2. See for example Figure 10. Thus we obtain the following Theorem.

Theorem 5 *[22] Let P be an arbitrary polygon. If no edge of P is parallel to one of the edges of $CH(P)$ then P can be held in form closure with one edge- and two point-contacts.*

However, it is not true that all polygons that do have parallel edges cannot be fixtured. We can even show that rectilinear polygons of which the convex-hull is not a rectangle are fixturable with an edge-fixel and two point-contacts.

The proof is again constructive, leading to an $O(n)$ algorithm for computing one form closure configuration.

Not all polygons can be fixtured in this model, for example a rectangle and some trapezoids cannot be

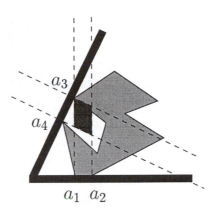

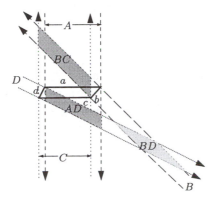

Figure 11: *A rectangle can not be fixtured in the model with one edge-contact and two point-contacts. We can always still translate it in horizontal or vertical direction.*

Figure 12: *Trapezoid abcd can not be fixtured in a model with one edge-fixel and two point-fixels, since the intersection of two slabs is never intersected by a third slab that is correctly oriented.*

Figure 13: *A polygon in contact with an angle-fixel, where the two contacts are both edge-contacts.*

fixtured as shown in Figures 11 and 12. The rectangle can always be translated in horizontal or vertical direction if we place the point-contacts at interiors of edges of P. The trapezoid with two parallel edges, a and c can not be fixtured. Consider all possibilities to place one edge of the trapezoid against the edge-fixel, then it can be seen that none of the correctly oriented contact normals will intersect in the slab defined by the edge-contact. Thus, according to Lemma 4 a form closure configuration does not exist.

5.3 One Angle-fixel and a Point-fixel

In a fixture model with one angle-fixel and a point-fixel, there are two possible sets of contacts. The angle-fixel can sometimes be placed such that both edges of the fixel make contact with an edge of the convex-hull of P. In this case the intersection between the two slabs defined by the edge-contacts is a parallelogram,

s. For an example see Figure 13. If we want to obtain a form-closure configuration the contact normal to the point-contact should intersect the parallelogram and be correctly oriented. Let a_1 be the the leftmost vertex contacting the first edge-contact and a_2 the rightmost vertex contacting this edge. Similarly a_3 is the leftmost vertex contacting the second edge and a_4 the rightmost contact vertex.

Lemma 6 *An object P is in form-closure with edge-contacts (a_1, a_2) and (a_3, a_4) and point-contact a_5 if and only if*

1. *The three vectors along $l(a_1), l(a_3), l(a_5)$ positively span $\mathbf{R}^2$, and*

2. *$l(a_5)$ should intersect the interior of $slab(a_1, a_2) \cap slab(a_3, a_4)$.*

In the other case one edge-fixel can be placed against an edge of $CH(P)$ (this contact will define a slab-constraint) and the other edge will contact P at a vertex of the convex hull (imposing an edge-vertex constraint as described in Chapter 3). If we intersect these two constraints (i.e. a slab and a line), we obtain a segment, s. In fact this is a special case of Lemma 6, where a_3 and a_4 coincide, Thus reducing the parallelogram to a segment. We will consider two different cases. One in which the two edge-fixels make a right angle and one in which the angle can be chosen arbitrarily.

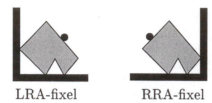

Figure 14: *A polygon in form closure with a LRA-fixel and a point contact and with a RRA-fixel and a point-contact.*

5.3.1 Right Angle Fixel and a Point

In this section we investigate fixturing an object with a right-angle-fixel and a point-fixel. A right-angle fixel offers two edge-contacts, perpendicular to each other. See Figure 14 for an example. We assume one of the edges to be horizontal. We will consider two possible orientations of the fixel that will give different form closure configurations. One way of placing the angle-fixel is such that the second (vertical) edge will be to the left of the object (LRA-fixel) and the other placement where the second edge is to the right of P (RRA-fixel). See Figure 14 for the two placements of the RA-fixel. An object is called fixturable in the model if it can be fixtured with at least one of the orientations of the RA-fixel. A rectangle can not be fixtured, since no triple of contact normals can positively span $\mathbf{R}^2$. Currently, we have no other examples of polygons that cannot be fixtured in this model.

We will now show that any convex polygon without parallel edges can be fixtured. For non-convex polygons this is still an open problem. The pair of boundary features (edges, vertex) of a convex polygon P intersected by the two closest parallel lines of support of P satisfies a property that turns out to be useful the proof of Theorem 8. A supporting line of P either intersects P in a vertex or along an edge. A pair of boundary features intersected by two parallel supporting lines is referred to as an antipodal pair. Note that in the case that the polygon P has no parallel edges, an antipodal pair consists of either two vertices or one edge and one vertex. The distance between the two closest parallel lines of support is referred to as the *width* of the polygon. The antipodal pair that determines the width

of P, i.e., the pair of features that are intersected by the two closest parallel supporting lines, consists of an edge e_w and a vertex v_w [15]. The pair (e_w, v_w) can be computed in time linear in the number of polygon vertices, Lemma 7 gives an interesting property of the antipodal pair determining the width.

Lemma 7 *Let P be a convex polygon without parallel edges and let v_w be the vertex and e_w be the edge intersected by the two closest parallel supporting lines of P. Then the line through v_w and perpendicular to e_w intersects the interior e_w.*

The lemma is true if and only if v_w lies in the open slab $sl(e_w)$ of all lines that perpendicularly intersect the edge e_w. Assume that the edge e_w and the closest parallel supporting lines are horizontal. We can show that if v_w is not in the slab $sl(e_w)$ (for example to the right of the slab), by a slight clockwise rotation of P about the right-endpoint of the edge e_w, the distance between the the horizontal parallel lines of support decreases. This contradicts the fact that the initial horizontal lines were the closest lines of support. Hence the vertex v_w can not be to the right of the slab. A similar arguments holds if we try placing v_w to the left of the slab. Informally, Lemma 7 tells us that we can put any convex polygon without parallel edges into an orientation where one of its edges is horizontal and the highest (top) vertex lies above that horizontal edge. This will be the orientation in which we will fixture the object.

Theorem 8 *Any convex polygon without parallel edges can be immobilized by a right angle fixel and a single point contact.*

Proof: Let e_w be the edge and v_w be the vertex intersected by the two closest parallel supporting lines of the polygon. Let ℓ_w be the line through v_w and perpendicular to the edge e_w, which, by Lemma 7, intersects the interior of e_w. We assume, without loss of generality, that the edge e_w and the two closest parallel supporting lines are horizontal. The line ℓ_w is vertical.

Consider the two lines of support that are perpendicular to the two closest parallel supporting lines (see Figure 15).

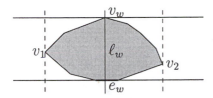

Figure 15: *The line ℓ_w through v_w perpendicularly intersects e_w. The vertical lines of support intersect the polygon in the vertices v_1 and v_2.*

Each of these lines intersects the polygon boundary in a vertex; in the accidental case that the line intersects the polygon along an edge, simply take its top endpoint to be the vertex of intersection. Note that the two resulting vertices v_1 and v_2 are the candidate contact points with the second fixel bar if one fixel bar is placed along e_w. Choose v_v to be the highest (in vertical direction) of the two vertices v_1 and v_2. We place the angle fixel in simultaneous contact with the full edge e_w and the vertex v_v. Let ℓ_v be the horizontal line through v_v, and let q be its intersection point with ℓ_w. The point q lies strictly below v_w. Note that the contact normal at v_v is directed along ℓ_v and pointing towards the interior of the polygon. The contact normals at e_w point upward.

It remains to show that there exists a point p on the polygon boundary such that the right angle fixel placed against e_w and v_v and a point contact at p immobilize the polygon. More specifically, we show that there exists a point p such that (i) the contact normal at p positively spans the plane with the contact normals at v_v and e_w and (ii) the supporting line $l(p)$ of the contact normal at p intersects the supporting line ℓ_v of the contact normal at v_v in q, which is contained in the slab induced by the edge contact at e_w. By Lemma 6, this will imply form closure.

Now assume that v_v is the vertex intersected by the left vertical supporting line. The contact normal at v_v points to the right. Figure 16 shows the situation. We claim that there is a point p satisfying (i) and (ii) on the convex polygonal chain C to the right of ℓ_w and above ℓ_v. The chain is strictly decreasing and contains no vertical edges by the observation that the vertex

of intersection with the right vertical supporting line of the polygon lies below ℓ_v. As a result, the contact normal at any point $p \in C$ satisfies condition (i). Let r be the endpoint of C on ℓ_v.

Let us consider the motion along ℓ_w of the intersection point s of $l(p)$ and ℓ_w when p moves from v_w to r (see Figure 16). By placing p sufficiently close to v_w on C, we can put the intersection point s arbitrarily close to v_w, and, hence, above q. By placing p at r or sufficiently close to r on C if r is a vertex - we can put s below q by the fact that r is on a non-vertical edge. If p moves towards r along some edge of C then the intersection point s moves downward along ℓ_w (see Figure 16 where s moves from s' to s'' as p moves from p' to p''); if p passes a vertex then s jumps back up. As a consequence, the point s will pass through q as p moves along one of the edges of C on its way from v_w to r. Hence, there exists a point p that satisfies conditions (i) and (ii). As a result, a point contact at p establishes form closure for the polygon placed against the angle fixel. Similar arguments apply when v_v is the vertex intersected by the right vertical supporting line. $\square$

This fixturability-proof can be generalized to angle-fixels with an arbitrary fixed angle, we will not give the proof here, but only state the corresponding theorem.

Theorem 9 *Any convex polygon without parallel edges can be immobilized by a fixed angle fixel and a single point contact.*

Since computation of the width of P takes $O(n)$ [15] and testing the edges on the chain C takes $O(1)$ per edge, a form closure configuration can be computed in linear time.

5.3.2 Adjustable Angle-fixel and a Point-fixel

In this section we consider fixtures with an adjustable angle fixel as shown in Figure 17 and a point contact. The angle-fixel is horizontally fixed, but the second edge can be adjusted by rotation about the joint that connects the two edges of the fixel. The second edge can be either to the left or to the right of the

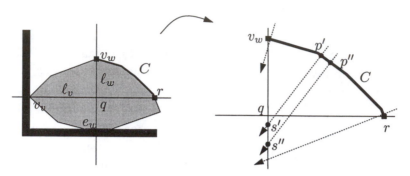

Figure 16: *The angle fixel is placed in simultaneous contact with e_w and v_v. A point contact p can be placed on the convex polygonal chain C such that p and the angle fixel hold the polygon in form closure. The closer view of C on the right shows that if p moves from p' to p'' along an edge of C, the intersection s of $l(p)$ and ℓ_w moves from s' to s''. Given the initial and final positions of s on ℓ_w, the point s must eventually pass through q.*

horizontally fixed edge. The most stable fixtures can be obtained if two edges of $CH(P)$ are in contact with the edge-fixels. We can prove that there is such a fixture for all polygons without parallel edges. In the case where we do have parallel edges (e.g. a rectangle) we may not be able to fixture P in this manner, but one of the contacts with the angle-fixel should be a point-edge-contact. We will prove fixturability of all polygons using these types of contacts. We use the following lemma.

Lemma 10 *Given a polygon P and a point c in the interior of P. If we grow a circle with center c, the first point that will be hit on the boundary of P is a point on an edge of P, tangent to the circle or a concave vertex of P. Similarly, if we grow a circle with center c in the exterior of P, we will first hit a convex vertex of P or an edge of P.*

We first deal with the case in which no two edges of the convex hull of P are parallel.

Theorem 11 *Let P be an arbitrary polygon. If no edges of $CH(P)$ are parallel then P can be held in form closure with an adjustable angle fixel and a point-contact.*

Proof: Let $MIC(CH(P))$ be the maximum inscribed circle of $CH(P)$. The center of $MIC(CH(P))$ is called c. Since P has no parallel edges, $MIC(CH(P))$ will

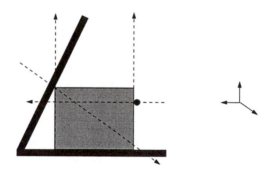

Figure 17: *A rectangle fixtured by an angle-fixel and a point-fixel.*

touch $CH(P)$ in at least three points: a_1, a_2, a_3. Let e_i be the convex hull edge containing a_i and let d_i be the direction of the contact normal at a_i $(1 \leq i \leq 3)$ The vectors (d_1, d_2, d_3) positively span $\mathbf{R}^2$.

If (at least) one of the edges e_1, e_2 or e_3 is an edge of P itself, say e_1, we can place the point contact at a_1 and the edge contacts at e_2 and e_3, it follows from Lemma 6 that we have a form closure configuration now. If none of the edges e_1, e_2 or e_3 is an edge of P, we have to find another contact point. We distinguish two cases.

Case 1. Point c is in the interior of P.
We grow a circle centered at c inside P, until we touch P either at a concave vertex or on an edge of P (Lemma 10). Let the contact point be a_4. The direction of the corresponding contact normal, d_4 will span $\mathbf{R}^2$ together

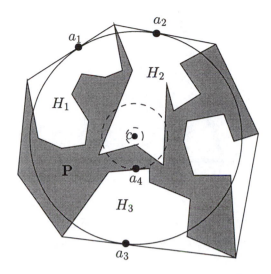

Figure 18: *Computation of the contact point if a_1, a_2 and a_3 are not on the boundary of P.*

with two vectors of V according to Lemma 2. Suppose these vectors are d_1 and d_2. Now we can position P such that e_1 and e_2 are the edge contacts and the point contact is placed at a_4. Since the contact normals intersect in c, this is a form closure configuration.

Case 2. Point c is in the exterior of P.

Since none of the distinct edges e_1, e_2 or e_3 belongs to P, each of the points a_1, a_2 and a_3 lie in cavities H_1, H_2 and H_3 of P (A cavity is a non-empty polygon bounded by the polygon edges connecting two adjacent convex hull vertices and the convex hull edge connecting these two vertices). Assume that H_3 is one of the cavities that does not contain c (as c can lie in only one cavity). Thus, by growing a circle from c, we will hit a convex vertex of H_3 (which is a concave vertex of P) or an edge of H_3 that is also an edge of P. The place where we hit H_3, a_4, is the position of the contact point. From d_1, d_2 or d_3 we can now select the two vectors that will positively span $\mathbf{R}^2$ together with the contact normal d_4 at a_4. The corresponding edges of $CH(P)$ will provide form closure by making them the edge contacts. □

Figure 17 suggests that polygons with parallel edges are fixturable in a model with an adjustable angle-fixel and a point-contact. We will prove this for convex polygons. In this case the fixel-edges cannot both be placed against convex-hull edges of P, because this can not always give us a form closure configuration as the example with the rectangle shows.

Theorem 12 *For every convex polygon P a form closure configuration exists in the model with one adjustable angle-fixel and a point-contact.*

Proof: We only give a sketch of the proof.
From the proof of Theorem 11 we can deduce that the problem with parallel edges arises when the maximal inscribed circle touches two parallel edges. Let us call these edges e_1 and e_2. The contact points are a_1 and a_2 resp. Observe that not all four angles of e_1 and e_2 with their adjacent edges can be acute angles (viewed from the interior of P). Otherwise P would not have been a convex polygon. Choose one of the edges (e_1, e_2) that has at least one obtuse angle with an adjacent edge. Without loss of generality suppose this edge is e_2. We can now show that there is always a form closure configuration such that e_1 is placed against the horizontal edge.

Let l be the line through a_1 and a_2, thus l also contains the center of $MIC(P)$, c. The line l is a vertical line, since we have placed e_1 horizontally, and passes through the center c of $MIC(P)$. Our goal is now to find two contact normals that are correctly oriented (they have to positively span $\mathbf{R}^2$ together with the upward pointing contact normal at e_1) and have their intersection point on l. Since l is contained in the slab defined by e_1, we then have a form closure configuration. More precisely, we want to find one contact normal with direction in $I_1 = (-\frac{1}{2}\pi, 0)$ and one with orientation in $I_2 = [\pi, 1\frac{1}{2}\pi)$, that intersect on l. Since P is convex, the contact normals associated with interval I_1 will originate from contacts to the left of l and those associated with I_2 from the right of l. We now proceed by distinguishing two cases:

Case 1. One of the edges adjacent to e_2 forms an acute angle with e_2, the other one an obtuse angle. An angle of $\frac{1}{2}\pi$ will be considered an obtuse angle. Without loss of generality we assume the acute angle to be to the left of e_2 and the obtuse angle to the right. For an example, see Figure 19.

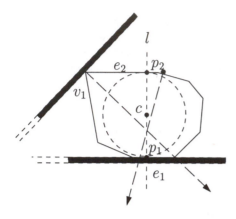

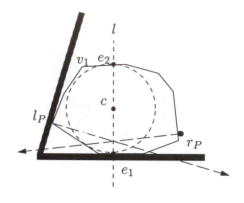

Figure 19: *A polygon with one obtuse and one acute angle adjacent to edge e_2 as in Case 1 of the proof of Lemma 12.*

Figure 20: *A polygon with two obtuse angles adjacent to edge e_2 as in Case 2 of the proof of Lemma 12.*

In this case the edge-fixel will be placed at the vertex of the acute angle adjacent to e_2. Let this vertex be v_1. By adjusting the angle of the edge-fixel from 0 to $\frac{1}{2}\pi$, we will rotate the edge of the fixel around v_1. Since the angle is acute and P convex, we can always do this without collisions with edges of P. The orientations of the contact normals will during the rotation increase from $-\frac{1}{2}\pi$ to 0. At the same time the intersection point of the contact normal at v_1 with l will travel (back and forth) along l from $y = -\infty$ to $y = y_{a_2}$, where y_{a_2} is the y-coordinate of a_2. In other words, for each point on l below a_2, there exists an orientation of the fixel edge such that the contact normal at v_1 passes through that point and has an orientation in the range I_1.

The point contact a_c will be placed somewhere at the interior of the edge adjacent to e_2 on the right side. Since the angle formed by this edge with e_2 is obtuse, we know that the contact normal at e_2 will be in the interval I_2. Because of this orientation, we know that the intersection point of the contact normal to a_c with l will have a y-coordinate y_c smaller than y_{a_2}.

Having fixed the point-contact, we now orient the angle-fixel such that the contact normal to v_1 intersects l in the same point as the contact normal to a_c, which is always possible. The orientation of the angle fixel will never be 0 or $\frac{1}{2}\pi$, because I_p does not contain the values $-\infty$ or y_{a_2}.

Case 2. Both edges adjacent to e_2 form an obtuse angle with e_2. See for example Figure 20.

Consider the y-coordinates of the highest left-most vertex l_P of P, y_l and of the lowest right-most point r_P of P, y_r. Without loss of generality assume that $y_l > y_r$. We will now prove that we can find a form closure configuration by placing the edge-fixel to the left of l and the point-fixel just above r_P.

Consider the edges with orientation in I_1 and their endpoints. This is a chain of edges starting from v_1 and ending with vertex l_P. The chain will at least contain one vertex (v_1). We now rotate the edge-fixel from angle 0 to $\frac{1}{2}\pi$. First the fixel will contact v_1 and an angle of $\frac{1}{2}\pi$ will result in a contact with l_P. During the rotation the edge will contact all vertices in the chain. If we look at the contact normals at the vertex of contact, during the rotation, we see that the intersection point with l will move continuously from y-coordinate $-\infty$ to y_{l_P}. During the process, this point might have a y-coordinate above y_{l_P}, but we do not need these points. We now have to find a point contact to the right of l with orientation in I_2 and intersection point with l somewhere below y_{l_P}. This is true for a point contact just above r_P. We can also find a form closure configuration in the case that $y_{l_P} = y_{r_P}$, but will not discuss this special case here. □

A fixturability-proof can also be given for arbitrary non-convex polygons. In this case we can always construct a form-closure-configuration if we make the adjustable angle extremely small (such that both edges are nearly parallel). The construction though is not very stable and therefore rather impractical.

The time to compute a form-closure configuration depends on the time needed to compute the maximal inscribed circle of P. This is $O(n)$ [7].

6   Modular Fixtures

In this section we consider fixturing polygonal 2-dimensional objects with modular fixturing systems. We are given a rectangular flat surface into which circular holes have been drilled to form a regular square lattice. We assume that the size of the fixture table is large enough to place the fixture elements such that all possible form closure configurations can be accomplished. Grid points are hole centers while grid lines are the vertical and horizontal lines through the grid points. Unit distance is defined as the distance between two adjacent grid points on a single grid line. Let P denote the given polygonal object with perimeter p and diameter d, and let P be defined by n edges. In this section we will consider algorithms that compute all possible form closure configurations on a grid in a specific model. The fixels are assumed to be of zero size, thus they can be placed anywhere without intersections with other fixels or the object. (This restriction is not realistic but can often be enforced by taking the Minkovsky sum of the polygon and the fixel, thus reducing the fixel to a point. Also solutions can be checked afterwards to see whether they are actually realisable.) We will use the following fixture elements:

Locators. A locator must be placed at a grid point. It provides a point-contact with P. The locator has zero radius and the point of contact is a grid point.

Clamps. The fixed part of the clamp must be placed at a grid point. The extension of the clamp is assumed to be at most one grid unit along a grid line. In our model, where we assume that fixels have no size, this means that a clamp can be placed anywhere along a grid line. A clamp provides a point-contact with P on a grid line.

Edge-fixels. An edge-fixel is assumed to be attached to the lattice on one of the grid-lines. We normally put the fixel along the x-axis of our reference frame with P in the positive y-half-plane. An edge-fixel provides an edge-edge contact with P.

Angle-fixels. An angle-fixel basically consists of two edge-fixels. One of the edge-fixels is placed horizontally along a grid-line, the other edge-fixel is attached to this fixel perpendicularly (Right-angle-fixel), such that it extends in the y-direction, or is attached by a joint, such that the angle with the horizontal edge-fixel is adjustable. The types of contacts with angle-fixels are discussed in the related subsections.

6.1   Four Point Contacts

We first discuss a number of different ways to fixture the polygon with four point-contacts.

6.1.1   3L/1C

Since achieving contact with four locators restricted to a grid is in general impossible, we need at least one fixel that takes care of the slack. So, the first model that we consider here is a model where we have three locators and one clamp (3L/1C). The existence of modular fixtures in this model was investigated by Zhuang et al. [30]. They proved a negative result (they identified a set of non-fixturable polygons). A complete algorithm to compute all possible form closure configurations was described by Brost and Goldberg [5]. We review both results briefly.

Non-fixturability

Intuitively one can see that a very small or long and skinny part might be unfixturable in the model, since three of the point-contacts (associated with the locators) must occur at a grid point. However, Zhuang et al. [30] proved the following theorem on convex polygons of arbitrary large diameter.

Theorem 13 *[30] For any given diameter there exists a convex polygon of greater diameter that can not be fixtured on a grid with three locators and one clamp.*

So, we can construct a convex part of arbitrary size that is unfixturable. The proof of the theorem consist of two main steps. First it is shown that for any given positive number M, a disk of radius $> M$ can be constructed that touches at most two grid points. The construction of this disk uses the finiteness of the set of all circles uniquely determined by all triplets of grid-points. It can then be shown that this disk can be transformed into a regular polygon while preserving this property. Since the part should contact three locators (placed at grid points) in order to obtain a form closure configuration, this regular polygon cannot be fixtured under the $3L/1C$ model.

Algorithm to compute all form closures

Brost and Goldberg [5] presented an algorithm that computes all possible form closure configurations in the $3L/1C$-model for polygonal objects. The algorithm is complete, since it outputs all possible form closure configurations. Furthermore, the algorithm can identify the optimal fixture according to an arbitrary quality metric. The main steps of the algorithm are the following.

1. For every combination of three (not all three identical) edges, the area swept out by the second edge by translating and rotating the first edge while maintaining contact with the origin of the grid is computed. This area is an annulus, of which only the part in the first quadrant has to be considered (to eliminate equivalent fixtures). Then for each grid point in this area, the possible positions for the third locator are in the intersection of the annulus defined by possible locations for the third edge with respect to the first and the annulus defined by the third edge with respect to the second locator position. This area can be refined by taking into account the angular limits of the part configurations that simultaneously contact the first and second locator.

2. Having found the triplets of locators and the corresponding contact edges, the set of consistent part configurations can be found by a so called configuration-space analysis. There may be two different poses of the part that permit simultaneous contact with the three

chosen locators. In this case two candidate locator setups are generated.

3. For each of the generated locator-set-ups in the previous step, the region of possible clamp-positions is identified, using force-sphere analysis. The analysis identifies all possible positions where form-closure is obtained. The force-sphere represents all planar forces by direction (x and y) and moment components of a line of force exerted in the plane. A locator-clamp set-up should be able to resist all forces applied in the plane. The set of correct contact normals obtained with the force-sphere analysis, is then mapped back onto the part perimeter and intersected with the horizontal and vertical grid-lines to obtain the clamp-positions that give a form closure configuration.

Brost and Goldberg prove the following theorem.

Theorem 14 *[5] All configurations holding P in form closure with three locators and one clamp can be computed in time $O(n^4 d^5)$. The maximum number of possible fixtures is $O(n^4 d^5)$.*

6.1.2   4C or T-slots

In order to be able to fixture more objects in a modular fixturing system with only four point-contacts than in the $3L/1C$-model, we consider here a model with four clamps that turns out to be equivalent to the T-slot model considered in [30]. In this model we have four continuous degrees of freedom, wheras for a locator the number of possible positions is finite and discrete. We assume that the clamp can have an extension of at most one grid unit. Thus, the extension of a clamp can increase continuously from 0 to 1 grid-unit. So, in the 4C-model, contacts that are made with the polygon can lie anywhere on a grid line. In this subsection we describe two classes of objects that were shown to be fixturable by Zhuang and Goldberg [30]. There is no algorithm yet that efficiently computes all form closure configurations in this model. It seems quite hard to do this, since we do not only have to deal with a discrete number of possible clamp positions, but also with four continuous degrees of freedom, namely the four extensions of the clamp plungers.

Fixturable polygons

Improving on a result of Mishra [19], Zhuang and Goldberg proved the following theorem. Remember that the distance between two adjacent grid points on a grid line is 1.

Theorem 15 *[30] Let P be a rectilinear polygon with all edges of length > 1, then there always exists a fixture using at most 4 clamps.*

The proof shows that a fixture always exists when P is aligned with the axes of the lattice. Because of the restriction on the minimal length of the edges, we can always embed a unit-length square into P. By extending the four sides of this unit-square, we intersect the boundary of P in eight points that are not vertices. It can then be shown that two sets of four of these points yield form closure for the part. The unit-square can be placed such that its vertices all lie on grid-points and thus the intended contact points are actually on grid-lines, which means that the four clamps can indeed be placed such that form closure is obtained.

Nguyen [21] showed how to find sets of 4 edge segments on an arbitrary polygon such that if we place a point contact somewhere on every of these four segments, the part is in form closure. Such segments are now called Nguyen segments. Zhuang and Goldberg also showed that a convex polygonal part is fixturable if there exists a set of at least three Nguyen segments of length $> \sqrt{2}$ [30]. Here we show another result, where existence of a fixture depends on the lengths of the edges of P itself rather than the lengths of the Nguyen segments.

Theorem 16 *Any polygon with all edges of length greater than $\sqrt{3}$ and without parallel edges is fixturable on a unit grid in the 4C model.*

Proof: Let P be a polygon with all edges of length > $\sqrt{3}$ and without parallel edges. The maximal inscribed circle $MIC(P)$ of P, which is assumed to be centered at c, touches P in three points. Assign the names a_1, a_2, and a_3 to these points in counterclockwise order such that $\angle a_1ca_2 \leq \angle a_2ca_3, \angle a_3ca_1$. The vectors $\vec{a_1c}$, $\vec{a_2c}$, and $\vec{a_3c}$ positively span the plane. Moreover, $\angle a_1ca_2 \leq 2\pi/3$.

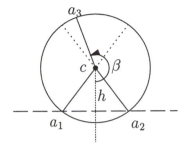

Figure 21: *The points a_1 and a_2 are placed on horizontal grid line. The half-line h bisects the angle between ca_1 and ca_2; β is the angle between h and ca_3.*

Each of the three points a_1, a_2, and a_3 is either a point in the interior of an edge of P or a concave vertex of P. In the case that a point a_i $(1 \leq i \leq 3)$ lies in the interior of an edge, a point contact at a_i induces a half-plane constraint along the supporting line of $\vec{a_ic}$. In the case that a_i is a concave vertex, a point contact at a_i induces two half-plane constraints whose conjunction implies the half-plane constraint along the supporting line of $\vec{a_ic}$. Moreover, the normals to the edges incident to a_i positively span $\vec{a_ic}$. As a result, the composite constraint induced when a_i is a concave vertex is stronger than the constraint induced when a_i is a point in the interior of an edge.

As a first step, we put the two points a_1 and a_2 on a horizontal grid line and hold them with horizontal clamps. We assume that the constraints imposed at a_1 and a_2 are the weakest constraints, namely half-plane constraints along the supporting line of $\vec{a_1c}$ and $\vec{a_2c}$. Assume that a_3 lies above the horizontal grid line and let h be the (vertical) half-line emanating from c and bisecting $\angle a_1ca_2$. Let β be the (counterclockwise) angle between h and ca_3. Figure 21 shows the half-line h and the angle β.

The fact that $\vec{a_1c}$, $\vec{a_2c}$, and $\vec{a_3c}$ positively span the plane and the bound $\angle a_1ca_2 \leq 2\pi/3$ jointly imply that $2\pi/3 \leq \beta \leq 4\pi/3$. As a consequence, an edge tangent to $MIC(P)$ at a_3 cannot be too steep. In fact, the projection on a horizontal grid line of any such edge of length greater than $\sqrt{3}$ will have more than unit length.

Simply assuming that the constraint imposed by the

contact at a_3 is a half-plane constraint would not give us form closure, as it will not exclude c as a center of both clockwise and counterclockwise rotation. A more thorough analysis of the contact at a_3 is necessary to exclude c. We disinguish three cases for a_3 (see Figure 22).

a_3 is a concave vertex and neither of the edges incident to it is tangent to $MIC(P)$. Instead of a single half-plane constraint at a_3, the point contact at a_3 induces a wedge of half-plane constraints (see Section 3), leaving a half-wedge of centers of clockwise rotation and a half-wedge of centers of counterclockwise rotation. By the assumption that neither of the edges incident to the vertex a_3 is tangent to $MIC(P)$, the supporting line of $\vec{a_3c}$ lies in the interior of the wedge of constraints. As a result, c is strictly in the interior of the wedge, excluding it as a center of both clockwise and counterclockwise rotation. Hence, P is in form closure. By sliding a_1 and a_2 along the horizontal grid line we can put a_3 on a vertical grid line, thereby establishing that all contact points lie on grid lines. Note that in this case we need only three clamps.

a_3 is a point in the interior of an edge. Since a_3 lies in the interior of an edge, we choose two points p and p' on either side of a_3 on the same edge. Point contacts at p and p' induce a slab constraint which has the supporting line of $\vec{a_3c}$ strictly in its interior. The slab constraint excludes c as a center of clockwise and counterclockwise rotation. Hence, the point contacts at a_1, a_2, p, and p' hold P in form closure. As each edge has length at least $\sqrt{3}$, we can choose p and p' such that the projection of the segment pp' onto a horizontal grid line has unit length. By sliding a_1 and a_2 along the horizontal grid line we can put p and p' on neighboring vertical grid lines, thereby establishing that all contact points lie on grid lines.

a_3 is a concave vertex and (exactly) one of the edges incident to it is tangent to $MIC(P)$. Let e be the edge tangent to $MIC(P)$ at a_3. We choose a point p in the interior of the edge p. Point contacts at p and a_3 induce a slab constraint and a wedge constraint which have the supporting line of $\vec{a_3c}$ as a common boundary. Careful analysis shows that c lies strictly in the interior of the union of the slab and the wedge: c

lies to the right of the interior of the slab and to the left if the interior of the wedge, or vice versa. As a consequence, no rotation about c is allowed. Hence, the point contacts at a_1, a_2, a_3, and p hold P in form closure. As each edge has length at least $\sqrt{3}$, we can choose p such that the projection of the segment a_3p onto a horizontal grid line has unit length. By sliding a_1 and a_2 along the horizontal grid line we can put a_3 and p on neighboring vertical grid lines, thereby establishing that all contact points lie on grid lines. $\square$

6.1.3 Vise

Wallack and Canny [28] described an algorithm to compute all form closure configurations using a fixture vise. A vise consists of two modular fixturing tables, of which one can translate horizontally relative to the other table. Locators can be placed at the grid points of both tables. See for an example of a vise Figure 4. Here we have in addition to the discrete number of placements for the locators, only one continuous degree of freedom, namely the translation of one vise table in x-direction with respect to the second table. The two tables are also referred to as the left and right jaw. The fixture elements that are used are simply locators. If one of the locators is placed on one table and the other three on the other table, this model is similar to the $3L/1C$-model (although there are some differences). A different case is encountered if two of the locators are placed on each fixture table. There are no known positive or negative results on existence of fixtures in this model.

Algorithm to compute all form closures
The algorithm of Wallack and Canny basically consists of two parts. In the first phase of the algorithm, all possible configurations of locators are generated, such that the locators are in contact with P. In the second phase of the algorithm, the corresponding orientation of the part is computed and then all configurations generated in the first phase are tested for form closure, using matrix computations (checking if the four contact points can resist arbitrary forces and torques). More specifically the first phase of the algorithm consist of three steps:

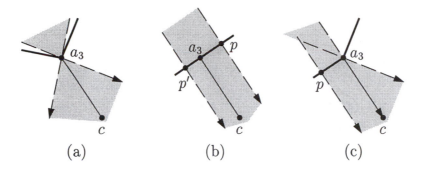

Figure 22: *Three cases: (a) c lies in the interior of the wedge induced by the point contact at a_3; (b) c lies in the interior of the slab induced by the point contacts at p and p'; (c) c lies in the interior of the union of the slab induced by the point contacts at a_3 and p and the wedge induced by the point contact at a_3.*

1. Enumerate all quadruple of edge-combinations of P, where every edge can occur more than once.

2. For each quadruple enumerate all different combinations of intended jaw-contacts, i.e. for every edge specify if it will contact a locator on the right jaw or on the left jaw. Seven situations should be considered for every quadruple of edges (four where three edge segments contact the left jaw and three where pairs of edge-segments contact both jaws).

3. For each quadruple obtained in the previous step, compute fixture configurations providing simultaneous contact. This is done by choosing the first locator at some origin and then examining all possible positions for the second, third and fourth locator consecutively.

It should be clear that the last step is rather complex, regions that are swept out while maintaining contact with the already fixed locators should be described. In the paper, a description of these regions is computed exactly, using two different types of configurations, one where there are two locators on each table and the other one where there are three locators on one table and one on the other. Analysis of their algorithm gives the following theorem.

Theorem 17 *[28] All configurations holding P in form closure on a vise with four locators can be computed in time $O(n^4 d^4 r^2)$, where r is the largest range of distance between points on two edges. The maximum number of form closure configurations is also $O(n^4 d^4 r^2)$.*

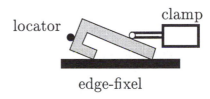

Figure 23: *A form closure configuration with an edge-fixel, a clamp and a locator.*

It should be noted that the algorithm uses this much time even when there are only a few resulting form closure configurations, since we only check for form closure after having generated all the possible contact positions of the locators.

6.2 One Edge-Fixel, a Locator and a Clamp

In this section we consider an algorithm to compute all fixtures in a model with one horizontal edge-fixel, a locator and a clamp. The edge-fixel is assumed to be attached to the grid parallel to one of the horizontal grid-lines. We can distinguish two possibilities for placing the clamp. One, where the clamp is placed such that its extension is along a horizontal grid line (horizontal clamp) and the other case where the extension is along a vertical grid line (vertical clamp). Here, we will only discuss the algorithm for a horizontal clamp in detail. In addition we will briefly describe how to apply similar ideas to vertical clamps. We will also give a theorem about a class of rectilinear objects that can

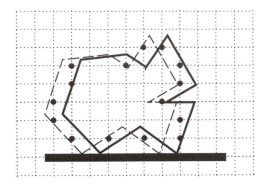

Figure 24: *Sliding the polygon one unit along the edge-fixel to enumerate all feasible locator positions.*

be fixtured on a modular grid.

Algorithm for computing all form closures

Since we know that at least four point contacts are necessary to fixture a polygon, we need to place P always such that at least two point contacts are made between P and the edge-fixel. The only assumption that we will make about the edge-fixel is that it is long enough, i.e. at least as long as the longest edge on the convex hull of P. P will always be placed with an edge of the convex hull against the edge-fixel. Hence, we only have $O(n)$ different orientations for P, which makes the algorithm more efficient than algorithms for fixturing with four point-contacts.

To get a better understanding of the problem, we first give a naive (non-output sensitive) algorithm for computing all form closure configurations.

Consider the polygon P shown bold in Figure 24 resting with some particular edge of its convex hull against the edge-fixel, the thick bar shown at the bottom. As P may not be touching a grid-point on its boundary, we slide it, say to the left, until it does. Now we have a possible position for the locator. To compute all possible locator positions, it is sufficient to slide the polygon only one unit to the left and mark all grid-points encountered during the slide. Let $P = P(0)$ indicate the initial position of the polygon and $P(t)$, the polygon when shifted by distance t to the left, $0 \leq t \leq 1$. We call t the *shift variable*. A locator position L exists only

at certain discrete values of the shift variable; at each one it contacts a certain edge e of P. Let us term each such contact as a *locator-edge combination*.

Clamp positions are those points on the boundary of P intersected by the grid lines at some $P(t)$. We distinguish two cases: those points intersected by the horizontal grid lines are called *horizontal clamp positions* while the others are *vertical clamp positions*. Notice that since the shift is horizontal, horizontal clamp positions remain constant for all shift values and can be computed from the unshifted polygon. On the other hand, vertical clamp positions vary with the shift value. A vertical grid line might intersect several edges e of P during the shift resulting in *(vertical) clamp-edge* combinations.

There are at most $O(p)$ grid points (locator positions) generated by this process (a more precise bound is $O(\min(d^2, p))$, where d is the diameter of P). Furthermore, the total number of locator-edge and clamp-edge combinations is $O(n + p)$ and can be computed in $O(n + p)$ time.

A naive procedure to generate all valid fixtures is therefore to consider each locator-edge combination and clamp-edge combination pair, check if they simultaneously exist (w.r.t. the shift variable; this is necessary only for locators with vertical clamps) and if so, test if they achieve form-closure. Iterating over all convex hull edges against the edge-fixel gives a $\Theta(n(p+n)^2)$ algorithm.

Horizontal clamp

Instead of considering the placements of the locator and clamp (on the grid), we can also look at the possible contact normals associated with these grid points. We can then abstract from the grid and polygon and only have to look at directed lines. We assume that P is placed with one of the edges of its convex hull $CH(P)$ against the edge-fixel. This edge-contact induces a slab as defined in Section 3. As mentioned before, horizontal clamp and locator contact points can be determined by intersecting the horizontal grid lines with the boundary of P. The contact points are always achievable, since we assume that the clamp can have

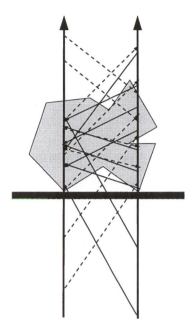

Figure 25: *The directed line segments (directed lines representing half-plane constraints restricted to the slab formed by the edge-fixel at the bottom) resulting from the object in Figure 24. The dotted line segments (from left to right) correspond to the locators while the solid lines (from right to left) are from the horizontal clamps.*

a one grid-unit extension. Let us assume for now that the locator will be placed to the left of P (meaning that by sliding P from right to left, we will obtain a contact with it) and the clamp is placed to the right. We obtain two sets of lines. The clamp lines are the lines normal to the points of contact (as obtained by intersection of P with the horizontal grid lines) directed into the object and having an orientation in the open interval $(\frac{1}{2}\pi, \frac{3}{2}\pi)$. This actually means that the lines cross the slab from right to left. There are $O(n + p)$ of these lines. The locator lines are the possible contact normals that cross the slab from left to right. The set of locator lines will will have $O(n + p)$ elements.

Thus we obtain two sets of directed lines defining half-plane constraints on P; one set of lines associated with possible clamp positions and one set of lines for the possible locator positions. We want to find all combinations of lines, one from each set, such that Condi-

tions 1 and 2 in Lemma 4 are satisfied. Intersecting the directed (infinite) lines with the slab corresponding to the fixed convex hull edge of P, we obtain two sets of directed line *segments*. Given a locator line segment, we wish to detect, in an output-sensitive manner, all clamp line segments that (properly) intersect it and such that the vectors along the two segments and the upward vertical vector positively span $\mathbf{R}^2$. See Figure 25. Formally, the basic query that we want to answer is:

QUERY Let S be a set of m directed line segments, each with the end-points along two vertical lines (the "slab"), and each with an orientation from the open interval $(-\frac{1}{2}\pi, \frac{1}{2}\pi)$. Store S such that for a query segment q also similarly anchored on the slab but with orientation in the interval $(\frac{1}{2}\pi, \frac{3}{2}\pi)$ one can report all segments $s \in S$ such that

1. the vectors along s, q and the upward vertical vector positively span $\mathbf{R}^2$, and

2. s and q intersect strictly inside the slab.

Using a two-level partition tree, such queries can be solved in time $O(\log^2 m + k)$, where k is the number of reported segments (see [22] for details). Since we have to perform this query for all $O(n + p)$ clamp segments and $O(n)$ possible orientations of P, we obtain a total time of $O(n(n + p) \log^2(n + p) + K)$, where K is the number of form closure configurations. Actually in contrast to querying the clamp lines one at a time, we can process them all together and improve the time complexity to $O(n(n + p) \log(n + p) + K)$.

Theorem 18 *[22] All configurations holding P in form closure using one edge-fixel, one locator and one horizontal clamp can be enumerated in time $O(n(n + p) \log^2(n + p) + K)$, where K is the number of form closure configurations.*

Vertical clamp

In the case of a vertical clamp we can apply a query algorithm that is similar to the one above. In this case, however, the place where the clamp will contact

P's boundary depends on the extension of the clamp. As described in the naive algorithm above, we have to deal with an extra 'shift'-variable. The actual clamp-line for a specific clamp position depends on this variable. Locator lines exist only at one time during the shift (when an edge of the polygon shifts over the corresponding grid-point) In order to solve this problem, we can add an extra dimension to the query problem, representing the shift. In this query structure a locator line is a segment inside a block and clamp lines are represented by parallelograms inside a block. Again, we query the structure with every parallelogram. Because of the three-dimensional nature of this problem this leads to the following result.

Theorem 19 *[22] All configurations holding P in form closure using one edge-fixel, one locator and one vertical clamp can be enumerated in time $O(n(n + p)^{\frac{4}{3}+\epsilon} + K)$, where K is the number of form closure configurations.*

Fixturable polygons

We can prove that a subset of rectilinear polygons is fixturable under the edge-locator-clamp-model. The result is comparable to [30]. Specifically, we are interested in simple polygons P that are rectilinear (the edges of P are parallel or perpendicular to each other), the convex hull of P is not a rectangle, and each edge is of length at least δ grid units. We term such a class of polygons as δ-*long rectilinear polygons*. We can prove the following theorem.

Theorem 20 $2 + \sqrt{2}$-*long rectilinear polygons are fixturable with an edge-fixel, a locator and a clamp.*

Such a fixture can be computed in $O(n)$ time. Note that as a consequence of Theorem 20, in the absence of a grid any rectilinear polygon for which the convex hull is not a rectangle can be fixtured with an edge and two points. This extends Theorem 5 in Section 5.

6.3   One Angle-fixel and a Clamp

Finally we discuss modular fixtures using an angle-fixel and one clamp. The three types of angle-fixels that we will consider are a so-called right-angle fixel, a

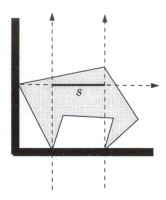

Figure 26: *The segment s that should be intersected by the contact normal to the clamp contact when P is placed against a right-angle fixel.*

fixed-angle fixel, where the angle is not $\frac{1}{2}\pi$ and the adjustable angle fixel. No results are known about what objects can be fixtured in this model. The algorithms described here are quite straightforward and follow directly from the geometric interpretation of the contacts.

Right angle fixel and a clamp

In order to obtain a form closure configuration P will always be placed with one of the edges of $CH(P)$ against one of the edges of the fixel. Assume this to be the horizontal edge. P is then slided along the horizontal edge, until it hits the second, vertical edge of the fixel. Usually this will yield a contact of a vertex of P with the vertical edge. Having done this, the orientation and position of P are fixed. For a fixed orientation of P, we can compute the intersection between the slab defined by the edge-contact with the horizontal edge of the angle-fixel and the contact normal to the point-contact with the vertical edge of the fixel (Lemma 6). The angle-fixel can be placed to the left or to the right of P. The computation of all form closure configurations is similar for both cases. Let us assume here that the right-angle fixel is placed to the left of the polygon as in Figure 26. The intersection of the contact normals give us a horizontal segment s. See Figure 26. The possible contact points for the clamp can be obtained by intersection of the boundary of P with the horizontal and vertical grid lines. This gives us $O(n+p)$ possible contact points and their corresponding contact normals. For every element in this set we now have to check if they have an orientation in the interval $(\pi, 1\frac{1}{2}\pi)$ and if they intersect s. Since we have to do this for every orientation of P, we get the following result.

Theorem 21 *All configurations holding P in form closure with a right-angle fixel and a clamp can be enumerated in time $O(n(n+p))$.*

Note that when P also touches the second, vertical, edge of the fixel with an edge of its convex hull, instead of a segment s, we obtain a rectangle that should be intersected by the contact normal of the clamp contact. This does not increase the time bound.

Fixed angle fixel and a clamp

The algorithm as described above for the case of a right-angle fixel can equally be applied to the case of an arbitrary fixed-angle fixel. The only things that we have to change are the check for the orientation of the contact normal to the clamp, if the angle of the fixel is α, the angle of the contact normal to the second, non-horizontal edge will have direction $(1\frac{1}{2}\pi + \alpha)$. This means that the orientation of the contact normals that will provide form closure must be somewhere in the interval $(\pi + \alpha, 1\frac{1}{2}\pi)$.

Theorem 22 *All configurations holding P in form closure with one fixed-angle fixel and a clamp, can be computed in time $O(n(n+p))$.*

Adjustable angle fixel and a clamp

If the angle of the angle-fixel is adjustable, the number of form closure configurations can be infinite, it might be possible to generate intervals of angles for the fixel together with a finite number of possible clamp positions that will provide form closure. No such algorithm currently exists. Another possibility is to restrict the form closure configurations to those configurations where both edges of the fixel are in contact with an edge of $CH(P)$. In this way though an existing solution might be overlooked, for example for a rectangle

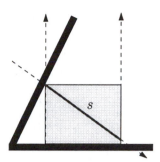

Figure 27: *The segment s that should be intersected by the contact normal to the clamp contact when P is placed against a fixed-angle fixel.*

we will not find a form closure configuration, although one exists.

7 Extensions and Open Problems

The algorithms described in this paper only give results for a basic situation (fixels of zero size, no quality metrics applied, only two-dimensional polygonal objects). Some of the algorithms can be extended or changed such that additional cases can be dealt with as well. Here we give a short overview of a number of known extensions.

Computation of Nguyen regions. Nguyen [21] introduced an algorithm to compute Nguyen regions of an object in a model with four point contacts. Nguyen regions define possible placements of the point contacts, such that form closure is obtained if we place at least one contact in every region. Since the Nguyen regions are independent of a grid, we can usually find more efficient algorithms to compute these regions than to compute all possible fixtures on a grid. Nguyen described an $O(n^4)$ algorithm to compute al these combinations of regions. In the case of a right-angled fixel and a point contact, all regions where we can place the point contact for all orientations of P can be computed in time $O(n \log^3 n + K)$ time, where K is the number of resulting regions. This can be achieved by adding an extra level in the partition tree. In the case of one edge and two point contacts we currently only have efficient algorithms for rectilinear polygons and c-oriented

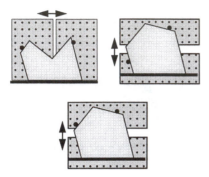

Figure 28: *Three different models for a vise combined with an edge-fixel.*

polygons.

Fixels with size > 0. Up to now we assumed the fixels to have 0 size. This is highly unrealistic. To deal with locators of non-zero size one can compute the Minkovsky sum of the polygon with the locator. This blows up the polygon and reduces the locator to a point. For the clamps we can either test for intersecting afterwards or use similar techniques. Finally, the size of the edge-fixels never causes a problem (it only shifts the polygon with respect to the grid). For example, the algorithm of Brost and Goldberg [5] thus needs time $O(n^5 d^5)$. The time complexity of the vise-algorithm of Wallack and Canny [28] (Theorem 17) does not change. The Minkovsky difference can also be embedded in the algorithm for one edge-fixel, a locator and a clamp [22] (Theorems 18 and 19), The time for a vertical clamp does not change (Theorem 19), for a horizontal clamp we get an $O(n(n+p)\log^3(n+p) + K)$ algorithm.

Vise with wall. Wallack and Canny [28] considered a vise with four locators, but what happens if we replace two of the locators by an edge-fixel? We can distinguish three cases. See Figure 28. In the first case the edge-fixel is placed in horizontal direction (the direction of the translation of one of the two tables of the vise). In addition one locator is placed on each of the tables of the vise. We can generate all form-closure configurations with the output-sensitive algorithm for an edge-fixel, a locator and a horizontal clamp (Theorem 18), since we can view the locator on the moving table as a horizontal clamp. If the edge-fixel is placed

perpendicular to the direction of movement of the vise, we can obtain form closure for two types of placements of the locators. In the first case the locators are both on different tables. This case is similar to a model with one edge-contact, one locator and one vertical clamp and we can use the algorithm for this model to compute all form closure configurations (Theorem 19). In the other case, the two locators are both placed on the top table (the moving table). It is an open problem how to compute all form closure configurations in an output-sensitive way in this case.

Curved edges. Wallack and Canny [29] also presented an algorithm to compute form closure configurations on a vise, where the objects to be fixtured are so called generalized polygons. These are two-dimensional objects with a boundary composed of linear edges and circular arcs. It is unclear how to generalize the other algorithms in this paper to polygons with curved edges.

Three dimensional fixturing. Markenscoff et al. showed that for three-dimensional polyhedra, seven frictionless point contacts are necessary and sufficient to fixture the part [18]. In addition Wagner et al. [27] proposed a three-dimensional modular fixturing system using seven struts that are mounted to the walls of a rectangular four-sided frame on which mounting-sockets are arranged in a regular grid. They also presented an algorithm that enumerates all fixtures for a polyhedron, whose pose is given. All possible seven-membered sets of simultaneously contacting strut-positions are enumerated and tested for form closure, leading to a $O(t^7)$ algorithm, where t is the number of lattice sites in the projection of the faces of the polyhedron on the fixture frame. Clearly, this bound is quite high and can possibly be improved. An open question is yet how to determine all possible poses of the polyhedron. It would be nice if there was a graphical representation of possible motions in 3-D similar to the 2-D representation of Reuleaux [24], since this would probably help in finding output-sensitive algorithms.

Quality metrics. After we have generated a list of all possible form closure configurations for a given part, we would like to rank them according to some quality metric. Several quality measures have been proposed in literature based on the smallest contact force necessary to resist applied forces [9, 17]. One such measure is the radius of the largest sphere that can be embedded inside the wrench convex [12, 16]. Brost and Goldberg [5] describe a quality metric that is used to rank the fixtures obtained by their algorithm for three locators and a clamp. The metric allows the user to specify a list of expected forces on the part. The quality metric scores each fixture by estimating the maximum contact reaction force required to resist the list of expected applied forces. However, any suitable quality metric can be applied, since ranking the fixtures is done as an extra final step of their algorithm. It would be nice if, rather than checking all solutions afterwards, one could incorporate the metric in the algorithm itself. Thus, hopefully, reducing the running time.

As can be seen from this paper much progress has been made in fixture design over the past few years but many questions are still open. For example, only partial results are known about which polygons can be fixtured in which model (either with or without a grid). Also many of the algorithms for modular fixturing seem to be far from optimal. Finally, the extensions to curved objects, three-dimensional fixtures and incorporation of the quality metric are far from being solved.

Acknowledgements

This work was partially supported by the Netherlands Organization for Scientific Research (N.W.O.) and by a NATO collaborative research grant number CRG.951224.

References

[1] H. Asada and A. B. By. Kinematic analysis of workpart fixturing for flexible assembly with automatically reconfigurable fixtures. *IEEE Journal of Robotics and Automation*, RA-1(2):86–94, June 1985.

[2] J. J. Bausch and K. Youcef-Toumi. Kinematic methods for automated fixture reconfiguration planning. In *International Conference on Robotics and Automation*, pages 1396–1401. IEEE, May 1990.

[3] A. Blake and M. Taylor. Planning planar grasps of smooth contours. In *International Conference on Robotics and Automation*, pages 834–839. IEEE, May 1993.

[4] W. Boyes, editor. *Handbook of Jig and Fixture Design, 2nd Edition*. Society of Manufacturing Engineers, 1989.

[5] R. C. Brost and K. Y. Goldberg. A complete algorithm for synthesizing modular fixtures for polygonal parts. In *International Conference on Robotics and Automation*, pages 535–549. IEEE, May 1994.

[6] R. C. Brost and M. T. Mason. Graphical analysis of planar rigidbody dynamics with multiple frictional contacts. In H. Miura and S. Arimoto, editors, *Robotics Research: The Fifth International Symposium*, pages 293–300. MIT Press, 1990.

[7] Francis Chin, Jack Snoeyink, and Cao-An Wang. Finding the medial axis of a simple polygon in linear time. In *Proc. 6th Annu. Internat. Sympos. Algorithms Comput. (ISAAC 95)*, volume 1004 of *Lecture Notes in Computer Science*, pages 382–391. Springer-Verlag, 1995.

[8] Y-C. Chou, V. Chandru, and M. M. Barash. A mathematical approach to automatic configuration of machining fixtures: Analysis and synthesis. *Journal of Engineering for Industry*, 111:299–306, November 1989.

[9] M. R. Cutkosky. *Grasping and fine manipulation for automated manufacturing*. PhD thesis, Carnegie-Mellon University, Dep. Mech. Eng., Jan 1985.

[10] J. Czyzowicz, I. Stojmenovic, and J. Urrutia. Immobilizing a polytope. In *Lecture Notes in Computer Science*, volume 519, pages 214–227. 1991.

[11] B. Faverjon and J. Ponce. On computing two-finger force-closure grasps of curved 2d objects. In *International Conference on Robotics and Automation.* IEEE, May 1991.

[12] C. Ferrari and J. Canny. Planning optimal grasps. In *International Conference on Robotics and Automation*, pages 2290–2295. IEEE, May 1992.

[13] M. V. Gandhi and B. S. Thompson. Automated design of modular fixtures for flexible manufacturing systems. *Journal of Manufacturing Systems*, 5(4):243–252, 1986.

[14] E. G. Hoffman. *Modular Fixturing.* Manufacturing Technology Press, Lake Geneva, Wisconsin, 1987.

[15] M. E. Houle and G. T. Toussaint. Computing the width of a set. *IEEE Trans. Pattern Anal. Mach. Intell.*, PAMI-10(5):761–765, 1988.

[16] D. G. Kirkpatrick, B. Mishra, and C-K Yap. Quantitative steinitz's theorems with applications to multifingered grasping. In *ACM Annual Symposium on Theory of Computing*, pages 341–351, 1990.

[17] Z. Li and S. S. Sastry. Task-oriented optimal grasping by multifingered robot hands. IEEE *Transactions on Robotics and Automation*, RA-4(1):32–43, 1988.

[18] X. Markenscoff, L. Ni, and C.H. Papadimitriou. The geometry of grasping. *International Journal of Robotics Research*, 9(1):61–74, 1990.

[19] B. Mishra. Workholding: Analysis and planning. In *International Conference on Intelligent Robots and Systems*, pages 53–56, July 1991.

[20] B. Mishra, J. T. Schwartz, and M. Sharir. On the existence and synthesis of multifinger positive grips. *Algorithmica*, 2(4):641–558, 1987.

[21] V-D. Nguyen. Constructing force-closure grasps. *International Journal of Robotics Research*, 7(3):3–16, 1988.

[22] M. Overmars, A. Rao, O. Schwarzkopf, and C. Wentink. Immobilizing polygons against a wall. In *ACM Symposium on Computational Geometry*, pages 29–38, June 1995.

[23] J. Ponce, S. Sullivan, J-D. Boissonnat, and J-P. Merlet. On characterizing and computing three- and four-finger force-closure grasps of polyhedral objects. In *IEEE Conference on Robotics and Automation*, pages 821–827, May 1993.

[24] F. Reuleaux. *The Kinematics of Machinery.* Macmilly and Company, 1876. Republished by Dover in 1963.

[25] E. Rimon and J. Burdick. On force and form closure for multiple finger grasps. In *IEEE Conference on Robotics and Automation*, 1996.

[26] J. C. Trinkle. On the stability and instantaneous velocity of grasped frictionless objects. *IEEE Transactions on Robotics and Automation*, 8(5):560–572, 1992.

[27] R. Wagner, Y. Zhuang, and K. Goldberg. Fixturing faceted parts with seven modular struts. In *IEEE International Symposium on Assembly and Task Planning*, pages 133–139, August 1995.

[28] A. Wallack and J. Canny. Planning for modular and hybrid fixtures. In *IEEE Conference on Robotics and Automation*, pages 520–527, May 1994.

[29] A. Wallack and J. Canny. Modular fixture design for generalized polyhedra. In *IEEE Conference on Robotics and Automation*, April 1996.

[30] Y. Zhuang and K. Goldberg. On the existence of solutions in modular fixturing. *International Journal of Robotics Research*, 1996. to appear.

Algorithms for Robot Grasp and Delivery

Prasad Chalasani, *Los Alamos National Laboratory, NM, USA*
Rajeev Motwani, *Stanford University, Stanford, CA, USA*
Anil S. Rao, *Qualcomm, Inc., San Diego, CA, USA*

We consider the following application from industrial automation: parts arrive on a conveyer belt and have to be grasped by a robot arm of finite capacity and then delivered to specified delivery points for packaging or other processing. An intelligent choice of the order of picking up the parts can significantly reduce the total distance traveled by the robot arm and, consequently, its cycle time. We model this application as problems of the form: Given n identical parts initially located on a conveyer belt, and a robot arm of capacity k parts, compute the shortest route for the robot arm to grasp and deliver the n parts, handling at most k at a time. We consider arbitrary finite k as well as unbounded k. We also allow the belt to be either motionless or moving with a known fixed velocity. For the moving belt problem, we require all parts to be delivered to a specified origin point, while for the static case we allow up to n different destinations for the parts. In either case, since the optimization problems that result are NP-hard, the focus is on devising efficient approximation algorithms. While the approaches taken for the static and the moving belt cases are very different, all our algorithms are simple to implement, efficient, and produce a solution within a small constant factor of the optimum.

1 Introduction

A major thrust of industrial automation today is towards rapidly configuring assembly lines to manufacture products described by a CAD model using parametrized modular components such as conveyer belts, low degrees of freedom robot arms, modular fixtures, flexible parts feeders, light beams, and simple 2D vision systems. The general scenario of rapidly configuring robots working in their cells to handle parts in an assembly line has been termed *rapid deployment automation* [3, 9, 28]. A number of challenging algorithmic problems arise in the efficient design, simulation, and implementation of such assembly lines.

We focus on the following specific issue related to the planning and scheduling of the motion of a robot arm in a typical industrial application. After manufacture, identical parts are dumped onto a conveyer belt in arbitrary positions. Prior to packaging or assembly, the parts must be collected by a four degree of freedom (x, y, z translation and z rotation) robot arm of capacity k and delivered to an empty pallet at a specified location. Once filled, the pallet is moved away by another conveyor belt, and a new empty pallet appears at the origin. An automated solution involves using an integrated vision system and a robot arm with k suction cups. The vision system identifies the positions of the n parts. The robot arm translates to k part positions, each time positioning an empty suction cup over a part and picking it up. The robot arm then drops off the collection of k parts onto the pallet before returning to pick up more parts. The goal is to plan the motion of the robot arm so that the total time to grasp and deliver the n parts is minimized.

The cost metric assumed in this paper is the total translation of the robot arm. We are neglecting the time for rotating the end-effector in order to correctly reorient itself over the part prior to grasping. This is done for simplicity, as we describe our algorithms for the planar configuration space; taking planar rotation into account would result in a three-dimensional configuration space. Our algorithms can be extended to higher-dimensional configuration spaces, with an exponential dependence on dimension. We also neglect the translation involved between the suction cups. This can be justified by a fast spinning suction cup. Finally, we ignore the vertical (z) distance traveled by the robot as this is fixed for a given n and k. These assumptions allow us to consider the robot arm and the parts to be points in the plane.

In this paper, we study two distinct scenarios: the *static* case where the conveyor belt is held fixed for the duration of the grasps/pick-ups, and the *dynamic* case where the conveyor belt continues to move at a fixed rate throughout. Since the minimization problems that result in either case are easily seen to be NP-hard, we focus on devising efficient approximation algorithms for these problems. Our algorithms return a solution whose cost is within a constant factor of the optimum. (For an instance of a minimization problem whose optimum solution has cost C^*, a k-approximation algorithm must return a solution of cost $C \leq kC^*$ [33].) While the approximation analysis (proving $C \leq kC^*$) is non-trivial, and is indeed the main contribution of this paper, the algorithms themselves are fairly simple and well-suited for practical implementations. The running-times of all our algorithms are bounded by low-degree polynomials in n and, wherever meaningful, in k.

The classical TSP problem requires finding a path of minimum length for a "salesman" who must start at the origin and visit n cities before returning to the origin. The problems we consider can be viewed as the variants of the classical traveling salesman problem (TSP) where the salesman's can visit at most k cities at a time, before returning to the origin. However, for the static case, we consider a generalization where we allow parts to be delivered to up to n different locations, leading to a problem that cannot be viewed as such a variant of TSP. We call this problem the **k-Delivery TSP**, which may be stated formally as: *Given n source points and n sink points in some metric space, with exactly one part placed at each source (all parts are identical), and a robot arm of capacity k, compute a minimum length route for the robot arm to deliver exactly one part to each sink.* The capacity k may be unbounded and the sources and sinks need not lie at distinct locations. The problem is easily seen to be NP-hard via a reduction from TSP: place a source and a sink very close to each point of the TSP problem, and set $k = 1$; now, an optimal solution to the 1-Delivery TSP is an optimal solution to the TSP instance. Similar reductions can be devised for arbitrary finite k as well as for unbounded k.

We provide what appear to be the first known polynomial runtime constant-factor approximation algorithms for the k-Delivery TSP. This problem can also be seen to be an instance of vehicle routing or schedul-

ing problems that have been the subject of intensive study in the literature [11, 21]. The problem of efficiently rearranging parts in the plane with a centrally placed gripper that can rotate and extend telescopically was studied by Atallah and Kosaraju [3] under different cost metrics.

We next consider the case of the moving belt and partially extend our previous results to this case. Formally, we define the following problem. **Dynamic k-Collect TSP**: *Given n point-objects moving in the Euclidean plane with fixed and identical velocities, and a robot arm with capacity k initially located at the origin, find a minimum length route to pick up and deliver these objects to the origin.* Note that the *static* version of this problem is the special case of k-Delivery TSP where the sinks are all at the origin. This problem has been shown to be NP-hard by Lenstra and Rinnooy Kan [30]. This problem also has some features of the *time-dependent* TSP [16, 31] wherein the distance function varies with time.

Although there does not appear to be any prior work on TSP with moving points, heuristics for related problems have been studied in the robotics literature. For example, Li and Latombe [28] consider a situation similar to ours: a vision system identifies parts on a moving belt; these parts must be grasped and shelved elsewhere. They consider heuristics for this on-line problem that plan the motions of two cooperating robots.

Two subtleties arise in the case of moving points leading to some complications. First, we lose symmetry in the distance matrix: given a robot initially at A, the time it requires to travel to B is not the same as the time required to travel to A starting from B initially. Fortunately, however, since the points are all moving at the same velocity, there is some structure in this asymmetry. The second problem is that the distance of points to the origin is time-dependent, although the distances between moving points are time-independent. The dynamic variants have some other interesting and counterintuitive aspects.

Consider a numerical example. Suppose $k = 1$ and there are two points, initially at $(10, 0)$ and $(15, 0)$. The robot is initially at the origin. Assuming that the robot can move at speed 1 and the points move at speed $1/2$ in the negative x direction, which point should be visited first? It is easy to check that visiting the *further* point first produces a smaller total time (20) than the other way round (roughly 22). In general, the more time we spend visiting points early in the tour, the closer the later points would have moved to the origin, and so we would spend less time visiting them from the origin.

Thus, the dynamic problem has some features of **Maximum Latency TSP:** *given a set of n points and a symmetric distance matrix* (d_{ij}) *satisfying the triangle inequality, find a path starting at* p_1 *and visiting all other points so as to* maximize *the total latency of the points, where the latency of a point is the length of the subpath up to that point in the tour.* We give an approximation algorithm for this problem and also study some variants that arise implicitly in the dynamic settings. While approximation algorithms for *Minimum Latency TSP* are known [6, 20], there does not appear to be any prior work on our version of the problem.

1.1 Organization and Results

In Section 2, we begin by considering 1-Delivery TSP. The 1-Delivery problem is to find an optimal tour that alternately visits the sources and sinks; therefore, we refer to it as the *Bipartite TSP*. This is closely related to the *Swapping Problem* for which Anily and Hassin [2] present a 2.5-approximation algorithm. We use matroid intersection to obtain a 2-approximation algorithm for this problem. In Section 2.1, based on the 1-Delivery approximation algorithm and additional lower bound arguments, we devise a 9.5-approximation algorithm for k-Delivery TSP with arbitrary finite k. Then, in Section 2.2, we present a 2-approximation algorithm for the case of unbounded k.

In Section 3, we give a 1/2-approximation algorithm for Maximum Latency TSP. We also study some variants that arise implicitly in the dynamic settings discussed in Section 4.

In Sections 4.1–4.3 we establish some basic properties of travel-times involving moving points (such as the triangle inequality) that are not as obvious as they might seem. In the rest of Section 4 we show that an optimal Dynamic 1-Collect tour must visit points in decreasing order of distances to the origin and that the case $k = \infty$ is an asymmetric TSP with bounded asymmetry. We then present a constant factor approximation for the Dynamic k-Collect problem for arbitrary finite k.

Finally, we summarize our work in Section 5. We discuss some problems left open in our work, as well as new directions for algorithmic research in the area of rapid deployment automation.

In the rest of the paper we will refer to the sources as blue points and the sinks as red points. We will also assume for convenience that n is a multiple of k; our results can be extended to the general case by introducing a few dummy sources and sinks close to one of the original points.

1.2 Relation to Previous Work

This paper contributes new results to the field of robot motion and task planning [26, 29, 38]. It can also be seen as contributing to the extensive Operations Research literature on capacitated delivery problems [11, 21]. Most prior research on such problems has focused on heuristics for finding the optimal solution (using sophisticated branch and bound techniques, for example) and not on finding provably good approximations. Examples of such work include the following. **(a)** *The Capacitated Vehicle Routing Problem (CVRP)* [1, 21, 24]: given n points in a metric space and an infinite fleet of vehicles of capacity k, find a collection of vehicle routes starting at origin such that every point is visited by exactly one vehicle. Typical objective functions to be minimized included number of vehicles, total distance traveled, maximum distance traveled by a vehicle, or some combination thereof. **(b)** *The Dial-a-Ride Problem* [5, 25, 35]: compute an optimal route for a k-capacity van to pick-up and drop-off n persons between different origin-destination pairs. Note that this problem differs from the k-Delivery TSP in that each person must be dropped off at a specific destination. **(c)** *The Precedence-Constrained TSP* [5]: for each vertex i there is a set $P(i)$ of vertices that must be visited before visiting i, and we are required to find an optimal TSP satisfying these constraints.

The *static* k-Collect TSP has received considerable attention in the literature and the best-known result is the 2.5-approximation algorithm of Altinkemer and Gavish [1]. Also, the k-Person TSP is a related problem and has a 1.5-approximation algorithm due to Frieze [18] (see also Frederickson, Hecht, and Kim [17]). Another problem with a similar flavor is the k-MST problem and a recent series of papers [4, 7, 19, 32, 36] has led to good approximation algorithms for it.

2   Static Conveyor Belts: The k-Delivery TSP

In this section we consider the case of a static belt and generalize possible part destinations to n in number; the robot arm, which can carry a maximum of k parts per cycle, must deliver one part to each sink. We begin by considering a unit-capacity robot, i.e., the case $k = 1$. When the robot arm has unit capacity, any delivery route must alternate between part pick-up and and part delivery. Viewing the 1-Delivery TSP as a graph problem, we can re-define it as follows:

> **Bipartite Traveling Salesman Problem.** Given an edge-weighted graph G satisfying the triangle inequality, with n blue vertices (sources) and n red vertices (sinks), find the optimal *bipartite* tour starting and ending at a designated blue vertex s and visiting all vertices. A tour is bipartite if every pair of adjacent vertices in it have different colors.

Anily and Hassin [2] have shown a 2.5-approximation algorithm for a generalization of this problem known as the Swapping Problem. Their algorithm finds a perfect matching M consisting of edges that connect red and blue vertices, and uses Christofides's heuristic [10] to find a tour T of the blue vertices. The final delivery route consists of visiting the blue vertices in the sequence specified by the tour T, using the matching edges in M to deliver an item to a sink and return to the blue vertex (or "short-cut" to the next blue vertex on T). If OPT is the optimal delivery tour, clearly $T \leq 1.5\text{OPT}$ and $M \leq 0.5\text{OPT}$, whereas the total length of the delivery tour is at most $T + 2M \leq 2.5\text{OPT}$. We exploit some combinatorial properties of bipartite spanning trees and matroid intersection to improve this factor to 2.

A naive approach towards a 2-approximation is to mimic the well-known 2-approximation algorithm for the TSP problem: pick a *bipartite* spanning tree of G, perform a depth-first traversal followed by short-cutting. A spanning tree of G is bipartite if each edge connects a red and blue vertex. Given a bipartite spanning tree T, we can think of it as a tree rooted at s, and do a depth-first traversal of T with short-cuts (there may in general be several ways to short-cut) and obtain a tour of G. However, such a tour may not be bipartite; there are bipartite spanning trees that do not yield a bipartite tour regardless of how we perform the depth-first traversal and short-cuts (see Figure 1).

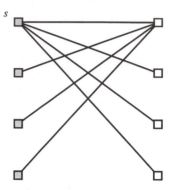

Figure 1: *A bipartite spanning tree for which no depth-first traversal yields a bipartite tour.*

Our 2-approximation algorithm is based on the following very simple observations.

> **Observations.** Let T be a bipartite spanning tree where each blue vertex has degree at most 2; then, T has exactly one blue vertex v_1 of degree 1. If T is rooted at v_1 then every blue vertex has exactly one (red) child. Clearly, if we traverse this rooted tree T in (any) depth-first order, then the sequence of vertices visited are of alternating color.

Clearly, the OPT bipartite tour contains a bipartite spanning tree where all blue vertices have degree at most 2. Therefore the weight of the minimum-weight bipartite spanning tree whose blue vertices have degree at most 2, is a lower bound on OPT. So if we can find (in polynomial time) the minimum-weight bipartite spanning tree T whose blue vertices have degree at most 2, then a depth first traversal of T (with short-cuts) will yield a tour whose length is at most twice OPT.

We now claim that the problem of finding T can be viewed as that of finding the minimum-weight, maximum-cardinality subset in the *intersection* of two matroids [27, 12, 13]. The two matroids in this case are: M_1, the matroid of all bipartite forests, and M_2, the matroid of all bipartite subgraphs whose blue vertices have degree at most 2. For completeness we review here the definition of a matroid, following the standard text [27].

Definition 1 *A matroid $M = (E, \mathcal{I})$ is a structure in which E a finite set of elements and $\mathcal{I}$ is a family of subsets (called* independent sets*) of E, such that: $\phi \in \mathcal{I}$ and all proper subsets of a set $I \in \mathcal{I}$ are in $\mathcal{I}$; and, if I_p and I_{p+1} are sets in $\mathcal{I}$ containing p and $p+1$ elements respectively, then there exists an element $e \in I_{p+1}$ such that $I_p \cup \{e\} \in \mathcal{I}$.*

An example of a matroid is the *graphic matroid* $M = (E, \mathcal{I})$ where E is the set of edges of an undirected graph, and a subset $I \subset E$ is in $\mathcal{I}$ if and only if I is cycle-free. Another example of a matroid is the *matrix matroid* $M = (C, \mathcal{I})$ where C is the set of columns of a fixed matrix A and a subset S of columns is in $\mathcal{I}$ if and only if the columns of S are linearly independent. A maximal-cardinality independent subset of a matroid is called a *base* of a matroid; all bases of a matroid have the same cardinality.

Returning to our problem, let E be the set of all edges that connect red vertices to blue vertices. Let $\mathcal{F}$ denote the collection of all subsets of E that are cycle-free, and let $\mathcal{D}$ denote the collection of subsets S of E such that no more than two edges of S are incident on any blue vertex. Then it is easily seen that $M_1 = (E, \mathcal{F})$ and $M_2 = (E, \mathcal{D})$ are matroids. In addition, the problem of finding *a minimum-weight bipartite spanning tree where the blue vertices have degree at most two*, is equivalent to the problem of finding *a minimum-weight common base of M_1 and M_2.*

This is a special case of the *matroid intersection* problem, which was first solved in polynomial time by Edmonds [12, 13]. Other authors [8] have exploited the special structure of problems such as ours to improve the running times. We obtain the following theorem.

Theorem 1 *There is a polynomial time algorithm that approximates the Bipartite TSP within ratio 2.*

2.1   Finite Capacity Robot Arms

We will now show how to obtain a constant-factor approximation for the case of arbitrary finite k using the algorithm for the unit capacity case. But first, we will establish some lower bounds on the optimal solution. Let C_k denote the (length of the) optimal k-Delivery tour, and let C_r and C_b denote the (length of the) optimal tours on the red and blue points respectively. Let A denote the weight of the minimum-weight perfect matching in the bipartite graph with red vertices on

one side and blue vertices on the other. To keep the notation simple, we will often use the same symbol to denote a graph and its weight, and the context will make it clear which one is intended.

Lemma 1 (a) $A \leq C_1/2$.   (b) $\frac{1}{2k}C_1 \leq C_k$.

Proof: Part (a) is easy to see since C_1 consists of two perfect matchings and the one with smaller weight is at least as heavy as A. To see part (b), start with an optimal k-Delivery tour C_k: this defines an ordering $r_1, r_2, \ldots, r_n$ on the red points and an ordering $b_1, b_2, \ldots, b_n$ on the blue points. We then construct a 1-Delivery tour T starting at the blue vertex b_1 as follows: Consider the blue vertices in the order imposed by C_k, connecting the ith blue vertex b_i to the earliest red vertex in the C_k-ordering that has not already been connected to a blue vertex; then add another edge connecting this red vertex to the next blue vertex b_{i+1} (if this red vertex is the last one, connect it to the starting blue vertex b_1). By the triangle inequality, each edge e of T is no longer than the sum of the C_k-edges connecting the endpoints of e; we can thus "charge off" each edge of T to a collection of edges of C_k. Since there is never a sequence of more than k consecutive red or consecutive blue points in the tour C_k, it follows that no edge of T is charged more than $2k$ times. Thus $T \leq 2kC_k$, from which the lemma follows since $C_1 \leq T$. $\square$

We omit the proof of the following lemma.

Lemma 2 (a) $C_r \leq C_k$.   (b) $C_b \leq C_k$.

We now use the lower bounds just presented to design a constant-factor approximation algorithm for the k-Delivery problem. We first use Christofides' heuristic to obtain a 1.5-approximate tour T_r of the red vertices and a 1.5-approximate tour T_b of the blue vertices. Next we (arbitrarily) break up T_r and T_b into paths of k vertices each, by deleting edges appropriately. It will be convenient to view each k-path as a "super-node" in the following. We now overlay the minimum-weight perfect matching of cost A on this graph. Note that any (red or blue) super-node now has degree exactly k, and that there may be several edges between two given super-nodes. Thus, we obtain a k-regular bipartite multi-graph. The following lemma is crucial to the design of our algorithm.

Lemma 3 *The edges of a d-regular bipartite multi-graph can be partitioned into d perfect matchings.*

Proof: First, note that the d-regularity implies there are an equal number of vertices on each side of the bipartition. Consider any subset S of the vertices on the left side of the bipartite graph and say $|S| = m$. Clearly the number of edges emanating from S is dm and by the Pigeonhole Principle if the set S has fewer than m neighbors on the right side then some vertex on the right side has degree greater than d, which is a contradiction. Thus any subset S of the left-vertices has at least $|S|$ neighbors. Clearly, this remains true even if replace each multi-edge by one of the edges. Then, by Hall's theorem [22, 27] for the existence of perfect matchings, it follows that the multi-graph contains a perfect matching. If we delete this perfect matching, we are left with a $(d-1)$-regular bipartite multi-graph, and the same argument can be repeated, each time removing a perfect matching. This proves the lemma. □

We can thus partition the perfect matching A into k perfect matchings on the super-nodes. We pick the least-weight matching M out of these and delete all other edges of T_1. Clearly $M \leq A/k \leq \frac{1}{2k}C_1$. At this stage we have a collection of n/k subgraphs $H_1, H_2, \ldots, H_{n/k}$, each consisting of a red supernode connected via an edge of M to a blue supernode. Now we re-introduce the edges of T_b that were removed when breaking T_b into k-paths; this imposes a cyclic ordering on the subgraphs H_i; let us relabel them $H_1, H_2, \ldots, H_{n/k}$ with this cyclic ordering, where H_1 contains the start blue vertex. We now traverse the subgraphs $H_1, H_2, \ldots, H_{n/k}$ in sequence as follows. Within each subgraph H_i first visit all the blue vertices, then use the edge of M to go to the red side and visit all the red vertices, and then return to the blue side, and go to the blue vertex that is connected via an edge e of T_b to the next subgraph H_{i+1}, and use the edge e to go to H_{i+1} (or H_1 if $i = n/k$).

We claim that this tour T is within a constant factor of the optimal k-Delivery tour. To verify this, notice that in short-cutting, by triangle inequality, we "charge" each edge of T_b no more than 3 times, each edge of T_r no more than 2 times and each edge of M at most 2 times. Thus, we obtain that

$$T \leq 3T_b + 2T_r + 2M$$

$$\leq 3 \times 1.5C_b + 2 \times 1.5C_r + \frac{2}{2k}C_1$$

$$\leq 9.5C_k.$$

Theorem 2 *The above algorithm gives approximates the k-Delivery TSP to within a factor 9.5.*

2.2 Unbounded Capacity Robot Arms

When the robot arm has unbounded capacity, the only restriction is that at any stage during the delivery, the number of items picked up so far be at least as many as the number of sinks visited so far. The corresponding graph problem is defined as follows:

> **Blue-Dominant TSP.** Given an edge-weighted graph G satisfying the triangle inequality, with n blue vertices (sources) and n red vertices find an optimal tour that starts at a blue vertex s and visits all the vertices (and returns to s) in *blue-dominant* order. An ordering of vertices is blue-dominant if in each prefix of the ordering, there are at least as many blue vertices as red vertices.

We show below a 2-approximation algorithm for this problem. It turns out that one can obtain a 2-approximation for this problem by first finding a minimum-weight spanning tree (not necessarily bipartite) and then using a *specific* depth first traversal and short-cutting scheme. As Figure 2 shows, for a fixed depth-first traversal there may be several ways to perform short-cuts.

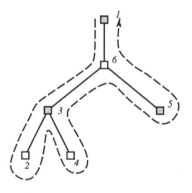

Figure 2: *A bipartite spanning tree that does not contain a perfect matching and yet there is a depth-first traversal that can be short-cut to yield a bipartite tour.*

Let us make precise the distinction between a depth-first traversal and a specific short-cutting scheme. Note that we can view a depth-first traversal as traversing edges rather than as visiting vertices. In the edge-traversing viewpoint, each edge is traversed exactly twice: first downward (away from the root) and later upward (toward the root). Thus each vertex may be "visited" several times by a depth-first traversal. The short-cutting scheme *marks* exactly one out of these several visits to a vertex. For a fixed depth-first traversal and short-cutting scheme, the marking order defines a tour of all the vertices. We will consider two types of short-cutting rules in particular: when the depth-first traversal reaches a subtree rooted at v, we may mark v either *before* or *after* visiting all its descendants. In the former case we say v is *pre-marked*, and in the latter case we say v is *post-marked*.

Theorem 3 *For any spanning tree T rooted at the blue vertex s, there is a depth-first traversal and a short-cutting scheme that marks the vertices in blue-dominant order.*

Proof: For any vertex v, let $\mathbf{r}(v)$ and $\mathbf{b}(v)$ be defined as before. For a marking sequence defined by a depth-first traversal and a short-cutting scheme, for any vertex v let $R(v)$ denote the number of red vertices marked just before the depth-first traversal *first* reaches v, and let $B(v)$ denote the number of blue vertices marked just before the depth-first traversal *first* reaches v.

Consider a depth-first traversal of T satisfying the following:

   Depth First Traversal Rule. After visiting a vertex v, if there are any children w of v such that $\mathbf{b}(w) \geq \mathbf{r}(w)$ then visit them (and their descendants) before visiting any remaining children.

The short-cutting scheme is:

   Short-Cutting Rule. Pre-mark blue-vertices (sources) and post-mark red vertices.

Figure 3 illustrates an application of these rules.

We would like to show that $B(v) \geq R(v)$ holds at all times. We will in fact claim that the following invariant holds whenever the depth-first traversal first reaches a vertex v:

$$B(v) \geq R(v), \quad \text{and}$$
$$B(v) + \mathbf{b}(v) \geq R(v) + \mathbf{r}(v).$$

The proof of the invariant is by structural induction on the rooted tree T. It is trivially true at the start

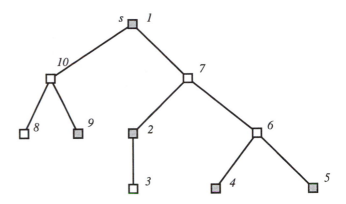

Figure 3: *When the vertices of this spanning tree are marked in depth-first order using the short-cutting scheme of Theorem 3, we get a valid delivery tour for an unbounded-capacity robot arm. The sources are shaded.*

vertex s since $B(s) = R(s) = 0$ and $\mathbf{b}(s) = \mathbf{r}(s) = n$. Assume inductively that the invariant holds just before the depth-first traversal reaches some vertex v. We argue that it will hold at each child of v, visited say in the order $w_1, w_2, \ldots, w_k$. For any vertex v, let $c(v)$ denote the quantity $\mathbf{r}(v) - \mathbf{b}(v)$, the "surplus of sinks", and let $D(v)$ denote $B(v) - R(v)$, the "source surplus". Our inductive assumption $D(v) \geq c(v)$ implies, if v is blue,

$$D(v) \geq \sum_{i=1}^{k} c(w_i) - 1,$$

and if v is red,

$$D(v) \geq \sum_{i=1}^{k} c(w_i) + 1.$$

If v is blue, it is marked before the depth-first traversal visits w_1, and therefore $D(w_1) = D(v) + 1$. If v is red, it is not marked before visiting w_1, so $D(w_1) = D(v)$. In either case we have $D(w_1) \geq 0$ and

$$D(w_1) \geq \sum_{i=1}^{k} c(w_i).$$

Note that for each $i = 1, 2, \ldots, k-1$,

$$\begin{aligned} D(w_{i+1}) &= D(w_i) - c(w_i) \\ &\geq c(w_{i+1}) + c(w_{i+2}) + \ldots + c(w_k). \end{aligned}$$

Our depth-first traversal rule specifies that children with $c(w_i) \leq 0$ must be visited first, and the above equation implies that the $D(.)$ value does not decrease after traversing the subtree rooted at such a child and therefore remains positive. So our invariant holds at each such child. After the subtree under the last such child (say w_j) has been traversed, we have $D(w_{j+1}) \geq 0$ and

$$D(w_{j+1}) \geq c(w_{j+1}) + c(w_{j+2}) + \ldots + c(w_k),$$

where each term on the right hand side is positive. It is now easy to see that $D(w_i) \geq 0$ and $D(w_i) \geq c(w_i)$ will hold for each $i = j + 1, \ldots, k$.   □

3 Maximum Latency TSP

This section is concerned with the latency measure for TSP tours, a minor digression from the family of robot grasp and delivery problems under consideration. A variant of the following latency TSP problem arises in the moving points case and is also of independent interest. Given a set of n points $\{p_1, \ldots, p_n\}$ in a metric space, find a path visiting all points, starting at a given point p_0, such that the total *latency* of the points is *maximized*. If in a given path P the length of the ith edge traversed is e_i then the latency of the jth point visited ($j > 0$) is $L_j = \sum_{i=1}^{j} e_i$ and the total latency $L(P)$ is

$$L(P) = \sum_{j=1}^{n} L_j = \sum_{i=1}^{n} (n - i + 1)e_i.$$

We would like to find a path P for which $L(P)$ is maximized. We can show that this problem is NP-hard, by reducing from the Maximum Hamiltonian Path (MaxHP) problem. A related problem, *minimum latency TSP*, has been addressed earlier [6, 20] and constant-factor approximation algorithms are known.

We show that the greedy strategy, *"At any stage, visit the farthest unvisited point from the current point,"* achieves a total latency at least half of the maximum latency path. In fact, the greedy path also has *length* at least half of the MaxHP from the starting point p_0. The key observation is the following lemma.

Lemma 4 *Let G_i be the length of the first i edges in the greedy path starting from p_0. Let P_i be the length of the maximum i-path, i.e., the longest path that visits i vertices from p_0. Then for $i \leq n$, $G_i \geq P_i/2$.*

Proof: For brevity, paths/edges and their lengths are denoted by the same symbols. Consider the maximum matching M in the maximum i-path P_i, i.e., the maximum-length collection M of independent edges from P_i. Let the (lengths of) edges of M be $m_1 \geq m_2 \geq \ldots \geq m_k$. Note that $P_i \leq 2(m_1 + m_2 + \ldots + m_k)$. We will argue that $G_i \geq (m_1 + m_2 + \ldots + m_k)$.

Consider the edges in G_i as being *directed* in the direction of travel, starting from p_0. Call an edge m of M an *anchor* if there is an edge of G_i that *starts* at an endpoint of m; the *earliest* such edge g of G_i is said to be *anchored* at m.

Claim (a): If g is anchored at m then $g \geq m$, since G_i is greedy and the other endpoint of m has not yet been visited at the time g was traversed by the greedy path.

Claim (b): If $m \in M$ is not an anchor, then *every* edge of G_i has length at least $m/2$. To see this, note that neither end-point of m is visited by G_i (except possibly at the end). By the triangle inequality, from any point p, at least one of the end-points of m is at distance at least $m/2$. Since G_i is greedy, its edges have length at least $m/2$.

Now suppose $m_1, m_2, \ldots, m_u$ are anchors, and m_{u+1} is not, i.e., m_{u+1} is the heaviest edge of M that is *not* an anchor. (Note that u could be 0.) Let $g_1, g_2, \ldots, g_u$ be the corresponding anchored edges of G_i. Then Claims (a) and (b) imply that $g_1 + g_2 + \ldots + g_u \geq m_1 + m_2 + \ldots + m_u$, and that every edge of G_i has length at least $m_{u+1}/2$. There are now two cases to be considered. Recall $k = |M|$.

Case 1. $[u = 0$ and $i = 2k - 1]$ If $u = 0$ then m_1 is not an anchor, so all edges of G_i have length at least $m_1/2$. Also $i = 2k - 1$ implies that the starting point p_0 must occur in the matching M. Thus the first edge of G_i must be anchored at an edge m' of M, and the total length of G_i is at least

$$\begin{aligned} m' + (i - 1)m_1/2 &\geq m' + (2k - 2)m_1/2 \\ &= m' + (k - 1)m_1 \\ &\geq m_1 + m_2 + \ldots + m_k. \end{aligned}$$

Case 2. $[u \geq 1$ or $i \geq 2k]$ Note that the total length L of unanchored edges of G_i is at least $(i - u)m_{u+1}/2$. If $u \geq 1$,

$$L \geq (2k - 1 - u - (u - 1))m_{u+1}/2 = (k - u)m_{u+1},$$

and if $i \geq 2k$,

$$L \geq (2k - u - u)m_{u+1}/2 = (k - u)m_{u+1}.$$

Thus, G_i is at least $(m_1 + m_2 + ... + m_u) + (k - u)m_{u+1} \geq m_1 + m_2 + ... + m_k$. $\qquad\square$

Clearly the latency of the ith point in the maximum latency path is at most P_i, so the lemma implies that the total latency of the greedy path is at least half that of the maximum latency path. Since P_n is the MaxHP, it also follows that the greedy path is a $\frac{1}{2}$-approximation of the MaxHP. Lemma 4 implies our result.

Theorem 4 *The greedy strategy of always visiting the furthest unvisited point achieves a total latency at least half that of the maximum latency path. The total length of the greedy path is at least half that of the maximum length path that visits all points exactly once (i.e., the MaxHP).*

We can also show that the greedy heuristic works well for a different "latency measure" that arises implicitly in Dynamic k-Collect TSP.

Theorem 5 *Let $\alpha < 1$ be a positive constant. For a Hamiltonian path P starting at p_0, define the cost $L_\alpha(P) = \sum_{i=1}^{n-1}(1 - \alpha^{n-i})e_i$. The greedy heuristic produces a path whose L_α cost is at least $(1 - \alpha)/2$ of the maximum-L_α path.*

Proof: Let G denote the greedy path, and H^* denote the Maximum Hamiltonian Path. Let P_α^* denote the path that maximizes $L_\alpha(.)$ and P^* be the path that maximizes $L(.)$. Since $(1 - \alpha^{n-i}) < 1$ for $i = 1, 2, ..., n - 1$, we have

$$L_\alpha(P_\alpha^*) \leq length(P_\alpha^*) \leq length(H^*),$$

and since $(1 - \alpha^{n-i}) \geq 1 - \alpha$ for $i = 1, 2, ..., n - 1$, we have

$$L_\alpha(G) \geq (1 - \alpha)length(G).$$

The desired result follows then follows from the fact that $length(G) \geq length(H^*)/2$. $\qquad\square$

4 Moving Conveyor Belts: Dynamic k-Collect TSP

In this section, we consider the case of parts on a moving conveyor belt. For convenience, we switch to the

L_1 metric to measure costs. That is, we assume that the robot only translates parallel to the x and y axes. While there are situations where this applies directly, it is also easy to see that this only causes an error of factor $\sqrt{2}$ with respect to the L_2 metric. We assume that the points $p_1, p_2, ..., p_n$ are always within the positive quadrant of the coordinate frame centered at the origin $p_0 = (0, 0)$ (see Figure 4). The robot moves with speed 1, and the belt (and each point p_i) moves with a velocity v directed in the negative-x direction. The y-axis represents the end of the conveyor belt, and to obtain meaningful results we must assume at the very least that v is suitably bounded below 1, since a slow robot may be unable to catch up with some points. In fact, in our application it turns out it is sufficient to have $v \leq \frac{k}{2n}$ to ensure that no p_i crosses the y-axis while the robot is in the process of executing the tour. We will assume this upper bound on v throughout the rest of this paper. Also, we define $\alpha = \frac{1-v}{1+v}$.

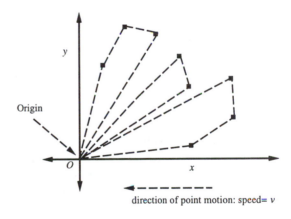

Figure 4: *Illustrating the moving points model. A 3-collect tour is shown.*

For clarity, we refer to a fixed point in space (such as the origin) as a *space-point*, to distinguish it from a *moving-point* p_i. Define (x_i, y_i) as the coordinates of the point p_i at time 0, and let $d_i = x_i + y_i$ denote the L_1 distance of p_i from the origin at time 0. Clearly the x-coordinate of p_i at time t is $x_i - vt$ and the y-coordinate doesn't change. The distance of p_i from the origin at time t is therefore $x_i + y_i - vt = d_i - vt$.

4.1 Shortest Paths and Triangle Inequality

We establish several basic properties of travel-times that are not obvious as they may seem. Henceforth, when we say the robot moves from a point A to a space-point B or meets a moving point p_i, we will assume that the robot takes the shortest-time path. What is the quickest way for a robot to meet a moving-point p? The following lemma characterizes such paths. A robot path is said to be *monotone* if no two points on the path have the same x-coordinate or the same y-coordinate.

Lemma 5 (Shortest Paths) *Suppose the robot is at a space-point A and meets moving-point p at the earliest possible time, say at space-point B. Then the robot's path from A to B is necessarily a shortest path between those points, and the robot never stops at any time before it meets p.*

Proof: At any time the robot may either "sit and wait," or move parallel to the x-axis, or move parallel to the y-axis. Suppose the robot meets point p after time T. If the robot's path to B is not the shortest path to B (from A) or the robot stopped at some time, this means that the robot could have arrived at B at an earlier time $T' < T$ (by either using a shorter path to B or not waiting along the way). At this time T' the moving-point p must be to the right of B a distance $(T - T')v$ away on the same y-coordinate, and the robot can meet p in time $(T - T')\frac{v}{1+v}$ by moving toward it. Thus, the total time to meet p could have been $T' + \frac{v}{1+v}(T - T')$, and the time saved would be $(T - T') - \frac{v}{1+v}(T - T') = \frac{1}{1+v}(T - T')$, which is positive. This contradicts our assumption that the robot met p at the earliest possible time. $\qquad\square$

We have the following important corollaries of the above lemma.

Corollary 1 (Monotone Path) *Suppose the robot moves from a space-point A to meet a moving point p. Then, the robot's path must be monotone. In particular a quickest way for the robot to meet p is to first move to the y-coordinate of p, then move toward p.*

Proof: The following fact will be useful in this and other proofs about shortest-paths:

Fact 1 *The shortest (and therefore least travel-time) path for the robot to move from a given space point A to a given space point B is a monotone path; in particular the monotone path that first moves to the y-coordinate of B and then to the x-coordinate of B is optimal.*

The monotonicity follows from Lemma 5 and Fact 1. Any monotone path that meets p can be replaced by one where all the x-motion is made after the y-motion, without changing the rendezvous time or coordinates. Clearly the y-motion consists simply of moving to the y-coordinate of p. At this time p may either be left or right of the robot. In case p is left of the robot, by monotonicity the quickest way to meet p is to move toward p. In case p is to the right of the robot, if the robot moves to the left then either the path becomes non-monotone before it meets p or it must stop and wait for p to catch up with it, both of which are not possible in a shortest path, by Lemma 5. Thus, in this case also the robot must move toward p. $\qquad\square$

Corollary 2 (Triangle Inequality 1) *If the moving points p_i and p_j are in the positive quadrant, and the robot is at p_i, then the time t_{ij} to travel directly to p_j is bounded by the time to travel to p_j via the origin.*

Proof: If the composition of the robot's path from p_i to the origin and the path from the origin to p_j is not monotone, then by the Monotone Path Corollary 1 it cannot be shorter than the shortest p_i-to-p_j path; if it is monotone then it cannot be shorter than the particular shortest p_i-p_j path described in Lemma 1. $\qquad\square$

Corollary 3 (Triangle Inequality 2) *Let t_{ij} denote the shortest time of travel from moving-point p_i to moving-point p_j. For any three points p_i, p_j, p_k, $t_{ij} + t_{jk} \geq t_{ik}$*

It is natural to wonder whether a given order of visiting the points takes less time when the points are moving than when they are at rest. We show that this is indeed the case (recall that $v \leq \frac{k}{2n}$).

Theorem 6 *Consider a tour T that starts at the origin, visits some set of moving-points, and returns to the origin, and let $T(v)$ be the time taken to do so. Then, for $v \geq 0$, $T(v) \leq T(0)$.*

Proof: Denote the origin by $p_0 = (x_0, y_0) = (0,0)$, and suppose T visits the points in the sequence $p_1, p_2, \ldots, p_m$ and returns to the origin. For $i = 0, 1, 2, \ldots, m - 1$, let $X_i = |x_i - x_{i+1}|$ and $Y_i = |y_i - y_{i+1}|$. Clearly then

$$T(0) = \sum_{i=0}^{m-1} (X_i + Y_i) \, + \, d_m.$$

For a positive speed v, from equation (2), the total time is given by

$$T(v) = (1 - v) \sum_{i=0}^{m-1} T_i + d_m,$$

where T_i is the time taken by T to go from p_i to p_{i+1}, and is given by one of the expressions (4), (5), (6). For each possible expression for T_i, it is easy to check that $(1 - v)T_i \leq (X_i + Y_i)$, so $T(v) \leq T(0)$. $\qquad \square$

4.2 Useful Time Expressions

The following time expressions are easily verified; we will use them in the proofs in this paper. In all of the following keep in mind that d_i denotes the distance of p_i from the origin at time 0.

- **Origin to Moving-Point.** The robot is at the origin at time τ, and meets moving point p_i. The earliest rendezvous time is

$$\tau + \frac{d_i - v\tau}{1 + v} = \frac{\tau + d_i}{1 + v}. \qquad (1)$$

- **Moving-point to Origin.** At time τ the robot is at a moving-point p_i and it then returns to the origin. The time of return to the origin is

$$\tau + d_i - v\tau = d_i + (1 - v)\tau. \qquad (2)$$

- **Moving-Point to Origin to Moving Point** At time τ the robot is at a moving-point p_i; it then returns to the origin and goes to moving-point p_j. From the previous expression, the time of arrival at the origin is $d_i + (1 - v)\tau$, and during this time the point p_j moves closer to the origin by a distance $v(d_i + \tau(1 - v))$. So the time of arrival at p_j is

$$
\begin{aligned}
d_i &+ \tau(1 - v) + \frac{d_j - v[d_i + \tau(1 - v)]}{1 + v} \\
&= \frac{d_i + d_j}{1 + v} + \tau \frac{1 - v}{1 + v}. \qquad (3)
\end{aligned}
$$

4.3 Optimal Rendezvous

We can use the above characterization to derive some expressions for the shortest time (or distance) to meet a moving point. Suppose at time t_0 the robot is at space-point $A = (x_1, y_1)$, and moving-point p is at (x_2, y_2). Let $x = |x_1 - x_2|$ and $y = |y_1 - y_2|$. If the robot moves

and meets moving-point p at the earliest possible time $t_0 + t$, then t equals one of the following. In view of Lemma 1 we may assume that the robot first moves to the y-coordinate of p, then meets p by moving parallel to the x-axis.

- If p is right of A at time t_0, and $vy \leq x$, then p remains on the right of the robot when it reaches the y-coordinate of p, so

$$t = y + \frac{x - vy}{1 + v} = \frac{x + y}{1 + v}. \qquad (4)$$

- If p is right of A at time t_0 and $vy > x$, the moving-point p will be left of the robot by the time the robot reaches the y-coordinate of p, so

$$t = y + \frac{vy - x}{1 - v} = \frac{y - x}{1 - v}. \qquad (5)$$

- If p is to the left of A at time t_0, then

$$t = y + \frac{vy + x}{1 - v} = \frac{x + y}{1 - v}. \qquad (6)$$

4.4 Robot Arms with Capacity 1

In the remainder of the paper we write C_k to denote the (length of the) optimal Dynamic k-Collect tour.

In Dynamic 1-Collect TSP, the robot must visit the points one at a time, returning to the origin after visiting each point. Suppose that the robot visits the points $p_1, \ldots, p_n$ in that order. We now derive an expression for the total time $T^{(1)}$ taken by a 1-Collect tour. Let T_m denote the time taken by the tour after it has visited m points and returned to the origin. We proceed by induction on m. The base case, $m = 1$, follows from an application of Equation 1 using $\tau = 0$ and $i = 1$; note that the factor of 2 comes from the requirement that the robot returns to the origin.

For the induction step, we assume that

$$T_{m-1} = \frac{2}{1 + v} \sum_{j=1}^{m-1} \alpha^{m-(i+1)} d_i.$$

Suppose that at time T_{m-1} the robot has just returned to the origin after visiting the point p_{m-1}, and now it is ready to go visit point p_m. The time taken to reach p_m is given by an application of Equation 1 with $\tau = T_{m-1}$ and $i = m$, and this equals

$$\frac{d_m}{1 + v} - \frac{vT_{m-1}}{1 + v}.$$

The total time required is given by

$$
\begin{aligned}
T_m &= T_{m-1} + 2\left(\frac{d_m}{1+v} - \frac{vT_{m-1}}{1+v}\right) \\
&= \left(1 - \frac{2v}{1+v}\right)T_{m-1} + \left(\frac{2}{1+v}d_m\right) \\
&= \alpha T_{m-1} + \frac{2}{1+v}d_m \\
&= \frac{2}{1+v}\left(\sum_{j=1}^{m-1}\alpha^{m-(i+1)+1}d_i\right) + \frac{2}{1+v}d_m \\
&= \frac{2}{1+v}\left(\sum_{j=1}^{m}\alpha^{m-i}d_i\right).
\end{aligned}
$$

So the total time $T^{(1)}$ taken by a 1-Collect tour is given by

$$
T^{(1)} = \frac{2}{1+v}\sum_{j=1}^{n}\alpha^{n-i}d_i. \tag{7}
$$

Consequently, we can easily obtain the following result:

Theorem 7 *For Dynamic 1-Collect TSP, the minimum time tour is one that visits the moving-points in decreasing order of their distance from the origin at time 0.*

Thus the optimal Dynamic 1-Collect tour has some aspects of a *Maximum Latency* tour. In fact the maximum latency problem is implicit in the Dynamic k-Collect problem for arbitrary finite k. We make this formal in Section 4.7 where we introduce the notion of *Geometric Latency* of a tour.

4.5 Robot Arms with Unbounded Capacity

In Dynamic k-Collect TSP with $k = \infty$, the robot must visit all n moving-points before returning to the origin. For the optimal tour suppose the last moving-point visited is p, and say p is at distance d from the origin at time 0. If the tour reaches p at time T_p, then by (2) the total time taken is $C_\infty = d + (1-v)T_p$.

Our strategy for approximating C_∞ is to "guess" the last point p visited by the optimal tour (there are only n possibilities for p) and approximate the minimum-length path T_p^* from the origin to p that visits every other moving-point before visiting p (i.e., the Minimum Hamiltonian Path from the origin to p).

The problem of approximating T_p^* can be set up as an Asymmetric TSP instance on a graph with $n+1$ vertices $v_0, v_1, \ldots, v_n$, where v_0 represents the origin and v_i the moving-point p_i. The directed distance $d(v_i, v_j)$ is defined to be the appropriate one out of the expressions (4), (5), and (6). Note that the ratio of the two directed distances between a given pair of vertices is bounded by either $(1+v)/(1-v)$ or $(y+x)/(y-x)$, where in the second case $vy > x$, i.e., $x/y < v < 1$. Thus the ratio never exceeds $1/\alpha$. The best-known approximation algorithm [23] for the Minimum Hamiltonian Path between two specified vertices for a *symmetric* distance matrix has a ratio $\frac{5}{3}$. Therefore, using this algorithm we can approximate T_p^* to within a factor $\frac{5}{3\alpha}$, and thereby approximate C_∞ to within the same ratio.

Theorem 8 *There is a $\frac{5}{3\alpha}$-approximation algorithm for Dynamic k-Collect TSP with $k = \infty$.*

4.6 Robot Arms with Finite Capacity k

We now consider Dynamic k-Collect TSP. We derive an expression for the time $T^{(k)}$ taken by a k-Collect tour, Let $m = n/k$ denote the number of returns to the origin. The sequence of edges traversed can be viewed as

$$
\begin{aligned}
(p_0, p_{v(0)}) \quad & L_1 \quad (p_{u(1)}, p_0), (p_0, p_{v(1)}) \\
& L_2 \quad (p_{u(2)}, p_0), (p_0, p_{v(2)}) \\
& L_3 \quad \ldots \quad L_m \quad (p_{u(m)}, p_0),
\end{aligned}
$$

where each L_i denotes the sequence of edges involving non-origin points in the ith excursion from the origin. We abuse notation and denote by L_i the time spent traversing the corresponding edges. Let $D_0 = d_{v(0)}$, $D_m = d_{u(m)}$, and for $2 \le i \le m-1$, $D_i = d_{u(i)} + d_{v(i)}$.

Notice first that the time till the end of L_1 is $T_1 = D_0/(1+v) + L_1$ and the time to the end of L_2 (from Equation 1) is

$$
T_2 = \alpha\left(\frac{D_0}{1+v} + L_1\right) + \frac{D_1}{1+v} + L_2
$$

If we were to return to the origin after L_2 (and d denotes the time-0 distance of the end of L_2 to the origin) the total time would be

$$
T' = d + (1-v)T_2 = D_0\alpha^2 + D_1\alpha + d + (1-v)[L_1\alpha + L_2].
$$

Generalizing this gives the following expression for $T^{(k)}$:

$$T^{(k)} = D_0\alpha^m + D_1\alpha^{m-1} + \ldots + D_{m-1}\alpha + D_m$$
$$+(1-v)[L_1\alpha^{m-1} + L_2\alpha^{m-2} + \ldots + L_m] \qquad (8)$$

It might appear from this expression that $T^{(k)} \to 0$ as $\alpha \to 0$. However, it must be kept in mind that the expression is only valid if at all times, all unvisited moving-points remain to the right of the origin (i.e., in the positive quadrant). In the moving belt scenario, the origin represents the end of the belt. A small α corresponds to a v close to 1, whereas the problem is meaningful only if v is small enough that the points do not cross the origin when the robot is in the process of grasping them. A reasonable restriction on v is that the time for an optimal k-Collect TSP tour should not suffice for any point to cross the origin. This implies that $C_1/k \leq d_{av}/v$, where d_{av} denotes the average of the distances d_i and C_1/k is a lower bound on the optimal in the static problem (see for instance [1]). Thus, we will assume that $v(2d_{av}n)/k \leq d_{av}$, i.e., $v \leq \frac{k}{2n}$.

Given that $v \leq \frac{k}{2n}$, in the expression (8) for $T^{(k)}$, the smallest coefficient on any term is

$$\alpha^m = \alpha^{n/k} \geq \frac{(1-\frac{k}{2n})^{n/k}}{(1+\frac{k}{2n})^{n/k}} \geq \frac{1}{e}$$

This implies that $T^{(k)} \geq \frac{1}{e}(D_0 + L_1 + D_1 + L_2 + D_2 + \ldots + L_m + D_m)$.

Let us denote the sum of terms in the parentheses by T'. Minimizing T' is a static k-Collect TSP problem on $n+1$ vertices $v_0, v_1, \ldots, v_n$ where the directed distances are defined as in the C_∞ approximation, except that for all $i > 0$, $d(v_0, v_i) = d_i = d(v_i, v_0)$. As we saw before, the ratio of the two directed distances between a pair of vertices never exceeds $1/\alpha$, so the k-Collect TSP approximation algorithm of [1] can be used to approximate T' to within a factor $2.5/\alpha$ of optimal. Thus, we can approximate the optimal k-Collect tour time C_k to within a factor $2.5e/\alpha$.

Theorem 9 *For $v \leq k/2n$, there is a $(2.5e/\alpha)$-approximation algorithm for Dynamic k-Collect TSP.*

The following lower bound, although not used in this paper, may lead to a different approximation algorithm for C_k than the one we presented above.

Theorem 10 *For any finite k, $C_\infty \leq C_k$.*

Proof: Consider the optimal tour C_k for capacity k. (We will abuse notation and use the same symbol to stand for a tour as well as its length). Let d be the time-0 distance from the origin, of the final moving-point p visited by C_k before returning to the origin, and let T_p be the time taken by T up to this final point p. From equation 2, the total time taken is $C_k = T_p(1-v) + d$. Note that before reaching the final point p, T may do the following several times: go from a moving-point p_i to the origin, then to moving-point p_j. However by the Triangle Inequality I (Corollary 2), a direct path from p_i to p_j is no longer than the path via the origin, so we can short-circuit each such indirect path by a direct path and gain time. By applying these short-circuits to every origin-return except the last one, we can modify the optimal tour C_k to an unbounded-capacity tour that reaches p at an earlier time $T'_p \leq T_p$, and the total time of this tour would be $(1-v)T'_p + d$ which is no larger than C_k. □

4.7 Geometric Latency

The somewhat unwieldy expression (8) for $T^{(k)}$ can be lower-bounded by a more pleasant cost expression which we call the *geometric latency* of a tour. We describe this below and show how a variant of the *maximum* latency problem arises implicitly in minimizing the geometric latency.

Consider the asymmetric k-Collect TSP problem with directed distances as defined in Section 4.6. Now consider a capacity-k tour on this graph, and let $e_1, e_2, \ldots, e_u$ be the sequence of (weights of) edges traversed. Fix some positive $\beta < 1$, and define the *geometric latency* G_β of this tour to be $G_\beta = \sum_{i=1}^{u} \beta^{u-i} e_i$. Assuming for convenience that n is a multiple of k, $u = m + n$.

Lemma 6 *Let $\beta = \alpha^{1/k}$ and fix a capacity-k tour of length $T^{(k)}$ for the moving points problem. Let G_β be the geometric latency of the corresponding (i.e., same order of visiting points) tour in the corresponding asymmetric k-Collect TSP instance as described above. Then $T^{(k)} \geq G_\beta$.*

Proof: Consider expression (8) for $T^{(k)}$. Since $(1-v) \geq \alpha$,

$$T^{(k)} \geq (D_0 + L_1)\alpha^m + (D_1 + L_2)\alpha^{m-1}$$
$$+ \ldots + (D_{m-1} + L_m)\alpha + D_m.$$

The lemma follows by comparing weights of corresponding terms in this expression and the one for G_β.
□

Thus, if we are able to find a constant-factor approximation to a capacity-k tour with minimum Geometric latency G_β, we would have a constant factor approximation for the dynamic k-Collect TSP problem. We know of no such algorithm that runs in polynomial time. It is worth noticing that G_β may be rewritten as

$$G_\beta = \sum_{i=1}^{u} e_i - \sum_{i=1}^{u-1}(1 - \beta^{u-i}e_i),$$

so minimizing G_β roughly involves simultaneously minimizing the total tour length and maximizing the second term, which is a variant of the linear latency. As we mentioned in Section 3, we can approximately maximize the second term, but we do not know how to simultaneously bound the length of the tour.

5 Conclusion

In this paper we have initiated a formal algorithmic study of problems arising in the context of rapid deployment automation. Specifically, we have considered problems related to rearranging parts by a robot in an industrial assembly line setting. Our work applies to parts both on a static as well as a moving conveyer belt. The static belt problems, which are variants of a problem we define as k-Delivery TSP, and the dynamic belt problems, variants of k-Collect TSP, are related to well known problems in the literature.

All the algorithms we have presented are, to the best of our knowledge, either the first polynomial time constant-factor approximation algorithms for the problems, or provide a better constant approximation than was previously known. The algorithms themselves are very simple and lend themselves to easy implementation. We briefly mention some easy extensions of the model and results for the moving belt scenario; the details are omitted.

- The first extension is to the case where all points are moving *away* from the origin at velocity v. For instance, this would be the case in mid-air refueling of a formation of planes. The results and analysis are similar to that presented above.

- The second extension is to the case where the conveyor belt is circular and is rotating around the origin. We can obtain a constant-factor approximation for the case where the rotation speed is bounded below the robot's speed.

A variety of open problems and new directions for algorithmic and robotic research arise out of our work. In general, there are a large number of problems related to the efficient design, simulation, and implementation of rapid deployment automation [9]. For instance: a typical design problem is the optimal placement of modules, e.g., given a fixed path, determine a location for the robot base so as to minimize cycle time; a typical simulation problem is that of finding space for dropped parts on a conveyor belt, i.e., find a pose where a new part can be placed without collision.

More specific to our work, it is important to experimentally study the performance of the algorithms presented here. Then, there is the natural question of whether the approximation results obtained in this paper can be improved, hopefully without sacrificing on the efficiency of the algorithms. In the case of a moving conveyor belt, there is the issue of weakening the assumption we made regarding the relative velocity of the conveyor belt. One approach to this would be to devise an efficient approximation algorithm for the Maximum Latency TSP problem with respect to the notion of geometric latency. It also remains to be seen whether any of our results can be extended to other vehicle routing and scheduling problems such as Dial-a-Ride and Precedence-Constrained TSP.

An important extension to our work would be to consider more complex cost metrics. The cost metric used in this paper is the total translational (x, y) distance, D, traveled by the end-effector to deliver all n parts. We neglected the cost of rotating the end-effector to orient itself correctly over each part prior to pick-up as well as the reorientation at delivery. Considering rotation would increase our configuration space by one dimension. As long as the rotational velocity of the end-effector is proportional to its (assumed constant) translational velocity, our algorithms can be extended to higher dimensions, with exponential dependence on dimension. We could also consider cost metrics of the form $D + c$, where c is a constant (or only dependent on n). Such a cost metric could model, for example, damping and the z translation by the end-effector (the total

z-translation depends only on the number of parts, n, and not on the actual configuration of the parts). Finally, an assumption that could be relaxed is that of uniform (x, y) velocity of the robot end-effector so as to model the non-zero accelerations of the end-effector as it starts its motion to a new part and as it gets there. This leads to a more complex bounded-velocity problem instead of our constant-velocity problem.

Acknowledgments

Rajeev Motwani was supported by the ARO MURI Grant DAAH04-96-1-0007. He is also partially supported by an Alfred P. Sloan Research Fellowship, an IBM Faculty Partnership Award, and NSF Young Investigator Award CCR-9357849, with matching funds from IBM, Schlumberger Foundation, Shell Foundation, and Xerox Corporation. This research was done when Anil Rao was with the Department of Computer Science, Utrecht University and supported by European Strategic Program for Research in Information Technology (ESPRIT) Basic Research Action No. 6546 (project PRoMotion).

We are deeply indebted to Ken Goldberg for introducing us to this family of problems, and for his valuable feedback and constant encouragement during the course of this work. We are also grateful to Alan Frieze for his help with the Bipartite Traveling Salesman Problem as well as for his encouragement. We thank Rafi Hassin for pointing out a simplification in our arguments in Section 2. Finally, we are grateful to an anonymous referee for suggestions that improved the presentation.

References

[1] K. Altinkemer and B. Gavish. Heuristics for delivery problems with constant error guarantees. *Transportation Science*, 24(4):294–297, 1990.

[2] S. Anily and R. Hassin. The swapping problem. *Networks*, 22:419–433, 1992.

[3] M.J. Atallah and S.R. Kosaraju. Efficient solutions to some transportation problems with applications to minimizing robot arm travel. *SIAM Journal on Computing*, 17:849–869 (1988).

[4] B. Awerbuch, Y. Azar, A. Blum, and S. Vempala. Improved Approximation Guarantees for Minimum-Weight k-Trees and Prize-Collecting Salesmen. In *Proceedings of the 27th Annual ACM Symposium on the Theory of Computing* (1995), pp. 277–283.

[5] L. Bianco, A. Mingozzi, S. Riccardelli, and M. Spadoni. Exact and heuristic procedures for the traveling salesman problem with precedence constraints, based on dynamic programming. *INFOR*, 32(1):19–32, 1994.

[6] A. Blum, P. Chalasani, D. Coppersmith, B. Pulleyblank, P. Raghavan, and M. Sudan. The Minimum Latency Problem. In *Proceedings of the 26th Annual ACM Symposium on the Theory of Computing* (1994), pp. 163–171.

[7] A. Blum, P. Chalasani, and S. Vempala. A Constant-Factor Approximation for the k-MST Problem in the Plane. In *Proceedings of the 27th Annual ACM Symposium on the Theory of Computing* (1995), pp. 294–302.

[8] C. Brezovec, G. Cornuejols, and F. Glover. A matroid algorithm and its application to the efficient solution of two optimization problems on graphs. *Mathematical Programming*, 42:471–487, 1988.

[9] B. Carlisle and K.Y. Goldberg. Report on TARDA. *Symposium on Theoretical Aspects of Rapid Deployment Automation*, Adept Technology, San Jose (1994).

[10] N. Christofides. Worst-case analysis for a new heuristic for the Traveling Salesman problem. In *Symposium on New Directions and Recent Results in Algorithms and Complexity*, (Ed: J.F. Traub), Academic Press (1976).

[11] N. Christofides. Vehicle routing. In *The Traveling Salesman Problem: A Guided Tour of Combinatorial Optimization* (Ed: E.L. Lawler, J.K. Lenstra, A.H.G. Rinnooy Kan, and D.B. Shmoys), John Wiley & Sons (1985), pp. 431–448.

[12] J. Edmonds. Submodular functions, matroids and certain polyhedra. In *Combinatorial Structures and their Applications, Proceedings of Calgary International Conference*, pages 69–87, 1970.

[13] J. Edmonds. Matroid intersection. *Annals of Discrete Mathematics*, 4:39–49, 1979.

[14] Flexible Assembly Systems. *Proceedings of the 3rd American Society of Mechanical Engineers*

(ASME) Conference on Flexible Assembly Systems. Block September 1991.

[15] B. Carlisle. Is U.S. Robotics Research of Any Use to U.S. Industry?. *IEEE Robotics and Automation Society Newsletter.* July, 1991.

[16] K.R. Fox, B. Gavish, and S.C. Graves. An n-constraint formulation of the (time-dependent) traveling salesman problem. *Operations Research* 28:1018–1021 (1980).

[17] G. N. Frederickson, M. Hecht, and C. Kim. Approximation algorithms for some routing problems. *SIAM Journal on Computing*, 7(2):178–193, 1978.

[18] A.M. Frieze. An Extension of Christofides' Heuristic to the k-person Traveling Salesman Problem. *Discrete Applied Mathematics*, 6:79–83, 1983.

[19] N. Garg and D.S. Hochbaum. An $O(\log k)$ approximation algorithm for the k minimum spanning tree problem in the plane. In *Proceedings of the 26th Annual ACM Symposium on the Theory of Computing* (1994), pp. 432–438.

[20] M.X. Goemans and J.M. Kleinberg. An Improved Approximation Ratio for the Minimum Latency Problem. In *Proceedings of the 7th Annual ACM-SIAM Symposium on Discrete Algorithms* (1996), to appear.

[21] B.L. Golden and A.A. Assad, Eds. *Vehicle Routing: Methods and Studies.* North-Holland, Amsterdam (1988).

[22] P. Hall. On representatives of subsets. *Journal of London Mathematics Society*, 10:26–30, 1935.

[23] J.A. Hoogeveen. Analysis of Christofides' heuristic: some paths are more difficult than cycles. *Operations Research Letters*, 10:291–295 (1991).

[24] K. Jansen. Bounds for the general capacitated routing problem. *Networks*, 23(3):165–173, 1993.

[25] M. Kubo and H. Kagusai. Heuristic algorithms for the single-vehicle dial-a-ride problem. *Journal of the Operations Research Society of Japan*, 30(4):354–365, 1990.

[26] J-C. Latombe. *Robot Motion Planning.* Kluwer Academic Press (1991).

[27] E. Lawler. *Combinatorial Optimization: Networks and Matroids.* Holt, Reinhart and Winston, New York, 1976.

[28] T-Y. Li and J-C. Latombe. On-line manipulation planning for two robot arms in a dynamic environment. In *Proceedings of the 12th Annual IEEE International Conference on Robotics and Automation* (1995), pp. 1048–1055.

[29] T. Lozano-Perez, J. Jones, E. Mazer, and P. O'Donnell. *Handey: A Robot Task Planner.* MIT Press, 1992.

[30] J. Lenstra and A. R. Kan. Complexity of Vehicle Routing and Scheduling Problems. *Networks* 11:221–227 (1981).

[31] A . Lucena. Time-dependent traveling salesman problem – the deliveryman case. *Networks* 20:753–763 (1990).

[32] J. Mitchell. Guillotine subdivisions approximate polygonal subdivisions: A simple new method for the geometric k-mst problem. In *7th Annual ACM-SIAM Symposium on Discrete Algorithms*, 1996.

[33] R. Motwani. *Approximation Algorithms*, Technical Report STAN-CS-92-1435, Department of Computer Science, Stanford University, 1992.

[34] *Proceedings of Robotic Industries Association (RIA) Workshop on Flexible Part Feeding.* October 1993.

[35] H. Psaraftis. Scheduling large-scale advance-request dial-a-ride systems. *American J. Math. Manage. Sci.*, 6:327–367, 1986.

[36] R. Ravi, R. Sundaram, M.V. Marathe, D.J. Rosenkrantz, and S.S. Ravi. Spanning trees short or small. In *Proceedings of the 5th Annual ACM-SIAM Symposium on Discrete Algorithms* (1994), pp. 546–555.

[37] D.J. Rosenkrantz, R.E. Stearns, and P.M. Lewis. An analysis of several heuristics for the traveling salesman problem . *SIAM Journal on Computing* 6:563–581 (1977).

[38] J.T. Schwartz, J. Hopcroft, and M. Sharir. *Planning, Geometry, and Complexity of Robot Motion.* Ablex, 1987.

Algorithms for Constructing Immobilizing Fixtures and Grasps of Three-Dimensional Objects

Attawith Sudsang, Jean Ponce and Narayan Srinivasa, *Beckman Institute, University of Illinois, Urbana, IL, USA*

We address the problem of computing immobilizing fixtures and grasps of three-dimensional objects, using simple fixturing devices and grippers with both discrete and continuous degrees of freedom. The proposed approach is based on the notion of second-order immobility introduced by Rimon and Burdick [48, 49, 50], which is used here to derive simple sufficient conditions for immobility and stability in the case of contacts between spherical locators and polyhedral objects. In turn, these conditions are the basis for efficient geometric algorithms that enumerate all of the stable immobilizing fixtures and grasps of polyhedra that can be achieved by various types of fixturing devices and grippers. Preliminary implementations of the proposed algorithms have been constructed, and examples are presented.

1 Introduction

We address the problem of immobilizing an object through a few contacts with simple modular fixturing elements, with applications in manufacturing (fixturing) and robotics (grasping). Our approach is based on the notion of second-order immobility introduced by Rimon and Burdick [48, 49, 50], and it is related to recent work in fixture planning by Wallack and Canny [57, 58], Brost and Goldberg [5], and Wagner, Zhuang, and Goldberg [56].

For concreteness, let us consider the fixturing device shown in Figure 1: it consists of two parallel plates with locator holes drilled along a rectangular grid, and of a set of spherical locators with integer height and radius. Four of these locators can be selected to form a fixture; either two of them are mounted on each plate (type I configuration, Figure 1(a)), or three locators are mounted on the first plate, the last one being mounted

on the second plate (type II configuration, Figure 1(b)). The distance between the plates is a continuously adjustable degree of freedom of the device. (This device is a generalization of the two-dimensional fixturing vise proposed by Wallack and Canny [57, 58].)

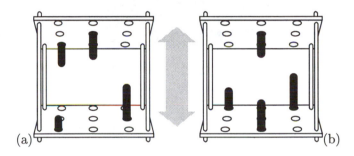

Figure 1: *The proposed fixturing device: (a) type I configuration, (b) type II configuration.*

Our goal is to compute the locator configurations (i.e., placements and heights) as well as the plate separation that will guarantee that a polyhedral part in *frictionless* contact with the locators is immobilized. To solve this problem, we must (1) formulate operational conditions for immobility, (2) enumerate all of the locator configurations that may achieve immobility, and (3) for each of these configurations, decide whether there exists a pose of the fixtured object that simultaneously achieves contact with the four locators and guarantees immobility. Our approach to step (1) is to specialize the conditions formulated by Rimon and Burdick to the class of fixturing elements and objects of interest. Step (2) can then be reduced to solving a combinatorial problem, so we can attack step (3) using numerical algebraic methods [28, 29, 37].

It should be noted that the conceptual design shown in Figure 1 can be implemented using standard modu-

lar fixturing elements such as the ones available in the
QU CO kit: for example, a type I configuration can be
constructed using two spherical locators mounted on a
plate and two additional locators mounted on a beam
clamp (Figure 2). A similar assembly with three locators and an adjustable vertical clamp can be used for
type II configurations.

Figure 2: *Implementing a type I configuration of the proposed fixturing device using standard modules from the QU
CO kit.*

Although we will emphasize the fixturing problem
in most of the paper, our approach applies to robotic
grasp planning and other related problems (see Section
4). In particular, we are also developing an automatically reconfigurable gripper (Figure 3) which consists of
two parallel plates whose distance can be adjusted by
a computer-controlled actuator. The top plate carries
a rectangular grid of individually-actuated locators,
which can translate vertically under computer control.
The bottom plate is a bare plane with three passive
planar degrees of freedom, which are used to simulate
the absence of friction and ensure grasp stability. This
design is reminiscent of Goldberg's sliding-jaw gripper
[17].

The rest of this paper is organized as follows. Previous work in fixture and grasp planning is reviewed in
Section 2. Our approach is described in Section 3: sufficient conditions for immobility and stability are derived
in Section 3.1; they are used in Section 3.2 to design
an efficient algorithm for enumerating all immobilizing
stable fixtures of a given polyhedral object using the

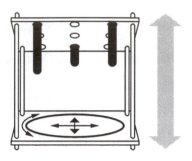

Figure 3: *A reconfigurable gripper. An actuated locator is
associated with each grid point. To avoid friction effects,
the bottom plate should have three passive planar degrees of
freedom.*

device shown in Figure 1. This algorithm is adapted
to the problem of planning immobilizing grasps for the
reconfigurable gripper of Figure 3 in Section 4, where
some other extensions are also described. Preliminary
experiments are presented in Section 5. Section 6 concludes with a brief discussion of our approach and future research directions.

2 Related Work

Modular fixturing systems consist of a kit of modules that can be reconfigured to fixture different parts.
They have the potential of avoiding the costs associated with the design of custom fixtures, but pose the
problem of automatically planning the module configurations adequate for a given part geometry. Traditionally, fixture designers have relied on heuristics such as
the *3:2:1 fixturing principle* [19, 53]: the object to be
fixtured is first positioned relative to a plane (*primary
datum*) defined by three contact points; it is then positioned relative to a line (*secondary datum*) defined by
two additional contact points, and finally positioned
relative to a last point contact (*tertiary datum*). When
the six points have been chosen correctly, the position
of the fixtured object is completely determined as long
as the contacts are maintained (*deterministic positioning* [1]). The object is then clamped into place by
one or several additional contacts (*total constraint* [1]).
Positioning is typically achieved through contact with
passive fixturing elements such as plates, vee blocks,

and locators, while clamping is achieved through contact with *active* fixturing elements, such as vises, toe clamps, or chucks.

The theoretical justification for such an approach finds its roots in the dual role of fixtures: immobilizing a part and resisting the forces and torques involved in manufacturing tasks such as assembly or machining.[1] Since screw theory [2, 22, 41] can be used to represented both displacements (*twists*) and forces and moments (*wrenches*), it is an appropriate tool for analyzing and designing fixtures. Indeed, it is known that six independent contact wrenches are necessary to prevent any infinitesimal displacement which maintains contact, and that a seventh one is required to ensure that contact cannot be broken (these correspond to the positioning and clamping contacts introduced above) [24, 54]. Such a fixture prevents any infinitesimal motion of the object, and it is said to achieve *form closure* [41, 47, 52]. A system of wrenches is said to achieve *force closure* when it can balance any external force and torque. Like wrenches and infinitesimal twists [51], force and form closure are dual notions and, as noted in [36, 39] for example, force closure implies form closure and vice versa.[2] In particular, fixtures achieving form/force closure also fulfill their second role as devices capable of resisting external forces and torques.

Past approaches to fixture planning have been based on expert systems [14, 18, 32], kinematic analysis and screw theory [1, 3, 8, 34], or a combination of both [12, 16]: Markus *et al.* have used a rule-based system to interactively design fixtures for box-type parts and to select appropriate fixture modules [32]. Ferreira

and Liu have used a generate-and-evaluate approach to determine the orientation of workpieces for machining operations [14]. Hayes and Wright have proposed *Machinist*, an expert-system-based process planner that incorporates fixturing information in the construction of a machining plan [18].

While expert systems are limited in their ability to generate fixture configurations based on analytical considerations, approaches based on screw theory can accurately predict the performance of fixture designs: Asada and By have proposed the *Automatically Reconfigured Fixturing (ARF)* system, which uses a detailed kinematic analysis to derive conditions for deterministic positioning, part accessibility and detachability, and total constraint [1]. Their approach has been integrated into a robotic assembly cell. Chou, Chandru, and Barash have developed a mathematical method based on screw theory for analyzing and synthesizing fixtures, and used linear programming to generate optimal clamp positions constrained to lie within convex contact polygons [8]. Bausch and Youcef-Toumi have introduced the notion of *motion stop* which represents the geometric resistance of a contact point to a given screw motion, and they have used it to compare fixture configurations. Their approach is integrated with the CATIA CAD system and is capable of synthesizing optimal fixture configurations from a discrete set of candidates.

Finally, it should be noted that some systems bridge the gap between expert systems and kinematic and force analysis: Gandhi and Thompson have proposed a methodology that relies on expert knowledge, force analysis, and geometric reasoning to synthesize and analyze modular fixture configurations [16]. Englert has also combined analytical considerations and knowledge-based methods to identify tradeoff relations between part production attributes and propose a control structure for planning part setup and clamping [12].

In the robotics community, the notions of form and force closure have been the basis of several approaches to grasp planning. Mishra, Schwartz, and Sharir [35] have proposed linear-time algorithms for computing a

---

[1]There are of course other issues involved in fixturing, for example the analysis of part deformation under clamping, see [10, 20] for approaches using finite-element methods.

[2]Note that there is unfortunately no general agreement on terminology in the grasping literature (see [55, 33] for discussions of this problem): for example, Reulaux [47], Salisbury [52], Ji [23], Markenscoff et al. [30] and Trinkle [55] use the expression form closure for what we call force closure, and reserve the expression force closure for grasps that can only balance certain external loads. Our definitions match the ones used by Mishra et al. [35], Nguyen [39], and Murray et al. [38].

finger configuration achieving force closure for friction-less polyhedral objects. Markenscoff and Papadim-itriou [31] and Mirtich and Canny [33] have proposed algorithms for planning grasps which are optimal ac-cording to various criteria [13]. In each of these works, the grasp-planning algorithm outputs a single grasp for a given set of contact faces. Assuming Coulomb friction [39], Nguyen has proposed instead a geomet-ric method for computing *maximal independent* two-finger grasps of polygons, i.e., segments of the polygo-nal boundary where the two fingers can be positioned independently while maintaining force closure, requir-ing as little positional accuracy from the robot as pos-sible. This approach has been generalized to handle various numbers of fingers and different object geome-tries in [4, 7, 42, 44, 45, 46]

Although robotic grasping and fixture planning are related (in both cases, the object grasped of fixtured must, after all, be held securely), their functional re-quirements are not the same: as remarked by Chou, Chandru, and Barash [8], machining a part requires much better positional accuracy than simply picking it up, and the range of forces exerted on the parts are very different. The role of friction forces is also differ-ent: in the grasping context, where fingers are often covered with rubber or other soft materials, friction effects can be used to lower the number of fingers re-quired to achieve form closure from seven to four; in the fixturing context, on the other hand, it is custom-ary to assume frictionless contact, partly due to the large magnitude and inherent dynamic nature of the forces involved [8] (see, however [27] for an approach to fixture planning with friction). Finally, the kine-matic constraints on the positions of the contacts are also quite different.

Nonetheless, as noted by Wallack [58], there has re-cently been a renewed interest in the academic robotics community for manufacturing problems in general and fixturing in particular. Mishra has studied the prob-lem of designing fixtures for rectilinear parts using toe clamps attached to a regular grid, and proven the ex-istence of fixtures using six clamps [34] (this result has since then been tightened to four clamps by Zhuang, Goldberg, and Wong [59]). In keeping with the idea of

Reduced Intricacy Sensing and Control (RISC) robotics of Canny and Goldberg [6], Wallack and Canny [57, 58] and Brost and Goldberg [5] have recently proposed very simple modular fixturing devices and efficient al-gorithms for constructing form-closure fixtures of two-dimensional polygonal and curved objects. Wagner, Zhuang, and Goldberg [56] have also proposed a three-dimensional seven-contact fixturing device and an al-gorithm for planning form-closure fixtures of a polyhe-dron with pre-specified pose.

All of the approaches discussed so far are based on the concepts of form and force closure. A different notion is *stability*: a part is said to be in stable equi-librium if it returns to its equilibrium position after having been subjected to a small displacement. Stabil-ity is very important in real mechanical systems which cannot be expected to have perfect accuracy. Nguyen has shown that force (or form) closure implies stability [40], but Donoghue, Howard and Kumar have shown that there exist stable grasps or fixtures which do not achieve form closure [11, 21]. Recently, Rimon and Burdick have introduced the notion of *second-order immobility* [48, 49, 50] and shown that certain equilib-rium grasps (or fixtures) of a part which do not achieve form closure effectively prevent any *finite* motion of this part through curvature effects in configuration space. They have given operational conditions for immobiliza-tion and proven the dynamic stability of immobilizing grasps under various deformation models [50]. An ad-ditional advantage of this theory is that second-order immobilization can be achieved with fewer fingers (four contacts for convex fingers) than form closure (seven contacts [24, 54]). As detailed in the next section, we propose to exploit second-order immobility in fixture planning for three-dimensional polyhedral objects.

3   Proposed Approach

We first derive simple sufficient conditions for immobil-ity and stability in the case of contacts between spher-ical locators and polyhedral objects (Section 3.1). We then use these conditions in Section 3.2 to design an efficient algorithm for planning stable immobilizing fix-tures of a polyhedral object with the device shown in

Figure 1.

3.1 Sufficient Conditions for Immobility and Stability

A sufficient Condition for Immobility. Let us consider a rigid object and the contacts between d locators and this object. Let us also denote by p_i ($i = 1, .., d$) the positions of the contacts in a coordinate frame attached to the object, and by n_i ($i = 1, .., d$) the unit inward normals to the corresponding faces.

We say that equilibrium is achieved when the contact wrenches balance each other, i.e.,

$$\sum_{i=1}^{d} \lambda_i \begin{pmatrix} n_i \\ p_i \times n_i \end{pmatrix} = 0, \tag{1}$$

for some $\lambda_i \geq 0$ ($i = 1, .., d$) with $\sum_{i=1}^{d} \lambda_i = 1$. Equilibrium is a necessary, but not sufficient, condition for force and form closure.

Czyzowicz, Stojmenovic and Urrutia have recently shown that three contacts in the plane and four contacts in the three-dimensional case are sufficient to immobilize (i.e., prevent any *finite* motion of) a polyhedron [9]. Rimon and Burdick have formalized the notion of immobilizing grasps and fixtures in terms of isolated points of the free configuration space [48, 49, 50]. They have shown that equilibrium fixtures that do not achieve form closure may still immobilize an object through second-order (curvature) effects in configuration space: a sufficient condition for immobility is that the relative curvature form associated with an essential equilibrium [3] grasp or fixture and defined by

$$\kappa_{\mathrm{rel}} = \sum_{i=1}^{d} \lambda_i |w_i| \kappa_i$$

be negative definite. Here the weights λ_i are the equilibrium weights of (1), $|w_i|$ is the magnitude of the wrench exerted by locator number i, and κ_i is the curvature form associated with the corresponding contact; this quadratic form is defined by:

$$\kappa_i = \frac{1}{|w_i|} (v^T, \omega^T)(\mathcal{C}_i^T \mathcal{L}_i \mathcal{C}_i + \mathcal{D}_i) \begin{pmatrix} v \\ \omega \end{pmatrix},$$

---

[3]Essential equilibrium is achieved when the coefficients λ_i in (1) are uniquely defined and strictly positive [49].

where v and ω denote the translational and rotational parts of an infinitesimal twist, $\mathcal{L}_i$ is a matrix related to the surface curvatures of the body and locator at the contact points, and the matrices $\mathcal{C}_i$ and $\mathcal{D}_i$ depend only on the position p_i of the contact point and the normal to n_i to the body's surface in p_i.

Specializing the above equations to equilibrium contacts between spherical locators and polyhedra yields:

$$\kappa_{\mathrm{rel}} = \sum_{i=1}^{d} \lambda_i \omega^T \mathcal{K}_i \omega, \quad \text{where} \tag{2}$$

$$\mathcal{K}_i = ([n_{i\times}]^T [p_{i\times}])^S - r_i [n_{i\times}]^T [n_{i\times}],$$

where r_i denotes the locator's radius, and by definition, $A^S = \frac{1}{2}(A + A^T)$

Note that there is no term involving the translation v in this case. It follows from (2) that a sufficient condition for immobility is that the 3×3 symmetric matrix

$$\mathcal{K} = \sum_{i=1}^{d} \lambda_i \mathcal{K}_i$$

is negative definite.

A Sufficient Condition for Stability. We prove a sufficient condition for the stability (in the sense of Nguyen [40], see also [11, 21]) of a fixture configuration, and show that it is equivalent to the immobility condition derived in the previous paragraph (see [50] for a more general statement of the dynamic stability of immobilizing grasps).

Each locator is modeled as a sphere of radius r_i attached to a linear spring whose axis is aligned with the inward normal to the corresponding contact face. As the solid moves, the sphere slides on the contact face and translates along the corresponding spring (Figure 4).

The potential energy of the fixture is the sum of the potential energies of the individual springs, i.e.,

$$U = \sum_{i=1}^{d} \frac{1}{2} \sigma_i^2,$$

where σ_i is the displacement of the spring associated with locator number i from its position at rest (we

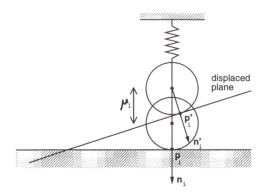

Figure 4: *Small displacement of a plane in contact with a sphere mounted on a spring.*

assume a unit spring constant for each locator). In general we can write $\sigma_i = -\lambda_i + \mu_i$, where λ_i is the compression at equilibrium (which is of course equal to the coefficient λ_i in (1)) and μ_i denotes the displacement of the sphere along n_i corresponding to a given displacement of the solid. An equilibrium fixture will be stable when it corresponds to a local minimum of the potential energy (as a function of small displacements of the object).

A rigid displacement is specified by a rotation $\mathcal{R}$ of axis a and angle θ, and a translation v. Following Nguyen [40], we use a second-order Taylor expansion of the exponential definition of rotations, and parameterize μ_i by the twist (v, ω), where $\omega = \theta a$.

The gradient and Hessian of the potential energy are respectively

$$\nabla U = \sum_{i=1}^{d}(-\lambda_i + \mu_i)\nabla\mu_i \quad \text{and} \qquad (3)$$

$$\nabla^2 U = \sum_{i=1}^{d} \nabla\mu_i \nabla\mu_i^T + (-\lambda_i + \mu_i)\nabla^2\mu_i.$$

A simple calculation shows that the gradient of μ_i at the origin is the wrench $(n_i^T, p_i \times n_i^T)^T$, and it follows as expected that the fixture is in equilibrium if and only if equation (1) is satisfied. To decide whether the equilibrium is stable, we must examine the Hessian of the potential function. Computing the Hessian of μ_i at the origin and substituting in (3) yields:

$$\nabla^2 U|_{0,0} = \mathcal{F} + \mathcal{S}, \quad \text{where}$$

$$\mathcal{F} = \sum_{i=1}^{d} \begin{pmatrix} n_i \\ p_i \times n_i \end{pmatrix} \begin{pmatrix} n_i \\ p_i \times n_i \end{pmatrix}^T, \quad \text{and}$$

$$\mathcal{S} = -\sum_{i=1}^{d} \lambda_i \begin{pmatrix} 0 & 0 \\ 0 & \mathcal{K}_i \end{pmatrix}.$$

The equilibrium is stable when $\mathcal{F} + \mathcal{S}$, the Hessian of the potential function, or *stiffness matrix*, is positive definite. The matrix $\mathcal{F}$ is of course positive semi-definite, its zeros being the twists reciprocal to the wrenches exerted by the locators (which are guaranteed to exist for frictionless fixtures when $d \leq 6$). These twists satisfy the equations

$$v \cdot n_i + \omega \cdot (p_i \times n_i) = 0 \quad \text{for} \quad i = 1, \ldots, d. \qquad (4)$$

For equilibrium fixtures, only three of the above equations are independent, and for any choice of ω there exists in general a vector v such that (4) is satisfied. Thus $\nabla^2 U$ is positive definite if and only if the matrix $\mathcal{K}$ is negative definite. This condition is the same as the sufficient condition for immobility derived earlier.

3.2 Planning Four-Locator Immobilizing Fixtures of Polyhedral Objects

In this section, we focus on the four-locator case and present an efficient algorithm for enumerating all immobilizing fixtures of a polyhedral object that can be achieved with the device of Figure 1. To simplify this planning process, we reduce the problem of achieving contact between a spherical locator and a plane to the problem of achieving point contact with a plane. This is done without loss of generality by growing the object to be fixtured by the locator radius and shrinking the spherical end of the locator into its center (see [5, 57, 58] for similar approaches in the two-dimensional case). For the sake of conciseness, we restrict our attention here to type II fixture configurations. Planning type I configurations involves analogous methods and has the same cost

The algorithm can be summarized as follows. For each quadruple of faces do:

1. Test whether they can be held in essential equilibrium.

2. Enumerate all locator configurations potentially achieving equilibrium through contacts with the selected faces.

3. For each such configuration, compute the pose of the object and test the immobilization condition.

Testing Essential Equilibrium For a polyhedral object, the normals n_i are fixed vectors. To ensure essential equilibrium, we restrict our attention to quadruples of faces such that no three of them have coplanar normals. This ensures that the coefficients λ_i in (1) are uniquely defined, and it allows us to compute them from the equation $\sum_{i=1}^4 \lambda_i n_i = 0$ and to test whether they all have the same sign. If they do not, the four candidate faces are rejected. If they do, we obtain three independent *linear* constraints on the positions of the locators on the faces:

$$\sum_{i=1}^4 (\lambda_i n_i) \times p_i = 0. \tag{5}$$

(Note that the coefficients λ_i are now constants depending only on the choice of faces.)

We can parameterize each contact p_i by two variables u_i, v_i. Assuming convex faces, the fact that the contact points actually belong to the faces can be written as a set of linear inequalities on u_i, v_i:

$$f_{ij}(u_i, v_i) \leq 0, \quad j = 1, .., k_i, \tag{6}$$

where k_i is the number of edges that bound face number i.

Given a choice of four faces, a necessary and sufficient condition for the existence of contact points within those faces which achieve equilibrium is that there exists a solution to (5) subject to the constraints (6). This can be tested using linear programming. If the test is negative, the quadruple of faces is rejected.

For quadruples passing this second test, there is only (in general) a subset of each face that can participate in an equilibrium configuration. The subset corresponding to face number i is determined by projecting

the five-dimensional polytope defined by (5) and (6) onto the plane (u_i, v_i). Several algorithms can be used to perform this projection, including Fourier's method [15], the convex hull and extreme point approaches of Lassez and Lassez [26, 25], and the Gaussian elimination and contour tracking techniques of Ponce *et al.* [46].

For faces with a bounded number of edges, all of these algorithms run in constant time, and they can be used to construct sub-faces that can be passed as input to the rest of the algorithm.

Enumerating Locator Configurations An exhaustive search of all possible grid coordinates would be extremely expensive: consider an object of diameter D (measured in units equal to the distance between successive plate holes); there are $O(D^8)$ type II possible configurations: one locator is at the origin with zero length, two locators have three integer coordinates, the last locator has only two. A similar line of reasoning also applies to type I configurations, and it yields the same order of complexity.

This has prompted us, like Wallack and Canny [57, 58] and Brost and Goldberg [5] in the two-dimensional case, to use bounds on the distance between two faces to restrict the set of grid coordinates under consideration. The minimum and maximum distances between pairs of points belonging to two given faces can be computed in constant time as follows: the maximum distance between two faces is always achieved for a pair of vertices. The minimum distance, on the other hand, may be achieved for any pair of face, edge, or vertex points (Figure 5). The first two cases shown in Figure 5 (face-face and edge-face pairs) only occur when two faces are parallel or when one edge is parallel to a face, and they reduce to computing the distance between a vertex and a face. Thus, there are only three non-trivial cases: the vertex-face, edge-edge, and vertex-edge pairs, and the corresponding distances are easily computed by constructing the unique straight line orthogonal to the pair of features of interest.

Let us position the first locator at the origin with zero length. The integer point corresponding to the second locator is then constrained to lie within the

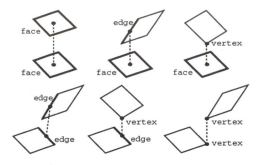

Figure 5: *A list of the feature combinations yielding the minimum distance between two faces.*

spherical shell centered at the origin with inner radius equal to the minimum distance between the two corresponding faces and outer radius equal to the maximum distance. Given the position of the second locator, the third locator is now constrained to lie within the region formed by the intersection of the two shells associated with the first and second locator. Finally, given the position of the third locator, the fourth locator is constrained to lie within a region formed by the intersection of three shells. The projection of this region onto the first plate yields the set of integer coordinates of the locator. Its last coordinate δ is determined at the next stage of the algorithm.

Computing the Pose Associated with a Given Locator Configuration. To avoid imposing a particular parameterization of the object's pose, we take advantage of the fact that a tetrahedron is completely determined by the lengths of its six edges.

We define the tetrahedron whose vertices are the four contacts by six quadratic equality constraints of the form

$$|\boldsymbol{p}_i - \boldsymbol{p}_j|^2 = l_{ij}^2, \text{ with } i = 1, 2, 3 \text{ and } i < j \leq 4, \quad (7)$$

which specify the lengths of the tetrahedron's edges.

At this point, the integer grid coordinates of the locators are fixed, and the coefficients l_{ij} are only functions of the variable δ. Thus the equalities (5) and (7) form a system of nine equations in the nine unknowns u_i, v_i ($i = 1, .., 4$) and δ. Since three of these equations are linear, and the remaining six are quadratic, this system admits at most $2^6 = 64$ solutions which can easily

be computed using homotopy continuation [37] or the toolkit for algebraic computation described in [28].

Once the solutions have been found, we can check whether they satisfy the linear inequalities (6) defining the contact faces, then check whether they achieve immobilization. Note that the object pose corresponding to a given locator configuration and plate distance is easily computed: since one of the locators is at the origin, we only need to compute the rotation mapping the fixturing device's coordinate frame onto the object's coordinate frame. Since we know the positions of the contacts in both coordinate systems, it is a simple matter to compute the corresponding rotation.

Algorithm Analysis. For each quadruple of faces, enumerating the locator configurations amounts to determining the integer positions falling in regions defined by the intersection of two, four, or six half-spaces bounded by spheres. A naive approach to that problem is to test every grid point against the constraints defining the regions of interest with cost $O(D^8)$, where D is as before the diameter of the object measured in units equal to the distance between two successive holes.

A better approach is to use a three-dimensional scan-line conversion algorithm to determine the integer points within a region in (optimal) time proportional to the number V of these points: scan-line conversion algorithms are used in computer graphics to render polygonal and curved shapes by enumerating pixels within these shapes one row at a time; they only require the ability to trace the shape boundaries and find their extrema in the horizontal and vertical dimension, and they have a time complexity linear in the size of their output. It is relatively straightforward to generalize these algorithms to the three-dimensional case: we can construct an explicit representation of the region boundaries by a procedure akin to boundary evaluation in constructive solid geometry. This process is simplified by the fact that in our case the boundary elements are sphere patches, circular arcs (intersections of two spheres), and vertices (intersections of three spheres). Constructing the boundary representation and its extrema in any direction can be done in constant time (given our bounded number of half-spaces), and scan-line conversion can then proceed, one plane at a time, in time proportional to V (Figure 6). Thus, the time complexity of our overall algorithm is $O(N^4 V)$ where N is the number of faces of the polyhedron. Of course, V is still, in the worst case, $O(D^8)$.

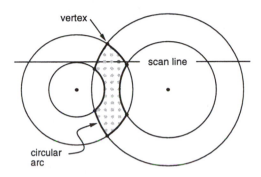

Figure 6: *Illustration of scan-line conversion in the 2D case: spans between consecutive boundary elements are filled one scan-line at a time.*

To obtain a more realistic estimate of the algorithm behavior, we can parameterize V by the diameter D of the object and the maximum range between the minimum and maximum distance between two faces, say r. If we assume that $r \ll D$ and that the distance between two faces is bounded below by some strictly positive number,[4] then the regions corresponding to the possible positions of the second, third, and fourth locator have respectively the "thick" spherical, circular, and punctual shapes shown in Figure 7. The corresponding volumes are respectively $O(D^2 r)$, $O(D r^2)$, and $O(r^3)$. The projection of the latter volume on the plate has an area of $O(r^2)$, and it follows that $V = O(D^3 r^5) \ll O(D^8)$. Experiments will allow us to conduct an empirical evaluation of this model.

4 Extensions

In this section, we show that our general approach to fixturing applies to a variety of related problems. For the sake of conciseness, we do not go into as much detail as in the previous section.

---

[4]This is of course not true in general for an arbitrary polyhedron; however in practice two contact points are separated by at least one unit of distance.

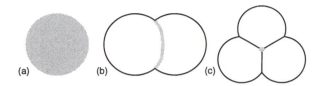

Figure 7: *The regions shown in grey correspond to the position of: (a) the second locator, (b) the third locator, (c) the fourth locator.*

4.1 Planning Immobilizing Grasps for a Reconfigurable Gripper

Consider the reconfigurable gripper shown in Figure 3: it is similar to the device of Figure 1, except that the bottom plate is just a bare plane, and that the top plate carries a rectangular array of individually-actuated locators. This gripper can be used to immobilize a polyhedral object through contacts with three of the top plate locators, and either a face, an edge-and-vertex, or a three-vertex contact with the bottom plate. Let us assume for the sake of simplicity that the faces of the polyhedron are triangular (convex faces can be handled in similar ways, see [8] for a related approach). Any wrench exerted at a contact point between a face and the bottom plate can be written as a positive combination of wrenches at the vertices. Likewise, the wrenches corresponding to an edge-and-vertex contact are positive combinations of wrenches exerted at the end-points of the line segment and at the vertex. Thus equilibrium configurations can be found, in general, by writing the equilibrium equation (1) for six elementary wrenches.

We detail the case of a contact between the bottom plate and a triangular face with unit normal $\boldsymbol{n}$ and vertices $\boldsymbol{v}_i$ ($i = 1, 2, 3$). Let $\boldsymbol{p}_i$ and $\boldsymbol{n}_i$ ($i = 1, 2, 3$) denote the remaining contact points and surface normals; we take advantage of the fact that the overall scale of the wrenches is irrelevant to rewrite (1) as

$$\sum_{i=1}^{3} \lambda_i \begin{pmatrix} \boldsymbol{n} \\ \boldsymbol{v}_i \times \boldsymbol{n} \end{pmatrix} + \sum_{i=1}^{3} \mu_i \begin{pmatrix} \boldsymbol{n}_i \\ \boldsymbol{p}_i \times \boldsymbol{n}_i \end{pmatrix} = 0, \quad \text{with} \quad (8)$$

$\lambda_1 + \lambda_2 + \lambda_3 = 1$, and $\lambda_i, \mu_i \geq 0$ ($i = 1, 2, 3$).

When the four surface normals are linearly independent, the equation $\boldsymbol{n} = -\sum_{i=1}^{3} \mu_i \boldsymbol{n}_i$ allows us to compute the coefficients μ_i and check whether they have the same sign. If they do not, the quadruple of faces under consideration is rejected. If they do, (8) provides four linear equations in the the nine unknowns λ_i, u_i, v_i $(i = 1, 2, 3)$, where u_i, v_i parameterize as before the position of the contact point p_i within the corresponding face. We test the existence of equilibrium configurations by using linear programming to determine whether the polytope defined by the constraints (6), (8), and $\lambda_i \geq 0$ $(i = 1, 2, 3)$ is empty. When this polytope is not empty, we determine as before the subset of each face that may participate in an equilibrium configuration by projecting it onto the plane u_i, v_i $(i = 1, 2, 3)$. Ideas similar to the ones used in the four-locator case can be used to reduce the subset of integer locator positions that needs to be considered.

We now show how to compute the pose of the object for a given configuration of the locators. We write that the difference in height between the contact points p_j and p_1 $(j = 1, 2)$ is equal to the difference in height δ_j between the corresponding locators. This yields two linear equations

$$(p_j - p_1) \cdot n = \delta_j, \quad j = 1, 2, \qquad (9)$$

in the unknowns u_i, v_i $(i = 1, 2, 3)$.

We still need three additional equations to compute the pose of the object. Let us denote by q_i $(i = 1, 2, 3)$ the projections of the three locators onto the plane of the bottom plate. The planar transformation mapping the vector $q_2 - q_1$ onto the vector $q_3 - q_1$ is the linear map $\mathcal{T}$ obtained by composing the rotation that aligns the two vectors with the scaling that gives them the same length. The transformation $\mathcal{T}$ is trivially determined from the points q_i.

Let us denote by $p'_i = p_i - (p_i \cdot n)n$ the projection of the point p_i $(i = 1, 2, 3)$ onto the plane of the bottom plate. We now write that the two triangles formed respectively by the points p'_i and q_i are similar. This can be expressed by the linear vector equation

$$(p'_3 - p'_1) = \mathcal{T}(p'_2 - p'_1), \qquad (10)$$

which expresses the fact that the two triangles are homotetic, and by the quadratic equation

$$|p'_2 - p'_1|^2 = |q_2 - q_1|^2, \qquad (11)$$

which expresses the fact that the two triangles have the same size.

Together, (8), (9) and (10) form a system of eight linear equations that can be used to solve for eight of the unknowns as a function of the remaining one. Substituting in (10) finally yields a univariate quadratic equation which is readily solved.

It follows that we can enumerate as before the equilibrium configurations, then test them for immobility. For planar contacts, equilibrium will guarantee immobilization, while for edge-and-vertex contacts a conservative condition will be to use point contact $(r_i = 0)$ in the expression of the relative curvature form. The complexity of the algorithm is $O(N^4 D^6)$. Interestingly, the solution of a six-contact fixturing problem has a lower complexity than the four-contact fixturing problem encountered before.

4.2 Planning Three-Locator Stable Equilibrium Fixtures

Consider the problem of computing "fixture" configurations in which a polyhedral object rests in a stable fashion on three locators with integer lengths attached to a single horizontal plate (Figure 8).[5] Here we take the effect of gravity into account, and assume that the vertical is aligned with one of the axes of the grid. The method is similar to the earlier one.

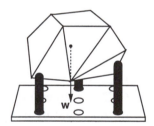

Figure 8: *A three-locator "fixturing" device.*

We choose the center of mass of the object as the origin. Writing that the weight w and the reaction forces achieve equilibrium yields

$$\binom{w}{0} + \sum_{i=1}^{3} \lambda_i \binom{n_i}{p_i \times n_i} = 0, \qquad (12)$$

---

[5]This problem was suggested by Ken Goldberg.

with $\lambda_i \geq 0$ for for $i = 1, 2, 3$.

We can assume unit weight, and by choosing one of the locators as the origin of the grid, we obtain six more quadratic constraints:

$$|w|^2 = 1, \quad \begin{cases} |p_1 - p_2|^2 = l_{12}^2 \\ |p_2 - p_3|^2 = l_{23}^2 \\ |p_3 - p_1|^2 = l_{31}^2 \end{cases}, \quad \text{and} \quad (13)$$

$$\begin{cases} (p_2 - p_1) \cdot w = k_1 \\ (p_3 - p_1) \cdot w = k_2 \end{cases}.$$

Note that the last two equations are necessary to ensure that the vertical is aligned with one of the grid axes. Together, equations (12) and (13) yield nine quadratic constraints in the nine variables λ_i, u_i, v_i ($i = 1, 2, 3$) (it is not necessary to explicitly use w as an unknown, since the vector $-\sum_{i=1}^{3} \lambda_i n_i$ can be used instead). The total degree of the system is $2^9 = 512$. Once the solutions of the equations have been found, they can be checked to see whether the corresponding contacts actually lie inside the faces and whether the equilibrium is stable (of course immobilization is not achieved anymore in this case). It should be noted that planning the position of three locators instead of four reduces the complexity of planning to $O(N^3 D^6)$. Another important point is that for the device of Figure 8 to work properly, it is very important that no friction occurs at the contacts. This can be achieved by mounting the locator caps on ball bearings, at the cost of some loss in positioning accuracy.

4.3   Four-Finger Immobilizing Grasp Planning

We now consider the problem of planning a four-finger immobilizing grasp of a polyhedral object. First, it should be obvious that when the positions of the contact points between fingers and object faces are unconstrained, equations (5) and (6) form a set of *linear* constraints defining the set of all equilibrium grasps. Thus this set can be completely characterized through linear programming. As shown earlier, an equilibrium grasp achieves immobilization when the matrix $\mathcal{K}$ is negative definite. For a given choice of faces, the only unknowns in the matrix $\mathcal{K}$ are the finger positions p_i.

However, testing for definite negativeness involves cubic constraints on the elements of $\mathcal{K}$.

As shown in [46], four-finger equilibrium grasps can be divided into three sub-classes: concurrent, two-pencil, and regulus grasps (Figure 9).[6] Thus we can examine each class separately: by specializing for each of them the expression of $\mathcal{K}_i$ given in (2) and, if appropriate, the twist reciprocity condition given in (4), we will simplify the form of the immobility condition with the goal of finding linear sufficient conditions for immobility.

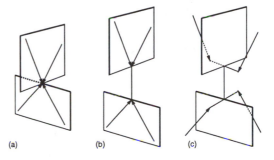

Figure 9: *Classes of grasps achieving equilibrium: (a) the contact forces intersect in one point; (b) they form two non-coplanar pencils; (c) they form a regulus. The contact faces are not shown.*

We illustrate this process with concurrent grasps, such that the lines of action of the four forces intersect in one point (Figure 9(a)). These grasps can be parameterized by

$$p_i = p_0 - \rho_i n_i, \quad (14)$$

where p_0 is the point where the lines of action of the forces intersect, and ρ_i is the signed distance between p_0 and p_i, such that $\rho_i > 0$ if p_0 is on the interior side of face number i. Substituting in (2) and taking advantage once again of the fact that, at equilibrium, $\sum_{i=1}^{4} \lambda_i n_i = 0$, we obtain immediately

$$\mathcal{K} = -\sum_{i=1}^{4} \lambda_i (\rho_i + r_i) [n_{i\times}]^T [n_{i\times}].$$

---

[6]Of course, four co-planar forces can also achieve equilibrium, but not essential equilibrium.

A sufficient (but not necessary) condition for $\mathcal{K}$ to be definite negative is of course that

$$\rho_i + r_i > 0, \quad \text{for} \quad i = 1, \ldots, 4, \qquad (15)$$

and it follows that the immobilizing grasps satisfying this condition can be completely characterized through linear programming under the constraints (5), (6), and (15). Since the stiffness matrix of the grasp is determined by $\mathcal{K}_i$, we can also compute optimal grasps by maximizing $\min_i(\rho_i + r_i)$ under these constraints.

We plan to seek similar linear sufficient conditions for the other two types of equilibrium grasps.

5   Implementation and Results

We have fully implemented the four-locator fixturing algorithm and the algorithm for planning immobilizing grasps using the reconfigurable gripper proposed in Section 4.1. The fixtures have been physically implemented using QU CO fixturing elements as shown in Figure 2. Because we are still in the process of constructing the reconfigurable gripper, we can only present simulated grasping experiments. Both implementations have been written in C, and they include the two pruning stages proposed in Section 3: the subsets of the faces that may participate in an immobilizing grasp are first found by projecting the polytope defined by the equilibrium constraints (5) and (6) onto the parameter space of the faces. Candidate configurations that satisfy the distance constraints associated with these subsets of the faces are then enumerated by scan-converting the volumes bounded by the corresponding spherical shells.

5.1   Four-Locator Fixture Planning

As shown in Figure 2, we have constructed the proposed four-locator fixturing device using modular fixturing elements from the QU CO kit. We have used an aluminum base plate with an array of threaded holes, compatible threaded bolts and nuts, removable spherical locator tips, and a horizontal beam. The bolts and spherical tips are used as locators, and different locator heights are implemented by attaching different number

of nuts to the bolts before screwing them through the threaded holes of the base plate. The horizontal beam is used as a support for the top locators.

Figure 10 shows some simulation results. The test object is a tetrahedron, and each result is shown from two different viewpoints. Figure 11 shows the fixturing device in the immobilizing configuration given in 10.a.

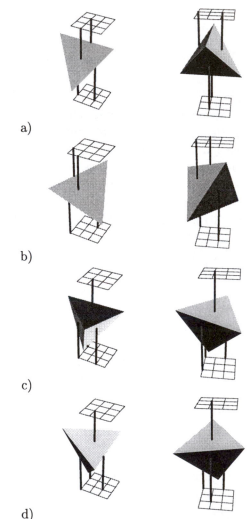

a)

b)

c)

d)

Figure 10: *Some solutions for the four-locator fixturing device.*

Table 1 shows some quantitative results for different grid resolutions. In our experiments we have used a $K \times K$ grid with various values of K, as well as locators whose height may take ten discrete values. Table 1

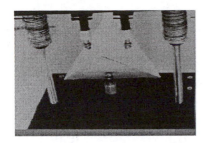

Figure 11: *The fixturing device immobilizing a tetrahedron.*

shows the results obtained without any pruning (N), using spherical shell pruning only (S), and combining the projection- and shell-pruning stages (P+S).

K	# Candidates		
	N	S	P+S
3	6,377,292	267,868	223,224
4	63,700,992	4,429,772	3,601,440
5	379,687,500	20,720,018	17,709,408
6	1,632,586,752	297,104,432	237,683,544

Table 1: *Quantitative results using a tetrahedron as a test object.*

We have used homotopy continuation [37] to solve the polynomial system of degree 64 that determines the poses of an object compatible with a given locator configuration. Our distributed implementation of continuation takes roughly 2.5 seconds on two networked four-processor SUN SPARCstations 10 to solve this system. Thus we have not been able in our actual experiments with moderate grid resolutions to compute in a reasonable time all of the achievable fixtures. Instead, we have stopped the computation once a few immobilizing fixtures had been found. The statistics given in Table 1 have been obtained by running only the part of the algorithm that enumerates all possible locator configurations. We have recently found a new pose parameterization that only requires solving a system of degree 32 and are in the process of implementing it.

5.2 Planning Grasps for the Reconfigurable Gripper

In the case of the gripper, we have recently found that the orientation of an immobilized object is independent of the heights of the three locators, and that this orientation can be computed by solving a univariate quadratic equation. This has allowed us to construct an efficient implementation of our grasp planning algorithm. Figures 12 and 13 show some of the grasps found for a tetrahedron for two different grid resolutions, and Table 2 shows some quantitative results. In this case, we have been able to compute all of the immobilizing grasps; the run times reported in Table 2 were measured on a SUN SPARCstation 10.

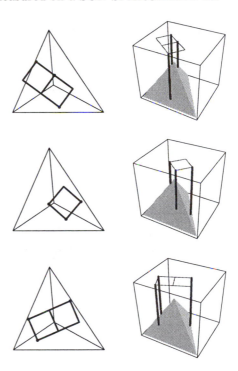

Figure 12: *Grasping a tetrahedron: some solutions for a* 5×5 *grid.*

Figures 14 amd 15 show some more simulation examples using a polyhedron with ten faces. As shown in Table 3, in this case pruning eliminates a much larger percentage of the possible configurations, corresponding to the fact that, for most choices of faces, the range between the minimum and maximum distances being

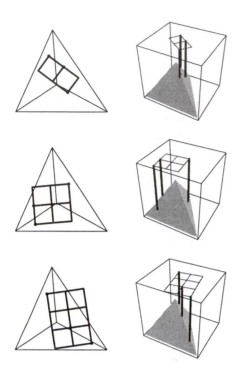

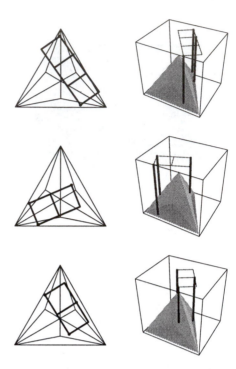

Figure 13: *Some solutions for a 6 × 6 grid.*

Figure 14: *Grasping a 10-face polyhedron: some solutions for a 5 × 5 grid.*

K	Number of Solutions	Run Time (s)			# Candidates		
		N	S	P+S	N	S	P+S
3	0	1	1	1	33	10	10
4	160	1	1	1	141	42	40
5	704	2	1	2	411	145	135
6	1,963	4	2	2	927	391	378
7	4,263	8	4	4	1,839	795	751

Table 2: *Quantitative results using a tetrahedron as a test object.*

smaller than in the previous case.

K	# Sol	Run Time (s)			# Candidates		
		N	S	P+S	N	S	P+S
3	0	20	1	2	2,772	750	712
4	189	47	3	4	11,844	2,213	2,102
5	794	72	9	9	34,524	3,819	3,537
6	2,326	142	20	20	77,868	7,811	7,125
7	5,046	341	43	41	154,476	16,259	14,951

Table 3: *Quantitative results using a 10-face polyhedron as a test object.*

6   Discussion

We have proposed and implemented various algorithms for fixturing and grasping three-dimensional polyhedra. We are in the process of constructing a prototype of the automatically reconfigurable gripper whose conceptual design is shown in Figure 3.

There are obviously polyhedral objects which cannot be fixtured with our device (a trivial example is an object whose diameter is smaller than the inter-locator distance). It would be interesting to characterize precisely the class of fixturable objects (see [59] for a discussion of the two-dimensional case).

Another interesting avenue of research would be to extend the proposed algorithm to parts bounded by algebraic patches (see [58, Chapter 6] for the two-dimensional case). Our overall approach extends to this case in a straightforward way, but working out the details of how to enumerate locator configurations and dealing with the very high degree of the equations in-

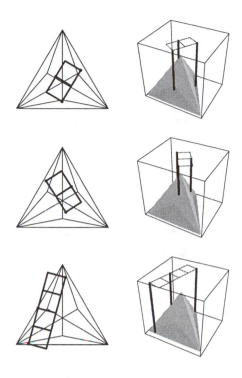

Figure 15: *Some solutions for a 6 × 6 grid.*

volved should prove quite challenging.

Finally, we plan to investigate in-hand manipulation using our reconfigurable gripper.

Acknowledgments

This research was supported in part by an equipment grant from the Beckman Institute for Advanced Science and Technology and by a Critical Research Initiative planning grant from the University of Illinois at Urbana-Champaign. Part of this work was done while the author was visiting the University of Southern California. Many thanks to G. Medioni and R. Nevatia for their hospitality, and to J. Burdick, K. Goldberg and his students, C. Hayes, S. Hutchinson, D. Kriegman, E. Rimon, S. Sullivan, and A. Wallack for useful discussions and comments.

References

[1] H. Asada and A. By. Kinematics of workpart fixturing. In *Proc. IEEE Int. Conf. on Robotics and Automation*, pages 337–345, St Louis, MI, 1985.

[2] R.S. Ball. *A treatise on the theory of screws.* Cambridge University Press, 1900.

[3] J. Bausch and K. Youcef-Toumi. Kinematics methods for automated fixture reconfiguration planning. In *Proc. IEEE Int. Conf. on Robotics and Automation*, pages 1396–1401, Cincinatti, OH, 1990.

[4] A. Blake. Computational modelling of hand-eye coordination. *Phil. Trans. R. Soc. Lond. B*, 337:351–360, 1992.

[5] R.C. Brost and K. Goldberg. A complete algorithm for synthesizing modular fixtures for polygonal parts. In *IEEE Int. Conf. on Robotics and Automation*, pages 535–542, San Diego, CA, May 1994.

[6] J.F. Canny and K.Y. Goldberg. "RISC" for industrial robotics: recent results and open problems. In *IEEE Int. Conf. on Robotics and Automation*, pages 1951–1958, San Diego, CA, 1994.

[7] I.M. Chen and J.W. Burdick. Finding antipodal point grasps on irregularly shaped objects. In *IEEE Int. Conf. on Robotics and Automation*, pages 2278–2283, Nice, France, June 1992.

[8] Y.C. Chou, V. Chandru, and M. Barash. A mathematical approach to automatic configuration of machining fixtures: analysis and synthesis. *ASME Journal of Engineering for Industry*, 111:299–306, 1989.

[9] J. Czyzowicz, I. Stojmenovic, and J. Urrutia. Immobilizing a polytope. volume 519 of *Lecture Notes in Computer Sciences*, pages 214–227. Springer-Verlag, 1991.

[10] M. Daimon and T. Yoshida. Study for designing fixtures considering dynamics of thin-walled-plate and box-like workpieces. *Annals of the CIRP*, 34(1):319–322, 1985.

[11] J. Donoghue, W.S. Howard, and V. Kumar. Stable workpiece fixturing. In 23rd *biennial ASME mechanics conference*, Minneapolis, Sept. 1994.

[12] P.J. Englert. *Principles for Part Setup and Workholding in Automated Manufacturing*. PhD thesis, Dept. of Mechanical Engineering and Robotics Institute, Carnegie-Mellon University, Pittsburgh, PA, 1987.

[13] C. Ferrari and J.F. Canny. Planning optimal grasps. In *IEEE Int. Conf. on Robotics and Automation*, pages 2290–2295, Nice, France, June 1992.

[14] P.M. Ferreira and C.R. Liu. Generation of workpiece orientation for machining using a rule-based system. *Journal of Robotics and Computer-Integrated Manufacturing*, 1988.

[15] J.B.J. Fourier. Reported in: Analyse des travaux de l'académie royale des sciences pendant l'année 1824. In *Partie mathématique, Histoire de l'Académie Royale des Sciences de l'Institut de France*, volume 7, xlvii-lv. 1827. Partial English translation in: D.A. Kohler, Translation of a Report by Fourier on his work on Linear Inequalities, *Opsearch* 10 (1973) 38-42.

[16] M.V. Gandhi and B.S. Thompson. Automated design of modular fixtures for flexible manufacturing systems. *Journal of Manufacturing Systems*, 5(4):243–252, 1986.

[17] K.Y. Goldberg. A kinematically-yielding gripper. In 22nd *Int. Symp. on Industrial Automation*, 1992. See U.S. Patent 5,098,145.

[18] C.C. Hayes and P.K. Wright. Automatic process planning: using feature interactions to guide search. *Journal of Manufacturing Systems*, 8(1):1–16, 1989.

[19] F.B. Hazen and P. Wright. Workholding automation: innovations in analysis. *Manufacturing Reviews*, 3(4):224–237, 1990.

[20] T. Hishi. Research for high-productivity machining in japan. In *Int. Conf. on High Productivity Machining, Materials and Processing*, pages 15–26, 1985.

[21] W.S. Howard and V. Kumar. Stability of planar grasps. In *Proc. IEEE Int. Conf. on Robotics and Automation*, pages 2822–2827, San Diego, CA, May 1994.

[22] K.H. Hunt. *Kinematic Geometry of Mechanisms*. Clarendon, Oxford, 1978.

[23] Z. Ji. *Dexterous hands: optimizing grasp by design and planning*. PhD thesis, Stanford University, Dept. of Mechanical Engineering, 1987.

[24] K. Lakshminarayana. Mechanics of form closure. Technical Report 78-DET-32, ASME, 1978.

[25] C. Lassez and J-L. Lassez. Quantifier elimination for conjunctions of linear constraints via a convex hull algorithm. In B. Donald, D. Kapur, and J. Mundy, editors, *Symbolic and Numerical Computation for Artificial Intelligence*, pages 103–122. Academic Press, 1992.

[26] J-L. Lassez. Querying constraints. In *ACM conference on Principles of Database Systems*, Nashville, 1990.

[27] S.H. lee and M.R. Cutkosky. Fixture planning with friction. *Trans. of the ASME*, 112:320–327, August 1991.

[28] D. Manocha. *Algebraic and Numeric Techniques for Modeling and Robotics*. PhD thesis, Computer Science Division, Univ. of California at Berkeley, 1992.

[29] D. Manosha. Computing selected solutions of polynomial equations. In *Int. Symp. on Symbolic and Algebraic Computation*, pages 1–8, Oxford, England, 1994.

[30] X. Markenscoff, L. Ni, and C.H. Papadimitriou. The geometry of grasping. *International Journal of Robotics Research*, 9(1):61–74, February 1990.

[31] X. Markenscoff and C.H. Papadimitriou. Optimum grip of a polygon. *International Journal of Robotics Research*, 8(2):17–29, April 1989.

[32] A. Markus, Z. Markusz, J. Farka, and J. Filemon. Fixture design using prolog: an expert system. *Robotics and Computer Integrated Manufacturing*, 1(2):167–172, 1984.

[33] B. Mirtich and J.F. Canny. Optimum force-closure grasps. Technical Report ESRC 93-11/RAMP 93-5, Robotics, Automation, and Manufacturing Program, University of California at Berkeley, July 1993.

[34] B. Mishra. Worksholding – analysis and planning. In *IEEE/RSJ Int. Workshop on Intelligent Robots and Systems*, pages 53–56, Osaka, Japan, 1991.

[35] B. Mishra, J.T. Schwartz, and M. Sharir. On the existence and synthesis of multifinger positive grips. *Algorithmica, Special Issue: Robotics*, 2(4):541–558, November 1987.

[36] B. Mishra and N. Silver. Some discussion of static gripping and its stability. *IEEE Systems, Man, and Cybernetics*, 19(4):783–796, 1989.

[37] A.P. Morgan. *Solving Polynomial Systems using Continuation for Engineering and Scientific Problems*. Prentice Hall, Englewood Cliffs, NJ, 1987.

[38] R.M. Murray, Z. Li, and S.S. Sastry. *A mathematical introduction to robotic manipulation*. CRC Press, 1994.

[39] V-D. Nguyen. Constructing force-closure grasps. *International Journal of Robotics Research*, 7(3):3–16, June 1988.

[40] V-D. Nguyen. Constructing stable grasps. *International Journal of Robotics Research*, 8(1):27–37, February 1989.

[41] M.S. Ohwovoriole. An extension of screw theory. *Journal of Mechanical Design*, 103:725–735, 1981.

[42] N.S. Pollard and T. Lozano-Pérez. Grasp stability and feasibility for an arm with an articulated hand. In *IEEE Int. Conf. on Robotics and Automation*, pages 1581–1585, Cincinatti, OH, 1990.

[43] J. Ponce. On planning immobilizing fixtures for three-dimensional polyhedral objects. In *IEEE Int. Conf. on Robotics and Automation*, pages 509–514, Minneapolis, MN, 1996.

[44] J. Ponce and B. Faverjon. On computing three-finger force-closure grasps of polygonal objects. *IEEE Transactions on Robotics and Automation*, 11(6), December 1995. In press.

[45] J. Ponce, D. Stam, and B. Faverjon. On computing force-closure grasps of curved two-dimensional objects. *International Journal of Robotics Research*, 12(3):263–273, June 1993.

[46] J. Ponce, S. Sullivan, A. Sudsang, J-D. Boissonnat, and J-P. Merlet. On computing four-finger equilibrium and force-closure grasps of polyhedral objects. *International Journal of Robotics Research*, 1995. In press. Also Beckman Institute Tech. Report UIUC-BI-AI-RCV-95-04, University of Illinois at Urbana-Champaign.

[47] F. Reulaux. *The kinematics of machinery*. MacMillan, NY, 1876. Reprint, Dover, NY, 1963.

[48] E. Rimon and J. W. Burdick. Towards planning with force constraints: On the mobility of bodies in contact. In *Proc. IEEE Int. Conf. on Robotics and Automation*, pages 994–1000, Atlanta, GA, May 1993.

[49] E. Rimon and J. W. Burdick. Mobility of bodies in contact-I: A new 2nd order mobility index for multi-finger grasps. *IEEE Transactions on Robotics and Automation*, 1995. Submitted. A preliminary version appeared in *IEEE Int. Conf. on Robotics and Automation*, pages 2329-2335, San Diego, CA, 1994.

[50] E. Rimon and J. W. Burdick. Mobility of bodies in contact-II: How forces are generated by curvature effects? *IEEE Transactions on Robotics and Automation*, 1995. Submitted. A preliminary version

appeared in *IEEE Int. Conf. on Robotics and Automation*, pages 2336-2341, San Diego, CA, 1994.

[51] B. Roth. Screws, motors, and wrenches that cannot be bought in a hardware store. In *Int. Symp. on Robotics Research*, pages 679–693. MIT Press, 1984.

[52] J.K. Salisbury. *Kinematic and force analysis of articulated hands*. PhD thesis, Stanford University, Stanford, CA, 1982.

[53] B. Shirinizadeh. Issues in the design of reconfigurable fixture modules for robotics. *Journal of Manufacturing Systems*, 12(1):1–14, 1993.

[54] P. Somov. Über Gebiete von Schraubengeschwindigkeiten eines starren Korpers bie verschiedener Zahl von Stutzflachen. *Zeitricht für Mathematik und Physik*, 45:245–306, 1900.

[55] J.C. Trinkle. On the stability and instantaneous velocity of grasped frictionless objects. *IEEE Transactions on Robotics and Automation*, 8(5):560–572, October 1992.

[56] R. Wagner, Y. Zhuang, and K. Goldberg. Fixturing faceted parts with seven modular struts. In *IEEE Int. Symp. on Assembly and Task Planning*, pages 133–139, Pittsburgh, PA, August 1995.

[57] A. Wallack and J.F. Canny. Planning for modular and hybrid fixtures. In *IEEE Int. Conf. on Robotics and Automation*, pages 520–527, San Diego, CA, 1994.

[58] A.S. Wallack. *Algorithms and Techniques for Manufacturing*. PhD thesis, Computer Science Division, Univ. of California at Berkeley, 1995.

[59] Y. Zhuang, K. Goldberg, and Y.C. Wong. On the existence of modular fixtures. In *IEEE Int. Conf. on Robotics and Automation*, pages 543–549, San Diego, CA, May 1994.

On Motion Planning of Polyhedra in Contact

Hirohisa Hirukawa, *Electrotechnical Laboratory, Tsukuba, Japan*

This paper studies motion planning of polyhedra in contact, which has a crucial role to automate mechanical assembly processes. We have been attacking this problem based on an algebraic formulation. We first revisit our complete and implemented algorithm for motion planning of convex polyhedra in contact to see that the complexity of the algebraic part heavily depends how the geometric problem is reduced to it. Then we present a predicate for nonconvex polyhedra in contact without overlapping, give algorithms to be applied to the nonconvex case, and investigate their geometric and algebraic complexity.

1   Introduction

Motion planning of objects in contact has a crucial role to automate mechanical assembly processes. It is also interesting from a theoretical viewpoint, since exact motion planning approaches seems to be more appropriate when the clearance between the objects is tight. This problem is closely related to that of finding the boundary of configuration space obstacles (C-obstacles in short), which is the image of fixed objects in the configuration space of a moving object, because a path on this boundary corresponds to the motion of the object in contact with the fixed ones.

Avnaim, Boissonnat and Faverjon proposed an algorithm to find the boundary of C-obstacles when a polygon moves within polygonal obstacles in 2-space[2], together with an algorithm for motion planning of polygons in contact [3]. The combinatorial complexity of the motion planning algorithm is $O(m^3 n^3 \log mn)$, where m and n is the complexity of the polygons in contact respectively. This is almost tight, because the lower bound of the complexity of the C-obstacle is known to be $\Omega(m^3 n^3)$[30]. Brost independently developed an algorithm for the same setting [4]. When a single connected component in the configuration space

is considered, its complexity is $O((mn)^{2+\epsilon})$, for any $\epsilon > 0$ [16], and motion of polygons in contact can be planned in the same running time.

When a polyhedron translates amidst polyhedral obstacles in 3-space, the complexity of the entire free configuration space is $\Theta(m^3 n^3)$ [30], where m and n are the complexity of the moving and fixed polyhedra respectively. If both the moving and fixed polyhedra are convex, the boundary of a C-obstacle can be found in $O(m+n+K)$, where K is the size of the C-obstacle and can be $O(mn)$ in the worst case[14]. Aronov and Sharir present an algorithm to plan motions in a single cell, and it runs in randomized expected time $O((mn)^{2+\epsilon})$, for any $\epsilon > 0$[1]. de Berg, Guibas and Halperin show an algorithm to find vertical decompositions for Triangles in 3-space, to which this motion planning problem can be reduced. It is a deterministic output-sensitive algorithm which runs in $O((mn)^2 \log mn + K \log mn)$, where K is the complexity of the decomposition.

In the case that a polyhedron translates and rotates in contact with another, the dimension of the configuration space becomes six, and it is very easy to imagine that the problem becomes much more difficult. In fact, the lower bound of the complexity of a C-obstacle in this case is $\Omega(m^6 n^6)$[15], and so the complexity is $\Theta(m^6 n^6)$. When both polyhedra are convex, the authors present an algorithm which runs in $O(mn)$[18]. Besides the combinatorial complexity, the bit complexity of the algorithm is not so poor and can be computed in a practical time, since the algorithm needs to solve quadratic equations in three variables rather than cubic equations in six variables. This becomes possible thanks to some astute geometric formulations. In the general case, we must face the awful combinatorial complexity as well as the unrealistic bit complexity.

Several related algorithms has been presented that are valid for high dimensional cases. Schwartz and Sharir present a general motion planning algorithm [29] based on the Collins decomposition[10], though both geometric and algebraic parts of the algorithm have double exponential complexity. Chazelle, Edelsbrunner, Guibas and Sharir give an algorithm for stratifying a real semi-algebraic set rather than producing a cell complex, and have improved the complexity of the geometric part to single exponential [8, 9]. Canny presents a general motion planning algorithm based on finding one-dimensional skeleton or "roadmap" of the stratified set in the configuration space, and both geometric and algebraic complexity of the algorithm is single exponential[5, 6, 7].

On the other hand, there is good news at least from the algebraic side. Manocha and Canny present multipolynomial resultant algorithms to solve algebraic equations[24], which run in the single exponential time in the number of the variables and so surpass the algorithms based on Groebner bases at least from the asymptotic complexity. For a sparse polynomial system which often arises in many applications, Emiris and Canny show a more practical method[13]. Manocha present an efficient algorithm to compute selected solutions of zero-dimensional equations in a real domain[25]. Most recently, Krishnan and Manocha give a numeric-symbolic algorithm for finding one-dimensional algebraic sets from polynomial equations [21]. These results have improved the running time for solving algebraic equations drastically, especially when we want to find zero or one dimensional sets of solutions. So our problem seems to be joining into a practical league at least from the algebraic aspect, if we employ some motion planning algorithms of the roadmap type [23], which need to find only points or curves from hypersurfaces.

This paper is organized as follows. In section 2, we revisit our algorithm for motion planning of convex polyhedra in contact to show that an astute geometric formulation induces an easier algebraic mission. In section 3, we study the planning of nonconvex polyhedra in contact. We present a predicate for nonconvex polyhedra in contact without overlapping, and investigate algorithms to be applied with their geometric and algebraic complexity. We conclude in section 4.

2 Convex Polyhedra in Contact

When a convex polyhedra moves in contact with a fixed convex one, each face on the boundary of a C-obstacle is connected, and the topology between the faces can be found only by the topology of these polyhedra. Besides, it can be determined by a local criteria, and so the complexity of the algorithm becomes $\Theta(mn)$, where m and n are the complexity of the polyhedra respectively. See [18] for details.

In this section, we describe how to plan a motion on a C-Obstacle boundary face. Though the motion planning is done in the working space essentially, but we formulate the problem in algebraic equations and find the curves in the configuration space by solving the equations, in order to examine the performance of algebraic tools and see how it depends on the geometric formulation.

At first, we define the configuration space and parameterize it, and describe a part of our algorithm for the above purpose. See also [18] for details.

2.1 Configuration space

We use the special unitary 2×2 matrix for parameterizing $SO(3)$, the Special Orthogonal group in 3-space, to have the algebraic representation of the constraints. Then a vector $\vec{X} = (x, y, z)$ is represented by an Hermitian 2×2 matrix,

$$X = xS_1 + yS_2 + zS_3. \tag{1}$$

Here S_j is the Pauli's spin matrix,

$$S_1 = \begin{pmatrix} 0 & 1 \\ 1 & 0 \end{pmatrix}, S_2 = \begin{pmatrix} 0 & -i \\ i & 0 \end{pmatrix}, S_3 = \begin{pmatrix} 1 & 0 \\ 0 & -1 \end{pmatrix}, \tag{2}$$

where $i^2 = -1$. In the following, we use a symbol such as X not only for representing the matrix but also for referring the corresponding geometric object. Let I be the unit 2×2 matrix and $e_0 = \cos(\frac{\phi}{2}), e_1 = u_x \sin(\frac{\phi}{2}), e_2 = u_y \sin(\frac{\phi}{2}), e_3 = u_z \sin(\frac{\phi}{2})$, then the rotation of a vector $\vec{X}$ along a unit vector (u_x, u_y, u_z) with an angle ϕ is given by

$$X' = QXQ^\dagger, \tag{3}$$

where

$$Q = e_0 I - i(e_1 S_1 + e_2 S_2 + e_3 S_3) \tag{4}$$

$$Q^\dagger = e_0 I + i(e_1 S_1 + e_2 S_2 + e_3 S_3). \tag{5}$$

Note that e_j is a real scalar such that $e_0^2 + e_1^2 + e_2^2 + e_3^2 = 1$, then we can forget that e_j comes from the trigonometric function and have an algebraic parameterization of $SO(3)$. We can also parameterize $SO(3)$ by three parameters respectively in the four cubes C_i, $i = 1, \cdots, 4$, where C_i is defined by the space of (e_0, e_1, e_2, e_3) in which e_i has the largest magnitude. That is, the rotation can be represented in C_i by

$$\vec{\Theta} \stackrel{def}{=} (\theta_j)_{j=0,\cdots,3;j\neq i}, \qquad \theta_j \stackrel{def}{=} \frac{e_j}{e_i} \qquad (6)$$

For an example, $\vec{\Theta} = (\frac{e_1}{e_0}, \frac{e_2}{e_0}, \frac{e_3}{e_0})$ in C_0.

Let the moving object be $\mathcal{A}$ and the fixed one $\mathcal{B}$. In this paper, we use the above parameterization for rotation and R^3 for translation to define the configuration space of a moving polyhedron, and represent the configuration of $\mathcal{A}$ by the translation after the rotation, which are specified by a Hermitian 2×2 matrix X and a special unitary 2×2 matrix Q respectively.

Let $V_{A_i}, i = 1, \cdots, m_V$ be an Hermitian 2×2 matrix representing the positions of the vertices of $\mathcal{A}$ when $\mathcal{A}$ is at a configuration $(X, Q) = (0, I)$, where m_V is the number of the vertices. Let $E_{A_j}, j = 1, \cdots, m_E$, be an Hermitian 2×2 matrix representing the directions of the edges, where the direction of an edge is defined to point from its starting vertex to its ending vertex on each face. Though each edge has two directions according to the corresponding faces, but its direction becomes unique when its starting vertex is specified. With a slight abuse of notation, we use E_{A_j} in this sense. Let $f_{A_k}, k = 1, \cdots, m_f$ the distances with signs between a point P and the face whose normal vector is N_{A_k} and the distance from the origin is $-d_{A_k}$. Then f_{A_k} is given by

$$f_{A_k} = \frac{1}{2}tr((-d_{A_k}I + N_{A_k})(I + P)), \qquad (7)$$

where $tr()$ is the trace of the matrix. The distance becomes positive when P is outside the object. Let $V_{B_p}, p = 1, \cdots, n_V$, $E_{B_q}, q = 1, \cdots, n_E$ and $f_{B_r} = \frac{1}{2}tr((-d_{B_r}I + N_{B_r})(I + P)), r = 1, \cdots, n_f$ be the notations for $\mathcal{B}$.

Let V'_{A_i} be the transformed matrices of V_{A_i} and f'_{A_k} that of f_{A_k} when $\mathcal{A}$ is at a configuration (X, Q). They can be given by

$$V'_{A_i} = QV_{A_i}Q^\dagger + X \qquad (8)$$

$$f'_{A_k} = \frac{1}{2}tr((-(d_{A_k} + \frac{1}{2}tr(QN_{A_k}Q^\dagger X))I + QN_{A_k}Q^\dagger) (I + P)). \qquad (9)$$

2.2 The Algorithm for motion planning

Let us consider the example whose initial configuration is shown in Fig.1(a) and final one in (b). The essen-

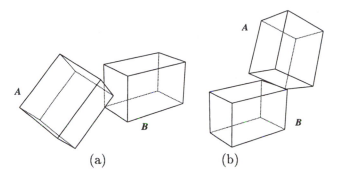

(a)   (b)

Figure 1: *Initial and final configurations*

tial idea of the algorithm is to plan robust motions for errors which may occur in the real world. For an example, motions are planned so that the middle point of an edge of the moving polyhedron is contact perpendicularly with that of the fixed one when the moving edge rotates around the fixed edge. See [18] for details.

Let us concentrate our attention to the planning of a motion which make edge $V_{A_1}V_{A_2}$ to be in contact with face F_{B_1} while keeping the position of V_{A_1}. See Fig.2.

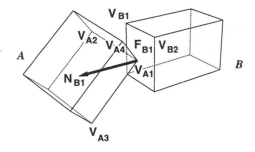

Figure 2: *Point contact between A's vertex and B's face*

This motion can be specified by

$$QV_{A_1}Q^\dagger + X = V_{A_1}^s, \qquad (10)$$

$$\frac{1}{2}tr(Q(V_{A_4} - V_{A_3})Q^\dagger N_{B_1}) = 0, \qquad (11)$$

$$\frac{1}{2i}tr(N_{B_1}(V_{A_2}^s - V_{A_1}^s)Q(V_{A_2} - V_{A_1})Q^\dagger) = 0. \qquad (12)$$

Eq.(10) keeps the position of V_{A_1} to be the initial position $V_{A_1}^s$, Eq.(11) does the direction of diagonal $V_{A_3}V_{A_4}$

to be parallel to F_{B_1}, and Eq.(12) determines the axis of rotation. The motion terminates when

$$\frac{1}{2}tr(Q(V_{A_2} - V_{A_1})Q^\dagger N_{B_1}) = 0, \qquad (13)$$

which means that edge $V_{A_1} V_{A_2}$ becomes parallel to F_{B_1}. These equations can be solved by the following algorithm.

Algorithm 2.1

1. *Solve Eqs.(11) and (12).*

2. *Substitute the solutions to Eq.(10).*

3. *Substitute the solutions to Eq.(13).*

A parametric representation of the curve in the configuration space is obtained by Step 1 and 2, and their end-points are determined by Step 3. That is, we can plan a motion only by solving quadratic equations in three variables. The solution of Eqs.(11) and (12) in cube C_0 is shown in Table 1. The planned motion is shown in Fig.3.

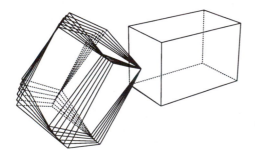

Figure 3: *Point contact to the edge contact*

Here is an important point to address. If we consider the edge contact as the conjunction of two point contacts, the resulting algebraic equations look like

$$\frac{1}{2}tr(-d_{B_1}QQ^\dagger + N_{B_1}(QV_{A_1}Q^\dagger + QQ^\dagger X)) = 0, \quad (14)$$

$$\frac{1}{2}tr(-d_{B_1}QQ^\dagger + N_{B_1}(QV_{A_2}Q^\dagger + QQ^\dagger X)) = 0, \quad (15)$$

instead of Eqs.(10) and (13), which consider the edge contact as the conjunction of one point contact and the orientation of the edge being parallel to the face. Then we must solve the equations with six variables, which should be much more difficult than our equations. Though Eqs.(14) and (15) have the linear forms

in X as pointed out in [12], but they can not be solved only in variables x, y, z. Because the coefficients of these variables come from the same N_{B_1}, so Eqs.(14) and (15) are linearly dependent with respect to x, y, z. The planned overall motion is illustrated in Fig.4. Our

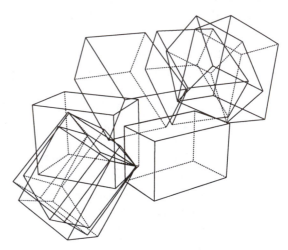

Figure 4: *Planned overall motion*

algorithm is complete, that is, it is guaranteed to find a solution.

2.3   Implementation

The geometrical part of the proposed algorithm is implemented in *Euslisp*, which is the subset of the CommonLisp with object oriented primitives[26], and its algebraic part is in *Mathematica*[31].

2.4   Complexity of the algebraic algorithms

The asymptotic time complexity of the algorithm to solve the algebraic equations increases single exponentially according to the number of variables when we use the multivariate resultant and doubly exponentially when we use the Groebner basis[22]. We have implemented the algebraic part of the proposed algorithm in *Mathematica*, which uses the Groebner basis computation. In spite of such difficulties, the proposed algorithm can be computed in the practical time generally, since our algorithm need to solve the quadratic equations only in three variables. In fact, we can compute the above example within several minutes using a Sparc Station 10/40. But it is unrealistic to apply such algebraic tools to solve qubic equations in six variables like

$$\theta_1 = \frac{778000. - 4.11\,10^6\,\theta_3 + 2.29\,10^6\,\theta_3{}^2 - 46000.\,\sqrt{841. - 7830.\,\theta_3 + 18200.\,\theta_3{}^2} + 9890.\,\theta_3\,\sqrt{841. - 7830.\,\theta_3 + 18200.\,\theta_3{}^2}}{-1.01\,10^6 + 4.69\,10^6\,\theta_3}$$

$$\theta_2 = 7.19\,10^{-6}\left(67900. - 107000.\,\theta_3 - 1.\,\sqrt{(-67900. + 107000.\,\theta_3)^2 - 278000.\left(10900. - 80500.\,\theta_3{}^2\right)}\right)$$

Table 1: *Solution in cube C_0*

Eqs.(14) and (15). We have tried to solve such equations by *Mathematica*, but it had run over a night and crashed due to out of memory. So we must be careful how to reduce geometric problems into algebraic ones.

It is usually assumed in the combinatorial complexity analysis that the algebraic computation can be done in $O(1)$ time, but, as we have seen, the complexity of the algebraic algorithms can vary drastically even for equivalent geometric problems. Then we need to design algorithms while taking both geometric and algebraic sides into account. We may call such a field *computational algebraic geometry*.

3 Nonconvex Polyhedra in Contact

3.1 Predicates for C-obstacles Boundary

3.1.1 C-obstacle Boundary

Two (possibly noncovex) polyhedra $\mathcal{A}$ and $\mathcal{B}$ overlap if and only if either a vertex of $\mathcal{A}$ is inside $\mathcal{B}$, or a vertex of $\mathcal{B}$ is inside $\mathcal{A}$, or an edge of $\mathcal{A}$ pierces a face of $\mathcal{B}$, or an edge of $\mathcal{B}$ pierces a face of $\mathcal{A}$[5]. The first and second cases correspond to the case that one polyhedron is fully included in another, and it is more troublesome to handle these cases when $\mathcal{A}$ or $\mathcal{B}$ is not convex. Here we have more interest in the C-obstacle's boundary rather than its inside. So we give a predicate representation of the C-obstacle boundary directly rather than that of the C-obstacle itself, to do without the inclusion cases.

Let us define a few terminologies first. We say that a vertex is *adherent* to an edge if the vertex is one of end points of the edge, and that an edge is *adherent* to a face if the edge is a boundary edge of the face. Note that the adherence is a transitive relation. When we consider an edge is directed in a face, we say the edge is directed from the starting vertex to the ending vertex. The *coboundary* of a vertex V is the set of all edges to which V is adherent. A vertex of a polyhedron is defined to be *convex* if the inside of the polyhedron

in the neighborhood of the vertex is convex, and *reflex* otherwise. A convex and reflex edges are defined in the same way. Based on some of these terminologies, a *primary point contact* is defined as follows.

Definition 1 *A primary point contact between two polyhedra $\mathcal{A}$ and $\mathcal{B}$ is a point contact between a face of $\mathcal{A}$ and a convex vertex of $\mathcal{B}$, a convex vertex of $\mathcal{A}$ and a face of $\mathcal{B}$, or a convex edge of $\mathcal{A}$ and a convex edge of $\mathcal{B}$ without overlapping of the insides of $\mathcal{A}$ and $\mathcal{B}$ in the neighborhood of the contact point.*

More formally, the non-overlapping of the insides of $\mathcal{A}$ and $\mathcal{B}$ in the neighborhood of a contact point X can be written by

$$\exists \epsilon > 0 \; s.t. \; int(\mathcal{A}) \cap int(\mathcal{B}) \cap \delta(X;\epsilon) = \emptyset \qquad (16)$$

where $\delta(X;\epsilon)$ is the open ball with the radius ϵ centered at X. Notice that the definition of a primary point contact is given only by local conditions and says nothing if some part of the boundary of $\mathcal{A}$ and that of $\mathcal{B}$ may intersect or not.

Then we claim the following theorem.

Theorem 1 *Let $\mathcal{A}$ and $\mathcal{B}$ be two regular polyhedra. $\mathcal{A}$ and $\mathcal{B}$ are in contact but do not overlap if and only if a primary point contact holds between $\mathcal{A}$ and $\mathcal{B}$, no edge of $\mathcal{A}$ pierces a face of $\mathcal{B}$, and no edge of $\mathcal{B}$ pierces a face of $\mathcal{A}$.*

Proof: Assume that $\mathcal{A}$ and $\mathcal{B}$ are in contact but do not overlap, then it is trivial that at least one primary point contact holds between $\mathcal{A}$ and $\mathcal{B}$, and that no edge of $\mathcal{A}$ pierces a face of $\mathcal{B}$ and no edge of $\mathcal{B}$ pierces a face of $\mathcal{A}$. Note, for an example, that a point contact between a convex vertex and a reflex vertex can be considered as a conjunction of primary point contacts between the convex vertex and the faces to which the reflex vertex is adherent.

Conversely, if no edge of $\mathcal{A}$ pierces a face of $\mathcal{B}$, and no edge of $\mathcal{B}$ pierces a face of $\mathcal{A}$, then only case for the overlapping of $\mathcal{A}$ and $\mathcal{B}$ is that one polyhedron is fully included by another polyhedron [5]. But this does not occur, since at least one primary point contact holds between $\mathcal{A}$ and $\mathcal{B}$ and so the insides of $\mathcal{A}$ and $\mathcal{B}$ in the neighborhood of the contact point do not overlap. It is clear that $\mathcal{A}$ and $\mathcal{B}$ are in contact. This completes the proof.   □

This theorem implies that we can tell when two polyhedra are in contact but do not overlap by the spacial relationships between pairs of the features of the polyhedra, which will lead to a predicate representation of the C-obstacle.

3.1.2 Parameterization of point contacts

Before exploiting the predicate, let us parameterize three predicates for point contacts between the affine hull of an $\mathcal{A}$'s face and a $\mathcal{B}$'s vertex, between an $\mathcal{A}$'s vertex and the affine hull of a $\mathcal{B}$'s face, and between supporting lines of $\mathcal{A}$ and $\mathcal{B}$'s edges. We denote each of them by $CS_{k,p}^{\alpha} = 0$, $CS_{i,r}^{\beta} = 0$ and $CS_{j,q}^{\gamma} = 0$ respectively, and parameterize $CS_{k,p}^{\alpha}$, $CS_{i,r}^{\beta}$ and $CS_{j,q}^{\gamma}$ by (X, Q) which has been defined in section 2. From equation (9), we can derive

$$CS_{k,p}^{\alpha} = \frac{1}{2}tr(-d_{A_k}QQ^{\dagger} + QN_{A_k}Q^{\dagger}(V_{B_p} - X)). \quad (17)$$

See Fig.5. In the same way, we can obtain

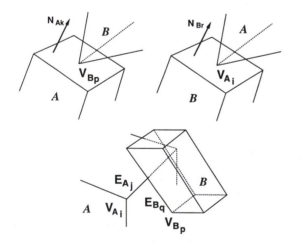

Figure 5: *Point contacts*

$$CS_{i,r}^{\beta} = \frac{1}{2}tr(-d_{B_r}QQ^{\dagger} + N_{B_r}(QV_{A_i}Q^{\dagger} + QQ^{\dagger}X)), \quad (18)$$

$$CS_{j,q}^{\gamma} = \frac{1}{2i}tr((V_{B_p} - X)QE_{A_j}Q^{\dagger}E_{B_q})$$
$$-QV_{A_i}E_{A_j}Q^{\dagger}E_{B_q})), \quad (19)$$

where V_{A_i} and V_{B_p} are the starting vertices of the E_{A_j} and E_{B_m} respectively.

3.1.3 Predicates Representation

The fact that $\mathcal{A}$ and/or $\mathcal{B}$ may be nonconvex implies that some faces of $\mathcal{A}$ and/or $\mathcal{B}$ may be noncovex. In the following, we assume that all nonconvex faces are decomposed into convex polygons without introducing Steiner points, and that f_{A_k}, E_{A_j}, f_{B_r} and E_{B_q} are renumbered according to the decomposition. Besides, m_f, m_E, n_f and n_E should be increased properly.

From Theorem 1, a predicate representation $\mathcal{CB}_{\mathcal{A},\mathcal{B}}$ of the C-obstacle boundary is given by

$$\mathcal{CB}_{\mathcal{A},\mathcal{B}} =$$
$$\left((\bigvee_{k=1}^{m_f} \bigvee_{p=1}^{n_v} \mathcal{CF}_{k,p}^{\alpha}) \bigvee (\bigvee_{i=1}^{m_v} \bigvee_{r=1}^{n_f} \mathcal{CF}_{i,r}^{\beta}) \bigvee (\bigvee_{j=1}^{m_E} \bigvee_{q=1}^{n_E} \mathcal{CF}_{j,q}^{\gamma}) \right)$$
$$\bigwedge \left(\bigwedge_{k=1}^{m_f} \bigwedge_{q=1}^{n_E} \mathcal{NO}_{k,q}^{\alpha} \right) \bigwedge \left(\bigwedge_{j=1}^{m_E} \bigwedge_{r=1}^{n_f} \mathcal{NO}_{j,r}^{\beta} \right),$$
$$(20)$$

where $\mathcal{CF}_{k,p}^{\alpha}$ is the predicate for a primary point contact between f_{A_k} and V_{B_p}, $\mathcal{CF}_{i,r}^{\beta}$ for that between V_{A_i} and f_{B_r}, $\mathcal{CF}_{j,q}^{\gamma}$ for that between E_{A_j} and E_{B_q}, $\mathcal{NO}_{k,q}^{\alpha}$ is the predicate for non-overlapping of f_{A_k} and E_{B_q}, and $\mathcal{NO}_{j,r}^{\beta}$ for that of E_{A_j} and f_{B_r}. Note that we must not include $\mathcal{CF}_{k,p}^{\alpha}$ in the disjunctions if V_{B_p} is a reflex vertex, $\mathcal{CF}_{i,r}^{\beta}$ if V_{A_i} is a reflex vertex, and $\mathcal{CF}_{j,q}^{\gamma}$ if E_{A_j} or E_{B_q} is a reflex edge. Note also that $\mathcal{CF}_{j,q}^{\gamma}$ need not be included in the disjunction if E_{A_j} or E_{B_q} was introduced by the decomposition of a concave face, since it is redundant.

These predicates can be broken down further as follows.

$$\mathcal{CF}_{k,p}^{\alpha} = (CS_{k,p}^{\alpha} = 0) \bigwedge (\bigwedge_{\xi} CS_{\xi,\lambda}^{\gamma} \leq 0) \bigwedge (\bigwedge_{\zeta} CS_{k,\zeta}^{\alpha} \geq 0),$$
$$(21)$$

where ξ is taken so that E_{A_ξ} should be every edge of f_{A_k}, E_{B_λ} is an arbitrary edge whose starting vertex is V_{B_p}, and ζ is taken so that V_{B_ζ} should be every ending vertex of the edge from the starting vertex V_{B_p}.

$CS^\alpha_{k,p} = 0$ and $\bigwedge_\xi CS^\gamma_{\xi,\lambda} \leq 0$ mean that V_{B_q} is on f_{A_k}, and $\bigwedge_\zeta CS^\alpha_{k,\zeta} \geq 0$ says that the coboundary of V_{B_q} is outside of f_{A_k}. In the same way,

$$CF^\beta_{i,r} = (CS^\beta_{i,r} = 0) \bigwedge (\bigwedge_\phi CS^\gamma_{\phi,\mu} \leq 0) \bigwedge (\bigwedge_\psi CS^\beta_{\psi,r} \geq 0), \tag{22}$$

where ϕ is taken so that E_{B_ϕ} should be every edge of f_{B_r}, E_{B_μ} is an arbitrary edge whose starting vertex is V_{A_i}, and ψ is taken so that V_{A_ψ} should be every ending vertex of the edge from the starting vertex V_{A_i}.

$$CF^\gamma_{j,q} = (CS^\gamma_{j,q} = 0) \bigwedge \left(CF^{\gamma_{out}}_{j,q} \bigvee CF^{\gamma_{in}}_{j,q}\right), \tag{23}$$

where

$$CF^{\gamma_{out}}_{j,q} = (CS^\alpha_{j_R,q_S} \geq 0) \bigwedge (CS^\alpha_{j_L,q_E} \geq 0) \bigwedge$$
$$(CS^\beta_{j_S,q_R} \geq 0) \bigwedge (CS^\beta_{j_E,q_L} \geq 0), \tag{24}$$

$$CF^{\gamma_{in}}_{j,q} = (CS^\alpha_{j_R,q_E} \geq 0) \bigwedge (CS^\alpha_{j_L,q_S} \geq 0) \bigwedge$$
$$(CS^\beta_{j_S,q_L} \geq 0) \bigwedge (CS^\beta_{j_E,q_R} \geq 0), \tag{25}$$

where $f_{A_{j_R}}$ and $f_{A_{j_L}}$ are the right and left face of E_{A_j} respectively, $V_{B_{q_S}}$ and $V_{B_{q_E}}$ are the starting and ending vertices of E_{B_q} respectively, and the others are defined similarly. $CF^{\gamma_{out}}_{j,q}$ and $CF^{\gamma_{in}}_{j,q}$ correspond to the cases that the vector $QE_{A_j}Q^\dagger E_{B_q}$ points to the neighborhoods of E_{B_q} and $QE_{A_j}Q^\dagger$ respectively. The non-overlap predicates are given in [5] as follows.

$$NO^\alpha_{k,q} = NO^{\alpha_{out}}_{k,q} \bigwedge NO^{\alpha_{in}}_{k,q}, \tag{26}$$

$$NO^\beta_{j,r} = NO^{\beta_{out}}_{j,r} \bigwedge NO^{\beta_{in}}_{j,r}, \tag{27}$$

where $NO^{\alpha_{out}}_{k,q}$ and $NO^{\alpha_{in}}_{k,q}$ correspond to the cases that E_{B_q} points outwards and inwards to f_{A_k} respectively. These predicates can be broken down further as

$$NO^{\alpha_{out}}_{k,q} = (CS^\alpha_{k,\mu} \geq 0) \bigvee (CS^\alpha_{k,\nu} \leq 0)$$
$$\bigvee \left(\bigvee_\tau CS^\gamma_{\tau,q} \geq 0\right), \tag{28}$$

$$NO^{\alpha_{in}}_{k,q} = (CS^\alpha_{k,\mu} \leq 0) \bigvee (CS^\alpha_{k,\mu} \geq 0)$$
$$\bigvee \left(\bigvee_\tau CS^\gamma_{\tau,q} \leq 0\right), \tag{29}$$

where V_{B_μ} is the starting vertex of E_{B_q}, V_{B_ν} its ending vertex, and τ is takes so that E_{A_τ} should be every edge

of f_{A_k}, and

$$NO^{\beta_{out}}_{j,r} = (CS^\beta_{\mu,r} \geq 0) \bigvee (CS^\beta_{\nu,r} \leq 0)$$
$$\bigvee \left(\bigvee_\tau CS^\gamma_{j,\tau} \geq 0\right), \tag{30}$$

$$NO^{\beta_{in}}_{j,r} = (CS^\beta_{\mu,r} \leq 0) \bigvee (CS^\beta_{\nu,r} \geq 0)$$
$$\bigvee \left(\bigvee_\tau CS^\tau_{j,\omega} \leq 0\right), \tag{31}$$

where V_{A_μ} is the starting vertex of E_{A_j}, V_{A_ν} its ending vertex, and τ is takes so that E_{B_τ} should be every edge of f_{B_r}.

Now the predicate $CB_{A,B}$ has been reduced to atomic formulae consisting of $CS^\alpha_{k,p}$, $CS^\beta_{i,r}$ and $CS^\gamma_{j,q}$ with $= 0$, ≥ 0 or ≤ 0.

3.2 The algorithm

It seems that $CB_{A,B}$ should define a compact semi-algebraic set, so we can apply the roadmap algorithm from [5, 6, 7] to plan motions of A in contact with B in principle. But the bit complexity of the roadmap algorithm is awful for our problem as well as the combinatorial complexity, mainly because it needs symbolic evaluation of the resultant matrix. So we consider to replace this part, corresponding to algorithm 3.1 and 3.2 in [5], by the numeric-symbolic algorithm for evaluating one-dimensional algebraic set from [21] and that for finding curve and surface intersections from [25] respectively. We clarify how these replacements can be done after describing some preparations.

3.2.1 Preparations

(1) Convex partitioning of concave faces

The definition of $CB_{A,B}$ has assumed that all non-convex faces of A and B have been decomposed into convex polygons. It is known that partitioning of a polygon with polygonal holes into the minimum number of convex pieces is NP-hard, even if Steiner points are disallowed[20]. Here we take Hertel and Mehlhorn algorithm[17] which is not worse than four-times optimal in the number of convex pieces and does not employ Steiner points, since Steiner points may cause numerical instability.

(2) Re-evaluation of the predicate tree

The depth of the predicate tree $CB_{A,B}$ should be no more than $O(\log N)$, N is the number of the leaves or the atomic formulae, to ensure the complexity bound of algorithm 4.6 in [5]. In other words, the time to re-evaluate $CB_{A,B}$ when the value of a single atomic formula changes should be no more than $O(\log N)$. Canny describes a depth compression algorithm to attain this [7].

In our case of $CB_{A,B}$, we can concentrate our attention to $O(1)$ number of CFs when we call algorithm 4.3 in [5], because CFs are combined by the disjunction. And each CF consists of $O(1)$ expected number of atomic formulae from an amortized analysis, so there is no need to apply the depth compression to the CFs' sub-tree. The NOs' sub-tree can be considered a conjunctive canonical form, which has an $\land$ vertex at the root and an $\lor$ vertex at each child of the root. So we can have a new NOs' sub-tree with $O(\log N)$ depth, by converting each sub-tree below every vertex of the canonical form to a balanced binary tree.

For simplicity, we will assume in the following that every $CS \leq 0$ is replaced by $-CS \geq 0$ in CFs and NOs, and write $CS \geq 0$ in the algorithm descriptions with a slight abuse of notation.

(3) Bounding boxes for the C-obstacles

We need a rectangular region on the projected plane including all of the silhouette curves of the C-obstacles to apply the algorithm from [21] to finding them. For translational axes, upper and lower boundaries of the region can be determined as follows.

Algorithm 3.1

1. *Find the bounding box of B.*

2. *Find the minimum sphere covering A and centered at its reference point.*

3. *Enlarge the bounding box by the radius r of the sphere in both directions along every axis.*

4. *Output the enlarged bounding box.*

Considering the degeneracy, it must be safer to use $r+\epsilon$ instead of r, where ϵ is a small positive constant. For rotational axes, θ_j varies between -1 and $+1$ within each cube C_i, so we can use these boundaries for our purpose.

3.2.2 The roadmap algorithm

A rough sketch of the algorithm[5, 6, 7] with the replacements of the algebraic part applied to our problem is the following.

Algorithm 3.2

1. *for $i = 1, \cdots 5$*

2.   *for each subset R of CFs s.t. $|R| = i$*

3. $M \leftarrow \cap_{r \in R}(CS = 0 \text{ in } r)$

4. $S_{local} \leftarrow \cup_{r \in R}(CS \geq 0 \text{ in } r)$

5. $S_{NO} \leftarrow \cup_{s \in NO}(CS \geq 0 \text{ in } s)$

6.

    Find the adjacency graph G_M of the silhouette curve $C = \Sigma(a|_M)$ of the manifold M under a generic linear map a. If C is not an empty set, find all the intersection points of this curve with the hypersurfaces $CS = 0 \in S_{local} \cup S_{NO}$, label each intersection point and curve segment with the signs of all the $CS \geq 0$, and store the set P of the critical points of $a_1|_{\Sigma(a|_M)}$.

7. *Construct the unified adjacency graph G from $G_M s$ obtained above.*

8. *for each critical point p of P*

9.   *Compute a link curve from p to a point on G.*

10.   *Add the link curve to graph G.*

11. *Output G.*

In Step 6, $\Sigma(a|_M)$ denotes the critical set of a with its domain restricted to M (See [5]), C is computed by the numeric-symbolic algorithm in [21] instead of algorithm 4.1 in [5], and the intersection points by the algorithm in [25]. As a side effect of the computation of C, we can also find the set P of the critical points of $a_1|_{\Sigma(a|_M)}$, where a_1 is a generic linear map to some axis. See [5] for details.

3.2.3 The motion planning algorithm

The motions of polyhedra in contact can be planned from the initial configuration, the final configuration, and graph G as inputs.

Algorithm 3.3

1. *Compute link curves from the initial configuration and the final configuration to G, and add them to G.*

2. *Search a path on G from the initial configuration to the final configuration.*

3. *If a path was found*

4.   *then output the path,*

5.   *else report "No path exists."*

Step 1 can be done in the same way as Step 9 and 10 in the previous algorithm. We can employ some complete algorithm for graph searching for Step 2.

3.2.4 Complexity of the algorithm

The combinatorial complexity of the algorithm is $O(m^6 n^6 \log mn)$, which comes directly from that of the roadmap algorithm[5]. This is close to the optimal, since the worst case lower bound of the C-obstacle complexity has been shown to be $\Omega(m^6 n^6)$ [15].

The constants hidden in this complexity is expected to be improved, since the algorithm does not include the symbolic evaluation procedures of the resultant matrices. But the algebraic problem imposed by the roadmap algorithm still seems too big to be solved by the numeric-symbolic algorithm, since the maximum Bezout number of the resulting equations is $2,187$. This number comes from the case when we try to find the silhouette curves for $i = 3$ in Algorithm 3.2, where the silhouette curves are the solutions of the equations with degrees $3, 3, 3, 9, 9$ respectively. On the other hand, the maximum Bezout number of the manifolds that appear in Step 6 in Algorithm 3.2 is $3^5 = 243$, which is significantly smaller than $2,187$. The number has been made bigger, because we try to find the silhouette of the manifolds. So, it would be nice to develop a motion planning algorithm that needs to solve equations with smaller Bezout number, especially when

we consider the experimental results that the numeric-symbolic algorithm can evaluate the intersection curves with Bezout number 324 from two surfaces in less than a second [21].

3.3 Possible performance improvements

We still face the high combinatorial complexity even if we can suppose that the complexity of the algebraic algorithms has been improved. The combinatorial complexity comes from the number of the subset R in Step 2 and the number of critical points p in Step 8 of algorithm 3.1. The number of the critical points is expected to be much smaller than the worst case in most practical cases, since the number of the critical points in the worst case comes from the Bezout bound of the corresponding equations, and the expected number of solutions is much smaller in general. But at least the number of the iterations for each R is $\Theta(m^5 n^5)$ in practice.

We present two ideas, one could enjoy simpler algebraic computations, and another could improve the expected running time in many practical cases. Note, unfortunately, that these are remaining just ideas, and neither their effectiveness nor completeness has been proved yet.

3.3.1 Applicability conditions

The silhouette curve C in Step 6 becomes an empty set, when $\mathcal{CF}s$ in R does not hold simultaneously. We could prune some of such R by checking a necessary condition for R to be an empty set as follows.

Let us re-examine the predicate

$$\mathcal{CF}_{k,p}^\alpha = (CS_{k,p}^\alpha = 0) \bigwedge (\bigwedge_\xi CS_{\xi,\lambda}^\gamma \le 0) \bigwedge (\bigwedge_\zeta CS_{k,\zeta}^\alpha \ge 0).$$
(32)

The predicate $\bigwedge_\zeta CS_{k,\zeta}^\alpha \ge 0$ means that the coboundary of V_{B_q} is outside of f_{A_k}, and is true if and only if

$$\bigwedge_\zeta \frac{1}{2} tr(Q N_{A_k} Q^\dagger (V_{B_\zeta} - V_{B_p})) \ge 0.$$
(33)

This condition is called the applicability condition for a primary point contact[12], and includes only rotational variables. Recall that Eqs.(10) and (13) are equivalent to Eqs.(14) and (15), but have less algebraic complexity. The applicability conditions construct an arrangement of quadratic surfaces in 3-space, which can be computed in $O(m^3 n^3 \beta(mn))$ time [30]. See [30] for the definition of function $\beta(\cdot)$. Our idea is to prune such R that yields an empty set by checking the conjunction of the applicability conditions for the primary point contacts corresponding to R. This query can be done quickly after the arrangement has been successfully constructed, and then we can do without the expensive procedure for finding the silhouette curve in 6-space.

3.3.2 Implicit construction of the graph

Algorithm 3.1 can yield the adjacency graph with $O(m^6 n^6)$ nodes and arcs in $O(m^6 n^6 \log mn)$ time. On the other hand, it is unrealistic to execute such motions by real robots, that correspond to paths going through hundreds nodes and arcs even if they have been successfully planned by the algorithm.

So, one practical way may be to specify the graph implicitly, and expand it from the initial or final node by a graph searching procedure while constructing it incrementally. Eq.(20) can be rewritten as

$$
CB_{A,B} = \left(\bigvee_{k=1}^{m_f} \bigvee_{p=1}^{n_V} (C\mathcal{F}_{k,p}^{\alpha} \bigwedge \mathcal{NO}_{A,B}) \right)
$$
$$
\bigvee \left(\bigvee_{i=1}^{m_V} \bigvee_{r=1}^{n_f} (C\mathcal{F}_{i,r}^{\beta} \bigwedge \mathcal{NO}_{A,B}) \right)
$$
$$
\bigvee \left(\bigvee_{j=1}^{m_E} \bigvee_{q=1}^{n_E} (C\mathcal{F}_{j,q}^{\gamma} \bigwedge \mathcal{NO}_{A,B}) \right),
$$
(34)

where

$$
\mathcal{NO}_{A,B} = \left(\bigwedge_{k=1}^{m_f} \bigwedge_{q=1}^{n_E} \mathcal{NO}_{k,q}^{\alpha} \right) \bigwedge \left(\bigwedge_{j=1}^{m_E} \bigwedge_{r=1}^{n_f} \mathcal{NO}_{j,r}^{\beta} \right).
$$
(35)

We can see in Eq.(34) that $CB_{A,B}$ is given by the disjunction of the elementary conjunctions, each of which is the predicate for a primary point contact without any intersection between the boundaries of A and B. So we can apply the roadmap algorithm to each combination of elementary conjunctions, which corresponds to each subset R at Step 2 in Algorithm 3.1, and construct the whole roadmap incrementally. We call the set of the configurations of the moving polyhedron at which a combination of elementary conjunctions holds

a **contact state**. This incremental construction can start from the contact state including the final configuration or the initial configuration. At an incremental step, we can find the neighboring contact states from the set of $CS = 0$ which pass through any node of the adjacency graph found for the current contact state. Note that we must not compute the adjacency graph again for the visited contact states. In other words, we should not revisit the subset R at Step 2 in Algorithm 3.1 to find the subset of the whole adjacency graph.

We can not improve the worst case running time by this idea, but the expected running time has some chance to be shorten.

4   Conclusions

This paper studied motion planning of polyhedra in contact. We first revisited our algorithm for the convex case to see that the complexity of the algebraic part heavily depends how the geometric problem is reduced to it.

The combinatorial complexity of the nonconvex case is very high as well as the algebraic complexity, but the gap between the current algorithms and practical ones has been getting smaller. We have shown several ideas for this purpose.

Open problems include the development of output sensitive algorithms for the combinatorial part, which need to solve algebraic equations with smaller Bezout number.

Acknowledgement

The author thanks Dan Halperin and Toshihiro Matsui for their help.

References

[1] B. ARONOV AND M. SHARIR, Castles in the air revisited, *Discrete Comput. Geom.*, Vol.12, No.2, pp. 119–150, 1994.

[2] F. AVNAIM AND J.-D. BOISSONNAT, Polygon placement under translation and rotation, *Proc. 5th Sympos. Theoret. Aspects Comput. Sci.*, in *Lecture Notes in Computer Science*, Vol.294, pp. 322–333, Springer-Verlag, 1988.

[3] F. AVNAIM, J.-D. BOISSONNAT AND B. FAVERJON, A practical exact motion planning algorithm for polygonal objects amidst polygonal obstacles, *Proc. 5th IEEE Internat. Conf. Robot. Autom.*, pp. 1656–1661, 1988.

[4] R. BROST, Computing metric and topological properties of configuration-space obstacles, *Proc. 6th IEEE Internat. Conf. Robot. Autom.*, pp. 170–176, 1989.

[5] J.CANNY, *The Complexity of Robot Motion Planning*, The MIT Press, 1988.

[6] J.CANNY, Some algebraic and geometric computation in PSPACE, *ACM Symp. on Theory of Computing*, pp.460–467, 1988.

[7] J.CANNY, Computing roadmaps of general semi-algebraic sets, *The Computer Journal*, Vol.36, No.5, pp. 504–514, 1993.

[8] B. CHAZELLE, H. EDELSBRUNNER, L. GUIBAS AND M. SHARIR, A singly-exponential stratification scheme for real semi-algebraic varieties and its applications, *Proc. 16th Internat. Colloq. Automata Lang. Program*, LNCS, Vol.372, pp. 179–192, 1989.

[9] B. CHAZELLE AND H. EDELSBRUNNER AND L. GUIBAS AND M. SHARIR, A singly-exponential stratification scheme for real semi-algebraic varieties and its applications, *Theoret. Comput. Sci.*, Vol.84, pp. 77–105, 1991.

[10] G.E.COLLINS, Quantifier elimination for real closed fields by cylindric algebraic decomposition, PROC. 2ND GI CONF. ON AUTOMATA THEORY AND FORMAL LANGUAGES, LNCS, Vol.33, pp. 134–183, 1975.

[11] M. DE BERG AND L. J. GUIBAS AND D. HALPERIN, Vertical decompositions for triangles in 3-space, *Proc. 10th Annu. ACM Sympos. Comput. Geom*, pp. 1–10, 1994.

[12] B. R. DONALD, A search algorithm for motion planning with six degrees of freedom, *Artificial Intelligence*, vol.31, no.3, pp.295–353, 1987.

[13] I.EMIRIS AND J.CANNY, A practical method for the sparse resultant, *Proc. ISSAC'93*, pp. 183–192, 1993.

[14] L. J. GUIBAS AND R. SEIDEL, Computing convolutions by reciprocal search, *Discrete Comput. Geom.*, Vol.2, pp. 175–193, 1987.

[15] D.HALPERIN AND O.SCHWARZKOPF, *Personal Communication*, 1995.

[16] D.HALPERIN AND M.SHARIR, Near-quadratic bounds for the motion planning problem for a polygon in a polygonal environment, *Discrete and Computational Geometry*, to appear.

[17] S. HERTEL AND K. MEHLHORN, Fast triangulation of simple polygons, *Proc. 4th Internat. Conf. Found. Comput. Theory*, in *Lecture Notes in Computer Science*, Vol.158, pp. 207–218, Springer-Verlag, 1983.

[18] H.HIRUKAWA, Y.PAPEGAY AND T.MATSUI, A motion planning algorithm for convex polyhedra in contact under translation and rotation, *Proc. IEEE Int. Conf. Robotics and Automation*, pp. 3020–3027, 1994.

[19] H. HIRUKAWA, T. MATSUI AND K. TAKASE, Automatic determination of possible velocity and applicable force of frictionless objects in contact, *IEEE Trans. Robotics and Automation*, Vol.10, No. 3, pp. 309–322, 1994.

[20] J. M. KEIL, Decomposing a polygon into simpler components, *SIAM J. Comput.*, No.14, pp. 799–817, 1985.

[21] S.KRISHNAN AND D.MANOCHA, Numeric-symbolic algorithms for evaluating one-dimensional algebraic sets, *ISSAC'95*, pp. 59–67, 1995.

[22] D.KAPUR AND Y.N.LAKSHMAN, Elimination methods: an introduction, in B.R. DONALD et al. (eds.), *Symbolic and Numerical Computation for Artificial Intelligence*, Academic Press Ltd., 1992.

[23] J.C.LATOMBE, *Robot Motion Planning*, Kluwer Academic, 1991.

[24] D.MANOCHA AND J.CANNY, Multipolynomial resultant algorithms, *Journal of Symbolic Computation*, Vol.15, No.2, pp. 99–122, 1993.

[25] D.MANOCHA, Computing selected solutions of polynomial equations, *Proc. ISSAC'94*, pp. 1–8, 1994.

[26] T.MATSUI AND M.INABA, An object-based implementation of Lisp, *Journal of Information Processing*, vol.13, no.3, 1990.

[27] K. MULMULEY, *Computational Geometry: An Introduction Through Randomized Algorithms*, Prentice Hall, New York, 1993.

[28] B.K. NATARAJAN, On planning assemblies, *Proc. 4th ACM Symposium on Computational Geometry*, 1988, pp. 299–308.

[29] J. T. SCHWARTZ AND M. SHARIR, On the "piano movers" problem II: general techniques for computing topological properties of real algebraic manifolds, *Adv. Appl. Math.*, Vol.4, pp. 298–351, 1983.

[30] M. SHARIR AND P. K. AGARWAL, *Davenport-Schinzel Sequences and Their Geometric Applications*, Cambridge University Press, 1995.

[31] S.WOLFRAM, *Mathematica*, Addison-Wesley Pub., 1991.

Issues in the Metrology of Geometric Tolerancing

Chee Yap, *Courant Institute of Mathematical Sciences, New York University*
Ee-Chien Chang, *Courant Institute of Mathematical Sciences, New York University*

The classification problem is fundamental in the metrology of Geometric Tolerancing: given a part B, is it within some tolerance Φ? The "standard methodology" for the classification problem comprises three steps: sampling σ, algorithmic step α and policy decision π. The issues we examine are posed around this methodology, using a simple 1-dimensional problem as case study.

1 Introduction

Geometric tolerancing is concerned with the specification of geometric shapes for use in the manufacture of mechanical parts. Since manufacturing processes are inherently imprecise, it is imperative that such geometric designs be accompanied by tolerance specifications. Figure 1 is a simple example of such a specification. Evidently, the ideal object indicated there is a 3×5

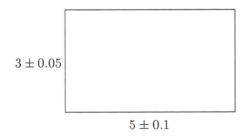

Figure 1: *Tolerancing a rectangle.*

rectangle, with some amount of deviation to be tolerated in its manufacture. There is a naive interpretation of this symbology, namely, "*any manufactured rectangle of dimension $L \times H$ where $|L - 5| \le 0.1$ and $|H - 3| \le 0.05$ is considered within tolerance*". This is the **parametric tolerancing** semantics. The problem with this approach is that no manufactured planar object[1] will be a rectangle or even a quadrilateral, but rather some planar shape with an uneven boundary. Following Requicha [9], we believe that a rigorous approach can be based on the **zone tolerancing semantics**. This interprets the symbology of figure 1 to specify some region called the **(tolerance) zone** and declare any body with boundary within this zone as "within tolerance".

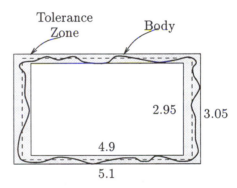

Figure 2: *Zone Semantics.*

Figure 2 illustrates one interpretation of the zone implicit in figure 1: it is a region bounded by two "aligned" rectangles with the indicated dimensions. We remark that the actual language for tolerance specifications used by professional engineers is highly intricate, involving both planar figures and textual symbols (e.g., [14, 8]) and requiring careful mathematical interpretation [13]. See [6] for an exposition. The other remark is that the semantics in the ANSI standard [14, 8]

---

[1]Parametric tolerancing may be useful in non-manufacturing applications. For instance, in a window-based graphical interface system, we might wish to tolerance the window sizes using this approach.

is basically a zone approach.

Having toleranced a geometric design, suppose we now hold in hand an actual (physical) object that is supposed to be an instance of the design. We need to check if this instance is within tolerance. The **metrology of geometric tolerancing** (or simply "tolerancing metrology") is the study of this measurement process[2]. Hopp [2] calls this subject "computational metrology". The use of **coordinate measuring systems** or **machines** (CMS or CMM) is considered the state-of-the-art in the practice of tolerancing metrology. Each CMS is basically a highly accurate mechanical probe, coupled to computational power. As such a CMS can be programmed to automatically take probes and perform computations.

We use the term "Geometric Tolerancing and Metrology" (GT&M) to refer to the combination of geometric tolerancing with computational metrology. The latter is by far the less understood half of GT&M. The present paper attempts to expose some basic metrological issues.

Importance of GT&M. Intellectually, GT&M is a cornerstone of modern high precision engineering. But this subject also has widespread implications in practice because achieving high precision in manufacturing is a major component of manufacturing costs. Studies have shown that manufacturing tolerances decrease linearly with time (on a semi-log scale), in every category of manufacturing, from large-scale manufacture such as shipbuilding to micron-level manufacture such as computer chips. Hence such issues are expected to become more acute with time.

A more comprehensive view of the "cost of a part" must look beyond the manufacturing cost. The "lifetime cost" of a manufactured part must include its reliability over its life-time. Reliability is positively correlated with tight tolerances, as is evident in the reliability of parts such as engine pistons. Hence we see a tradeoff between reliability versus manufacturing cost.

---

[2]Note that "metrology" is the science of measurement and as such is much wider than its applications to geometric tolerancing.

Just as manufacturing is inherently imprecise, metrology is also inherently imprecise. This has implications be making policies for accepting or rejecting manufactured parts: there is a cost for rejecting good parts as well as a cost for accepting bad parts. For instance, machining a high-precision part such as a gear for a jet engine may cost up to $50K. Without solid metrological principles, we may be hard pressed between making a $50K mistake (if we wrongly reject such a gear) versus making a graver mistake (if we wrongly accept such a gear).

Current tolerancing metrology practice often appears ad hoc. For instance, a subcontractor for a car manufacturer is to produce aluminum space-frames which are basically tubes that are bent to some specified shape. The contractual agreement calls for coordinate measuring machine (CMM) measurements at 64 locations in each space-frame, with the criteria is that each measurement q_i is within 1mm from given ideal measurement p_i: $\|q_i - p_i\| \leq 1$ $(i = 1, \ldots, 64)$. This turns out to be remarkably hard to achieve. It is also unclear if this contractual bases of performance is the most appropriate.

For overviews of current GT&M, we refer to [12, 10, 1]. A recent report [4] describes the economic impact of CMM research carried out at the National Institute of Standards and Technology (NIST) from 1975-85; it is also a good background for CMM research. See [15] for the connection to computational geometry and exact computation.

2 Standard Methodology in Classification

In metrology, our original datum is a **body** $B \in \mathbb{R}^n$ $(n = 1, 2, 3)$. For our purposes, define a body B to be any compact, simply connected subset of $\mathbb{R}^n$. The boundary of B is denoted ∂B. For each toleranced geometric design D, we obtain a corresponding set Φ_D comprising all bodies B that are "within tolerance". Figure 1(a) is an example of D. Viewing $\Phi_D(B)$ as a predicate, we have the corresponding **classification problem** for D: *given B, does $\Phi_D(B)$ hold?* The following three steps constitute the **standard methodology** for classification problems:

I. Measurement Step: Make some measurements S on the given body B. Let μ denote this measurement procedure,

$$\mu : B \mapsto S.$$

II. Algorithmic Step: From S, we compute a small set V of "characteristic values". Let α denote the algorithm for this,

$$\alpha : S \mapsto V.$$

III. Policy Step: Using V, we finally make a decision to accept or reject the body B. Let this decision be based on a policy π,

$$\pi : V \mapsto \{accept, reject\}.$$

Let us illustrate this methodology for the well-known "roundness problem" (e.g., [11]). The **(annular) width** of a body $B \subseteq \mathbb{R}^2$,

$$\texttt{width}(B)$$

is defined to be the minimum width $R - r$ of an annulus with radii (R, r) that contains the boundary of B. If S is a finite set, $\texttt{width}(S)$ is similarly defined. Suppose that we are interested in the classification predicate

$$\Phi(B) : \text{``}\texttt{width}(B) \leq 0.1\text{''}.$$

The standard methodology (μ, α, π) for this classification problem is typically as follows:

μ: Probe B to get n sample points $S \subseteq \partial B$.
α: Compute $\texttt{width}(S)$.
π: Accept iff $\texttt{width}(S) \leq 0.1$.

The following questions arise:

(i) Accuracy: How shall n be chosen? Intuitively, it is clear that as n grows, the accuracy improves.

(ii) Sampling strategy: How shall S be obtained? In practice, the sampling strategies are ad hoc and unanalyzed.

(iii) Validity of substitution: The above policy apparently treats $\texttt{width}(S)$ as a substitute for $\texttt{width}(B)$. Why is this appropriate — what is their relation?

(iv) Policy analysis: Wrong decisions are inevitable in any policy (recall that metrology is inherently imprecise). How should we hedge our bets concerning making **erroneous rejections** versus **erroneous acceptances**? Borrowing a terminology of hypothesis testing in statistics, we may refer to these two kinds of errors as **type I** and **type II** errors, respectively.

(v) Integration: The connections between μ, α, π are often not exploited. In [7], an integrated approach to (μ, α, π) is given for the problem of classifying roundness of convex polygons.

We remark that often computational geometers simply address the algorithmic part (α) of this methodology. But from the metrological perspective, this is not the most significant aspect. We believe that computational geometry can contribute to the entire process.

3   A Simple Probe Model

To expose some issues, consider a toy problem in one dimension. Let $B \subseteq \mathbb{R}^1$ be a closed line segment of some unknown length L. The classification problem we address is this: does L satisfy the bounds

$$1 - \delta \leq L \leq 1 + \delta \tag{1}$$

where $0 < \delta < 1$ is fixed. We also write $L \in [1 \pm \delta]$ instead of (1). The probe model we consider is the Δ-**grid** for some $\Delta > 0$. It is comprised of infinitely many point "sensors" that are uniformly spaced Δ-apart along the entire real line. Figure 3 illustrates the Δ-grid. To make a probe, we simply place B on the grid and we obtain as **probe result** the number n_0 of sensors that are covered by B. Note that we have no control over how to place B, so that n_0 is not a deterministic function of L. In figure 3, the value of n_0 is 5.

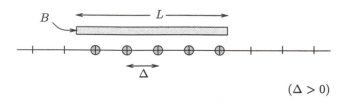

$(\Delta > 0)$

Figure 3: Δ-*grid*.

Nevertheless, we shall only consider[3] **deterministic** policies based on n_0.

E.g., A simple policy might be as follows: accept B iff $n_0 = \lfloor 1/\Delta \rceil$. Here $\lfloor 1/\Delta \rceil$ denotes the rounding of $1/\Delta$ to the closest integer. How good is this policy?

We classify policies into three basic types: (1) A policy is **conservative** if it never accepts a bad L. (2) It is **liberal** if it never rejects a good L. A conservative policy is **trivial** if it rejects every input; a liberal policy is likewise trivial if it accepts every input. The basic deduction from our probe result n_0 is the following:

$$(n_0 - 1)\Delta \le L < (n_0 + 1)\Delta.$$

It follows that a non-trivial conservative policy exists iff there exist $n_*, n^* \in \mathbb{N}$, $1 \le n_* \le n^*$, such that

$$[(n_* - 1)\Delta, (n^* + 1)\Delta] \subseteq [1 \pm \delta].$$

Choosing $n^* - n_*$ as large as possible, we have the "best" conservative policy: *accept iff $n_0 \in [n_*, n^*]$.* As corollary:

(i) If $\Delta > \delta$, no non-trivial conservative policy exists.

(ii) If $\Delta \le \delta/2$, non-trivial conservative policies exist.

(iii) If $\Delta \le \delta$ and Δ is an Egyptian fraction, non-trivial conservative policies exist.

Recall that an Egyptian fraction is one whose reciprocal is an integer. On the other hand, note that non-trivial liberal policies always exist. Indeed, (i)-(iii) are subsumed by the following:

---

[3]But our policies are necessarily nondeterministic when viewed as functions of L.

Lemma 1 *Let* $\varepsilon_0 = \lceil 1/\Delta \rceil - (1/\Delta)$, $\varepsilon_1 = (1/\Delta) - \lfloor 1/\Delta \rfloor$, *and* $\varepsilon = \min\{\varepsilon_0, \varepsilon_1\}$. *Then a non-trivial conservative policy exists iff* $\delta \ge (1 + \varepsilon)\Delta$.

Except for the "perfect policy" (which does not exist under our probe model), note that conservative and liberal policies are mutually incompatible. Conservative (resp., liberal) policies may be useful when the cost of wrong acceptance (resp., rejection) is extremely high. But usually, a tradeoff is more useful. We therefore formulate another class of policies: let $0 \le \delta_* \le \delta^* \le \infty$. A policy is δ_*-**neoliberal** if $|L - 1| < \delta_*$ implies that it surely accepts. A policy is δ^*-**neoconservative** if $|L - 1| > \delta^*$ implies that it surely rejects. Say a policy is a (δ_*, δ^*)-**policy** if it is δ_*-neoliberal and δ^*-neoconservative. We say the (δ_*, δ^*)-policy is **compatible** with the classification of equation (1) if

$$\delta_* \le \delta \le \delta^*.$$

Note that a compatible (δ_*, δ^*)-policy is not constrained to behave in any particular way on inputs that satisfy $\delta_* \le |L - 1| \le \delta^*$: it may behave conservatively or liberally. Also, a non-trivial conservative policy may not be δ_*-neoliberal for any $\delta_* > 0$.

Lemma 2 (δ_*, δ^*)-
Policies exist iff (a) $\lceil (1 - \delta^*)/\Delta \rceil < \lfloor (1 - \delta_*)/\Delta \rfloor$, *and (b)* $\lceil (1 + \delta_*)/\Delta \rceil < \lfloor (1 + \delta^*)/\Delta \rfloor$. *Equivalently, such policies exist iff* $\delta^* - \delta_* \ge (1 + \varepsilon)\Delta$ *where*

$$\varepsilon = \max\{\frac{1 - \delta_*}{\Delta} - \lfloor\frac{1 - \delta_*}{\Delta}\rfloor, \lceil\frac{1 + \delta_*}{\Delta}\rceil - \frac{1 + \delta_*}{\Delta}\}$$

EXPLICIT POLICIES. For $\alpha, \beta \ge 0$, the **policy** (α, β) is as follows: *Accept iff* $1 \in [(n_0 - \alpha)\Delta, (n_0 + \beta)\Delta]$. Not all (δ_*, δ^*)-policies can be achieved by explicit policies. The following shows some attainable bounds.

Lemma 3 *Fix* $\alpha, \beta \ge 0$. *For all* Δ, *the policy* (α, β) *is a* (δ_*, δ^*)-*policy where*

$$\begin{aligned} \delta_* &= \Delta \cdot \min_0\{\alpha - 1, \beta - 1\}, \\ \delta^* &= \Delta \cdot \max\{\alpha + 1, \beta + 1\}. \end{aligned} \qquad (2)$$

The notation $\min_0\{x, y\}$ *here refers to* $\max\{0, \min\{x, y\}\}$. *This is the best possible* in the sense that there exists Δ_* (resp., Δ^*) for which this choice of δ_* (resp., δ^*) is optimal.

Proof. We know that $L \in [(n_0 \pm 1)\Delta]$ and we accept iff $1 \in [(n_0 - \alpha)\Delta, (n_0 + \beta)\Delta]$. We want to determine a suitable δ_*, δ^*. Suppose $\alpha < 1$. We must show that $\delta_* = 0$ is "best possible" choice. For any $\varepsilon > 0$, there exists Δ such that the policy (α, β) rejects an input of length L where $|L - 1| = \varepsilon$: we simply choose Δ to satisfy $1 = (n_0 - \alpha)\Delta - \varepsilon$. Hence $\delta_* = 0$ is best possible in the sense of the lemma. Similarly if $\beta < 1$ then $\delta_* = 0$ is the best possible. On the other hand, if $\alpha \geq 1$ and $\beta \geq 1$, then $|L - 1| \leq \Delta \cdot \min\{\alpha - 1, \beta - 1\}$ implies that policy (α, β) will surely accept L. This proves the choice $\delta_* = \Delta \cdot \min\{\alpha - 1, \beta - 1\}$ is valid. It is the best possible in the sense that for any $\varepsilon > 0$, if $\delta_* = \varepsilon + \Delta \cdot \min\{\alpha - 1, \beta - 1\}$ then we can choose Δ so that policy (α, β) may reject an input of length L where $|L - 1| < \delta_*$. Finally, it is easy to see that if $|L - 1| > \Delta \cdot \max\{1 + \alpha, 1 + \beta\}$ then policy (α, β) will surely reject. Hence the choice of δ^* is valid. Again, it is the best possible. **Q.E.D.**

For instance, policy $(1/2, 1/2)$ is a $(0, 3\Delta/2)$-policy, and policy $(3/2, 2)$ is a $(\Delta/2, 3\Delta)$-policy.

POLICY FOR FIXED δ, Δ. Given δ, Δ, we want to choose α, β such that policy (α, β) is a (δ_*, δ^*)-policy that is compatible with equation (1) and $\delta^* - \delta_*$ is minimized. If we use the bounds (2) from the above lemma, then we see that $\delta^* - \delta_* \geq 2\Delta$, with equality iff $\alpha = \beta$. When $\alpha = \beta$, "policy (α, β)" will be simply called "policy α". Clearly we can choose a compatible (δ_*, δ^*)-policy with $\alpha = \beta$. For compatibility, we can choose any $\alpha > 1$ such that $(\alpha - 1)\Delta < \delta < (\alpha + 1)\Delta$. However, we can choose δ_*, δ^* that are sharper than (2) for every fixed δ, Δ. The following is a "natural choice" for α, β:

$$\alpha_0 = \left\lfloor \frac{1 + \delta}{\Delta} \right\rfloor - \frac{1}{\Delta}, \quad \beta_0 = \frac{1}{\Delta} - \left\lceil \frac{1 - \delta}{\Delta} \right\rceil. \quad (3)$$

If we set $m_0 := \left\lceil \frac{1-\delta}{\Delta} \right\rceil$ and $m_1 := \left\lfloor \frac{1+\delta}{\Delta} \right\rfloor$, then policy (α_0, β_0) amounts to "accept iff $n_0 \in [m_0, m_1]$". Let us assume that $\Delta \leq \delta$. Then $m_0 < m_1$. We may verify that

$$1 - \delta \in [(m_0 - 1)\Delta, m_0 \Delta], \quad 1 + \delta \in [m_1 \Delta, (m_1 + 1)\Delta].$$

Thus, for this is a (δ_*, δ^*)-policy for some $0 < \delta_* < \delta^* < \infty$. Indeed, $\delta_* = \min\{1 - m_0\Delta, m_1\Delta - 1\}$ and $\delta^* = \max\{1 - (m_0 - 1) \cdot \Delta, (m_1 + 1) \cdot \Delta - 1\}$. Moreover,

$$\Delta \leq \delta^* - \delta_* < 2\Delta.$$

PENALTY MODELS. To distinguish among the possible explicit policies, we may introduce a penalty model. Consider a simple linear model parametrized by the real numbers $p_A, p_R \geq 0$: if we wrongly accept then we pay a penalty of $p_A \cdot \max\{L - 1 - \delta, 1 - \delta - L\}$, and if we wrongly reject then we pay a penalty of $p_R \cdot \min\{1 + \delta - L, L - 1 + \delta\}$. See figure 4.

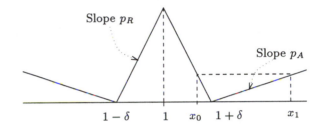

Figure 4: *Linear Penalty Model*

We define a policy to be **optimal** if it minimizes the maximum penalty over all decisions. In general, we want to choose a (δ_*, δ^*)-policy in which the maximum penalty for wrong acceptance equals the maximum penalty for right acceptance. In case Δ is Egyptian, this amounts to the choice of $\delta_* = x_0$ and $\delta^* = x_1$ such that $(\delta - x_0) \cdot p_R = (x_1 - \delta) \cdot p_A$, as illustrated in figure 4. This is not enough to pin down x_0, x_1. But using a formula such as (2), the additional constraint $\delta^* - \delta_* = 2\Delta$ will pin down x_0, x_1 uniquely. Of course, such x_0, x_1 may not exist in case the optimum policy is not δ_*-neoliberal for any $\delta_* > 0$.

Note that the "natural" policy (3) may not be optimal if p_A, p_R are very different. However, it is easy to see that the optimal policy is only small modification of the natural policy. That is, the optimal policy is to accept iff $n_0 \in [m'_0, m'_1]$ for some m'_0, m'_1 satisfying $|m'_i - m_i| \leq 1$ $(i = 0, 1)$.

4 Statistical Analysis

Evaluation of policies is a staple subject of mathematical statistics (policies are usually called "tests" or "decision rules" in that setting). To deploy statistical methods, we must first introduce probabilistic assumptions. Indeed, there is a very natural one in our context: note that the probe value n_0 is[4] either $\lfloor L/\Delta \rfloor$ or $\lceil L/\Delta \rceil$. Assuming that every position of B on the Δ-grid is equally likely, then the probability of getting the value $\lfloor L/\Delta \rfloor$ is

$$p := \frac{L}{\Delta} \bmod 1$$

where $x \bmod 1$ denotes the fractional part of a real number x. Let us now generalize our decision problem by performing m probes, for some given $m \geq 1$. Let the average value of the m probes be the random variable X_m. Clearly, $\lfloor L/\Delta \rfloor \leq X_m \leq \lceil L/\Delta \rceil$. We note a slight problem in case X_m is an integral value n_0: this happens iff all our m probe values are equal to a single value n_0, and we have no way of telling whether this n_0 is $\lfloor L/\Delta \rfloor$ or $\lceil L/\Delta \rceil$. To get around this, we will assume that we do know the value of $\lfloor L/\Delta \rfloor$ through some prior experiment, and let X denote the "shifted" random variable $X_m - \lfloor L/\Delta \rfloor$. Thus

$$X \in \{0, 1/m, \ldots, (m-1)/m, 1\}$$

and the probability $\Pr\{X = i/m\}$ ($i = 0, \ldots, m$) has the binomial distribution $B(m, p)$.

Returning to the classification problem (1), we may now think of this as the problem of testing the **null hypothesis** $H : L \in [1 \pm \delta]$ against its **complementary hypothesis** $K : L \notin [1 \pm \delta]$. The standard tool here is the Neyman-Pearson theory [5, 3] for testing a pair (H, K) of hypothesis. Let us now describe this setting. We need a randomized test (or policy)

$$\phi : X \mapsto [0, 1]$$

where $\phi(X)$ corresponds to the probability of rejecting the hypothesis. Let X have probability distribution F_θ

---

[4]If L/Δ is integer, we must interpret $\lceil L/\Delta \rceil$ to mean $1 + \lfloor L/\Delta \rfloor$. We shall employ this interpretation throughout this discussion.

where $\theta \in \Theta \subseteq \mathbb{R}$ is an unknown real parameter. Suppose H (resp., K) correspond to the hypothesis that $\theta \in [\theta_0, \theta_1]$ (resp., $\theta \notin [\theta_0, \theta_1]$). A **type I error** occurs when we wrongly reject H and a **type II error** occurs when we wrongly accept H. For any $0 < \alpha < 1$, we call ϕ a **most powerful test** (MPT) of **significance** α for (H, K) if the probability of type II error is minimized by ϕ, subject to the probability of **type I error** being at most α. The Neyman-Pearson theory studies conditions for the existence of MPT's.

Let us now assume that $\{F_\theta : \theta \in \Theta\}$ is a "one-parameter exponential family" of distributions, i.e., F_θ has the form

$$C(\theta)e^{Q(\theta)T(x)}h(x)$$

where $Q(\theta)$ is strictly increasing. A generalization of the Neyman-Pearson lemma [5, theorem 6, p.101] says that for each α, there is a MPT of significance α for the pair (K, H) (not the pair (H, K)).

We may apply this generalized lemma for our problem where $\Theta = [0, 1]$ and F_θ is the binomial distribution $B(m, \theta)$. (The reader may verify that this is a one-parameter exponential family.) For any chosen significance α, this lemma tells us how to construct a corresponding MPT. We note some shortcomings in this approach: (i) There does not seem to be a convenient algorithm for constructing the MPT ϕ. (ii) Although we freely choose the significance α, we have no *a priori* bound on the type II error. (iii) Finally, this lemma only applies when K is the null hypothesis, but not when H is the null hypothesis.

5 Multiprobe Strategies

In the preceding study, we focused on the policy π since the sampling σ and algorithm α are trivial. We now make the algorithm non-trivial by allowing "multiprobes". For any $k \geq 0$, a k-**multiprobe** is a sequence of $k + 1$ probes in which:

(a) The initial (zeroth) probe is, as before, an arbitrary placement of B on the Δ-grid.

(b) Each of the subsequent k probes are obtained by user-determined "shifts" of B. The shifts can depend on the results of previous shifts.

A "shift" is determined by a real number σ where $0 < |\sigma| < \Delta$ and amounts to translating B by the distance σ. The **outcome** of a probe is, instead of being just an integer, the set of grid points that are covered by B. The choice of σ_i $(i = 1, \ldots, k)$ in the ith shift may be chosen to depend on the outcomes of the previous i probes.

E.g., a 1-multiprobe allows us to shift B once after its initial placement on the grid. Intuitively, it is clear that the best choice for this shift is $\sigma = \pm\Delta/2$.

Note that there is an uncertainty of Δ at each end of the segment B after the initial (zeroth) probe. The total uncertainty is therefore 2Δ. With each additional probe, we reduce the uncertainty at both ends of the segment B. For any $k \geq 0$, let U_k denote the minimax total uncertainty (in units of Δ) at the two ends of B after k additional probes.

Clearly $U_0 = 2$. For $k = 1$, with $\sigma = 1/2$ for the shift (which is clearly optimal), we achieve the bound $U_1 = 1$. For $k = 2$, the choice $\sigma = 1/3$ for the first shift leads to the bound $U_2 \leq 2/3$. It is also not hard to see that this is optimal. In general, let $0 < \sigma_k \leq 1/2$ denote the first shift in an optimal k-multiprobe. The key remark is this: *once the first shift in a k-multiprobe is given, the optimal strategy for the remaining shifts can essentially be solved by induction on k.* The initial values of σ_k and U_k can be calculated using this remark; they are illustrated in the following table.

k	1	2	3	4	5	6	7	8
σ_k	$\frac{1}{2}$	$\frac{1}{3}$	$\frac{5}{17}$	$\frac{15}{49}$	$\frac{79}{275}$	$\frac{237}{787}$	$\frac{1261}{4409}$	$\frac{3783}{12601}$
U_k	1	$\frac{2}{3}$	$\frac{8}{17}$	$\frac{16}{49}$	$\frac{64}{275}$	$\frac{128}{787}$	$\frac{512}{4409}$	$\frac{1024}{12601}$

The asymptotic behavior of U_k is not immediate from these initial values. However, we may note that

$$2^{1-k} \leq U_k \leq 2^{1-\lfloor k/2 \rfloor}.$$

The first inequality follows from the fact that in the best case, each probe may simultaneously reduce both uncertainty intervals in half. The second inequality follows from the fact that if we treat the two uncertainty intervals independently, then we need two probes to reduce both intervals by half. In a separate paper [16], we determine the asymptotic behavior of U_k.

6 Final Remarks

1. The case study here is intended to raise issues that should be of greater practical interest in 2 or 3 dimensions. E.g., the probe model can clearly be generalized to higher dimensions and applied to other classification problems.

2. Our probe model is "RISC-like" (following Canny and Goldberg). It has the useful property that there is no corresponding "localization problem". In contrast, many probe methods require that certain **localization problems** be solved before the measurement proper can begin. E.g., the method may require that the body be first put in some restricted placement, or it may require estimating the Euclidean transformation from the actual placement of the object to be measure to some special placement.

3. We have looked at deterministic policies, except when we discussed statistical approaches. One expects non-deterministic policies to be more powerful in general. It may also be possible to introduce "uncertainty" into the model, independent of any probabilistic assumptions. We explored a possible statistical approach and this seems to leave interesting unresolved questions.

References

[1] Shaw C. Feng and Theodore H. Hopp. A review of current geometric tolerancing theories and inspection data analysis algorithms. Technical Report NISTIR-4509, National Institute of Standards and Technology, U.S. Department of Commerce. Factory Automation Systems Division, Gaithersburg, MD 20899, February 1991.

[2] Theodore H. Hopp. Computational metrology. In *Proc. 1993 International Forum on Dimensional Tolerancing and Metrology*, pages 207–217, New York, NY, 1993. The American Society of Mechanical Engineers. CRTD-Vol.27.

[3] Maurice G. Kendall and Alan Stuart. *The advanced theory of statistics: volume 2, inference*

and relationship. Hafner Publishing Company, New York, second edition, 1973.

[4] David P. Leech and Albert N. Link. The economic impacts of NIST's software error compensation research. NIST Planning Report 96-2, National Institute of Standards and Technology, TASC, Inc., 1101 Wilson Boulevard, Suite 1500, Arlington, Virginia 22209, June, 1996.

[5] E.L. Lehmann. *Testing statistical hypotheses.* John Wiley and Sons, New York, second edition, 1986.

[6] James D. Meadows. *Geometric Dimensioning and Tolerancing.* Marcel Dekker, Inc, 1995.

[7] K. Mehlhorn, T.C. Shermer, and C.K. Yap. Probing for near-centers and approximate roundness, to appear, 1996. Preliminary version was presented at the ASME Workshop on Tolerancing & Metrology, University of North Carolina at Charlotte, June 21-23, 1995. See URL http://cs.nyu.edu/cs/faculty/yap.

[8] Alvin G. Neumann. The new ASME Y14.5M standard on dimensioning and tolerancing. *Manufacturing Review*, 7(1):16–23, March 1994.

[9] Aristides A. G. Requicha. Toward a theory of geometric tolerances. *International Journal of Robotics Research*, 2(4):45–60, 1983.

[10] U. Roy, C.R. Liu, and T.C. Woo. Review of dimensioning and tolerancing: representation and processing. *Computer-aided Design*, 23(7):466–483, 1991.

[11] U. Roy and X. Zhang. Establishment of a pair of concentric circles with the minimum radial separation for assessing rounding error. *Computer Aided Design*, 24(3):161–168, 1992.

[12] Vijay Srinivasan and Herbert B. Voelcker, editors. *Dimensional Tolerancing and Metrology*, 345 East 47th Street, New York, NY 10017, 1993. The American Society of Mechanical Engineers. CRTD-Vol. 27.

[13] Richard K. Walker and Vijay Srinivasan. Creation and evolution of the ASME Y14.5.1M standard. *Manufacturing Review*, 7(1):16–23, March 1994.

[14] ANSI Y14.5M-1982. *Dimensioning and tolerancing.* American Society of Mechanical Engineers, New York, NY, 1982.

[15] Chee K. Yap. Exact computational geometry and tolerancing metrology. In David Avis and Jit Bose, editors, *Snapshots of Computational and Discrete Geometry, Vol.3.* McGill School of Comp.Sci, Tech.Rep. No.SOCS-94.50, 1994. A Volume Dedicated to Godfried Toussaint.

[16] Chee K. Yap and Ee-Chien Chang. A simultaneous searching problem, August, 1996. pickup site, URL http://cs.nyu.edu/cs/faculty/yap.

Kinematic Tolerance Analysis *

Leo Joskowicz, *The Hebrew University, Jerusalem, Israel*
Elisha Sacks, *Purdue University, West Lafayette, IN, USA*
Vijay Srinivasan, *IBM T.J. Watson Research Center, Yorktown Heights, NY, USA*
E-mail: josko@cs.huji.ac.il, eps@cs.purdue.edu, vijay@watson.ibm.com

We present a general method for worst-case limit kinematic tolerance analysis: computing the range of variation in the kinematic function of a mechanism from its part tolerance specifications. The method covers fixed and multiple contact mechanisms with parametric or geometric part tolerances. We develop a new model of kinematic variation, called kinematic tolerance space, that generalizes the configuration space representation of nominal kinematic function. Kinematic tolerance space captures quantitative and qualitative variations in kinematic function due to variations in part shape and part configuration. We derive properties of kinematic tolerance space that express the relationship between the nominal kinematics of mechanisms and their kinematic variations. Using these properties, we develop a practical kinematic tolerance space computation algorithm for planar pairs with two degrees of freedom.

1 Introduction

We present a new method of kinematic tolerance analysis based on configuration spaces. Kinematic tolerance analysis studies the variation in the kinematic function of mechanisms resulting from manufacturing variation in the shapes and configurations of their parts. The results support the synthesis of designs that reduce manufacturing cost by maintaining kinematic function under increased part variation. They help designers predict performance, uncover design flaws, and optimize tolerance allocation. Kinematic tolerance analysis complements tolerancing for assembly and for other design functions.

The kinematic function of a mechanism is the rela-

tionship among the motions of its parts imposed by contacts between them. For example, the kinematic function of an ideal gear pair is to transform an input rotation into an output rotation with a constant angular velocity ratio. The shapes and configurations of the parts determine the kinematic function under the assumption that they are rigid, hence cannot deform or overlap. Kinematic analysis computes the kinematic function by identifying the part contacts and deriving the resulting motion constraints. Variation in the part shapes and part configurations produces variation in the kinematic function. For example, variation in the gear profiles and in the configurations of the rotation axes causes backlash and transmission ratio variation. Kinematic tolerance analysis computes the kinematic variation from the part variations.

Kinematic models for tolerancing must account for multiple, changing contacts between irregular parts that generate complex, discontinuous kinematic functions. Nominal kinematic models are of limited use because they assume idealized contacts between nominal parts in permanent contact. For example, nominal linkage models assume that the links are permanently connected by ideal joints, nominal cam models assume that the cam and the follower have smooth profiles with a single, permanent contact, and nominal gear models assume that the teeth have perfectly meshed involute profiles. Part tolerances invalidate the idealized shape and contact assumptions.

We illustrate the effect of tolerances on the kinematic function of a Geneva mechanism (Figure 1). The driver consists of a driving pin and a locking arc segment mounted on a cylindrical base (not shown). The wheel consists of four locking arc segments and four slots. The driver rotates around axis O_d and the wheel

---
* This paper first appeared in *Computer-Aided Design*, 1996. It is reproduced with permission from the publisher.

rotates around axis O_w. In the nominal model, each rotation of the driver causes a nonuniform, intermittent rotation of the wheel with four drive periods where the driver pin engages the wheel slots and with four dwell periods where the driver locking segment engages the wheel locking segments. The pin fits perfectly into the slots, thus producing positive drive without play. Tolerancing the pin slightly smaller than the slots introduces play due to contact changes between the pin and the sides of the slots. The analysis must account for these contacts to measure the play and to determine the transmission ratio variation. Undercutting, interference, and jamming provide further examples of how part tolerances lead to multiple contacts that invalidate the idealized contact assumption.

Current kinematic tolerance analysis methods [5] are limited by intrinsic factors and by their reliance upon predefined kinematic models, which are available only for linkages and for a few higher pairs. Specialized methods are available for specific mechanisms with specific part tolerances, for example linkages whose links vary in length. The most common general methods are statistical sampling and sensitivity analysis. Statistical sampling estimates the kinematic variation by generating mechanism instances based on conjectured part tolerance distributions and comparing their kinematic functions with the nominal function. The method requires many instances to provide an accurate picture of the variation over the entire kinematic function and can miss important cases. Sensitivity analysis computes the kinematic variation under the assumption that the kinematic function is a smooth function of some tolerance parameters. It linearizes the function around the nominal parameter values and computes the variations in the linearized function. The method is limited to small deviations and cannot compute the effects of changing contacts because they violate the smoothness assumption.

In this paper, we present a general method for worst-case limit kinematic tolerance analysis: computing the range of variation in kinematic function from the part tolerance specifications. The method covers all mechanisms: planar and spatial, fixed topology and varying topology mechanisms consisting of linkages, gears, cams, and other higher pairs. It applies to parametric and geometric part tolerance specifications. We define a new model of kinematic variation, called kinematic tolerance space, that generalizes our configuration space model of nominal kinematics to cover part variations. Kinematic tolerance space encodes the quantitative effect of part variations on kinematic function along with possible qualitative effects, such as changes in operating mode or unintended functions. It generalizes kinematic function sensitivity analysis to large scale quantitative and qualitative variations by capturing the interplay between part variations and multiple, changing contacts. We derive the basic properties of kinematic tolerance space and describe an implemented kinematic tolerance space computation algorithm for planar pairs with two degrees of freedom. We conclude with a discussion of applications and of future work.

2 Nominal kinematics

We set the stage for kinematic tolerance analysis with a brief review of the configuration space method for nominal kinematic analysis. Configuration space provides a uniform geometrical model of kinematic function that is concise, complete, and explicit. It simplifies and systematizes kinematic analysis by reducing it to computational geometry.

2.1 Configuration space

The configuration space of a mechanism is the space of configurations (positions and orientations) of its parts. The dimension of the configuration space equals the number of degrees of freedom of the parts. For example, a nominal gear pair has a two-dimensional configuration space because each gear has one rotational degree of freedom. The gear orientations provide a natural coordinate system. Configuration space partitions into free space where the parts do not touch and into blocked space where some parts overlap. The common boundary, called contact space, contains the configurations where some parts touch without overlap and the rest do not touch. Only free space and contact space are physically realizable.

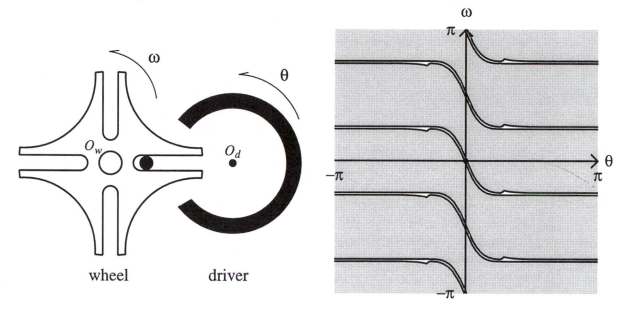

Figure 1: *Geneva mechanism and its configuration space. The mechanism is displayed in configuration* $\theta = 0$, $\omega = 0$, *marked by the dot at the configuration space origin.*

We illustrate these concepts on the Geneva mechanism (Figure 1). The configuration space is two-dimensional (each part has one degree of freedom), with coordinates the orientations θ and ω of the driver and the wheel. The shaded region is the blocked space where the driver and the wheel overlap. The white region is the free space. It forms a single channel that wraps around the horizontal and vertical boundaries, since the configurations at $\pm\pi$ coincide. The width of the channel measures the potential backlash of the pair. The curves that bound the free and blocked regions form the contact space. They encode the contact relations between the wheel and the driver. The horizontal segments represent contacts between the locking arc segments, which hold the wheel stationary. The diagonal segments represent contacts between the pin and the slots, which rotate the wheel.

The configuration space topology reflects the semantics of rigid body kinematics. Free space is an open set because free parts remain free under small motions. Blocked space is open because overlapping parts remain overlapping under small motions. Contact space is closed because it is the complement of the union

of free space and blocked space. It forms the common boundary of free space and blocked space because touching parts overlap when they move closer and become free when they move farther apart. Contact space partitions configuration space into connected components.

The configuration space encodes the space of kinematic functions under all driving motions. It represents the motion constraints induced by part contacts and the configurations where contacts change. The kinematic functions under specific driving motions are paths in configuration space that consist of contact and free segments separated by contact change configurations. For example, clockwise rotation of the driver produces a path that follows the bottom of the free space from right to left. The kinematic function consists of four horizontal segments alternating with four diagonal segments. The pin makes contact with the slot at the start of the diagonal segments and breaks contact at the end.

Another important property, called compositionality, is that the configuration space of a mechanism is determined by the configuration spaces of its pairs of parts [9]. We embed the pairwise configuration spaces in the mechanism configuration space by inverse projection. Each pairwise configuration (a, b) maps to the set of configurations $(a, b, \mathbf{x})$ where $\mathbf{x}$ varies over all values of the other coordinates. The mechanism free space equals the intersection of the embedded pairwise free spaces because a mechanism configuration is free when every pair of parts is free. The blocked space equals the union of the embedded pairwise blocked spaces because a mechanism configuration is blocked when at least two parts overlap.

Configuration space computation can be formulated in terms of algebraic geometry. The formulation requires that part shapes be specified as algebraic curves and surfaces. The kinematic condition that the parts cannot overlap is expressed by multivariate polynomial inequalities in the configuration space coordinates. The configurations that satisfy the constraints are the free and contact spaces. Computing this set takes time polynomial in the geometric complexity of the parts and exponential in the number of degrees of freedom with large constant factors [12].

2.2 The HIPAIR mechanism configuration space computation program

We [9, 10, 16, 17] have developed an efficient configuration space computation program, called HIPAIR, for planar mechanisms composed of linkages and of higher pairs with two degrees of freedom, such as gears and cams. HIPAIR covers 80% of higher pairs and most mechanisms based on a survey of 2500 mechanisms in Artobolevsky's [2] encyclopedia of mechanisms and on an informal survey of modern mechanisms, such as VCR's and photocopiers. Other researchers have developed algorithms for some higher pairs that HIPAIR does not cover, including planar polygonal pairs with three degrees of freedom [3, 4] and a polyhedral body with six degrees of freedom amidst polyhedral obstacles [6, 11].

HIPAIR computes the configuration space of a mechanism by composing the configuration spaces of its higher pairs and linkages. The correctness of this procedure follows from the compositionality of configuration space.

HIPAIR computes the configuration space of a higher pair from the contacts between the part features (vertices and curves on its boundary). The configuration space is two dimensional because the pair has two degrees of freedom, each of which is translation along a planar axis or rotation around an orthogonal axis. The contacts occur along curves in configuration space. The contact curves partition configuration space into connected components that form the free and blocked spaces. The component boundaries are sequences of contact curve segments that meet at curve intersection points where two contacts occur. The component that contains the initial configuration is the realizable space. HIPAIR enumerates the feature pairs, generates the contact curves, computes the planar partition with a line sweep, and retrieves the realizable component. The curves come from a table with entries for all combinations of part features and degrees of freedom. Figure 2 shows the contact curves, intersection points, and components in a portion of the Geneva mechanism configuration space where the driver locking segment disengages from a wheel locking segment and the driver pin engages in the adjacent wheel slot.

HIPAIR computes linkage configuration spaces by homotopy continuation. It composes the pairwise and linkage configuration spaces by linearizing the contact zone boundaries and intersecting them with the simplex algorithm. The result is a partition of the mechanism configuration space into free and blocked regions defined by linear inequalities in the motion coordinates. We visualize the configuration space, which can have any dimension, by projecting it onto the pairwise configuration spaces.

HIPAIR simulates the kinematic function of a mechanism by propagating driving motions through part contacts. The result is a configuration space path. HIPAIR constructs the mechanism region that contains the initial configuration, computes the segment of the motion path that lies in the region, replaces the initial configuration with the endpoint of the segment, and repeats the process. It computes the motion path in a region by combining the input motion with the contact constraints. In free configurations, the motion path is a line tangent to the input motion. In contact configu-

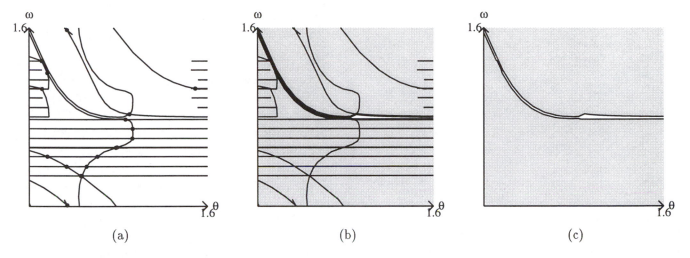

Figure 2: *Detail of configuration space computation for the Geneva mechanism: (a) contact curves and intersection points; (b) connected components and realizable space; (c) final configuration space.*

rations, the motion path is the projection of the input motion onto the tangent space of the contact surface (in 2D, onto the tangent to the contact curve).

HIPAIR is written in Common Lisp with a C graphics interface and runs on Unix workstations. It has been tested on over 1,000 parametric variations of 50 higher pairs with up to 10,000 contacts, and on a dozen mechanisms with up to ten moving parts. It computes the higher pair configuration spaces in under one second and the mechanism configuration spaces in under ten seconds. All the figures in this paper are annotated HIPAIR output.

3   Tolerance specifications

Tolerance specifications define the allowable variation in the shapes and configurations of the parts of mechanisms. The most common are parametric and geometric tolerance specifications [20]. Parametric specifications restrict shape and configuration parameters of part models to intervals of values. For example, a tolerance of $r = 1 \pm 0.1$ restricts the radius r of a disk to the interval $[0.9, 1.1]$. Geometric specifications restrict part features to zones around the nominal features, typically to fixed-width bands, called uniform profile tolerance zones, whose boundaries are the geometric inset and offset of the nominal features. For

example, a uniform geometric profile tolerance of 0.1 on a disk of radius 1 constrains its surface to lie inside an annulus with outer radius 1.1 and inner radius 0.9.

We define the variational class of a mechanism relative to some tolerance specifications as the set of mechanisms whose parts satisfy the tolerance specifications. This definition generalizes the standard definition of the variational class of a single part [15]. We define the kinematic variational class of a mechanism as the set of kinematic functions of the mechanisms in its variational class. The properties of the variational class of a mechanism determine its kinematic variational class. We define two properties, monotonicity and independence, that we will use in computing kinematic variational classes.

The variational class of a mechanism is monotone if it contains a maximal and a minimal instance. In every instance, each part is a subset of the corresponding maximal part and is a superset of the corresponding minimal part. Most geometric tolerance specifications generate monotone variational classes, including profile offsets, sweeps, and tolerance zones [1]. Parametric tolerance specifications rarely produce monotone variational classes because the values that maximize some features need not maximize others. In particular, position tolerances never lead to monotone instances because every instance has the same shape.

The variational class of a mechanism is independent if each part tolerance is specified with respect to its own reference datum. It is equal to the cross product of the part variational classes. Independent tolerance specifications facilitate the engineering goal of designing parts that can be manufactured and gauged independently [1]. Shape tolerances are usually independent for manufacturing reasons. Assembly tolerance specifications, which express functional relations between parts, are sometimes dependent. Position tolerances are independent when they refer to a common external datum or a single reference part. They are dependent when they refer to multiple parts or external datums. Tolerance specifications for pairs of parts are always independent, as one of the parts serves as the reference object. One-dimensional chains of parametric tolerances and vectorial tolerances are also independent.

4 Kinematic tolerance space

We now describe a new method for kinematic tolerance analysis. We perform kinematic tolerance analysis by generalizing our nominal kinematic analysis to variational classes of mechanisms. We generalize configuration space to kinematic tolerance space, a uniform geometrical model of kinematic variation that is concise, complete, and explicit. Kinematic tolerance space simplifies and systematizes kinematic tolerance analysis by reducing it to computational geometry.

We model the kinematic variational class of a mechanism with a kinematic tolerance space. The kinematic tolerance space is a partition of configuration space into free, blocked, and contact zones that model the range of variation from the free, blocked, and contact spaces of the nominal mechanism. The free zone, defined as the intersection of the free spaces of the variational class, is the set of configurations for which part instances in the variational class are free. It is the portion of the nominal free space that is guaranteed to persist. The blocked zone, defined as the intersection of the blocked spaces, is the set of configurations for which part instances in the variational class are blocked. It is the portion of the nominal blocked space that is guaran-

teed to persist. The contact zone, defined as the union of the nominal contact spaces, bounds the deviation from the nominal kinematic function.

Figure 3 shows a detail of the kinematic tolerance space of the Geneva mechanism with a uniform geometric profile tolerance of 0.05 units (0.5% of the diameter of the pair). The figure shows the region where the driver locking segment disengages from the wheel locking segment and the driver pin engages the slot of the wheel. In the detail of the pair, the dashed lines mark the nominal part shapes and the shaded bands represent their variational classes. In the kinematic tolerance space, the free zone is white, the contact zone is light grey, and the blocked zone is dark grey. The dashed curves mark the nominal contact space. Its outside boundary corresponds to the kinematic function of the minimal part shapes, while its inside boundary corresponds to the kinematic function of the maximal part shapes.

The structure of the kinematic tolerance space of a mechanism depends on the structure of its variational class, which is determined by the tolerance specifications. For example, Figure 4 shows the kinematic tolerance space for a parametric tolerance of 0.05 units on the horizontal and vertical distances between the centers of rotation of the wheel and the driver (parameters a and b). Comparing it with Figure 3, we see that the uniform geometric profile tolerance produces a narrow, uniform-width free zone with a smooth boundary, whereas the parametric tolerances produce a broader, irregular free zone with a jagged boundary that marks changes in sensitivity due to contact changes.

The topology of the kinematic tolerance space reflects the semantics of kinematic tolerances. The free and blocked zones are subsets of the free and blocked spaces of the nominal mechanism because configurations that are free or blocked for the entire variational class are free or blocked for every instance. They are open sets because configurations that are free or blocked for the class remain so under small motions. The contact zone equals the complement of the union of the free and blocked zones, hence is closed. (The proofs of the following facts appear in the Appendix.)

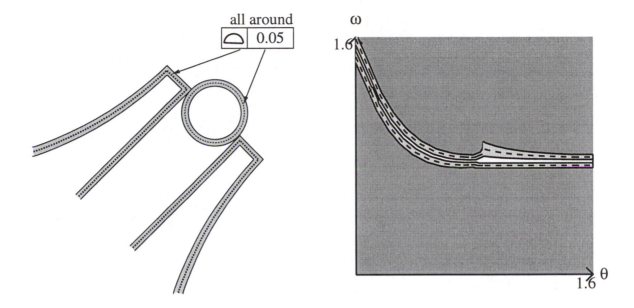

Figure 3: *Detail of the Geneva mechanism with uniform geometric profile tolerance of 0.05 units and a detail of its kinematic tolerance space.*

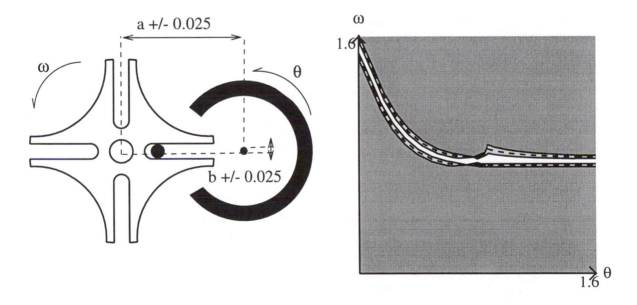

Figure 4: *Geneva mechanism with a parametric tolerance specification of 0.05 units and a detail of its kinematic tolerance space.*

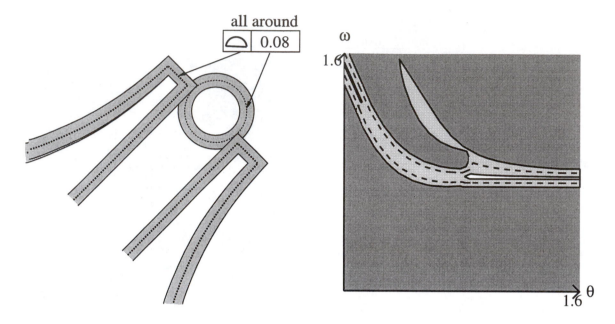

Figure 5: *Detail of the Geneva mechanism with uniform geometric profile tolerance of* 0.08 *units and a detail of its kinematic tolerance space.*

The contact zone is a superset of the nominal contact space because nominal contact configurations do not belong to the free or blocked zones. The contact zone boundary bounds the free zone, the blocked zone, and the (possibly empty) interior of the contact zone. It partitions kinematic tolerance space into connected components just as contact space partitions configuration space.

The kinematic tolerance space models the variations from the nominal kinematics. With small tolerances, the free and blocked zones are shrunken versions of the nominal free and blocked spaces, whereas the contact zone is a fattened version of the nominal contact space. The contact zone bounds the quantitative variation from the nominal kinematics. Qualitative changes in kinematics do not occur because the free zone, hence every free space in the variational class, is homeomorphic to the nominal free space. With larger tolerances, the contact zone can grow fat enough to alter the free zone topology, thus producing qualitative changes in kinematics.

We observe quantitative and qualitative variation in the Geneva mechanism. A uniform geometric profile tolerance of 0.05 units produces quantitative variation in kinematic function because the contact zone is narrower than the nominal free space (Figure 3). A profile tolerance of 0.08 units produces qualitative changes because the contact zone pinches closed the nominal free space and carves necks in the nominal blocked space (Figure 5). The qualitative changes are most pronounced in the mechanisms with maximal and minimal parts (Figure 6). The maximal free space shrinks to four partial channels. The driver pin cannot fit in the wheel slots, hence cannot rotate the wheel. The minimal free space grows to include necks at the junctions of the horizontal and diagonal segments. The driver pin can slip out of the wheel slots and jam against the locking segments.

Kinematic tolerance space has a simple structure when the variational classes of the parts are monotone. The free zone equals the free space of the maximal parts because it is a subset of every free space. The blocked zone equals the blocked space of the minimal parts because it is a subset of every blocked space. We

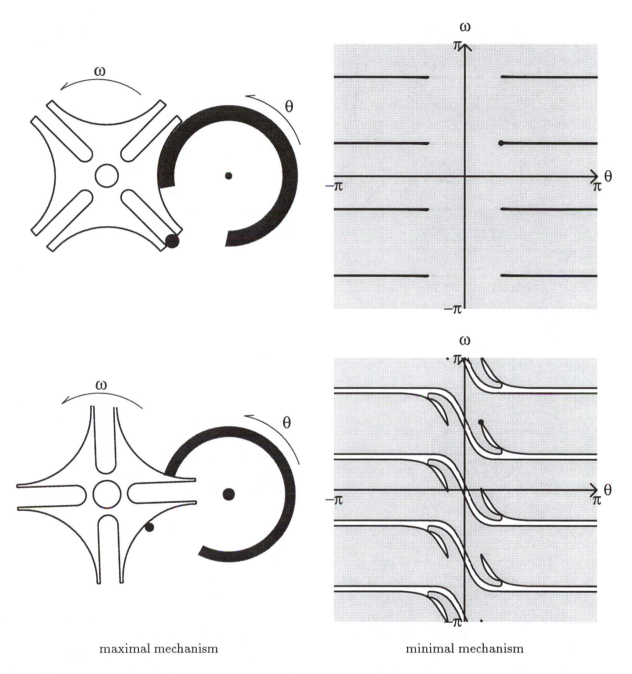

maximal mechanism   minimal mechanism

Figure 6: *Maximal and minimal Geneva mechanisms with a 0.08 unit uniform geometric profile tolerance and their configuration spaces. The mechanisms are displayed in jamming configurations, which correspond to the dots in the configuration spaces.*

illustrate this property on the Geneva mechanism with the profile tolerance of 0.08 units. Comparing Figure 5 with Figure 6, we see that the free zone equals the free space of the maximal mechanism and that the blocked zone equals the blocked space of the minimal mechanism.

Kinematic tolerance space has a compositional structure akin to that of configuration space for mechanisms with independent part tolerances. The free zone equals the intersection of the embedded pairwise free zones because a mechanism configuration is free when every instance of every pair of parts is free. The blocked zone equals the union of the embedded pairwise blocked zones because a mechanism configuration is blocked when some instance of some pair of parts is blocked.

5   Kinematic tolerance space computation

Computing kinematic tolerance spaces is at least as difficult as computing configuration spaces. We have developed a kinematic tolerance space computation program for planar pairs with two degrees of freedom. We follow the HIPAIR strategy of exploiting domain properties to manage the worst-case complexity of the computation. The program generalizes HIPAIR and reuses several subroutines. We discuss only those steps that differ from the HIPAIR algorithm.

We exploit the topological property that the contact zone boundary bounds the free zone, the blocked zone, and the contact zone interior. We compute the contact zone boundary, derive the kinematic tolerance space partition that it induces with HIPAIR, and classify the components. The classification algorithm employs point/region containment and region adjacency. The components that contain endpoints of nominal contact curves form the contact zone. The free zone consists of those remaining components in which a single, arbitrary configuration belongs to the nominal free space. The rest of the components form the blocked zone.

We compute the exact contact zone boundary for uniform geometric profile tolerances. The contact zone equals the complement of the union of the free and

blocked zones, as proved in the Appendix. Its boundary equals the union of their boundaries because they are open. The free and blocked zones equal the free space of the maximal pair and the blocked space of the minimal pair by monotonicity. Their boundaries are the maximal and minimal contact spaces. We compute the contact spaces of the pairs with HIPAIR and construct the contact, free, and blocked zones from them with a line sweep algorithm. For example, Figure 6 shows the contact spaces of the maximal and minimal Geneva mechanisms, which form the contact zone boundary shown in Figure 5.

Computing the exact contact zone boundary for parametric tolerances is more difficult because the pairs are not monotone. The general solution is to compute the extremal contact curves with respect to the parameters, which is a computationally expensive functional optimization. Instead, we approximate the boundary under the assumption that the tolerances are small with respect to the kinematic function. The method generalizes sensitivity analysis to varying contacts. We denote the tolerance vector by $\mathbf{p}$, its nominal value by $\overline{\mathbf{p}}$, its lower bound by $\mathbf{l}$, and its upper bound by $\mathbf{u}$. The nominal contact space consists of contact curves of the form $y = f(x, \mathbf{p})$ with x and y the kinematic tolerance space coordinates. We approximate the contact zone boundary by linearizing the contact curves around $\overline{\mathbf{p}}$ and offsetting them by the variation in y. The y variation is defined by:

$$\delta y = \sum_i \frac{\partial f}{\partial p_i}(w_i - \overline{p}_i) \text{ with } w_i = \begin{cases} u_i & \text{if } \partial f/\partial p_i > 0 \\ l_i & \text{otherwise.} \end{cases}$$

The ith term is the maximal linear variation in y induced by p_i variations. It occurs when p_i is maximal or minimal depending on whether y increases or decreases with p_i. We compute the contact curves with HIPAIR, discretize them with respect to x, offset each sample point by $\pm \delta y$, and connect the offset points into upper and lower curves that approximate the contact zone boundary. For example, Figure 4 shows the kinematic tolerance space of the Geneva mechanism with the tolerance vector $\mathbf{p} = (a, b)$.

Computing the kinematic tolerance space for geometric tolerance specifications requires roughly three

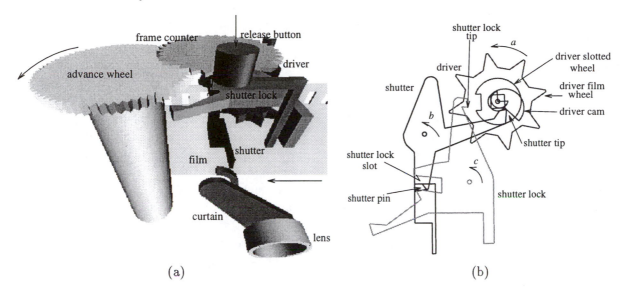

Figure 7: *Disposable camera: (a) shutter mechanism; (b) top view of driver, shutter, and shutter lock assembly.*

times as much time as computing the nominal configuration spaces: one for the maximal contact space, one for the minimal contact space, and one to construct the zones. The kinematic tolerance spaces in Figures 3 and 5 were computed in four seconds apiece on an Iris Indigo 2 workstation. The computation time for parametric tolerances depends on the number of parameters and on the linearization accuracy. The kinematic tolerance space in Figure 4 was computed to an accuracy of 0.01% in one second.

6 Kinematic tolerance analysis

Kinematic tolerance space provides the information for systematic kinematic tolerance analysis. Free zones whose topology differs from that of the nominal free space indicate possible failure modes, such as undercutting, interference, and jamming. Contact zones describe the variability of the contact function. The width of the contact zone bounds the quantitative deviation in kinematic function. Contact zones that contain horizontal or vertical segments indicate that the slope of the nominal contact space can change, which interchanges locking and driving functions. Contact relations that hold in the contact zone hold in every instance, whereas ones that fail in some contact

zone configurations may fail in some instances. The linearized contact equations, which define the contact zone, specify the kinematic variation in terms of the part variations, thus supporting sensitivity analysis of higher pair assemblies with changing contacts.

We demonstrate the role of kinematic tolerance space in kinematic tolerance analysis by means of an extended example involving the shutter mechanism of a disposable camera (Figure 7a). The shutter mechanism consists of 10 higher pairs, none of which have standard kinematic or tolerance models. The advance wheel moves the film forward by one frame and rotates the driver cam, which engages the spring-loaded shutter in the shutter lock. Pressing the release button rotates the shutter lock, thus releasing the shutter. The shutter trips the spring-loaded curtain, which briefly rotates away from the lens and exposes the film.

We focus on the loading sequence of the driver, shutter, and shutter lock, which is the most complex kinematic function (Figure 7b). The driver consists of three planar pieces: a cam, a slotted wheel, and a film wheel mounted on a shaft. The shutter consists of two planar pieces, a tip and a pin, and is spring-loaded counterclockwise. The shutter lock is planar and is spring-loaded clockwise. The driver cam interacts with the shutter tip. The driver slotted wheel interacts with the shutter lock tip. The shutter pin interacts with the shutter lock slot.

The film advance continuously rotates the driver counterclockwise via the film wheel. In the initial configuration (snapshot 1 of Figure 8), the shutter tip lies on the driver cam, the shutter pin lies in the shutter lock slot, and the shutter lock does not touch the slotted wheel. As the driver rotates counterclockwise, the driver cam rotates the shutter clockwise by pushing the shutter tip (snapshot 2). The shutter pin leaves the slot in the shutter lock (snapshot 3). The shutter lock spring rotates the shutter lock clockwise until the tip touches the driver slotted wheel (snapshot 4). The tip then follows the wheel contour. When the shutter tip passes the highest point of the driver cam, it breaks contact with the cam. The shutter spring forces the shutter to rotate counterclockwise, causing the pin to engage the shutter lock on the surface below the slot (snapshot 5). The loading sequence ends when the shutter lock tip drops into the driver wheel slot and blocks further rotation (snapshot 6).

We analyze the kinematic variation under a uniform geometric profile tolerance of 1 unit for all parts (0.5% of the diameter of the smallest part). The kinematic tolerance spaces of the kinematic pairs reveal the sensitivity of the pairwise kinematic functions to part deviations (Figure 9). (The three spaces were computed in between 0.75 and 1.25 seconds apiece.) The driver/shutter function is insensitive to the part tolerances. The shutter tip follows the driver cam profile with a small deviation from its nominal path. The driver/shutter lock space also shows insensitivity to the part tolerances since the vertical channel is roughly the same as the nominal channel. The shutter lock tip follows the contour of the driver slotted wheel until it drops into the driver wheel slot. A larger tolerance can eliminate the channel and prevent blocking. The shutter/shutter lock space shows that small deviations in the pin and the slot can cause the shutter lock to release the shutter spontaneously under the action of the spring. This happens when the lower boundary of the horizontal slot in the blocked space has a positive slope. Shape deviations of 0.1 units can make the slope positive because the nominal value is a small negative number.

The tolerance spaces reveal the sensitivity of the mechanism function to part deviations. Coupling between part interactions leads to a global failure mode in which the driver cam does not push the shutter tip far enough for the shutter pin to clear the slot in the shutter lock. The failure occurs in mechanism instances where the value of b, b_1, at the lowest point of the driver/shutter contact space is greater than the value b_2 at the bottom, left corner of the horizontal slot in the shutter/shutter lock contact space. Figure 10 shows the projections of the nominal and failure modes in the driver/shutter configuration space. In the nominal mode, the configuration follows the lower segment of the horizontal slot from right to left, drops into free space at the left end of the horizontal slot when the shutter pin clears the shutter lock slot, and moves right and down as the shutter pin engages below the shutter lock slot. In the failure mode, the configuration follows the horizontal slot from right to left, does not reach the left end of the horizontal slot, and returns from left to right without engaging the pin in the slot.

The tolerance spaces show that the variational class of mechanisms may contain a failing instance where $b_1 > b_2$ because $-0.26 \leq b_1 \leq -0.22$ and $-0.24 \leq b_2 \leq -0.15$. This does not guarantee the existence of an instance in which $b_1 > b_2$ because the shutter effects the values in opposite ways. Increasing the shutter size decreases b_1 by shifting the contact between its tip and the driver cam, but also decreases b_2 by increasing the size of the pin, which must clear the shutter lock slot. We compute the maximum of $b_1 - b_2$ over the variational class by binary search in the shutter offset, using the monotonicity in the other parts. (The kinematic tolerance spaces contain the equations for more sophisticated optimization algorithms.) The maximum occurs when the driver is minimal, the shutter lock is maximal, and the shutter offset is 0.72 units. The mechanism works correctly for this value, but fails when we increase the shutter lock offset from 1 unit to 1.25 units. Manual analysis can easily miss this kind of failure.

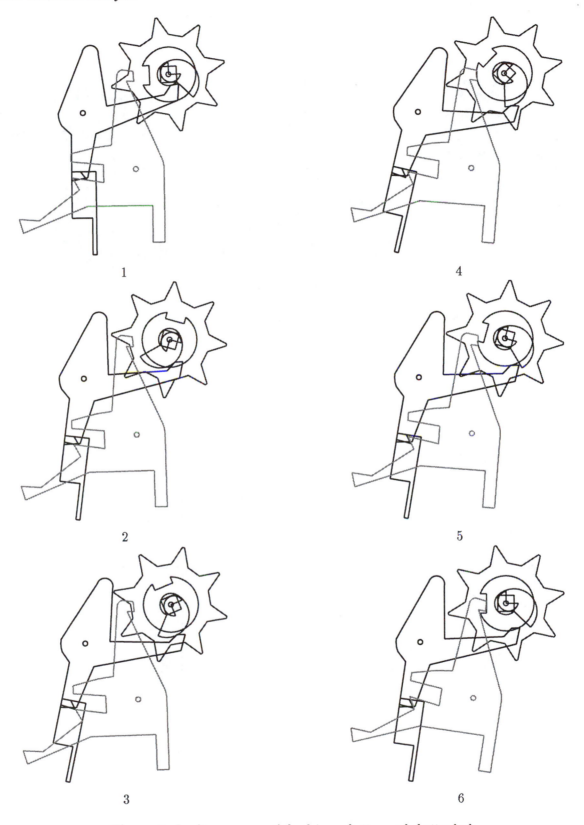

Figure 8: *Loading sequence of the driver, shutter, and shutter lock.*

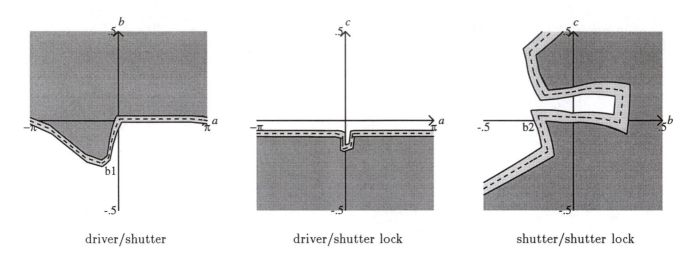

driver/shutter   driver/shutter lock   shutter/shutter lock

Figure 9: *Shutter mechanism pairwise tolerance spaces.*

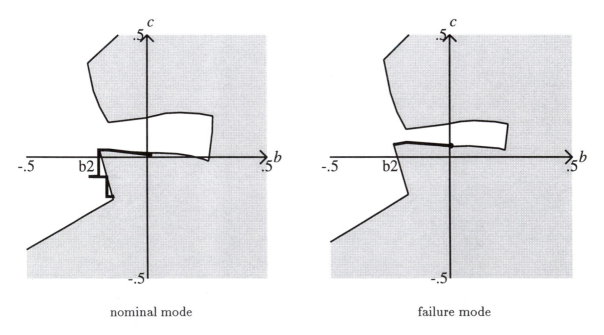

nominal mode   failure mode

Figure 10: *Projections of kinematic function onto the driver/shutter lock configuration space.*

7 Conclusion

We present a general method for worst-case limit kinematic tolerance analysis: computing the range of kinematic variation of a mechanism from its tolerance specifications. The method covers all types of mechanisms with parametric or geometric part tolerances. We develop a model of kinematic variation, called kinematic tolerance space, that generalizes the configuration space representation of kinematics. We derive properties of kinematic tolerance space that express the relationship between the nominal kinematics of mechanisms and their kinematic variations. Using these properties, we develop an efficient kinematic tolerance space computation program for planar pairs with two degrees of freedom.

Our work advances the state of the art in kinematic modeling for tolerancing. Kinematic tolerance space is the first general representation for variational classes of kinematic functions. It represents fixed and changing contacts and quantitative and qualitative variations in kinematic function. It represents the nominal kinematics and its variations in a uniform manner, thus eliminating the need for a separate tolerance model. The kinematic tolerance space computation program automates a significant part of kinematic tolerance modeling. It is also useful for related tasks in mechanism analysis and design. For example, we can model part wear with geometric tolerances and compute the kinematic tolerance space to analyze the consequences of the wear. We can fine-tune designs by specifying a range of variations and selecting the best parameter value combination as the nominal specification. Tolerancing for assembly can also be studied with kinematic tolerance spaces. The task is to compute whether every mechanism in the variational class can be assembled, which involves reasoning about the variational class of kinematic function under assembly motions [7, 21, 13].

Our analysis focuses on the effects of part variations on the nominal degrees of freedom of mechanisms. As a consequence, the dimension of the configuration space and the kinematic tolerance space are equal. A full analysis would require computing the effects on all six degrees of freedom of every part, including those that are fixed in the nominal model. We could compute high-dimensional kinematic tolerance spaces to perform this analysis. A more practical approach is to model the added degrees of freedom as linear perturbations of the nominal degrees of freedom, as is common in linkage tolerancing [14]. For example, the Geneva mechanism has two nominal degrees of freedom because the other ten are fixed by the perfect fits between the wheel and the driver and their mounting shafts. We compute a two-dimensional kinematic tolerance space that neglects imperfect fit. We can model play due to imperfect fit with a 12-dimensional space or as a linear perturbation of the two-dimensional space.

We [?] have recently developed a kinematic tolerance analysis algorithm for assemblies of planar, two degree of freedom pairs with independent part tolerances. For example, the algorithm computes the sensitivity of the camera shutter function to simultaneous variations in the driver, shutter, and shutter lock. We avoid the exponential cost of computing high dimensional kinematic tolerance spaces by computing only the one-dimensional subspace traversed by the actual mechanism function.

We see several directions for future work. We plan to test the practicality of kinematic tolerance space computation on industrial tolerancing tasks. This will necessitate extensions to the kinematic tolerance space computation algorithm. We must compute kinematic tolerance spaces for pairs with mixed parametric and geometric tolerance specifications, which arise when part shapes and configurations vary simultaneously. We must automate kinematic tolerance space interpretation tasks, such as detecting qualitative changes in kinematic function or measuring the maximum play. We would like to extend the coverage to spatial pairs and to pairs with three or more nominal degrees of freedom.

8 Update

One year after writing this paper, we have developed a comprehensive, efficient algorithm for kinematic tolerance analysis of mechanisms with parametric part tolerances. The algorithm covers multi-pair assemblies with many parameters. It is orders of magnitude faster than the algorithm reported here and produces both worst-case and statistical analysis results. The extensions are reported in [19] (submitted for publication).

The new algorithm for pairs computes the exact kinematic variation and runs in time proportional to the number of parameters times the number of contacts, whereas the old running time also depends on the desired accuracy. It analyzes pairs with tens of parameters in well under a minute, enabling the interactive analysis of detailed functional models of complicated mechanisms.

The multi-pair algorithm computes the kinematic variation under a given driving motion, rather than computing the entire kinematic tolerance space. This analysis is appropriate for most mechanisms, since designers are usually interested in quantifying deviations in a few operating modes. The driving motion defines a path in the nominal configuration space of the mechanism that represents the operating mode. We compute the nominal path and the portion of the contact zone that surrounds the path. The result is a sensitivity analysis along the path with discontinuities at contact change configurations. The computation is simple and fast because the variation occurs on a one-dimensional set instead of on the entire nominal contact space.

The statistical kinematic tolerance analysis algorithms cover pairs and multi-pair assemblies. The inputs are the pairwise contact zones, the nominal motion path, and the joint distribution of the tolerance parameters. The outputs are the distributions of the kinematic variation in the contact zones and along the motion path. The central task is to compute the distribution of a kinematic function $y = f(\mathbf{x}, \mathbf{p})$ from the joint distribution $\mathbf{p}$ of the tolerance parameters at a given $\mathbf{x}$ value. In pairwise analysis, we compute a distribution for each contact zone region, while in mechanism analysis we compute a distribution for each motion path segment. We approximate the distribution as a linear combination of the parameter distributions and analyze it by standard probabilistic methods. We ignore nonlinear kinematic variations, as is standard in the field, because the contact curves are smooth and the parameter variation is small.

We demonstrate the algorithms on detailed functional models of the Geneva pair and of a camera shutter mechanism. The examples show that the program provides a comprehensive description of the kinematic variation and helps identify and quantify subtle failures. These results subsume the analyses in this paper because the new part models are more detailed and more general than the old models.

Appendix: Proof of topological properties

We prove that the free zone of a kinematic tolerance space is open, the blocked zone is closed, and the contact zone equals the complement of the free and blocked zones. We treat the cases of uniform geometric profile tolerances and of parametric tolerances where the parameters are defined on closed intervals and the part shapes and configurations are continuous functions of the parameters.

The first two results are immediate for monotone tolerance specifications, which include uniform geometric profile tolerances. The free zone equals the free space of the minimal mechanism and the blocked zone equals the blocked space of the maximal mechanism. Both spaces are open because free and blocked configurations are preserved by sufficiently small motions. We prove the parametric result via properties of continuous functions on compact spaces. The relevant function for the free zone, called the Hausdorff metric, is the distance between the closest pair of points on two parts of the mechanism. Let $d(x, \mathbf{p})$ denote the Hausdorff distance in configuration x with $\mathbf{p}$ the tolerance parameters. The function d is continuous because the parts depend continuously on $\mathbf{p}$ by hypothesis. Let P denote the domain of $\mathbf{p}$, X denote the domain of x, and x_0 be a point in x. The set $x_0 \times P$ is compact because P is a cross-product of closed intervals by hypothesis. We use these properties to prove that the free zone is open. If x_0 belongs to the free zone, d is positive on $x_0 \times P$ by definition. Each point in $x_0 \times P$ has a positive neighborhood in $X \times P$ by the continuity of d. The union of these neighborhoods is an open covering of $x_0 \times P$, which has a finite subcovering by compactness. The projection into X of the intersection of the subcovering is a neighborhood of x_0 on which d is positive. An analogous proof applies to the blocked zone with the overlap area of the parts replacing the Hausdorff distance.

We prove that the contact zone equals the complement of the free and blocked zones for parametric tolerances. The proof covers offsets where the parts are continuous functions of the offset radii, which includes uniform geometric profile tolerances. Given a configuration, we define a function $f(\mathbf{p})$ whose value is the Hausdorff distance between the parts if $\mathbf{p}$ defines a free instance, 0 if $\mathbf{p}$ defines a contact instance, and the negative overlap area of the parts if $\mathbf{p}$ defines a blocked instance. The function f is continuous at free and blocked $\mathbf{p}$ values by elementary calculus. It is continuous at contact values because the Hausdorff distance and the overlap area are both 0. It is positive at free values, zero at contact values, and negative at blocked values. Given a configuration that is not in the free zone or the blocked zone, f cannot be positive for all $\mathbf{p}$ or negative for all $\mathbf{p}$. It must be zero for some $\mathbf{p}$ by the intermediate value theorem. The configuration belongs to the contact zone because this $\mathbf{p}$ value yields a contact instance.

Acknowledgments

We thank M. Jakiela and R. Gupta of the MIT Mechanical Engineering Department for providing preliminary parametric models of the camera parts. Elisha Sacks is in part supported by NSF grant CCR-9505745 and by the Purdue Center for Computational Image Analysis and Scientific Visualization.

References

[1] The American Society of Mechanical Engineers, New York. *ASME Y14.5M-1994 Dimensioning and Tolerancing Standard*, 1994.

[2] Artobolevsky, I. *Mechanisms in Modern Engineering Design*, volume 1–4. (MIR Publishers, Moscow, 1979). English translation.

[3] Brost, R. C. Computing metric and topological properties of configuration-space obstacles. in: *Proceedings IEEE Conference on Robotics and Automation*, pages 170–176, 1989.

[4] Caine, M. E. The design of shape interactions using motion constraints. in: *Proceedings of the IEEE International Conference on Robotics and Automation*, pages 366–371, 1994.

[5] Chase, K. W. and Parkinson, A. R. A survey of research in the application of tolerance analysis to the design of mechanical assemblies. *Research in Engineering Design* 3 (1991) 23–37.

[6] Donald, B. R. A search algorithm for motion planning with six degrees of freedom. *Artificial Intelligence* 31 (1987) 295–353.

[7] Giordano, M. and Duret, D. Clearance space and deviation space. in: *Proceedings of the 3rd CIRP Seminar on Computer-Aided Tolerancing*, pages 184–196, 1993.

[8] Goldberg, K., Halperin, D., Latombe, J., et al. (Eds.). *The Algorithmic Foundations of Robotics*. (A. K. Peters, Boston, MA, 1995).

[9] Joskowicz, L. and Sacks, E. Computational kinematics. *Artificial Intelligence* 51 (1991) 381–416. reprinted in [8].

[10] Joskowicz, L. and Sacks, E. Configuration space computation for mechanism design. in: *Proceedings of the 1994 IEEE International Conference on Robotics and Automation*. IEEE Computer Society Press, 1994.

[11] Joskowicz, L. and Taylor, R. H. Interference-Free Insertion of a Solid Body into a Cavity: An Algorithm and a Medical Application. *International Journal of Robotics Research*, June 1996, Vol. 15 No. 3, MIT Press, pp 1-17.

[12] Latombe, J.-C. *Robot Motion Planning*. (Kluwer Academic Publishers, 1991).

[13] Latombe, J.-C. and Wilson, R. Assembly sequencing with toleranced parts. in: *Third Symposium on Solid Modeling and Applications*, 1995.

[14] Lee, S. J. and Gilmore, B. J. Determination of the probabilistic properties of velocities and accelerations in kinematic chains with uncertainty. *Journal of Mechanical Design* **113** (1991).

[15] Requicha, A. A. G. Mathematical definition of tolerance specifications. *Manufacturing Review* **6** (1993).

[16] Sacks, E. and Joskowicz, L. Automated modeling and kinematic simulation of mechanisms. *Computer-Aided Design* **25** (1993) 106–118.

[17] Sacks, E. and Joskowicz, L. Computational kinematic analysis of higher pairs with multiple contacts. *Journal of Mechanical Design* **117** (1995) 269–277.

[18] Sacks, E. and Joskowicz, L. Parametric kinematic tolerance analysis of plannar pairs with multiple contacts. *Proc. of the 8th International ASME Design Automation Conference*, Irvine, CA 1996.

[19] Sacks, E. and Joskowicz, L. Parametric kinematic tolerance analysis of plannar mechanisms. Submitted for publication, 1996.

[20] Voelcker, H. A current perspective on tolerancing and metrology. *Manufacturing Review* **6** (1993).

[21] Wilson, R. and Latombe, J.-C. Geometric reasoning about mechanical assembly. *Artificial Intelligence* **71** (1994) 371–396.

Dynamic Simulation:
Model, Basic Algorithms, and Optimization

Ammar Joukhadar, *INRIA Rhône Alpes & GRAVIR, 38330 Montbonnot Saint-Martin, FRANCE*
Christian Laugier, *INRIA Rhône Alpes & GRAVIR, 38330 Montbonnot Saint-Martin, FRANCE*

We describe models and algorithms designed to produce efficient and physically consistent dynamic simulations. These models and algorithms have been implemented within the RobotΦ system[13] which can potentially be configured for a large variety of intervention-style tasks such as dextrous manipulations with a robot hand; manipulation of non-rigid objects; teleprogramming of the motions of an all-terrain vehicle; and some robot assisted surgery tasks (e.g. positioning of an artificial ligament in knee surgery). The approach uses a novel physically based modeling technique to produce dynamic simulations which are both efficient and consistent according to the laws of the Physics. The main advances over previous works in Robotics and Computer Graphics fields are twofold: the development of a unique framework for simultaneously processing motions, deformations, and physical interactions; and the incorporation of appropriate models and algorithms for obtaining efficient processing times while insuring consistent physical behaviors.

1 Introduction

Some robotic tasks involve complex contact interactions whose characteristics may strongly modify the behavior of the robot (and consequently the probability of success of the task). Such situations occur, for instance, when manipulating a rigid or a deformable object using a dextrous hand [1], when moving an all-terrain vehicle on a hilly terrain [5], or when evaluating the influence of the ligament graft position and initial tension on the reconstructed knee kinematics[17]. On one hand, it is seldom easy for a human operator to guess what the robot behavior will be when the com-

plexities of robot kinematics, environmental collisions and contact, motions and deformations of interacting objects are combined with the task constraints. On the other hand, it is, obviously, out of the state of art to include all these considerations in a planner. Consequently, powerful dynamic simulation tools are required.

Classical geometrical models developed in the field of CAD-Robotics and of motion planning are obviously not adapted to the processing of such interactions. Indeed, the purpose of geometrical models is to represent spatial and shape properties of objects, while complex contact interactions and object behaviors mainly depend on physical properties such as mass, mass distribution, rigidity/elasticity factors, viscosity, collision forces. Dealing with these problems requires the use of appropriate models which have the capability to represent motions, deformations, and objects interactions in an unified framework consistent with the laws of Dynamics. This is why we have developed a system (the RobotΦ system [2]) which exhibits such characteristics. This system uses a novel physically based modeling technique to produce dynamic simulations which are both efficient and consistent with the laws of the Physics. The main advances over previous works in the Robotics and Computer Graphics fields are twofold: the development of an unified framework for processing simultaneously motions, deformations, and physical interactions; and the incorporation of appropriate models and algorithms for obtaining efficient processing time while insuring consistent physical behaviors. The next sections respectively describe the problem to be solved, the main characteristics of the Robotφ system, the *inertia-based adaptive discretization* principle, the *energy-based adaptive time step* principle, the fast contact detection between deformable polyhedra, and some experimental results.

2 Modeling problems and related works

Usually, Robotics systems combine geometric models and partial physical models[1] in order to study contacts [6], to simulate collision effects [19], to avoid obstacles [21], or to study stability properties[24]. These approaches are generally efficient when solving the considered sub-problems under some restrictions such as, completely rigid objects[23], or ballistic motion[19]. Constructing a complete dynamic simulation system, requires the combination of three partial physical models: motion, deformation, and contact interaction. Combining existing efficient partial models is not always possible, and even when it is possible, it may give a non efficient model. For instance, the efficient motion model "rigid body dynamic" and the deformation model "finite element" can not be automatically combined because they use paradoxical conditions (rigid and deformable). Also, the "finite element" is not an efficient model, because it needs at least a squared time $O(n^2)$ to calculate deformations.

Several physically based models have been developed in the field of Computer Graphics in order to produce behavioral animation [4, 8, 18, 15]. Among these models, the "Spring based approaches", have been widely used [8, 15]. This model is initially developed in the mechanics domain to study the traction [11] and flexion (Timoshenko-beam model[9]) behavior of objects.

We have already shown in [5] how this type of model could be applied with some appropriate improvements for solving robotics applications involving strong physical interaction constraints; we have also shown in [2] that some computation problems have to be solved when dealing with robotics applications. Indeed, robotics applications differ from graphics applications by the fact that graphic systems try to generate "looks right" animated pictures (even if some basic

principles of Physics are violated), while Robotics systems have to control "real" mechanical systems (and consequently, the dynamics laws governing it, have to be satisfied at all time). The reason for some inconsistent behaviors exhibited by the Particles based systems comes from the large number of particles of the models and the associated numerical and algorithmic problems (time and space discretization, error accumulations). Such problems generate numerical divergences which may lead the system to violate some basic principles of the Physics (for instance, the mechanical energy of an object increases without any physical reason). This is why appropriate models and algorithms have to be integrated in these approaches in order to be able to consider Robotics applications.

The $Robot\Phi$ system, which generalize the "particles models", has been devised in order to meet these requirements [2]. In this paper, we will focus on the modeling and algorithmic problems we have recently solved for making the system more efficient and consistent at any time with the basic principles Physics. We will show how we have drastically reduced the required computing time, and how the consistency properties can be guaranteed by controlling the incurred errors and the numerical convergence of the differential equations solver:

- Energy-based adaptive time step : this approach allows us to estimate the incurred error, to minimize the computational time and to avoid numerical divergence.

- Inertia-based adaptive discretization: this approach allows us to minimize the number of particles needed to represent an object.

- surface/surface interaction: this approach allows us to detect and localize contact between deformable polyhedra in linear time $O(n)$.

3 Outline of the $Robot\Phi$ system

The $Robot\Phi$ system basically allows the construction of complex physical scenes involving various interacting objects and robots (i.e. articulated quasi-rigid ob-

---

[1]Partial physical model represent the motion, the deformation, or the interaction but not the three aspects at the same time

jects), and the "control" of these scenes using appropriate force-based operators. In practice, the user can: (1) construct the physical models of a set of objects (rigid, deformable or articulated objects), (2) control the motions of these objects by applying selected external forces, (3) simulate the dynamic behavior of the involved objects, and (4) define and record the motion parameters which are relevant for programming the Robotic application considered. The main features of our approach are the following:

- *Robot*Φ is a generalization of the spring/mass system. An object is represented by a network of interacting primitive connected by appropriate damper/ spring connectors. A primitive is an object whose motion and deformations, between two time step and under the effect of a given force, can be given by an existing dynamic model (dynamics of point, dynamics of rigid object, Lagrangian formulation, etc..). At the moment, the available models are the point mechanics and the rigid body mechanics.

- The geometrical form of an object is represented by a set of convex polyhedra. One polyhedron can be associated to a part of a primitive (the case of a concave rigid object), to a primitive (the case of convex rigid object) or to more than one primitive (a deformable object represented by a set of point masses). These polyhedra help to get the same geometrical model for all primitives, to simplify the semi-automatic construction of the physical representation from the geometrical one and to simplify the calculation of the interaction between two primitives from the same object or two primitives from two different objects.

- Primitives have to be connected by means of a set of connectors in order to make a solid object. This is why, each primitive has to have the same physical representation. Each primitive is discretized into a set of punctual masses (particles), which respect locally and globally the inertial properties of the primitive. Connectors can relate two particles from two primitives but not two particles from the same primitive, ie, that when a deformable object, which is represented by a set of particles, each particle is considered as a primitive.

- Three types of connectors are considered in our model:

  - *linear connector LS* which connects a pair of particles. This connector specifies the distance to maintain between these two particles. The force which is associated to this linear constraint can be expressed by $F = -\lambda\Delta d - \mu\dot{d}$, where λ and μ are constants, d is the distance between the two particles and $\dot{d}$ is the relative speed of the two particles.

  - *torsion connector TS* which connects three particles. This connector specifies the angle to maintain between these three particles. The force which is associated to this angular constraint can be expressed by $F = -\lambda\Delta\theta - \mu\dot{\theta}$, where θ is the angle formed by the three particles and $\dot{\theta}$ is the angular speed. The force given by this connector depends only on the angle between particles and not to the distance between them. If we represent an angular constraints by a set of three linear connectors instead of an angular one, we get a wrong behavior, because we relate the flexion force with the traction on, and consequently the length of an object may depend to its curvature.

  - *The joint connector JS* which connect a set of four particles. This connector specifies the angle to maintain between three of these particles (like the case of the torsion connector), and the rotation axis which is defined using the fourth particle. This type of connector is useful for modeling revolute joints.

To simulate the plastic behavior of an object, we allow that the rest position of these connectors be variable. It changes in function to the internal tension of the connector.

- Three types of physical interactions are considered for simulating the dynamic behavior of objects:

viscous-elastic collision, static & kinetic friction, and viscosity.

- The complexity of the algorithm which computes at each time step the position/ deformation/ interaction of the object is $O(n)$, where n is the number of particles of the object model.

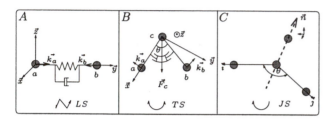

Figure 1: *LS, TS and JS connectors.*

4   Object modeling

Objects have to be discretized into particles, to be modeled. The level of this discretization mainly depends on the deformability characteristics of the object. Usually, a very fine and uniform discretization of objects is used for modeling deformable objects [20]. But this straightforward solution generates prohibitive execution times and numerical problems. Non-uniform discretization techniques, as the one used in [15], may also generate inconsistent behaviors if the inertia properties of the objects are not preserved. In order to solve these problems, we have developed a technique consisting in subdividing each object into a set of components having the same inertia properties of the object (inertia matrix and inertia center) where each component is not subjected to large deformations. The discretization is made using a wave propagation technique (figure 2) applied from the inertia center of the object to model [13]. The number and sizes of these components obviously depends on the deformability properties of the object to model (i.e. deformable parts will be more deeply subdivided, and almost rigid parts will be represented by a few components). If the object is rigid the resulting set of particles will form only one primitive. When it is deformable, each particle is considered as a primitive and appropriate connectors will be

added between theme (see [13]). The second modeling step consists in giving a "volumetric" dimension to the objects, by adding the implicit facets delimited by the external particles of the previous network. These facets are necessary to process collision and contact phenomenon (see § 7). The last step is to determine

Figure 2: *Discretization of a convex polyhedron*

the mass distribution (i.e. the mass m_i of each particle of the model) which preserve the inertia properties of the object. Theoretically, this can be done by solving the following system:

$$\begin{pmatrix} A & -F & -E \\ -F & B & -D \\ -E & -D & C \end{pmatrix} =$$

$$\begin{pmatrix} \sum (y_i{}^2 + z_i{}^2) * m_i & -\sum x_i * y_i * m_i & -\sum x_i * z_i * m_i \\ -\sum x_i * y_i * m_i & \sum (x_i{}^2 + z_i{}^2) * m_i & -\sum y_i * z_i * m_i \\ -\sum x_i * z_i * m_i & -\sum y_i * z_i * m_i & \sum (x_i{}^2 + y_i{}^2) * m_i \end{pmatrix}$$

$$\sum_{i=1}^{n} m_i * (P_i - P_c) = 0$$

where A, B, C, D, E, F are the components of the inertia matrix of the real object (it normally known), the second matrix represent the inertia matrix of a set of n particles, P_c is the position of the inertial center of m_i, P_i the position of the particle i and x_i, y_i, z_i are the coordinates of P_i. If we make the assumption that $\{P_i \ i \in [1..n]\}$ have been determined by the previous discretization technique, then the system has 9 equations and n unknown variables (the m_i parameters), Consequently, it will have an infinite number of solutions in the general case. Moreover, the inertia properties of an object change when the object deforms, which is why we need to apply a technique which preserves both globally and locally the inertia properties. This technique consists in decomposing the object into a set of tetrahedra. The number of these tetrahedra has to be chosen that one tetrahedron does not deforms a lot during the deformation of the object. Each tetrahedron being modeled by 5 particles having the following masses (see [13] for more details): $\frac{1}{20}M$ for the particles located on the vertices of the tetrahedron, and $\frac{4}{5} * M$ for the particle located onto the inertia center of the tetrahedron (M is the mass of the tetrahedron considered). As the inertia matrix of the object is the sum of the inertia matrices of these tetrahedra, the inertia matrix of the object is also preserved.

5   Motion and deformation generation

5.1   Basic idea

The motion and the deformation of a physical object are the result of the motion of its primitives (in our case these are rigid objects or particles which can be considered as a particular case of a rigid object[2]).

The motion of a primitive can be divided into two phases: translation and rotation. The translation of a primitive is characterized by the Newton's law $\vec{F} = m\vec{\gamma}$, where $\vec{F}$ is the sum of external forces, m is the mass of the primitive, and γ its acceleration which is equal to the second derivative of the position of the inertia center $\vec{\gamma} = \ddot{\vec{P}}$. The rotation is characterized by the differential equation: $\vec{N} = I\dot{\vec{W}} + \vec{W} \wedge (I.\vec{W})$. where $\vec{N}$ is the sum of external torques, I is the inertial matrix of the object and $\vec{W}$ is the rotational velocity vector, if the orientation the object is given by a vector $\Theta = (\theta_x, \theta_y, \theta_z)$, then $|\vec{W}| = |\dot{\Theta}|$. If the primitive is a single particle then $\vec{W}$ always null. As the primitives are related by spring/dampers connectors, the last differential equations are not independent and give the following differential equation system:

$$\vec{F}_i = m_i\vec{\gamma}_i \qquad (1)$$

$$\vec{N}_i = I_i\dot{\vec{W}}_i + \vec{W}_i \wedge (I_i.\vec{W}_i) \qquad (2)$$

$$\vec{F}_i = f(\vec{P}_j, \dot{\vec{P}}_j, \Theta_j, \vec{W}_j, \vec{FI}_i, \vec{FC}_i)|i, j \in [1..n] \qquad (3)$$

$$\vec{N}_i = g(\vec{F}_i) \qquad (4)$$

where n is the number of primitives, $\vec{P}_i$ is the position of the inertia center of the primitive i at time t, m_i is its mass, $\dot{\vec{P}}_i$ is the velocity of its inertia center, Θ_i is the orientation of the primitive, $\vec{W}_i$ is the rotation velocity around its inertia center, $\vec{F}_i$ is the sum of the forces applied on the particle i (this force is a function f of $\vec{P}_j, \dot{\vec{P}}_j, \Theta_j$ and $\vec{W}_j$, where j is a primitive within the n primitives in the scene; it also depends on the interaction forces $\vec{FI}$ and the control forces $\vec{FC}$ applied on the primitive i), and $\vec{N}_i$ is the sum of the external torques applied on the primitive i, this torque is a function g of the external forces.

The object motion and deformation are given by the solution of this system.

Unfortunately, the system (1) is *neither continuous nor linear*, because it involves collision forces and control forces that can be applied at any time by the user on the environment. Consequently, it is difficult to solve analytically this system or to use an implicit[3] numerical method to solve it. A classical way to solve such a system is to use an explicit method based on a second order limited development:

$$\vec{V}_{t+\tau} = \vec{V}_t + \tau\,\gamma_t + O(\frac{\tau^2}{2}\dot{\gamma}_t) \qquad (5)$$

$$\vec{P}_{t+\tau} = \vec{P}_t + \tau\,\vec{V}_t + \frac{\tau^2}{2}\,\gamma_t + O(\frac{\tau^3}{6}\dot{\gamma}_t) \qquad (6)$$

$$\vec{W}_{t+\tau} = \vec{W}_t + I^{-1}(\vec{N}_t - \vec{W}_t \wedge (I.\vec{W}_t))\tau \qquad (7)$$

Some well known drawbacks of such methods for dynamical simulation are the followings:

- Such a process is very sensitive to the chosen time step: if it is too large, they may lead to numerical divergence generated by the error term (figure 3); if it is too small, then the execution time is prohibitive. Moreover, it is difficult to choose a time step which insures numerical stability, because it depends on many factors like collision characteristic or deformability properties. In practice, this

---

[2]Adding other types of primitives does not change the general form of the dynamical equations.

[3]If $P_{t+\tau} = f(P_x) \mid x <= t$ the approach is explicit (incremental), else it is implicit.

time step is empirically chosen after several attempts of executing the simulation process.

- The error term can't be estimated, because the system has a non continuous behavior. This is a major drawback for robotics applications.

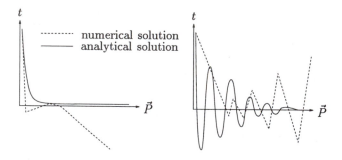

Figure 3: *Without a suitable time step the numerical solution may diverge*

5.2 Adaptive time step and error estimation

5.2.1 Problem and approach

As shown by the equations (5), the error term associated to the computation of $\vec{V}_{t+\tau}$, $\vec{P}_{t+\tau}$, and $\vec{W}_{t+\tau}$ increases when τ and/or the derivative of the acceleration $\dot{\gamma}$ increase. Since the value of $\dot{\gamma}$ can not be controlled, the only way to decrease the error term is to decrease the time step, and consequently to increase the computational time. An efficient strategy is to use an adaptive time step τ (figure 4), where the sampling rate varies according to the variation of $|\dot{\gamma}|$ in order to keep the error term less than a given threshold while using a few samples as possible.

The main problem comes from the fact that the value of $\dot{\gamma}$ is not known, and consequently can not be used to estimate the error term. Moreover the discontinuity of the force function make it impossible to have a numerical estimation of the error term. In order to bypass this problem, we have decided to make use of the physical meaning of our system, and to estimate the error variations by evaluating its consequence onto the variation of the mechanical energy. Let E_m be the mechanical energy of the physical system S considered

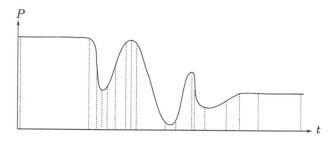

Figure 4: *A good strategy to avoid divergence and reduce the computational time is to use an adaptive time step τ which allows the system to be integrated using a minimum number of samples.*

($E_m = E_p + E_k$, where E_p is the potential energy and E_k is the kinetic energy). E_m depends on $\vec{P}$ and $\vec{V}$:

$$E_m = E_p + E_k = -\int_P^{P+\Delta P} \vec{F}.\delta P + \frac{1}{2}m\Delta V^2$$

Three cases have to be considered when reasoning about the variation of E_m:

- Case 1: S in an isolated system without energy dissipation. This situation arises when an object (or a set of objects) is not subjected to an external force, nor friction phenomenon. In this case, E_m is constant.

- Case 2: S is an isolated system with energy dissipation. This situation arises when a friction force appears between objects. In this case E_m decreases.

- Case 3: S is a Non-isolated system. This situation arises when S is subjected to an external forces (eg. a control force or a collision with another object in the environment). In this case E_m increases.

5.2.2 Time step determination

The time step determination algorithm is basically applied during the time intervals where the object is considered as an isolated system (figure 5). The extension of the domain of validity of this algorithm is discussed below (§5.2.3). As previously explained, the

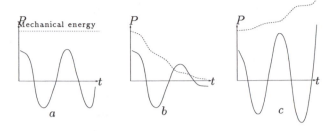

Figure 5: *(a) an isolated object without energy dissipation:* E_m *is constant. (b) An isolated object with energy dissipation:* E_m *decreases. (b) Non isolated object:* E_m *increases.*

mechanical energy E_m is **constant** when the system is isolated: $\Delta E_m = 0$. If ΔE_m increases in the numerical process, this means that this "non physical" variation is the result of the error term. Then, the basic idea is to monitor the value of $|\Delta E_m|$ at each time step and to choose the greatest time step which satisfies $|\Delta E_m| < \epsilon_e$ for a small ϵ_e. The following algorithm gives a practical implementation of this idea:

> **Let $\vec{P_t}$ be the position of the object.**
> **Let $\vec{V_t}$ be its velocity.**
> **Let $\vec{F_t}$ be the sum of the forces applied on it.**
> **LOOP**
>   **Calculate $\vec{P_{t+\tau}}$ and $\vec{V_{t+\tau}}$**
>   **Calculate ΔE.**
>   **IF $|\Delta E| > \epsilon_e$ THEN**
>   **LOOP**
> $\tau = \frac{\tau}{2}$
>   **Calculate $\vec{P_{t+\tau}}$ and $\vec{V_{t+\tau}}$**
>   **Calculate ΔE**
>   **UNTIL $|\Delta E| < \epsilon_e$**
>   **ELSE**
> $\tau = \frac{3}{2}\tau$
> **UNTIL ∞**

Justification: $|\Delta E_m| > \epsilon_e$ means that the time step is too large, and consequently the error term is too large; $|\Delta E_m| < \epsilon_e$ means that the probability that the state of the system is "correct" is very heigh. An intuitive justification of this assumption relies on the fact that two different states of the system having approx-

imatively the same numerical value for E_m can't be associated to two consecutive time step, because of the continuity of the object motions. For instance, when two objects penetrate each other, the mechanical energy increases continuously until these two object leave each other. So the two objects have to move in a single time step a distance which is greater than the sum of their dimensions in order to reach a false state which verifies the conservation of the energy.

In the following section we show that the previous idea still applies when an external perturbation occurs.

5.2.3 Dealing with external perturbations

When an external force F_e (a control force or a collision force) is applied on the system, E_m is no longer constant during a very short interval of time. After this event, we can take into account the potential energy generated by F_e and the system will returns to a state where E_m is constant. A safe and blind iteration (using a very small time step ϵ_τ) of the algorithm is necessary to evaluate the value of the new mechanical energy E_m. A good way to choose ϵ_τ, which is consistent with the algorithm described above, is to search for the time step ϵ_τ for which ΔE_m between t and $t + \tau$ is smaller than ϵ. It is always possible to find such time step because of the **continuity** of E_m. Consequently, the previous algorithm still applies where an external force is applied on the system.

It also can be shown that the validity of the algorithm can be extended to the case when a force F_e not derived from potential (eg. the friction force where $\int_P^{P+\Delta P} \vec{F}.\delta\vec{P} \neq 0$): The algorithm is applied between two time steps, where forces are approximately constant (this assumption comes from the explicit approach figure 6 illustrates) and the variation of the force can be represented by a stepwise function.

The figure 7 shows the evolution of τ when a ball falls down on the ground.

5.2.4 Main characteristics of our approach:

Let T_a be the average time required for an iteration (passing from t to $t + \Delta t$) using an adaptive time step, T_c the time required for an iteration using a constant

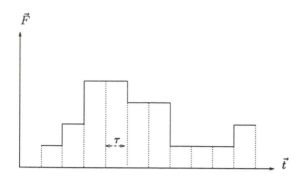

Figure 6: *Using an explicit approach, The force is, in practice, constant during each time step.*

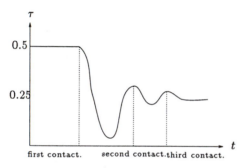

Figure 7: *The evolution of τ when a ball falls down on the ground. Note that τ decreases when the ball hits the ground and increases when it leaves the ground. Observe also that the decrement value depend on the force of the collision which is greater when the ball is faster.*

time step, τ the current time step, τ_g the time step verifying the energy constraint, and $N_{lookfor}$ the number of iterations needed to reach τ_g.

- Because one divides τ by 2 at each iteration, $N_{lookfor}$ has a very small value:

$$N_{lookfor} = \log_2 \tau - \log_2 \tau_g$$

In practice, and for a big number of test, we have remarked that

$$T_a \simeq 1.5 * T_c$$

This can be explained by the continuity of the motion, which make that τ_g and τ are very close, so $N_{lookfor} = 1$ when $\tau_g < \tau$ and $N_{lookfor} = 0$ when $\tau < \tau_g$. So the average value of $N_{lookfor}$ is 0.5.

- The gain in computational time during t second is proportional to the relationship between the average time step and the smallest time step used during the simulation:

$$Gain_t = \frac{\text{average time step}}{\text{smallest time step}} > 1$$

Experimentally, the gain is very large (for instance, it can reach a value as high as 10000 for a stiff collision).

- We do not use the same factor when increasing (one multiplies by 1.5) and when decreasing (one divides by 2) the time step in order not to obtain the same values and to be as close as possible to the greatest value of time step which respect our constraint.

- To estimate the velocity error term at each iteration, one can assume that in the worst case, the energy variation ϵ_e is fully transformed into kinetic energy; then one can write:

$$\Delta E_k = \frac{1}{2}m\Delta V^2 \Rightarrow \varepsilon_v = \sqrt{\frac{2\varepsilon_e}{m}}$$

- Since $\Delta P \simeq V\tau$, the position error is approximately $\varepsilon_p \simeq \varepsilon_v * \tau$.

- Experimentally, the cumulative position error is proportional to the value of the used ϵ (figure 8).

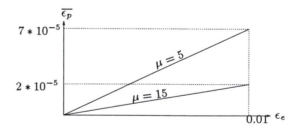

Figure 8: *The average value of the position error ϵ_p as a function of the energy error ϵ_e during 10 second and for two different values of μ (μ is a damping factor). This result is obtained using MAPPLE.*

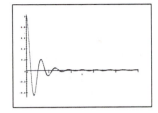 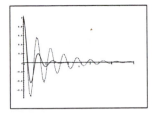

Figure 9: *The variation of the relative position of two particles as a function of the time. At the left, one uses an adaptive time step, the incurred error is 0.00006. At the right one uses a constant time step which is equal to the average time step obtained by the adaptive approach, the incurred error is 0.003. The used parameters are $\lambda = 100$, $\mu = 5$, $\tau_m = 0.065$, $\epsilon_e = 0.01$.*

6 Collision detection

It is well known that collision detection is a bottle neck for a large number of geometric based algorithms, and in particular for dynamic based simulation (a large percentage of computational time is devoted to collision detection). Several interesting results have been obtained for processing collision between rigid polyhedra: Lin & Canny[6] have proposed an incremental algorithm of complexity $O(1)$ for calculating the positive distance between rigid and convex polyhedra in motion; this algorithm has been extended by Kotoku [22] in order to localize the contact point when the distance between the two polyhedra is negative. Gilbert et al[7] have developed an algorithm which calculates, in linear time $O(n)$, the positive distance between the convex-hulls of two sets of points, and gives an approximation of the negative distance when "small" interpenetrations occur[7].

Concave polyhedra can also be considered by decomposing them into a set of convex components. When these components are rigid or when they are subjected to a first order deformation [3] one can use on of these algorithms to detect collision between theme. However, the previous algorithms can not be applied when deformable polyhedra are involved: such polyhedra may alternatively be concave or convex during the simulation and hence can not be statically divided into convex polyhedra. Also, it is very expensive in time consuming

to try to divide them dynamically during the motion ($O(n^2)$).

6.1 Outline of our approach

Given two deformable polyhedra A and B, A is composed of N_A facets and N_a vertices, B is composed of N_B facets and N_b vertices. Let C_X be the convex hull of the polyhedron X and F_X^i be the *ith* facet of the polyhedron X. Our approach consists in detecting collision between each pair of facets (F_A^i, F_B^j). This requires $N_A * N_B$ operations which is very expensive. So, before considering these pair of facets, we begin by eliminating from each polyhedron some facets that can not be in contact, in order to determine two minimal sets of facets that can be a potential candidates. This elimination of facets can be made in linear time $O(N_A + N_B)$ using some parameters. These parameters can be obtained by extending the Gilbert algorithm in order to calculate the real value of the negative distance.

Our algorithm returns two sets of facets that can be in collision, it returns also the amount of maximum interpenetration (between C_A and C_B) and the direction of this collision. Our algorithm assumes that, although there is many contact points between the two polyhedra, there is only one collision direction (figure 10). So one polyhedron cannot be in contact with two opposite sides from another polyhedron. This restriction comes from the fact that our polyhedra are initially convex and they become concave because of the collision.

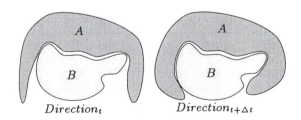

Figure 10: *Our algorithm is restricted to the case where there is only on collision direction. The collision direction between the two polyhedra A and B changes with the time but it is fixed at a given time*

6.2 Determining potential candidate of facets

Eliminating facets needs to use the following parameters which characterize the contact(See figure 11):

- The negative distance d_n: it is the smallest amount of displacement that can separate the two convex hulls C_A and C_B.

- The contact direction $\vec{n}_{A/B}$: it is the direction in which the smallest displacement has to be performed. It goes from A to B.

- The contact plane $P_{A/B}$: it is the plane which the director vector is $\vec{n}_{A/B}$ and passes by the closest point of C_A to C_B.

- The impact zone $Z_{A/B}$: it is the projection of the part of A, which exists in the negative side of $P_{B/A}$, on $P_{B/A}$.

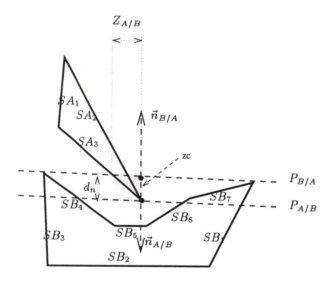

Figure 11: *One can eliminate, in linear time, all facets from the polyhedron A (resp. B) that can not be in contact with the polyhedron B (resp. A).*

Given these three factors, one can eliminate, in linear time, all facets from the polyhedron A (resp. B) that can not be in contact with the polyhedron B (resp. A). Two criteria are evaluated for eliminating such facets:

- if $\vec{n}_{A/B} \times \vec{N_{SA_i}} < 0$ (where $\vec{N_{SA_i}}$ is the external normal on the facet SA_i of A), then the facet is invisible to the polyhedron B and can be eliminated. This is the case of the facets SA_1, SA_2, SB_1, SB_2, SB_3 in the figure 11. This criterion can be applied only under the above-mentioned restriction, if we are working in the general case where polyhedra can be initially concave, we don't have to apply this type of elimination.

- If $Distance(SA_i, P_{B/A}) > 0$, then the facet is very far from the other polyhedron and can be eliminated. This is the case of the facets SA_1, SB_2, SB_5.

- If the projection of SA_i on $P_{A/B}$ is out of $Z_{B/A}$, then the facet can be eliminated. This is the case of SB_1, SB_3, SB_6, SB_7.

After applying these three tests on each facet, the algorithm will return the two sets of facets that still candidates for the collision (in our case they are $\{SA_3\}$ and $\{SB_4\}$). These three tests are atomic and they can be made in constant time for each facet. So the time needed to perform facet elimination depends on the number of facets $nf1 + nf2$ only. So, to obtain a linear complexity we need to be able to obtain these parameters in linear time.

6.3 Determining contact parameters in extending Gilbert algorithm

Gilbert algorithm is designed to calculate the Euclidean distance between the convex hulls of two polyhedra A and B (without calculating these convex hulls) defined only by their vertices N_a and N_b in m-dimensional space. The Euclidean distance between C_A and C_B is defined by:

$$d(C_A, C_B) = \min\{|x - y| : x \in C_A, y \in C_B\}$$

$$= \min\{|z| : z \in C_A \ominus C_B\}$$

$$C_A \ominus C_B = \{z : z = x - y, x \in C_A, y \in C_B\}$$

So the distance between C_A and C_B is equal to the distance of their Minkowski set difference (the

Minkowski set difference of two convex polyhedra is always a convex one) $C_A \ominus C_B$ from the origin O. The algorithm of Gilbert finds the closest point $\vec{P} = \vec{P}_A - \vec{P}_B$ from $C_A \ominus C_B$ to the origin and returns $\vec{P}_A$ and $\vec{P}_B$, so $d = |P_A - P_B|$ and $\vec{n}_{A/B} = \frac{\vec{P}_B - \vec{P}_A}{|\vec{P}_B - \vec{P}_A|}$. Although this algorithm uses $C_A \ominus C_B$, but it does not calculate all the vertices of $C_A \ominus C_B$, which is why it can make use of the continuity of the motion of A and B in order to converge in linear time $O(N_a + N_b)$.

In the following, we show how this algorithm can be used to find the contact parameters[14].

Finding the contact direction: In the space of Minkowski set difference $C_A \ominus C_B$, the contact direction $\vec{n}$ is the direction in which the origin O has to be moved in order to bring it out of $C_A \ominus C_B$ with a minimum displacement. For simplification purposes, the algorithm will be justified in the space of the Minkowski set difference. But in practice, we do not calculate this difference.

We can remark that there is a compact zone $Z = \vec{z}_i$ around $n_{A/B}^\rightarrow$ verifying that for every displacement according to $\vec{z}_i \in Z$ (which separates C_A and C_B) the new contact direction (which is positive and obtained by the Gilbert algorithm) is closer to $n_{A/B}^\rightarrow$ than to $\vec{z}_i$: this is clear in the space of the Minkowski set difference (figure 12), that the closet point of $C_A \ominus C_B$ to the origin belongs always to a facet f and the contact direction is the external normal on this facet. In this case the zone Z is defined by the origin and the extremities of f (see?? for more details).

Given an initial value $\vec{n}_0 \in Z$ of the contact direction, the two above properties allows us to find $\vec{n}_{A/B}$ using the following algorithm:

VECTOR Find_Contact_Direction($\vec{n}_0$)
{
    Separate C_A and C_B according to $\vec{n}^0$.
    Apply the Gilbert algorithm to get the new positive contact direction $\vec{n}_{A/B}$.
    IF $\vec{n}_0 = \vec{n}_{A/B}$ THEN
        return $\vec{n}_{A/B}$

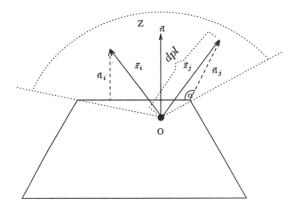

Figure 12: *Inside the zone Z any displacement $(\vec{z}_i, \vec{z}_j, ..)$ of O will give a new positive contact direction $(\vec{n}_i, \vec{n}_j, ..)$ which is closer to the exact contact direction $\vec{n}$.*

    ELSE
        return Find_Contact_Direction ($\vec{n}_{A/B}$)
}

The behavior of this algorithm depends on two parameters: the initial direction $\vec{n}_0$ and the amount of displacement dpl. Because of the continuity of the motion of two polyhedra, the contact direction obtained at the instant t is a very good approximation to initialize the algorithms at the instant $t + \Delta t$. The displacement value dpl: We have to choose the smallest amount of displacement that can separate the two polyhedra, because, as much as dpl is small, the contact direction obtained after the the displacement is closer to the real contact direction. Figure 13 illustrates this. This is why we choose this value to be proportional to the negative distance obtained in the last iteration. If this value is not large enough, the Gilbert algorithm will detect a negative value, then our algorithm will duplicates this displacement and repeats the process.

The choice of $\vec{n}_{A/B}^0$ and dpl explained above, allowed us to converge practically very fast. In our experimentation the average value of I was very close to 1.

Finding the negative distance d_n: To find the value of the negative distance $d_n(t)$, we apply a displacement dpl on the polyhedron A in the direction $\vec{n}_{B/A}(t)$ and we apply the Gilbert algorithm to get the

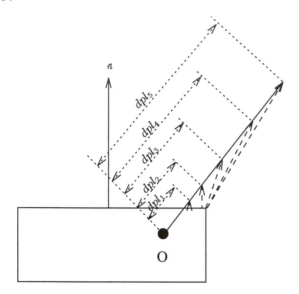

Figure 13:

positive distance d_p. Then negative distance d_n is equal
to $dpl - d_p$.

Finding the impact zone $Z_{A/B}$: The impact zone
$Z_{A/B}$ is the projection of the part of A, which exists
in the negative side of $P_{B/A}$, on $P_{B/A}$. The vertices
of this projection can be obtained in projecting each
vertex of A (which is in the negative side of $P_{B/A}$) on
$P_{B/A}$, and in calculating the intersections of all the
edges of A with $P_{B/A}$. Then the impact zone can be
simplified by the smallest circle which contains these
vertices.

7   Objects interactions

When objects move in a real environment they may
interact with each other. Three types of interactions
have to be considered: collision, friction and external
viscosity.

7.1   Collisions response

When two objects collide with each other, they locally deform. Consequently, they store potential energy which will be transformed into kinematic energy
when these objects return to their initial shapes. This
means that force is, in some sense, proportional to the

amount of deformation. If these two objects are completely rigid, we can make the assumption that the
duration of the collision is negligible, and that "impulse based method" can give an estimation of the collision force[23, 19]. When these objects are deformables,
the duration of the collision cannot be considered as
negligible and consequently the so-called e "penalty
method"- in which the collision force is "proportional"
to the amount of "penetrating volume" of the two objects and presented by a spring/damper parallel combination. A non linear damper is used[10, 16] in order to get a restitution coefficient which depends only
on the relative velocity of the two objects and not to
their masses. The restitution coefficient e is defined by
$v_0 = -ev_i$, where, v_0 is the relative velocity before the
collision and v_i is the relative velocity after the collision. So the collision force is given by:

$$\vec{F_c} = \begin{cases} (-\lambda v - \mu \dot{v} v)\vec{k} & \text{if } v < 0 \\ \vec{0} & \text{else} \end{cases} \qquad (8)$$

where λ is the rigidity factor of the collision, μ is
a damping factor (which represent the dissipation of
the energy), v the volume of interpenetration between
these two object, and $\vec{k}$ is the contact direction. This
force is supposed to act on the inertia center $\vec{P_c}$ of the
interpenetration volume.

In practice, it is very difficult to calculate quickly
the interpenetration volume between two polyhedra.
A practical approach to calculate this volume is to
calculate the contribution of each pair of facets; $v = \sum_{i,j} v(SA_i, SB_j)$, where SA_i, SB_j are the candidates
that have not been eliminated in the previous step(§6).
$v(SA_i, SB_j)$ is by definition the volume between the
projection of SA_i on SB_j and the projection of SB_j
on SA_i after eliminating the part of SA_i which is in
front of SB_j and the part of SB_j which is in front
of SA_i . These two projections are parallel to $\vec{n}_{A/B}$.
Figure 14 shows a $2D$ example.

Only the particles have masses (triangular facets
used in our system are implicit), so the collision force
between two facets has to be distributed between the
three particles (P_1, P_2, P_3) which define the three vertices of the triangular facet. So the collision force $\vec{F_c}$ is

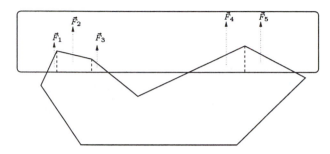

Figure 14: *the interpenetration volume is the sum of the contribution of each pair of facets in this volume.*

distributed between P_i according to the following criteria:

$$\vec{F_c} = \vec{F_1} + \vec{F_2} + \vec{F_3}$$

$$\vec{F_c}\vec{P_c} = \vec{F_1}\vec{P_1} + \vec{F_2}\vec{P_2} + \vec{F_3}\vec{P_3}$$

Where F_i is the force to apply on the particle i. As the direction of the forces F_i is known and equal to $\vec{n}_{A/B}$, the previous system has a solution.

When a particle i is subjected to a collision force it can loss energy (the case of plastic collision). This energy dissipation can be represented by a force $\vec{F_d}$ which is proportional to its velocity:

$$\vec{F_d^i} = -\mu \vec{V}_{relative}^i$$

where μ is the energy dissipation factor, and $\vec{V}_{relative}^i$ is the relative velocity of the particle i relatively to the other facet.

7.2 Friction

Two objects in contact are subjected to a friction force. This friction force is defined by the Coulomb's law which show that the magnitude of the friction force $|\vec{F_f}|$ is proportional to the magnitude of the normal reaction $|\vec{F_n}|$ [4] :

$$\vec{F_f} = \begin{cases} -c\,|\vec{F_n}|\frac{\vec{V}}{|\vec{V}|} & \text{If } \vec{V} \neq \vec{0} \quad (sliding) \\ -s\,|\vec{F_n}|\frac{\vec{F_t}}{|\vec{F_t}|} & \text{If } |\vec{F_t}| > s|\,\vec{F_n}| \quad (intermediate) \\ -\vec{F_t} & \text{Else} \quad (sticking) \end{cases}$$

---

[4] An other approach consists in representing the static friction using a spring connecting the two colliding objects[12].

where c is the kinetic friction parameter, s the static friction parameter $(s > c)$, $\vec{V}$ the relative speed of the colliding facets, and $\vec{F_t}$ the tangential force applied by one facet on the other.

7.3 Viscosity

When an object is moving, it also interacts with the environment (air, water). This interaction can be neglected if the motion is slow enough and if the viscosity of the environment is not very high. Otherwise the interaction (which can be seen as a succession of micro-collision between the object and the molecules of the gaze or the liquid) can generates forces which may modify the motion of the object. For example, the rotation of a fallen object is the result of its interaction with the air. Representing this interaction force (called viscosity), requires to associate a physical representation to the environment. But a representation of fluid involving a very fine decomposition is not applicable in practice; this why we use a macroscopic representation of the resulting phenomena by considering that the generated force $\vec{F_v}$ is proportional to the speed of the particle[5]: $\vec{F_v} = -k\vec{V}$, where k is the viscosity factor and $\vec{V}$ the velocity. For an infinitesimal surface δs; $\vec{F_v^s} = -k\vec{V_s}$; for a surface S the expression is modified as follow:

$$\vec{F_v} = -k\int_S \vec{V_s}\delta s = -k\int_S \frac{\delta \vec{P_s}}{\delta t}\delta s = -k\frac{\delta}{\delta t}\int_S \vec{P_s}\delta s$$

$$\vec{F_v} = -k\,S\frac{\delta \vec{P}_{Ic}}{\delta t} = -k\,S\,\vec{V}_{Ic}$$

where Ic denote the inertia center of the facet. Obviously, only those facets which are colliding with the oncoming fluid are subjected to this force. The viscosity force is maximum when $\vec{V} \times \vec{n} = |\vec{V}|$, and it is null when $\vec{V} \times \vec{n} < 0$, where $\vec{n}$ is the external normal vector to the facet. Consequently, the force $\vec{F_v}$ is defined as follow:

$$(9) \qquad \vec{F_v} = \begin{cases} -k\,S\,(\vec{V}_{Ic} \times \vec{n}).\frac{\vec{v}}{|\vec{v}|} & \text{if } \vec{V}_{Ic} \times \vec{n} > 0 \\ 0 & \text{else} \end{cases} \qquad (10)$$

---

[5] This approximation is true only for an object moving in the air with a small velocity (this is the case for actual robots). For the fluid, this force is of the form $\vec{F_v} = -k\,\vec{V}^2$

In our approach, objects are composed of triangular
facets. Then each facet (P_1, P_2, P_3) has the following
velocity for I_c:

$$\vec{V}_{Ic} = \frac{1}{3} \sum_{i=1}^{3} \vec{V}_i$$

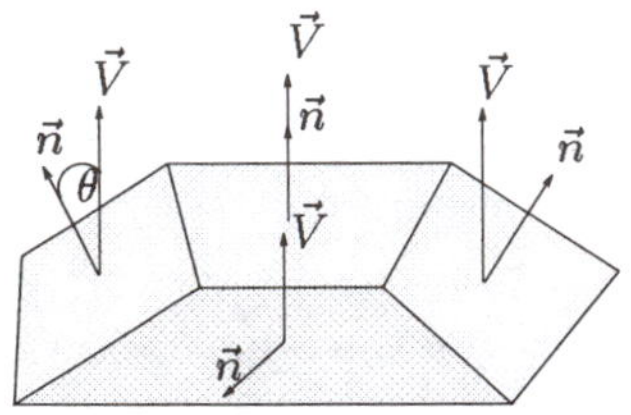

Figure 15: *The viscosity force applied on a facet, depends
on the area of this facet and on the angle between the normal
on this facet and its velocity.*

8   Implementation and experimental results

The approaches described in this paper have been integrated within the $Robot\Phi$ system ($Robot\Phi$ uses the
$3D$ visualization system $Geomview$). We have simulated a wide range of physical systems, from the three
basic mechanical behaviors of a stiff bar (traction, flexion, torsion) to complex robotic and biological systems.
Figure 16 shows a manipulation of a cylinder by an
articulated robot. The robot is represented by 160
particles and 4 articulations. The cylinder is represented by 36. The robot moves toward the cylinder
because of a tracking PD force applied on its end-
effector; during this motion, the robot's joints are pas-
sives and they "follow" the motion of its end-effector.
Two stiff interactions are considered; the interaction
$Robot/Cylinder$ and the interaction $cylinder/earth$.
The average time step was 0.0003 second. A very stiff
collision ($lambda = 2000$) between an object and a table has also been simulated. The sampling frequency
needed varies between 300 and 600 Hertz.

The correctness of the mathematical approach used
to find the trajectory of an object has been tested using MAPLE. For a case which can be solved analytically (an object of two particles), we have compared

Figure 16: *The system $Robot\Phi$ helps to simulate the behavior of robots.*

the adaptive time step approach ATS with the constant time step approach CTS (figure 9). Using ATS,
the numerical solution converges and is much more accurate than the solution obtained using CTS. The position is found to be proportional to the energy error
level ϵ.

The stability of a power grasp, in which the object is
grasped by a dextrous hand, can also be evaluated using our approach: this stability is the result of a large
number of contact interactions between the hand and
the object, as shown in the figure 17. This large number of contact interactions can easily be taken into account using our physical model (including the involved
friction force). There was 200 particles, the average
time step was 0.004 second, and the execution time
needed to simulate the behavior of the hand for 50 second, was 48 second.

This system has also been used to simulate the behavior of the knee ligament during the passive motion[6]
(figure 18). To demonstrate the accuracy of the model

---

[6]In co-operation with the TIMC laboratory at
Grenoble/FRANCE, and the instituto ortopedico Rizzoli
Bologna/Italy

and its ability to describe the real ACL dynamical behavior during passive motion, we have imposed to the mathematical model of ACL the recorded real trajectory, computing the resulting ACL position, cross sections, volume, twist, fibers elongation and forces on attachment sites during the passive motion. We have verified that the position of ACL anterior fibers match the recorded real position within 5% relative error, and that the elongation of fibers near the 3 ACL bundles are comparable with data reported in literature by other investigators. Moreover we are able to compute the other fibers stress/strain curves at all flexion angle, and to compute the complete $3D$ changes of shape, volume and surfaces of the ligament during the whole range of passive motion[17].

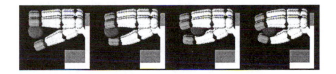

Figure 17: *Power grasp simulation.*

These examples showed that in the case of objects in motion we need only one iteration to update the value of the contact direction, because it changes very slowly during the simulation. So, practically, we obtain the same complexity when computing the negative and the positive distance between two polyhedra.

Figure 18: *Simulation of the behavior of the knee ligament during the passive motion.*

9 Summary and Conclusions

In this paper we have described some algorithms and approaches, which allows to construct and optimize

a dynamic simulation system. These algorithms are integrated in the *Robot*Φ system. They allow us to model robots and their environment in order to simulate their motions, their deformations, and their interaction with the environment. They allow us also to semi-automatically create a physical representation of an object from its geometrical model, which respects the inertial proprieties of the real object and it optimizes the needed particle number. It uses an energy-based adaptive time step in order to avoid the numerical divergence, to estimate incurred errors and to reduce the computation time. The motion complexity at each time step $O(n)$, where n is the number of particles in the environment. Contact detection between deformable polyhedra is made in linear time $O(n)$ so the total complexity of the system in $O(n)$ operation at each time step.

Acknowledgments

The work presented in this paper has been partly supported by the CNES (Centre National des Etudes Spatiales) through the RISP national project, and by the Rhône-Alpes Region through the IMAG/INRIA Robotics project SHARP.

References

[1] A.Joukhadar, C.Bard, and C.Laugier. Planning dextrous operations using physical models. *IEEE/ICRA*, May 1994.

[2] A.Joukhadar and C.Laugier. Dynamic modeling of rigid and deformable objects for robotic task: motions, deformations, and collisions. *International conference ORIA*, Decembre 1994.

[3] D. Baraff and A. Witkin. Dynamic simulation of non-penetrating flexible bodies. *Computer Graphics, volume 26, Number 2*, july 1992.

[4] David Baraff. Analytical methods for dynamic simulation of non-penetrating rigid bodies. *Computer Graphics, volume 23, Number 3*, july 1989.

[5] C.Laugier, C.Bard, M.Cherif, and A.Joukhadar. Solving complex motion planning problems by

combining geometric and physical models: the case of a rover and of a dextrous hand. *Workshop on the Algortihmic Fondations of Robotics (WAFR)*, February 1994.

[6] Ming C.Lin and John F.Canny. A fast algorithm for incremental distance calculation. *IEEE/ICRA*, April 1991.

[7] E.G.Gilbert, D.W.Johnson, and S.S.Keerthi. A fast procedure for computing the distance between objects in three-dimentional space. *IEEE/ICRA*, 1988.

[8] J.-P. Gourret. *Modélisation d'images fixes et animées*. MASSON, 1994.

[9] Stephane H.Carandall, Dean C.Karnopp, Edward F.Krtz, and David C.Pridmore-Brown. *Dynamics of mechanical and electronical systems*. Krieger publishing company, Malabar, Florida, 1968.

[10] K. H Hunt and F. R. E. Crossley. Coefficient of restitution interpreted as damping in vibroimpact. *ACME Journal of Applied Mechanics*, 1975.

[11] J.Dorlot, J.Batlon, and J.Masounve. *DES MATRIAUX*. Ecole polytechnique de Montereal, 1986.

[12] S. Jimenez and A. Luciani. Animation of interacting objects with collision and prolonged contacts. *Modeling in computer graphic*, 1993.

[13] Ammar Joukhadar. *robotφ*: Dynamic modeling system for robotics applications. research report RR-2543, INRIA, May 1995.

[14] Ammar Joukhadar, Ahmad Wabbi, and Christian Laugier. Fast contact localisation between deformable polyhedra in motion. *IEEE/Computer Animation*, June 1996.

[15] A. Luciani and al. An unified view of multiple behaviour, flexibility, plasticity and fractures: balls, bubbles and agglomerates. In *IFIP WG 5.10 on Modeling in Computer Graphics*, pages 55–74. Springer Verlag, 1991.

[16] Duane W. Marhefka and Davide E. Orin. Simulation of contact using a nonlinear damping model. In *International Conference on Robotics and Automation (ICRA)*, Minneapolis, Minnesota, April 1996.

[17] S. Martelli, A. Joukhadar, S. Lvallee, G. Champleboux, and M. Marcacci. Acl mathematical model for implant simulation in animal. *To be presented in the 7th Congress of the European Society of Sports Traumatology, Knee Surgery and Arthroscopy (ESSKA)*, May 1996.

[18] D. Metaxas and D. Terzopoulos. Dynamic deformation of solid primitives with constraints. *Computer Graphics, volume 21, Number 2*, july 1992.

[19] Brian Mirtich and John F.Canny. Impulse-based, real time dynamic simulation. *Workshop on the Algortihmic Fondations of Robotics (WAFR)*, February 1994.

[20] D. Terzopoulos, J. Platt, A. Barr, and K. Fleischer. Elastically deformables models. *Computer Graphics, volume 21, Number 4*, july 1987.

[21] C. Thibout, P. Even, and R. Fournier. Virtual reality for teleoperated robot control. *International conference ORIA*, Decembre 1994.

[22] K.Komoriya T.Kotoku and K.Tanie. A force display system for virtual environments and its evaluation. *IEEE/International Workshop on Robot and Human Communication (RoMan'92)*, September 1992.

[23] Yu Wang and Matthew T.Mason. Modeling impact dynamics for robotic operations. *IEEE/ International Conference on Robotics and Automation (ICRA)*, 1987.

[24] D. Williams and O. Khatib. The virtual linkage: A model for internal forces in multy-grasp manipulation. *IEEE*, 1993.

Geometry and the Discovery of New Ligands

Lydia E. Kavraki* *Stanford University, Stanford, CA 94305, USA*

Computer-aided drug design is a significant component of the rational approach to pharmaceutical drug design. Chemists now consider the geometric and chemical characteristics of molecules early in the design process in an effort to quickly identify ligands that have good chances of becoming potent pharmaceutical drugs. Computer assistance is not only helpful but also necessary to narrow down the search for potential ligands. Depending on the level of accuracy desired to model drug action, detailed quantum mechanical methods or approximate molecular mechanics methods are used. Even when simple approximations are made, efficient approaches are needed to compute, among other things, molecular surfaces and molecular volume, models of receptor active sites, reasonable dockings of ligands inside protein cavities, and geometric invariants among different ligands that exhibit similar activity. This paper surveys several problems and approaches in the area of computer-aided pharmaceutical drug design and draws analogies with problems from robotics and computational geometry.

1 Introduction

The design of pharmaceutical drugs is an extremely complex and still not completely understood process [2]. Computational chemists combine their knowledge of molecular interactions and drug activity together with visualization techniques, detailed energy calculations, geometric considerations, and data filtered out of huge databases, in an effort to narrow down the search for potent pharmaceutical drugs. Computer-aided drug design is a significant component of rational

*Author's current address: Department of Computer Science, Rice University, Houston, TX 77005, USA

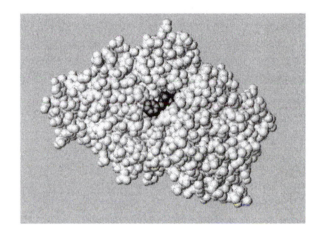

Figure 1: *The protease thermolysin with one of its known inhibitors (1TMN)*

drug design [6], and is becoming more relevant as the understanding of molecular activity improves and the amount of available experimental data that requires processing increases.

A fundamental assumption for rational drug design is that drug activity, or pharmacophoric activity, is obtained through the molecular recognition and binding of one molecule (the ligand) to the pocket of another, usually larger, molecule (the receptor). In their active, or binding, conformations, the molecules exhibit geometric and chemical complementarity, both of which are essential for successful drug activity [2, 49]. There is no simple way to explain how drugs achieve their desired effects. It is known, however, that several pharmaceutical drugs are inhibitors, i.e. inhibit reactions that would take place without their presence. For example, if a cavity of a molecule provides a favorable environment for a reaction, a ligand that fills that cavity in an energetically stable conformation can prevent this reaction from happening. Figure 1 shows the protease thermolysin and one of its inhibitors. Thermolysin is the large molecule shown in the picture, while the inhibitor (1TMN) is drawn in a darker color near the center of that picture. The 3D structure of the complex has been obtained by X-ray crystallography and can be retrieved from the Brookhaven protein data bank.

The modeling of molecular structure is a complex task. Quantum mechanics provide a detailed description of molecules in terms of atomic nuclei and electron distribution among them. However, quantum mechanical calculations cannot be used to treat large systems because of high computational demands. The modeling of the binding process is also a difficult task. The characteristics of the receptor, the ligand, and the solvent in which these are found have to be taken into account. Although chemists strive to obtain models that are as accurate as possible, several approximations have to be made in practice. Molecules are thus visualized to have surfaces and volume similar to our perception of surfaces and volume of macroscopic objects, or are considered under ideal conditions (i.e. in vacuum). It is clear that the more accurate the model used, the better the chances chemists stand in predicting molecular interactions. Nevertheless, there is a large number of predictions made with approximate models which have been confirmed with experimental observations [6, 34]. This has encouraged researchers to build tools that use approximate models and investigate the extent to which these tools can be useful. More accurate molecular modeling, gained through better understanding of drug activity or increased computational power, can only improve the techniques developed with simpler models.

Depending on whether the chemical and geometric structure of the receptor is known or not, the problems arising can be classified in two broad categories. If the receptor is known, chemists are interested in finding if a ligand can be placed inside the binding pocket of the receptor in a conformation that results in low energy for the complex. This problem is referred to as the *docking problem*. It has several variations: an accurate description of the binding may be desired, or an approximate estimate of which ligands from a huge database are likely to fit inside a receptor may be sought. Very often the binding pocket is not known. In fact, the 3D structure of few large molecules (or macromolecules) has been determined by X-ray crystallography or NMR techniques. When the receptor is not known, what is usually known is a number of ligands that interact with that specific receptor. These ligands have been discovered mainly by experiments. Using the geometric structure and the chemical characteristics of these molecules, chemists attempt to infer information about the receptor. In particular, chemists are interested in identifying the *pharmacophore* present in these ligands. The pharmacophore is a set of features at a specific 3D arrangement contained in all the active conformations of the considered molecules. A prevailing hypothesis is that the pharmacophore is the part of the molecule that is responsible for any observed drug activity, while the rest of the molecule is a scaffold for the pharmacophore's features. If the pharmacophore is isolated, chemists can use it to design a more potent pharmaceutical drug by examining the different activities, relative shapes, and chemical structures of the initial molecules [27].

The techniques that have been used so far in computer-aided drug design include geometric calculations (surface computation), numerical methods (energy minimization), randomized algorithms (conformational search), and a variety of other techniques like genetic algorithms and simulated annealing (docking). The machines used for these calculations range from desktop workstations to supercomputers. It is only recently that chemists have tools for complex geometric and energy calculations and the success of these computer-aided methods is currently being evaluated [2, 6].

This paper describes some of the computational problems arising in rational drug design. It surveys recent work on surface and volume calculations, conformational search, docking, pharmacophore generation, and database searching. The discussion reveals the wealth and diversity of the problems that arise in the domain of computer-aided pharmaceutical drug design. Analogies with problems from robotics and computational geometry are also drawn.

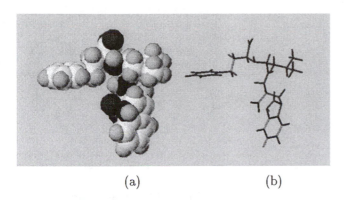

(a)   (b)

Figure 2: *The hard-sphere model and stick diagram of 1TMN*

2 Molecular Modeling

The *hard-sphere* model of 1TMN, the inhibitor of thermolysin of Figure 1, is drawn in Figure 2(a). This model is an abstraction frequently used by chemists to approximate the volume of a molecule. A sphere is drawn around the center of every atom of the molecule. The radius of each sphere reflects the space requirements of the corresponding atom and has been determined by a combination of experimental observations and quantum mechanical calculations. A set of radii that are commonly used are the *van der Waals radii* [7]. If the van der Waals radii are used, the envelope surface of the hard-sphere model is called the *van der Waals surface*.

The *stick diagram* of a molecule (Figure 2(b)) draws a line segment for each chemical bond. The angle between two consecutive bonds is called the *bond angle* and the angle formed by the first and the third of three consecutive bonds, when one looks along the axis of the second bond, is called the *dihedral* or *torsional angle*.

A priori, all bond lengths, bond angles, and torsional angles are degrees of freedom (DOF) of the molecule. Because of their chemical characteristics certain bonds cannot rotate about themselves and, as a result, all the torsions in which they participate as middle bonds are fixed. Bond lengths and bond angles tend not to exhibit large variations in their values. It is fairly common to consider bond lengths and bond angles con-

stant in calculations [28, 34]. Torsional angles, however, vary significantly and this affects the 3D shape of the molecule. When bond lengths and bond angles are considered fixed and only torsions vary, a molecular chain with n torsions can be viewed as an articulated mechanism with n revolute joints.

Standard geometries are commonly used to construct reasonable models of molecules. For example, there exist tables that show "preferred values" for bond lengths and these depend on the kind of atoms participating in these bonds [7]. Preferred values have also been calculated for bond angles and torsional angles, and again depend on the types of atoms linked by the corresponding bonds. The exact values used are obtained from statistical analysis of structural data in X-ray databases, like the Brookhaven or the Cambridge databases. Although it is true that there is variability in the geometric data in these depositories, the information gathered provides a reasonable approximation of reality [7, 34].

As far as calculations of energy are concerned, empirical force fields are used in practice instead of more detailed methods like quantum mechanics. A typical empirical force field includes terms for bond-stretch, bond-angle and torsional-angle deformations, and terms for van der Waals and Coulomb potentials [52]. Frequently, terms that model solvation effects are also included. Interaction of the molecule with the solvent in which it is dissolved is very important but also difficult to model accurately [7, 34]. An example of how the energy of conformation c can be calculated with empirical force fields when the molecule is considered in vacuum is given below:

$$
\begin{aligned}
E(\mathbf{c}) = \quad & \sum_{bonds} \tfrac{1}{2} K_b (R - R_0)^2 + \\
& \sum_{angles} \tfrac{1}{2} K_a (\theta - \theta_0)^2 + \\
& \sum_{torsions} K_d [1 + \cos(n\phi - \gamma)] + \\
& \sum_{i,j} \left\{ 4\epsilon_{ij} \left[\left(\frac{\sigma_{ij}}{r_{ij}}\right)^{12} - \left(\frac{\sigma_{ij}}{r_{ij}}\right)^6 \right] + \frac{q_i q_j}{\epsilon r_{ij}} \right\}.
\end{aligned}
$$

In the above $K_b, K_a,$ and K_d are force constants, ϵ is the dielectric constant, and n is a periodicity constant. R, θ, and ϕ are the measured values of the bond lengths, bond angles, and torsional angles in conformation **c**, while R_0, θ_0, and γ are equilibrium (or preferred) values for these bond lengths, bond angles, and torsional angles. r_{ij} measures the distance of atom centers in **c**. The parameters $\sigma_{ij}, \epsilon_{ij}$, and q_i are the Lennard-Jones radii, well depth, and partial charge for each atom in the system. All parameters and constants above are derived by a combination of quantum mechanics, vibrational methods, and experimental data. Once the values of bond lengths, bond angles and torsional angles of a conformation are known, obtaining the energy of a molecule with an empirical force field is a straightforward task. Minimization of this energy is not easy however, since force fields are non-linear functions and may contain a large number of local minima.

Calculations of energy are very important in the molecular world. In nature, molecules are usually found in low-energy conformations. Protein-ligand complexes are stable when the binding energy of the system is low. It should be emphasized that the exact calculation of molecular and binding energies is by no means a simple task [52], and that empirical force fields offer only an approximation. Nevertheless, as noted above, there are several cases where reasoning with these approximations has produced meaningful results [5, 7, 34].

Before describing specific problems let us also define the concept of molecular *features*. Chemists group atoms according to their chemical characteristics and use a label to refer to these groups. Given a molecule there are rules that identify the hydrophilic and hydrophobic parts of that molecule, the hydrogen-bond donors and acceptors, the charged centers, etc. These features are used, for example, to define pharmacophores, or to specify database queries that will retrieve ligands with certain characteristics. The accurate definition of features is a difficult task for the chemist, but is out of the scope of this paper [17, 32].

3   Molecular Surfaces & Volume

Computing surfaces and volume of molecules analogous to our perception of macroscopic surfaces and volume

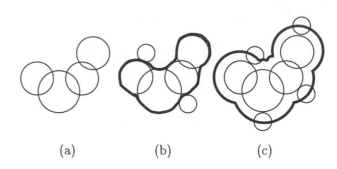

Figure 3: *(a) van der Waals, (b) molecular, and (c) solvent accessible surface*

has attracted considerable attention [58]. This information is useful in calculations for molecular recognition and docking [12], or computations of the energy of a molecule in solution [24]. Surface computation is also useful in pharmacophore identification, since atoms that are buried or little exposed are not likely to participate in a pharmacophore.

3.1   Types of Surfaces

Surfaces that are of interest to chemists include *the van der Waals* surface (defined in Section 2), the *molecular* surface and the *solvent accessible* surface [48, 66]. Figure 3 illustrates these different surfaces.

The molecular and the solvent accessible surfaces are defined with the help of a solvent atom which is a sphere of radius r. In particular, the molecular surface is defined by the front of the solvent sphere when this is rolling around the van der Waals surface. The solvent accessible surface is defined by the center of the solvent when this is rolling around the van der Waals surface of the molecule. In other words, the solvent accessible surface is the boundary of the free placements of the center of the sphere of the solvent, when this is moving among the atom spheres of the molecule. It can thus be computed, using configuration space techniques from robotics, as the union of the Minkowski sums of each molecular atom sphere and the sphere of the solvent [36].

3.2   Methods

Approximation and analytical methods have been used for the computation of surfaces of molecules. A survey of early techniques is given in [58]. Two widely used methods are Richards' method [65], which approximates the volume of a molecule by polyhedra, and Connolly's approach [11], which analytically computes molecular surface patches. More recently, methods from computational geometry are being employed to efficiently compute surfaces and volume without the limitations of previous approaches [19, 36, 50, 70].

Halperin and Overmars [36] observed that the complexity of the arrangement defined by n atomic spheres of a molecule is $\Theta(n)$, as opposed to $O(n^3)$ for a general arrangement of spheres in space. The complexity of an arrangement is defined as the overall number of cells in that arrangement. In the same paper it is shown that the arrangement of atomic spheres can be decomposed into an arrangement of simple cells whose total complexity is $O(n)$. As a result, it is possible to construct a hashing data structure that uses $O(n)$ space and can answer intersection queries for spheres of comparable radii to the atomic spheres in constant time. Computation of surfaces and volume follows nicely from this data structure. In particular, the van der Waals surface of a molecule can be constructed in $O(n \log n)$ time. Similar results can be obtained for the solvent accessible and the molecular surface.

Edelsbrunner uses alpha shape theory to accurately compute the surface and volume of molecules [18]. The alpha shape is the space occupied by the simplices of an alpha complex. These simplices are constructed in such a way that they are always a subset of the simplices defined by the weighted Delaunay triangulation of the molecule. In the above model, α is a parameter that regulates the radius $d = \sqrt{w^2 + \alpha}$ of atomic spheres, where w denotes the van der Waals sphere of an atom. If α is increased from its least possible value (a negative value) to zero, the shape of a molecule grows from a set of points to its van der Waals shape. Appropriate simplices are maintained as α changes, and when $\alpha = 0$ the set of constructed simplices, the alpha complex, contains important information about atom inter-

sections and the topology of the molecule. The alpha complex can be computed in $O(n \log n)$ time and then it is possible to quickly identify the atoms on the surface of the molecule, and compute the van der Waals, molecular, and solvent accessible surfaces. The volume of the alpha complex can be combined with the volume of the surface atoms to compute the volume of the molecule. Furthermore, the topological structure of the apha complex permits the identification of voids and canyons in the molecule [19, 20, 50]. Alpha shapes have also been used by Varshney el al [70] for molecular modeling. This work has produced a parallelizable algorithm that scales linearly with the number of atoms in a molecule for computing molecular surfaces.

3.3   Dynamic Maintenance

Although algorithms that compute molecular surfaces have been widely investigated, little has been done for their dynamic maintenance. For calculation of binding energies, to give an example, it is interesting to know how the surface that one particular atom contributes to the outer van der Waals surface changes, as the shape of the molecule changes. Work on dynamic data structures is useful in this respect [35].

4   Conformational Search

Conformational search is a fundamental problem in molecular biology. Perhaps the most well-known conformational search problem is the protein folding problem. It is believed that proteins have "unique" 3D shapes which correspond to global minima of their total energy and which are specified only by the chemical composition of the molecules. Finding these conformations is by no means an easy task and involves several hundreds of DOF [16].

For small ligands, finding the conformation with the minimum energy is of little interest. What is interesting is to find a set of conformations whose energy is below a threshold and which are geometrically distinct [46]. Such conformations are used in docking [59] and pharmacophore identification [55]. Low-energy conformations of a molecule that also respect certain "distance constraints" (i.e. have certain features at specific positions in 3D space) are also of interest to computational chemists. Tools that can produce such conformations have applications in database screening [56].

Several approximations are made during conformational search depending on the level of detail required. For example, it is usual to consider bond lengths and bond angles almost fixed, choose torsional angle values from predefined distributions, and simplify the energy model considered [28, 34]. Frequently the molecule is assumed to be in vacuum. An external potential can be considered with most conformational search methods but may result in longer computation times. Depending on whether distance constraints are imposed when conformational search is performed, we distinguish conformational methods into unconstrained and constrained techniques.

4.1 Unconstrained Search

A wide variety of methods for searching conformational space have been described in recent years (for a survey see [46]). Systematic search methods sample each torsional DOF of the ligand at regularly spaced intervals and were among the first to be developed and used [51]. The discretization of the torsional values is typically as coarse as 30° or 60° [46]. Even with such a resolution the number of conformations that are generated with systematic search can be very large. Typically the energy of all generated conformations is minimized which is an expensive operation. Several heuristics have been used to quickly prune down conformations that are close to previously generated conformations [68] in an effort to enhance the diversity of the sample.

A variety of randomized methods are also under investigation: conformations are obtained by applying random increments to the torsional DOF of the molecule starting from a user-specified initial conformation [26], or from a previously found low-energy conformation [9]. Recent articles, which attempt to compare different methods, emphasize the superior quality of the results obtained with stochastic methods [26].

Randomized techniques have been proven useful for high-dimensional search problems [38] and this direc-

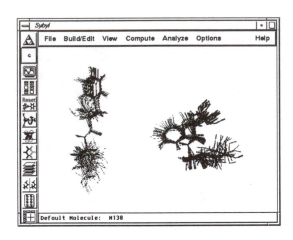

Figure 4: *Two clusters of 1TMN*

tion deserves to be further explored in the context of conformational search. A random sampling method for exploring the conformation space of small molecules has been recently developed in [22]. This method borrows ideas from randomized techniques for planning in high-dimensional configuration spaces [38]. The approach is divided into three steps: generation of random conformations, minimization of these conformations, and grouping or clustering of the minimized conformations. Initially, a large number of conformations, frequently tens of thousands, are generated at random over the conformational space of the molecule. The generation of these conformations is done by selecting each torsional DOF of the molecule uniformly from its allowed range. The selection can also be done according to a distribution that reflects preferred values for each torsional DOF, if such information is available. The resulting structure is stored only if it avoids intersections of the spheres of non-bonded atoms. Subsequently an efficient minimizer is used to obtain a conformation at a low energy minimum. At this step only conformations below a user-defined energy threshold are retained. Experimental observations have shown that the number of these conformations can be very large. Since only conformations that are geometrically distinct are interesting, it is necessary to partition the low-energy conformations into clusters of similar conformations. Figure 4 shows two of the clusters obtained with the randomized approach of [22] for 1TMN. At the end of the clustering step, a representative per cluster can be retained.

Conformational search raises a number of interesting issues. Many open questions remain on what are good minimization techniques for the energy models that are available for small molecules, what are reasonable similarity measures for conformations, and how partitioning can be done efficiently. Improvements in each of these domains can affect the performance of conformational search software and the quality of solutions obtained for the problems where these conformations are actually used (i.e. docking and pharmacophore identification).

4.2 Constrained Search

Most of the techniques described in the previous section will produce poor results when distance constraints are imposed in the structure of the molecule. Distance constraints arise frequently in practice. For example, chemists may be interested in conformations that keep two atoms of the molecule at specific positions in space because these two atoms belong to a pharmacophore. Ring structures impose distance constraints by their own nature: maintaining ring closure when a torsional angle in the ring changes, requires the atom at the beginning and the atom at the end of the chain to be at a bond's length distance from each other.

The constrained conformational search problem has a direct analog in robotics, namely the problem of *inverse kinematics*. If the bond lengths and bond angles in a single molecular chain are considered fixed, then the chain can be viewed as a serial manipulator with revolute joints (these joints correspond to the torsional DOF of the chain).

Manocha et al [54] exploit the work done in robotics for computing inverse kinematics of manipulators to find valid conformations for small molecular chains. In particular, the case of a serial manipulator with 6 revolute DOF has been extensively studied (6 is the minimum number of DOF for a robot to be able to span a full-rank subset of $SE(3)$ [13]). Symbolic manipulation

of the equations of Raghavan and Roth [63] transforms the inverse kinematics problem into one of computing the eigenvalues and eigenvectors of a matrix, which in turn can be done efficiently [53]. In a similar way, the inverse kinematics of a serial molecular chain with 6 torsional DOF can be computed by finding the eigenvalues and eigenvectors of appropriately defined matrices. For chains with $n > 6$ torsions, 6 torsions are considered free while the rest $n - 6$ are assigned discrete values and this procedure is repeated for different values of the $n - 6$ "fixed" torsions. The techniques in [54] are very efficient when computing conformations that maintain ring closure and for local deformations of small protein chains. It is worth mentioning that algebraic equations in 6 unknowns were also derived in [28] for finding the permissible conformations of a single-loop molecule when only 6 torsional angles are considered free, and solutions in limited cases were obtained.

Other kinds of methods, like distance geometry [14], are also being tested for constrained conformational search problems. Distance geometry exploits the fact that lower and upper values on interatomic distances can be derived from the restriction that atoms belong to a 3D chemical structure. These distances are used to refine 3D models of molecules by a variety of constraint propagation and "bounds smoothing" techniques. Distance geometry can also deal with large scale constrained conformational search problems like the ones arising from NMR data [14]. NMR produces distances between atoms of a macromolecule and chemists seek to reconstruct the 3D conformation of the molecule that produced these distances. The drawback of distance geometry methods is that they may fail to converge to a solution and can be relatively slow in practice [54].

Note finally that constrained optimization techniques, which minimize the energy of a conformation while observing distance constraints, can be used to obtain more stable conformations starting with conformations produced by algebraic or distance geometry techniques. The speed of constrained conformational procedures is crucial if these procedures are used to screen large databases [46].

5 Receptor is Known: Docking

Surface calculations and results of conformational search are used when trying to find a "reasonable" docked position of a ligand inside a known receptor. Information about the geometry of the receptor is obtained by X-ray crystallography or NMR techniques. For docking, it is generally assumed that the receptor molecule is rigid [47]. This approximation is justified by experimental data i.e. crystals of the molecule with and without the ligand, but exceptions have also been noted [60]. For the ligand however, it is essential to address its flexibility.

A central question for the docking problem is how to represent the geometry of the cavity, and how to compare it to the geometry of the ligand. The computation of the binding energy of the complex is a very important issue to be addressed in docking, but his problem is out of the scope of the present paper.

5.1 Rigid Ligand

If the ligand is considered rigid, it is possible to systematically search its six-dimensional configuration space for possible placements inside the binding pocket, but such a process can be time consuming. Several recent methods adopt a different approach: they try to match points (features) of the binding pocket to points (features) of the ligand. The points inside the pocket are referred to as "hot spots" [30], "essential points" [59], or "matchprobes" [71].

The definition of matching points in the receptor and the ligand varies widely with the method used. Some approaches use energy calculations to define these points. They describe, for example, the chemical environment of the pocket using a 3D grid, and define matching points as energetically favorable sites for certain functional groups [30, 59]. When the ligand molecule is placed in the grid region, the interaction energy can be efficiently calculated using precomputed data. Other docking approaches use only the geometry of the receptor and the ligand to define matching points. DOCK [43, 67], one of the earliest methods for docking, generates spheres inside the binding site in a way that they touch the surface of the pocket in two

points and have their centers along the surface normal at one of these points. The centers of these typically overlapping spheres are the receptor's matching points. Spheres are created in a similar way inside the ligand and their centers are the matching points of the ligand. The description of the binding pocket by the spheres described above is not unique and may seem arbitrary, but several successful predictions have been reported [67].

After essential points have been identified in the pocket and the ligand, the docking problem reduces to a matching problem. All possible combinations of ligand-receptor points can be tried if their number is small [45]. Simple heuristics can be used to narrow the search. DOCK, for example, selects a pair of receptor points and measures their distance. Then a pair of ligand points that are at approximately the same distance with the receptor points is found. A third receptor point is chosen its distances with the previously selected receptor points are used to identify a third point of the ligand. This process continues until a specified number of pairs is found or until no possible matches can be found. In that case the algorithm backtracks. At least four points are necessary to define an unambiguous orientation of a ligand inside a receptor. Other approaches [42] build a "docking graph" using the receptor and ligand matching points. The graph has a node for all pairs of receptor-ligand points, and an edge between two nodes, if the pairs corresponding to the nodes can be matched at the same time. A maximal clique in this graph will produce a maximal matching between the receptor and the ligand. It is well known that this problem is NP-hard [25] but the method is reported to work well in practice [42].

The matching problem that arises in docking, has analogies with the geometric matching performed for model-based shape recognition [21]. These analogies are extensively discussed in [62]. In geometric matching, a 3D model of an object is known. Given a set of 3D points which may lie on the surface of that object, a rigid transformation is sought to align these points to the model. In the context of molecular docking the ligand provides the model, and the receptor provides the set of 3D points that are checked against the model.

Techniques developed for model-based recognition, like interpretation trees [33] or geometric hashing [44], are thus applicable to the docking problem. In fact, geometric hashing has already been used for molecular docking in protein-ligand and protein-protein studies [1, 61]. In geometric hashing, a hash table for the ligand is computed and this is a transformation invariant representation of the molecule. Given a set of points in the receptor, matches can be detected through a voting scheme. An advantage of this approach is that the hash table for the ligand can be computed off-line, and after that it is possible to dock the ligand to multiple receptors.

5.2 Flexible Ligand

To address conformational flexibility, a widely used approach has been to consider different low-energy conformations of the ligand. These conformations, which are frequently obtained by a conformational search procedure, are tried against the receptor cavity using a technique developed for docking a rigid ligand to a rigid receptor [59]. To facilitate such docking approaches, several molecular databases now store a set of geometrically distinct conformations per ligand [39]. It is clear that if the active conformation is not one of the conformations considered, these methods will fail to produce the optimal docking.

Conformational flexibility has also been addressed directly by simulated annealing techniques. In that case, the torsional DOF of the molecule are changed inside the receptor's cavity [31]. One could also imagine using randomized sampling techniques instead of simulated annealing to find low-energy conformations of the ligand inside the binding pocket. Matching points defined inside the binding pocket are again useful when flexible ligands are considered. In this case however, fragments of the ligand are docked independently and the fragments are later joined into conformations which are in turn refined and ranked with appropriate scoring functions [15, 64, 71]. The idea of "building" a ligand inside a binding pocket is also popular with methods that suggest unsynthesized compounds or add functionality to a known inhibitor [40].

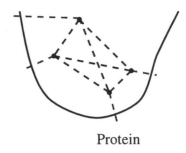

Protein

Figure 5: *The features of the pharmacophore interact with features of the receptor cavity*

Allowing for ligand flexibility is a challenging and still unsolved problem in protein-ligand docking. Efficient geometric techniques that can exclude placements of fragments that are in collision with the rest of pocket, or can suggest reasonable placements for these fragments may help prune the number of placements that are subjected to rigorous energy calculations. Researchers have also stressed the need for more accurate scoring functions for characterizing the energy of the binding. The development of such functions remains a difficult and poorly understood problem [5].

6 Receptor is Unknown: Pharmacophores

When the 3D structure of the target macromolecule is not known, the identification of a pharmacophore is key to the development of new pharmaceutical drugs [55]. A prevailing assumption in rational drug design is that if different ligands exhibit similar activity with a receptor, this activity is due largely to the interaction of the features of the pharmacophore to "complementary" features of the receptor (see Figure 5). Thus, if a pharmacophore has been isolated, chemists can use it as a template to build more potent drugs [27]. Given 5-10 ligands that are very flexible, finding a set of features that is present in the same 3D arrangement in the active conformation of these ligands is by no means a simple task. Figure 6 shows 4 different inhibitors of the protease thermolysin which was drawn in Figure 1. These ligands have 5 to 11 torsional DOF and each of these molecules can assume a large number of distinct low-energy conformations when these torsions are

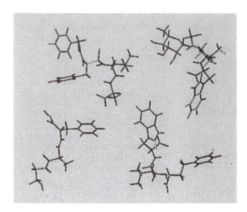

Figure 6: *Four different inhibitors of thermolysin*

varied [22, 46].

Initial approaches to pharmacophore identification searched simultaneously, in a systematic way, the conformational space of all molecules [57]. These approaches are now abandoned due to their prohibitive computational cost. The most popular recent algorithms start with a collection of distinct low-energy conformations per molecule, obtained by a conformational search procedure. They search for an invariant present in at least one conformation of most of the given molecules. Requiring that the invariant be present in all molecules may unnecessarily exclude solutions, since conformational search methods do not guarantee that all distinct low-energy conformations have been produced.

DISCO [55], one of the most popular algorithms for pharmacophore identification, uses clique detection to identify invariants. Initially the program considers a pair of conformations c_1 and c_2 belonging to different molecules. A "correspondence graph" G is constructed and this graph is similar to the "docking graph" described in Section 5.1. The nodes of G are again all node pairs of c_1 and c_2. An edge in G is created if the pairs in each of the connected nodes can be matched simultaneously. The Bron-Kerbosch clique detection algorithm [8] is then used to find cliques in G. These correspond to invariants in c_1 and c_2 and thus to candidate pharmacophores. The algorithm seems to work well in practice [55, 72]. Generalization of the above approach to n conformations is straightforward by considering one of the conformations as a reference and comparing it with all other $n - 1$ conformations. Common parts of all pairwise invariants need to be computed in the end.

If a large number of conformations per molecule are considered, there can be a combinatorial explosion in the number of basic operations performed by algorithms like DISCO [3]. This is the main reason why different approaches are under development. One idea is to start with small invariants (2-3 features) and gradually expand them [3]. Another idea is to use randomized techniques when searching for invariants. When conformations c_1 and c_2 are compared in [22], a randomized sampling scheme is used to select atoms (features) in conformation c_1, and a hashing structure is built to find possible matchings of these atoms (features) in c_2. This process is repeated for all pairs of conformations of two molecules and produces several invariants. It is then checked if these invariants are present in the rest of the considered molecules with an elaborate hashing scheme.

7   Database Searching

Searching databases of 3D chemical structures for ligands with specific characteristics is becoming a basic tool in rational drug design [56, 72]. Although, it is fairly simple to do an initial screening of a database with one million compounds, it is difficult to narrow down the results at later stages [72]. Ligand flexibility can increase dramatically the number of cases that need to be examined before it is decided that a molecule does not match a query.

Queries in current database systems are usually specified by a 3D graph whose nodes correspond to specific features and whose edges correspond to diatomic distances. Formulating a query in this way is consistent with the definition of a pharmacophore. To find ligands with a specific pharmacophoric pattern in a database, a combination of the techniques described in this paper can be used. The efficiency requirements for these techniques are however increased considerably. For example, algorithms developed for surface computation or conformational search may need to be reevaluated in the context of database queries: it may be possible to find if a feature is on the surface of a conformation without computing the whole surface, or to produce a conformation which is very different from a given one without performing a large scale conformational search.

Many database queries result in a constrained conformational search problem which is currently poorly addressed [72]. Distance geometry, systematic/randomized search, and genetic algorithms have been tried but have produced slow algorithms [4, 9, 10, 23]. One of the most efficient existing techniques for flexible searching is the "Directed Tweak Method" [37, 69]. The method minimizes a pseudoenergy function which combines the energy of the molecule and the sum of the squares of the deviations of the distances found in the molecular structure to the distances expressed in the database query. Unfortunately the pseudoenergy function contains a large number of local minima and conformations having high energy are frequently returned [10]. Techniques that can produce low-energy geometries that avoid these local minima are clearly needed [72].

8   Discussion

Computed-assisted methods for rational drug design are likely to combine a number of different techniques like randomized search methods, efficient indexing schemes, algebraic techniques, constrained optimization, etc. Undoubtedly, the geometry of the ligands is only one part of the picture of rational drug design, the other being the energy and chemical properties of the molecules involved. Software tools that consider molecular geometries and perform simple energy calculations can help in the early stages of drug development [2, 5, 6, 72]. The increased use of such tools may also contribute to an improved understanding of drug action and to the development of models that can better explain drug activity [29, 41]. Last but not least, the amount of data that is now available in molecular databases makes such tools indispensable to medicinal chemists. From a computational point of view, the geometric problems that arise in drug design,

even when simple energy models are assumed, are truly challenging.

Acknowledgment

The author is partially supported by a grant from Pfizer Central Research. Many of the problems and ideas described in this paper have been formulated in the course of a joint project between the Robotics Laboratory at Stanford University and Pfizer Central Research. The author is grateful to Prof. Jean-Claude Latombe of Stanford and Dr. Paul Finn of Pfizer Central Research who initiated and developed the project and to the rest of the people involved in this effort: Dan Halperin, Rajeev Motwani, Christian Shelton, and Suresh Venkatasubramanian.

References

[1] O. Bachar, D. Fischer, R. Nussinov, and H. Wolfson. A computer-vision based technique for 3d sequence independent structural comparison of proteins. *Protein Engineering*, 6(3):279–288, 1993.

[2] L. Balbes, S. Mascarella, and D. Boyd. A perspective of modern methods in computer-aided drug design. In K. Lipkowitz and D. B. Boyd, editors, *Reviews in Computational Chemistry*, volume 5, pages 337–370. VCH Publishers, 1994.

[3] D. Barnum, J. Greene, A. Smellie, and P. Sprague. Identification of common functional components among molecules. To appear in J. Chem. Inf. Comput. Sci., 1996.

[4] J. Blaney, G. Crippen, A. Dearing, and J. Dixon. Dgeom: Distance geometry. Quantum Chemistry Program Exchange, 590, Dept. of Chemistry, Indiana Univ., IN.

[5] J. Blaney and S. Dixon. A good ligand is hard to find: Automated docking methods. *Perspectives in Drug Discovery and Design*, 1:301–319, 1993.

[6] B. Boyd. Successes of computer-assisted molecular design. In K. Lipkowitz and D. B. Boyd, editors, *Reviews in Computational Chemistry*, volume 1, pages 355–371. VCH Publishers, 1990.

[7] D. B. Boyd. Aspects of molecular modeling. In K. Lipkowitz and D. B. Boyd, editors, *Reviews in Computational Chemistry*, volume 1, pages 321–351. VCH Publishers, 1990.

[8] C. Bron and J. Kerbosch. Finding all cliques of an undirected subgraph. *Commun. ACM*, 16:575–577, 1973.

[9] G. Chang, W. Guida, and W. Still. An internal coordinate monte-carlo method for searching conformational space. *J. Am. Chem. Soc.*, 111:4379–4386, 1989.

[10] D. Clark, G. Jones, P. Willet, P. Kenny, and R. Glen. Pharmacophoric pattern matching in files of three-dimensional chemical structures: Comparison of conformational searching algorithms for flexible searching. *J. of Chem. Inf. Comput. Sci.*, 34:197–206, 1994.

[11] M. Connolly. Analytical molecular surface calculation. *J. of Applied Crystallography*, 16:548–558, 1983.

[12] M. Connolly. Shape complementarity at the hemoglobin alpha1-beta1 subunit surface. *Biopolymers*, 25:1229–1247, 1986.

[13] J. Craig. *Introduction to Robotics*. Addison-Wesley, Reading, MA, 1986.

[14] G. Crippen and T. Havel. *Distance Geometry and Molecular Conformation*. Research Studies Press, Letchworth, U.K., 1988.

[15] R. DesJarlais, R. Sheridan, J. Dixon, I. Kuntz, and R. Venkatarghavan. Docking flexible ligands to macromolecular receptors by molecular shape. *J. of Medicinal Chemistry*, 29:2149–2153, 1986.

[16] K. Dill. Folding proteins: Finding a needle in a haystack. *Current Opinion in Structural Biology*, 3:99–103, 1993.

[17] J. V. Drie, D. Weininger, and Y. Martin. Alladin: An integrated tool for computer-assisted molecular design and pharmacophore recognition, from geometric steric and substructure searching of three-dimensional molecular structures. *J. of Computer-Aided Molecular Design*, 3:225–251, 1989.

[18] H. Edelsbrunner. The union of balls and its dual shape. In *Proc. of the 9th Annual Symposium on Computational Geometry*, pages 218–231, 1993.

[19] H. Edelsbrunner, M. Facello, P. Fu, and J. Liang. Measuring proteins and voids in proteins. In *Proc. of the 28 Hawaii International Conf. on Systems Sciences*, pages 256–264, Wailea, Hawaii, 1995.

[20] H. Edelsbrunner, M. Facello, and J. Liang. On the definition and the construction of pockets in macromolecules. In *DIMACS Workshop on Computational Biology*, Rutgers, NJ, 1995.

[21] O. Faugeras. *Three-Dimensional Computer Vision*. MIT Press, Cambridge, MA, 1993.

[22] P. Finn, D. Halperin, L. Kavraki, J.-C. Latombe, R. Motwani, C. Shelton, and S. Venkatasubramanian. Geometric manipulation of flexible ligands. In M. Lin and D. Manocha, editors, *LNCS Series - 1996 ACM Workshop on Applied Computational Geometry*. Springer-Verlag, 1996.

[23] E. Fontain. Applications of genetic algorithms in the field of constitutional similarity. *J. Chem. Inf. Comput. Sci.*, 32:748–752, 1992.

[24] B. Freyberg, T. Richmond, and W. Braum. Surface area effects on energy refinement of proteins: a comparative study on atomic solvation parameters. *J. Molecular Biology*, 233:275–292, 1993.

[25] M. Garey and D. Johnson. *Computers and Intractability: A Guide to the Theory of NP-Completeness*. Freeman, San Francisco, 1980.

[26] A. Ghose, J. Kowalczyk, M. Peterson, and A. Treasurywala. Conformational searching methods for small molecules: I. study of the sybyl search method. *J. of Computational Chemistry*, 14(9):1050–1065, 1993.

[27] R. Glen, G. Martin, A. Hill, R. Hyde, P. Wollard, J. Salmon, J. Buckingham, and A. Robertson. Computer-aided design and synthesis of 5-substituted tryptamines and their pharmacology at the $5-HT_{10}$ receptor: Discovery of compounds with potential anti-migraine properties. *J. of Medicial Chemistry*, 38:3566–3580, 1995.

[28] N. Go and H. Scherga. Ring closure and local conformational deformations of chain molecules. *Macromolecules*, 3(2):178–187, 1970.

[29] V. Golender and E. Vorpagel. Computer-assisted pharmacophore identification. In H. Kubinyi, editor, *3D QSAR in Drug Design*, pages 137–149. ESCOM, Leiden, 1993.

[30] P. Goodford. A computational procedure for determining energetically favored binding sites on biologically important macromolecules. *J. of Medicinal Chemistry*, 28:849–857, 1985.

[31] D. Goodsell and A. Olson. Simulated annealing and docking. *Proteins*, 8:195–202, 1990.

[32] J. Greene, S. Kahn, H. Savoj, P. Sprangue, and S. Teig. Chemical function queries for 3D database search. *J. Chem. Inf. Comput. Sci.*, 34:1297–1308, 1994.

[33] W. Grimson and T. Lozano-Pérez. Model-based recognition and localization from sparse range and tactile data. *The International Journal of Robotics Research*, 3(3):3–35, 1984.

[34] W. Guida, R. Bohacek, and M. Erion. Probing the conformational space available to inhibitors in the thermolysin active site using monter carlo/energy minimization techniques. *J. of Computational Chemistry*, 13(2):214–228, 1992.

[35] D. Halperin, J.-C. Latombe, and R. Motwani. Dynamic maintenance of kinematic structures. In J.-P. Laumond and M. Overmars, editors, *Algorithmic Foundations of Robotics*. A K Peters, MA, 1996.

[36] D. Halperin and M. Overmars. Spheres, molecules and hidden surface removal. In *Proc. 10th ACM Symposium on Computational Geometry*, pages 113–122, Stony Brook, 1994.

[37] T. Hurst. Flexible 3D searching: The directed tweak method. *J. Chem. Ing. Comp. Sci.*, 34:190–196, 1994.

[38] L. Kavraki. *Random Networks in Configuration Space for Fast Path Planning*. PhD thesis, Stanford University, 1995.

[39] S. Kearsley, D. Underwood, R. Sheridan, and M. Miller. Flexibases: A way to enhance the use of molecular docking methods. *J. of Computer-Aided Molecular Design*, 8:565–582, 1994.

[40] G. Klebe and T. Mietzener. A fast and efficient method to generate biologically relevant conformations. *J. of Computer-Aided Molecular Design*, 8:583–606, 1994.

[41] H. Kubinyi. *3D QSAR in Drug Design*. ESCOM, Leiden, 1993.

[42] G. Kuhl, G. Crippen, and D. Friesen. A combinatorial algorithm for calculating ligand binding. *J. of Computational Chemistry*, 5:24–34, 1984.

[43] I. Kuntz, J. Blaney, S. Oatley, R. Langridge, and T. Ferrin. A geometric approach to macromolecular-ligand interactions. *J. of Molecular Biology*, 161:269–288, 1982.

[44] Y. Lamdan and H. Wolfson. Geometric hashing: A general and efficient model-based recognition scheme. In *IEEE International Conference on Computer Vision*, pages 238–249, Tampa, FL, 1988.

[45] M. Lawrence and P. Davis. CLIX: A search algorithm for finding novel ligands capable of binding proteins of known three-dimensional structure. *Proteins*, 12:31–41, 1992.

[46] A. Leach. A survey of methods for searching the conformational space of small and medium sized molecules. In K. Lipkowitz and D. Boyd, editors, *Reviews in Computational Chemistry*, volume 2, pages 1–47. VCH Publishers, 1991.

[47] A. Leach and I. Kuntz. Conformational analysis of flexible ligands in macromolecular receptor sites. *J. of Computational Chemistry*, 13:730–748, 1992.

[48] B. Lee and F. Richards. The interpretation of protein structures: Estimation of static accessibility. *J. of Molecular Biology*, 55:379–400, 1971.

[49] T. Lengauer. Algorithmic research problems in molecular bioinformatics. In *IEEE Proc. of the 2nd Israeli Symposium on the Theory of Computing and Systems*, pages 177–192, 1993.

[50] J. Liang, P. Sudhakar, H. Edelsbrunner, P. Fu, and S. Subramanian. Analytical shape computing of macromolecules: Molecular area and volume through alpha-shapes. In preparation.

[51] M. Lipton and W. Still. The multiple minimum problem in molecular modeling: Tree searching internal coordinate conformational space. *J. of Computational Chemistry*, 9(4):343–355, 1988.

[52] T. Lybrand. Computer simulation of biomelecular systems using molecular dynamics and free energy perturbation methods. In K. Lipkowitz and D. B. Boyd, editors, *Reviews in Computational Chemistry*, volume 1, pages 295–320. VCH Publishers, 1990.

[53] D. Manocha. *Algebraic and Numeric Techniques for Modeling and Robotics*. PhD thesis, University of California, Berkeley, 1992.

[54] D. Manocha, Y. Zhu, and W. Wright. Conformational analysis of molecular chains using nanokinematics. *Computer Application of Biological Sciences (CABIOS)*, 11(1):71–86, 1995.

[55] Y. Martin, M. Bures, E. Danaher, J. DeLazzer, and I. Lico. A fast new approach to pharmacophore mapping and its application to dopaminergic and benzodiazepine agonists. In *J. of Computer-Aided Molecular Design*, volume 7, pages 83–102, 1993.

[56] Y. C. Martin, M. G. Bures, and P. Willet. Searching databases of three-dimensional structures. In K. Lipkowitz and D. B. Boyd, editors, *Reviews in Computational Chemistry*, volume 1, pages 213–256. VCH Publishers, 1990.

[57] D. Mayer, C. Naylor, L. Motoc, and G. Marshall. A unique geometry of the active site of angiotensin-converting-enzyme consistent with structure activities studies. *J. of Computer-Aided Molecular Design*, 1:3–16, 1989.

[58] P. G. Mezey. Molecular surfaces. In K. Lipkowitz and D. B. Boyd, editors, *Reviews in Computational Chemistry*, volume 1, pages 265–289. VCH Publishers, 1990.

[59] M. Miller, S. Kearsley, D. Underwood, and R. Sheridan. Flog: A system to select 'quasi-flexible' ligands complementary to a receptor of known three-dimensional structure. *J. of Computer-Aided Molecular Design*, 8:153–174, 1994.

[60] M. Nicklaus, S. Wang, J. Driscoll, and G. Milne. Conformational changes of small molecules binding to proteins. *Bioorganic and Medicinal Chemistry*, 3(4):411–4128, 1995.

[61] R. Norel, D. Fischer, H. Wolfson, and R. Nussinov. Molecular surface recognition by a computer-based technique. *Protein Engineering*, 7(1):39–46, 1994.

[62] D. Parsons and J. Canny. Geometric problems in molecular biology and robotics. In *Intelligent Systems for Molecular Biology*, pages 322–330, Palo Alto, CA, 1994.

[63] M. Raghavan and B. Roth. Kinematic analysis of the 6r manipulator of general geometry. In *International Symposium of Robotics Research*, pages 314–320, Tokyo, 1989.

[64] M. Rarey, B. Kramer, and T. Lengauer. Time efficient docking of flexible ligands into active sites of proteins. In *International Conference on Intelligent Systems for Molecular Biology*, Cambridge, 1995.

[65] F. Richards. The interpretation of protein structures: Total volume, group volume distributions and packing density. *J. of Molecular Biology*, 82:1–14, 1974.

[66] F. Richards. Areas, volumes, packing, and protein structures. *Ann. Rev. Biophys. Bioeng.*, 6:151–176, 1977.

[67] B. Shoichet, D. Bodian, and I. Kuntz. Molecular docking using shape descriptors. *J. of Computational Chemistry*, 13(3):380–397, 1992.

[68] A. Smellie, S. Kahn, and S. Teig. Analysis of conformational coverage: 1. validation and estimation of coverage. *J. Chem. Inf. Comput. Sci.*, 35:285–294, 1995.

[69] Tripos. *UNITY*. St. Louis, MO.

[70] A. Varshney, F. P. Brooks, Jr., and W. V. Wright. Computing smooth molecular surfaces. *IEEE Computer Graphics & Applications*, 15(5):19–25, September 1994.

[71] W. Welsh and A. Jain. Hammerhead: Fast fully automated docking of flexible ligands to protein binding sites. In preparation.

[72] P. Willet. Searching for pharmacophoric patterns in databases of three-dimensional chemical structures. *J. of Molecular Recognition*, 8:290–303, 1995.

Facial Analysis and Synthesis Using Image-Based Models

Tony Ezzat, *MIT Artificial Intelligence Laboratory, Cambridge, MA, USA*
Tomaso Poggio, *MIT Artificial Intelligence Laboratory, Cambridge, MA, USA*

In this paper, we describe image-based modeling techniques that make possible the creation of photo-realistic computer models of real human faces. The image-based model is built using example views of the face, bypassing the need for any three-dimensional computer graphics models. A learning network is trained to associate each of the example images with a set of pose and expression parameters. For a novel set of parameters, the network synthesizes a novel, intermediate view using a view morphing approach. This image-based synthesis paradigm can adequately model both rigid and non-rigid facial movements.

We also describe an analysis-by-synthesis algorithm, which is capable of extracting a set of high-level parameters from an image sequence involving facial movement using embedded image-based models. The parameters of the models are perturbed in a local and independent manner for each image until a correspondence-based error metric is minimized.

A small sample of experimental results is presented.

1   Introduction

Facial analysis and synthesis have emerged to be two important requirements for a vast array of vision-based applications. Facial analysis refers to the extraction from video sequences of information concerning the location of the head, its pose, and the movement of facial features such as the eyes and the mouth. Facial synthesis refers to the reverse process of animating a facial model using a set of high-level parameters that control the face's gaze, mouth orientation, and pose. Facial analysis would be useful for such applications as eye-tracking, facial expression recognition, and visual speech understanding. Facial synthesis would be useful for animating cartoon characters or digital actors. Together, facial analysis and facial synthesis *in tandem* would be useful for model-based coding applications such as video email and video-teleconferencing, as well as interactive animation of cartoon characters using facial motions.

Many of the attempts at facial analysis and synthesis involve modeling the human face in three dimensions using computer graphics techniques. In this work, we adopt an *image-based model*, whose basis is to completely forego any underlying computer graphics models, and instead model the face using *example images*. Within the image-based synthesis literature, a number of researchers ([4], [7], [15], [16]) have noticed the viability of a *view interpolation* approach to image synthesis, where novel, intermediate images of a scene are synthesized from example endpoints using a morphing technique. In this work, we adopt the particular approach of Beymer, Shashua, Poggio [4], who cast the view interpolation approach in a *learning-by-example* framework: each example image is associated with a position in a high-level, multi-dimensional parameter space denoting pose and expression. By training on the examples, a learning network can then generalize, and generate suitable novel images that lie at intermediate points in the example space. The trained network, in essence, becomes a *synthesis network*, which generates images as output, for suitable parameters as input. Beymer, Shashua, Poggio [4], in fact, showed that this technique is capable of modeling rigid facial transformations such as pose changes, as well as non-rigid transformations such as smiles.

From the analysis standpoint, we are motivated in particular by the work of Jones and Poggio who constructed models of line drawings [11] and faces [12], and used a stochastic gradient descent algorithm to match the models to novel line drawings or faces input by the user. The models themselves consisted of a linear combination of prototypes [14], and the error metric which the gradient descent algorithm tried to minimize was the pixel-wise error between the novel drawing and the current guess for the closest model image. At every iteration, the algorithm would compute the gradient of this error metric with respect to the model parameters, and proceed to a new guess for a set of parameters that would produce a new model image closer to the novel image.

The first contribution of this work is to extend the *synthesis network* paradigm of Beymer, Shashua, Poggio [4] into a *synthesis module* paradigm more suitable for analysis: Firstly, each synthesis network is additionally parameterized with a set of affine parameters, such as translation, rotation, and scale. Secondly, a flexible mask-based segmentation scheme is incorporated into the synthesis module that is capable of segmenting the head in any of the images output by the network. Thus, from an input-output perspective, the synthesis module is capable, for the appropriate input parameters, of producing images of segmented faces at various scales, rotations, positions, poses, and expressions.

The second contribution of this work is to embed the synthesis modules mentioned previously in an analysis-by-synthesis algorithm similar to that of Jones and Poggio [11]. In our case, however, we define a *correspondence-based error metric* instead of a pixel-based error metric, in an attempt to make the analysis algorithm more robust to changes in lighting, position, scale, rotation, and hairstyle. Essentially, the parameters of the embedded synthesis modules are perturbed in a local and independent manner for each image in the sequence until the correspondence-based error metric is minimized.

In Section 2, we describe the construction of the basic synthesis networks to be used for analysis and synthesis. In Section 3, we describe additional techniques that allow for more complicated synthesis networks to be constructed. In Section 4, we transition from the synthesis network to the synthesis module, and sketch an outline of our analysis-by-synthesis algorithm. In Section 5, we describe and depict a small sample of experimental results. Finally, in Section 6, we briefly critique our approach, and discuss future work.

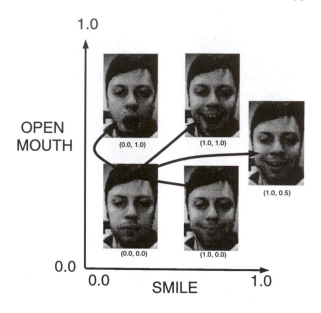

Figure 1: *A 5-example, 2-dimensional example set in a smile/open-mouth configuration.*

2 Building the Synthesis Networks

2.1 Choosing the Example Set and the Parameter Space

The first step in the creation of the synthesis network is the selection of the example images and the association of each example with a point in a hand-crafted parameter space x. Figure 1 depicts five example images arranged in a two-dimensional parameter space where each axis is limited to values between 0.0 and 1.0. One axis denotes degree of *smile*, while the other denotes degree of *mouth openness*. The top-right example image, for instance, would be associated with the position in parameter space $x = (1.0, 1.0)$.

2.2 Learning the Map from Parameters to Correspondences

Given the example images and the associated parameters, the desired task is to generate novel intermediate images lying in the space spanned by the examples. Beymer, Shashua, and Poggio [4] re-cast this task as a *learning problem*, in which it is necessary to learn an unknown function $y = f(x)$ that maps between the parameter space, x, and the example space, y, given a set of N training samples (x_i, y_i) of the function $f(x)$.

Learning such a function would allow one to generalize the function at points other than the example points, and hence synthesize appropriate novel intermediate images.

Poggio and Brunelli [13], however, made the crucial observation that trying to approximate a function $y = f(x)$ that maps between the parameter space x and an example space y *of images* would probably not work due to the discontinuous nature of the underlying map. Instead, Poggio and Brunelli [13] argued that it is better to try to learn a map between a parameter space x and an example space y *of correspondence vectors* that define corresponding features across the example images. The underlying intuition is that such a map is easier to learn because the correspondence vectors factor out lighting effects, and also because they undergo reasonably smooth change during motion of the underlying object to be modeled.

A helpful, and often important, way (discussed in Beymer [3]) to think about the distinction between images and correspondences is to view correspondence as a way to sample the *motion* (or shape) of an object between two views, and to view images as a way to sample the object's *texture* for those views. Synthesizing novel intermediate correspondences is thus equivalent to synthesizing novel intermediate motions (or shapes) of the face.

2.3 Defining and Obtaining the Correspondence

In this work, a *dense, pixel-wise* correspondence is defined between two images: for a pixel in image A at position (i, j), the corresponding pixel in image B lies at position $(i + \Delta x(i, j), j + \Delta y(i, j))$, where Δx and Δy are arrays or matrices that contain the x and y components of the correspondence vectors, respectively. In the rest of this paper, we use the symbol y to refer to both the x- and the y-components of the correspondences, and the reserve symbol x to refer to the imposed multidimensional parameter space. Equations involving the use of y imply that they are performed twice: once on the x-components and once on the y-components.

From the standpoint of synthesis network design, in which more than two images may be involved, a *reference example* image is designated, and correspondence between it and the rest of the images in the example set

is found. For example, in Figure 1, the bottom-left image is the reference example, and four correspondence vectors y_i are obtained between it and the other examples. A fifth, and null, correspondence vector, y_0, is designated to represent the correspondence between the reference example and itself.

To obtain such a dense, pixel-wise correspondence between the example images, optical flow algorithms borrowed from the computer vision literature are utilized. We specifically use the coarse-to-fine, gradient-based optical flow algorithms developed by Bergen and Hingorani [2], which have yielded good results in practice. In cases where they have not yielded good results, such as in cases when there is significant movement or occlusion, we have found [9] that concatenating optical flow between a set of intermediate images improves the final correspondences dramatically.

2.4 Constructing the Mapping Function

We approximate the unknown function $y = f(x)$ which maps from parameters to correspondences given the samples $(y_i, x_i)_{i=1}^N$, using a *radial basis function with Gaussian centers*:

$$f(x) = \sum_{\alpha=1}^{n} c_\alpha G(\|x - t_\alpha\|) \qquad (1)$$

where

$$G(x) = e^{-\frac{x^2}{\sigma^2}} \qquad (2)$$

and the $t'_\alpha s$ are arbitrary parameters termed *centers*.

A radial basis function as defined in Equation 1 was shown in [10] to be a type of *regularization network* which incorporates *a priori* knowledge about the smoothness of the function that can be learned from a set of samples. Such a smoothness prior is necessary because the problem of generalizing a function from a set of samples is inherently ill-posed, and many functions may be found which pass through the sample points. The approximating function in Equation 1 is chosen from among all the possible solutions because it is simultaneously close to the data samples and the smoothness constraints.

The learning stage of a radial basis function consists of the specification of three sets of variables: the centers t_α, the σ's of the Gaussians, and, most importantly, the coefficients c_i.

In this work, we associate one example parameter x_i with each center t_α, so the approximating function $f(x)$ may be rewritten as

$$f(x) = \sum_{i=1}^{N} c_i G(\|x - x_i\|) \tag{3}$$

where N denotes the total number of examples used to train the network. One can now visualize a Gaussian center associated with each example parameter x_i used to train the network.

The sigmas of the Gaussians, which denote the width of their influence, are determined using an *average inter-example distance strategy*. For each Gaussian, the average distance between its associated example parameter and all the other example parameters is found. The final sigma value for that Gaussian is chosen to be some fixed constant times the resulting average.

Finally, the c_i coefficients are chosen in a manner that minimizes the empirical error between the approximating function $f(x)$ and the sample points $(y_i, x_i)_{i=1}^{N}$. If we substitute all the sample pairs into the Equation 3 we obtain the equation

$$Y = CG \tag{4}$$

where

$$Y = [\; y_1 \quad y_2 \quad \cdots \quad y_N \;], \tag{5}$$

$$C = [\; c_1 \quad c_2 \quad \cdots \quad c_N \;], \tag{6}$$

and

$$G = \begin{bmatrix} G(\|x_1 - x_1\|) & G(\|x_2 - x_1\|) & \cdots & G(\|x_N - x_1\|) \\ G(\|x_1 - x_2\|) & G(\|x_2 - x_2\|) & \cdots & G(\|x_N - x_2\|) \\ \vdots & & \ddots & \vdots \\ G(\|x_1 - x_N\|) & G(\|x_2 - x_N\|) & \cdots & G(\|x_N - x_N\|) \end{bmatrix} \tag{7}$$

The coefficients C are then determined by computing

$$C = YG^+ \tag{8}$$

where G^+ is the pseudo-inverse of G.

2.5 The Dual Representation for Synthesis

It is extremely helpful to rewrite the approximation function in Equation 3 into its *dual representation*, which illustrates the nature of its interpolative properties. Continuing from Equation 3, we have

$$y(x) = Cg(x) \tag{9}$$

where

$$g(x) = [\; G(\|x - x_1\|) \quad G(\|x - x_2\|) \quad \cdots \quad G(\|x - x_N\|) \;]. \tag{10}$$

Substituting Equation 8 into Equation 9 we obtain

$$y(x) = YG^+g(x). \tag{11}$$

Gathering the terms not related to Y together, we have

$$y(x) = \sum_{i=1}^{N} b_i(x)y_i \tag{12}$$

where

$$b_i(x) = (G^+)_i g(x). \tag{13}$$

Equation 12, which represents the dual representation of Equation 3, is arguably the most central equation for synthesis. Equation 12 represents any novel intermediate correspondence vector y as a *linear combination* of the N example correspondence vectors y_i. The coefficients of the combination, $b_i(x)$, depend nonlinearly on the parameter x. The learning stage defines the structure of the b_i kernels, which are typically Gaussian-like in nature, centered around each of the example parameters x_i used for training. This is not surprising given the approximation of $y = f(x)$ using a radial basis function with Gaussian centers.

2.6 Warping

A new correspondence vector y synthesized from Equation 12 defines a *position* in correspondence space that we would like the novel, intermediate image to be located at. A simple *forward warp* operation that pushes the pixels of the reference example image along the synthesized correspondence vector is sufficient to generate a novel intermediate image, but such an approach would not utilize the image texture from *all* the examples in the network. To utilize the image texture from all the examples, we adopt a correspondence re-orientation procedure, described in [4], that re-orients the synthesized correspondence vector from Equation 12 so that it originates from each of the other example images and points to the same position as the original synthesized correspondence. This allows us to subsequently forward warp all the examples along their respective re-oriented correspondence vectors.

The forward warp algorithm used does not explicitly treat *pixel overlaps*, or *folds*, in any special way, since there is no a-priori visibility model [16] built into the algorithm, unlike [6] and [7]. Hence, the order of the warp is a simple *top-down, left-to-right* order. Pixel destination values are rounded to the nearest integer location.

2.7 Hole-filling

Since the correspondences produced by the optical flow algorithms are not strictly one-to-one mappings, forward warping usually exposes regions in the warped image that are unfilled. These regions, called *holes*, must be explicitly treated, since they lead to noticeable degradations in the quality of the final images that are synthesized. In particular, holes due to local image expansion and inaccuracies in the optical flow algorithms due to lack of discriminating texture usually lead to small *specks* in the warped image. These holes are identified and eliminated as in [7], by filling the warped image with a reserved "background" color prior to warping. For those pixels which retain the background color after the warp, new colors are computed by interpolating the colors of the adjacent non-background colors.

2.8 Blending

Warping all the example images along the re-oriented correspondence vectors results in a set of warped images that need to be combined to produce the final image, which is done using *blending*. Blending refers to multiplying each image with a blending coefficient, and then adding all the scaled images together to form the final image. The blending coefficients chosen are the same as the coefficients $b_i(x)$ from Equation 12. Intuitively, the blending coefficient associated with a particular example image decreases with the distance of the novel synthesis image from the example image. Conse-

Figure 2: *Two sets of intermediate, novel images generated from 1-dimensional, 2-example synthesis networks for smile (top) and rightwards pose (bottom). The original images are the leftmost and rightmost images for each synthesis segment.*

quently, more weight in the blending stage is given to the warped images from the closer examples.

2.9 Results and Discussion

Figure 2 illustrates a number of novel images synthesized from two one-dimensional networks. The original images for both networks are the leftmost and rightmost images. The correspondences were obtained using direct optical flow estimation between both endpoint images.

Figure 3 illustrates a large number of novel images synthesized from the two-dimensional, 5-example network shown in Figure 1. The correspondences were obtained by concatenating optical flow between a number of intermediate images.

It is important to note that the combination of warping and blending, also known as morphing, is more powerful than either technique on its own. Blending alone can generate intermediate images, but a sense of *movement* between images will be lacking. Warping alone will expose the deficiencies of the optical flow algorithms, and particularly in our case their linearization errors: the flow estimates at each pixel are only a *linear* approximation to the actual flow. As a result, warping from one image by itself will lead to suitable intermediate images *only* when the parameters are close to the parameters of the example image, but will lead to incorrect images as the parameters move farther away. Warping from all the examples combined

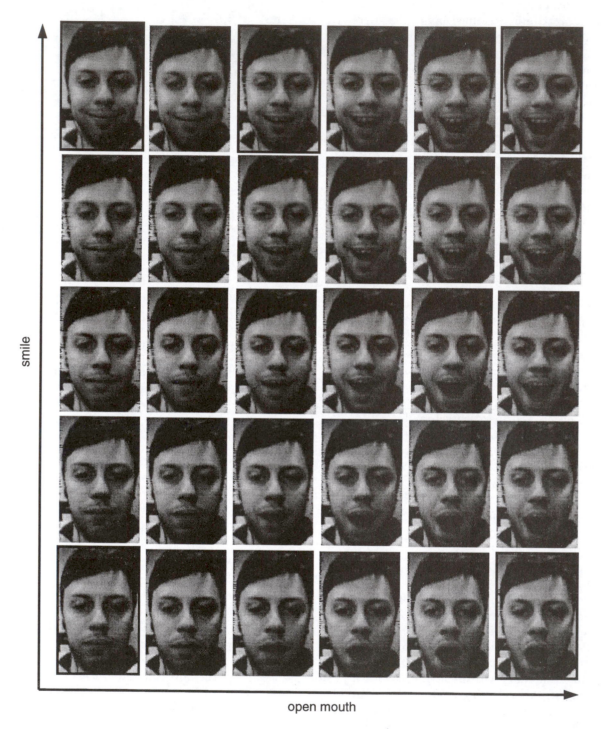

smile

open mouth

Figure 3: *Examples of the intermediate, novel images generated from the 2-dimensional, 5-example synthesis network for facial expressions shown in Figure 1. The original images are high-lighted with darker borders.*

with weighted blending, however, eases the linearization errors because, as the parameters move farther away from one example image, the pixel motion and pixel values of another example image begin to take effect.

It is also extremely heartening that a technique that combines warping, blending, and concatenated optical flow can lead to results that are good for cases in which large occlusions are present, as was the case in Figure 3. This is considerably surprising especially in light of the fact that our optical flow algorithms have no a-priori visibility model [16] built into them, as mentioned earlier.

3 Other Types of Networks

Although the synthesis network paradigm described above is adequate for modeling a large number of facial motions, it is necessary to augment it with additional techniques to address certain issues. In this section, we describe those issues, and the techniques adopted to address them.

3.1 Regional Networks

One of the problems associated with the example-based synthesis paradigm explored in this work is that a large number of example images are needed whenever a new dimension is added to a synthesis network. For example, suppose we are modeling 6 eye positions and 4 mouth positions. Modeling a fifth mouth position would require 6 additional examples, one for each separate modeled eye position.

One approach to alleviate the need for such a large number of example images is to create separate, *regional* networks for different parts of the face that move independently of each other. Such an approach would decorrellate the eye-mouth network described above into two separate, regional networks: one regional eye network composed of 6 example images modeling the various eye positions, and one regional mouth network composed of 4 images modeling the various mouth positions. Modeling a fifth mouth position would thus require only one additional example image.

Regional decomposition needs to address two issues: how to specify which regions each network controls, and how to combine the synthesized outputs of all the regional networks back together again.

A *mask-based* approach was adopted to specify which regions of the face each network controls. At the outset of the example set selection, the example set designer uses a special tool to "mask out" which region of the face each network controls. The mask produced by the tool is essentially a binarized image. During synthesis, a navigational mechanism first determines which parameters have changed relative to previous parameters, and identifies which regional network is activated. The parameters associated with that regional network are then used to synthesize an image. The mask associated with that regional network is then used to extract the appropriate portion of the synthesized image.

To combine the masked regions back together again, a simple *paste* approach was adopted, where the regions are pasted on top of a *base image* of the face. This approach works extremely well if the motion is contained *within* the regions themselves. Ideally, one would want to *blend* the regions onto the base image using more sophisticated techniques.

As an example, a regional network was constructed for left eye motions, right eye motions, and mouth motions, as shown in Figure 4. The regional left and right eye networks were composed of the *same* six images placed in a 2-dimensional arrangement, with different masks for each eye regional network. The mouth regional network consisted only of two examples to model an opening mouth. The mask for the mouth network consisted of all the pixels not contained in the left and right eye regional networks; this approach enables one to avoid creating a mask with a more explicit segmentation of the mouth region, which is hard to do because mouth movements affect a large portion of the face. The masked outputs of each regional network are pasted onto the base image shown in the center of the figure.

The gain in possible eye-mouth configurations given the number of example images is now much higher than in a standard approach not involving regional networks. Using only 6 original eye images (since the left and right eye regional networks use the same images) and 1 additional open-mouth image (since the reference image is the same for all the regional networks), various combinations of eye-mouth positions may be synthesized, as shown in Figure 5. This added flexibility is also a boon for the example set designer, who needs fewer example images to build the desired synthesis model. On the other hand, the example set designer now needs to specify the mask regions.

3.2   Composed Networks

Another problem with the example-based synthesis network paradigm described in the Section 2 is that it requires warping and blending of *all* the example images within the network. Such an approach does not take advantage of the inherent *locality* of the image space. For example, suppose we wanted to model vertical and horizontal head pose movements using a two-dimensional, 3-by-3 network such as the one shown in Figure 6.

Such a 3-by-3 network, however, may be viewed as *four 2-by-2* networks that share a common set of example images along the adjacent edges. Instead of traversing one large network space, smaller, *local* network spaces are traversed, and a navigational mechanism is utilized to determine which local network is currently activated. Experiments were performed with exactly such a set of 4 composed local networks denoting horizontal and vertical pose movement, and some of the synthesized results are shown in Figure 7. The navigational mechanism used in this case performs a horizontal and vertical threshold check based on the input parameters to check which network is activated.

There are several major advantages of using such a network composition technique. Firstly, *composition is natural*, at least within a synthesis framework based on morphing. If the 2-dimensional space is large, chances are that the intermediate examples should only be determined from the example images that are the closest.

Secondly, *composition maintains constant computation complexity*. No matter how large an example set space becomes, if one synthesizes only from the four closest examples, then the computational complexity remains constant. The only price to be paid is the price of having to decide which network to activate, which is not as computationally intensive as having to warp and blend from a large number of examples.

Thirdly, *composition improves final image quality*. By using only the most relevant images for synthesis, the final image quality is improved. Image quality tends to decrease when a large number of examples are warped and blended together, due to the accumulated errors.

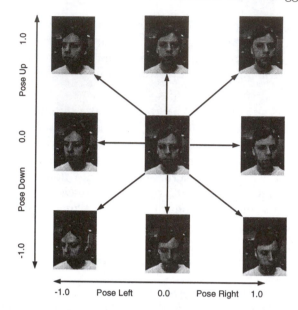

Figure 6: *The examples for a 3-by-3 network involving pose movements in all directions.*

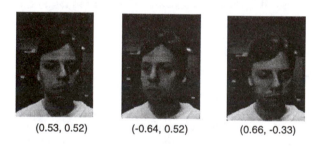

(0.53, 0.52)   (-0.64, 0.52)   (0.66, -0.33)

Figure 7: *Some intermediate examples generated from the synthesis network of Figure 6, and their positions in the imposed parameter space.*

3.3   Hierarchical Networks

Another modification to the general example-based synthesis paradigm introduced in the second section emerged in the course of attempting to model eye, mouth, and pose movements simultaneously. It became apparent that there is an *inherent hierarchical relationship* between certain facial motions. For example, eye motions and mouth motions are subordinate to pose motions: changing pose necessarily affects the appearance of the eyes and the mouth, while movements of the mouth and the eyes do not change the overall pose of the head. Consequently, there was a need for a modified synthesis approach which attempted to en-

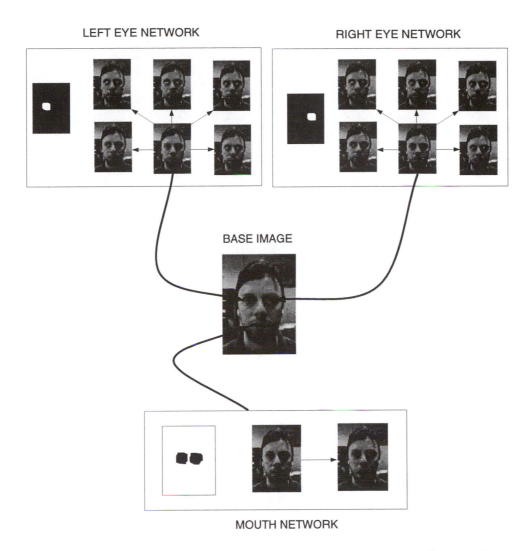

Figure 4: *Construction of a 7-example, 5-dimensional regional synthesis network controlling mouth movement and eye movement.*

Figure 5: *Synthesized images generated from the network in the Figure 4.*

code this new notion of hierarchy between networks.

The modified hierarchical synthesis paradigm was applied to a 14-example, 4-dimensional network that involved eye, mouth, and pose movements, shown in Figure 8. Firstly, a 7-example eye-mouth network was constructed for eye and mouth movements *at a single head pose*. The eye-mouth network was composed of two regional networks for the eyes and mouth, as described in the previous section on regional networks. A similar 7-example network was also created for the eye and mouth movements *at a second rightwards pose*. The two 7-example eye-mouth networks, shown in Figure 8 a), thus constitute *subnetworks* to be placed within a larger pose network.

The next step in the creation of the eyes-mouth-pose network, shown in Figure 8 b), is to compute a set of *cross-flows* linking the images between the two subnetworks. The cross-flows may be obtained using any one of various methods described in [9], and essentially allow the two eye-mouth subnetworks to be placed in correspondence, in the same manner as two images would be placed in correspondence.

The third and most important step in the hierarchical synthesis paradigm, shown in Figure 8 c), is to *synthesize an intermediate eye-mouth subnetwork for a change in pose*. Synthesizing such a new intermediate eye-mouth subnetwork consists of two steps:

- The first involves synthesizing the new, intermediate *images* that belong in the new, intermediate subnetwork. The synthesis of the new images proceeds along the respective cross-flow vectors. Essentially, temporary 1-dimensional synthesis networks are created, where the corner images are the corresponding images in the mouth subnetworks, and the correspondence vector is the cross-flow vector. Synthesis of the intermediate images proceeds in the standard manner described in Section 2.

- The second step, involves the synthesis of new, intermediate *correspondences* tying the images within the new, intermediate subnetwork together. In this case, a temporary network is created *in which the endpoints are not images, but the two correspondences from the corner mouth subnetworks*. These correspondences are warped to produce the intermediate correspondences that tie

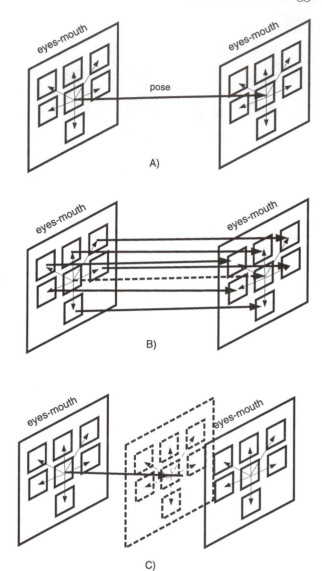

Figure 8: *The stages of the new hierarchical synthesis approach for the 14-example, 4-dimensional, eyes-mouth-pose network.*

the images within the intermediate subnetwork together.

Experiments were performed with such a hierarchical eyes-mouth-pose network, and Figure 9 shows two sequences of images generated from the same eyes-pose-mouth synthesis network. In the top row, the mouth is kept closed while the eyes and the pose are changed. In the bottom row, all three facial features are changed.

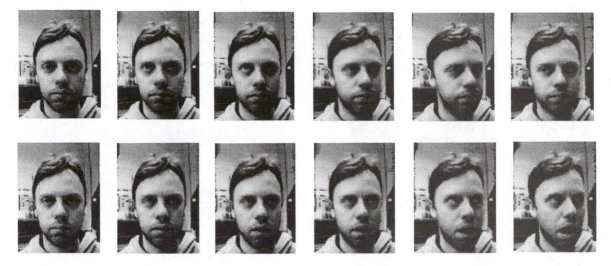

Figure 9: *Two separate image sequences synthesized from the 14-example, 4-dimensional eyes-mouth-pose network in Figure 8.*

It is interesting to point out that the modified hierarchical synthesis approach is not a new paradigm at all, but a generalization. In the old synthesis method, *images* were warped and blended together to achieve novel, intermediate images. In the new method, this notion of warping and blending is extended to include not only *images*, but also *flows*, and hence *entire networks*. One can alternatively think of the synthesis technique as warping and blending *nodes*, where a node can be an image, a network, a network of networks, and so on.

4 Analysis

4.1 Overview

In this section, a model-based analysis algorithm is outlined which is capable of extracting a set of high-level parameters from novel image sequences. The analysis approach is, in fact, an *analysis-by-synthesis* approach, where the synthesis networks created in the previous section are themselves used for analysis. An important and useful consequence of this approach is that the only parameters that may be extracted from the novel sequence are those that are encoded by the synthesis networks themselves.

4.2 Affine Parameters

Before analyzing with respect to novel image sequences, the synthesis networks must be additionally parameterized with a set of affine parameters. This is needed because novel sequences involve movements of the head that are at scales, positions, and rotation angles that are different from those in the network. Augmenting the synthesis networks with a set of four affine parameters (two translation parameters, an angle parameter, and a scale parameter), is straightforward. Essentially, the network first synthesizes the head at the intrinsic parameters imposed by the user, and then it performs an affine transformation according to the desired translation, scale, and rotation. Ideally, we would also like to augment the synthesis network with a set of projective parameters, but this was beyond the scope of this work.

4.3 Segmentation

In addition to augmenting the synthesis network with a set of affine parameters, it is also necessary to incorporate segmentation. This is needed because, in its effort to match the synthesis network with a novel sequence, the analysis algorithm needs to match only on the region in the synthesized network that corresponds to the face. This will allow the algorithm to be less sensitive to background changes, as well as hairstyle and clothing changes.

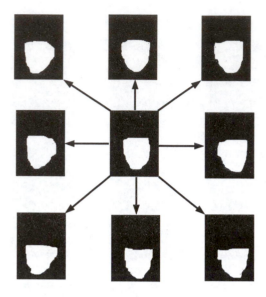

Figure 11: *Various segmented and affine-perturbed images synthesized from a 3-by-3 pose network similar to the one shown in Figure 6.*

Figure 10: *The masks associated with the 3-by-3 pose network in Figure 6.*

In attempting to segment the head in a *network*, as opposed to segmenting the head in just an *image*, there is a need for a *flexible* segmentation scheme, because the outline of the head changes shape as the head changes pose, position, rotation, and scale. One rigid mask is thus not capable of segmenting the head properly.

Consequently, a *network scheme* for flexible segmentation was adopted, where a network of the same dimensions and orientation as the corresponding image synthesis network is created, except that instead of images, the examples are masks. Each mask example serves to segment the head for the corresponding image example, and the correspondence flows relating the masks together are the same as those within the image synthesis network. The masks are defined by hand, although it is possible to use other automatic techniques. Whenever the synthesis network synthesizes a new image, it also synthesizes a new mask appropriate for the same image using the same warping and blending technique described in Section 2, with minor modifications to preserve the black-and-white pixel integrity of the mask.

Figure 10 depicts the masks that would be associated with the 3-by-3 pose network in Figure 6. Figure 11 shows various affine-perturbed, segmented images which are synthesized from a network similar to the

3-by-3 pose network of Figure 6.

We can thus begin to conceptualize a *synthesis module* that, from an input-output perspective, can generate images of a face at a various positions, rotations, scales, poses, expressions, etc., for the appropriate set of input parameters. It is important to note, in light of the forthcoming description of our analysis algorithm, that in addition to images, the synthesis module can also output correspondences and masks.

4.4 A Correspondence-Based Error Metric

A key feature of our analysis algorithm is that instead of using the embedded synthesis module to synthesize *images* to match to the novel images, and thereby have to rely on an *image-based* error metric, as in [11], the algorithm instead tries to match novel *correspondence*. For every iteration, the algorithm computes the optical flow between two consecutive novel frames, and then attempts to find the best matching correspondence from within its embedded synthesis module. The rationale for using a correspondence-based metric, as opposed to an image-based metric, is that trying to minimize a correspondence-based error metric is less susceptible to noise, local minima, and lighting changes. Both Beymer, Shashua, and Poggio [4] and Essa and Pentland [8] also found that comparing novel incoming motion with stored motion (or motion energy) templates led to good facial analysis results.

It is important to note that the set of correspondences that can be synthesized by the network are *only* those correspondences involved with the facial motions that the user chose to model, in addition to the correspondences involved with the affine movements. Consequently, even though the novel sequences will generate arbitrary types of correspondences in general, we are constrained, through this synthesis module paradigm, of matching them with only a certain repertoire of acceptable correspondences. Our matching approach may be viewed as a form of *motion regularization*, in which unconstrained optical flow is regularized with the flows from our synthesis module. Other motion regularization approaches were made by Basu, Essa, and Pentland [1] and Black and Yacoob [5]. Basu, Essa, and Pentland regularized the unconstrained optical flow with the motion of a three-dimensional ellipsoid for head-tracking purposes. Black and Yacoob regularized *local* regions of optical flow using projective planar models and curve models.

4.5 Parameter Perturbation Strategy

The analysis-by-synthesis algorithm is based on *iterative, local, independent perturbations of the synthesis parameters*. A sketch of the steps of the algorithm are as follows:

1. For a novel correspondence obtained from two consecutive novel images (say images A and B) in the sequence, the parameters of the embedded synthesis model are perturbed. The perturbations include the affine parameters, and vary each parameter independently in the positive and negative directions by a small *delta* factor.

2. For each set of perturbed parameters, the algorithm then synthesizes a correspondence from the module that corresponds to the perturbation. For reasons described in [9], we have opted to obtain the correspondence associated with the perturbation by synthesizing the two perturbed images first, and then computing optical flow between them.

3. The algorithm then computes the Euclidean distance between each perturbed correspondence and the novel correspondence, and finds the closest synthesized correspondence of the set. All distances are computed only in the regions specified by the masks associated with the perturbed correspondences.

4. The algorithm then repeats steps 1 through 3, iteratively perturbing around the set of parameters associated with the closest synthesized correspondence found in step 3.

5. For each iteration, the synthesized correspondence that yielded the overall smallest distance with respect to the novel correspondence is preserved; if a set of perturbations do not yield any new correspondences that reduce the overall minimum, the delta factors are halved and the iterations proceed once again. Thus when the algorithm gets close to the optimum synthesized correspondence, it proceeds with smaller and smaller perturbations to achieve a better match. The iterations terminate when the delta factors have been reduced to a degree where perturbations made using those factors do not make any significant changes in the synthesized correspondences.

6. Once a parameter estimate is obtained for the given novel flow, the algorithm computes the next consecutive novel correspondence in the sequence (say, between images B and C), and starts to perturb around the set of parameters found in the previous iteration. This whole process is performed across the entire sequence.

The first image in the novel sequence needs to be treated differently from the other images, since there is no prior flow within the sequence itself against which to match. Consequently, we compute the correspondence from the reference image in the network to the image, and then apply the iterative parameter perturbation technique to find the closest synthesized correspondence. This strategy suffers from the weakness that if the optical flow fails due to the fact that the heads are too far away from each other, then the extracted parameters for the first image will be incorrect. Consequently, in the novel sequences that we used to test the analysis algorithm on, the head in the initial frame was not placed too far away from the head in the reference image of the embedded synthesis module, although, of course, significant deviations in translation, rotation, scale, pose, and other variables did exist nevertheless.

5   Results

A varied but limited set of experiments were performed to test our analysis-by-synthesis technique in estimating different facial movements such as pose movements, eye movements, mouth movements, as well as head translations, rotations, and scales. The novel sequences involved changes in lighting, position, scale, rotation, background, clothing, and hairstyle. It should be noted that to change his hairstyle, the author shaved his head! In each illustration, a few frames from the novel sequence are juxtaposed against a few frames from the synthesized sequence. The analysis parameters extracted by the algorithm are also shown.

The synthesis modules embedded within the analysis algorithm were based on the networks described in Sections 2 and 3 of this paper. Specifically, the two-dimensional, 3-by-3 pose network shown in Figure 6 was used to analyze various novel pose movements of the head. Figures 12 through 18 on the following pages depict the results from the experiments. In addition, the 4-dimensional, 14-example eye-pose-mouth network shown in Figure 8 was used to analyze various combinations of mouth, eye, and pose movements. The results from these experiments are shown in Figures 20 through 24. Finally, the 2-dimensional, 5-example expression network shown in Figure 1 was used to analyze both mouth expression movements and a variety of affine head movements. The results from two such experiments are shown in Figures 26 and 28.

6   Discussion and Future Work

Our analysis experiments are still very preliminary, and more thorough testing is needed on a larger database of facial expressions and head movements. On the other hand, the results are extremely encouraging.

At present, the most salient difficulty with the analysis-by-synthesis algorithm presented in this work is that, like many iterative nonlinear optimization techniques, it is computationally inefficient. Formal timing tests were not performed, but it takes between a few minutes to half an hour to analyze one frame, depending on the complexity of the underlying synthesis module. Future work definitely needs to explore improving the efficiency of the algorithm.

It also seems that the analysis-by-synthesis paradigm as presented is also strongly *user-dependent*, although

formal tests were not performed to confirm this: the example-based models can only extract analysis parameters *reliably* from faces whose examples were used to build the model. Further work is needed to determine the limitations of the example-based models in this respect, and to overcome those limitations.

On the positive side, however, it seems that our decision to use a correspondence-based metric, in addition to the incorporation of affine perturbation and segmentation, have allowed us to achieve very good analysis in spite of changes in background, lighting, hairstyle, position, rotation, and scale.

Furthermore, it seems that the analysis-by-synthesis technique is fairly general, and can serve to analyze a wide variety of rigid and non-rigid facial movements, which would be useful for many tasks such as eye-tracking, facial expression recognition, visual speech understanding, and pose estimation.

Acknowledgments

The authors would like to thank and acknowledge David Beymer, who provided code which formed the basis for this work, as well as David Sarnoff Research Labs for their optical flow code and image libraries. Also, the authors would like to thank Mike Jones, Steve Lines, Federico Girosi, and Theodoros Evgeniou for many thoughtful discussions and criticisms.

References

[1] Sumit Basu, Irfan Essa, and Alex Pentland, "Motion Regularization for Model-based Head Tracking", MIT Media Laboratory Perceptual Computing Section Technical Report No. 362, 1996.

[2] J.R. Bergen and R. Hingorani, "Hierarchical Motion-Based Frame Rate Conversion", David Sarnoff Research Center Technical Memo, April, 1990.

[3] David Beymer, "Vectorizing Face Images by Interleaving Shape and Texture Computations", MIT AI Lab memo, No. 1537, September, 1995.

[4] D. Beymer, A. Shashua, and T. Poggio, "Example Based Image Analysis and Synthesis", MIT AI Lab Memo, No. 1431, 1993.

[5] Michael Black and Yaser Yacoob, "Tracking and Recognizing Rigid and Non-rigid Facial Motions using Local Parametric Models of Image Motion", International Conference on Computer Vision, Cambridge, Massachusetts, pp. 374–381, June, 1995.

[6] Leonard McMillan and Gary Bishop, "Plenoptic Modeling: An Image-Based Rendering System", SIGGRAPH '95 Proceedings, Los Angeles, CA, 1995.

[7] S.E. Chen and L. Williams, "View Interpolation for Image Synthesis", SIGGRAPH '93 Proceedings, Anaheim, California, pp. 279–288, August, 1993.

[8] Irfan A. Essa and Alex Pentland, "Facial Expression Recognition Using a Dynamic Model and Motion Energy", MIT Media Lab Perceptual Computing Section Technical Report No. 307, 1995.

[9] T. Ezzat, "Example-Based Analysis and Synthesis for Images of Human Faces", Master's Thesis, School of Electrical Engineering and Computer Science, Massachusetts Institute of Technology, February 1996.

[10] F. Girosi, M. Jones, and T. Poggio, "Priors, Stabilizers, and Basis Functions: From Regularization to Radial, Tensor, and Additive Splines", MIT AI Lab Memo, No. 1430, June, 1993.

[11] M. Jones and T. Poggio, "Model-Based Matching of Line Drawings by Linear Combinations of Prototypes", International Conference on Computer Vision, Cambridge, Massachusetts, pp. 531–536, June, 1995.

[12] M. Jones and T. Poggio, "Model-Based Matching by Linear Combinations of Prototypes", unpublished MIT AI memo.

[13] T. Poggio and R. Brunelli, "A Novel Approach to Graphics", MIT AI Memo No. 1354, 1992.

[14] T. Poggio and T. Vetter, "Recognition and Structure from One 2D Model View: Observations on Prototypes, Object Classes, and Symmetries", MIT AI Memo, No. 1347, 1992.

[15] Steve M. Seitz and Charles R. Dyer, "Physically-Valid View Synthesis by Image Interpolation", Proc. IEEE Workshop on the Representation of Visual Scenes, pp. 18–25, June, 1995.

[16] T. Werner, R. D. Hersch, and V. Hlaváč, "Rendering Real-World Objects Using View Interpolation", International Conference on Computer Vision, Cambridge, Massachusetts, pp. 957–962, June, 1995.

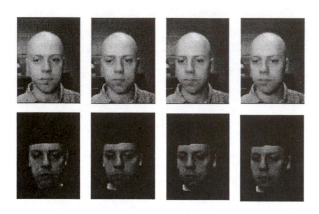

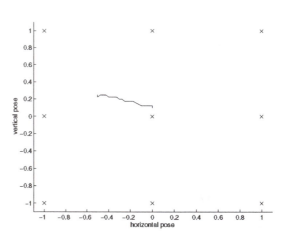

Figure 12: *A novel sequence with leftwards pose movement (top), juxtaposed along with the synthesized sequence (bottom). The synthesis module is the 9-example 3-by-3 pose network shown in Figure 6 in this paper.*

Figure 13: *The pose parameters extracted from the sequence in Figure 12. The analysis algorithm extracts a set of four affine parameters as well as 2 pose parameters, but only pose parameters are shown here for illustration. The "x" marks denote the positions of the 9 examples in pose space.*

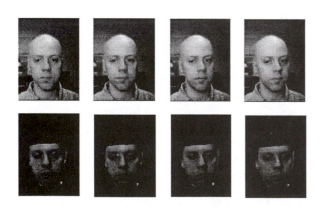

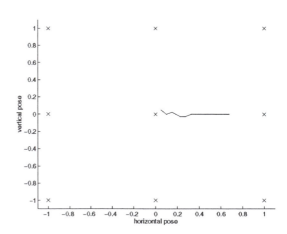

Figure 14: *A novel sequence with rightwards pose movement (top), juxtaposed along with the synthesized sequence (bottom). The synthesis module is the 9-example 3-by-3 pose network shown in Figure 6 in this paper.*

Figure 15: *The pose parameters extracted from the sequence in Figure 14. The analysis algorithm extracts a set of four affine parameters as well as 2 pose parameters, but only pose parameters are shown here for illustration. The "x" marks denote the positions of the 9 examples in pose space.*

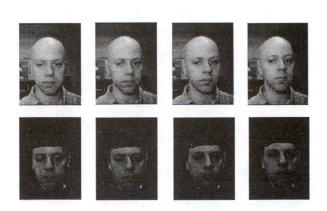

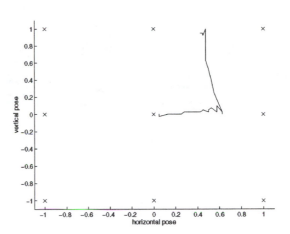

Figure 16: *A novel sequence with top-rightwards pose movement (top), juxtaposed along with the synthesized sequence (bottom). The synthesis module is the 9-example 3-by-3 pose network shown in Figure 6 in this paper.*

Figure 17: *The pose parameters extracted from the sequence in Figure 16. The analysis algorithm extracts a set of four affine parameters as well as 2 pose parameters, but only pose parameters are shown here for illustration. The "x" marks denote the positions of the 9 examples in pose space.*

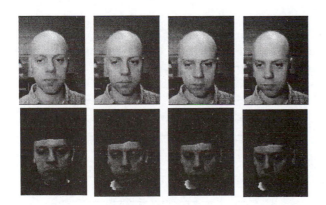

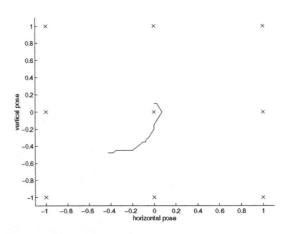

Figure 18: *A novel sequence with bottom-leftwards pose movement (top), juxtaposed along with the synthesized sequence (bottom). The synthesis module is the 9-example 3-by-3 pose network shown in Figure 6 in this paper.*

Figure 19: *The pose parameters extracted from the sequence in Figure 18. The analysis algorithm extracts a set of four affine parameters as well as 2 pose parameters, but only pose parameters are shown here for illustration. The "x" marks denote the positions of the 9 examples in pose space.*

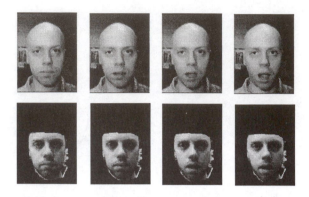

Figure 20: *A novel sequence with mouth movement (top), juxtaposed along with the synthesized sequence (bottom). The synthesis module is the 4-dimensional, 14-example network shown in Figure 8.*

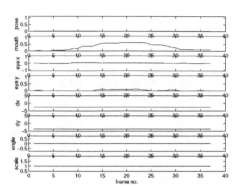

Figure 21: *The complete set of parameters extracted from the sequence in Figure 20. All the activity occurs in the mouth parameter, which denotes degree of openness.*

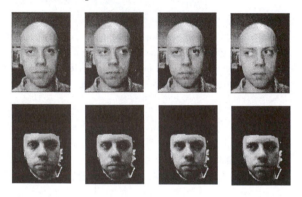

Figure 22: *A novel sequence with eye movement (top), juxtaposed along with the synthesized sequence (bottom). The synthesis module is the same 4-dimensional, 14-example network shown in Figure 8.*

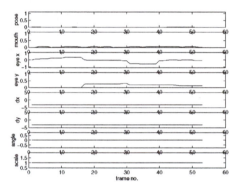

Figure 23: *The complete set of parameters extracted from the sequence in Figure 22. All the activity occurs in the eyes x and y parameters.*

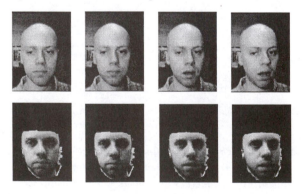

Figure 24: *A novel sequence with eye, pose, and mouth movement (top), juxtaposed along with the synthesized sequence (bottom). The synthesis module is the same 4-dimensional, 14-example network shown in Figure 8.*

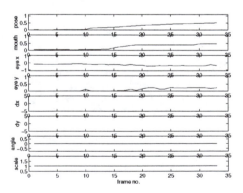

Figure 25: *The complete set of parameters extracted from the sequence in Figure 24. All the activity occurs in the eye, mouth, and pose parameters.*

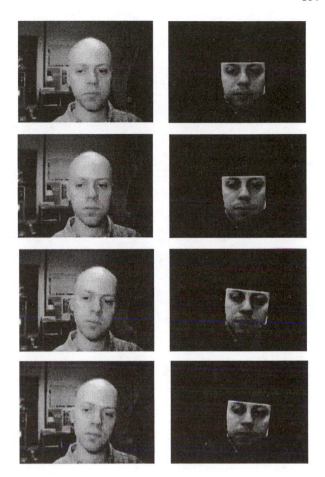

Figure 26: *A novel sequence with mouth movement (left), juxtaposed along with the synthesized sequence (right). The synthesis module is the 2-dimensional, 5-example network shown in Figure 1.*

Figure 28: *A novel sequence with on-plane rotation of the head (left), juxtaposed along with the synthesized sequence (right). The synthesis module is the same 2-dimensional, 5-example expression network shown in Figure 1.*

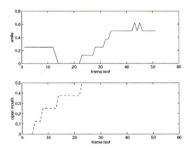

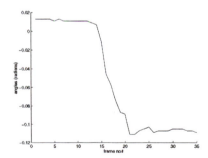

Figure 27: *The smile and open-mouth parameters extracted from the sequence in Figure 26. The analysis algorithm also extracts a set of four affine parameters as well, but only the expression parameters are shown here for illustration.*

Figure 29: *The on-plane rotation parameter extracted from the sequence in Figure 28. Another set of three affine parameters, as well as two expression parameters are also extracted, but only the rotation parameters are shown here for illustration.*

Author Index